Advances in Delivery Science and Technology

Series Editor
Michael J. Rathbone

For further volumes:
http://www.springer.com/series/8875

Navnit Shah • Harpreet Sandhu • Duk Soon Choi
Hitesh Chokshi • A. Waseem Malick
Editors

Amorphous Solid Dispersions

Theory and Practice

Springer

Editors
Navnit Shah
Kashiv Pharma LLC
Bridgewater
New Jersey
USA

Hitesh Chokshi
Roche Pharma Research & Early Development
Roche Innovation Center
New York
New York
USA

Harpreet Sandhu
Merck & Co., Inc.
Summit
New Jersey
USA

Duk Soon Choi
Kashiv Pharma LLC
Bridgewater
New Jersey
USA

A. Waseem Malick
Pharmaceutical and Analytical R&D
Hoffmann-La Roche Ltd.
Nutley
New Jersey
USA

ISSN 2192-6204 ISSN 2192-6212 (electronic)
ISBN 978-1-4939-1597-2 ISBN 978-1-4939-1598-9 (eBook)
DOI 10.1007/978-1-4939-1598-9
Springer New York Heidelberg Dordrecht London

Library of Congress Control Number: 2014947686

Printed on acid-free paper

Springer is part of Springer Science+Business Media (www.springer.com)

To extraordinary scientists at Hoffmann-La Roche who advanced the field of amorphous science to transform poorly soluble "sand-like" compounds into important medicines

Preface

The idea of writing this book was triggered by the development of ASD utilizing microprecipitated bulk powder (MBP) technology at Hoffmann-La Roche and the successful application of this technology to poorly soluble molecules, such as vemurafenib. This technology was instrumental in transforming this novel molecule into a medicine (Zelboraf®) for malignant melanoma patients. It was a gratifying and fulfilling experience for all of us when Zelboraf® became a key drug for this deadly disease and made a difference in the lives of many patients. We believe that many pharmaceutical scientists face such a challenge, and a book covering the theory and practice of amorphous solid dispersion technologies would be very useful to industrial and academic scientists as well as students in understanding and handling the challenges associated with developing such molecules.

Poorly water soluble drug molecules emerging from contemporary discovery programs often have inadequate and/or variable in vivo exposure, presenting pharmaceutical scientists with considerable challenges during development. Drugs with poor and variable oral absorption often have suboptimal therapeutic performance and significant food effect, thereby raising safety concerns, particularly for narrow therapeutic window drugs. As a result, promising molecules can be terminated prematurely if these issues are not adequately addressed. A number of formulation strategies have been developed to enhance the bio-performance of such molecules. Among these technologies, particle size reduction by micronization or nano milling improves the rate of dissolution; however, this strategy has resulted in limited success for poorly water soluble molecules having a solubility of less than 10 mcg/mL. Solubilization in lipid vehicles and self-emulsifying delivery systems have certainly added value, but their utility has been limited by drug loading, which remains a major issue. Similarly, salts of weak acids and bases have met with limited success due to precipitation of these salts in physiological fluids resulting in significant variability. Co-crystallization has been recently explored, but its utility has yet to be realized for poorly soluble molecules.

The amorphous form of a drug offers high free energy and therefore higher kinetic solubility, which provides an opportunity for overcoming solubility-related absorption and bioavailability challenges. The amorphous form, however, is thermodynamically unstable, and stabilization of molecules in this physical state still remains a formidable task. A greater understanding of the scientific principles governing these systems and the development of amorphous solid dispersion (ASD) formulations for stabilizing amorphous molecules have created tremendous opportunities for the pharmaceutical scientist to address issues relating to the bioavailability of poorly soluble molecules. ASD technology has become one of the most powerful and versatile technology platforms in recent years. The design and development of successful ASD formulations requires the integration of scientific, technological and biopharmaceutical aspects to arrive at a robust drug product. Amorphous formulation technologies and our understanding of amorphous systems have advanced significantly in the last decade. A greater appreciation of the underlying physical science and thermodynamics, the emergence of newer technologies for the preparation of amorphous formulations, and the availability of newer excipients and polymers for stabilizing ASD have vastly expanded the opportunities for pharmaceutical scientists to establish stabilization strategies for these systems. The interest in developing amorphous formulations has increased more than ever due to the successful market introduction of such products over the last decade.

Written by experts from industry, academia and government, this book provides an excellent reference for pharmaceutical research scientists in the understanding, preparation and stabilization of ASD. In this book, we present the three primary factors for the stabilization and successful development of ASD, namely (a) the physical and chemical properties of the drug substance, (b) polymers and their impact on the stability of the final product, and (c) processing technologies to put ASD into practice. These aspects are extensively covered by the inclusion of case studies.

The first few chapters of the book cover the fundamentals and theoretical aspects of amorphous systems, an overview of ASD technologies, and details on excipients and polymers used in ASD, along with their safety aspects. "Fundamentals of Amorphous Systems" discusses the theoretical aspects of thermodynamics and kinetics with respect to the energy barrier. Also addressed are the active pharmaceutical ingredient (API) properties and polymer characteristics necessary for preparing stable ASD, involving solubility and miscibility, interaction parameters and drug loading impact. "Overview of Amorphous Solid Dispersion Technologies" provides a detailed presentation of each technology and its limitations. The chapter on excipients presents different classes of excipients, their physico-chemical properties and their interrelationship with different processes; the safety and stability of excipients are also described at length.

Later chapters present details of ASD manufacturing technologies, including spray drying, hot melt extrusion, and a breakthrough novel solvent-controlled microprecipitation technology (MBP). Each technology is illustrated with processing fundamentals and scale up factors along with specific case studies, which provide the scientist with approaches for handling challenges presented by different types of molecules as well as building process flexibility. In addition, a dedicated section covers the miniaturization of technologies for screening polymers and processes with small amounts of API, particularly during the discovery and early development phases addressing preclinical needs. Since all of the technologies used in preparing ASD systems require downstream processing for developing viable drug products, the chapter on downstream processing covers the physical and mechanical factors impacting product performance. The analytical tools for the characterization of amorphous solid dispersions, prediction of long term stability, evolving suitable dissolution methods particularly addressing supersaturation kinetics, as well as regulatory aspects germane to amorphous solid dispersion formulations and technologies are also extensively covered.

This volume explores technologies on the horizon, such as supercritical fluid processing, mesoporous silica, KinetiSol®, and the use of non-salt forming organic acids and amino acids for the stabilization of amorphous systems. It presents a comprehensive overview of the theory and practice of amorphous solid dispersions in overcoming the challenges associated with poorly soluble drugs, and it includes practical examples based on commercially successful products using different manufacturing technologies and stabilization strategies. *Amorphous Solid Dispersions* provides pharmaceutical scientists with up-to-date knowledge on amorphous solid dispersions that will further enhance their ability to handle more challenging molecules and will pave the way for future innovation to bring cutting-edge therapeutics to patients in need.

Sincerely
Navnit Shah
Harpreet Sandhu
Duk Soon Choi
Hitesh Chokshi
A.Waseem Malick

Acknowledgement

The editors want to thank all the individuals who provided scientific input and critique as well as offered valuable changes and suggestions. These contributions truly enhanced the quality of the book. We acknowledge and express our sincere and deep appreciation for their efforts.

It is hard to express our gratitude in words to Hoffmann-La Roche Inc. for supporting high quality research and creating an atmosphere that was conducive to exploring new ideas and innovation. The inspirational and collaborative environment enabled us to pursue original research and contribute to the advancement of amorphous systems. These efforts led to cutting edge innovations that in turn enabled development of effective new medicines. None of this would have been possible if it were not for the dedication and enormous effort of many outstanding scientists from the Pharmaceutical and Analytical R&D Department, strong partnership of scientists from other disciplines as well as unwavering management support. In reality, getting these differentiated medicines to the patients in need is the true inspiration for writing this book.

We do not have enough words to express our earnest thanks to all the authors and co-authors for accepting our request for contributing to this effort and most importantly for providing high quality contents to enable the timely completion of the book. We are highly appreciative of their patience in responding to our numerous requests throughout this process. It indeed has been a privilege to work with people of such high scientific caliber and integrity.

We want to express our sincere thanks to Springer for the invitation for writing this book. Our special thanks to Ms. Carolyn Honour and Ms. Sarah McCabe for their valuable suggestions, helpful comments and especially for putting up with our slight tardiness in completion of the book.

The editors want to acknowledge the valuable contributions of Mr. Dinesh Shah and Dr. Martin Infeld for reviewing documents and providing constructive comments for enhancing the quality of the book. Our thanks to Ms. Lisa Mitchell for handling the logistics and all the support for arranging the book contents, formatting, preparing tables, and valuable edits. A special thank you to Ms. Vicky Pacholec for keeping us together during challenging times.

Most importantly, our heartfelt thanks are owed to our families. Our spouses, Gita, Farooq, Kwanghee, Sonal and Aneeza and our children for their patience and sacrifice during the last two years, for tolerating our long hours away from them and encouraging us to undertake this mission. Without their support, the timely completion of this book would not have been possible.

Sincerely
Navnit Shah
Harpreet Sandhu
Duk Soon Choi
Hitesh Chokshi
A.Waseem Malick

Contents

Contributors

Jean-Rene Authelin Pharmaceutical Sciences Operation, Sanofi R&D, Vitry Sur Seine, France

Robert A. Bellantone Division of Pharmaceutical Sciences, Long Island University, Brooklyn, NY, USA

Murali Mohan Bommana Kashiv Pharma LLC, Bridgewater, NJ, USA

Chad Brown Merck & Co., Inc., Whitehouse Station, NJ, USA

Duk Soon Choi Kashiv Pharma LLC, Bridgewater, NJ, USA

Hitesh Chokshi Roche Pharma Research & Early Development, Roche Innovation Center, New York, NY, USA

Yogesh Choudhari W. R. Grace and Company, Columbia, MD, USA

Dipen Desai Kashiv Pharma LLC, Bridgewater, NJ, USA

James DiNunzio Pharmaceutical Sciences & Clinical Supplies, Merck & Co., Inc., Summit, NJ, USA

Ralph Diodone Pharmaceutical Technical Development Chemical Actives, F. Hoffmann-La Roche Ltd., Basel, Switzerland

Jelena Djordjevic Kashiv Pharma LLC, Bridgewater, NJ, USA

Michael Eglesia Merck & Co., Inc., Whitehouse Station, NJ, USA

Adam Feiler Nanologica AB, Stockholm, Sweden

Seth Forster Merck & Co., Inc., Whitehouse Station, NJ, USA

Nikoletta Fotaki Department of Pharmacy and Pharmacology, University of Bath, Bath, UK

Alfonso Garcia-Bennett Department of Materials and Environmental Chemistry, Arrhenius Laboratory, Stockholm University, Stockholm, Sweden

Filipe Gaspar Particle Engineering Services, Hovione, Lisbon, Portugal

Kirsten Graeser pRED, Roche Innovation Center, F. Hoffmann-La Roche Ltd., Basel, Switzerland

Holger Grohganz Department of Pharmacy, University of Copenhagen, Copenhagen, Denmark

Abhay Gupta Division of Product Quality and Research, Center of Drug Evaluation and Research, U.S. Food and Drug Administration, MD, USA

Hans Hoefer W. R. Grace and Company, Columbia, Maryland, USA

Qingyan Hu Formulation Development, Regeneron Pharmaceuticals, Inc., Tarrytown, NY, USA

Raman Iyer Novartis, East Hanover, NJ, USA

Katrine Tarp Jensen Department of Pharmacy, University of Copenhagen, Copenhagen, Denmark

Justin M. Keen DisperSol Technologies LLC, Georgetown, TX, USA

Mansoor Khan Division of Product Quality and Research, Center of Drug Evaluation and Research, U.S. Food and Drug Administration, MD, USA

Xiang Kou Chemical and Pharmaceutical Profiling, Novartis Pharmaceuticals, Shanghai, China

Riikka Laitinen School of Pharmacy, University of Eastern Finland, Kuopio, Finland

Matthew Lamm Merck & Co., Inc., Whitehouse Station, NJ, USA

Cristian Libanati W. R. Grace and Company, Columbia, MD, USA

Korbinian Löbmann Department of Pharmacy, University of Copenhagen, Copenhagen, Denmark

Chiau Ming Long Department of Pharmacy and Pharmacology, University of Bath, Bath, UK

Michael Lowinger Merck & Co., Inc., Whitehouse Station, NJ, USA

Hans J. Mair Pharmaceutical Technical Development Chemical Actives, F. Hoffmann-La Roche Ltd., Basel, Switzerland

A. Waseem Malick Pharmaceutical and Analytical R&D, Hoffmann-La Roche Ltd., Nutley, NJ, USA

Patrick Marsac Merck & Co., Inc., Whitehouse Station, NJ, USA

Reto Maurer Formulation Research and Development, F. Hoffmann-La Roche Ltd., Basel, Switzerland

William McCarthy W. R. Grace and Company, Columbia, MD, USA

Craig McKelvey Merck & Co., Inc., Whitehouse Station, NJ, USA

Joke Meeus Drug Delivery and Disposition, University of Leuven, Leuven, Belgium

Robert Meyer Merck & Co., Inc., Whitehouse Station, NJ, USA

Dave A. Miller DisperSol Technologies LLC, Georgetown, TX, USA

Fred Monsuur W. R. Grace and Company, Columbia, MD, USA

Guy Van den Mooter Drug Delivery and Disposition, University of Leuven, Leuven, Belgium

Michael Morgen Bend Research, Inc., Bend, OR, USA

Filipe Neves R&D Drug Product Development, Hovione, Loures, Portugal

Susanne Page Formulation Research and Development, F. Hoffmann-La Roche Ltd., Basel, Switzerland

Amrit Paudel Drug Delivery and Disposition, University of Leuven, Leuven, Belgium

Research Center Pharmaceutical Engineering GmbH (RCPE), Graz, Austria

Wantanee Phuapradit Kashiv Pharma LLC, Bridgewater, NJ, USA

Petra A. Priemel School of Pharmacy, University of Otago, Dunedin, New Zealand

Thomas Rades Pharmaceutical Design and Drug Delivery, Faculty of Health and Medical Sciences, Department of Pharmacy, University of Copenhagen, Copenhagen, Denmark

Ziyaur Rahman Division of Product Quality and Research, Center for Drug Evaluation and Research, U.S. Food and Drug Administration, MD, USA

Harpreet Sandhu Merck & Co., Inc., Summit, NJ, USA

Luke Schenck Merck & Co., Inc., Whitehouse Station, NJ, USA

Abu T. M. Serajuddin Department of Pharmaceutical Sciences, College of Pharmacy and Health Sciences, St. John's University, Queens, NY, USA

Ankita Shah Department of Pharmaceutical Sciences, College of Pharmacy and Health Sciences, St. John's University, Queens, NY, USA

Navnit Shah Kashiv Pharma LLC, Bridgewater, NJ, USA

Pratik Sheth Forum Pharmaceuticals, Inc., North Grafton, MA, USA

Koji Shiraki Research division, Chugai Pharmaceutical Co., Ltd., Gotemba, Shizuoka, Japan

Akhtar Siddiqui Division of Product Quality and Research, Center of Drug Evaluation and Research, U.S. Food and Drug Administration, MD, USA

Brandye Smith-Goettler Merck & Co., Inc., Whitehouse Station, NJ, USA

Cindy Starbuck Merck & Co., Inc., Whitehouse Station, NJ, USA

Clare J. Strachan Division of Pharmaceutical Chemistry and Technology, University of Helsinki, Helsinki, Finland

Sachin Surwase Division of Pharmaceutical Chemistry and Technology, University of Helsinki, Helsinki, Finland

Kin Tang Pharma Technical Regulatory, Genentech Inc., South San Francisco, CA, USA

Graciela Terife Merck & Co., Inc., Whitehouse Station, NJ, USA

Gregory Troup Merck & Co., Inc., Whitehouse Station, NJ, USA

Siva Ram Kiran Vaka Kashiv Pharma LLC, Bridgewater, NJ, USA

Joao Vicente R&D Drug Product Development, Hovione, Loures, Portugal

David T. Vodak Bend Research, Inc., Bend, OR, USA

Nicole Wyttenbach pRED, Roche Innovation Center, F. Hoffmann-La Roche Ltd., Basel, Switzerland

Liping Zhou Ipsen Biosciences, Cambridge, MA, USA

About the Editors

Dr. Navnit Shah is president and CSO of Kashiv Pharma LLC in Bridgewater, New Jersey. Prior to joining Kashiv, he was a distinguished scientist at Hoffmann-La Roche Inc., where he headed the oral dosage form development group for many years. He received his Ph.D. in pharmaceutics from St. John's University in Queens, New York, and he has published 120 abstracts and more than eighty scientific papers in the drug delivery area, particularly in the field of amorphous solid dispersion. He is inventor and co-inventor on nineteen issued patents and thirteen patent applications. He is the recipient of numerous awards, including AAPS Fellow, Thomas Alva Edison Patent Award, St John's University Distinguished Alumni Award, New Jersey Biomedical Research Leadership Award, and the New Jersey Inventor of the Year award by New Jersey Inventors Hall of Fame (NJIHoF). He is an adjunct professor at the University of Rhode Island in Kingston and has mentored several graduate students for their M.S. & Ph.D. research.

Dr. Harpreet Sandhu Sandhu is a principal scientist at Merck, Inc. in the formulation sciences group. Prior to this, she was at Hoffmann-La Roche Inc., where she held positions of increasing responsibility in the area of oral formulation development. She received her Ph.D. in physical pharmacy from the University of Connecticut in Storrs and an M.B.A. from Rutgers University in Newark, New Jersey. Her research interests have been shaped by her ardent desire to meet the growing challenges of poor aqueous solubility. During her tenure at Roche, she led the development of multiple programs covering simple formulations to the most complex delivery systems and mentored junior scientists, graduate students and postdocs.

Dr. Duk Soon Choi is a vice president at Kashiv Pharma LLC in Bridgewater, New Jersey. His responsibilities include the leadership of the preformulation and chemistry teams. Dr. Choi's research focuses on active pharmaceutical ingredient (API) modifications and drug delivery systems to overcome the challenges presented by molecules with poor drug-like properties. He received his Ph.D. in Chemistry from Louisiana State University in Baton Rouge, and prior to joining Kashiv, he was the head of the Preformulation and Solid State Group at Hoffmann-La Roche in Nutley, New Jersey, where he played a key role in transitioning a number of discovery compounds into clinical studies.

Dr. Hitesh Chokshi is a senior leader in Roche Pharma Research & Early Development at Roche Innovation Center, New York. He joined Roche in Nutley as a preformulation scientist after completing his Ph.D. and postdoctoral research at the University of Georgia in Athens, Georgia, and University of Kansas in Lawrence, respectively. Hitesh has held diverse scientific and leadership responsibilities at Roche for the development of new medicines: small molecules, peptides, and biologics. His research includes pharmaceutical profiling, development and characterization of fit-for-purpose dosage forms for toxicological and clinical studies as well as for the market. He has a special interest in amorphous solid dispersion (ASD) design and characterization, establishing predictive in-vitro models to de-risk bioequivalence, and in quality by design (QbD) solid state characterization of drug substances and dosage forms.

Dr. A. Waseem Malick is a retired pharmaceutical R&D executive who has had an extensive career in the industry, previously serving as vice president and global head of pharmaceutical and analytical R&D at Hoffmann-La Roche. He received his Ph.D. in pharmaceutics from the University of Michigan, Ann Arbor, and an M.S. in pharmaceutics from Columbia University, New York. His research interests include drug product development, formulation and drug delivery research. He has published 110 research papers and 169 abstracts and has presented extensively in the area of drug delivery research. He was elected as fellow of the American Association of Pharmaceutical Scientists (AAPS) and has received several honors and awards. He serves on the board of directors of several non-profit organizations and on the Science Board of Advisors at New York University. He also has an adjunct professor appointment at the University of Rhode Island in Kingston.

Part I
ASD Introduction

Chapter 1
Fundamentals of Amorphous Systems: Thermodynamic Aspects

Robert A. Bellantone

1.1 Introduction

Drugs with poor aqueous solubility present a major challenge to pharmaceutical scientists because they tend to show low oral bioavailability. This has been one of the most critical issues in pharmaceutical industry for many decades. In 1995, the biopharmaceutical classification system (BCS) was introduced to facilitate drug development, classifying drug molecules according to their solubility and permeability (Amidon et al. 1995). It was estimated that 60–70 % of compounds in development are poorly water soluble. Moreover, it has recently been pointed out that the percentage can be even higher, as high as 90 %, for certain drug categories (Williams et al. 2012).

Pharmaceutical companies screen thousands of new chemical entities (NCEs) every year in the hope to find cures for diseases, but many of these face ill-fated discontinuation partially due to inadequate exposures owing to poor aqueous solubility. In the current environment where only 20–40 NCEs make it to the market (USA) per year (Herschler and Humer 2012), it will be beneficial to the patient and society, even if one or two more compounds are developed annually as medicines by overcoming solubility related issues.

For an orally administered drug to have a therapeutic effect, the drug molecules must be dissolved in the gastrointestinal (GI) fluids, pass through GI membrane to the circulatory system, and reach the target in sufficient quantity. That is, drug molecules must be dissolved in aqueous-based GI fluids in sufficient quantity to have any therapeutic effect. If the solubility of the drug in GI fluids is not sufficient, the bioavailability will be compromised as the absorption will be "solubility limited." It is quite common for poorly soluble compounds to also dissolve slowly. If the dissolution

R. A. Bellantone (✉)
Division of Pharmaceutical Sciences,
Long Island University, Brooklyn, NY, USA
e-mail: Robert.Bellantone@liu.edu

N. Shah et al. (eds.), *Amorphous Solid Dispersions*,
Advances in Delivery Science and Technology, DOI 10.1007/978-1-4939-1598-9_1

rate of the drug is too slow in the timeframe of absorption, the bioavailability will again be compromised as absorption will be "dissolution limited."

It is noteworthy that the mechanism of absorption requires that the drug should be in solution. In this chapter, the term "dissolved" or "in solution" refers to the state in which individual molecules are dispersed in the solvent medium. Complexed or bound molecules such as in micelles, emulsions, drug polymer complexes, or inclusion complexes are excluded from this definition. With this definition, the dissolved form corresponds to the form that is absorbable in vivo. This distinction is important because complexation or other means of solubilization can also increase the apparent dissolved concentration of the drug, but these complexes are not really absorbable "as is" unless they dissociate into individual components.

One way to increase the oral bioavailability is to increase the concentration of the dissolved drug in the GI fluids. This can be achieved by increasing the dissolution rate, increasing the drug solubility, or the combination of both. One of the most remarkable approaches to achieve faster dissolution and higher apparent solubility is converting crystalline drug to amorphous drug (Leuner and Dressman 2002). This amorphous approach has been extensively studied over the past several decades (Simonelli et al. 1969; Chiou and Riegelman 1971; Hancock and Zografi 1997).

Two types of amorphous solids are relevant to the pharmaceutical sciences—pure amorphous material (referred to as neat active pharmaceutical ingredient, or neat API) and solid solutions/dispersions (referred to as amorphous solid dispersions, or ASDs). Both can increase the solubility and dissolution rate, but they are very different microscopically. For a neat API, the material is pure drug and the molecular packing is altered in a manner that weakens the average attractive energy between drug molecules, which in turn lowers the energy barrier for drug molecules to go into solution. On the other hand, in an ASD, the molecular packing is disrupted by dispersing the drug molecules in a solid medium or carrier. Neat amorphous forms will be the subject of this chapter. (ASDs are considered in the next chapter.)

1.1.1 Function of "Dissolved Drugs" in Absorption

The dissolution rate of a drug can be described by the Nernst–Bruner equation

$$\frac{dM}{dt} = k_d(C_S - C) \quad C = \frac{M}{V}, \tag{1.1}$$

where M is the mass of drug dissolved in a volume V of GI fluid, C_s is the solubility of the drug, C is the concentration of the dissolved drug at time t, and k_d is the parameter that depends on factors such as the diffusion coefficient, total surface area, agitation rate, etc. Poorly soluble drugs produce a low level of "dissolved" drug because C_S is inherently low. In addition, dissolution rates tend to be slow for poorly soluble drugs. These factors, as mentioned earlier, make the poorly soluble drug also poorly bioavailable. Taking the concentration C in Eq. (1.1) as the "dissolved" drug concentration in GI fluids, the absorption rate is given by

$$\begin{bmatrix} \text{absorption} \\ \text{rate} \end{bmatrix} = k_a C, \tag{1.2}$$

where k_a is an absorption rate constant that depends on the surface area, membrane permeability, etc. Although Eqs. (1.1) and (1.2) represent an overly simplified version of dissolution of a drug in the GI track, they state that an increasing apparent saturation solubility C_s will improve oral bioavailability by improving the dissolution rate as well.

Despite the aforementioned advantages of amorphous forms, only a few products have been developed using neat API amorphous forms. These include Accolate® (zafirlukast), Accupril® (quinapril hydrochloride), Ceftin® (cefuroxime axetil), and Viracept® (nelfinavir mesylate).

1.2 Structural Aspects

An amorphous solid is characterized by the lack of long-range order symmetry operators (translational, orientational, and conformational order) found in crystalline solid. The absence of long-range order can be ascribed to a random distribution of molecular units. Individual molecules are randomly oriented to one another and exist in a variety of conformational states. The molecular pattern of an amorphous solid is often depicted as that of a frozen liquid with the viscosity of a solid having many internal degrees of freedoms and conformational diversities (disorder). An amorphous solid, at the molecular level, has properties similar to liquids; but at the macroscopic level, it has properties of solids.

The lack of symmetry operators are commonly manifested by the lack of X-ray diffraction peaks typically found in crystalline solid, but may exhibit a characteristic broad amorphous "halo" (Fig. 1.1). In addition, an amorphous solid is further characterized by the lack of distinct melting point and birefringence properties.

The internal molecular arrangement of a solid in general can be projected as a continuum between well-ordered crystalline state and completely disordered amorphous state (Fig. 1.2). A crystalline material is depicted as having three-dimensional long-range symmetry operators over a domain of at least 1000 individual molecules. A mesophase material (liquid crystals, plastic crystals) is depicted as having intermediate symmetry operators, and an amorphous state has no symmetry operators (Klug and Alexander 1974).

Such diverse packing arrangements explain different physicochemical properties of solids, such as differences in density, hardness, thermal properties, conductivity, solubility, etc. In practice, perfect crystals without any defects are not seen, nor are perfect amorphous systems. Recent studies suggest that amorphous solid of small molecules may not be in truly amorphous state but may contain certain structural elements. In fact, amorphous solids can exhibit short-range order over domains that are too small to show crystalline properties (Gavezzotti 2007).

Amorphous solids are considered as glasses, which have rheological properties of solids and molecular properties of liquids (Kittel 1986). The behavior of amorphous glasses can be explained by heat content or molar volume changes with the changes of temperature. When the heat content or molar volume of a sample is plotted against

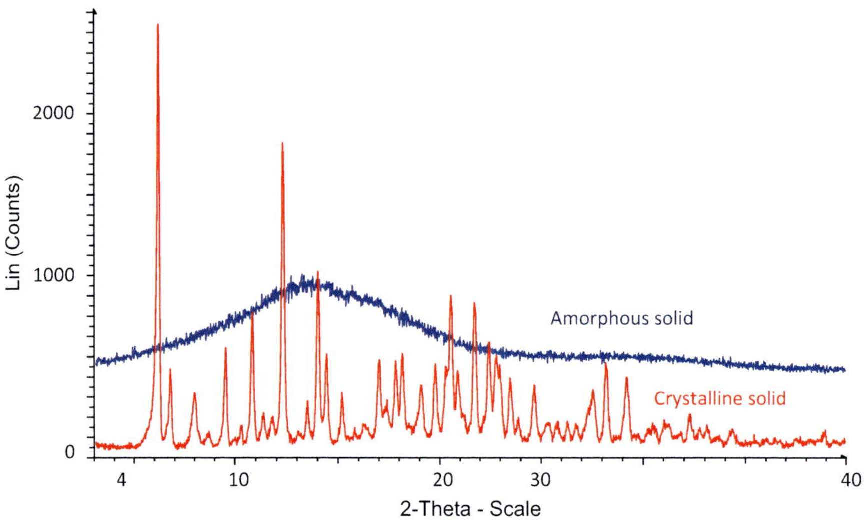

Fig. 1.1 Illustration of X-ray diffraction pattern of crystalline solid (*red*) and amorphous solid (*blue*). Owing to periodic lattice planes, crystalline solid scatters X-ray beam constructively in well-defined directions producing characteristic X-ray diffraction pattern whereas amorphous solid is anisotropic, scatters X-ray beam in many directions producing broad amorphous halos instead of high-intensity narrow peaks

the temperature, these variables change smoothly until it comes to the region known as glass transition temperature, where the variables change abruptly. The temperature region below the glass transition temperature is known as "glassy state" and above the glass transition temperature is known as "rubbery state." The physicochemical and physicomechanical properties of the materials are starkly different between these two regions.

1.3 Thermodynamic Aspects

1.3.1 Two Approaches to Understanding Neat Amorphous Forms

There are two general approaches to understand the behavior of amorphous materials, one based on macroscopic thermodynamic arguments and the other based on microscopic molecular arguments. While these approaches should lead to the same conclusions, each view provides unique ways of explaining material behavior.

Thermodynamics is based on macroscopic observations that characterize average behavior of material based on energy contents. The most fruitful approach is to apply equilibrium thermodynamics, which can be applied even if the system is not in an equilibrium state as long as the intensive variables are uniform within the material. This approach is useful to describe the solubility of the material, in which

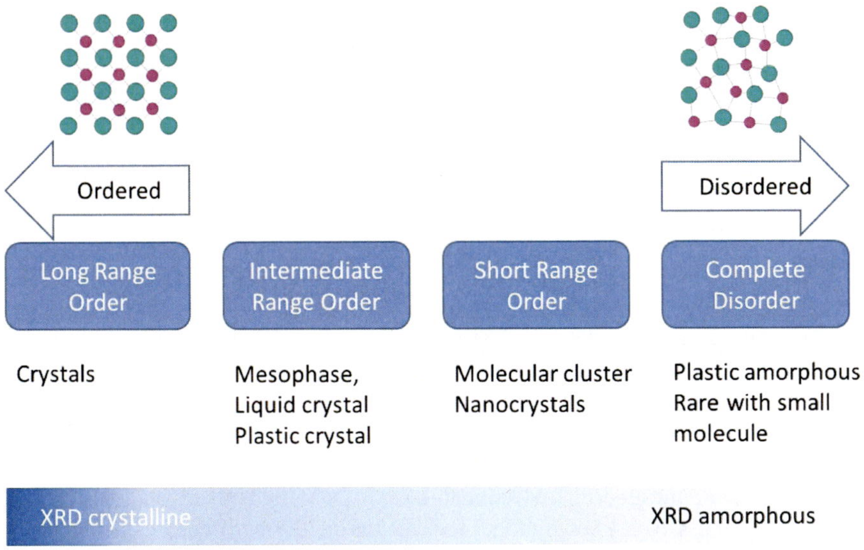

Fig. 1.2 Solid form system illustrating long-range ordered (translational, orientational, and conformational) crystals on one end and completely disordered amorphous material on the other end. Solid forms can assume various length scale of order (long range, medium range, short range) and/or mesomorphic states (smectic, nematic)

the properties of the drug in the dissolved and undissolved forms are be taken into the consideration together.

The material behavior of solid forms is ultimately attributed to the strength of interactions between molecules, which vary significantly depending on the chemical structure and molecular separations. The latter factor is a function of the molecular packing, which sets the distance between molecules. Thus, a microscopic point of view is well suited for a fundamental interpretation of the solid form in terms of molecular interactions and packing.

1.3.2 Description of Forming a Solution

The dissolution process involves removing drug molecules from the undissolved solid and placing them into the holes generated in the solvent system in such a way that molecules are dispersed in the solvent matrix uniformly. In this sense, "undissolved" means the drug is in a solid form and "dissolved" means the drug is molecularly dispersed in a solvent medium. In the undissolved state, the drug molecules interact with each other via various intramolecular cohesive forces to form a condensed solid phase, whereas in the dissolved state, the drug molecules interact with the surrounding solvent molecules.

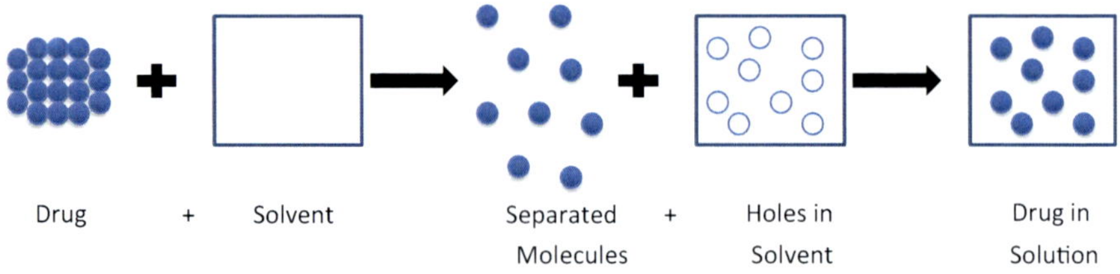

Fig. 1.3 Schematic of dissolution for pure API

Dissolution has been described by a number of models. The ideal solution has been described as a consequence of solute being melted and mixed with the solvent (Atkins 1998). This simplistic view is generalized by adding specific solvent effects in the regular solution theory (Hildebrand and Scott 1950). In particular, the Hildebrand–Scott model includes the differences in interaction energies between the "unmixed" state (drug–drug and solvent–solvent) and the "mixed" or solution state (drug–solvent). A well-known model for calculating these differences is the "hole" model (Martin et al. 1983; Hill 1986). In this model, the drug molecules are separated from the solid and "holes" are generated in the solvent. The drug molecules are then placed in the holes and allowed to migrate until they are uniformly dispersed throughout the solvent as illustrated in Fig. 1.3.

This "hole" model can be depicted as a sequence of conceptual steps as shown below. Because the Gibbs energy is a thermodynamic state function, this sequence of steps does not have to occur in reality, but it provides a means to calculate the changes in Gibbs free energy due to dissolution (Klotz and Rosenberg 1974). As the changes in Gibbs energy only depend on the final state and the initial states, this is analogous to "Hess' law" of thermodynamics (Atkins 1998).

Assuming the temperature and pressure (T, P) remain constant, dissolving a solid drug in a liquid solvent can be modeled by the following sequence.

1. Break up the lattice structure of the solid to supercooled liquid at (T, P).
2. Separate the molecules from the supercooled liquid at (T, P).
3. Make uniformly distributed "holes," one for each drug molecule, in the solvent at (T, P).
4. Put the drug molecules in the holes at (T, P).

Steps 1–3 require an input of energy to overcome the drug–drug molecular attractions to separate them, and to overcome solvent–solvent molecular attractions to create holes within the solvent. In step 4, energy is given back as a result of drug–solvent molecular attractions.

Figure 1.4 illustrates a diagram of the Gibbs energy versus the dissolution process at a given (T, P). The diagram includes the initial raw material state of undissolved drug and solvent (A), the final solution state of the drug dissolved in the solvent (D), and two hypothetical intermediate states B and C. The changes of Gibbs energy by forming the solution, ΔG_S, are the Gibbs energy of state D minus state A. For state

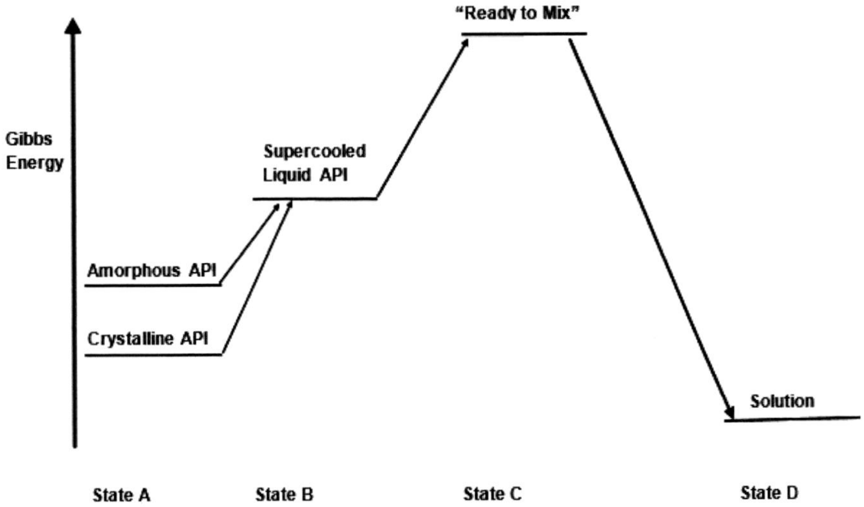

Fig. 1.4 Gibbs energy diagram for the dissolution scheme. *State A* represents the initial state consisting of undissolved solid API and empty solvent. *State B* represents a supercooled liquid API plus empty solvent. *State C* represents separated drug molecules and holes in the solvent. *State D* represents the final solution in which the drug is dissolved in the solvent

A, two initial states are shown—one in which the undissolved API is crystalline, and another in the amorphous form.

The intermediate state B represents the "empty" solvent and "supercooled" liquid API. For a given amount of API at (*T, P*), all liquid forms of API (including supercooled) have the same Gibbs energy regardless of the starting solid form in state A. State C is another intermediate state in which the drug molecules have been separated and uniformly distributed holes have been made in the solvent, in which the number of holes equals to the number of separated drug molecules. This can be thought of as a "ready to mix" (RTM) state, and the energy level is also the same regardless of the starting undissolved drug forms. The RTM state also represents the highest Gibbs energy level in the overall dissolution processes.

With the assumption that all drug molecules in state A dissolve, the Gibbs energy level in the final solution is independent of the starting solid forms. Thus, the energy differences between states B, C, and D are independent of the starting undissolved solid form of state A . However, the energy required to pass from state A to state B does depend on the starting solid form, and is less for the amorphous than the crystalline form. In this regard, state B serves as a reference point, especially with regard to the energy of state A, which reflects different energies for different starting forms.

As depicted in Fig. 1.4, the difference between the initial and final Gibbs energy levels (state D minus state A) ΔG_S dictates the solubility of the drug in the solvent. The difference between the initial energy levels and the intermediate RTM state energy level (state C minus state A) represents the energy required to bring the system to the dissolved state. This can be thought of as analogous to an energy barrier or activation energy for dissolution, which is related to the kinetics of dissolution. It should be noted that this is not the only factor affecting dissolution rates. Thus, because of smaller activation energy, material with higher initial energy levels will result in higher solubility and faster dissolution. Since amorphous systems start at a higher level than crystalline forms (level A), the reduced energy barrier (activation energy) partially explains why the kinetics of dissolution is faster for amorphous systems.

1.3.3 Calculating the Gibbs Energy Change ΔG_S of Forming a Solution

The change in Gibbs energy due to dissolving the solid, ΔG_S, is the sum of changes for each step. If the starting and ending (*T, P*) are the same, it can be shown that (Bellantone et al. 2012)

$$\Delta G_S = \sum \Delta H - T \sum \Delta S, \tag{1.3}$$

Thus, ΔG_S can be calculated by finding the changes in enthalpy and entropy from steps 1–4 above and inserting them into Eq. (1.3), as illustrated below.

Step 1 can be modeled as follows. At constant pressure, the ΔH is given by (1) the heat required to bring the solid from T to its melting temperature T_{M1}, (2) melting the solid at that temperature, and then (3) cool the melt as a supercooled liquid back to the original temperature:

$$\Delta H(1) = \int_T^{T_{M1}} n_1 C_{P1}^{(S)} dT + n_1 \Delta h_{M1} + \int_{T_{M1}}^{T} n_1 C_{P1}^{(L)} dT, \tag{1.4}$$

where n_1 represents the moles of drug, $C_{P1}^{(S)}$ and $C_{P1}^{(L)}$ represent the molar heat capacities of the solid and supercooled liquid melt, respectively, and Δh_{M1} represents the molar heat of melting of the drug. In general, $C_{P1}^{(S)}$, $C_{P1}^{(L)}$, and ΔC_{P1} are functions of the temperature, so they are brought into the integral. Noting that $dS = (C_P/T)dT$ and $\Delta S_{M1} = \Delta H_{M1}/T_{M1}$, and defining $\Delta C_{P1} = C_{P1}^{(S)} - C_{P1}^{(L)}$, the enthalpy and entropy changes associated with step 1 are:

$$\Delta H(1) = n_1 \Delta h_{M1} + n_1 \int_T^{T_{M1}} \Delta C_{P1} dT \quad \Delta S(1) = \frac{n_1 \Delta h_{M1}}{T_{M1}} + \int_T^{T_{M1}} \frac{n_1 \Delta C_{P1}}{T} dT, \tag{1.5}$$

The enthalpic contributions from steps 2–4 are typically combined to give a total enthalpy of mixing ΔH_{mix}, which represents the net total of the energy input to separate the drug molecules and to make holes in the solvent, minus the energy released due to drug–solvent interactions. For regular solutions, this is often modeled as

$$\Delta H_{mix} = n_1 x_0 \chi RT, \tag{1.6}$$

where $x_0 = n_0/(n_1+n_0)$ is the mole fraction of the solvent. The interaction parameter χ has been modeled theoretically in terms of solubility parameters as

$$\chi = \frac{V_{m,1}\phi_0^2}{RT}(\delta_1 - \delta_0)^2, \tag{1.7}$$

where δ_1 and δ_0 represent the solubility parameters of the drug and solvent, respectively, $V_{m,1}$ is the molar volume of the drug, and ϕ_0 represents the volume fraction of the solvent (which is very close to 1 for dilute solutions). Equation (1.7) has limitations because it requires the heat of solution to be always positive, which conflicts with observations that they can be positive (endothermic) or negative (exothermic). Extensions have been proposed to correct for this, such as (Adjei et al. 1980)

$$\chi = \frac{V_{m,1}\phi_0^2}{RT}\left(\delta_1^2 + \delta_0^2 - 2W\right), \tag{1.8}$$

in which the parameter W can allow for negative values of the interaction parameter.

In step 3, choosing the holes to be uniformly distributed in the solvent ensures that the molecules spread out to achieve a uniform average concentration. This is part of the entropy of mixing ΔS_{mix}, which accounts for the spreading out of the drug molecules when they go from the condensed phase (liquid or solid) into solution, and is given by (Atkins 1998)

$$\Delta S_{mix} = -R[n_1 \ln x_1 + n_0 \ln x_0], \tag{1.9}$$

where $x_1 = n_1/(n_1 + n_0)$ denotes the mole fraction of the drug in solution. As $x_1 + x_0 = 1$, each log term is less than one and $\Delta S_{mix} > 0$. It is assumed that this is the only entropy effect due to mixing. (Although not considered for the purposes of this discussion, there can be other sources of entropy change on mixing. For instance, water molecules orient around drugs that are polar or ionic, which would result in a loss of entropy that would be taken into account along with an adjusted the entropy of mixing.)

The change in Gibbs energy that results from dissolving a drug in a solvent is

$$\Delta G_S = G(\text{solution}) - G(\text{unmixed components}), \tag{1.10}$$

Adding the contributions from steps 1–4 gives

$$\Delta G_S = n_1 \Delta h_{M1}\left(1 - \frac{T}{T_{M1}}\right) + n_1 \int_T^{T_{M1}} \Delta C_{P1} dT \tag{1.11}$$

$$-n_1 T \int_T^{T_{M1}} \frac{\Delta C_{P1}}{T} dT + n_1 x_0 \chi RT + RT[n_1 \ \ln \ x_1 + n_0 \ \ln \ x_0].$$

For drugs dissolving in liquid solvents, the heat capacity terms in Eq. (1.11) are relatively small compared to the heat of melting terms and can be neglected. In some models, they are not neglected but the $\Delta C_{P,1}$ term is taken as constant. While both of these approximations can introduce errors, the heat capacity terms will be neglected in the discussion that follows in order to simplify the equations and highlight the principles, so Eq. (1.11) will be replaced by the simpler form:

$$\Delta G_S = n_1 \Delta h_{M1} \left(1 - \frac{T}{T_{M1}}\right) + n_1 x_0 \chi RT + RT[n_1 \ln x_1 + n_0 \ln x_0]. \quad (1.12)$$

It should be noted that the effects of the heat capacities can be especially important when modeling solid solution ASDs (Bellantone et al. 2012).

1.3.4 *Solubility and Chemical Potential*

From thermodynamics, it is well established that materials will convert from a higher to lower chemical potential form. In this context, the chemical potentials of the drug in its dissolved and undissolved states are of primary interest. The chemical potential of a drug is the change in the Gibbs energy per changes in drug amount, holding all other factors (remaining chemical composition, temperature, and pressure) constant. This is written as (Atkins 1998)

$$\mu_1 = \left(\frac{\partial G}{\partial n_1}\right)_{T,P,n_0}. \quad (1.13)$$

The change in chemical potential for the drug due to dissolution is given by

$$\Delta\mu_1 = \begin{bmatrix}\text{chemical potential of}\\ \text{dissolved drug}\end{bmatrix} - \begin{bmatrix}\text{chemical potential}\\ \text{of undissolved drug}\end{bmatrix}. \quad (1.14)$$

From Eq. (1.13), this is given by (Bellantone et al. 2012)

$$\Delta\mu_1 = \left(\frac{\partial \Delta G_S}{\partial n_1}\right)_{T,P,n_0}. \quad (1.15)$$

When chemical potential of the undissolved drug is higher than that of the dissolved drug, $\Delta\mu_1 < 0$, the solution is below saturation, and more drug can dissolve if present. On the other hand, if the chemical potential of the dissolved form is higher than the undissolved form, $\Delta\mu_1 > 0$ and the solution is supersaturated, and precipitation is favored. The solubility of the drug in a solvent can be taken as the dissolved concentration at which $\Delta\mu_1 = 0$.

Equation (1.15) is important because it shows that $\Delta\mu_1$ represents the slope of ΔG_S versus the moles of drug "dissolved" (both per constant amount of solvent), which is shown in Fig. 1.5. The solubility criterion of $\Delta\mu_1 = 0$ corresponds to

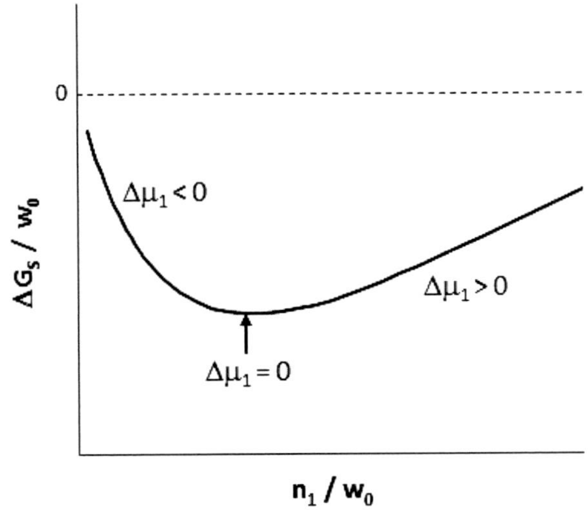

Fig. 1.5 Change in Gibbs energy versus moles of dissolved drug (both per gram of solvent)

the minimum of the plot, where the slope is zero, and the solubility is taken as n_1/w_0 (which can then be converted to other units). To the left, corresponding to lower amounts of dissolved drug, the dissolved concentration is below the solubility ($\Delta\mu_1 < 0$), and to the right it is supersaturated ($\Delta\mu_1 > 0$).

Noting that the mole fractions of the dissolved drug and solvent are given by $x_1 = n_1/(n_1 + n_0)$ and $x_0 = n_0/(n_1 + n_0)$, respectively, applying the derivative in Eq. (1.15) to Eq. (1.12) leads to

$$\Delta\mu_1 = \Delta h_{M1}\left(1 - \frac{T}{T_{M1}}\right) + x_0^2\chi RT + RT\ \ln\ x_1. \tag{1.16}$$

Since $0 < x_1 < 1$, the left-hand side of Eq. (1.16) is negative when the "dissolved" concentration x_1 is low, but becomes less negative as the dissolved drug concentration increases. Figure 1.6 shows a plot of Eq. (1.16), in which $\Delta\mu_1$ is negative for dissolved concentrations below the solubility, equals zero when the "dissolved" concentration equals the solubility, and is positive for supersaturated solutions. Equation (1.16) can be simplified for dilute solutions by noting that x_0^2 is very close to 1. Doing this and setting $\Delta\mu_1 = 0$, the drug solubility x_{1S} is obtained as

$$\ln\ x_{1S} = -\frac{\Delta h_{M1}}{RT} - \chi + \frac{\Delta h_{M1}}{RT_{M1}}. \tag{1.17}$$

Equation (1.17) can be equivalently written in terms of an activity coefficient γ_1 instead of the interaction parameter χ as

$$\ln\ x_{1S} + \chi = \ln\ \gamma_1 x_{1S} = \frac{\Delta h_{M1}}{R}\left(\frac{1}{T_{M1}} - \frac{1}{T}\right), \quad \text{where } \ln\ \gamma_1 = \chi. \tag{1.18}$$

Figure 1.6 shows the effect of the chemical potential change on solubility. It can be seen that the solubility increases exponentially with the increase in chemical

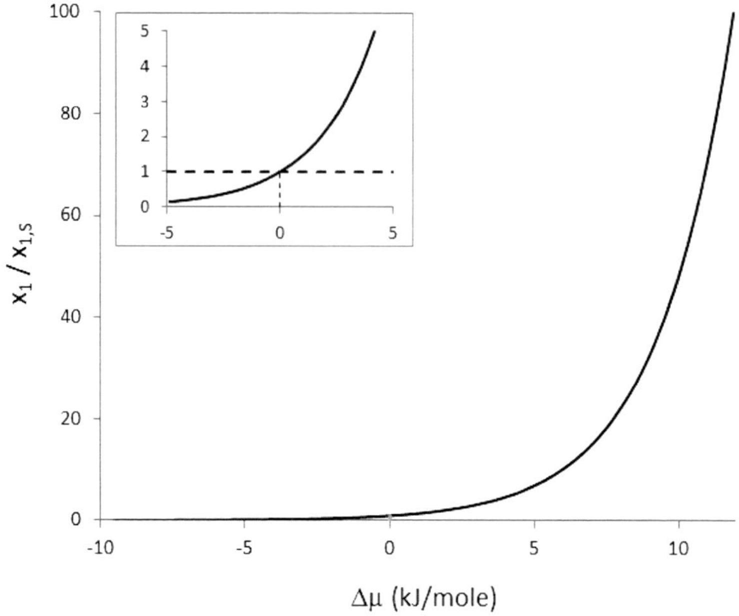

Fig. 1.6 Concentration versus chemical potential change

potential, and for $\Delta\mu_1 > 0$, even small increases in the chemical potential result in very large increases in the solubility.

The dotted line in the insert represents the solubility of the stable crystalline drug form.

1.3.5 *Amorphous Versus Crystalline Form Solubility*

For a given drug and solvent at (T, P), the chemical potential of the dissolved drug is determined by the concentrations of the components. On the other hand, the chemical potential of the undissolved drug is a function of a number of parameters that can be altered, including the molecular arrangement.

The solubility of the drug in terms of mole fraction is given by Eq. (1.17). Any factor that makes the right-hand side of Eq. (1.17) more positive (or less negative) will work to increase the solubility of the drug. For a given drug and solvent, physical factors that influence the right-hand side include the heat of melting Δh_{M1} and the melting temperature T_{M1} of the "undissolved" solid form, both of which are different for the amorphous versus crystalline forms. The effect can be seen as follows. The solubility (mole fraction), heat of melting, and melting temperature for the amorphous form will be denoted as x_S^A, Δh_{M1}^A, and T_{M1}^A, respectively, and the analogous parameters for the most stable crystalline form will be denoted using the superscript

"C." The solubility with respect to each form is given by Eq. (1.17), using the appropriate mole fraction, heat of melting, and melting temperature. Thus, subtracting $\ln x_{1S}^{(A)} - \ln x_{1S}^{(C)}$ gives

$$\ln \frac{x_{1S}^{A}}{x_{1S}^{C}} = -\frac{\left(\Delta h_{M1}^{A} - \Delta h_{M1}^{C}\right)}{RT} + \left(\frac{\Delta h_{M1}^{A}}{RT_{M1}^{A}} - \frac{\Delta h_{M1}^{C}}{RT_{M1}^{C}}\right) = -\frac{\delta h_{M1}}{RT} + \frac{\delta s_{M1}}{R}, \quad (1.19)$$

where $\delta h_{M1} = \left(\Delta h_{M1}^{A} - \Delta h_{M1}^{C}\right)$ and $\delta s_{M1} = \left(\Delta h_{M1}^{A}/T_{M1}^{A} - \Delta h_{M1}^{C}/T_{M1}^{C}\right)$. It is noteworthy that χ does not appear in Eq. (1.19), which reflects the fact that the "dissolved" drug molecules interact with the solvent in the same way regardless of the "undissolved" solid form. From Eq. (1.19), any effect that makes the right-hand side more positive will result in an increase in the relative solubility $x_{1S}^{(A)}$ vs. $x_{1S}^{(C)}$. Since the left-hand side of Eq. (1.19) is positive, a constraint is imposed on the right-hand side in that $\delta s_{M1} > 0$ is required for all amorphous systems, which says that the amorphous form will have higher entropy than the crystalline form.

Solubility information can be applied to the solid-state forms as well, since the relative solubilities of two forms will also give the difference in chemical potential between them in the solid state. Denoting the difference in the chemical potentials between forms as $\delta\mu = \mu_{1S}^{A} - \mu_{1S}^{C}$, Eq. (1.19) leads to

$$\delta\mu = RT \ln \frac{x_{1S}^{A}}{x_{1S}^{C}}. \quad (1.20)$$

Thus, a solid form that displays a higher solubility than the crystalline form must also display a higher chemical potential. Since material tends to move or convert from higher chemical potential to lower chemical potential forms, the form with the lower chemical potential is more stable, verifying that the more stable form will also have a lower solubility.

It is possible to show that the amorphous form is less stable than any crystalline form over all temperatures by considering the following. From Eqs. (1.19) and (1.20), the relative stability of the amorphous or any polymorph compared to the more stable form, as given by $\delta\mu$, depends on the temperature. Polymorph pairs can be classified as either monotropic or enantiotropic. A monotropic pair is one in which one form is more stable at all temperatures up to the melting temperature of the less stable form. An enantiotropic pair is one in which there is a crossover or transition temperature T_X such that for temperatures lower than T_X one form is more stable while above T_X the other form is more stable. An implied requirement for enantiotropic pairs is that neither form melts below the crossover temperature, otherwise, the pair would be monotropic. (Once melted, there is no solid structure, so there can be no amorphous or polymorphic properties.)

Several rules of thumb have been identified (Burger and Ramberger 1979) to help predict whether a pair is monotropic or enantiotropic. One can be illustrated in terms of the melting temperatures and melting enthalpies as follows. Consider two forms, A and C, where form C has the higher melting temperature. The transition temperature

T_X is defined as the temperature at which the chemical potential of the two forms is equal, so ln $\left(x_{1S}^A/x_{1S}^C\right) = 0$ and Eq. (1.19) leads to

$$T_X = \frac{\left(\Delta h_{M1}^A - \Delta h_{M1}^C\right)}{\left(\frac{\Delta h_m^A}{T_{M1}^A} - \frac{\Delta h_m^C}{T_{M1}^C}\right)} = \frac{\left(\Delta h_{M1}^A - \Delta h_{M1}^C\right) T_{M1}^A T_{M1}^C}{\left(\Delta h_{M1}^A T_{M1}^C - \Delta h_{M1}^C T_{M1}^A\right)}. \tag{1.21}$$

Noting that T_{M1}^C denotes the higher melting point and defining $\delta T_{M1} = T_{M1}^A - T_{M1}^C < 0$, Eq. (1.21) can be rewritten as

$$\frac{T_X}{T_{M1}^A} = \frac{\delta h_{M1} T_{M1}^C}{\delta h_{M1} T_{M1}^C - \Delta h_{M1}^C \delta T_{M1}} = \left(1 - \frac{\Delta h_{M1}^C}{\delta h_{M1}} \frac{\delta T_{M1}}{T_{M1}^C}\right)^{-1}. \tag{1.22}$$

Since δT_{M1} is negative by convention while the heat of melting and melting temperature Δh_{M1}^C and T_{M1}^C are positive, $T_X/T_{M1}^A < 1$ when δh_{M1} is positive, and $T_X/T_{M1}^A > 1$ when δh_{M1} is negative. In other words, if $\Delta h_{M1}^C < \Delta h_{M1}^A$ (so $\delta h_{M1} > 0$) it is predicted that the transition temperature is below the melting point of the less stable form, or $T_X < T_{M1}^A$, which is characteristic of an enantiotropic pair. On the other hand, if $\Delta h_{M1}^C > \Delta h_{M1}^A$ (so $\delta h_{M1} < 0$) it is predicted that $T_X > T_{M1}^B$, so form C is more stable at all temperatures below the melting point of form A and the pair is monotropic. This leads to what Burger and Ramberger (1979) term the heat of fusion (HFR) rule: When the form with the higher melting temperature has the lower heat of melting, the forms are usually enantiotropic, otherwise they are monotropic. Thus, since amorphous solids display lower melting temperatures and heats of melting than the crystalline forms, amorphous solids are less stable at all temperatures than the crystalline forms.

1.4 The Microscopic View

In addition to thermodynamic and macroscopic models, the behavior of amorphous systems can be viewed in terms of microscopic and molecular arguments. While these are equivalent in theory, each view provides different insights and advantages for explaining certain behaviors. In particular, the microscopic viewpoint allows a fundamental interpretation in terms of molecular interactions and packing, while thermodynamics allows material-independent equations to be developed based on macroscopic energy content arguments.

From the standpoint of dissolution, the microscopic viewpoint allows a simple but useful picture to be constructed that gives a clear and intuitive way to think of the properties of amorphous systems with regard to dissolution. When a solid drug dissolves, molecules leave the solid matrix and disperse in the solvent, which can be thought of as resulting from several processes. (1) Solvent molecules interact and associate with drug molecules on the solid surface. (2) Energy is released in the form of heat due to the solvation. (3) The heat energy that is released helps the molecule leave the solid pack and migrate into the solvent, where solvent molecules surround it.

From this point of view, a number of factors can influence the dissolution rate. (1) Stronger attractive forces between the solvent and drug molecules result in releasing more heat of solvation, which will increase the rate at which undissolved molecules can pass into the solvent. (2) For a given drug and solvent, physical forms of undissolved solids with reduced drug–drug attractions will make it easier to remove drug molecules from the undissolved form. This is the dominant effect in explaining the increased dissolution and solubility of amorphous forms compared with crystalline forms. Presumably, molecules in the amorphous form show weaker attractive forces and more easily give up molecules from the solid surface to the solvent medium. (3) The addition of heat will increase the dissolution or solubility by making it easier for the solvent to remove drug molecules from the undissolved environment. These effects have been described thermodynamically in the previous section, but can also be described in molecular terms based on separations and intermolecular attractions and the total kinetic energy of the molecules.

Another property of amorphous forms is that they are not physically stable, and tend to transform into crystalline forms over time. This is almost always accompanied by an increase in density, indicating that the molecules in the crystalline form are arranged more densely on average than in the amorphous form.

All of these can be explained using molecular interaction models, and these are in fact the basis of computational research done today. The basic ideas and how they explain observed behaviors are discussed below.

1.4.1 Interaction Versus Total Energy

The properties of amorphous materials can be explained to a great extent by considering intermolecular interactions such as van der Waals and dipolar interactions, and kinetic energy associated with molecular motion. In the discussions that follow, the standard convention for energies and forces are used. Negative energies denote situations in which energy must be added to separate molecules. Negative forces denote attraction, so molecules move closer to reduce their separation, while positive forces values are repulsive.

Molecular interactions have been described for both neutral and ionic cases. The former case includes hydrogen bonding and weaker van der Waals interactions (dipole/dipole, dipole/induced-dipole, and induced-dipole/induced dipole). In all of these cases, the interactions are attractive unless the molecules are forced so close that electron cloud overlap creates a repulsion. For poorly soluble drugs, the neutral interactions are typically considered most important, but it has recently been shown that coulombic (ionic) interactions also contribute to the total intermolecular interactions (Gavezzotti 2007). However, to illustrate the molecular concepts, this discussion will focus on neutral interactions.

Neutral interactions can be qualitatively described as a function of distance between molecules. While this concept is straightforward for atoms and simple

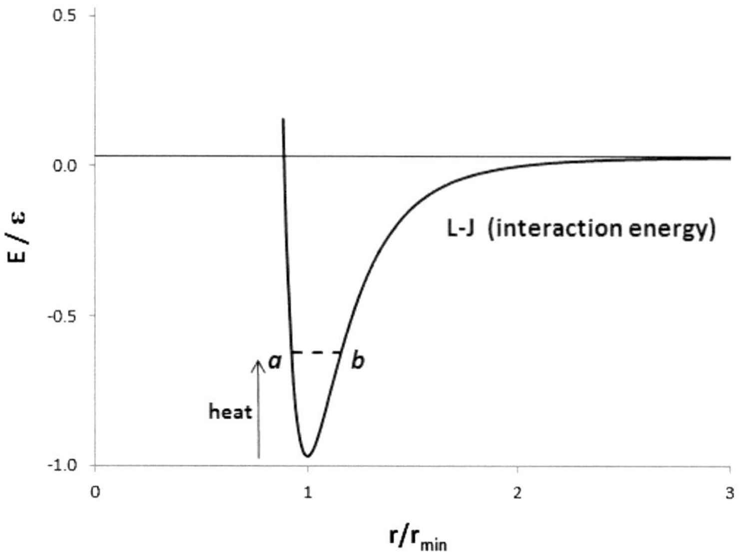

Fig. 1.7 Typical dependence of pairwise energy versus separation

molecules, it is less so for organic molecules owing to the complexity of the molecule, which requires that orientation be taken into account. In this case, the distance between molecules can be considered an average quantity naturally related to the number of molecules per unit volume, which is the density of a substance. The distance between molecules can be thought of as roughly a center-to-center distance for simple molecules, but it is understood that this is a crude approximation for more complicated organic molecules.

It is well established that molecules repel each other when they are brought close together and attract when separated by larger distances. A number of models have been proposed to describe these interactions. For the purposes of this discussion, the exact equation is not of primary concern, but rather the general features of the interaction plot. However, it is often the case for pairs of simple molecules separated by a distance r that the repulsive interaction energy can be described as a/r^m, and the attractive energies can be described by $-b/r^n$. The total interaction energy E_{int} is the sum of the two, resulting in positive interaction energy at small separations, a separation at which the interaction energy is most negative, and zero interaction when molecules are separated by large distance. This behavior is shown in Fig. 1.7 for the Lennard-Jones (L-J) model (Atkins 1998; Hill 1986), which is one of the most studied interaction models. The L-J model gives the interaction energy as

$$E_{\text{int}} = \varepsilon\left[\left(\frac{r_{\text{min}}}{r}\right)^{12} - 2\left(\frac{r_{\text{min}}}{r}\right)^{6}\right]. \tag{1.23}$$

The greatest attractive energy occurs at the minimum of the plot, where $E_{\text{int}} = -\varepsilon$ and the separation is $r = r_{\text{min}}$. The separation at which $E_{\text{int}} = 0$, which occurs at the intersection of the L-J plot and the horizontal zero line, is denoted by r_0, so the interaction energy for the molecule pair is negative for $r > r_0$. The interaction force is $F_{\text{int}} = -\partial E_{\text{int}}/\partial r$, which is the negative of the slope of the E_{int} versus r plot. Since a negative slope corresponds to a repulsive force, the force between the molecules is repulsive when the separation is less than r_{min} and attractive when it is greater. At $r = r_{\text{min}}$, the slope is zero and there is no net force due to interactions between the pair of molecules.

The L-J curve shown in Fig. 1.7 denotes only the potential energy between two molecules because the interaction energy does not include kinetic energy, so it represents the total energy of the pair only when there is no heat content, which occurs at absolute zero ($T = 0$ K). In that case, the molecules can be thought of as being at constant positions and the separation distances between molecules do not change.

When heat is added, the total energy of the pair will increase by a quantity equal to the heat added, so the molecules acquire kinetic energy and move relative to each other. Figure 1.7 shows an example in which heat is added to a molecular pair initially separated by r_{min}. The arrow shows the increase in energy to the level denoted by line segment *a–b,* and the corresponding energy will be denoted by E_{ab}. Since $E_{ab} < 0$, the separation between molecules will remain finite and the pair will oscillate at that energy along the line segment between points *a* and *b,* which correspond to the minimum and maximum separations r_a and r_b. As the molecules approach each other, they will slow as they approach and stop when they reach separation r_a, then begin to move apart because of a repulsive force at that distance. Similarly, as they are moving apart, they will slow as they approach and stop at r_b, then begin to move closer because of the attractive force at that separation.

As depicted in Fig. 1.7, the potential energy graph for real systems is not typically symmetric about r_{min}, so at energy E_{ab} the magnitude of displacement from the minimum is smaller for $r_{\text{min}} - r_a$ than $r_b - r_{\text{min}}$. An important consequence of this asymmetry is that the average separation increases when heat is added, which results in thermal expansion. (It should be noted that this is the average position over time, which is greater than the simple average of r_a and r_b.)

1.4.2 Extension to Macroscopic Systems

Organic molecules do not follow the L-J model exactly, but typically behave in a qualitatively similar manner. For instance, the L-J model describes interactions between pairs of molecules, so each molecule experiences the effects from other molecule as a function of the distance *r*. In bulk solids, individual molecules are surrounded by multiple neighboring molecules at specific locations with thermal motion (for $T > 0$). Thus, a distance of a molecule "*j*" to adjacent molecules "*k*" may vary depending on location, which are time dependent and denoted as $r_{jk}(t)$. However, because the interactions decrease as $1/r^6$, their magnitude drops sharply with even small increases in distance, so the sum is typically required only over the nearest neighbor molecules (i.e., the first "layer" or so of neighbors surrounding the molecule).

Despite qualitative similarities, organic molecules deviate from the L-J model as a consequence of molecular arrangements. A large number of adjacent molecules mean that the total interaction energy will be complex and of larger magnitude. In addition, the L-J model is one-dimensional (1-d), while molecular packing of real materials occur in three dimensions (even though on surfaces may not be surrounded from all directions). Since energies and forces are additive, the total interaction energy and force felt by molecule "*j*" are additive and summed over all nearest neighbors "*k*," as

$$E_{\text{int}}^{Tot}(j) = \sum_{k} E_{\text{int}}(r_{jk}) \quad \text{and} \quad F_{\text{int}}^{Tot}(i) = \sum_{j} F_{\text{int}}(r_{ij}). \tag{1.24}$$

At temperature above absolute zero, molecules have kinetic energy. Thus, their positions and corresponding distance $r_{jk}(t)$ vary with time, so the sums in Eq. (1.24) are extremely complicated and must be done using computational models such as molecular dynamics simulations (Gavezzotti 2007). In addition, when the molecules are moving relative to one another, the average vibrational amplitude becomes important. For instance, at low temperatures, the amplitudes are small, and the molecules are vibrating about average locations that are approximately fixed. However, when the average amplitude becomes similar to the average molecular separation, molecules can begin changing places so their "average" position changes with time. When this occurs for a sufficient fraction of the molecules, the solid transforms into a melt. It is significant that this is a consequence of three-dimensional packing and motion. (In the L-J model, which is one-dimensional in *r*, switching locations would require molecules going through zero separation, which would require infinite energy in theory. Thus, there is no corresponding concept in 1-d.)

With these considerations in mind, a hypothetical behavior of a neat API sample is shown in Fig. 1.8, which plots the potential energy versus the average distance as a dotted-solid line combination (through *e-a-b-d-f*). This line will not be exactly proportional to any L-J curve because the relative positions of the molecules in 3-d will change as a result of thermal expansion. The curved line containing points *b* and *c* represents the time-average molecular distance $\langle r \rangle$, which increases with increasing heat content as a reflection of thermal expansion. Along the vertical axis, level *A* corresponds to the energy content at absolute zero and level *B* corresponds to the energy after adding some quantity of heat.

The line segment *a–d* represents the average minimum-to-maximum separations between molecules, which approximately corresponds to the average vibrational amplitude in 3-d for molecules in a sample with that heat content. Of particular interest is the length of *a–d* compared to the average separation $\langle r \rangle = c$. In this example, the energy level *B* was chosen so the length of *a–d* is similar to the value of *c,* so the average vibrational amplitude is comparable to the average separation. Thus, level *B* corresponds to where the solid transitions to the melt phase, and the difference between levels *B* and *A* is the heat of melting.

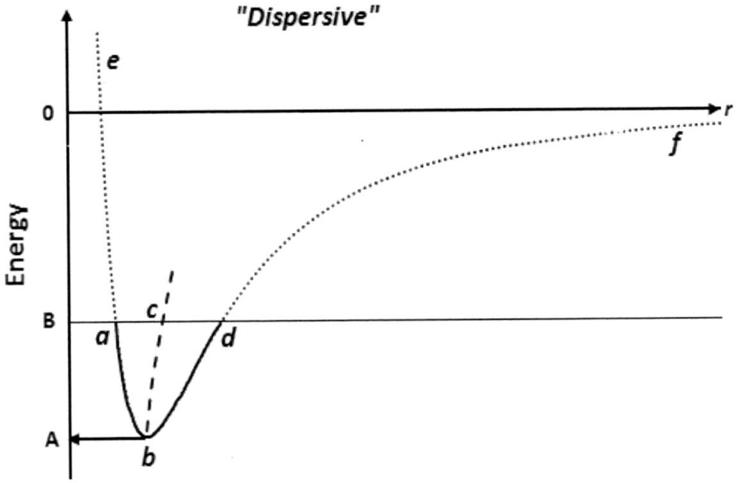

Fig. 1.8 Typical dependence of energy versus the average separation. The *dotted-solid line* (*e-a-b-d-f*) represents the average total attractive interaction energy experienced by a molecule potential versus the average separation between molecules. The *dashed line* going through the points *b* and *c* represents the average molecular separation at each energy level. The *line segment a–d* represents the average vibrational amplitude of molecules with total energy *B*, and point *c* represents the average separation. The energy difference between levels *B* and *A* represents the heat added to melt the material

Several comments are warranted. First, the heat of melting in this example was taken relative to a reference state of absolute zero. The heat of melting for samples at higher temperatures with energy levels higher than *A* is represented by the difference between level *B* and the new starting level. Second, although the example shown here assigned the melting energy at the level for which the length of *a–d* equals the average separation *c,* this is a crude approximation for illustration only. Although the molecular mobility increases dramatically when the vibrational amplitude becomes similar to the average separation, the exact ratio is difficult to determine and will require the use of molecular dynamics or similar approaches.

Third, unless the melt is supercooled, for a given temperature and heat content, the energy level *B* becomes a reference point. This is because the melt has no long-range structure and should contain the same energy content regardless of the starting solid form. Thus, different solid forms will show different energy shapes along points *e–a–b–d–f* and different lengths *a–d*, with more stable forms showing level *A* more negative than for less stable forms, but with the same level *B* energy. In fact, the amorphous form, which would be least stable, would have the highest (least negative) energy value at absolute zero.

The difference in energy between level *B* and the zero line corresponds to the energy that must be added to reach the dispersive region, in which the total energy for the molecules is greater than zero. This corresponds to the case in which the kinetic energy is sufficient to overcome the negative interaction energy due to molecular attractions, so the molecules are free to "spread out." Since the energy content of the melted form at a given temperature and pressure does not depend on the starting form, the heat energy required to go from level *B* into the dispersive region is constant for a given (*T, P*).

Finally, as more heat is added to solids with energy below level *B*, more modes of vibration become available and the average amplitude of the vibrations increases, so the kinetic energy increases from two sources. Thus, the change in kinetic energy with change in temperature is not constant. This is reflected by the fact that heat capacities tend to increase with increased temperature.

1.5 Glasses and Amorphous Forms

An amorphous solid form is often considered to be equivalent to glass form. A glass can be thought of as a supercooled liquid lacking long-range molecular packing order but with rheological properties of a solid. The viscosity η of liquid increases as the temperature is decreased. Depending on the cooling rate and compound properties, some melts crystallize when cooled below melting temperature, while others may remain as supercooled liquids. For those that crystallize, the viscosity increases drastically due to the phase transition. For those that survive as supercooled liquids, the viscosity increase is in continuum with the decreasing temperature over some temperature range till it reaches the glass transition temperature, at which the viscosity increases typically by 6 orders of magnitude or more over temperature range of typically 5–10°C. This is shown in Fig. 1.9 for supercooled liquids, which shows log η versus the temperature.

The viscosity is strongly related to timeframes for molecular motions or relaxations, which are characterized by an average relaxation time τ that is temperature dependent (Donth 2001; Yu 2001). In particular, systems that respond quickly to deviations from equilibrium, such as an applied stress, are characterized by short relaxation times and low viscosities, while systems that respond slowly are characterized by long relaxation times and high viscosities. The relationships are shown by the dual vertical axes in Fig. 1.9, in which larger viscosities correspond to larger values of τ, indicating that decreasing the temperature increases relaxation times. Since glasses are amorphous and supercooled liquids, they are unstable and the relaxation times are of interest in terms of glass formation and kinetics of conversion to crystalline forms.

As seen in Fig. 1.9, the relaxation time or viscosity change drastically at glass transition temperature. As discussed below, the values of T_M, T_g, and the width of T_g range are important parameters in characterizing glasses.

For supercooled liquids, two competing effects are of particular interest. As the temperature is decreased below freezing point, the supercooled liquid becomes thermodynamically less stable with respect to the crystalline form, and the relaxation times become longer as a result of slow diffusion. A major factor that determines

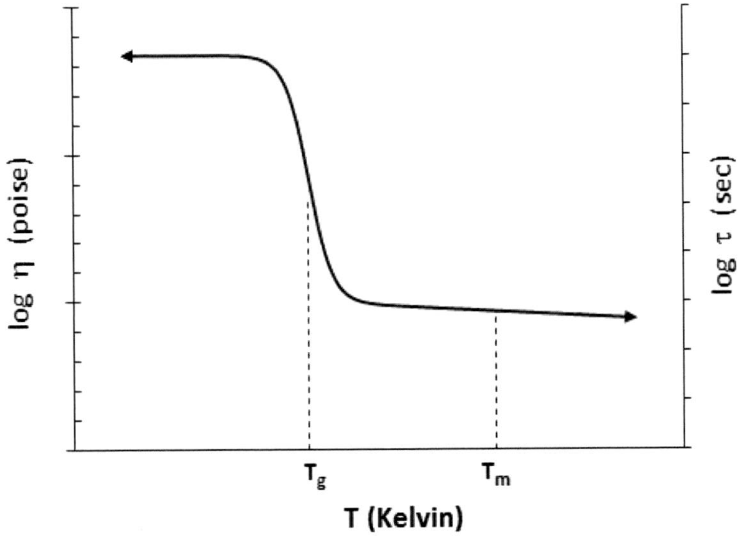

Fig. 1.9 Log η and log τ versus temperature and the glass transition

these effects is the temperature dependency of the relaxation times for the supercooled liquid, and specifically the magnitude of the change in τ with the change in temperature. If the change in relaxation time per change in the temperature is small, the thermodynamic instability of the supercooled melt will result in crystallization over the timeframe of observation. In that case, further cooling is likely to incur crystallization in the observed timeframe, and the supercooled form is less likely to survive. On the other hand, if cooling results in large increases in the relaxation time, crystallization is likely to be slow within the observation timeframe (from minutes to years), and further cooling is unlikely to incur crystallization during that timeframe. These competing effects are further complicated when the cooling rate is considered, since rapid cooling will result in less time for crystallization to occur before the relaxation times are increased. This is discussed further below.

The relationship between the relaxation time and the temperature between glass transition and melting is described by several empirical equations. The one of the most commonly employed Vogel–Fulcher–Tamman (VFT) equation is given by (Mauro et al. 2009; Yu 2001)

$$\tau = \tau_0 \exp\left[\frac{DT_0}{T - T_0}\right] \log \tau = \log \tau_0 + \frac{DT_0}{2.303(T - T_0)}, \tag{1.25}$$

where D is referred to as the strength parameter, T_0 is referred to as the Vogel temperature, and τ_0 is a reference relaxation time corresponding theoretical temperatures approaching infinity. In practice, all three parameters are obtained from fits of the data between T_g and T_M using viscosity or dielectric relaxation time, etc. For many materials, the relaxation times at the glass transition temperature are approximately

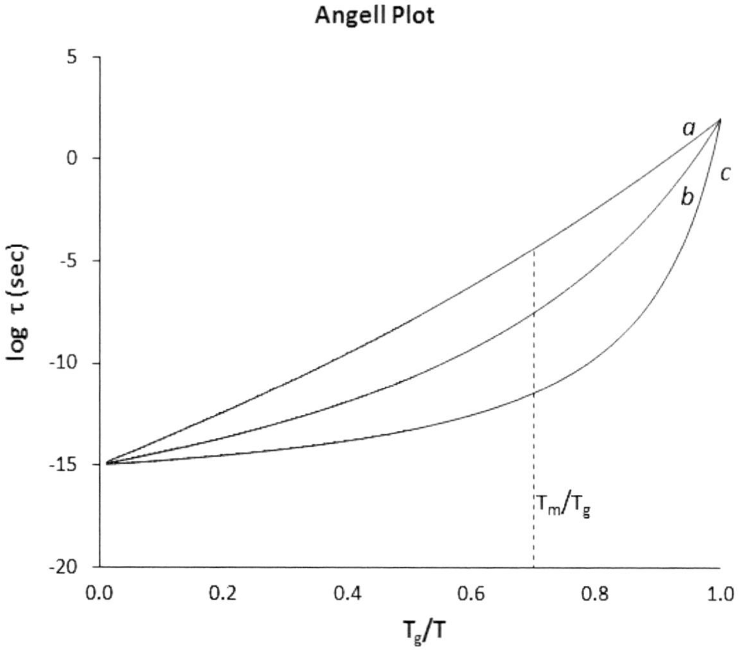

Fig. 1.10 Example of an Angell plot for typical glasses. *Line a* represents a strong glass former ($m = 25$), *b* represents a medium glass former ($m = 50$), and *c* represents a weak glass former ($m = 150$)

$\tau(T_g) \approx 100\ s$, which sets the reference relaxation time $\log \tau_0 \approx 15$. For many materials, D has been related to the Vogel and glass transition temperatures by the empirical equation

$$T_g/T_0 \approx 1 + D/39.1. \tag{1.26}$$

Using this equation, it is possible to correlate the relaxation time and the temperature between T_g and T_M by constructing an Angell plot (Angell 1995; Donth 2001), which plots log τ versus T_g/T. Angell plots give information about the temperature sensitivity of the relaxation time near or above the glass transition temperature. Figure 1.10 shows hypothetical data for three glass formers, each with glass transitions and melting temperatures of 45 and 175 °C. Thus, T_g/T (in Kelvin) ranges from approximately 0.7 to 1.0 in the plot. As the temperature is decreased (moving to the right in the plot), the material represented by line *a* increases its relaxation times much more slowly than the material represented by line *c*. Line *b* represents an intermediate case. Of special interest is the slope of the plot as T_g/T approaches 1, which corresponds to the temperature range in which vitrification occurs. The slope of line at $T_g/T = 1$ is called the fragility *m*, which is formally defined as (Donth 2001):

$$m = \left. \frac{d \log \tau}{d(T_g/T)} \right|_{T=T_g}. \tag{1.27}$$

From Fig. 1.10, line *a* has lowest fragility ($m = 25$) while line *c* has the largest ($m = 150$). Glasses are termed as strong if $m < 40$ or so, and weak if $m > 75$ or so. The significance of the terms is apparent from the Angell plot. As noted above, supercooled melts are thermodynamically unstable, but the crystallization kinetics is affected by the relaxation time. For weak glass formers with large fragility values, the relaxation time remains short until the temperature is lowered close to the T_g so more molecular mobility is retained at lower temperatures, thus making crystallization more likely before the glass transition is reached. In other words, larger values of m are associated with melts that are kinetically more likely to crystallize and less likely to form glasses or amorphous solids.

The fragility can be related to the strength factor and glass transition for typical materials by applying the derivative in Eq. (1.27) to Eq. (1.25), which gives

$$m = \frac{D(T_g/T_0)}{2.303(T_g/T_0 - 1)^2}. \tag{1.28}$$

Applying Eq. (1.26) leads to

$$m \approx 684 \left(\frac{1}{D} + \frac{1}{39.1} \right), \tag{1.29}$$

which can be applied to many typical materials. From Eq. (1.29), it can be said that strong glass formers are associated with strength factors $D > 25$–30 and weak ones with $D < 10$–12.

Numerous publications refer to the association between strong glass formers and Arrhenius-like behavior, as opposed to weak glass formers and non-Arrhenius behaviors (Angell 1995; Yu 2001). While this distinction is not critical for understanding the temperature effects, it is of interest and warrants a short discussion. Arrhenius behavior is associated with systems in which a plot of $\log \tau$ versus $1/T$ (or equivalently, T_g/T) is linear. In the context of Angell plots, the temperature range is from T_g to 50–100°C above T_g. Equation (1.25) shows that this linearity approximately occurs when $1/(T - T_0) \sim 1/T$, or $T/T_0 > 2 - 3$. This is the case for strong glass formers, since for $D > 40$ the value of $T_g/T > 2$, and is higher for any temperature above the glass transition. On the other hand, for weak glass formers with $D < 10$, $T_g/T < 1.25$, and the plots will show curvature (non-Arrhenius behavior).

Consistent with the above discussions, the ability to form glasses depends on the experimental conditions. For instance, while it is possible to supercool a melt for most materials, many materials tend to crystallize near the melting temperature, and will form supercooled melts only if relatively high cooling rates are employed. This is because the conversion of supercooled melt to more stable crystalline forms occurs as a function of relaxation times, in which shorter relaxation times make crystallization

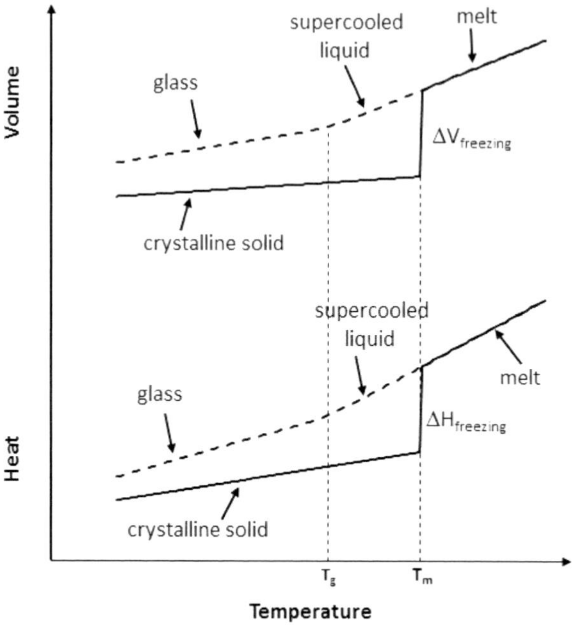

Fig. 1.11 Relationship between the temperature, volume, and heat for glasses and crystalline materials

more likely to occur. If a melt is cooled slowly, the relaxation times also increases slowly and the supercooled melt will have longer timeframes to crystallize. On the other hand, rapid cooling will increase the relaxation times faster than the material can crystallize. If cooled fast enough, the relaxation time of the supercooled melt will become very long before crystallization occurs, so vitrification occurs and a glass will form.

An interesting observation with most glasses is that the behavior on cooling is different from heating. This is a consequence of the fact that nucleation is favored at lower temperatures, while growth is more favored at higher temperatures, but both are very slow at low enough temperatures. Thus, on cooling, it is possible for the temperatures where nucleation is most favored to be below the temperatures where growth is favored, so nucleation without growth may occur. Conversely, on heating the nucleation phase is encountered before the growth phase, so nuclei are present with the growth is most favored. Thus, cooling tends to favor formation of glasses and amorphous forms for systems that behave in this manner.

1.5.1 Thermodynamic Implications of the Glass Transition

Although the glass transition is not a thermodynamic transition in a sense of phase transitions such as crystallization or melting, it provides important thermodynamic implications. Experimentally, there is an abrupt change in the slope of such properties as molar volume or heat content at the glass transition temperature, while the properties themselves are continuous. Examples of the changes in volume and heat are shown in Fig. 1.11. In the top half, the solid line represents a melting phase transition for a crystalline solid, which shows an abrupt change in the volume at the melting temperature T_M. The dashed plot above it represents the behavior of a melt that forms a supercooled liquid when the temperature is cooled below T_M. It shows a decrease in the magnitude of the slope of V vs. T at the glass transition temperature T_g. Directly below the volume plot is a plot of the heat content vs. T. Again, there is a drop in the heat content at the phase transition temperature, but at the glass transition temperature the slope changes while the heat content is continuous.

This behavior has important energetic consequences. When cooled to below melting temperature, there is an abrupt decrease in the volume and loss of heat if the material undergoes crystallization. If supercooled without crystallization, there is no abrupt decrease in the volume or the heat, which are thus higher than the crystalline solid. The significance of this observation is that the average distance between molecules is larger in a glassy state than in a crystalline state at the same temperature and pressure. A similar analysis can be made for the heat content. Since the supercooled liquid and glassy state do not experience abrupt decreases in the heat content on freezing, the heat content is higher in supercooled form than in the crystalline form at the same (T, P).

These observations can be interpreted in terms the energy versus separation plot of Fig. 1.8. At a given temperature below the melting point, the average separation between molecules and the heat content are both greater in the glassy state than in crystalline forms. As a result, the total energy is higher for the amorphous form than the crystalline form at a given temperature, and the energy needed to reach the dispersive region from the amorphous form is less. As discussed in the next section, this indicates that glassy forms have higher energy and are expected to dissolve more readily than crystalline forms.

1.6 Implications for Solubility and Dissolution

In drug delivery, the primary purpose for considering amorphous drug forms is to increase the solubility and/or dissolution rate of a drug in aqueous media. For amorphous solids, the lack of long-range order has the consequence that the molecular packing is less efficient than crystalline forms, resulting in a larger average molecular separation and weaker attractive forces between molecules. Thus, for a given temperature or heat content, the total energy per amount of material is higher for amorphous than crystalline forms, and the energy required to reach the melt state (level B in Fig. 1.8) and the dispersive region is less for amorphous than crystalline forms.

The lower energy requirements for amorphous solids to reach the dispersive region result in faster dissolution and higher solubility. At a constant temperature, the source of heat in dissolution is the solvation energy. A simple picture of solvation is that solvent molecules "attach" to undissolved drug molecules, and kinetic energy from the solvent molecules is transferred to a drug–solvent complex. The transfer of energy is a function of the attraction between the drug and solvent, and the number of solvent molecules that transfer energy to the undissolved drug. If the transferred kinetic energy is sufficient to overcome the drug–drug attractions, the drug molecule will leave the undissolved environment and move into the solvent, where it will be surrounded by more solvent molecules and go into solution. Once in solution, the drug molecules can disperse in the solvent, and they are in the dispersive region shown in Fig. 1.8.

With this picture in mind, any modification to the undissolved form that reduces the drug–drug attractive forces reduces the energy needed for "undissolved" molecules to become dissolved forms. At a given temperature, this can be achieved either by increasing the average distance between molecules, which will reduce the magnitude of the attractive forces, or by reducing the number of nearest neighbors. Both will make the total energy experienced by a molecule less negative and reduce the energy needed from the solvent to reach the dispersion region.

The first can be achieved by creating amorphous forms. By increasing the average distance between molecules, the attractive energy is less negative and the total energy increases. In addition, different polymorphic forms represent different molecular packing arrangements, and can reduce the number of nearest neighbor molecules. Since the total attractive energy is additive, creating arrangements with fewer nearest neighbors reduces the number of interactions and makes the total energy less negative.

Another way to reduce the number of nearest neighbors is by increasing the total surface area per volume, since molecules on surface lack the molecules above the surface. In addition, for extremely small particles (for instance, nuclei with radii of curvature less than 10 nm), the number of molecules per volume is reduced due to steric hindrance. This is one source of the Kelvin effect, in which the solubility of a substance increases with decreasing radius of curvature (Adamson and Gast 1997).

A third method to reduce the energy needed to reach the dispersive region is by inserting excipient molecules between the drug molecules, in which the drug–excipient interactions are weaker than the drug–drug interactions, or excipient–water interactions are stronger than drug–water interactions. The first case reduces the energy required to displace the drug and reach the dispersive region, and the second increases the energy obtained from the water–excipient interactions to dissolve the drug molecules. This is the basis for ASDs, which will be covered in subsequent chapters.

For a given drug and solvent, a kinetic view of dissolution can be formed. The "undissolved" solid with weak drug–drug attractions will allow the drug to leave from the solid more readily than one with stronger interaction. Kinetically, this corresponds to a lower energy barrier, which can approximately be thought of as the energy required to remove a drug molecule from its "undissolved" solid. Since dissolution occurs when drugs leave a solid surface, the specific energy that must be considered is the surface energy, which is related to the bulk energy by relationships such as the Scapski–Turnbull rule (Adamson and Gast 1997). Since the rate at which molecules leave the surface often follows an Arrhenius relationship, the dissolution rate can be related to the energy barrier ε_a as

$$\text{Dissolution rate} \propto \exp\left(-\frac{\varepsilon_a}{k_B T}\right). \tag{1.30}$$

Of course, the dissolution rate also depends on other factors such as the diffusion coefficient of the drug in the solvent, stirring, etc. However, if the energy barrier is large, then the rate-limiting step for dissolution is the removal of the drug from the surface of the undissolved solid. In that case, Eq. (1.30) can describe surface limited dissolution, as opposed to diffusion-limited dissolution. (This is a topic of some debate, as some pharmaceutical scientists claim that surface limited dissolution does not occur in practice.)

1.7 Implications for Physical Stability

Amorphous forms show a tendency toward crystallization as the means to reduce the total energy content. The rate of conversion must be slow enough to provide an intended shelf life for amorphous forms to be commercially viable. In a pure substance, crystallization may occur after short translational motions; whereas in ASD, longer diffusion motions are required. In this sense, pure amorphous forms convert to crystalline forms more rapidly than ASD's. The following illustrates the order of magnitude of the molecular motions on crystallization.

For a neat substance with a true density of ρ_0 (g/cm) and molecular weight M (g/mole), the average distance between drug molecules $\langle r \rangle$ can be estimated as

$$\langle r \rangle = \left(\frac{M}{\rho_0 N_A}\right)^{1/3}, \tag{1.31}$$

where N_A represents Avogadro's number (6.02×10^{23} molecules per mole). Also, Eq. (1.31) shows $\langle r \rangle \propto (1/\rho_0)^{1/3}$. As examples, the average distance between water molecules in the liquid state is $\langle r \rangle \sim 0.31nm$, while for ice ($\rho_0 = 0.917g/cm^3$) it is $\langle r \rangle \sim 0.32nm$, or about 3 % larger than for liquid water. For solid ibuprofen, ρ_0 is close to 1 g/cm^3 (D'Arcy and Persoons 2011) and the calculated average separation is $\langle r \rangle \sim 0.7nm$. Since the relative density differences between polymorphs and amorphous forms are small, the relative changes in average molecular separations are small as well. Thus, crystallization of amorphous forms involves short range motions or rearrangements of molecules that are typically shorter than 0.1 nm.

The main energy barrier to crystallization for pure amorphous materials is rotational, not diffusional. The kinetics of crystallization depend on the ability of the molecules to come into proximity plus to overcome energy barriers associated with going through intermediate arrangements to achieve the optimal structural alignment (the crystalline arrangement). For neat amorphous forms, the molecules are already in close proximity, so the second step is more important. In comparison, the average distance between drug molecules is larger in ASD's. Thus, the motions required for crystallization to occur in ASD's are significantly larger. In addition, in a solid solution ASD, the drug molecules are separated both by a large distance and have excipient molecules (polymers) between them. Thus, crystallization must occur after diffusion, which slows the overall process of losing the amorphous nature.

1.8 Kinetic Considerations

As discussed earlier in Sect. 1.5, the relaxation time of amorphous material is increased as the temperature is decreased. The Angell plot (Fig. 1.10) gives details of the temperature sensitivity for supercooled liquids between the melt and the glass transition temperatures, but in general it does not extend to lower temperatures for which $T_g/T > 1$. In fact, the extrapolations below the glass transition temperature often fail and the methods to determine the exact forms of equations for extrapolation are still being studied actively (Garca-Coln et al. 1989; Trachenko 2008; Mauro et al. 2009). Still, as would be expected, the relaxation time will continue to increase in some manner. As noted earlier, the relaxation time at the glass transition temperature for many materials is approximately $\tau(T_g) \approx 100s(\log \tau = 2)$, which is long enough for most organic molecules to undergo some crystallization within reasonably short timeframes.

It will be useful to estimate how far from the glass transition temperature an amorphous material can be stored to ensure that the material is kinetically stable over the duration of shelf life. Several rules of thumb have been proposed, of which probably the best known is the " $T_g - 50$ rule" for estimating the required storage temperature (Hancock et al. 1995). If this temperature is far below room temperature, storage of the amorphous drug at that temperature will not be practical, and alternative means of stabilizing the amorphous drug will be needed. (It should be noted that the " $T_g - 50$ rule" is a generalization based on concepts such as those detailed above. While many materials do not follow this rule, it is still a useful concept.)

The kinetics of crystallization has been simulated by many models, including the Avrami model (House 2007). The rate of conversion from amorphous to crystalline states can be measured by using thermal analysis (differential scanning calorimetry, DSC) and/or X-ray diffraction. The rate of conversion from amorphous to crystalline form depends on a number of factors. The process occurs in two steps, nucleation and growth (Mullin 2001), which are affected by various factors and occur at different rates. Specifically, for crystallization to occur, a seed or nucleus must form, on which subsequent growth will occur. Thus, the rate of nucleation is of primary interest. By analogy with Arrhenius-type processes, the nucleation rate can be written as

$$\frac{dN_{\text{nuc}}}{dt} = A \exp\left[-\frac{\Delta g_{\text{cr}}}{k_B T}\right], \tag{1.32}$$

where $\mathrm{d}N_{\text{nuc}}/\mathrm{d}t$ denotes the number of nuclei formed per volume per time, and Δg_{cr} reflects an activation energy for nucleation. This activation energy is related to the interfacial energy between a crystallizing nucleus and surrounding amorphous medium, and thus depends on the particle size via its surface area. For the idealized case of a spherical nucleus, the change in Gibbs energy with formation of a particle depends on the environment. For instance, when nucleation occurs from a supersaturated solution of a drug in a liquid solvent, Δg_{cr} and the nucleation rate are given by

$$\Delta g_{cr} = \frac{16\pi v^2\gamma^3}{3(k_B T \ln S)^2} \text{ and } \frac{dN_{nuc}}{dt} = A\exp\left[-\frac{16\pi v^2\gamma^3}{3k_B^3 T^3(\ln S)^2}\right], \tag{1.33}$$

where v is the molecular volume, γ is the interfacial tension of the nuclei in the amorphous environment, and S is the degree of supersaturation. On the other hand, if nucleation occurs from the melt of a pure substance, these are given by (Mullin 2001; Adamson and Gast 1997):

$$\Delta g_{cr} = \frac{16\pi v^2\gamma^3 T_M^2}{3\Delta h_M^2(T_M - T)^2} \text{ and } \frac{dN_{nuc}}{dt} = A\exp\left[-\frac{16\pi v^2\gamma^3 T_M^2}{3k_B T\Delta h_M^2(T_M - T)^2}\right], \tag{1.34}$$

where, γ represents the interfacial tension between the crystalline form and the melt. In an amorphous solid, it can be expected that nucleation will follow a pattern analogous to Eq. (1.34), in which γ would represent the interfacial tension between the crystalline and amorphous forms of the drug. This reflects the notion that a pure API is not microscopically homogeneous when nucleation and crystallization are occurring, so on a microscopic scale, there are regions with different molecular packing, density, and energy patterns. These regions create interfaces and interfacial tensions, based on the different molecular packing of the regions.

Because of the interfacial terms in Eq. (1.34), nucleation is sometimes modeled as occurring randomly in pure amorphous materials. This is one of the central assumptions of models for crystallization kinetics, and has led to solid-state kinetic models such as the Prout–Tompkins and Avrami models (House 2007), in which crystallization is modeled as random nucleation events in location and time, followed by growth by a definable mechanism. For instance, in 2-d and 3-d growth models, nucleation is treated as occurring at a constant rate per unconverted (amorphous) fraction of the sample volume, followed by a growth rate that is approximated as being constant along a given dimension such as the radius. This leads to an expression of a general Avrami equation (also known as the Avrami-Erofe'ev, or Johnson-Mehl-Avrami equations), given by

$$\alpha = 1 - \exp(-kt^n), \tag{1.35}$$

where α represents the fraction of a sample that has been converted to crystalline form, k is a function of several physical variables, and n is a function of the geometry (plates, spheres, etc.) for the growing crystals. The parameters k and n can be obtained from nonlinear fits of the data, or classic double log plots such as

$$\ln[-\ln(1-\alpha)] = \ln k + n\ln t. \tag{1.36}$$

An example calculation will illustrate the approach. If the number of nuclei formed per time per unconverted volume is denoted by a constant rate $\dot{N}_0$, the number of nuclei that form in a sample between times τ and $\tau + d\tau$ is $\dot{N}_0 V_T(1-\alpha)d\tau$. Here, V_T denotes the total sample volume, V is the volume that has converted to the crystalline form, and $\alpha = V/V_T$. The crystals are assumed to form perfect spheres of volume $4\pi r^3/3$, and the interfacial energy between the crystals and surrounding amorphous material is assumed to be constant (which is reasonable once crystals grow to larger than several nanometers in radius), so the radius of each sphere can be assumed to increase at the same constant rate $\dot{r}_0$. With these assumptions, the size of a sphere at a time t that formed at time between τ and $\tau + d\tau$ would be $4\pi \dot{r}_0^3(t-\tau)^3/3$. Also, at time t the total volume occupied by all spheres formed volume occupied by all spheres formed between τ and $\tau + d\tau$, denoted by $dV(t,\tau)$, would be the volume of each times the number that formed in that time interval, or

$$dV(t,\tau) = \frac{4\pi \dot{r}_0^3}{3}(t-\tau)^3 \times V_T(1-\alpha) \times \dot{N}_0 d\tau, \tag{1.37}$$

which can be rewritten as

$$\frac{d\alpha}{(1-\alpha)} = \frac{4\pi \dot{r}_0^3 \dot{N}_0}{3}(t-\tau)^3 d\tau. \tag{1.38}$$

At any time, $\alpha(t)$ can be found by integrating both sides. Assuming no converted fraction initially, the left-hand side is integrated from 0 to α and the right-hand side is integrated over all formation times τ from 0 to the observation time t, which leads to

$$\ln(1-\alpha) = \frac{\pi \dot{r}_0^3 \dot{N}_0 t^4}{3} \tag{1.39}$$

or

$$\alpha = 1 - exp(-kt^4) \quad k = \frac{\pi \dot{r}_0^3 \dot{N}_0}{3}. \tag{1.40}$$

This form of the Avrami equation describes a uniform 3-d growth of spherical particles whose nuclei randomly form at a constant average rate per unconverted volume, and is of a mathematical form referred to as the Avrami (A4) nucleation model (Khawam and Flanagan 2006). Other forms for other growth patterns have been derived, such as for crystals that grow as disks as (House 2007)

$$\alpha = 1 - exp(-kt^3) \quad k = \frac{\pi \dot{r}_0^3 \dot{N}_0}{3}. \tag{1.41}$$

These equations have been successfully used in practice to model crystallization data. Since equations have been derived for a number of Avrami models, this approach has the advantage that the model does not have to be known a priori, but instead can be chosen based on which equations fit the data (along with some physical insight). It is also possible to perform experiments at different temperatures or under nonisothermal conditions to facilitate further analyses such as obtaining growth activation energies, and the reader is referred to other works for detailed treatments (Khawam and Flanagan 2006).

1.9 Summary and Closing Remarks

Amorphous solids have been of great interest to pharmaceutical scientists for decades. Their potential for increasing bioavailability is of great importance, although they have not been utilized to the full potential to date. In this chapter, the thermodynamic and molecular foundation of neat amorphous APIs were presented with the goal of relating these properties to possible dissolution and solubility advantages, as well as to shed light on physical instability. Some kinetic considerations on amorphous materials were also presented because of their practical importance.

It is the opinion of this author that neat amorphous API and ASD formulations show great promise for future applications, but their utilization has not fully materialized due to a lack of understanding of how the systems work. The fact that the approach of using amorphous forms for drug delivery has been studied for nearly five decades and still yet not used in more drug products is indicative of the nature of the problem. Studying amorphous forms for drug delivery is interdisciplinary in nature, and requires high levels of expertise in fields ranging from pharmacy to chemistry, biology, physics and computer science. We are in exciting times because experimental equipment and computational platforms are rapidly increasing our fundamental knowledge of material science. Looking forward, it seems inevitable that these forms will find more commercial applications in drug delivery in the not-too-distant future.

Acknowledgments I would like to thank Drs. Navnit Shah, Harpreet Sandhu, and Duk Soon Choi for their invaluable interactions over the past several years. I would also like to especially thank Drs. Kosha Shah and Piyush Patel, with whom I have had countless scientific conversations and who have kept me on track when I was overwhelmed with commitments.

References

Adamson AW, Gast AP (1997) Physical chemistry of surfaces, 6th edn. Wiley, New York

Adjei A, Newburger J, Martin A (1980) Extended Hildebrand approach: solubility of caffeine in dioxane-water mixtures. J Pharm Sci 69:659–661

Amidon GL, Lennernäs H, Shah VP, Crison JR (1995) A theoretical basis for a biopharmaceutic drug classification: the correlation of in vitro drug product dissolution and in vivo bioavailability. Pharm Res 12:413–420

Angell CA (1995) Formation of glasses from liquids and biopolymers. Science 267:1924–1935

Atkins PW (1998) Physical chemistry, 6th edn. Freeman, New York

Bellantone RA, Patel P, Sandhu H, Choi D, Singhal D, Chokshi H, Malick AW, Shah NH (2012) A method to predict the equilibrium solubility of drugs in solid polymers near room temperature using thermal analysis. J Pharm Sci 101:4549–4558

Burger A, Ramberger R (1979) On the polymorphism of pharmaceutical and other molecular crystals I. Theory of thermodynamic rules. Mikrochimica Acta 11:259–281

Chiou WL, Riegelman S (1971) Pharmaceutical applications of solid dispersion systems. J Pharm Sci 60:1281–1302

D'Arcy DM, Persoons T (2011) Mechanistic modelling and mechanistic monitoring: simulation and shadowgraph imaging of particulate dissolution in the flow-through apparatus. J Pharm Sci 100:1102–1115

Donth E (2001) The glass transition: relaxation dynamics in liquids and disordered materials. Springer, Berlin
Garca-Coln LS, del Castillo LF, Goldstein P (1989) Theoretical basis for the Vogel-Fulcher-Tammann equation. Phys Rev B 40:7040–7044
Gavezzotti A (2007) Molecular aggregation: structure analysis and molecular simulation of crystals and liquids. Oxford University Press, New York (Chap. 14)
Hancock BC, Zografi G (1997) Characteristics and significance of the amorphous state in pharmaceutical systems. J Pharm Sci 86:1–12
Hancock BC, Shamblin SL, Zografi G (1995) Molecular mobility of amorphous pharmaceutical solids below their glass transition temperatures. Pharm Res 12:799–806
Herschler B, Humer C (2012) FDA new drug approvals hit 16-year high in 2012. Reuters Online. 2012. http://www.reuters.com/article/2012/12/31/us-pharmaceuticals-fda-approvals-idUSBRE8BU0EK20121231. Accessed 31 Dec 2012
Hildebrand J, Scott R (1950) Solubility of non-electrolytes, 3rd edn. Rheinhold, New York
Hill TL (1986) An introduction to statistical thermodynamics. Dover, New York
House JE (2007) Principles of chemical kinetics, 2nd edn. Elsevier, Amsterdam
Khawam A, Flanagan DR (2006) Basics and applications of solid state kinetics: a pharmaceutical perspective. J Pharm Sci 95:472–498
Kittel C (1986) Introduction to solid state physics, 6th edn. Wiley, New York
Klotz IM, Rosenberg RM (1974) Chemical thermodynamics—basic theory and methods, 3rd edn. Benjamin-Cummings, New York
Klug HP, Alexander LE (1974) X-ray diffraction procedures for polycrystalline and amorphous materials. Wiley, New York
Leuner C, Dressman J (2002) Improving drug solubility for oral delivery using solid dispersions. Eur J Pharm Biopharm 50:47–60
Martin A, Swarbrick J, Cammarata A (1983) Physical pharmacy, 3rd edn. Lea & Febiger, Philadelphia
Mauro JC, Yue Y, Ellison AJ, Gupta PK, Allan DC (2009) Viscosity of glass-forming liquids. Proc Natl Acad Sci 106:19780–19784
Mullin JW (2001) Crystallization, 4th edn. Butterworth-Heinemann, Boston
Simonelli AP, Mehta SC, Higuchi WI (1969) Dissolution rates of high energy polyvinyl pyrrolidone (PVP)-sulfathiazole coprecipitates. J Pharm Sci 58:538–549
Trachenko K (2008) The Vogel–Fulcher–Tammann law in the elastic theory of glass transition. J Non-cryst Solids 354:3903–3906
Williams RO III, Watts AB, Miller DA (eds) (2012) Formulating poorly water soluble drugs. Springer, New York (Preface)
Yu L (2001) Amorphous pharmaceutical solids: preparation, characterization and stabilization. Adv Drug Del Rev 48:27–42

Chapter 2
Theoretical Considerations in Developing Amorphous Solid Dispersions

Riikka Laitinen, Petra A. Priemel, Sachin Surwase, Kirsten Graeser, Clare J. Strachan, Holger Grohganz and Thomas Rades

2.1 Introduction

The term "solid dispersion" was introduced by Chiou and Riegelmann in 1971 who defined solid dispersions as "a dispersion of one or more active ingredients in an inert carrier at the solid state, prepared by either the melting, the solvent or the melting solvent method" (Chiou and Riegelman 1971). Although the concept of melting an active ingredient and carrier together had previously been used, Chiou and Riegelmann were the first to introduce a classification system for solid dispersions that was based on the possible physical states of the active pharmaceutical ingredient (API) and the carrier. Unfortunately, it is often found that the term "solid dispersion" is used somewhat inconsistently in the pharmaceutical literature; therefore, this first section briefly gives an overview over the nomenclature of different types of solid

T. Rades (✉)
Pharmaceutical Design and Drug Delivery, Faculty of Health and Medical Sciences, Department of Pharmacy, University of Copenhagen, Universitetsparken 2, 2100 Copenhagen, Denmark
Tel.: +45 35 33 60 01
e-mail: thomas.rades@sund.ku.dk

R. Laitinen
School of Pharmacy, University of Eastern Finland, Kuopio, Finland

P. A. Priemel
School of Pharmacy, University of Otago, Dunedin, New Zealand

S. Surwase · C. J. Strachan
Division of Pharmaceutical Chemistry and Technology, University of Helsinki, Helsinki, Finland

K. Graeser
pRED, Roche Innovation Center, F. Hoffmann-La Roche Ltd., Basel, Switzerland

H. Grohganz
Department of Pharmacy, University of Copenhagen, Copenhagen, Denmark

N. Shah et al. (eds.), *Amorphous Solid Dispersions*,
Advances in Delivery Science and Technology, DOI 10.1007/978-1-4939-1598-9_2

Table 2.1 Classification of solid dispersions

State of API	Number of phases	
	1	2
Crystalline	Solid solution	Eutectic mixture
Amorphous	Glass solution	Glass suspension

API active pharmaceutical ingredient

dispersions (Table 2.1). The number of components is not limited to two; however, in the following chapter, we restrict ourselves to binary systems, consisting of API and carrier. We use the term "solubility" to describe the solubility of the crystalline drug in the carrier and the term "miscibility" to describe the miscibility of the amorphous drug with the polymer.

The advantage of using solid dispersions as drug delivery systems to increase the dissolution behavior and apparent solubility of a drug is discussed in more detail later in this chapter. In brief, molecularly dispersing a poorly water-soluble API in a hydrophilic carrier (amorphous or crystalline) often leads to increased dissolution behavior and supersaturation of the drug when this system is exposed to water. This is attributed to a number of factors such as improved wettability of the drug by the polymer, minimal particle size of the drug, separation of individual drug particles by polymer particles, and subsequent prevention of drug precipitation upon contact with aqueous media.

2.2 Classification of Solid Dispersions

2.2.1 Eutectic Mixtures

A simple eutectic mixture consists of two compounds that are completely miscible in the liquid state (melt) but only show limited miscibility in the solid state. At a specific composition (E in Fig. 2.1), the two components crystallize simultaneously when the temperature is reduced (Fig. 2.1). If mixtures with different compositions to the eutectic composition of A and B are cooled, one component will start to crystallize before the other, which initially leads to a mixture of pure solid compound and liquid. Therefore, a true eutectic only exists for a defined composition of A and B. The microstructure of a eutectic mixture is different from the microstructure of either components, and this property may be used to differentiate the eutectic mixture from other forms of crystalline mixtures. A theoretical method to determine the eutectic composition of a binary mixture and the temperature at which it crystallizes has been suggested by Karunakaran (1981).

Eutectics of poorly soluble compounds and water-soluble inert carriers have been shown to enhance the dissolution rate of the poorly soluble compound. When the eutectic is exposed to water or the gastrointestinal (GI) fluids, the carrier will dissolve

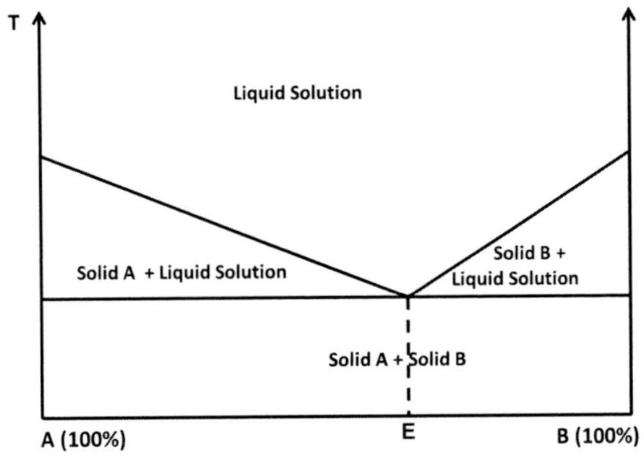

Fig. 2.1 Phase diagram of a simple eutectic system

rapidly and release fine crystals of the drug. Through the large surface area and the improved wettability from the carrier, the dissolution rate of the API will be enhanced.

2.2.2 *Solid Solutions*

Solid solutions are formed when a solute is nonstoichiometrically incorporated into the crystal lattice of the solvent (Moore and Wildfong 2009). Solid solutions can be classified according to the solubility of the solute in the crystal lattice (continuous vs. discontinuous) or according to the way in which the solute molecules are distributed. In general, the term "solid solution" refers to systems that contain a crystalline carrier.

Continuous Solid Solutions In continuous solid solutions, the components are miscible in all proportions. This occurs if the strength of the bonds between the two different molecules is higher than that of the bonds of the molecules of the same species. Organic molecules do not tend to form this kind of solid solutions and therefore, they are not of great importance in the pharmaceutical field (Leuner and Dressman 2000).

Discontinuous Solid Solutions The term "discontinuous" refers to the fact that solid solubility only exists at specific compositions of the mixture, not over the entire compositional range. Figure 2.2 represents a phase diagram of a discontinuous solid solution. Each component is capable of completely dissolving the other component in a specific compositional region (regions α and β in Fig. 2.2) whereby the solubilization capability of the components is temperature dependent. It is maximal at the eutectic temperature and decreases when the temperature is reduced (Leuner and Dressman 2000). In reality, limited solid solubility most likely exists for all, or at least very many, binary systems. Goldberg et al. (1965) therefore proposed to use the term "solid solution" only if the mutual solubility of the two components exceeds 5 %. In their work, they could show that the postulated eutectic mixture of sulfathiazole in urea by Sekiguchi and Obi (1961), which was shown to increase the

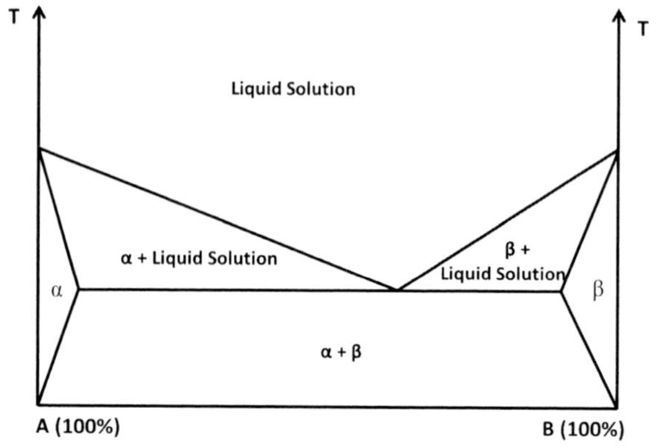

Fig. 2.2 Phase diagram of a discontinuous solid solution

drug absorption rate in man, should indeed be regarded as a physical mixture of two solid solutions.

Substitutional Solid Solutions In typical solid solutions with a crystalline carrier, a solute molecule can substitute for a carrier molecule in the crystal lattice (illustrated in Fig. 2.3a). Substitution is only possible if the size of the solute molecule is approximately similar to the size of the carrier molecule. Substitutional solid solutions can be continuous or discontinuous.

Interstitial Solid Solutions If the solute molecules are smaller than the solvent molecules, it is possible for them to occupy the interstitial spaces in the crystalline lattice (illustrated in Fig. 2.3b). The diameter of the solute molecules should not exceed 0.59 times the diameter of the solvent. Interstitial solid solutions can only form solid solutions of the discontinuous type (Khachaturyan 1978).

2.2.3 Glass Solutions

In glass solutions, the carrier is amorphous and the solute molecules are dispersed molecularly in the amorphous carrier. Glass solutions therefore are homogeneous one-phase systems. However, due to the much higher viscosity of glass solutions compared to liquid solutions, the distribution of solute molecules in the carrier may be irregular and a homogeneous distribution within the glass solution needs to be ensured by mixing. Chiou and Riegelmann (1969) formulated a glass solution of griseofulvin in citric acid to improve the dissolution rate of griseofulvin. In the past, sugars and urea were used as amorphous carriers, but more recently, organic polymers such as poly(vinylpyrrolidone) (PVP) and cellulose derivatives are commonly used. These polymers often exhibit amorphous regions (or are fully amorphous) and can be tailor-made for specific purposes. If the solubility of the drug in the carrier is not exceeded, the glass solution is thermodynamically stable. However, the concentration of the solute is often supersaturated to achieve higher drug loads and

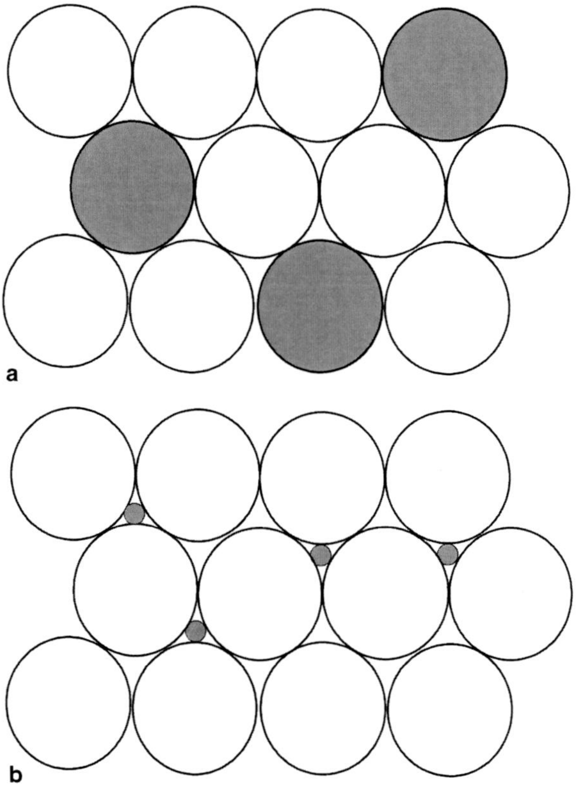

Fig. 2.3 **a** Substitutional crystalline solid solution and **b** interstitial crystalline solid solution

therefore recrystallization and precipitation may occur. Recrystallization may be retarded by kinetic stabilization. Phase separation, as the first step to recrystallization, requires a certain degree of molecular mobility within the system, and storing glass solutions at temperatures well below the glass transition temperature (T_g) will slow down mobility and may increase the stability of supersaturated glass solutions.

2.2.4 Glass Suspensions

As stated above, it is often observed that the miscibility of an amorphous drug in an amorphous carrier is limited, and as the drug content increases, phase separation can occur. If the drug forms a separate amorphous phase (or a drug-rich amorphous phase), the glass solution has been converted to a glass suspension. In this form, the drug is still present in the amorphous form, so it will still show increased dissolution behavior compared to the crystalline form; however, these amorphous precipitates have a high likelihood for recrystallization of the amorphous drug (usually the T_g of the drug or drug-rich phase will be lower than the T_g of the polymer or polymer-rich phase).

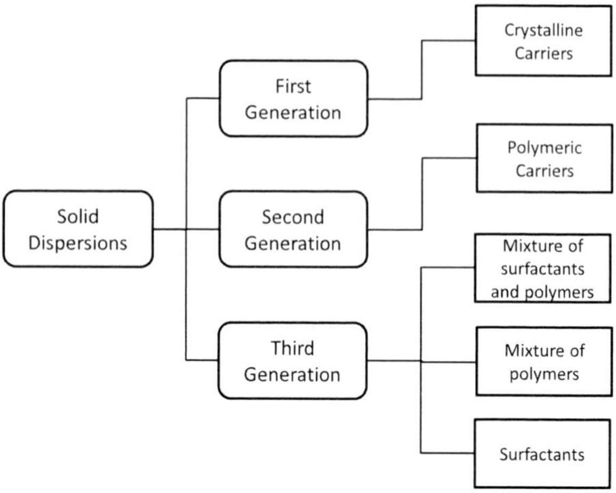

Fig. 2.4 Classification of solid dispersions. (Adapted from Vasconcelos et al. 2007)

Recently, a number of publications have presented a classification scheme of solid dispersions which is not based on the molecular structure but rather based on the advancement of knowledge and complexity of these systems. The authors categorize solid dispersions as first-, second-, and third-generation solid dispersions (Vasconcelos et al. 2007). This classification can be regarded as a kind of timeline showing the evolution of solid dispersion development and their increasing complexity as drug delivery systems (Fig. 2.4). This chapter predominantly covers second-generation solid dispersions.

2.3 Theoretical Considerations Regarding Solid Dispersions

2.3.1 Solubility Advantage of the Amorphous Form

Amorphous forms show excess free energy, enthalpy, and entropy compared to the corresponding crystalline state(s) and therefore their solubility in the GI tract may be higher, which results in potentially higher bioavailability of the drug.

However, before pursuing the laborious route of amorphous formulation development, the formulation scientist would benefit from a priori knowledge of whether the amorphous route is viable and how much solubility improvement, and hence potential increase in bioavailability, can be expected. For a crystalline material, simple solubility measurements and (intrinsic) dissolution testing are the most common methods for comparing solubility. Solubility and dissolution testing of amorphous compounds, however, is not as straightforward, and there are numerous reports in the literature on the difficulties associated with solubility and dissolution rate determination of an amorphous form, as amorphous material tends to recrystallize upon contact with the dissolution media (Alonzo et al. 2010; Babu and Nangia 2011; Egawa et al. 1992; Greco and Bogner 2010; Imaizumi et al. 1980).

For different polymorphs, the solubility advantage of one polymorph over the other can easily be calculated using the thermodynamic parameters of each polymorph. In a method proposed by Parks et al. (1928; 1934), the solubility difference between different polymorphs is estimated using the solubility ratio, σ^1/σ^2 The solubility ratio is directly related to the free energy difference, $\Delta G^{1,2}$, between polymorphs 1 and 2:

$$\Delta G_{\mathrm{T}}^{1,2} = -RT \ln\left(\sigma_{\mathrm{T}}^{1}/\sigma_{\mathrm{T}}^{2}\right), \tag{2.1}$$

where $\Delta G_{\mathrm{T}}^{1,2}$ is the free energy difference between polymorphs 1 and 2 at any given temperature (T) and R is the gas constant. The difference in free energy can be estimated from the enthalpy (ΔH) and entropy (ΔS) differences of both forms:

$$\Delta G_{\mathrm{T}}^{1,2} = \Delta H_{\mathrm{T}}^{1,2} - (T\Delta S_{\mathrm{T}}^{1,2}). \tag{2.2}$$

For crystalline compounds, the calculated solubility ratio gives accurate results. It has therefore been proposed to apply this equation to amorphous systems:

$$\Delta G_{\mathrm{T}}^{a,c} = \Delta H_{\mathrm{T}}^{a,c} - (T\Delta S_{\mathrm{T}}^{a,c}) \tag{2.3}$$

whereby the entropy and enthalpy differences can be calculated from the entropy and enthalpy of fusion and the heat capacities (C_{p}):

$$\Delta H_{\mathrm{T}}^{\mathrm{a,c}} = \Delta H_{\mathrm{f}}^{\mathrm{c}} - (C_{\mathrm{p}}^{\mathrm{a}} - C_{\mathrm{f}}^{\mathrm{c}})(T_{\mathrm{f}}^{\mathrm{c}} - T) \tag{2.4}$$

$$\Delta S_{\mathrm{T}}^{\mathrm{a,c}} = \Delta S_{\mathrm{f}}^{\mathrm{s}} - (C_{\mathrm{p}}^{\mathrm{a}} - C_{\mathrm{p}}^{\mathrm{c}})(\ln T_{\mathrm{f}}^{\mathrm{c}} - T) \tag{2.5}$$

$$\Delta S_{\mathrm{f}}^{\mathrm{c}} = \Delta H_{\mathrm{f}}^{\mathrm{c}}/T_{\mathrm{f}}^{\mathrm{c}}. \tag{2.6}$$

In this simplified approach, the amorphous form is treated as a pseudo-equilibrium state. Solubility advantages calculated by this method have predicted solubility advantages of up to 1600-fold for amorphous systems compared to their crystalline counterparts (Graeser et al. 2010; Hancock and Parks 2000). When compared to experimentally determined solubility data, the observed increase in solubility, however, was considerably lower. In the past, this has always been attributed to the difficulty in measuring the solubility of amorphous systems in aqueous media due to recrystallization and the nonequilibrium nature of the amorphous form (Egawa et al. 1992; Imaizumi et al. 1980). In a recent publication, however, some shortcomings of the proposed simplified calculations were highlighted. It was postulated that the large discrepancies between the theoretical and experimentally determined solubility increase can be attributed to inaccurate assumptions regarding the C_{p} differences, changes in the amorphous free energy due to water sorption, and a reduced fraction ionized in saturated solutions of the amorphous form (Murdande et al. 2010a). By including correction terms for these three considerations, the authors were able to calculate the solubility advantage for amorphous indomethacin to be 7-fold instead of the previously determined 25–104-fold higher solubility, compared to the crystalline form of the drug. This was in closer agreement to the experimentally observed value of 4.9. In a second study, the authors reported that the calculated and observed solubility advantages were in agreement only if recrystallization in the medium was slow (Murdande et al. 2010b). This limitation highlights once again the difficulties in determining the solubility and dissolution behavior of amorphous compounds.

Determination of the solubility advantage using this modified method, however, gives useful estimates of the expected solubility increase and can thus serve as a basis to decide whether the amorphous route should be pursued in development of a specific poorly water-soluble drug. Only drugs for which the solubility increase is considered sufficient should be selected for further testing and development.

2.3.2 Glass-Forming Ability (GFA) and Glass Stability (GS)

Research over the past decades has shown that different crystalline drugs have different tendencies to be converted into and remain in the amorphous form. Although several technologies exist to convert a crystalline drug into the amorphous form, but in practice, not every drug is susceptible to amorphization. Thermolabile drugs may not be transformed using heat-based methods, while poor solubility in organic solvents may prevent the use of spray drying or precipitation methods. Additionally, drugs that show a poor tendency to amorphization regardless of the technology used exist.

To assess whether a compound is capable of being developed into an amorphous dosage form, the glass-forming ability (GFA) of the drug can be estimated. The GFA has been defined as the ease of vitrification of a liquid upon cooling, and there is no shortage of structural and kinetic theories behind it (Avramov et al. 2003). Once successfully converted to the amorphous form, drugs may exhibit different tendencies to revert back to the energetically favored crystalline state. This recrystallization process may occur in a time frame of several seconds up to several years, depending on the drug used and the conditions at which it is stored.

Whereas the GFA describes the ease of vitrification of a compound, the glass stability (GS) describes its resistance to recrystallization. In the past decades, researchers not only in the pharmaceutical field but also in the field of inorganic chemistry have investigated the GFA and GS of numerous glasses. However, despite such research, the fundamental understanding, and thus prediction models, of the recrystallization process is still lacking. For the pharmaceutical scientist, prediction models for the GFA and GS would be of great benefit, as real-time preparation and storage experiments would potentially become redundant. This would be a time- and cost-saving improvement in the development process for new drugs.

After assessing whether a poorly soluble drug would benefit from being formulated in the amorphous form in terms of solubility increase, the GFA should be the subsequent property to determine. The most common parameter to estimate the GFA of a compound is to assess its minimum cooling rate, i.e., the slowest cooling rate from the melt at which the material can still be transformed to the amorphous form. Drugs which show poor GFA are thought to exhibit a high degree of mobility in

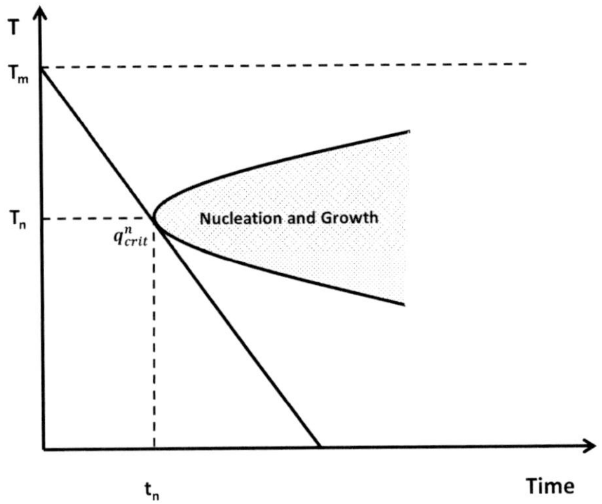

Fig. 2.5 Annotated time–temperature transformation diagram, showing the minimum cooling rate to avoid crystallization. T_m is the melting temperature, T_g is the glass transition temperature, and T_n and t_n are the temperature and time point at the locus, respectively. (Adapted from Karmwar et al. 2011a)

the melt, so that the melt can only solidify to an amorphous form if the temperature change (cooling rate) is sufficiently rapid. In contrast, the amorphous form can be attained by slow cooling rates for drugs with high GFA.

This critical cooling rate ($q^{\mathrm{n}}_{\mathrm{crit}}$) has been estimated by use of isothermal time–temperature transformation (TTT) diagrams (Uhlmann 1972) or continuous cooling transformation (CT) curves Onorato and Uhlmann 1976).

Applying the TTT method, the critical cooling rate is given by Eq. 2.7:

$$q^{\mathrm{n}}_{\mathrm{crit}} = \frac{T_{\mathrm{m}} - T_{\mathrm{n}}}{t_{\mathrm{n}}} \tag{2.7}$$

whereby T_{m} is the melting temperature and T_{n} and t_{n} are the minimum temperature and time where an amorphous form can still be achieved (Fig. 2.5).

However, it has been discussed that the critical cooling rate calculated by the TTT method often differs by up to one order of magnitude from the experimentally determined values (Huang et al. 1986). As the measurement of $q^{\mathrm{n}}_{\mathrm{crit}}$ is time and material consuming and the measurements are often not straightforward, a variety of thermal observations have been proposed as surrogates for the critical cooling rate such as, the melting temperature T_{m}, the glass transition temperature T_{g}, the crystallization temperature T_{c}, and combinations thereof. These thermal events can easily be obtained by differential scanning calorimetry (DSC) measurements. Barandiaran and Colmenero (1981) were among the first to develop a DSC-based method in which the crystallization temperature is determined while a liquid is being cooled in a DSC instrument at different rates, and they established the following relationship (Eq. 2.8), which was later refined to Eq. 2.10 (Cabral et al. 2003):

$$\mathrm{In}q = A - \frac{B}{(\Delta T_c^c)^2} \tag{2.8}$$

$$\Delta G(T_c^c) = S_m \Delta T_c^c \tag{2.9}$$

$$q_{\text{crit}}^{\text{n}} = \exp\left(A - \frac{B}{T_{\text{m}}^{2}}\right), \tag{2.10}$$

where q is the cooling rate, A and B are empirical constants obtained from linear regression, T_c^c is the crystallization peak temperature on cooling, and ΔT_c^c is the difference between T_{m} and T_c^c. In essence, the equation states that compounds which have a low thermodynamic driving force for recrystallization, $\Delta G(T_c^c)$, and high melting entropy (S_{m}) should be good glass formers (Eq. 2.9). Only few researchers have calculated the minimum cooling rate using this method; however, the calculated and experimentally determined values agreed (Whichard and Day 1984). A number of other researchers have investigated and proposed alternative methods for calculating the critical cooling rate, making small alterations to already existing equations. These different approaches are beyond the scope of this chapter and the interested reader is referred to the literature for further details (Gutzow and Schmelzer 2013; Whichard and Day 1984).

The concept of the GFA was developed for inorganic materials, and studies on the estimation of the GFA through measurement of the critical cooling rate for small organic molecules are scarce. For pharmaceutically relevant systems, DSC-based methods are of particular interest as they offer a rapid and simple way of estimating the GFA of an unknown drug. These methods use the thermal events from heating experiments in a DSC to assess the GFA. In the 1940s, Kauzmann introduced a ratio which was later termed the reduced glass transition temperature (K_{T}) by Turnbull (Kauzmann 1948; Turnbull 1969).

$$K_{\text{T}} = \frac{T_{\text{g}}}{T_{\text{m}}} \tag{2.11}$$

with T_{m} being the melting temperature and T_{g} the glass transition temperature. K_{T} is considered a predictor for the resistance to crystallization. The higher the value of K_{T}, the higher the GFA should be. The theory behind this simple equation relies on the assumption that the viscosity of compounds is equal at T_{g} and therefore materials with a higher value for K_{T} are expected to have a higher viscosity between T_{g} and T_{m}. Hence, the closer the value of K_{T} is to 1, the higher is the GFA of the compound.

Weinberg (1994) used an approach in which not only the T_{g} but the difference between T_{g} and the crystallization temperature was used to estimate the GS. It is assumed that larger differences between these two temperatures reflect a higher GFA:

$$K_{\text{W}} = \frac{T_{\text{x}}^{\text{h}} - T_{\text{g}}}{T_{\text{m}}}, \tag{2.12}$$

where K_{W} is the "Weinberg parameter" and T_{x}^{h} is the crystallization onset temperature.

In the past, there has been no shortage of authors and equations, all intending to establish the optimal equation in order to estimate the GFA of a drug (Eqs. 2.13, 2.14

and 2.15; Duan et al. 1998; Hrubý 1972; Lu and Liu 2002, 2003; Ota et al. 1995; Saad and Poulain 1987).

$$K_{\mathrm{H}} = \frac{T_{\mathrm{x}}^{\mathrm{h}} - T_{\mathrm{g}}}{T_{\mathrm{m}} - T_{\mathrm{x}}^{\mathrm{h}}} \tag{2.13}$$

$$K_{\mathrm{LL}} = \frac{T_{\mathrm{x}}^{\mathrm{h}}}{T_{\mathrm{m}} + T_{\mathrm{g}}} \tag{2.14}$$

$$K_{\mathrm{SP}} = \frac{(T_{\mathrm{x}}^{\mathrm{h}} - T_{\mathrm{g}})(T_{\mathrm{c}}^{\mathrm{h}} - T_{\mathrm{x}}^{\mathrm{h}})}{T_{\mathrm{g}}}, \tag{2.15}$$

where $T_{\mathrm{c}}^{\mathrm{h}}$ is the crystallization peak temperature.

The general consensus is that compounds which show a high GFA also exhibit a high degree of GS. In some instances, the abovementioned equations are used to determine the GFA, but other authors have employed them to estimate GS (Nascimento et al. 2005). It is mentioned here that "the definitions of GS are somewhat arbitrary," and in their study, the authors calculated the GS for 6 inorganic compounds using 14 different stability parameters by employing the previously mentioned equations. As an estimate of the GFA, the authors used the critical cooling rate which was determined experimentally. They compared their calculated GFAs and GS and could show an inverse relationship between GFA, determined as the critical cooling rate, and GS for a number of compounds. It can, therefore, be concluded that the equations can be used to estimate either GFA or GS for inorganic materials.

At this stage, it has to be mentioned that the majority of these equations and theories stem from the world of inorganic chemistry and have been applied mainly in this field. In recent years, however, interest in the pharmaceutical field has increased and an attempt has been made to translate these concepts to small organic molecules. Another parameter that has been introduced to the pharmaceutical sciences is the reduced crystallization temperature, here termed K_{Tcred}:

$$K_{\mathrm{Tcred}} = \frac{(T_{\mathrm{x}}^{\mathrm{h}} - T_{\mathrm{g}})}{(T_{\mathrm{m}} - T_{\mathrm{g}})} \tag{2.16}$$

This value describes how far above the T_{g} a compound must be heated before spontaneous crystallization occurs. The authors concluded that the higher the value of K_{Tcred}, the higher is the GFA (Zhou et al. 2002).

It should be noted that when dealing with amorphous drugs, it is the stability of the glass (and not the supercooled melt) that is of interest as this is the preferred state for manufacturing and storage of amorphous drug products. To calculate the GFA or GS of amorphous compounds, however, the equations use thermal characteristics that are obtained from heating the glass above its glass transition. So, in reality, it is not the stability of the glass but the stability of the supercooled melt that is calculated from these equations (Baird et al. 2010). In the remainder of this chapter, the term "GS" is only be used when the stability of the actual glass is concerned. Otherwise, the term "liquid GS" will be used.

In one of the first studies that investigated the relationship between GFA and GS experimentally, a TTT diagram using indomethacin as the model compound was generated (Karmwar et al. 2011a). These authors were the first to verify for small organic molecules that the critical cooling rate, as calculated from Eq. 2.10, of 1.0 K/min was in very good agreement with the experimentally determined value of 1.2 K/min. Furthermore, differently cooled indomethacin samples showed differences in stability upon storage. Correlation of the experimentally observed stability with the GFA parameters K_T and K_{Tcred} showed that for differently cooled indomethacin samples, these GFA parameters could be used as a surrogate for stability of the glass.

In order to test whether this attempt would be successful when different compounds are analyzed, the group of Lynne Taylor evaluated a potential correlation between the GFA, the GS, and the "liquid GS" for a set of 51 drugs (Baird et al. 2010). Compounds were separated into three different classes according to their physical state and behavior after the melt quenching. Class I compounds were not able to be transformed to the amorphous form through melting as they recrystallized during cooling; class II compounds were transformed to an amorphous form, but recrystallized upon reheating above the T_g. Class III molecules were made amorphous and did not recrystallize upon heating above the T_g. GFA and "liquid GS" parameters were calculated from eight different equations and compared to results from storage experiments. Class I compounds recrystallized directly upon cooling and were thus classified as having a low GFA. These compounds were cooled using different cooling rates (5–20 °C/min), and additionally, the critical cooling rate, q^n_{crit}, was determined using Eq. 2.10. It was discovered that calculation of the critical cooling rate, q^n_{crit}, was inaccurate or contradictory to the experimental observations for a number of molecules, and this was attributed to the relatively small differences in cooling rates. The q^n_{crit} could not be calculated for class II and III compounds as they did not crystallize upon cooling. This showed another drawback of this equation when used for small organic molecules, as not every drug recrystallizes on cooling. As a surrogate for GFA, the reduced temperature, K_T, was calculated and compared to the other "liquid GS" and GFA parameters. The overall results from their comparative study were that there was no reliable correlation between any of the calculated parameters and the behavior upon heating. They concluded that "[· · ·] in contrast to inorganic systems where there is some evidence [. . .], these parameters may not be ideal to predict the GFA or 'liquid GS' of organic molecules" (Baird et al. 2010).

There was, however, an indication that there may be a link between GFA and "liquid GS" for crystallization behavior above the T_g, with class I molecules (low GFA) also showing the lowest GS and class III molecules (high GFA) showing the highest GS for this selection of drugs. Class II compounds (high GFA) could be distinguished from class III compounds (high GFA) as class II compounds recrystallized upon storage (low GS). In a follow-up study, additionally using principal component analysis (PCA) on thermal properties such as T_m, T_g, ΔH, ΔS, ΔG, and molecular parameters, this was further investigated, and it could be shown that PCA in addition to some compound properties may provide predictive capabilities of recrystallization behavior (Van Eerdenbrugh et al. 2010).

In general, it has been shown that the applicability of the simple calculated GFA and "liquid GS" parameters in the field of small organic molecules is limited and that conclusions drawn from the inorganic sciences do not hold true for pharmaceuticals. As the physical stability is of great importance in the pharmaceutical field, other methods of attempting to predict the stability are required to circumvent time- and material-consuming experiments.

2.3.3 Calculation of Physical Stability

The observation that recrystallization is a complex process involving more than one or two parameters and that PCA with thermal and molecular input parameters may show some degree of predictability of stability was already noted by Graeser et al. (2009b) who attempted to correlate the observed stability below the T_g with a number of thermodynamic and kinetic factors they calculated from DSC experiments for a set of 12 drugs. Unsatisfactory linear correlation prompted the authors to suggest that multivariate analysis with the same parameters may be more successful.

In order to successfully develop an amorphous drug, a priori prediction of the GS would be beneficial. This section does not deal with the underlying physics and kinetics of recrystallization from the amorphous form; the details are covered elsewhere in this book. Rather, it briefly highlights the attempts in the pharmaceutical world to find predictive parameters of physical stability.

It is generally understood that the amorphous form possesses excess entropy, enthalpy, free energy (ΔS, ΔH, ΔG), and increased molecular mobility compared to their crystalline counterparts and that these factors should, therefore, play a crucial role in the recrystallization process. To date, the factors governing the recrystallization process are poorly understood, making general rules challenging to apply. Initially, it was considered that storing an amorphous compound at 50 °C below its glass transition temperature should ensure sufficient stability, as molecular mobility at this temperature was considered negligible. However, it could be shown that this rule of thumb does not always hold true and recrystallization still occurs at temperatures well below the T_g (Yoshioka et al. 1994). Research in the past has focused on molecular mobility and thermodynamic properties as the main factors governing the physical stability.

A number of equations exist to calculate the molecular mobility or its reciprocal, the relaxation time, τ (Andronis and Zografi 1997; Di Martino et al. 2000; Mao et al. 2007; Yoshioka et al. 1994). The two most commonly used equations that give an indication on the mobility within an amorphous sample are the Kohlrausch–Williams–Watts (KWW) equation and the Adam–Gibbs (AG) equation.

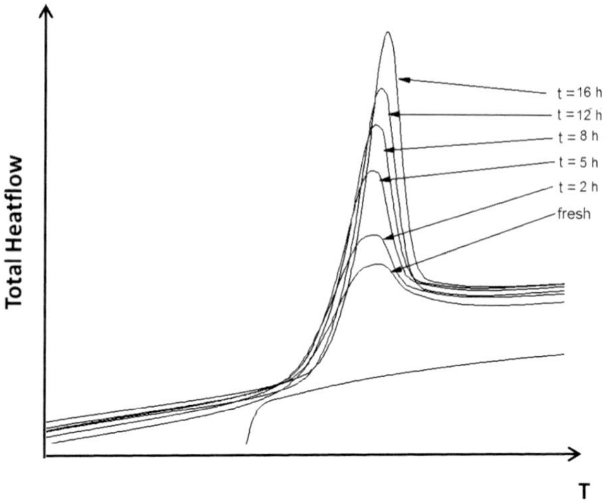

Fig. 2.6 Relaxation endotherms of the glass transition temperature for quench-cooled simvastatin annealed for different lengths of time. (Adapted from Graeser et al. 2009a)

2.3.3.1 Kohlrausch–Williams–Watts (KWW) Equation

The underlying assumption for the empirical KWW equation is that the relaxation time of a sample can be determined by measuring the enthalpy lost during annealing (Graeser et al. 2009a; Liu et al. 2002; Van den Mooter et al. 1999). An amorphous compound will relax, thereby losing some of its excess thermodynamic properties, e.g., enthalpy. Upon reheating of the sample, the enthalpy is recovered and can be visualized in the DSC as an enthalpic overshoot at the T_g (see Fig. 2.6).

In Eq. 2.17, the enthalpic relaxation, ΔH_{relax}, is related to the relaxation function φ and the maximal theoretical enthalpic relaxation. Samples are aged for various lengths of time and φ is obtained for various time points.

$$\varphi = 1 - \frac{\Delta H_{relax}}{\Delta H_\infty} \tag{2.17}$$

$$\Delta H_\infty = \Delta C_p(T_g - T), \tag{2.18}$$

where φ is the relaxation function, ΔH_{relax} is the enthalpic relaxation, ΔH_∞ is the maximal theoretical enthalpic relaxation, ΔC_p is the heat capacity change at T_g, and T is the annealing temperature.

The calculated relaxation functions are then fitted to the KWW equation (Eq. 2.19) using nonlinear regression and the KWW parameters, relaxation time τ, and the exponential parameter β are obtained.

$$\varphi = \exp\left[-(\tfrac{t}{\tau})^\beta\right], \tag{2.19}$$

where t is the time, τ is the relaxation time, and β is the stretched exponential parameter ($0 < \beta < 1$).

The KWW equation has been widely used to estimate the average relaxation time of single-component systems (Kawakami and Pikal 2005; Mao et al. 2006b; Surana et al. 2005; Van den Mooter et al. 1999), as well as binary systems; however, it was shown that the predictive capability from the KWW equation for physical stability is limited. Limitations of the equation include:

- KWW equation assumes τ does not change significantly during relaxation; however, studies have shown this not to be the case.
- The KWW value of τ is accompanied by the stretched exponential parameter β, which makes direct comparison of τ values imprecise unless the β values are close (± 0.1).

In small-scale studies with a limited number of drugs (i.e., two to three), the calculated relaxation time showed some degree of correlation with stability of the amorphous form; however, when the sample set was increased, this correlation could not be supported (Graeser et al. 2009b). The predictive potential of the KWW equation in terms of physical stability is small.

2.3.3.2 AG Equation

Limitations of the widely used KWW equation have been discussed and addressed by employing the Vogel–Tammann–Fulcher (VTF) and AG equations in order to calculate the relaxation time above and below the glass transition temperature more precisely (Shamblin et al. 1999).

In brief, the VTF equation is commonly used to describe the temperature dependence of relaxation time of a fluid:

$$\tau(T) = \tau_0 \exp\left(\frac{DT_0}{T - T_0}\right), \tag{2.20}$$

where τ_0 denotes a constant, taken as the lifetime of atomic vibrations, 10^{-14} s, D is Angell's strength parameter, and T_0 is the temperature where no structural mobility occurs.

As the temperature is decreased below T_g, free rotation of the individual molecules decreases and molecular motions become restricted. In this temperature region, the collective movement of cooperatively rearranging regions (CRR) of the molecules dominates. Decrease of temperature leads to an increase in the size of the CRRs and hence to the decrease of configurational entropy below T_g (Metatla and Soldera 2007). Any description of relaxation time below T_g therefore has to consider the contribution of configurational entropy (Eq. 2.21):

$$\tau(T) = \tau_0 \exp\left(\frac{C}{T S_{conf}(T)}\right) \tag{2.21}$$

Here, C denotes a material-dependent constant and S_{conf} is the configurational entropy.

Once created, a glass will always strive towards a lower energetic state, hence reducing a portion of its excess enthalpy and entropy. This results in the configurational entropy being not only temperature dependent but also time dependent. A convenient way of expressing the time and temperature dependence of S_{conf} is by introducing the fictive temperature T_{f}. The fictive temperature is the temperature at which the equilibrium system has the same thermodynamic properties as the real (nonequilibrium) system at temperature T and time t, thus relating the nonequilibrium state (glass) to the equilibrium state (supercooled liquid) and enabling the use of the AG equation. Introducing T_{f} into Eq. 2.21 leads to the modified AG equation that allows description of relaxation times below the glass transition temperature:

$$\tau = \tau_0 \exp\left(\frac{DT_0}{T\left(1\frac{T_0}{T_{\mathrm{f}}}\right)}\right). \tag{2.22}$$

A straightforward way to assess the relaxation time from easily obtainable DSC data is outlined in the literature (Mao et al. 2006a). Despite the improvement of the AG equation over the KWW equation, relaxation times calculated from the AG equation also did not show the capability of predicting the physical stability for a set of drugs.

In studies where the authors have calculated both, relaxation times from the KWW and the AG approach, they found that these values differed considerably (Graeser et al. 2009a, b; Karmwar et al. 2011b). This was not unexpected as both equations have different underlying assumptions. The KWW equation is an empirical equation, whereas the AG equation accounts for the nonlinearity of relaxation processes and includes the time dependence of the configurational entropy. The AG equation also suggests that that the relaxation time is controlled partly by the properties of the glass-forming liquid. The shortcomings of the AG equation are the non-exponentiality of the relaxation and neglecting other entropic contributions than configurational entropy. In both cases, the equations are simplified to make them usable in a laboratory environment. Despite these shortcomings, there seems to be some indication that the relaxation time calculated from either the KWW or AG equation may predict the stability correctly for the same drug which is prepared as an amorphous form by different methods (Graeser et al. 2009a; Karmwar et al. 2011b)—at least qualitatively.

2.3.3.3 Thermodynamic Properties

Other researchers have focused on the thermodynamic contributions to physical stability and have investigated if these parameters could be exploited for stability predictions. The configurational heat capacity (Cp_{conf}) is defined as the difference between the heat capacity of the amorphous and crystalline forms of a material (Eq. 2.23):

$$Cp_{\mathrm{conf}} = Cp^{\mathrm{amorph}} - Cp^{\mathrm{crystal}}. \tag{2.23}$$

The values for Cp_{conf} are close but not identical to the heat capacity change at T_{g} for amorphous compounds (ΔCp) as shown in Fig. 2.7. Although this assumption

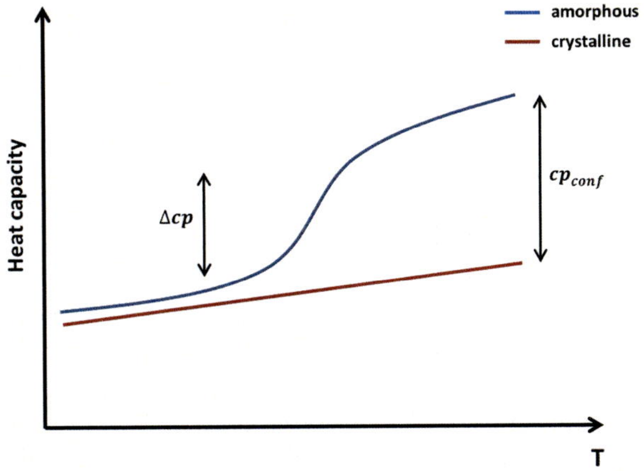

Fig. 2.7 Heat capacity difference (ΔCp) and configurational heat capacity (Cp_{conf}) of indomethacin. (Adapted from Graeser et al. 2010)

has often been made in order to simplify calculations of stability, the excess heat capacity of the glass over the crystal has to be considered as it adds to the overall excess configurational entropy.

The change in free energy is regarded as a driving factor for recrystallization from the amorphous form: the larger the difference in free energy between the amorphous and crystalline state, the more thermodynamically favorable is the situation upon recrystallization.

Calculation of the free energy difference (equivalent to the thermodynamic driving force for recrystallization) can be achieved by taking into account the temperature dependencies of the heat capacities for the amorphous and crystalline material. Upon measurement of the heat capacities and calculation of the configurational heat capacity, the configurational enthalpy (H_{conf}), configurational entropy (S_{conf}), and the configurational free energy (G_{conf}) may be calculated:

$$G_{\text{conf}}(T) = H_{\text{conf}}(T) - S_{\text{conf}}(T). \tag{2.24}$$

The configurational enthalpy and entropy are the differences between the amorphous and the crystalline forms (Eqs. 2.25 and 2.26), and the values may be obtained from their relationship with the heat capacity (Eqs. 2.27 and 2.28).

$$H_{\text{conf}}(T) = H^{\text{amorph}}(T) - H^{\text{crystal}}(T) \tag{2.25}$$

$$S_{\text{conf}}(T) = S^{\text{amorph}}(T) - S^{\text{crystal}}(T) \tag{2.26}$$

$$H_{\text{conf}} = \Delta H_{\text{m}} + \int_{T_{\text{m}}}^{T} Cp_{\text{conf}}\,\text{d}T \tag{2.27}$$

$$S_{\text{conf}} = \Delta S_{\text{m}} + \int_{T_{\text{m}}}^{T} \frac{Cp_{\text{conf}}}{T}\text{d}T. \tag{2.28}$$

The melting entropy can be obtained from the following relationship:

$$\Delta S_{\mathrm{m}} = \frac{\Delta H_{\mathrm{m}}}{T_{\mathrm{m}}}. \tag{2.29}$$

These configurational thermodynamic values give an indication of the relationship between the amorphous form and the crystalline state of a compound. The larger the configurational values, the greater are the differences between the crystalline and the amorphous forms.

A large value for H_{conf} indicates that the amorphous form holds a large amount of excess enthalpy compared to the crystalline form. H_{conf} determines the enthalpic driving force for crystallization, and as a consequence, compounds which show large values of H_{conf} should show a higher tendency of recrystallization. The entropic barrier of recrystallization is given by S_{conf}. For compounds with a high value, this indicates high structural diversity of the amorphous and the crystalline states. It has been suggested that the higher the entropic barrier, the more difficult it is for the molecule to arrange itself in the correct conformation for recrystallization. Compounds with a high S_{conf} value should therefore be more stable towards crystallization.

Investigation of this theory led to varying results. In some cases, a relationship between H_{conf} and recrystallization above the T_{g} was suggested (Marsac et al. 2006a), and in other studies, researchers commented on the relationship of S_{conf} and recrystallization (Graeser et al. 2009b; Karmwar et al. 2011b; Zhou et al. 2002). It should be noted, however, that these correlations were drawn for the stability of the supercooled liquid, rather than the glass itself.

The procedure of evaluating thermodynamic parameters in terms of their configurational properties is an attempt to incorporate not only the properties of the amorphous form but also of the original crystalline state, as the driving force for recrystallization is the energy difference between these two states.

Calculation of the thermodynamic quantities requires the system under investigation to be in equilibrium. For an amorphous system, this is achieved in the temperature range above the T_{g}, in the supercooled liquid state. Theoretically, the calculations of the configurational thermodynamic properties are not valid for the glass; however, the behavior of the amorphous form above the T_{g} may give an indication on the stability behavior below the T_{g} and calculation of these parameters provides further information on the behavior of the amorphous form.

To summarize this section, the formulation scientist will consider an amorphous approach for formulating a poorly water-soluble drug if sufficient solubility enhancement compared to the crystalline form can be expected. In addition to a superior solubility profile, the drug under consideration should be prepared easily in the amorphous form (high GFA) and it should show resistance against recrystallization (high GS). These two factors can to date only be roughly estimated, despite the abundance of equations available. In reality, an amorphous drug will rarely be manufactured on its own; rather, it will usually be formulated with a polymer as a glass solution. The polymer should be chosen to increase the physical stability in the solid state and in the aqueous media of the GI tract and also maintain a degree of supersaturation.

2.4 Drug–Polymer Miscibility: Theoretical Approaches

Solid polymer dispersions have become a preferred method to enhance drug dissolution and to stabilize the amorphous form of a drug (Sekiguchi and Obi 1961; Leuner and Dressman 2000; Newman et al. 2012; Serajuddin 1999; Sethia and Squillante 2003; Williams et al. 2013; Zheng et al. 2012). As pointed out earlier, the term "solid dispersion" is used for various solid mixtures of a drug with excipients; however, the focus of this chapter is on amorphous single-phase dispersions, i.e., glass solutions. There is an extensive amount of research on the mechanisms of how glass solutions lead to increased physical stability of the amorphous drug, but in spite of this, there is still a lack of sufficient understanding of the issue. However, some commonly accepted generalizations can be made.

A high glass transition temperature (T_g) of the carrier polymers increases the T_g of the glass solution compared to the pure amorphous drug, lowers the mobility of drug molecules, and kinetically acts as a crystallization inhibitor (Hancock et al. 1995; Janssens and Van den Mooter 2009). Furthermore, intermolecular drug–polymer interactions have been shown to be responsible for the stabilization of the drug in solid dispersions, even with relatively small polymer amounts (Janssens and Van den Mooter 2009; Matsumoto and Zografi 1999). Most importantly, however, the drug and the polymer should be mixed homogeneously at the molecular level during processing, and optimally, miscibility will also be maintained at the storage conditions of the solid dispersions. In case the drug is supersaturated beyond its solubility/miscibility limit in a particular polymer, thermodynamics dictates that the systems will tend to phase separate, but due to slow dynamics, the blend may still be kinetically stable enough for the intended use (Marsac et al. 2006b). Thus, knowledge of the phase behavior of a drug–polymer system is needed to avoid potential phase separation driven by supersaturation of the drug substance, and is essential regarding estimation of correct drug loading for manufacturing stable solid dispersion formulations. It should be noted that miscibility describes the tendency of the supercooled liquid/glassy form of the drug to mix with a polymer, whereas solubility refers to the ability of the polymer to act as a solvent and dissolve a crystalline drug (Qian et al. 2010a). Physical instability has been observed to be more common in systems where the mixing is heterogeneous; drug–polymer miscibility is poor and drug loading exceeds the drug–polymer miscibility limit (Marsac et al. 2009; Paudel et al. 2010; Qian et al. 2010a).

Hence, drug–polymer miscibility is an essential prerequisite for the successful formulation of a physically stable glass solution. In addition, glass solutions provide an environment in which the solid state of the drug is altered to give rise to enhanced solubility and dissolution rate. Owing to the potential for the successful formulation of a poorly water-soluble drug, the study of miscibility of drug with polymer has increasingly become a topic of interest in both academic and industrial research (Djuris et al. 2013; Wyttenbach et al. 2013). However, many real pharmaceutical formulation cases are particularly challenging because the drug loading in a single-phase amorphous glass solution needs to be relatively high for the required drug dose to be achieved. Often, this means that the drug is present in a supersaturated form relative to the solubility of the crystalline form in the system (Newman et al. 2012). This supersaturation is maintained by the apparent miscibility of the components in the amorphous form. However, also in these systems, a thermodynamic tendency for phase separation and subsequent drug crystallization is present. True miscibility of amorphous components would represent a thermodynamically stable single-phase system, similarly to miscibility of liquids. The concept of miscibility is well defined for polymer blends, in which the polymers are usually in a stable amorphous form. However, in the case of miscibility of a low molecular weight amorphous drug with a polymer in a solid dispersion, the drug is in an unstable form relative to the crystalline form and has the tendency to crystallize. The measurable miscibility in the case of a drug–polymer dispersion is therefore associated with a metastable equilibrium which is different to that defined in polymer science. Thus, a miscible drug–polymer system would eventually reach equilibrium with regard to the crystalline drug, and this equilibrium composition would be the solubility of the crystalline drug in the polymer (see Sect. 2.5). Miscibility at temperatures near and below T_g is only apparent and involves the kinetics of phase separation and structural relaxation. The theoretical equilibrium miscibility may practically only be estimated from extrapolation and modeling (Qian et al. 2010a).

It is of interest to experimentally determine if miscible binary drug–polymer systems were achieved upon processing, such as melt extrusion, spray drying, or freeze drying. This is routinely done by DSC, where only a single T_g should appear for miscible systems (Forster et al. 2001; Matsumoto and Zografi 1999). In an ideally mixed system (assuming no interaction between the components), the single T_g lies between the values of the pure component, depending on composition. However, problems may arise if the drug and the polymer have similar T_gs, as it may not be possible to resolve whether one or two thermal transitions are occurring. For such situations, the aging method may be helpful (ten Brinke et al. 1994). In this method, the sample is aged at a temperature somewhat below the lowest T_g and two distinct endothermic peaks may be observed on heating through the glass transition region, as the enthalpy lost during aging is recovered. Furthermore, the phase-separated regions have to be large enough for a DSC to be able to detect phase separation. It has been shown with amorphous solid dispersions consisting of BMS-A (a poorly water-soluble drug) and PVP-VA that phase separation at the scale of tens of microns may still be undetectable by DSC (Qian et al. 2010a).

The T_g of an ideally mixed solid solution can be predicted from the sum of the weight fractions (w) and T_g values of the individual components of the mixture by using the Gordon–Taylor equation (Gordon and Taylor 1952):

$$T_{\mathrm{gmix}} = \frac{w_{\mathrm{d}} T_{\mathrm{gd}} + K w_{\mathrm{p}} T_{\mathrm{gp}}}{w_{\mathrm{d}} + K w_{\mathrm{p}}}, \tag{2.30}$$

where $T_{\mathrm{g\,mix}}$ is the theoretical glass transition temperature of the drug–polymer blend and w_{d} and w_{p} and T_{gd} and T_{gp} are the weight fractions and glass transition temperatures (in Kelvin) of the pure drug and polymer, respectively. K is a constant which can be estimated from the densities of the components by the Simha–Boyer rule (Eq. 2.31) or from the heat capacities (Eq. 2.32; Couchman and Karasz 1978; Simha and Boyer 1962):

$$K \approx \frac{\rho_d T_{gd}}{\rho_p T_{gp}} \tag{2.31}$$

$$K \approx \frac{\Delta C_{pp}}{\Delta C_{pd}}, \tag{2.32}$$

where ρ_d and ρ_p are the true densities (g/cm^3) and ΔC_{pd} and ΔC_{pp} are the changes in heat capacity at T_g of the amorphous drug and polymer, respectively. Analogous equations can be derived for mixtures containing more than two components (Gupta and Bansal 2005).

In solid dispersion systems for which the T_g can be accurately predicted, the two components may be ideally mixed in the liquid form and be fully miscible at the molecular level (Baird and Taylor 2012). Deviations of the experimentally determined T_g's from the theoretical T_g's are generally considered as indicative of differences in the strengths of the intermolecular interactions between the individual components and components in the blend (Nair et al. 2001). A positive deviation from theoretical T_g would suggest that the amount and strength of bonding between the components in the mixture would be higher than interactions existing in the individual components due to, e.g., intermolecular hydrogen bonding (Gupta and Bansal 2005). In contrast, a negative deviation has been explained by an overall loss in the number and/or strength of interactions upon mixing (Shamblin et al. 1996; Taylor and Zografi 1998). Related to this, an increase in free volume due to mixing can be the cause for a lower T_g than predicted (Shamblin et al. 1998).

In addition to the measurement of the T_g, there are several other methods that can provide information about miscibility in drug–polymer solid dispersions. These include solubility parameter calculations (Forster et al. 2001; Greenhalgh et al. 1999; Hancock et al. 1997; Marsac et al. 2009), the use of partition coefficients (Yoo et al. 2009), rheological methods (Liu et al. 2012), melting point depression measurements (Marsac et al. 2006b; Paudel et al. 2010; Wiranidchapong et al. 2008), and computational methods based on X-ray diffraction (XRD) data (Ivanisevic et al. 2009). An essential element in the solubility parameter and melting point depression approaches is the determination of the Flory–Huggins (F–H) interaction parameter as an estimate of miscibility, and thus these methods are described in detail later in this chapter.

The extent of miscibility between a given compound and polymer is sensitive to different variables, e.g., environmental factors (such as moisture) may all affect the outcome. Whether the system forms a stable single-phase mixture or separates into coexisting phases is determined by the thermodynamic miscibility of these two components at the specific condition (temperature, pressure, and composition). In many pharmaceutical circumstances, a change in the experimental conditions causes phase separation to occur in a homogeneous one-phase system. This issue is also covered later in this chapter.

2.4.1 General Thermodynamics of Mixing

Amorphous miscibility is defined as the level of molecular mixing which is adequate to result in macroscopic properties expected for single-phase amorphous material. A miscible system in the amorphous form is represented by a single glass transition which is not identical to the glass transition of any of the components included in the system (Forster et al. 2001; Marsac et al. 2006b; Matsumoto and Zografi 1999). However, as stated above, a single T_g is not necessarily a definite indicator of the miscibility of a system (Qian et al. 2010b) and it gives no information on the thermodynamics of mixing.

Whether the formation of a one-phase binary amorphous system is thermodynamically favorable is determined by different factors. In any case, for a system to be thermodynamically miscible, the free energy, given by Eq. 2.33, must be negative:

$$\Delta G_{mix} = \Delta H_{mix} - T\Delta S_{mix}, \tag{2.33}$$

where ΔH_{mix} and ΔS_{mix} are the enthalpy and entropy of mixing, respectively, which are both temperature and composition dependent. The combinatorial entropy of mixing for the system of a drug and polymer can be represented mathematically by Eq. 2.34:

$$\Delta S_{mix} = -R(n_d \ln \phi_d + n_p \ln \phi_p), \tag{2.34}$$

where ΔS_{mix} is the entropy of mixing, R is the gas constant, and $n_{d,p}$ and $\varphi_{d,p}$ are the number of moles and volume fraction of the drug and polymer, respectively. Since $\varphi_{d,p} < 1$, their logarithms will be negative and ΔS_{mix} will be positive, as expected, since mixing increases disorder of the system. Thus, a positive ΔS_{mix} makes a negative contribution to ΔG_{mix}, as can be seen from Eq. 2.33, meaning that entropy always favors mixing, and that amorphous–amorphous phase separation in solid dispersions will be largely driven by the positive enthalpy values (ΔH_{mix}), which can be represented as:

$$\Delta H_{mix} = (H_{dd} + H_{pp}) - H_{dp}. \tag{2.35}$$

This means that the enthalpy of mixing (ΔH_{mix}) between two components is equal to the difference between the enthalpies of the pure components and the mixture.

The significance of enthalpic interactions to determine the miscibility of a mixture is known for polymer blends. Compared to enthalpy, the entropy of mixing of two polymeric materials, although typically favorable, is very small due to the low number of possible configurations for the two components in the binary mixture. However, the combinatorial entropy is more favorable for mixing in the case of a small molecule with a polymer, and as a consequence, the system can tolerate a larger unfavorable enthalpy of mixing and still achieve a negative ΔG_{mix}. The enthalpy value is positive in the case of endothermic mixing, negative in the case of exothermic mixing, or zero in the case of athermal mixing (Baird and Taylor 2012; Marsac et al. 2006b, 2009).

Exothermic mixing is achieved when adhesive interactions formed between the two components are stronger and/or more numerous than cohesive interactions in the pure components, as, for example, in the case of indomethacin–Eudragit mixtures, where disruption of the indomethacin dimer due to strong drug–polymer associations was observed upon formation of the solid dispersion (Liu et al. 2012). In contrast, even if strong adhesive interactions may be formed between two components in a binary mixture, these interactions come at the cost of breaking strong cohesive interactions in the individual components and a positive ΔH_{mix} may result, as, for example, in the case of mixtures of citric acid and indomethacin (Lu and Zografi 1998). Athermal mixing implies that adhesive interactions are formed between the two components; however, these are similar in magnitude and/or extent to the cohesive interactions. Therefore, in order to predict miscibility between a drug and polymer, it would be informative to estimate the magnitude of the interactions.

2.4.2 *Prediction of Drug–Polymer Miscibility*

Predictive models for drug–polymer miscibility have been introduced, and they are largely derived from solution thermodynamics. Lattice-based solution models, such as the F–H theory, can be used to assess miscibility in drug–polymer blends, for which the F–H interaction parameter can be considered as a measure of miscibility. In addition, solubility parameter models can be used for this purpose. The methods used to estimate interaction parameters include melting point depression and the determination of solubility parameters using group contribution theory.

2.4.2.1 Flory–Huggins (F–H) Lattice Theory

F–H theory is a classical theory of phase separation based on the Gibbs free energy in polymer–solvent systems (Flory 1953). A small molecule drug–polymer pair can be considered as analogous to a solvent–polymer system and thus can be described by the F–H theory. The F–H lattice theory takes into account the nonideal entropy of mixing of a large molecule with a small molecule, as well as any enthalpy of mixing contributions (Baird and Taylor 2012; Zhao et al. 2011). It describes a lattice model in which the structural units of a polymer and the solvent molecules are placed. Changes in entropy and enthalpy can be calculated based on the location of the molecule units in the lattice, and interactions between the structural units of the polymer and the solvent molecules, respectively. In a solid dispersion, the free energy of mixing (ΔG_{mix}) can be described as follows:

$$\frac{\Delta G_{mix}}{RT} = n_d \ln \varphi_d + n_p \ln \varphi_p + \chi_{dp} n_d \varphi_p, \tag{2.36}$$

where R is the gas constant, T is the absolute temperature (K), $n_{d,p}$ are the number of moles of the drug and the polymer, respectively, $\varphi_{d,p}$ are the volume fractions of the drug and the polymer, respectively, and χ is the F–H interaction parameter. Using volume fractions instead of mole fractions in the equation allows accounting for the comparatively large volume occupied by a polymer chain compared to the drug in the binary system (Thakral and Thakral 2013). The first two terms on the right-hand side describe the entropic contribution (combinatorial entropy), and the last term describes the enthalpic contribution to the total free energy of mixing of the binary system. As discussed above, entropy will always favor mixing by always being positive. Thus, the enthalpic part of the equation will determine whether ΔG_{mix} will become negative. The determining factor for this would be χ, which can consequently be considered as an indicator for drug–polymer miscibility. A negative (or slightly positive) χ value would lead to a negative (favorable) enthalpy of mixing between the components and an overall negative free energy of mixing. This would be an indication of adhesive forces between the drug and polymer, which have to be greater than the cohesive (drug–drug and polymer–polymer) forces (Pajula et al. 2010). However, the F–H theory only considers nonspecific dispersion force interactions and the inability of the theory to take specific interactions, such as hydrogen bonds, into account is a limitation. This should be kept in mind when determining χ for a particular system in which drug–polymer hydrogen bonding exists (Janssens et al. 2010). It has been suggested that for polymers with hydrogen bonding and other interactions, an extra free energy term ($\Delta G_{\text{H}}/RT$) should be added to the Eq. 2.36. This is referred to as the modified F–H theory (Coleman and Painter 1995; Li and Chiappetta 2008). Thus, in this case, the enthalpy of mixing (which is proportional to χ) and specific interactions (ΔG_{H}) would determine the miscibility of two polymers. Other limitations of the F–H theory include the mean-field approximation to facilitate the calculation for placement of a polymer molecule in a partly filled lattice, which is only sufficient when the volume fraction of the polymer is high. In addition, in F–H theory, the energy for breaking the crystal lattice is not taken into account and therefore the theory can only describe the miscibility of amorphous drugs with amorphous polymers (Zhao et al. 2011).

It is also important to note that χ is dependent on temperature, composition, and polymer chain length. The temperature dependence can be described by Eq. 2.37 (Rubinstein and Colby 2003):

$$\chi = A + \frac{B}{T} C_1\varphi + C_2\varphi^2, \tag{2.37}$$

where A is a temperature-independent constant (entropic contribution), B is a temperature-dependent constant (enthalpic contribution), and C_1 and C_2 are fitting constants of χ with respect to volume fraction. For a particular drug–polymer combination, a simplification of Eq. 2.37 can be made:

$$\chi = A + \frac{B}{T}. \tag{2.38}$$

Knowledge of the relationship between χ and temperature allows one to draw temperature-composition phase diagrams, as described later.

2.4.2.2 Estimation of the Interaction Parameter by Solubility Parameter Approach

Calculation of solubility parameters (δ) for a drug and a polymer have been used as a method for predicting miscibility in amorphous solid dispersions (Hancock et al. 1997), using, for example, the Hildebrand solubility parameter (Greenhalgh et al. 1999), which was calculated from the cohesive energy density (CED) by:

$$\delta = (\mathrm{CED})^{0.5} = \left(\frac{\Delta E_{\mathrm{v}}}{V_{\mathrm{m}}}\right)^{0.5}, \quad (2.39)$$

where ΔE_{v} is the energy of vaporization and V_{m} the molar volume.

CED is the cohesive energy per unit volume and it can be used to predict the solubility of one material in another. Solubility parameters can be evaluated by conducting solubility studies for the test materials in solvents of known cohesive energies (Reuteler-Faoro et al. 1988), by using inverse gas chromatography (IGC; Adamska et al. 2008; Phuoc et al. 1986), from heat of vaporization data (not suitable for many polymers), or by calculation using group contribution methods. The Hansen solubility parameters are calculated from the chemical structures of compounds using the approaches of Hoftyzer/Van Krevelen. In this method, the dispersive forces, interactions between polar groups, and hydrogen bonding groups are taken into account:

$$\delta_{\mathrm{coh}}^2 = \delta_{\mathrm{d}}^2 + \delta_{\mathrm{p}}^2 + \delta_{\mathrm{h}}^2, \quad (2.40)$$

where δ_{d}, δ_{p}, and δ_{h} are the dispersive, polar, and hydrogen bonding solubility parameter components, respectively, which in turn can be calculated as:

$$\delta_{\mathrm{d}} = \frac{\sum F_{\mathrm{di}}}{V}, \quad \delta_{\mathrm{p}} = \frac{\sqrt{\sum F_{\mathrm{pi}}^2}}{V}, \quad \delta_{\mathrm{h}} = \sqrt{\frac{\sum E_{\mathrm{hi}}}{V}}, \quad (2.41)$$

where F_{di}, F_{pi}, and E_{hi} are the group contributions for different components of structural groups and V is the group contribution to molar volume. The group contribution values are listed in the literature (Van Krevelen 1997).

Compounds with similar δ values are generally considered miscible since the energy required for mixing the components is compensated by the energy released by interactions between the components. For the evaluation of miscibility, it has been suggested that if the difference between the solubility parameters of the components to be mixed ($\Delta\delta$) is smaller than 7 $\mathrm{MPa}^{1/2}$, then the components are likely to be miscible, and if $\Delta\delta$ is smaller than 2 $\mathrm{MPa}^{1/2}$, the components might form a solid solution. $\Delta\delta$ values larger than 10 $\mathrm{MPa}^{1/2}$ denote immiscibility. Forster et al. (2001) used this approach in order to predict formation of glass solutions upon melt extrusion of 2 model drugs and 11 different excipients. Miscibility of the drugs with these excipients was determined experimentally by DSC and hot-stage microscopy (HSM), and the experimental results agreed with solubility parameter predictions. The drug/excipient combinations predicted to be generally immiscible exhibited more than one T_{g} upon reheating in the DSC. Melt extrusion of miscible components resulted in amorphous glass solution formation, whereas extrusion of immiscible components led to amorphous drug dispersed in a crystalline excipient. In solid dispersions prepared by solvent evaporation, the correlation between amorphous miscibility/physical stability and physicochemical properties, such as the Hildebrand solubility parameters, was investigated (Yoo et al. 2009). It was found that $\Delta\delta$ values smaller than (or equal to) 2.8 for drug–excipient systems indicated miscibility and formation of an amorphous dispersion. However, when an acid–base interaction was present, amorphous miscible systems were formed regardless of the other parameters (hydrogen bonding energy and log P). No correlation between $\Delta\delta$ and physical stability could be established.

The F–H interaction parameter can be estimated from solubility parameters as follows. First, the enthalpy of mixing for a drug–polymer system can be described by Eq. 2.42:

$$\Delta H_{\mathrm{m}} = V_{\mathrm{dp}}\varphi_{\mathrm{d}}\varphi_{\mathrm{p}}(\delta_{\mathrm{d}} - \delta_{\mathrm{p}})^2, \tag{2.42}$$

where φ_{d} and φ_{p} are the volume fractions of the drug and the polymer, respectively, δ_{d} and δ_{p} are the solubility parameters of drug and polymer, respectively, and V_{dp} is the volume of mixture. According to the F–H theory, ΔH_{m} can be given by the van Laar expression:

$$\Delta H_{\mathrm{m}} = \chi_{\mathrm{dp}} RT \varphi_{\mathrm{d}}\varphi_{\mathrm{p}}. \tag{2.43}$$

Thus, when combining Eqs. 2.41 and 2.42, the interaction parameter between a drug and polymer, χ_{dp}, can be calculated as follows:

$$\chi_{\mathrm{dp}} = \frac{V_{\mathrm{dp}}(\delta_{\mathrm{d}} - \delta_{\mathrm{p}})^2}{RT}. \tag{2.44}$$

Solubility parameters, determined by group contribution theory, were used for the estimation of interaction parameters using Eq. 2.44 for felodipine–PVP and nifedipine–PVP systems. χ values of 0.5 and 2.0 were obtained, respectively, which predicted that mixing in these two systems would be endothermic and that the nifedipine–PVP system would be immiscible. However, this was found to be in conflict with the experimental results which have shown that both nifedipine and felodipine are miscible with PVP at all compositions. The authors explained that Eq. 2.44 assumes the enthalpic interactions between the different components to be equal to the geometric mean of the enthalpic interactions of similar components. This assumption would generally be sufficient for systems with van der Waals interactions, but not for systems with specific, directional interactions such as hydrogen bonding. It has been shown that strong hydrogen bonding exists between nifedipine and PVP. Thus, miscibility estimated by the interaction parameter, obtained through the solubility parameter approach, is likely to be an underestimation for systems like nifedipine–PVP (Marsac et al. 2006a, b).

When the interaction parameter is known, a phase diagram can be constructed for a binary system, and several examples of this can be found in the literature (Thakral

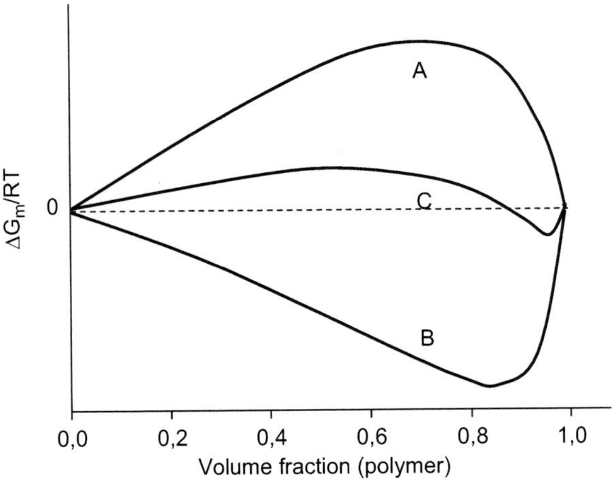

Fig. 2.8 Composition dependence of the total free energy of mixing calculated for PEG 6000 and **a** sucrose ($\chi = 22.313$), **b** phenylbutazone ($\chi = 0.063$), and **c** chlorampenicol ($\chi = 2.45$). (Schematic drawing adapted from Thakral and Thakral 2013)

and Thakral 2013; Tian et al. 2013; Zhao et al. 2011). The phase diagram shows the free energy of mixing as a function of different compositions, based on F–H theory. Thakral and Thakral (2013) investigated polyethylene glycol (PEG) 6000 as a model polymer and a dataset of 83 drugs and some commonly used excipients. Molar volumes for each of the molecules in the dataset and volume fractions of polymer and drug for each binary mixture were calculated. Subsequently, solubility parameters for the polymer and all the drugs/excipients were calculated, and these values were used to calculate χ according to Eq. 2.44. The phase diagrams were then constructed for each drug/excipient combination by using Eq. 2.36. On the basis of the shapes of the phase diagrams, the authors categorized the drugs in three different groups. Type I compounds showed a negative value of total free energy of mixing with all combinations (phase diagram for phenylbutazone/PEG 6000 shown in Fig. 2.8b). Thus, drugs in this category can be considered miscible with PEG 6000 at all mixture ratios. The values for χ in this group were found to be < 0.98 (at 298 K). In contrast, type II compounds showed a positive total free energy of mixing regardless of composition, and these compounds can be considered as immiscible with the polymer at all mixture ratios (phase diagram for sucrose/PEG 6000 shown in Fig. 2.8a). The χ values for the compounds in this category were found to vary between 5.19 and 28.27 (at 298 K). For type III compounds, the total free energy of mixing for compositions containing a low amount of polymer was found to be positive, whereas increasing the polymer fraction turned the energy of mixing to negative values (phase diagram for chloramphenicol/PEG 6000 shown in Fig. 2.8c). Thus, these drug/excipient combinations form a biphasic system at low polymer concentrations and a single phase at high polymer concentrations. In the study, this kind of behavior was observed for compounds exhibiting χ values in the range of 1.09–4.19 (at 298 K). However, it should be noted that, as Eq. 2.38 shows, χ is temperature dependent, and it is expected to be reduced at higher temperatures. Thus, an immiscible system (at 298 K) becomes miscible when the temperature is raised. Vice versa, a single-phase binary system could be expected to undergo phase separation when the temperature of the binary system is reduced. When the authors studied miscibility of the model drugs with PEG experimentally by DSC, they found that PEG-immiscible drugs (type II) were distinguished well from the compounds showing complete miscibility and composition-dependent miscibility with PEG (types I and III), but the differentiation between type I and type III was less clear. Nevertheless, the authors concluded that the knowledge of the interaction parameters could be helpful in drug development, in initial screening of polymers that may yield a stable binary mixture with a particular drug.

2.4.2.3 Estimation of the Interaction Parameter Using Melting Point Depression

The presence of a polymer changes the melting behavior of a drug in pharmaceutical systems if the drug is miscible with the polymer. Thus, a reduction in the melting temperature of the crystalline drug is observed, as, for example, for the piroxicam–PVP system (Tantishaiyakul et al. 1999). This is due to the fact that there is a negative free energy of mixing associated with the spontaneous mixing of the polymer with the liquid phase of the drug and the chemical potential of the drug in the mixture is decreased compared to that of the pure liquid drug. Thus, the melting process becomes thermodynamically more favorable which leads to melting point depression (Baird and Taylor 2012). If there is an amorphous proportion of the drug present, it further lowers the initial chemical potential of the drug and thus lowers the melting point additionally. When a polymer is above its T_g, it can act as a solvent for the drug and the end point of the melting endotherm is the temperature at which all the drug has dissolved in the polymer. Accordingly, if the temperature scanning rate of the experiment is slow enough to allow the equilibrium to be reached, the solubility of the crystalline drug in the supercooled polymer can be obtained, as will be discussed in more detail later. The melting point depression is larger in the case of strongly exothermic mixing, less for weakly endothermal or athermal mixing, and if the drug and polymer are immiscible with each other, melting point depression does not occur (Marsac et al. 2009). This phenomenon has been utilized for studying miscibility of polymer–polymer and polymer–solvent systems. Reduction of melting temperature of the crystalline phase as a function of composition and polymer–polymer interaction has been analyzed using the Nishi–Wang equation (Nishi and Wang 1975) based on the F–H theory:

$$T_m^{pure} - T_m^{mix} = \frac{-T_m^{pure} B V_2 \varphi_1^2}{\Delta H_{fus}}, \tag{2.45}$$

where T_m^{pure} is the melting temperature of the pure crystalline component, T_m^{mix} is the melting temperature of the crystalline component of the mixture, B is the interaction energy density between blend components, V_2 is the molar volume of the repeating unit of the crystalline component, φ_1 is the volume fraction of the amorphous component in the blend, and ΔH_{fus} is the heat (enthalpy) of fusion of the crystalline component per mole of the repeating unit. The value of B is independent of composition, and it represents the intensity of molecular interaction during mixing (Paudel et al. 2010). The more negative the value of B, the larger the interaction between the components.

The Nishi–Wang equation was found to predict T_m^{mix}, miscibility, and the existence of a specific interaction between the components in the molten state when investigating the miscibility between 17β-estradiol and Eudragit RS (Wiranidchapong et al. 2008). The B value obtained from curve fitting was -0.28 ± 0.0094 J/g cm^3, indicating some degree of interaction between 17β-estradiol and Eudragit RS in the system. The interaction-related B value was found to depend on the molecular weights of the mixing components. Paudel et al. (2010) obtained B values of −89.17, −118.03, and −68.05 for naproxen mixed with PVP K12, PVP K25, and PVP K90, respectively. This indicated that the interaction potential with the particular drug was highest with PVP K25. By extending the equations for polymer–solvent systems (Hoei et al. 1992), the melting point depression data from DSC measurements was related to the F–H interaction parameter for drug–polymer systems by Eq. 2.46 (Marsac et al. 2006b):

$$\left(\frac{1}{T_m^{mix}} - \frac{1}{T_m^{pure}}\right) = \frac{-R}{\Delta H_{fus}}\left[\ln \varphi_d + \left(1 - \frac{1}{m}\right)\varphi_p + \chi\varphi_p^2\right], \tag{2.46}$$

where T_m^{mix} is the melting temperature of a drug in the presence of a polymer, T_m^{pure} is the melting temperature of the pure drug without a polymer, ΔH_{fus} is the heat of fusion of the pure drug, $\varphi_{d,p}$ are the volume fractions of the drug or polymer, respectively, and m is the ratio of the volume of the polymer to the volume of the lattice site. Estimation of the interaction parameters from melting point depression data for nifedipine and felodipine mixed with PVP K12 was performed by rearranging Eq. 2.46 in order to establish a linear relationship as a function of φ_p^2 from which χ was determined as a slope of the curve. In addition, the visualization of the melting point depression by DSC required a reduced scanning rate and controlled particle size, due to slow mixing kinetics. For a polymer volume fraction up to 0.25, the curve was linear for both felodipine and nifedipine, but at higher concentrations, linearity was lost. This was explained to originate from the composition dependence of the interaction parameter and the increasingly unfavorable kinetics of drug–polymer interaction as the melting point is depressed closer to the T_g of the polymer. Thus, only the linear part for low polymer concentrations was used to obtain χ values of −3.8 and −4.2 for nifedipine–PVP and felodipine–PVP, respectively. When compared to the values obtained by the solubility parameter approach (2.0 and 0.5, respectively), it could be concluded that the melting point depression method resulted in values that were better in agreement with the experimental observations (Marsac et al. 2006a). In another study, the interaction parameters for felodipine with Soluplus and hydroxypropyl methyl cellulose acetate succinate (HPMCAS) were calculated using the solubility parameter and melting point depression methods. Similar values were obtained, indicating that either method is suitable in this case. The resulting values were in the range between 2.8 and 7.5 $MPa^{1/2}$ which predicts the felodipine—Soluplus and felodipine—HPMCAS systems to show limited miscibility (Tian et al. 2013).

The application of the melting point depression method for the evaluation of the miscibility for several drug–polymer systems showed that systems identified as miscible exhibited melting point depression, while systems identified as immiscible or only partially miscible showed only slight or no melting point depression (Marsac et al. 2009). Applying the melting point data to Eq. 2.46, χ values for the systems could be obtained. Negative or close to zero values were estimated for all PVP systems that were miscible, and systems previously known to be immiscible gave large positive χ values. Although the theoretical interaction parameters have often been found to be in reasonable agreement with the experimental results, there are limitations associated with the melting point depression method. First, application of the method to pharmaceutical systems requires that the drug and polymer are stable over the temperature range of interest and that there are sufficient physical interactions between the components for the melting point depression to be observed. In addition, the melting point of the drug should be sufficiently high for the polymer to exist in the supercooled liquid state, allowing mixing and interaction with the drug. Thus, the method is best suited for systems where the polymer has a T_g significantly lower than the T_m of the drug. Furthermore, Eq. 2.46 is linear only at comparatively low polymer concentrations, which probably is due to the kinetics of mixing during the experiment. It should also be noted that the method does not provide a universal value for χ but an estimation close to the melting point of the drug.

Zhao et al. (2011) attempted to correlate χ with temperature and construct a temperature phase diagram for indomethacin–PVP-VA systems. This would allow an estimation of the miscibility behavior of a drug polymer pair at different temperatures and compositions. In general, phase diagrams are useful in describing the compatibility of binary mixtures (Lin and Huang 2010). Phase diagrams give information on the temperature ranges in which the particular mixture would be miscible and/or unstable. An unstable binary mixture will separate into two phases, the compositions of which can be seen from the phase diagram. Two different sets of χ values, $\chi_1(T_1)$ and $\chi_2(T_2)$, were calculated, the first set being obtained from melting point depression experiments at T_m and the other set from solubility parameter calculations at room temperature. This allowed the authors to obtain values for A and B in Eq. 2.38. By substituting Eq. 2.38 into Eq. 2.36 and taking into account the fact that the component volume fractions equal 1, they were able to obtain the free energy versus composition phase diagram for the indomethacin–PVP-VA system at different temperatures (Fig. 2.9a). Subsequently, this relation was transformed to a temperature-composition phase diagram which summarizes the phase behavior of the mixture by showing regions of stability, instability, and metastability (Fig. 2.9b). The first phase boundary was determined by the tangent of the free energy curve where the first derivative of the free energy of mixing with respect to volume fraction is set equal to zero (Fan et al. 1992; Zhao et al. 2011):

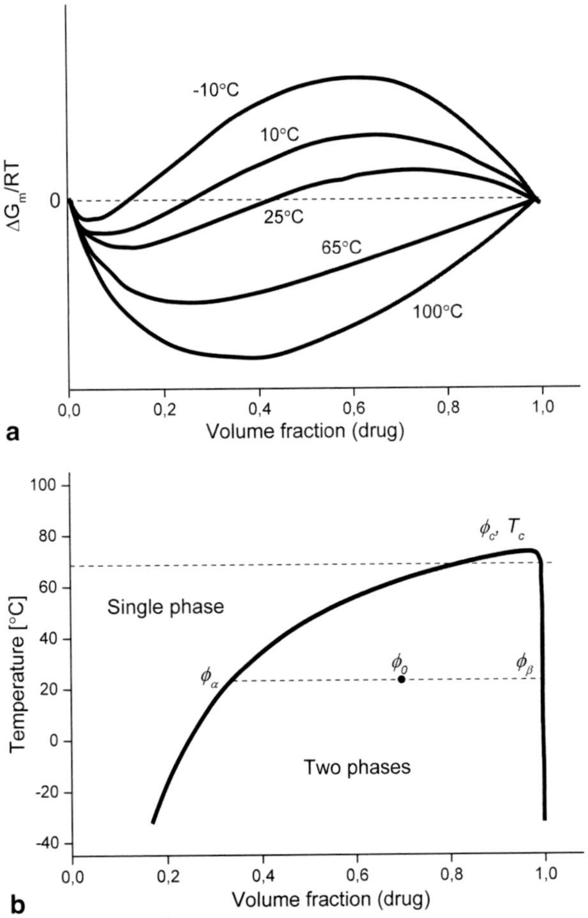

Fig. 2.9 **a** Free energy versus composition phase diagram for the indomethacin and PVP-VA system and **b** temperature versus composition phase diagram for the indomethacin and PVP-VA system showing the spinodal curve (the binodal curve was not determined in the study). (Adapted from Zhao et al. 2011)

$$\frac{\partial \Delta G_{\text{mix}}}{\partial \varphi_{\text{drug}}} = \ln \varphi_{\text{drug}} + 1 - \frac{1}{m_{\text{poly}}} - \frac{1}{m_{\text{poly}}} \ln (1 - \varphi_{\text{drug}}) + (1 - 2\varphi_{\text{drug}}) \chi_{\text{drug-poly}} = 0, \tag{2.47}$$

where ΔG_{mix} is free energy of mixing, φ_{drug} and φ_{poly} are the volume fractions of the drug and the polymer, respectively, m_{poly} is the degree of polymerization of the amorphous polymer, and $\chi_{\text{drug-poly}}$ is the F–H interaction parameter.

The numerical solution of Eq. 2.47 is possible by combining it with Eq. 2.36 for the interaction parameter at different temperatures. The phase boundary obtained corresponds to the boundary between the stable and metastable region and is known as the *binodal curve*. In addition, the *spinodal curve* that represents the boundary between the unstable and metastable region can be obtained by setting the second derivative of the free energy to zero:

$$\frac{\partial^2 \Delta G_{\text{mix}}}{\partial^2 \varphi_{\text{drug}}} = \frac{1}{\varphi_{\text{drug}}} + \frac{1}{m_{\text{poly}}} \frac{1}{(1 - \varphi_{\text{drug}})} - 2\chi_{\text{drug-poly}} = 0. \quad (2.48)$$

The highest or lowest point of the spinodal curve is the *critical point* (φ_c), a point where the spinodal and binodal curves meet. It can be calculated by setting the third derivative of the free energy to zero:

$$\frac{\partial^3 \Delta G_{\text{mix}}}{\partial^3 \varphi_{\text{drug}}} = \frac{1}{\varphi_c^2} - \frac{1}{m_{\text{poly}}} \frac{1}{(1 - \varphi_c)^2} = 0. \quad (2.49)$$

The critical interaction parameter χ_c and the critical temperature T_c can be obtained from substituting back to Eqs. 2.36 and 2.48. Binodal and spinodal curves do not exist if the mixture has no critical point. This is usually due to χ being negative or very small over the entire temperature range, i.e., the system is miscible at all conditions.

From Fig. 2.9, it can be observed that the ΔG_{mix} curve is negative and convex at 100 °Cf. At this temperature, thermodynamically stable single-phase mixtures are obtained at all indomethacin–PVP-VA compositions. The temperature at which ΔG_{mix} becomes positive depends on the volume fraction of the polymer, i.e., the bigger the polymer fraction, the lower the temperature. The critical temperature for the indomethacin–PVP-VA system was found to be 73 °C (Fig. 2.9). In addition, the critical point is at a very low polymer concentration, meaning that a high amount of polymer is required to prevent phase separation. The spinodal curve is the boundary between unstable and metastable regions, meaning that below this curve, the system would spontaneously phase separate into two phases with the compositions $\varphi\alpha$ and φ_β at the temperature in question with any given homogenously mixed state φ_0 (Fig. 2.9). After formation of the bicontinuous polymer-rich and drug-rich domains, phase separation would proceed through a coarsening process. Generally, the binodal curve (which was not studied in the publication in question) would be located above (and adjacent to) the spinodal curve. Together, these give boundaries for a region where the mixture is metastable, in the sense that the mixture will only start to phase separate after a sufficiently large fluctuation in concentration or temperature. For systems between the binodal and spinodal boundaries, amorphous phase separation would occur through nucleation and growth, i.e., the drug-rich domains first appear as small droplets which subsequently grow in size (Fan et al. 1992; Lin and Huang 2010).

2.4.2.4 In Silico Estimation of Miscibility

Advances in the computational field, such as molecular mechanics and dynamics, are enabling a mechanistic understanding of different systems, thus allowing prediction of the thermodynamic behavior for a system that is not well characterized experimentally (Cui 2011). Molecular simulation tools in combination with the F–H

Table 2.2 Solubility parameters for indomethacin (*IND*), polyethylene oxide (*PEO*), glucose (*GLU*), and sucrose (*SUC*) obtained from molecular dynamics simulations and group contribution methods (Van Krevelen and Hoy, $MPa^{0.5}$). (Values obtained from Gupta et al. 2011)

Compound	$\delta_{VanKrevelen}$	δ_{Hoy}	$\delta_{Simulations}$
IND	23.3	21.9	23.9 ± 0.3
PEO	22.9	20.0	22.2 ± 0.2
GLU	39.0	37.8	34.8 ± 0.2
SUC	36.0	33.5	29.9 ± 0.5

IND indomethacin, *GLU* glucose, *PEO* polyethylene oxide, and *SUC* sucrose

theory were employed to estimate the free energy of mixing, which was subsequently used to estimate the drug solubility in lipid excipients (Huynh et al. 2008). The miscibility behavior of three different binary mixtures (solvent with solvent, polymer with solvent, and polymer with polymer) was studied by using a combination of the F–H theory and molecular simulation techniques (Fan et al. 1992). The temperature dependence of the interaction parameter could be calculated by the extension to the F–H theory, which also considers molecules that can be arranged irregularly, rather than regularly, in the lattice (i.e., the off-lattice model). In all mixture types, the calculated and experimentally observed upper critical miscibility temperatures were in agreement. The approach provided an opportunity to test the F–H theory for a number of model binary systems and estimates their miscibility behavior without possessing specific knowledge or experimental data of the system under investigation. However, the simulation approach used in the study also possessed some of the general deficiencies of F–H theory, namely the inability to predict the existence of the lower critical solution temperature and composition dependence of χ, and to take the entropic contributions other than simply mixing the components into account. Nevertheless, it was possible to provide the molecular basis of the temperature dependence of χ and thus calculate this dependence. In addition, the study supported the previous observation that the F–H theory might strongly overestimate values of the critical temperature.

A computational model based on molecular dynamics was developed to predict the miscibility of indomethacin in the carriers polyethylene oxide (PEO), glucose, and sucrose (Gupta et al. 2011). The cohesive energy density and the solubility parameters were determined by simulations using the condensed-phase optimized molecular potentials for atomistic simulation studies (COMPASS) force field. The simulations predicted miscibility for indomethacin with PEO ($\Delta\delta < 2$), borderline miscibility with sucrose ($\Delta\delta < 7$), and immiscibility with glucose ($\Delta\delta > 10$; Table 2.2).

The solubility parameter values obtained from the simulations were found to be in reasonable agreement with those calculated using the traditional group contribution methods (Table 2.2). Furthermore, the miscibility/immiscibility prediction was experimentally confirmed, since DSC showed melting point depression of PEO with increasing levels of indomethacin and thermal analysis of blends of indomethacin with sucrose and glucose verified general immiscibility. The solubility parameter values for glucose and sucrose obtained from the simulations were somewhat lower than the calculated ones. The difference was explained by the strong hydrogen bonding in glucose and sucrose, which cannot be described adequately using the group contribution methods. The advantage of molecular dynamics is that it takes such directional-dependent interactions into account during computations. The study showed that molecular simulations are a powerful tool for screening and determining solubility parameters of pharmaceutical compounds and gaining molecular level insight into the nature of interactions within complex systems. An additional advantage of the in silico analysis is that it can be performed at relevant temperatures and overcomes the temperature restrictions of experimental methods.

As an attempt to create an in silico model for sophisticated selection of stabilizing excipients for amorphous dispersions, experimental characterization techniques (DSC and the pair distribution function (PDF)) have been coupled with different molecular descriptors to provide insight into miscibility of drug–polymer (PVP-VA) mixtures using a materials informatics approach (Moore and Wildfong 2011). It was assumed that there is an underlying molecular property responsible for the ability of a drug molecule to form a single phase upon mixing with a carrier. Molecular descriptors were calculated for each of the 12 model compounds and tested for correlation to solid dispersion potential using logistic regression. A univariate model was created that predicted the solid dispersion potential from a single-molecular descriptor and the model was challenged using three compounds not included in the calibration set. Six of the model drug–PVP-VA mixtures in the calibration set were experimentally classified as miscible, and the most promising molecular descriptor correlating with miscibility was the R3m (the atomic mass-weighted third-order *R* autocorrelation) index. When R3m increased, the probability for a miscible dispersion was found to increase. This parameter predicted the dispersion potential probabilities well for 10 of the 12 compounds, while the remaining 2 compounds only slightly deviated. The relationship between the R3m index and the mechanism of drug–polymer miscibility could be attributable to multiple molecular features and, for example, in the case of structurally related nifedipine and felodipine, differences in substitutes of the benzene ring in the molecular structure caused an increase in the R3m index from 0.579 for nifedipine to 0.813 for felodipine, making felodipine completely miscible with the polymer whereas nifedipine phase separated. The formation of a single-phase mixture for two external and a phase-separated mixture for one external compound was successfully predicted by the R3m model. However, it should be noted that this computational model is not generally applicable to all systems; rather, it serves as an attempt to use in silico tools to create models that may provide the means for intelligent selection of polymers in the design of amorphous molecular solid dispersions.

A similar concept was tested with the aim to develop a rapid screening method for the miscibility of mixtures of two small molecules (Pajula et al. 2010). In their approach, the F–H interaction parameter was determined in silico by using software which combines the modified F–H model and molecular modeling techniques. The temperature-dependent χ was calculated by generating a large number of pair configurations and calculating binding energies by Monte Carlo simulation. This approach was more rapid than molecular dynamic simulations. This was followed by temperature averaging of the results by using the Boltzmann distribution factor. Arranging molecules off-lattice leads to the coordination number, Z, which is the number of screen molecules A that can be packed around a single base molecule B. By calculating an average of temperature-dependent binding energies and coordination numbers of all possible pair configurations of molecules A and B, the energy of mixing (ΔE_{mix}) at temperature T can be calculated by:

$$\Delta E_{\text{mix}} = \frac{1}{2}[Z_{\text{bs}}(E_{\text{bs}})_{\text{T}} + Z_{\text{sb}}(E_{\text{sb}})_{\text{T}} - Z_{\text{bb}}(E_{\text{bb}})_{\text{T}} - Z_{\text{ss}}(E_{\text{ss}})_{\text{T}}], \tag{2.50}$$

where Z is the coordination number, E is the binding energy, and s and b indicate the base and screen molecules, respectively. With these extensions, the determination of the temperature-dependent interaction parameter, temperature-and composition-dependent free energy of mixing, the phase diagram, the critical point, as well as the coexisting and stability curves of the binary system were obtained. Based on DSC experiments, the 39 drug molecules in the study were classified into three different categories according to their crystallization tendency, i.e., highly crystallizing, moderately crystallizing, and noncrystallizing compounds. Subsequently, the authors calculated interaction parameters for 1122 compound pairs in the study and verified the miscibility/immiscibility experimentally by polarized light microscopy for 26 binary mixtures randomly selected from the highly crystallizing component category. The F–H interaction parameter was found to be a suitable predictor for thermodynamic miscibility and phase stability, and predicted either miscibility or immiscibility with 88 % confidence. As predicted, all of the analyzed immiscible pairs crystallized during the experiment. However, 27 % of the miscible pairs crystallized despite exhibiting favorable interaction parameters. The authors explained this by a marginal difference between the calculated and critical interaction parameter, meaning that the composition of crystallized binary mixtures may have been too close to the thermodynamic line that separates miscible and immiscible mixtures.

2.4.3 *Impact of Temperature, Pressure, and Moisture on Miscibility*

As discussed above, the stability of amorphous solid dispersions is influenced by the miscibility of the components, which in turn is determined by the thermodynamics of mixing, specifically by the mixing enthalpy. The mixing enthalpy in turn is determined by the relative strength of the cohesive (drug–drug) and adhesive interactions (drug–polymer). However, it is known that unfavorable conditions during processing and storage, such as high temperature and/or humidity, can induce amorphous–amorphous phase separation by affecting amorphous solid dispersion structure on a molecular level.

2.4.3.1 Impact of Temperature

The temperature dependence of kinetic factors and thermodynamic driving force for nucleation have opposite signs, with thermodynamics and increased viscosity favoring nucleation at low temperatures and kinetics and greater molecular motion favoring nucleation at higher temperatures (Bhugra and Pikal 2008; Craig et al. 1999; Hancock and Zograf 1997). As shown by the phase diagrams, temperature also influences miscibility of binary mixtures, meaning that they can undergo mixing and demixing upon temperature changes if a thermodynamic driving force exists and the process kinetics are not too slow. Variable-temperature Fourier transform infrared (FTIR) was used to study the miscibility behavior of a felodipine–PVP system at different temperatures (Marsac et al. 2010). This system has previously been shown to be miscible at room temperature, promoted by intermolecular interactions between the drug and polymer. When studying the mixture in the temperature range of 0–180 °C, it was observed that miscibility of the system and the drug–polymer interactions were maintained until the melting point of the drug at all compositions was studied. However, the hydrogen bonding interactions were found to weaken and/or become less numerous upon increasing the temperature, but these changes were reversible, indicating that the system was well mixed at all temperatures studied. In another study, melting point depression data to construct plots of the free energy of mixing (ΔG_{mix}) as a function of drug volume fraction for felodipine systems with HPMCAS and Soluplus® were used (Tian et al. 2013). The temperature at which the ΔG_{mix} became positive, and thus unfavorable for mixing, depended on the polymer and the volume fraction of the drug. This was ≤ 120 °C for felodipine–HPMCAS systems at high drug loadings. For felodipine–Soluplus systems, temperatures lower than 80 °C resulted in positive ΔG_{mix} at drug volume fractions larger than 0.3. Plotting drug solubility and miscibility as a function of drug weight fraction and temperature and combination with the T_g values led to phase diagrams describing the stability of the system.

2.4.3.2 Impact of Moisture

Amorphous materials can absorb water from the environment into their structure, which can increase the molecular mobility significantly and thus plasticize the material sufficiently to lower the T_g of the system to that of the storage temperature or lower (Hancock and Zografi 1994). Absorbed moisture can be considered as a third component in a drug–polymer mixture, addition of which alters the thermodynamic properties of the system. Thus, for an initially miscible drug–polymer system, the presence of water may cause the system to become immiscible, inducing amorphous–amorphous phase separation and subsequent crystallization (Konno and Taylor 2008; Marsac et al. 2010; Vasanthavada et al. 2005). Systems particularly prone to this phenomenon are those consisting of a hydrophobic drug and hydrophilic polymer. In amorphous solid dispersions consisting of different hydrophobic drugs and a hydrophilic polymer (PVP), the moisture-induced phase separation occurred in some cases at relative humidity (RH) values as low as 54 %, while other systems remained miscible even at 94 % RH (Rumondor et al. 2009). Stronger hydrogen bonding, which affects the mixing enthalpy, might lead to a better resistance against moisture-induced immiscibility. Furthermore, it was suggested that drug crystallization in hydrophobic drug–hydrophilic polymer systems can occur through different routes: either from a plasticized one-phase solid dispersion or from a plasticized drug-rich amorphous phase in a two-phase solid dispersion. In the former case, the polymer can inhibit the crystallization more effectively, being in the same phase with the drug. In addition to strong drug–polymer interactions, using a less hydrophilic polymer, such as PVP-VA or HPMCAS, might protect from moisture-induced phase separation. Compounds undergoing moisture-induced phase separation with PVP did the same when mixed with PVP-VA, but to a lesser extent, whereas systems containing the least hydrophilic HPMCAS seemed to be least susceptible to phase separation (Rumondor and Taylor 2010).

In the case of absorbed moisture, the F–H model (Eq. 2.36) can be extended to involve water as a third component in the mixture to calculate the Gibbs free energy of mixing:

$$\frac{\Delta G_{\mathrm{mix}}}{RT} = n_{\mathrm{w}} \ln \varphi_{\mathrm{w}} + n_{\mathrm{d}} \ln \varphi_{\mathrm{d}} + n_{\mathrm{p}} \ln \varphi_{\mathrm{p}} + \chi_{\mathrm{wd}} n_{\mathrm{w}} \varphi_{\mathrm{d}} + \chi_{\mathrm{wp}} n_{\mathrm{w}} \varphi_{\mathrm{p}} + \chi_{\mathrm{dp}} n_{\mathrm{d}} \varphi_{\mathrm{p}}, \tag{2.51}$$

where ΔG_{mix} is the free energy of mixing for the ternary system and the subscripts w, d, and p refer to the components water, drug, and polymer, respectively. The first three terms of Eq. 2.51 represent the combinatorial entropy of mixing (ΔS_{mix}) and the remainder of the terms form the enthalpy of mixing (ΔH_{mix}). Assuming that adding the third component to the system is favorable to the mixing entropy, the amorphous–amorphous phase separation should only occur when ΔH_{mix} becomes sufficiently positive, which in turn depends on the relative values of the interaction parameters (χ_{wd}, χ_{wp}, and χ_{dp}). Thus, drug–polymer interactions can be considered important in preventing moisture-induced phase separation. In the case of hydrophobic drugs, the interaction parameter between drug and water (χ_{wd}) can be assumed to be positive (unfavorable). However, the amount of the absorbed water (n_{w}) also affects the enthalpic contribution. The extent of water sorption in turn depends on the hygroscopicity of the mixture components. At the same drug-to-polymer weight ratio, the amount of water absorbed by solid dispersions containing PVP was observed to be higher than in the case of HPMCAS. This resulted in less favorable enthalpic contributions and may explain moisture-induced demixing in some drug–polymer systems. In addition, hydrophobic drugs, with larger χ_{wd}, and thus an unfavorable effect on ΔH_{mix}, would favor drug–polymer phase separation even at low moisture levels (Rumondor and Taylor 2010).

In conclusion, miscibility of a ternary system (drug, polymer, and water) depends on the balance between the thermodynamic factors, which in turn are affected by the amount of water absorbed by the system. The extent of water sorption depends on the hygroscopicity of both polymer and drug, in addition to the strength of the intermolecular interactions. Thus, a system that has strong drug–polymer interactions, low hygroscopicity of the amorphous solid dispersion, and a less hydrophobic drug may lead to the best resistance against moisture-induced phase separation (Rumondor et al. 2011).

2.4.3.3 Impact of Pressure

Compression can induce phase transformations in metastable materials (Chan and Doelker 1985) and thus it might also induce phase separation in amorphous systems. The effect of compression pressure on miscibility of naproxen–PVP spray-dried solid dispersions with different compositions (20, 30 and 40 % of naproxen) was studied at different compression pressures and their miscibility was evaluated prior to and after compression by modulated DSC and FTIR (Ayenew et al. 2012). It was found that a solid dispersion containing 20 % of naproxen remained intact after compression at 565.1 MPa, as demonstrated by the single T_g in DSC measurements and an unaltered spectral profile. In contrast, a solid dispersion containing 30 % of naproxen showed a single, but expanded, T_g after compression at 188.4 and 376.7 MPa, which was considered as an indication for induction of inhomogeneity in the mixture. With higher compression pressures (565.1 and 1130.1 MPa), two T_gs corresponding to drug-rich and polymer-rich phases appeared in the thermogram. In addition, the FTIR spectra indicated disruption of the hydrogen bonding between the drug and polymer, induced by the higher compression pressures. The behavior of a solid dispersion containing 40 % of naproxen was similar to that of the 30 % dispersion, with even more pronounced changes in T_g, suggesting formation of a higher fraction of the drug-rich domain. The authors concluded that the composition of 30 % naproxen and a compression pressure of 376.7 MPa was the threshold condition for the compression-induced demixing in this case.

2.5 Drug (Crystalline) Solubility Within a Polymer Matrix

The preceding section on miscibility evaluated the theoretical miscibility of a drug with a polymer. The aforementioned approaches, such as the F–H lattice theory and the interaction parameter, allow prediction whether or not a certain drug and polymer are miscible. However, they do not contain quantitative information about how much of a drug is miscible or soluble in the polymer. From a pharmaceutical perspective, this information is of importance for developing a glass solution with the maximum drug load, but avoiding recrystallization of the drug due to supersaturation within the polymer. This is of special interest since glass solutions are a means to overcome poor solubility as new drug candidates increasingly belong to biopharmaceutics classification system (BCS) classes II and IV. Knowledge of drug–polymer solubility is advantageous for formulation and production of stable glass solutions.

Solubility of a drug in liquid media can be determined directly, simply by dispersion of excess solid drug into the liquid and measurement of drug concentration once equilibrium has been established (Roy and Flynn 1989). However, determining drug solubility in a polymer is not as straightforward as both components are normally solid at room temperature. Different approaches have been applied to account for the much slower kinetics of mixing, dissolution, and reaching equilibrium in the solid state.

Drug–polymer solubility has been determined using:

- A low molecular weight polymer analog, e.g., methyl- or ethylpyrrolidone (EP) for PVP (Marsac et al. 2006b, 2009; Paudel et al. 2010)
- Melting point depression of the drug in the respective polymer (Marsac et al. 2006b, 2009; Sun et al. 2010; Tao et al. 2009)
- Recrystallization from supersaturated glass solutions (Mahieu et al. 2013; Ma et al. 1996)
- The F–H interaction parameter (Tian et al. 2013)
- The changes in the Gibbs energy (Bellantone et al. 2012)

2.5.1 *Determination of Drug Solubility Within a Polymer Matrix*

2.5.1.1 Melting Point Depression Methods

Polymer-induced melting point depression or drug dissolution into a polymer can be used to experimentally determine drug–polymer miscibility (as discussed in Sect. 2.4.2.3) and drug–polymer solubility by DSC. A physical mixture of drug and polymer is scanned at a rate varying from 0.1 to 10 K/min (Marsac et al. 2006b, 2009; Sun et al. 2010). The end of the melting point endotherm (T_{end}) is determined as the intersection of the end of the dissolution endotherm and the baseline after dissolution. T_{end} can be used directly for extrapolation of solubility at lower temperatures than the measured melting point or for further calculations, e.g., to relate the melting point depression to F–H lattice theory.

Problems in the accuracy of the measured T_{end} values arise due to the polymer's high viscosity that slows the mixing and dissolution process down. This is most prominent at temperatures close to or below the glass transition temperature (T_g) of the drug–polymer mixture as molecular mobility is reduced. The mixing and dissolution processes at temperatures close to the T_g may become so slow that equilibrium is not reached during the time frame of a DSC scan. These measurements lead to a wrong assumption of the equilibrium solubility. Therefore, solubility measurements close to or below the T_g of the mixture are not feasible with DSC measurements due to the time frame of the experiments.

During dissolution, a mass transfer from the crystalline drug (solute phase) to the glass solution (solution phase) takes place. Particle size reduction of the raw material and mixing of the drug and polymer before the DSC scan can facilitate dissolution and reduce DSC measurement times. Cryo-milling of the drug–polymer mixture with liquid nitrogen reduces plastic deformation and favors particle fracturing, reducing the particle size while at the same time blending the components (Tao et al. 2009).

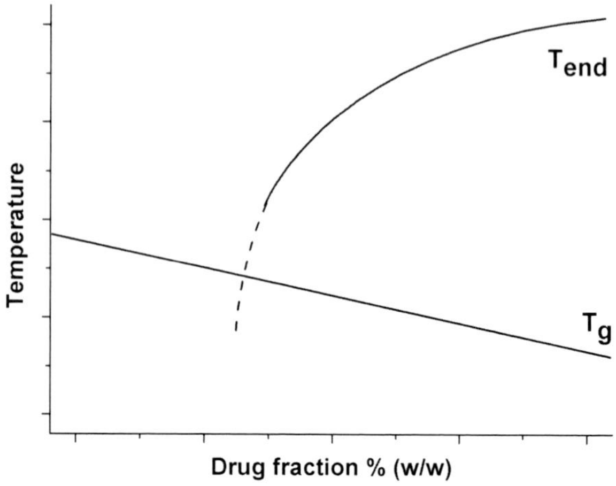

Fig. 2.10 Schematic of a temperature versus weight fraction diagram: *the straight line* represents the T_g of the drug–polymer mixture, *the curved solid line* represents the measured T_{end} values, and *the dashed curve line* represents the extrapolated T_{end} values of drug in polymer. The intersection marks the solubility at T_g for a saturated solution. (Adapted from Tao et al. 2009)

However, some drug–polymer combinations can be converted to the amorphous form during the milling process as has been shown for piroxicam with either PVP, methyl cellulose, or Soluplus® (Kogermann et al. 2013), indomethacin with PVP (Ke et al. 2012), and sulfathiazole or sulfadimidine with PVP (Caron et al. 2011). To prevent solid-state changes, the milling time has to be optimized for each drug–polymer pair.

In the approach developed by Tao et al., the cryo-milled drug–polymer mixtures were subjected to DSC runs and T_{end} was measured (Tao et al. 2009). DSC runs were performed with different heating rates and the respective T_{end} values were extrapolated to 0 K heating rate. Using this method, T_{end} values were obtained for drug–polymer mixtures of different composition. However, with a high polymer fraction, viscosity and T_g of the samples increase and the detection limit of the drug melting point may be challenging. Hence, T_{end} cannot be determined over the entire range of possible drug–polymer compositions but has to be extrapolated for low drug loadings.

Graphically, this can be shown in a temperature versus weight fraction diagram depicting the T_{end} and the T_g curve (Fig. 2.10).

The disadvantage of this method is that many DSC runs are necessary to determine solubility and those runs require a very slow heating rate (e.g., 0.1 K/min). Furthermore, solubility measurements are only possible above a certain drug content due to increased viscosity and the detection limit at low drug concentration.

Once solubility values were obtained, their accuracy can be further improved, by annealing the drug–polymer mixtures close to their equilibrium solubility for 4–10 h (giving the drug–polymer mixture additional time to equilibrate), followed by a DSC scan with a standard heating rate (e.g., 10 K/min) to determine if any crystals remain in the annealed sample (Sun et al. 2010). If the drug–polymer mixture is below its equilibrium solubility, the drug should have completely dissolved in the polymeric matrix and no melting event can be detected. In contrast, if the drug–polymer mixture is above the equilibrium solubility, some drug crystals will remain independent of the annealing time and give a melting signal in the standard DSC run. Generally, two approaches are possible, either to change the composition of the drug–polymer mixture and anneal at a set temperature or to use one set composition and change the temperature range. With the second approach, an upper and lower solubility of a certain drug–polymer mixture can be determined with regard to the temperature.

When the equilibrium solubility values obtained with this improved method were compared to Tao's method described above, they were very consistent at higher drug fractions. However, at lower drug fractions, T_{end} values were about 7 K lower with the new method. Hence, the T_{end} curve in Fig. 2.10 would be lower. As outlined above, viscosity in a drug–polymer mixture with a low drug content increases close to the mixture's T_g and at a higher polymer content. This may lead to measurements where equilibrium has not been established during the DSC run, and equilibrium solubility would be underestimated. The additional annealing step provides a higher certainty of the equilibrium being reached and therefore improves the accuracy. However, this method requires that the solubility of drug in polymer is roughly known beforehand and that neither drug nor polymer degrade or undergo polymorphic changes during annealing.

2.5.1.2 Recrystallization from Supersaturated Glass Solutions

Supersaturated glass solutions are thermodynamically unstable and are expected to phase separate and the drug to recrystallize until the drug–polymer mixture has reached equilibrium solubility. Glass solutions prepared by, e.g., solvent evaporation can be stored at temperature and humidity conditions of interest, and the existence and growth of crystals can be observed using polarized light microscopy (Ma et al. 1996; Reismann and Lee 2012). Solubility can easily be assessed with a light microscope at storage conditions of interest, under the prerequisite that no crystallinity is induced by impurities. On the downside, the time until crystallization starts may be very long (months or even years). This approach has been used for solubility determination of drugs in films intended for transdermal application.

In another approach, a supersaturated glass solution (e.g., 85:15 (w/w) drug-to-polymer ratio) can be created via milling of the drug–polymer mixture (Mahieu et al. 2013). The glass solution is then annealed at a defined temperature above its T_g where molecular mobility is higher and excess drug can crystallize. At the equilibrium solubility, the excess drug has completely crystallized and a glass solution of the soluble drug fraction and the polymer has formed. The T_g of the annealed sample is analyzed and the composition of the sample can be determined with the Gordon–Taylor equation. This way the equilibrium solubility of the drug in the polymer at the annealing temperature is assessed and a solubility curve can be built up using several annealing temperatures. In the case of indomethacin–PVP mixtures, annealing temperatures between 120 and 160 °C for 2 h were sufficient to reach equilibrium. The approach is faster than the melting point depression methods of Tao and Sun. The kinetics of demixing and crystallization from supersaturation are supposedly faster than dissolution of drug into polymer and the T_g of the demixing drug–polymer mixture is lower than that of the same composition during the mixing process. However, with increasing demixing and crystallization of the excess drug, the T_g of the drug–polymer mixture increases. Hence, solubility can only be determined in reasonable time frames at annealing temperatures that are above the T_g of the resulting glass solution. Solubility at lower temperatures has to be estimated by extrapolation.

2.5.1.3 F–H Interaction Parameter

The F–H lattice theory was described earlier in connection with miscibility and molecular interactions. For solubility calculations, a monomer/oligomer of the polymer (e.g., EP) in combination with the activity coefficient (Marsac et al. 2009) can be used as well as calculated data such as the van Krevelen solubility parameter (Tian et al. 2013).

The mole fraction solubility of drug, x_{drug}, can be calculated by Eq. 2.52 (Marsac et al. 2009):

$$\ln \gamma_{drug}\chi_{drug} = -\frac{\Delta G_{fus}}{RT} = -\frac{\Delta H_{fus}(T_m)}{RT}\left[1 - \frac{T}{T_m}\right] - \frac{1}{RT}\int_{T_m}^{T} T\Delta C_p \mathrm{d}T + \frac{1}{R}\int_{T_m}^{T} T\frac{\Delta C_p}{T}\mathrm{d}T, \quad (2.52)$$

where γ_{drug} is the drug's activity coefficient in the drug–polymer mixture at the solubility limit, ΔG_{fus} is the difference in the free energy between crystal and supercooled liquid, and R and T are the universal gas constant and the absolute temperature, respectively. Experimentally accessible are the heat of fusion (ΔH_{fus}), the melting temperature (T_m), and the difference in heat capacity between the supercooled liquid and the crystal, ΔC_p. The activity coefficient can be estimated either from solubility measurements in EP or by usage of the F–H theory. For the EP solubility-based approach, the ratio of ideal to experimental solubility (in EP) is adjusted assuming that physicochemical properties, molecular mobility, and, hence, solubility are the same in the liquid (EP) and the polymer (PVP). The activity coefficient and the F–H interaction parameter can be related to each other via Eq. 2.53:

$$\ln \gamma_{drug}^{PVP} = \frac{MV_{drug}}{MV_{lattice}}\left[\frac{1}{m_{drug}}\ln\frac{\phi_{drug}}{x_{drug}} + \left(\frac{1}{m_{drug}} - \frac{1}{m_{PVP}}\right)\phi_{PVP} + \chi\phi_{PVP}^2\right], \quad (2.53)$$

where γ_{drug}^{PVP} is the activity coefficient of drug in PVP, MV is the molecular volume, $\phi_{drug/polymer}$ is the volume fraction of drug or polymer, respectively, $m_{drug/polymer}$ is the ratio of the drug or polymer volume to that of the lattice site, and X is the F–H interaction parameter determined via melting point depression. The same approach was also used to determine the solubility of itraconazole in Eudragit® E 100 using a 2:1:1 w/w/w 2-dimethylaminoethyl methacrylate, methyl methacrylate, and butyl methacrylate mixture as a polymer analog (Janssens et al. 2010).

2.5.1.4 Changes in Gibbs Energy

This approach divides the formation of solid solutions into four different steps (Bellantone et al. 2012):

- Heating of the pure drug from the starting temperature to its melting point, T_{m}
- heating of the pure polymer from the starting temperature to the T_{m} of the drug
- Mixing molten drug and polymer at T_{m} to form a solution
- Cooling of the solution down to the starting temperature

As the temperature at the start and end of the process is the same, the total change in Gibbs energy can be described by the changes in the total entropy and enthalpy as:

$$\Delta G_{\mathrm{SS}} = \Delta H_{\mathrm{SS}} - T\Delta S_{\mathrm{SS}}, \tag{2.54}$$

where SS in ΔG_{SS} stands for solid solution; it can be further divided into three components where the first two use heat capacity values obtained via thermal analysis and the third component uses the calculated F–H interaction parameter:

$$\Delta G_1 = \int_T^{T_{M11}} [C_{\mathrm{P1}} + C_{\mathrm{P2}} - C_{\mathrm{P12}}]\, \mathrm{d}T - T\int_T^{T_{\mathrm{M1}}} \left[\frac{C_{\mathrm{P1}} + C_{\mathrm{P2}} - C_{\mathrm{P12}}}{T}\right] \mathrm{d}T \tag{2.55}$$

$$\Delta G_2 = n_1 \Delta h_{\mathrm{M1}} \left(1 - \frac{T}{T_{\mathrm{M1}}}\right) \tag{2.56}$$

$$\Delta G_3 = RTn_1\phi_2\chi(T) + RT(n_1 \ln \phi_1 + n_2 \ln \phi_2), \tag{2.57}$$

where C_{P1} represents the heat capacity of unmixed drug at constant pressure and heat capacity C_{P2} and C_{P12} represent the heat capacity of the polymer and the solid solution, respectively. Δh_{M1} is the molar enthalpy of melting of the drug's melting point (T_{M1}). Φ_1 and Φ_2 are the respective volume fractions of drug and polymer in the solid solution and n denotes the moles of drug or polymer. The solubility is determined as the minimum in the plot of ΔG_{SS} divided by the weight of the polymer versus the drug weight fraction.

Dissolution of a poorly soluble drug in a liquid has only a small effect on the heat capacity. However, the difference in heat capacity of unmixed components and the solid solution also contributes to the changes in Gibbs free energy. Hence, the difference in heat capacity is an important aspect for the determination of solubility of drugs in polymers, especially at room temperature.

2.5.1.5 Solubility Determination in PEG

PEG differs from other polymers commonly used in glass solutions. Depending on the chain length, it may be liquid or solid at room temperatures, thus allowing different approaches for solubility determination. Thermal analysis together with the Fox equation was used to determine the solubility of six drugs in PEG 400, which is a viscous liquid at room temperature (Haddadin et al. 2009). DSC measurements were performed in three cycles; drug and polymer were kept at predefined temperatures below the melting point of the drug for 5 min. The drug dissolved into the polymer and equilibrium should be reached a lot faster in liquid polymers compared to solid polymers. In a second step, the solutions were quench cooled and scanned again with modulated DSC to determine the T_g of the glass solution formed in the first and second cycle.

The T_gs of the pure components (T_{g1} and T_{g2}) are known and the T_g of the drug–polymer mixtures, $T_{g(total)}$, was measured in cycle 3 of the DSC scan. With the Fox equation (Eq. 2.58), the weight fractions, w, of drug and polymer can be determined:

$$\frac{1}{T_{g(total)}} = \frac{w_1}{T_{g1}} + \frac{1 - w_1}{T_{g2}}. \tag{2.58}$$

2.5.2 Summary of Solubility Determination Methods

Neither of the abovementioned methods to determine the solubility of a drug in a polymer is straightforward. Solubility measurements in low molecular weight polymer analogs assume that the solubility is the same in a monomer as in the polymer, even though their physicochemical properties and molecular mobility differ. Approaches using melting point depression and recrystallization from supersaturated glass solutions are only applicable over a certain range of drug–polymer compositions and at elevated temperatures. This may be very useful to predict drug–polymer solubility during manufacturing of a glass solution (e.g., via hot-melt extrusion), rather than being a stability indicator for the storage time of a pharmaceutical formulation. Using the F–H theory, the interaction parameter is calculated from melting point depression, but is extrapolated to room temperature, although a linear relationship only was found for certain drug–polymer composition ranges.

2.5.3 Effect of Drug Solubility on the Physical Stability of the Glass Solution

According to published data (Table 2.3), drug solubility in a polymeric matrix is relatively low and often the drug load would be insufficient to reach therapeutic concentrations.

However, that does not mean that glass solutions with drug content above equilibrium solubility cannot be used for pharmaceutical applications. Even though such glass solutions are not thermodynamically stable, they may be kinetically stable when recrystallization is prevented over pharmaceutically relevant storage time periods. Phase diagrams built up with data from drug solubility, miscibility, and T_g values of the glass solutions (as indicated in Sect. 2.4) may function as a useful guide with regard to thermodynamic and kinetic stability of a glass solution at a given composition.

Table 2.3 Overview of determined solubility values of various drugs in polymers as reported in literature

Drug	Polymer	Method	Solubility value (%)	Refs.
Felodipine	PVP K12 PVP K29/32 or K90	Interaction parameter	7.2 6.5	Marsac et al. (2009)
Ketoprofen	PVP K12 PVP K29/32 or K90		1.1 1.0	
Indomethacin	PVP K12 PVP K29/32 or K90		14.4 13.4	
Nifedipine	PVP K12 PVP K29/32 or K90		5.4 4.9	
Acetaminophen	PEO N10	Interaction parameter	12.1	Yang et al. (2001)
Itraconazole	Eudragit® E 100	Interaction parameter	0.01	Janssens et al. (2010)
Indomethacin Nifedipine	PVP/VA	Melting point depression	50–60 (110°C) 30–40 (123°C)	Sun et al. (2010)
Indomethacin Nifedipine	PVP/VA	Melting point depression	28 (85°C) 12 (95°C)	Tao et al. (2009)
Indomethacin	PVP K12	Recrystallization from supersaturation	67 (120°C)	Mahieu et al. (2013)
Felodipine	HPMCAS Soluplus	Calculated Flory–Huggins interaction parameter using a or b	0.001[a] (25°C) 0.6[b] (25°C) 0.02[a] (25°C) 3.69[b] (25°C)	Tian et al. (2013)
Indomethacin Griseofulvin Itraconazol	PVP K30 Eudragit E 100 Eudragit L 100 PVP K30 Eudragit E 100 Eudragit L 100 PVP K30 Eudragit E 100 Eudragit L 100	Change in Gibbs free energy	~8 (25°C for all) ~9 ~10 ~9 ~9 ~12 ~9 ~8 ~13	Bellantone et al. (2012)
Lidocain	Polyacrylate	Flux	20.8	Jasti et al. (2004)

PVP poly(vinylpyrrolidone), *VA* vinyl acetate, *PEO* polyethylene oxide, *HPMCAS* hydroxypropyl methyl cellulose acetate succinate

[a]Melting point data

[b]Van Krevelen solubility parameter

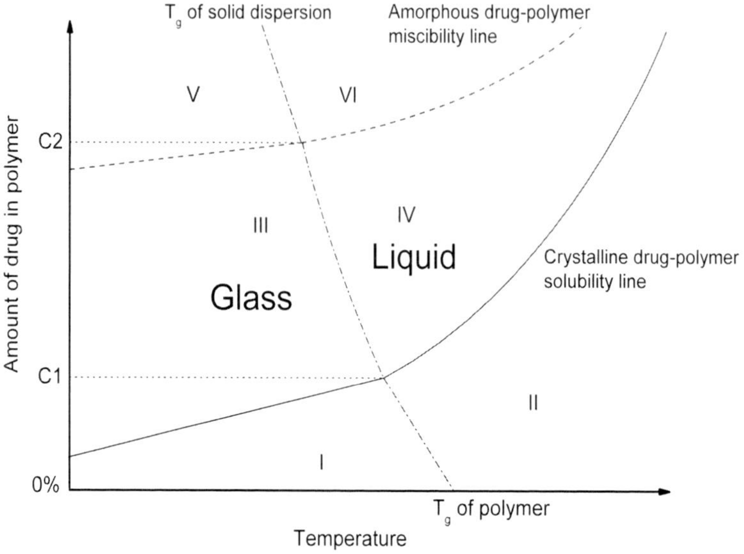

Fig. 2.11 Schematic representing a solid dispersion system with different drug loadings and temperatures. The regions (**I–VI**) are divided by the glass transition (T_g) of the solid dispersion, crystalline drug–polymer solubility limit, and amorphous drug–polymer miscibility limit. The zones (**I–VI**) are **I**: thermodynamically stable glass; **II**: thermodynamically stable liquid; **III**: supersaturated glass; **IV**: supersaturated liquid; **V**: supersaturated and immiscible glass; and **VI**: supersaturated and immiscible liquid. **C1**: drug solubility in the polymer at T_g and **C2:** miscibility of the drug in the polymer at T_g. (Adapted from Qian et al. 2010a)

Several authors, including some of those mentioned above and others (Qian et al. 2010a; Tao et al. 2009; Tian et al. 2013), have provided schematics to identify suitable drug–polymer compositions according to solubility and miscibility in combination with the T_g. In Fig. 2.11, temperature is plotted versus drug fraction, showing the T_g of the glass solution, the drug–polymer solubility, and miscibility curves. These curves build up six areas (indicated I–VI) that can help to choose a suitable drug–polymer composition. The zones V and VI in the Fig. 2.11 are located above the drug–polymer miscibility line and are areas where phase separation is thermodynamically favored. Thus, at equilibrium, a drug-rich and a polymer-rich phase coexist and two separate T_gs are observed for the system (Tian et al. 2013; Vasanthavada et al. 2005). The amorphous drug–polymer miscibility line is always higher than the crystalline drug polymer solubility line due to the fact that the amorphous drug has a higher chemical potential than its crystalline counterpart. Equilibrium liquid zones (II, IV, VI) and nonequilibrium glass (I, II, V) are separated by the T_g values of the solid dispersion, which can either be measured by DSC or estimated from the Gordon–Taylor equation. In the equilibrium zone, structural relaxation occurs rapidly and miscibility (and solubility) can be measured. In contrast, in the nonequilibrium glassy region, molecular mobility is low and structural relaxation occurs slowly which leads to equilibrium solubility not being strictly defined or experimentally measured. The solubility line is the limit between the thermodynamically stable (I, II) and unstable (III–VI) regions. This defines the maximum limit for drug loading in the polymer, below which any drug concentration or temperature fluctuation will not cause phase separation or crystallization of the drug. Between the miscibility and solubility lines (zones III, IV), a solid dispersion might be able to maintain its thermodynamically metastable status depending on the crystallization behavior of the drug. Especially in zone III, a solid dispersion may be stabilized both thermodynamically and kinetically. In contrast, zone VI has no thermodynamic barrier to prevent crystallization and it should be avoided, but zone V can be stabilized kinetically (i.e., by storage at temperature sufficiently below the T_g). Furthermore, the drug solubility in the polymer at T_g (C1 in Fig. 2.11) and miscibility of the drug in the polymer at T_g (C2 in Fig. 2.11) are illustrated. A solid dispersion at C1 would be thermodynamically stable at temperatures above T_g and may be kinetically stable at temperatures below T_g. Thus, a solid dispersion with a drug loading not higher than C1 could be able to resist the stress coming from temperature fluctuation across T_g during processing and storage, and thus remain stable. However, C2 may serve as a practical limit for solid dispersion formulation design, if the drug can remain amorphous during the time frame of formulation processing and storage, especially when a high drug loading is needed to satisfy the dose requirement.

2.5.4 Maintenance of Supersaturated Conditions: The "Spring and Parachute Effect"

The ultimate goal of an amorphous formulation is to increase the drug concentration in solution (extent and time) after administration. Thus, the dissolution and crystallization behavior of the drug after amorphous solid dispersions have been introduced to physiologically relevant aqueous media requires careful consideration.

It is worth first considering the dissolution behavior of pure amorphous and crystalline drug (without polymer). The theoretical solution concentration profiles of a drug that is added to an aqueous medium in its stable crystalline form (with respect to the aqueous medium conditions) and pure amorphous form are shown in Fig. 2.12. The crystalline drug dissolves until the solution concentration reaches the thermodynamic solubility of the drug, after which the concentration remains constant. The amorphous form, on the other hand, initially dissolves more rapidly than the crystalline form and reaches a higher concentration. This is often termed the "spring" effect and is a direct function of the higher apparent solubility of the amorphous form. However, the higher concentration is typically short-lived, as the solution is supersaturated with respect to the crystalline form and therefore precipitation into the crystalline form occurs until the concentration equals the solubility of the crystalline form.

The exact supersaturation profile depends on the interplay between the dissolution and crystallization rates of the drug. When considering crystallization, there are two potential recrystallization routes in aqueous media: direct solid–solid (same

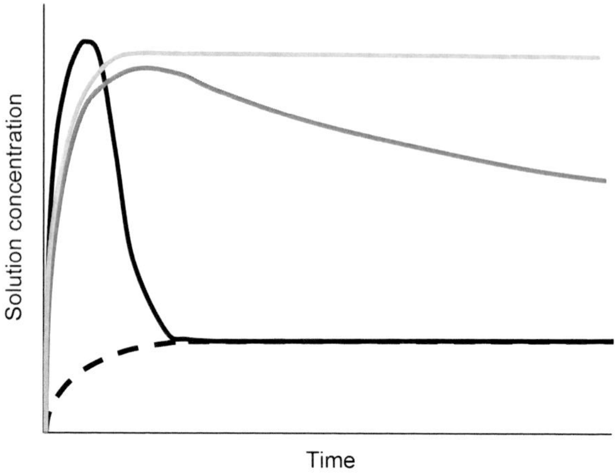

Fig. 2.12 Schematic showing of theoretical solution time profiles of **a** crystalline drug (*black, dashed*) and **b** pure amorphous drug without polymer (*black*), and solid dispersion with two different polymers: A (*medium gray*) and B (*light gray*). Note that all three amorphous formulations have a "spring effect," while the polymer A exerts a parachute effect while polymer B completely inhibits recrystallization and maintains the same degree of supersaturation

as during storage) and solution-mediated crystallization. Direct solid–solid crystallization can be accelerated in an aqueous medium. In such a medium, there is a greater driving force for water penetration into amorphous solid than in humid (or dry) conditions, and, as a result, a greater degree of plasticization of an amorphous drug may occur, facilitating crystallization within the solid (direct solid–solid transformation). If this occurs during dissolution, the spring effect (dissolution rate or maximum concentration) may be reduced due to crystalline material replacing the amorphous form available for dissolution. However, in a liquid medium, arguably the more important crystallization mechanism is solution-mediated crystallization, and this is the mechanism responsible for the drop in solution concentration until the solubility of the crystalline material is reached (Bhugra and Pikal 2008; Ozaki et al. 2012; Savolainen et al. 2009).

When a polymer is introduced in the form of an amorphous solid dispersion, a spring effect is also observed, but the subsequent drop in drug concentration may be slower (known as the parachute effect) or even completely inhibited. This is due to the polymer inhibiting solution-mediated crystallization of the drug. The theoretical spring and parachute effects of two different polymers (A and B) are shown in Fig. 2.12. Both the spring and parachute effects can be affected by polymer presence and type. There are several mechanisms contributing to the concentration profile of the drug during both the spring and parachute phases.

While the spring effect is due to the drug being in the amorphous form, the rate and extent of dissolution is also affected by the polymer in the solid dispersion. Dissolution of the polymer can promote more rapid dissolution of the drug by helping to liberate drug molecules (or particles), increase the drug solubility in the solvent (cosolvency effect), and inhibit direct solid–solid crystallization. On the other hand, the drug's chemical potential is lower for the solid dispersion than pure amorphous drug (if solubility changes due to dissolved polymer are disregarded; Kawakami and Pikal 2005; Ozaki et al. 2012). It is quite common that the spring effect is more pronounced for the pure drug than for a glass solution.

The parachute effect of the polymer is also due to a combination of mechanisms (Murdande et al. 2011; Raghavan et al. 2001; Sekikawa et al. 1978; Yokoi et al. 2005). First, polymers can elevate the equilibrium solubility of the drug (cosolvency effect) and therefore, can reduce the degree of supersaturation and hence, the thermodynamic driving force for solution-mediated crystallization (Warren et al. 2010). The drug molecules and polymer may form complexes in solution via electrostatic bonds, van der Waals forces, or hydrogen bonding. Even the addition of low amounts (e.g., 0.1–0.25 % w/w) of polymers such as PVP and hydroxypropyl methyl cellulose to solution have been shown to result in significant improvements in aqueous solubility (Loftsson et al. 1996). Second, polymers dissolved in the dissolution media can become adsorbed onto the surface of any crystallites that start to form, which may block the interaction of drug molecules with crystal surfaces. Electrostatic bonds, van der Waals forces, or hydrogen bonding may affect the degree and strength of interaction of the polymer and crystal faces, and therefore the degree of crystal growth inhibition. Third, the viscosity of the polymer solution may also inhibit the diffusion of the molecules during nucleation and crystal growth (Warren et al. 2010).

Many studies have shown that polymeric solid dispersions can be very effective at promoting spring and parachute effects. In order to promote these effects, both solid–solid and solution-mediated crystallization should be inhibited (Boersen et al. 2012). However, the partial or complete inhibition of crystallization alone does not guarantee a sufficient or, indeed, any supersaturation. While there are many, sometimes competing, mechanisms affecting the dissolution and crystallization of drug, there is evidence that the strength of polymer–drug intermolecular interactions is crucial. If the interactions are very strong, crystallization may be completely inhibited (during storage and administration), but both the drug and polymer may not dissolve sufficiently in the aqueous medium. Further research is needed to help rationalize polymer selection with respect to promoting both supersaturation generation and maintenance.

2.6 Summary

In the field of amorphous solid dispersion research and development, the elucidation of the theoretical basis for this increasingly important group of drug delivery systems is still an active and intensely pursued area. In this chapter, we have attempted to give an overview of the current status of the field. While it is obvious that much more fundamental work needs to be done, the current body of knowledge on the theoretical considerations in the development of amorphous solid dispersions lays a foundation for the applied formulation scientist concerned with the rational development of these drug delivery systems.

References

Adamska K, Bellinghausen R, Voelkel A (2008) New procedure for the determination of Hansen solubility parameters by means of inverse gas chromatography.J Chromatogr A 1195(1–2):146–149
Alonzo D, Zhang GZ, Zhou D, Gao Y, Taylor L (2010) Understanding the behavior of amorphous pharmaceutical systems during dissolution. Pharm Res 27(4):608–618
Andronis V, Zografi G (1997) Molecular mobility of supercooled amorphous indomethacin, determined by dynamic mechanical analysis. Pharm Res 14(4):410–414
Avramov I, Zanotto ED, Prado MO (2003) Glass-forming ability versus stability of silicate glasses. II. Theoretical demonstration. J Non-Cryst Solids 320(1–3):9–20
Ayenew Z, Paudel A, Van den Mooter G (2012) Can compression induce demixing in amorphous solid dispersions? A case study of naproxen-PVP K25. Eur J Pharm Biopharm 81(1):207–213
Babu NJ, Nangia A (2011) Solubility advantage of amorphous drugs and pharmaceutical cocrystals. Cryst Growth Design 11(7):2662–2679
Baird JA, Taylor LS (2012) Evaluation of amorphous solid dispersion properties using thermal analysis techniques. Adv Drug Deliv Rev 64(5):396–421
Baird JA, Van Eerdenbrugh B, Taylor LS (2010) A classification system to assess the crystallization tendency of organic molecules from undercooled melts. J Pharm Sci 99(9):3787–3806
Barandiaran JM, Colmenero J (1981) Continuous cooling approximation for the formation of a glass. J Non-Cryst Solids 46(3):277–287
Bellantone RA, Patel P, Sandhu H, Choi DS, Singhal D, Chokshi H, Malick AW, Shah N (2012) A method to predict the equilibrium solubility of drugs in solid polymers near room temperature using thermal analysis. J Pharm Sci 101(12):4549–4558
Bhugra C, Pikal MJ (2008) Role of thermodynamic, molecular, and kinetic factors in crystallization from the amorphous state. J Pharm Sci 97(4):1329–1349
Boersen N, Lee T, Hui H-W (2012) Development of preclinical formulations for toxicology studies. In: Faqi AS (ed) A comprehensive guide to toxicology in preclinical drug development. Academic Press, London, pp 69–86
Bøtker JP, Karmwar P, Strachan CJ, Cornett C, Tian F, Zujovic Z, Rantanen J, Rades T (2011) Assessment of crystalline disorder in cryo-milled samples of indomethacin using atomic pair-wise distribution functions. Int J Pharm 417(1–2):112–119
Branham ML, Moyo T, Govender T (2012) Preparation and solid-state characterization of ball milled saquinavir mesylate for solubility enhancement. Eur J Pharm Biopharm 80(1):194–202
Cabral AA, Cardoso AAD, Zanotto ED (2003) Glass-forming ability versus stability of silicate glasses. I. Experimental test. J Non-Cryst Solids 320(1–3):1–8
Caron V, Tajber L, Corrigan OI, Healy AM (2011) A comparison of spray drying and milling in the production of amorphous dispersions of sulfathiazole/polyvinylpyrrolidone and sulfadimidine/polyvinylpyrrolidone. Mol Pharm 8(2):532–542
Chan HK, Doelker E (1985) Polymorphic transformation of some drugs under compression. Drug Dev Ind Pharm 11(2–3):315–332
Chiou WL, Riegelman S (1969) Preparation and dissolution characteristics of several fast-release solid dispersions of griseofulvin. J Pharm Sci 58(12):1505–1510
Chiou WL, Riegelman S (1971) Pharmaceutical applications of solid dispersion systems. J Pharm Sci 60(9):1281–1302
Coleman MM, Painter PC (1995) Hydrogen-bonded polymer blends. Prog Polym Sci 20(1):1–59
Couchman PR, Karasz FE (1978) Classical thermodynamic discussion of effect of composition on glass-transition temperatures. Macromolecules 11(1):117–119
Craig DQM, Royall PG, Kett VL, Hopton ML (1999) The relevance of the amorphous state to pharmaceutical dosage forms: glassy drugs and freeze dried systems. Int J Pharm 179(2):179–207
Cui Y (2011) Using molecular simulations to probe pharmaceutical materials. J Pharm Sci 100(6):2000–2019
Di Martino P, Palmieri GF, Martelli S (2000) Molecular mobility of the paracetamol amorphous form. Chem Pharm Bull 48(8):1105–1108

Djuris J, Nikolakakis I, Ibric S, Djuric Z, Kachrimanis K (2013) Preparation of carbamazepine-Soluplus ® solid dispersions by hot-melt extrusion, and prediction of drug-polymer miscibility by thermodynamic model fitting. Eur J Pharm Biopharm 84(1):228–237

Duan RG, Liang KM, Gu SR (1998) Effect of changing TiO_2 content on structure and crystallization of CaO-Al2O_3-SiO_2 system glasses. J Eur Ceram Soc 18(12):1729–1735

Egawa H, Maeda S, Yonemochi E, Oguchi T, Yamamoto K, Nakai Y (1992) Solubility parameter and dissolution behavior of cefalexin powders with different crystallinity. Chem Pharm Bull 40(3):819–820

Fan CF, Olafson BD, Blanco M, Hsu SL (1992) Application of molecular simulation to derive phase-diagrams of binary-mixtures. Macromolecules 25(14):3667–3676

Flory PJ (1953) Principles of polymer chemistry. Cornell University Press, Ithaca

Forster A, Hempenstall J, Tucker I, Rades T (2001) Selection of excipients for melt extrusion with two poorly water-soluble drugs by solubility parameter calculation and thermal analysis. Int J Pharm 226(1–2):147–161

Goldberg AH, Gibaldi M, Kanig JL (1965) Increasing dissolution rates and gastrointestinal absorption of drugs via solid solutions and eutectic mixtures. I. Theoretical considerations and discussion of the literature. J Pharm Sci 54(8):1145–1148

Graeser KA, Patterson JE, Rades T (2009a) Applying thermodynamic and kinetic parameters to predict the physical stability of two differently prepared amorphous forms of simvastatin. Curr Drug Deliv 6:374–382

Graeser KA, Patterson JE, Zeitler JA, Gordon KC, Rades T (2009b) Correlating thermodynamic and kinetic parameters with amorphous stability. Eur J Pharm Sci 37(3–4):492–498

Graeser KA, Patterson JE, Zeitler JA, Rades T (2010) The role of configurational entropy in amorphous systems. Pharmaceutics 2:224–244

Greco K, Bogner R (2010) Crystallization of amorphous indomethacin during dissolution: effect of processing and annealing. Mol Pharm 7(5):1406–1418

Greenhalgh DJ, Williams AC, Timmins P, York P (1999) Solubility parameters as predictors of miscibility in solid dispersions. J Pharm Sci 88(11):1182–1190

Gupta P, Bansal AK (2005) Molecular interactions in celecoxib-PVP-meglumine amorphous system. J Pharm Pharmacol 57(3):303–310

Gupta J, Nunes C, Vyas S, Jonnalagadda S (2011) Prediction of solubility parameters and miscibility of pharmaceutical compounds by molecular dynamics simulations. J Phys Chem B 115(9):2014–2023

Gutzow IS, Schmelzer JWP (2013) The vitreous state: thermodynamics, structure, rheology, and crystallization. Springer, Berlin

Haddadin R, Qian F, Desikan S, Hussain M, Smith RL (2009) Estimation of drug solubility in polymers via differential scanning calorimetry and utilization of the fox equation. Pharm Dev Technol 14(1):18–26

Hancock BC, Parks M (2000) What is the true solubility advantage for amorphous pharmaceuticals? Pharm Res 17(4):397–404

Hancock BC, Zografi G (1994) The relationship between the glass-transition temperature and the water-content of amorphous pharmaceutical solids. Pharm Res 11(4):471–477

Hancock BC, Zograf G (1997) Characteristics and significance of the amorphous state in pharmaceutical systems. J Pharm Sci 86(1):1–12

Hancock BC, Shamblin SL, Zografi G (1995) Molecular mobility of amorphous pharmaceutical solids below their glass-transition temperatures. Pharm Res 12(6):799–806

Hancock BC, York P, Rowe RC (1997) The use of solubility parameters in pharmaceutical dosage form design. Int J Pharm 148(1):1–21

Hoei Y, Yamaura K, Matsuzawa S (1992) A lattice treatment of crystalline solvent-amorphous polymer mixtures on melting-point depression. J Phys Chem 96(26):10584–10586

Hrubý A (1972) Evaluation of glass-forming tendency by means of DTA. Czechoslov J Phys B 22(11):1187–1193

Huang W, Ray CS, Day DE (1986) Dependence of the critical cooling rate for lithium silicate glass on nucleating-agents. J Non-Cryst Solids 86(1–2):204–212
Huynh L, Grant J, Leroux JC, Delmas P, Allen C (2008) Predicting the solubility of the anti-cancer agent docetaxel in small molecule excipients using computational methods. Pharm Res 25(1):147–157
Imaizumi H, Nambu N, Nagai T (1980) Pharmaceutical interaction in dosage forms and processing.18. Stability and several physical-properties of amorphous and crystalline forms of indomethacin. Chem Pharm Bull 28(9):2565–2569
Ivanisevic I, Bates S, Chen P (2009) Novel methods for the assessment of miscibility of amorphous drug-polymer dispersions. J Pharm Sci 98(9):3373–3386
Janssens S, Van den Mooter G (2009) Review: physical chemistry of solid dispersions. J Pharm Pharmacol 61(12):1571–1586
Janssens S, De Zeure A, Paudel A, Van Humbeeck J, Rombaut P, Van Den Mooter G (2010) Influence of preparation methods on solid state supersaturation of amorphous solid dispersions: a case study with itraconazole and eudragit E100. Pharm Res 27(5):775–785
Jasti BR, Berner B, Zhou SL, Li X (2004) A novel method for determination of drug solubility in polymeric matrices. J Pharm Sci 93(8):2135–2141
Karmwar P, Boetker JP, Graeser KA, Strachan CJ, Rantanen J, Rades T (2011a) Investigations on the effect of different cooling rates on the stability of amorphous indomethacin. Eur J Pharm Sci 44(3):341–350
Karmwar P, Graeser K, Gordon KC, Strachan CJ, Rades T (2011b) Investigation of properties and recrystallisation behaviour of amorphous indomethacin samples prepared by different methods. Int J Pharm 417(1–2):94–100
Karunakaran K (1981) Theoretical prediction of eutectic temperature and composition. J Solut Chem 10(6):431–435
Kauzmann W (1948) The nature of the glassy state and the behavior of liquids at low temperatures. Chem Rev 43(2):219–256
Kawakami K, Pikal MJ (2005) Calorimetric investigation of the structural relaxation of amorphous materials: evaluating validity of the methodologies. J Pharm Sci 94(5):948–965
Ke P, Hasegawa S, Al-Obaidi H, Buckton G (2012) Investigation of preparation methods on surface/bulk structural relaxation and glass fragility of amorphous solid dispersions. Int J Pharm 422(1–2):170–178
Khachaturyan AG (1978) Ordering in substitutional and interstitial solid-solutions. Prog Mater Sci 22(1–2):1–150
Kogermann K, Penkina A, Predbannikova K, Jeeger K, Veski P, Rantanen J, Naelapää K (2013) Dissolution testing of amorphous solid dispersions. Int J Pharm 444(1–2):40–46
Konno H, Taylor LS (2008) Ability of different polymers to inhibit the crystallization of amorphous felodipine in the presence of moisture. Pharm Res 25(4):969–978
Leuner C, Dressman J (2000) Improving drug solubility for oral delivery using solid dispersions. Eur J Pharm Biopharm 50(1):47–60
Li JJ, Chiappetta D (2008) An investigation of the thermodynamic miscibility between VeTPGS and polymers. Int J Pharm 350(1–2):212–219
Lin DX, Huang YB (2010) A thermal analysis method to predict the complete phase diagram of drug-polymer solid dispersions. Int J Pharm 399(1–2):109–115
Liu JS, Rigsbee DR, Stotz C, Pikal MJ (2002) Dynamics of pharmaceutical amorphous solids: the study of enthalpy relaxation by isothermal microcalorimetry. J Pharm Sci 91(8):1853–1862
Liu HJ, Zhang XY, Suwardie H, Wang P, Gogos CG (2012) Miscibility studies of indomethacin and Eudragit ® E PO by thermal, rheological, and spectroscopic analysis. J Pharm Sci 101(6):2204–2212
Loftsson T, Fridriksdottir H, Gudmundsdottir K (1996) The effect of water-soluble polymers on aqueous, solubility of drugs. Int J Pharm 127(2):293–296
Lu ZP, Liu CT (2002) A new glass-forming ability criterion for bulk metallic glasses. Acta Mater 50(13):3501–3512

Lu ZP, Liu CT (2003) Glass formation criterion for various glass-forming systems. Phys Rev Lett 91 (11)
Lu Q, Zografi G (1998) Phase behavior of binary and ternary amorphous mixtures containing indomethacin, citric acid, and PVP. Pharm Res 15(8):1202–1206
Ma X, Taw J, Chiang C-M (1996) Control of drug crystallization in transdermal matrix system. Int J Pharm 142(1):115–119
Mahieu A, Willart JF, Dudognon E, Danède F, Descamps M (2013) A new protocol to determine the solubility of drugs into polymer matrixes. Mol Pharm 10(2):560–566
Mao C, Chamarthy SP, Byrn SR, Pinal R (2006a) A calorimetric method to estimate molecular mobility of amorphous solids at relatively low temperatures. Pharm Res 23(10):2269–2276
Mao C, Chamarthy SP, Pinal R (2006b) Time-dependence of molecular mobility during structural relaxation and its impact on organic amorphous solids: an investigation based on a calorimetric approach. Pharm Res 23(8):1906–1917
Mao C, Chamarthy SP, Pinal R (2007) Calorimetric study and modeling of molecular mobility in amorphous organic pharmaceutical compounds using a modified Adam-Gibbs approach. J Phys Chem B 111(46):13243–13252
Marsac PJ, Konno H, Taylor LS (2006a) A comparison of the physical stability of amorphous felodipine and nifedipine systems. Pharm Res 23(10):2306–2316
Marsac PJ, Shamblin SL, Taylor LS (2006b) Theoretical and practical approaches for prediction of drug-polymer miscibility and solubility. Pharm Res 23(10):2417–2426
Marsac PJ, Li T, Taylor LS (2009) Estimation of drug-polymer miscibility and solubility in amorphous solid dispersions using experimentally determined interaction parameters. Pharm Res 26(1):139–151
Marsac PJ, Rumondor ACF, Nivens DE, Kestur US, Stanciu L, Taylor LS (2010) Effect of temperature and moisture on the miscibility of amorphous dispersions of felodipine and poly(vinyl pyrrolidone). J Pharm Sci 99(1):169–185
Matsumoto T, Zografi G (1999) Physical properties of solid molecular dispersions of indomethacin with poly(vinylpyrrolidone) and poly(vinylpyrrolidone-co-vinylacetate) in relation to indomethacin crystallization. Pharm Res 16(11):1722–1728
Metatla N, Soldera A (2007) The Vogel-Fulcher-Tamman equation investigated by atomistic simulation with regard to the Adam-Gibbs model. Macromolecules 40(26):9680–9685
Moore MD, Wildfong PLD (2009) Aqueous solubility enhancement through engineering of binary solid composites: pharmaceutical applications. J Pharm Innov 4(1):36–49
Moore MD, Wildfong PLD (2011) Informatics calibration of a molecular descriptors database to predict solid dispersion potential of small molecule organic solids. Int J Pharm 418(2):217–226
Murdande SB, Pikal MJ, Shanker RM, Bogner RH (2010a) Solubility advantage of amorphous pharmaceuticals: I. A thermodynamic analysis. J Pharm Sci 99(3):1254–1264
Murdande SB, Pikal MJ, Shanker RM, Bogner RH (2010b) Solubility advantage of amorphous pharmaceuticals: II. Application of quantitative thermodynamic relationships for prediction of solubility enhancement in structurally diverse insoluble pharmaceuticals. Pharm Res 27(12):2704–2714
Murdande SB, Pikal MJ, Shanker RM, Bogner RH (2011) Solubility advantage of amorphous pharmaceuticals, part 3: is maximum solubility advantage experimentally attainable and sustainable? J Pharm Sci 100(10):4349–4356
Nair R, Nyamweya N, Gonen S, Martinez-Miranda LJ, Hoag SW (2001) Influence of various drugs on the glass transition temperature of poly(vinylpyrrolidone): a thermodynamic and spectroscopic investigation. Int J Pharm 225(1–2):83–96
Nascimento MLF, Souza LA, Ferreira EB, Zanotto ED (2005) Can glass stability parameters infer glass forming ability? J Non-Cryst Solids 351(40–42):3296–3308
Newman A, Knipp G, Zografi G (2012) Assessing the performance of amorphous solid dispersions. J Pharm Sci 101(4):1355–1377
Nishi T, Wang TT (1975) Melting-point depression and kinetic effects of cooling on crystallization in poly(vinylidene fluoride) poly(methyl methacrylate) mixtures. Macromolecules 8(6):909–915

Onorato PIK, Uhlmann DR (1976) Nucleating heterogeneities and glass formation. J Non-Cryst Solids 22(2):367–378
Ota R, Wakasugi T, Kawamura W, Tuchiya B, Fukunaga J (1995) Glass-formation and crystallization in Li_2O-Na_2O-K_2O-SiO_2. J Non-Cryst Solids 188(1–2):136–146
Ozaki S, Kushida I, Yamashita T, Hasebe T, Shirai O, Kano K (2012) Evaluation of drug supersaturation by thermodynamic and kinetic approaches for the prediction of oral absorbability in amorphous pharmaceuticals. J Pharm Sci 101(11):4220–4230
Pajula K, Taskinen M, Lehto VP, Ketolainen J, Korhonen O (2010) Predicting the formation and stability of amorphous small molecule binary mixtures from computationally determined Flory-Huggins interaction parameter and phase diagram. Mol Pharm 7(3):795–804
Parks GS, Huffman HM, Cattor FR (1928) Studies on glass. II: the transition between the glassy and liquid states in the case of glucose. J Phys Chem 32:1366–1379
Parks GS, Snyder LJ, Cattoir FR (1934) Studies on glass. XI: some thermodynamic relations of glassy and alpha-crystalline glucose. J Chem Phys 56:595–598
Paudel A, Van Humbeeck J, Van den Mooter G (2010) Theoretical and experimental investigation on the solid solubility and miscibility of naproxen in poly(vinylpyrrolidone). Mol Pharm 7(4):1133–1148
Phuoc NH, Luu RPT, Munafo A, Ruelle P, Namtran H, Buchmann M, Kesselring UW (1986) Determination of partial solubility parameters of lactose by gas solid chromatography. J Pharm Sci 75(1):68–72
Qian F, Huang J, Hussain MA (2010a) Drug–polymer solubility and miscibility: stability consideration and practical challenges in amorphous solid dispersion development. J Pharm Sci 99(7):2941–2947
Qian F, Huang J, Zhu Q, Haddadin R, Gawel J, Garmise R, Hussain M (2010b) Is a distinctive single T-g a reliable indicator for the homogeneity of amorphous solid dispersion? Int J Pharm 395(1–2):232–235
Raghavan SL, Trividic A, Davis AF, Hadgraft J (2001) Crystallization of hydrocortisone acetate: influence of polymers. Int J Pharm 212(2):213–221
Reismann S, Lee G (2012) Assessment of a five-layer laminate technique to measure the saturation solubility of drug in pressure-sensitive adhesive film. J Pharm Sci 101(7):2428–2438
Reuteler-Faoro D, Ruelle P, Namtran H, Dereyff C, Buchmann M, Negre JC, Kesselring UW (1988) A new equation for calculating partial cohesion parameters of solid substances from solubilities. J Phys Chem 92(21):6144–6148
Roy SD, Flynn GL (1989) Solubility behavior of narcotic analgesics in aqueous media: solubilities and dissociation constants of morphine, fentanyl and sufentanil. Pharm Res 6(2):147–151
Rubinstein M, Colby RH (2003) Polymer physics. Oxford University Press, New York
Rumondor ACF, Taylor LS (2010) Effect of polymer hygroscopicity on the phase behavior of amorphous solid dispersions in the presence of moisture. Mol Pharm 7(2):477–490
Rumondor ACF, Marsac PJ, Stanford LA, Taylor LS (2009) Phase behavior of poly(vinylpyrrolidone) containing amorphous solid dispersions in the presence of moisture. Mol Pharm 6(5):1492–1505
Rumondor ACF, Wikstrom H, Van Eerdenbrugh B, Taylor LS (2011) Understanding the tendency of amorphous solid dispersions to undergo amorphous-amorphous phase separation in the presence of absorbed moisture. Aaps Pharm 12(4):1209–1219
Savolainen M, Kogermann K, Heinz A, Aaltonen J, Peltonen L, Strachan C, Yliruusi J (2009) Better understanding of dissolution behaviour of amorphous drugs by in situ solid-state analysis using Raman spectroscopy. Eur J Pharm Biopharm 71(1):71–79
Sekiguchi K, Obi N (1961) Studies on absorption of eutectic mixture. I. A comparison of the behavior of eutectic mixture of sulfathiazole and that of ordinary sulfathiazole in man. Chem Pharm Bull 9(11):866–872
Sekikawa H, Nakano M, Arita T (1978) Inhibitory effect of polyvinylpyrrolidone on crystallization of drugs. Chem Pharm Bull 26(1):118–126

Serajuddin ATM (1999) Solid dispersion of poorly water-soluble drugs: early promises, subsequent problems, and recent breakthroughs. J Pharm Sci 88(10):1058–1066

Sethia S, Squillante E (2003) Solid dispersions: revival with greater possibilities and applications in oral drug delivery. Critic Rev Ther Drug Carrier Syst 20(2–3):215–247

Shamblin SL, Huang EY, Zografi G (1996) The effects of co-lyophilized polymeric additives on the glass transition temperature and crystallization of amorphous sucrose. J Therm Anal 47(5): 1567–1579

Shamblin SL, Taylor LS, Zografi G (1998) Mixing behavior of colyophilized binary systems. J Pharm Sci 87(6):694–701

Shamblin SL, Tang XL, Chang LQ, Hancock BC, Pikal MJ (1999) Characterization of the time scales of molecular motion in pharmaceutically important glasses. J Phys Chem B 103(20):4113–4121

Sheng Q, Weuts I, De Cort S, Stokbroekx S, Leemans R, Reading M, Belton P, Craig DQM (2010) An investigation into the crystallisation behaviour of an amorphous cryomilled pharmaceutical material above and below the glass transition temperature. J Pharm Sci 99(1):196–208

Simha R, Boyer R (1962) On a general relation involving the glass temperature and coefficients of expansion of polymers. J Chem Phys 37:185–192

Sun YE, Tao J, Zhang GGZ, Yu L (2010) Solubilities of crystalline drugs in polymers: an improved analytical method and comparison of solubilities of indomethacin and nifedipine in PVP, PVP/VA, and PVAc. J Pharm Sci 99(9):4023–4031

Surana R, Pyne A, Rani M, Suryanarayanan R (2005) Measurement of enthalpic relaxation by differential scanning calorimetry—effect of experimental conditions. Thermochim Acta 433(1–2): 173–182

Saad M, Poulain M (1987) Glass forming ability criterion. Mater Sci Forum 19(20):11–18

Tantishaiyakul V, Kaewnopparat N, Ingkatawornwong S (1999) Properties of solid dispersions of piroxicam in polyvinylpyrrolidone. Int J Pharm 181(2):143–151

Tao J, Sun Y, Zhang GGZ, Yu L (2009) Solubility of small-molecule crystals in polymers: D-mannitol in PVP, indomethacin in PVP/VA, and nifedipine in PVP/VA. Pharm Res 26(4):855–864

Taylor LS, Zografi G (1998) Sugar-polymer hydrogen bond interactions in lyophilized amorphous mixtures. J Pharm Sci 87(12):1615–1621

ten Brinke G, Oudhuis L, Ellis TS (1994) The thermal characterization of multicomponent systems by enthalpy relaxation. Thermochim Acta 238:75–98

Thakral S, Thakral NK (2013) Prediction of drug-polymer miscibility through the use of solubility parameter based Flory-Huggins interaction parameter and the experimental validation: PEG as model polymer. J Pharm Sci 102(7):2254–2263

Tian Y, Booth J, Meehan E, Jones DS, Li S, Andrews GP (2013) Construction of drug-polymer thermodynamic phase diagrams using flory-huggins interaction theory: identifying the relevance of temperature and drug weight fraction to phase separation within solid dispersions. Mol Pharm 10(1):236–248

Turnbull D (1969) Under what conditions can a glass be formed? Contemp Phy 10(5):473–488

Uhlmann DR (1972) A kinetic treatment of glass formation. J Non-Cryst Solids 7(4):337–348

Van den Mooter G, Augustijns P, Kinget R (1999) Stability prediction of amorphous benzodiazepines by calculation of the mean relaxation time constant using the Williams-Watts decay function. Eur J Pharm Biopharm 48(1):43–48

Van Eerdenbrugh B, Baird JA, Taylor LS (2010) Crystallization tendency of active pharmaceutical ingredients following rapid solvent evaporation-classification and comparison with crystallization tendency from undercooled melts. J Pharm Sci 99(9):3826–3838

Van Krevelen DW (1997) Cohesive properties and solubility. In: Van Krevelen DW (ed) Properties of polymers. Their correlation with chemical structure: their numerical estimation and prediction from additive group contributions. Elsevier, Amsterdam, pp 189–225

Vasanthavada M, Tong WQ, Joshi Y, Kislalioglu MS (2005) Phase behavior of amorphous molecular dispersions—II: role of hydrogen bonding in solid solubility and phase separation kinetics. Pharm Res 22(3):440–448

Vasconcelos T, Sarmento B, Costa P (2007) Solid dispersions as strategy to improve oral bioavailability of poor water soluble drugs. Drug Discov Today 12(23–24):1068–1075
Warren DB, Benameur H, Porter CJH, Pouton CW (2010) Using polymeric precipitation inhibitors to improve the absorption of poorly water-soluble drugs: a mechanistic basis for utility. J Drug Target 18(10):704–731
Weinberg MC (1994) Glass-forming ability and glass stability in simple systems. J Non-Cryst Solids 167(1–2):81–88
Whichard G, Day DE (1984) Glass-formation and properties in the gallia-calcia system. J Non-Cryst Solids 66(3):477–487
Williams HD, Trevaskis NL, Charman SA, Shanker RM, Charman WN, Pouton CW, Porter CJH (2013) Strategies to address low drug solubility in discovery and development. Pharmacol Rev 65(1):315–499
Wiranidchapong C, Tucker IG, Rades T, Kulvanich P (2008) Miscibility and Interactions between 17-estradiol and Eudragit ® RS in Solid Dispersion. J Pharm Sci 97(11):4879–4888
Wyttenbach N, Janas C, Siam M, Lauer ME, Jacob L, Scheubel E, Page S (2013) Miniaturized screening of polymers for amorphous drug stabilization (SPADS): rapid assessment of solid dispersion systems. Eur J Pharm Biopharm 84(3):583–598
Yang M, Gogos CG, Wang P (2001) A new systematic methodology to determine drug's solubility in polymer. pp 1354–1359
Yokoi Y, Yonemochi E, Terada K (2005) Effects of sugar ester and hydroxypropyl methylcellulose on the physicochemical stability of amorphous cefditoren pivoxil in aqueous suspension. Int J Pharm 290(1–2):91–99
Yoo SU, Krill SL, Wang Z, Telang C (2009) Miscibility/stability considerations in binary solid dispersion systems composed of functional excipients towards the design of multi-component amorphous systems. J Pharm Sci 98(12):4711–4723
Yoshioka M, Hancock BC, Zografi G (1994) Crystallization of indomethacin from the amorphous state below and above its glass-transition temperature. J Pharm Sci 83(12):1700–1705
Zhao YY, Inbar P, Chokshi HP, Malick AW, Choi DS (2011) Prediction of the thermal phase diagram of amorphous solid dispersions by flory-huggins theory. J Pharm Sci 100(8):3196–3207
Zheng WJ, Jain A, Papoutsakis D, Dannenfelser RM, Panicucci R, Garad S (2012) Selection of oral bioavailability enhancing formulations during drug discovery. Drug Dev Ind Pharm 38(2): 235–247
Zhou DL, Zhang GGZ, Law D, Grant DJW, Schmitt EA (2002) Physical stability of amorphous pharmaceuticals: importance of configurational thermodynamic quantities and molecular mobility. J Pharm Sci 91(8):1863–1872
Zimper U, Aaltonen J, McGoverin C, Gordon K, Krauel-Goellner K, Rades T (2010) Quantification of process induced disorder in milled samples using different analytical techniques. Pharmaceutics 2(1):30–49

Chapter 3
Overview of Amorphous Solid Dispersion Technologies

Harpreet Sandhu, Navnit Shah, Hitesh Chokshi and A. Waseem Malick

3.1 Introduction and Background

A survey of recent literature shows considerable growth in the application of amorphous solid dispersion (ASD) to solve solubility-related challenges in product development (Williams et al. 2010, 2013; Repka et al. 2013). This growth is primarily driven by three factors:

a. development and expansion of acceptable excipients especially at the dose level that is needed for solid dispersion,
b. application of newer technologies, and
c. enhanced understanding of amorphous systems using predictive analytical tools for stability and dissolution.

The earlier developments in ASD were hindered by the lack of scientific understanding of the metastable high-energy form and the availability of suitable technologies (Sekiguchi et al. 1964). For the purpose of this chapter, the processing technologies are classified into two main classes primarily, i.e., solvent based or fusion based. A schematic of this classification is shown in Fig. 3.1 to help orient the readers (Miller 2012). Based on their maturity, selected technologies are covered in this chapter with a goal to provide the necessary tools to help select an appropriate technology for a specific application.

H. Sandhu (✉)
Merck & Co., Inc., 556 Morris Avenue, Summit, NJ 07901, USA
e-mail: harpreet.sandhu@merck.com

N. Shah
Kashiv Pharma LLC, Bridgewater, NJ, USA

H. Chokshi
Roche Pharma Research & Early Development, Roche Innovation Center, New York, NY, USA

A. W. Malick
Pharmaceutical and Analytical R&D, Hoffmann-La Roche Ltd., Nutley, NJ, USA

N. Shah et al. (eds.), *Amorphous Solid Dispersions*,
Advances in Delivery Science and Technology, DOI 10.1007/978-1-4939-1598-9_3

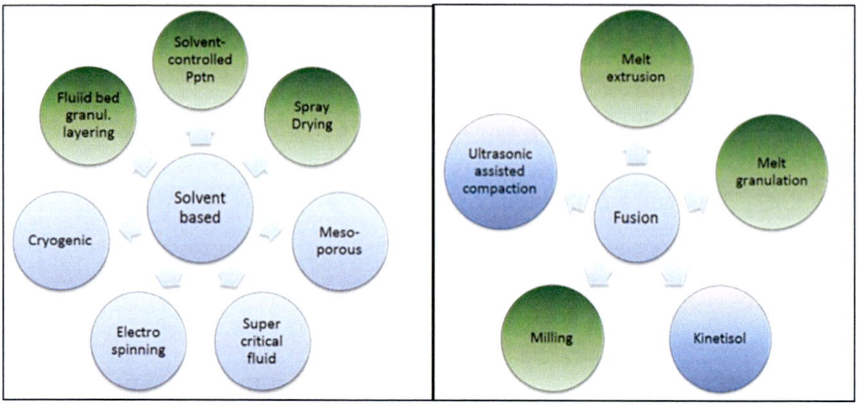

Fig. 3.1 Commonly used processing technologies in the manufacture of amorphous solid dispersion (ASD)

- **Solvent-based technologies** listed below rely on the preparation of a solution of the drug together with the stabilizing component:
  - Spray drying: Rapid removal of the solvent in a controlled environment (temperature and pressure) that is accelerated by generating high surface area
  - Fluid bed granulation/layering/film coating: Removal of solvent in various conventional pharmaceutical equipments
  - Coprecipitation: Solvent-controlled precipitation technologies, e.g., microprecipitated bulk powder (MBP), evaporative precipitation into aqueous solution (EPAS), Nanomorph, and flash precipitation, etc.
  - Supercritical fluid- based technologies and its variations such as FormulDisp®
  - Cryogenic processing, e.g., spray freeze drying (SFD) and thin film freezing (TFF)
  - Electrospinning: Drawing nanofibers from solution or molten material under high electrostatic voltage
  - Rotating jet spinning: Combination of centrifugation and pinning to produce nanofibers
- Fusion-based technologies where the drug and the stabilizing component are heated and mixed:
  - Melt granulation
  - Melt extrusion
  - KinetiSol: High-shear mixing combined with high temperature
  - Milling: High-shear milling/cryogrinding with and without excipients, e.g., Biorise®
  - Deposition of molten material on a carrier by hot-melt coating in a fluid bed process, e.g., Meltdose®

3.2 Solvent Evaporation

A key prerequisite for ASD is the elimination of drug's crystallinity and the best means to achieve that state is by dissolving the crystalline drug in a suitable solvent. In some cases, it may be possible to obtain pure amorphous drug but due to stability considerations the drug is generally processed with a polymer that stabilizes the amorphous form through mechanical and physicochemical interactions. An ideal means to achieve coprecipitation involves the solubilization of active pharmaceutical ingredient (API) and polymer in a common solvent followed by solvent removal. Typically, solubility of a drug in organic solvents drives the selection of stabilizing polymer and the process. The design of a formulation using solvent evaporation process generally consists of the following sequential steps:

- Solvent selection
- Selection of polymer and additives
- Selection of an evaporation method that produces ASD with acceptable residual solvent levels

To get a better insight into these processes, each of these steps is discussed in the following sections.

3.2.1 *Solvent Selection*

For successful application of solvent-based techniques, adequate solubility in organic solvents is critical for generating an ASD. In most cases, the solubility screen conducted during preformulation studies forms the basis for selecting a solvent. The criteria for solvent selection include solubility of API and polymer in a common solvent, drying efficiency of the solvent, acceptable level of residual solvents (based on International Conference on Harmonization classification), and desired shelf-life stability. From thermodynamic perspective, the drying of solvent involves complex interplay of heat and mass transfer and depends primarily on the supply of heat and efficiency of vapor removal. On a process level, the drying efficiency depends on the solvent evaporation rate that in turn depends on the boiling point, specific heat of solution, heat of vaporization, surface area, vapor pressure, percent solid content, and solution viscosity (Abeysena and Darrington, 2013).

Drying is an energy-intensive process that requires careful selection of a solvent that can provide adequate solubility of drug and polymer and is easy to remove. From the thermal perspective, the amount of heat required to remove a solvent represents the sum of latent heat of vaporization (ΔH_{vap}), heat required to raise the temperature to the boiling point, and losses in the process (Murugesan et al. 2011). Assuming that energy loss is an equipment factor and will be similar for different solvents, the heat required (Q_H) to remove a solvent can be estimated by ΔH_{vap} and $C_p \Delta T$:

$$Q_H = C_{P*}(T_b - T_{RT}) + \Delta H_{vap} \times \frac{1000}{\text{Mol weight}},$$

Table 3.1 Drying-related properties of some commonly used solvents

Solvent	Mol wt (g/mol)	Heat capacity (J/g °C)	Heat of vapor-ization (kj/mol)	Boiling point (°C)	Vapor pressure @20 °C (kpa)	Heat energy required to evaporate 1 kg solvent (J)
Water	18	4.18	40.7	100	2.3	2596
Ethanol	46	2.44	38.7	78	5.8	983
Acetone	58	2.17	29.1	70	24	610
Dimethylsulfoxide	78	1.96	52.9	189	0.06	1009
Dimethylacetamide	87	2.0	46.2	165	0.3	828
N-methylpyr-rolidone	99	1.7	54.5	204	0.04	846

where C_p is specific heat capacity, T_b is the boiling point, and T_{RT} is room temperature.

A summary of the relevant thermophysical properties of commonly used solvents is provided in Table 3.1. In addition to boiling point and heat of vaporization, vapor pressure of the solvent is also critical in assessing the drying efficiency as that determines the surface renewal efficiency. Understanding the temperature-dependent changes in vapor pressure can provide useful insights into the means of improving drying efficiency. It has been shown that solvents such as water, toluene, n-heptane, and N-methylpyrrolidone (NMP) are difficult to remove because the increase in vapor pressure as a function of temperature is very slow. Furthermore, it is also important to note that the properties summarized in Table 3.1 are for pure solvents and can vary significantly depending on the additives and their interactions with the solvents.

Driven by the desire to maximize API solubility and for optimization of drying efficiency, on many occasions the formulation scientists resort to using mixed solvents. Although considered an annoyance from the perspective of purification, the azeotropes are preferred for ASD in the event a pure solvent cannot be used. In the absence of azeotropes, the differences in the evaporation rates of binary solvents may result in variable supersaturation of the precipitating material thus potentially resulting in phase separation. A list of some commonly used solvents that can form azeotrope is provided in Table 3.2 for reference. When using mixed solvents for ASD, it is likely that much more extensive work will be required to optimize the right combination and ascertain its impact on product quality to derisk potential problems during manufacturing and scale-up.

In the course of selecting a suitable solvent for ASD preparation, it is important to ensure that material is chemically and physically stable in the solvent. The intent of solvent selection is to convert crystalline material into amorphous form, however, some solids may form solvates or the residual solvent may lower the glass transition temperature (T_g) of the material resulting in unfavorable stability. The stability needs

Table 3.2 Listing of commonly used solvents with respect to their ability to form azeotrope (http://en.wikipedia.org/wiki/Azeotrope_(data), Accessed 10 Dec 2013)

Solvents	Boiling Point of Solvent (°K)	Azeotropic Temp (°K)	Azeotropic Composition (%w/w)
Ethanol:Water	352/373	352	96:4
Acetone:Water	330/373	Zeotropic	Zeotropic
Water:Acetone:Chloroform	373/330/334	334	4:38:58
Ethanol:Tetrahydrofuran	352/339	339	97:3
Ethanol:Ethyl acetate	352/350	345	69:31
Dichloromethane:Water	313/373	312	99:1
Dichloromethane:Ethanol	313/352	313	95:5

to be established to ensure that sufficient hold time can be achieved especially during scale-up where the run times can extend over days.

3.2.2 *Selection of Polymer and Other Additives*

Primary criterion for the selection of a polymer for ASD by solvent evaporation method depends on its solubility in the solvent. The other criteria which are also important include miscibility with API in the solid state, ability to yield high-drug loading, supportive toxicological data package, and its impact on achieving and maintaining high supersaturation. These additional criteria are covered in details in the other chapters. A brief summary of different polymers and the solvents that have been used for various applications is provided in Table 3.3. Solvent-based processes provide options to include other additives, such as surfactant or secondary stabilizers, to augment product quality. Feed solution ranging from solution to suspension can be processed by spray drying process; however, for ASD manufacture it is desirable that all components are in the dissolved state. From a downstream processing perspective, most spray-dried intermediates require densification to improve the density and flow properties prior to manufacturing the final dosage form. The predominant consolidation mechanism for amorphous materials especially with relatively large proportion of polymer is plastic deformation (Iyer et al. 2013). Choice of the polymer, any additives and their relative amounts in the feed solution, and the characteristics of the final amorphous intermediate may impact critical properties of the material such as particle size and density that could have significant effect on downstream processing.

Table 3.3 Relevant properties of commonly used pharmaceutical polymers

Polymer	Commercial product/late-stage experience (ASD)	Thermal properties T_g (°C)	Solubility in solvent
Hypromellose	Sporanox®, Fluid bed coating on non-pareil seeds	170–180	Ethanol:dichlormethane (1:1, 2:1) Methyl acetate:methanol (1:1)
Hydroxypropylcellulose acetate succinate	InCivek®and Kalydeco® by spray drying, Zelboraf®by coprecipitation, Noxafil DR® (HME)	100–110	Ethanol, methanol, dichloromethane, chloroform
Methacrylic acid copolymers (Eudragit L100, S100)	No known product as ASD, used in controlled relase product	> 150	Acetone, ethanol, methanol Ethanol:dichlormethane (1:1)
Amino methacrylate copolymer (Eudragit EPO)	Taste-masking application	48	Acetone, ethanol, methanol, ethyl acetate, methyl ethyl ketone, dichloromethane, tetrahydrofuran
Povidone (PVP)	Cesamet®) granulation with ethanol (Conine 1980), Rezulin®(HME)	175 (K30) 180 (K90)	Acetone, ethanol, methanol, ethyl acetate, methyl ethyl ketone, dichloromethane, tetrahydrofuran
Copovidone (PVP/VA)	Kaletra® and Norvir® made by HME	106	Acetone, ethanol, methanol, ethyl acetate, methyl ethyl ketone, dichloromethane, tetrahydrofuran
Polyethylene glycol	GrisPEG® melt granulation	Tm = 55–63 (PEG6000)	Acetone, ethanol, methanol, ethyl acetate, methyl ethyl ketone, dichloromethane, tetrahydrofuran
Poloxamers	Late-stage experience as nanoparticles and crystalline dispersion	Tm = 52–57 (P188)	Acetone, ethanol, methanol, ethyl acetate, methyl ethyl ketone, dichloromethane, tetrahydrofuran
Polyvinyl acetate phthalate	Primary use in tablet coating	NA	Ethanol, ethyl acetate, methanol Ethanol:dichloromethane (1:1)
Cellulose acetate phthalate	NA	160–170	Acetone, methyl ethyl ketone, ethyl acetate
Hypromellose phthalate	NA	133–137	Acetone, ethanol:dichlormethane (1:1) methanol, ethyl acetate, methyl ethyl ketone, dichloromethane, tetrahydrofuran
Soluplus®	NA, primarily used in melt extrusion	70	Acetone, methanol, ethanol, dichloromethane

ASD amorphous solid dispersion, *HME* hot-melt extrusion

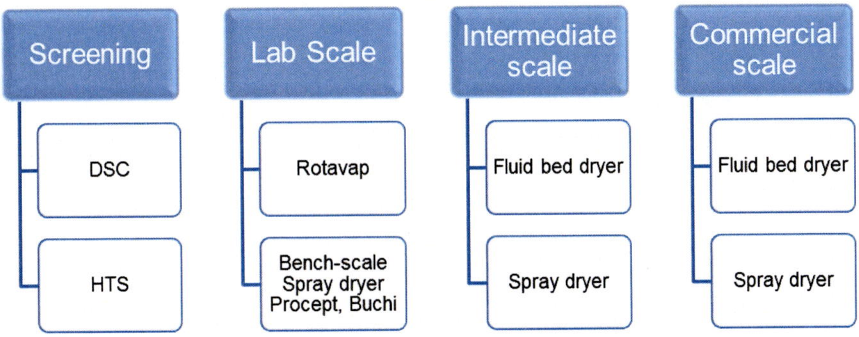

Fig. 3.2 Selection of solvent removal process

3.2.3 Selection of Solvent Evaporation Process

There are several literature reports demonstrating the role of solvent removal process in the development of ASD (Joe et al. 2010; Miller and Gil 2010). The choice of solvent evaporation process is influenced by the scale, the stability of the formulation, and the availability of equipment. Commonly used solvent removal processes in the pharmaceutical industry are shown in Fig. 3.2. Even though spray drying is the most efficient, well understood, and established process for ASD, other methods are also frequently used. Fluid bed drying includes either spray granulation or fluid bed layering on inert beads. The granulated product can be converted to tablets or capsules although the multiparticulate pellets produced by fluid bed processes are generally more suitable for encapsulation.

Owing to its suitability for high-throughput screening, solvent evaporation is the most widely used process during preformulation screening for optimal selection of solvent, polymer, and drug loading. Because of the small sample volume (typically few microliters) and the efficiency of solvent removal process, the screening studies tend to simulate the spray drying process fairly well. However, in the chronicles of ASD development, the weakest link between preformulation screening and the manufacture of small-scale batches has been the availability of suitable laboratory-scale equipment. Rotary evaporators that are used in early development may lead to false negatives for compounds with high crystallization tendency and the small-scale spray dryers suffer from low yield. It is generally recognized that compounds with low tendency for crystallization can be manufactured by any solvent evaporation process but the rate of solvent removal and the long exposure time to high-temperature conditions pose serious challenges for compounds with high crystallization tendency. Due to the solvent removal efficiency and single-stream continuous processing, spray drying offers the most favorable conditions for manufacture of ASD. With recent developments in the design of spray dryers, spray drying can now be realized across all scales ranging from laboratory to commercial. The laboratory-scale spray dryer supplied by ProCepT® can work with volumes as low as 1 mL to 24 L with more than 90 % yield (ProCepT 2014). One of the challenges with all solvent-based techniques

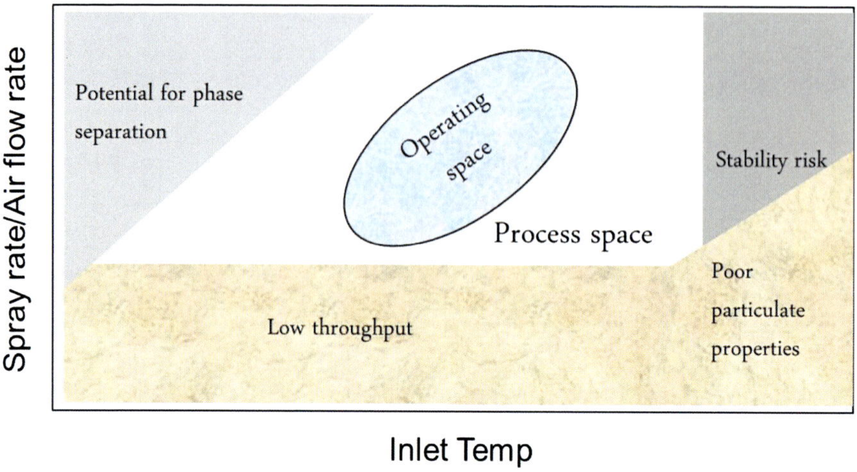

Fig. 3.3 Spray drying design space

is the complete removal of solvent. In addition to the safety concerns, the residual solvent can have a detrimental effect on the stability of the product. Therefore, spray drying is usually followed by secondary drying. Processes ranging from tray drying to fluid bed have been used to achieve the desired level of residual solvent. Among the various modes of solvent removal, spray drying has become the most widely adopted process. The key features of spray drying processes that are relevant for design and development of an ASD product are listed below:

- Design of spray dryer: Closed-loop versus open-loop systems
- Atomization and nozzle design: Rotary, multi-fluid pneumatic (two to four fluids), pressure, and ultrasonic nozzles
- Drying gas: Type (cocurrent versus current orientation) and air volume
- Feed material: Solid content, foaming, viscosity, solvent system, T_g, and stability
- Collection system: Cyclone, filter bags, and electrostatic precipitators
- Secondary drying: Tray drying, fluid bed drying, rotary, agitated dryer, and fluidized spray drying
- Downstream processing: Densification, compaction, agglomeration, dissolution, and stability

The product quality and particulate properties can be controlled by optimizing the process variables. Types of equipment setups that can be used to support the development of ASD product from early screening phase to commercial scale are shown in Fig. 3.2 and the key processing variables are shown in Fig. 3.3 (Appel 2009).

3.3 Hot-Melt Extrusion

From the discovery of ASD, the two methods that have dominated the literature are solvent evaporation and melt extrusion (fusion-based method). Although spray drying continues to be an important technology, the commercial success achieved with melt extrusion has placed hot-melt extrusion (HME) at the top of the technology list. This stems from the specific advantages of the HME process that provides solvent-free continuous processing, modularity, and ability to produce a close-to-final product. Comprehensive discourses focusing on the application of melt extrusion in the pharmaceutical industry have been the subject of several research-based textbooks that have become available in the recent past (Ghebre-Selassie et al. 2003; Douroumis 2012 and Repka et al. 2013). The following section provides an overview of the formulation and process considerations in the development of the HME process. The key areas that need special consideration are listed below and elaborated further in the text:

- Selection of polymer, additives, and drug loading
- Selection of extruder and the processing conditions
- Downstream processing and performance optimization

3.3.1 Selection of Polymer, Additives (Plasticizer, Flow Aid and Surfactant), and Drug Loading in HME

The use of a polymer in ASD development is primarily for stabilizing the amorphous form, but in the case of the HME it is critical for processing as well. The molten polymer provides a medium in which the drug is either solubilized or dispersed. Therefore, in addition to improving the performance (dissolution and stability) of the product, the polymer also serves as an enabler for processing. Key characteristics of the polymer and the overall composition that are suitable for melt extrusion can be summarized as:

- Melting point and/or T_g of the drug
- Melting point and/or T_g of the polymer
- Molecular weight and melt viscosity of the polymer
- Specific interactions between drug and polymer leading to plasticization or antiplasticization, especially in the molten state
- Thermal stability of the components at the processing temperature
- Properties of additives such as physical state, melting point, miscibility, and stability
- Particulate properties of the polymer

A systematic analysis of potential drug:polymer blends may provide insight into the selection of a suitable polymer, e.g.:

- Solubility parameter estimation and differential scanning calorimetry (DSC) help assess the drug:polymer miscibility and determine drug loading.
- Rheological studies provide key insights into the viscoelastic properties and potential torque-limited extrusion.
- Assessment of the plasticizer to improve processability (lowering processing temperature or reducing torque).
- Microscopic investigation, especially atomic force microscopic and light microscopic methods, in characterization of the extrudates.
- Dissolution studies to monitor the rate and extent of solubility enhancement as well as to determine the need for surfactants.

Utilizing the melting point depression data from DSC, it is possible to calculate the Flory–Huggins interaction parameter that can then be used to construct the temperature–composition phase diagram for a binary system. The maximum drug loading that can be achieved in the solid dispersion that provides acceptable dissolution performance depends on the thermal (T_m/T_g ratio) and hydrophobic properties of the compound (log*P*). Based on the trend analysis of the available data, an empirical relationship has been proposed that demonstrate that drug substance with log*P* less than 6 and T_m/T_g ratio less than 1.3 may accommodate payloads as high as 50 % w/w (Friesen et al. 2008; DiNunzio 2013).

Plasticizer Plasticizers are low molecular weight additives that may be used in the HME process to help lower the processing temperature or reduce the melt viscosity of formulations containing high-melting actives or high molecular weight polymers. The processing of ASD by HME has been envisioned to occur in either the solubility regime or miscibility regime. In most cases, it is the molten API that is mixed with the molten polymer to produce an ASD. For some challenging compounds that do not have adequate solubility in the molten polymer, plasticizers are added to the formulation to aid in the process. Since plasticizers can have a negative impact on other aspects of the product such as dissolution, physical stability, T_g, hygroscopicity, chemical stability, appearance, and milling, their use in the formulation should be based on balancing and optimizing their effect on both processing and performance of the ASD.

A list of commonly used plasticizers is summarized in Table 3.4. Selection of the plasticizer is based on its intended functionality in the formulation such as reducing the processing temperature or reducing the melt viscosity. An ideal plasticizer is a temporary plasticizer that imparts the desired processing advantage but is removed from the formulation before final processing to minimize its negative impact. Supercritical carbon dioxide (CO_2), low boiling solvents, and reagents that can evaporate or sublime are all being evaluated for this purpose (Verreck et al. 2005; Desai 2007). In some cases, drug itself may provide adequate plasticization of the polymer (Zhu et al. 2002).

Table 3.4 Commonly used processing aids: plasticizers, surfactants, and flow aids in HME processing

Class	Type	Name/functionality	Mode of addition
Plasticizer	Liquid	Triethyl citrate Tributyl citrate Triacetin Polyethylene glycol 400 Acetyl tributyl citrate Dibutyl sebacate Solutol HS15 Cremophor EL and\RH40	Preblending could be challenging and may require preprocessing either by high-shear mixing or crude HME processing Through liquid addition port
	Solid	Low-melting drugs (does not belong because of title of table) Methyl paraben Citric acid PEG 8000 Stearic acid Glyceryl behenate Soluplus	Preblending feasible May be added through a different port
			May require large amount 5–10 % to achieve desired benefit May also act as flow aid
	Gaseous	Supercritical fluid (CO_2)	Are added via liquid addition port
		Low-boiling solvents (Acetone/ethanol/ethyl acetate)	Requires optimal screw design to prevent back-flow
		Camphor	Engineering controls required in the facility

Table 3.4 (continued)

Class	Type	Name/functionality	Mode of addition
			Mostly act as temporary plasticizers, but residual solvent control may be needed
Surfactants	SPANS	Sorbitan esters of long-chain fatty acids	Mostly added in preblend but can be metered via dedicated feed port
	Poloxamer	Block co-polymer PEO-PPO	May impact T_g and other quality attributes
	Tween	Polyoxyethylene derivatives of sorbitan esters (C12-C18)	Pure docusate sodium is semisolid but a powder form is available that contains sodium benzoate
	Labrasol	PEG-8 Caprylic/Capric Glycerides	May require lower processing temperatures due to chemical stability
	Gelucire 44/14, 50/13	Lauroyl polyoxylglycerides Stearoyl polyoxylglycerides	Usage amount should consider permissible daily intake
	Docusate sodium	Dioctyl sodium sulfosuccinate	
	Sodium lauryl sulfate	Strong surfactant	
	Vitamin E TPGS	–	
Flow aid	Mannitol/isomalt		
	Starch/maltodextrin	Commonly used pharmaceutical excipients	They may be added in powder blend to improve feeding efficiency of poor-flowing powders
	Colloidal silica	Flow aid	

HME hot-melt extrusion, *TPGS* tocopheryl polyethylene glycol

Surfactant Despite having successfully converted the crystalline drug to amorphous form, the HME product may not always provide the desired dissolution advantage. This is attributed to the poor wetting of the extrudate caused by hydrophobicity of the drug and the low porosity of the extrudates. Inclusion of a surfactant in the formulation improves the dissolution properties resulting in improved bioavailability of the product (Rosenberg et al. 2005; Mosquera-Giraldo et al. 2014). Listing of the commonly used surfactants is provided in Table 3.4. The key considerations in terms of selection criteria include impact on stability, daily usage limit, and processing feasibility.

Flow Aids A key consideration in the development of HME process is being able to uniformly feed the extruder. The consistent feed rate depends on the flow properties of the material. Depending on the number of feeders used, the drug and the polymer can be fed either as a common blend or separately through different feeders. To ensure the uniformity of blend, it is important to closely match the particulate properties of the drug and the polymer. A milling step may be required to ensure that drug and polymer are adequately mixed prior to extrusion.

To aid in the dissolution of API in the molten polymer, micronized API is frequently used in the extrusion process. However, this poses challenges in terms of poor flow and electrostatic charges that may limit the feeding of the API:polymer blend to the extruder. The low bulk density of the powder blend may further compromise the feeding efficiency giving rise to feed rate fluctuations and process instability. Commonly used pharmaceutical excipients shown in Table 3.4 can be included in the HME formulation to aid in the flow of material. Since some of these materials are crystalline in nature, they may affect the miscibility of drug in the polymer or simply increase the analytical complexity.

Thus, formulation design requires judicious selection of each component while considering their impact on the desired and undesired attributes of the product.

3.3.2 Selection of Extruder and the Processing Conditions

From the early days of introduction of melt extrusion processing in the pharmaceutical world, co-rotating twin-screw extruders have dominated this technology owing to their superior efficiency of mixing and self-wiping action ensuring first-in-first-out material flow. Several extruder types and sizes are available to achieve the desired product attributes that meet the phase-dependent needs of the product. Small-scale extruders provide an API-sparing option to support early studies such as pharmacokinetic (PK) feasibility or range-finding toxicology. These, however, may not always reflect the actual shear stress that the product will be subjected to during intermediate to large-scale manufacturing. Some of the challenges faced during small-scale manufacturing using a laboratory-scale extruder (degradation or incomplete conversion to amorphous form) may be resolved with larger extruders due to more efficient material flow and controlled residence time and residence time distribution. Typically, extruders greater than 12 mm provide representative extrusion conditions for scale-up with respect to the geometric similarity between total length, screw geometry, shear conditions, feeding mechanism, temperature of zones, and die dimension. The key equipment considerations in the development of an extrusion process include:

- Selection of extruder type (corotating versus counter rotating, motor power, and gear box)
- Length/diameter ratio (L/D ratio)
- Die design, size, and number of openings
- Feeding mechanism, number, and type of feeders including liquid injection port
- Optimization of screw geometry (distribution of kneading and conveying zones across the screw length)
- Temperature of each zone
- Online processing of extrudates: Cooling belt, pelletization, milling, and chillers
- Calendaring or direct shaping of materials such as films, implants, or tablets
- Downstream processing of the extrudates

A key consideration in the development of scalable process requires maintaining geometric similarity, i.e., L/D ratio and the degree of fill. Similar L/D ratios along with temperature and screw design across the barrel length provide comparable temperature and shear stress profiles. And the comparable degree of fill ensures consistent residence time and residence time distribution. This ensures that product is exposed to similar energy as given by the following equation:

$$\text{Specific energy}(SE) = \frac{K_{wm} EG_{\%} TS_{\%} \frac{RPM\text{run}}{RPM\text{max}}}{Q_n},$$

where SE is kw/h/kg, K_{wm} is motor power in kw (horsepower/1.34), $EG_{\%}$ efficiency of the gear system (95 %), $TS_{\%}$ (percent of torque and is formulation specific), RPM_{run} is the screw speed during the run and RPM_{max} the maximum feasible for the machine, Q_h is the feed rate (kg/h). Since most of the parameters are equipment specific, the two process variables are screw speed and feed rate.

Owing to its direct impact on the performance and efficiency of the process, feed rate is an important factor to consider during development. Representative feed rates that can be geometrically scaled up ensure reproducibility of the process and product. Multiple feeders can be used to improve the throughput as long as the product robustness has been established in that feed rate range.

Screw Design (Screw Elements and Shaft) The unique feature of melt extrusion process is its modularity and the prime illustration of that is in the design of screw configuration. In most pharmaceutical operations, the screw design consists of three elements: conveying, mixing, and zoning. Each of these regions can be moved, lengthened, or shortened with relative ease to achieve desired product characteristics. Conveying elements are low-shear elements, however, mixing elements depending on the design can generate significant shear and result in distributive mixing whereas zoning elements are primarily included to block the backflow especially in case of gas or supercritical fluid addition. Screw design can be optimized to accomplish uniform mixing, modify residence time, and/or to improve the chemical stability of thermally labile compounds. The screw elements are assembled on the shaft that controls the amount of torque being transferred from the motor to the product. Optimal design of screw shaft can further improve the extrusion efficiency by ensuring that the extruder power is effectively transferred to move the screws especially for high-viscosity products.

3.3.3 Downstream Processing and Performance Optimization

The most common type of output from pharmaceutical extrusion process is a spaghetti-shaped extrudate that may appear as transparent glass for pure amorphous material and has characteristically high density. For manufacture of standard oral dosage forms, after adequate cooling, these extrudates are generally milled to obtain granules. The granules are mixed with other excipients such as disintegrant, compression aid, and lubricant for either encapsulation or compression into final dosage form. Contrary to spray-dried material, HME granules possess excellent flow properties requiring minimal lubrication. However, HME granules generally have very poor compaction characteristics that are attributed to low porosity of the extrudates and ductile properties of the polymeric systems. Process modification, such as inclusion of supercritical fluids in the extrusion, increases the porosity of the extrudates that has favorable effect on the compaction properties. Thus, process and material properties play an important role in achieving the desired quality attributes ranging from appearance, integrity, and dissolution. Some of the issues encountered during development and possible means of resolution are summarized below:

- **Low T_g product and milling**: Ideally, selection of polymer and drug loading takes into consideration the T_g, specifically for physical stability purposes, however, in some instances it may not be possible to improve the T_g. Products with low T_g ($T_g < 50\,°C$), may not be suitable for conventional milling by impact mills such as hammer mill due to the potential of melting and blinding of the screen. In such cases, lowering the density of extrudate with inclusion of volatile solvents or supercritical CO_2 may improve the milling behavior. Alternatively, air jet milling or cryo-milling may provide viable options to address the milling issues with the extrudates. Particle size reduction may also improve the porosity of the granules thus helping with compaction.
- **Slow dissolution**: Despite using the same formulation composition, HME products when compared to ASD manufactured by other techniques, such as spray drying or microprecipitation, may provide slower dissolution rate (Dong et al. 2008). The slow dissolution rate is attributed to low surface area of the particle (low porosity surface and particle size). Several examples have been cited in the literature that uses surfactants in the formulation to overcome the dissolution problem. High hydrophilic–lipophilic balance (HLB) surfactants such as docusate sodium, d-α-tocopheryl polyethylene glycol 1000 succinate (vitamin E TPGS),

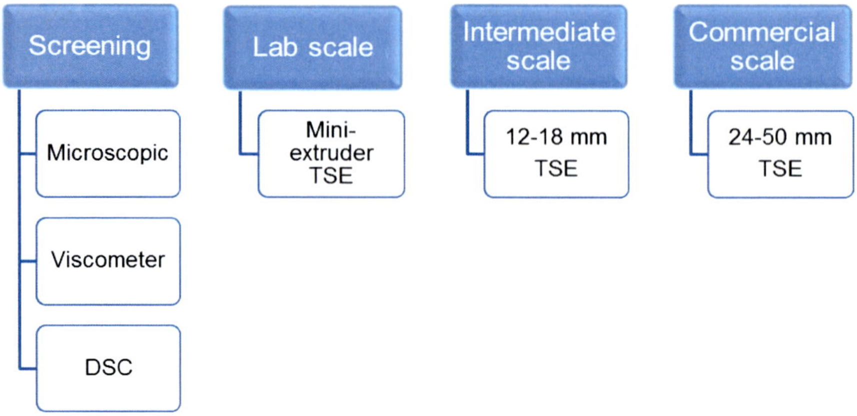

Fig. 3.4 Types of extruders used during product development

spans, tweens, cremophor, sodium laurylsulfate, and poloxamer are frequently used to improve the dissolution rate. As mentioned earlier, the selection of surfactant requires a systematic assessment of the allowable use limit, thermal stability, physical stability, and dissolution. The processing factors such as particle size reduction, use of superdisintegrants, and foaming agents in the extrusion can also help in improving the release rate by increasing the surface area and porosity.

- **Poor compaction**: Although milling of the extrudates may produce fine particles but due to low inherent porosity and ductility of polymers, for most part HME granules result in tablets of low tensile strength. This can be overcome by increasing the porosity of the extrudates either by the use of foaming agents or adding materials with brittle fracture characteristics during extrusion or prior to compression.

Figure 3.4 shows the equipment train that has been commonly used in the industry during different stages of development and Fig. 3.5 shows the key processing considerations during the development of HME process (Schenck et al. 2011).

3.4 Microprecipitation: MBP

The microprecipitation technology is especially suited for APIs that do not have adequate solubility in volatile organic solvents, and/or are thermally labile either due to high melting point or poor stability. According to Yalkowski, solubility of a compound can be estimated by its crystal structure (melting point and heat of fusion) and hydrophobicity (Yang et al. 2002). It has been observed that some compounds with high crystal lattice energy present solubility challenges in all types of solvents, i.e., aqueous as well as pharmaceutically acceptable cosolvents and vehicles. These brick dusts-like molecules have been shown to dissolve in polar solvents like dimethylacetamide (DMA), dimethylformamide, dimethylsulfoxide, and NMP.

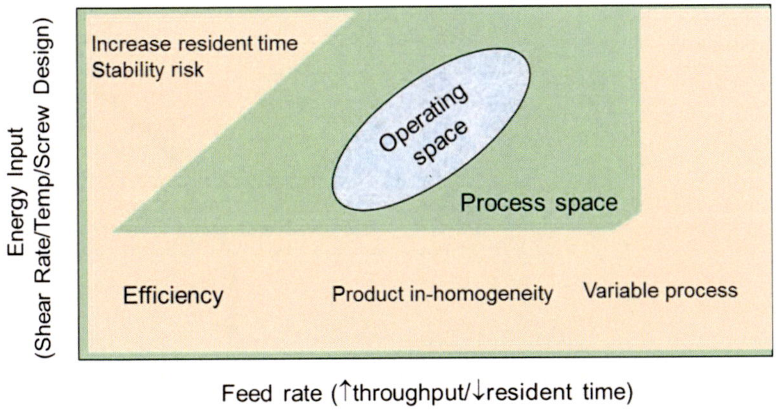

Fig. 3.5 Design space consideration during development of the hot-melt extrusion (HME) formulation

Such compounds are not suitable for either melt-based processes because of thermal stability or spray drying due to of the high boiling points of these solvents. Microprecipitation takes advantage of the solubility of the API and polymer in these polar solvents to produce amorphous form of API by solvent-controlled precipitation (Shah et al. 2012, 2013).

A schematic of the process is shown in Fig. 3.6 where a solution of the drug and polymer (ionic) is slowly added into a large volume of antisolvent to induce precipitation. The rapid precipitation conditions achieved due to insolubility of drug and polymer in the antisolvent as well as low processing temperature help preserve the amorphous form. From conceptual perspective, it can be visualized that the particle formation in microprecipitation occurs by extraction of solvent by the antisolvent. Because of high solubility of DMA in aqueous fluid, the extraction process is highly efficient resulting in amorphous particles with high porosity and superior wetting characteristics compared to spray drying. Although some work has been done using organic solvents as antisolvents, the most advanced systems use aqueous-based antisolvents to induce precipitation (Kadir 2012). Figure 3.7 shows a hypothetical scheme proposing the mechanism of particle formation during spray drying as well as the microprecipitation process. It appears that due to the formation of a skin on the surface of the particle, the rate of solvent removal could drop substantially in spray drying whereas this is not a concern in microprecipitation where the porous structure produced due to solvent removal is filled with aqueous fluid (antisolvent) which further promotes the solvent exclusion.

The salient features of the microprecipitation technology include:

Advantages:

- Suitable for challenging compounds (low solubility in volatile organic solvents and high melting point).
- Low temperature processing.

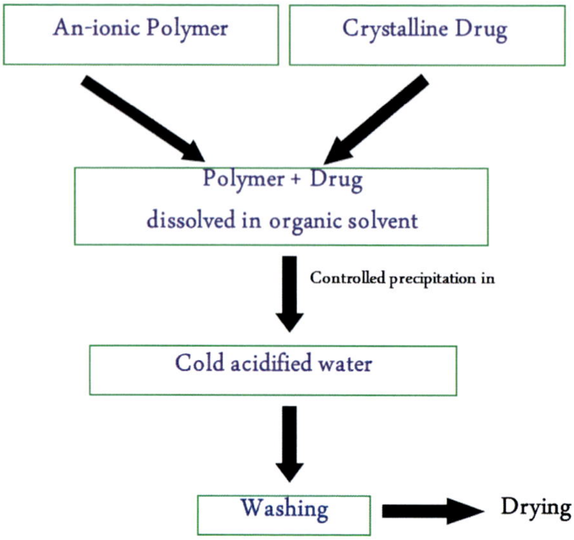

Fig. 3.6 Schematic of a microprecipated bulk powder (MBP) process

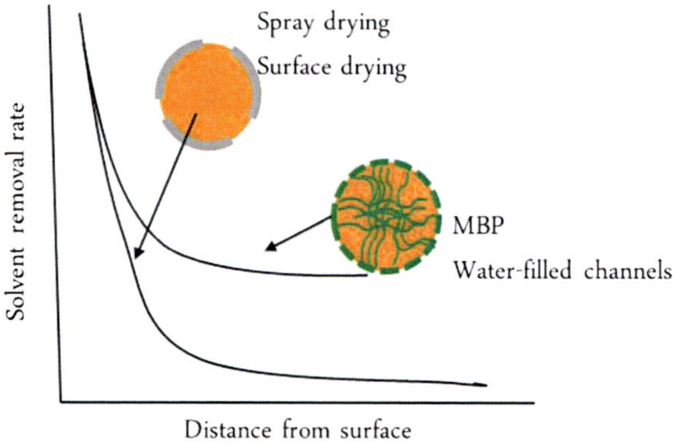

Fig. 3.7 Particle formation during solvent removal process by spray drying versus microprecipitation

- Suitable across different scales with high yield (few milligrams to thousand kilos).
- Superior particulate properties enable compaction and dissolution with least amount of external additives.
- Reduction in the need for plasticizers or surfactants.
- Rapid rate of quenching may provide higher drug loading.
- Ionic polymers used in creating MBP may impart superior stability (ionic interactions and low water activity).

Limitations:

- Some pH-sensitive compounds may not have an adequate window for processing due to pH-dependent solubility and stability.
- Ionic polymers release drugs in certain region of the gastrointestinal tract that may limit the applicability for drugs with narrow window of absorption.

- Removal of the nonaqueous solvent is by extraction but the final drying of the material containing water is generally performed in forced-air oven or a fluid bed dryer. Heat and moisture during final drying may promote recrystallization.

3.4.1 MBP Methodology

The key components of MBP technology involve two main aspects: preparation of amorphous dispersion and downstream processing to make the final product.

- Preparation of ASD:
  - Dissolution of API and polymer in a common solvent
  - Selection of antisolvent: Solubility and stability of API and polymer in solvent and solvent-rich-antisolvent phase
  - Precipitation conditions (pH, temperature, shear, solvent to antisolvent ratio, and time)
  - Mode of addition
  - Batch versus continuous processing
  - Washing of the precipitate to remove the residual solvents
  - Isolation of the precipitate
  - Drying of the precipitate
- Downstream processing:
  - Milling/sizing
  - Encapsulation or compaction
  - Coating

3.4.2 Preparation of ASD

Even though it is counter-intuitive to use an aqueous phase as antisolvent for the preparation of ASD, appropriate conditions can be generated that provide adequate supersaturation for both the polymer and API to induce rapid precipitation. The current literature is primarily based on using the pH condition that allows the precipitation of ionic polymers. Commonly used polymers include hypromellose acetate succinate, L, M, H grades, cellulose acetate pthalate, cellulose acetate butyrate, polyvinyl phthalate, hypromellose pthalate, polymethacrylates (Eudragit L100–55, Eudragit L100, Eudragit S-100, and Eudragit EPO). Use of low temperature, low solvent–antisolvent ratio, and appropriate shear help in maximizing the precipitation efficiency. As shown in Fig. 3.7, due to the differences in the mechanism of solvent removal process the surface properties of the two materials are also different. The MBP material produced by solvent exchange process has high porosity and better wetting compared to spray-dried or melt-extruded material that imparts better compaction and dissolution, thus reducing the need for additives such as compaction and wetting agents. Furthermore, due to the rapid quenching of the solution phase

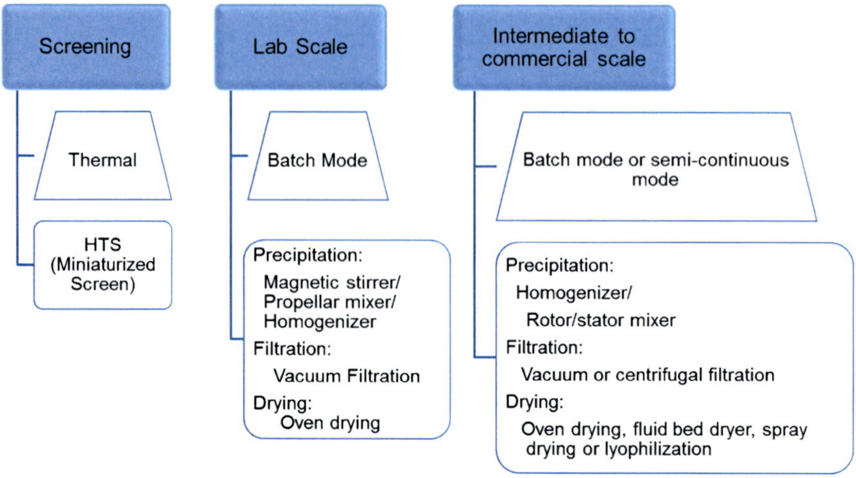

Fig. 3.8 Selection of precipitation, filtration, and drying methodologies at different scales

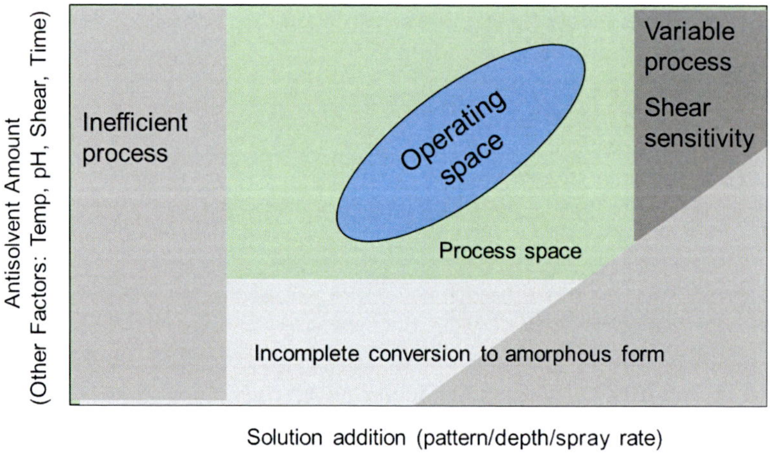

Fig. 3.9 Design space considerations during microprecipitation process design (solvent to antisolvent ratio and mixing). Other relevant factors related to supersaturation include pH, temperature, time, and shear

in solid state, it is also possible to increase the drug loading to as high as 70 % for some compounds. A general overview of the various processing options that can be used during different stages of development is summarized in Fig. 3.8 and the key processing aspects are shown in Fig. 3.9.

3.4.3 Downstream Processing

Depending on the particulate properties, the material may be of low bulk density (~0.1–0.3 g/cc) and densification may be required for further processing. Although MBP is primarily produced from aqueous media, once isolated and dried, the product requires appropriate protection from humidity and water due to physical stability of the amorphous form. Therefore, dry granulation is the preferred method to achieve the desired attributes of the granulates.

Several variations of the solvent-controlled precipitation have been evaluated to produce ASD, e.g., EPAS, Nanomorph, flash nanoprecipitation, and controlled precipitation (CP). In EPAS, the solution of drug and polymer is atomized into a heated aqueous solution, where the solvent (generally dichloromethane) is evaporated by the heated antisolvent (Vaughn et al. 2005). Due to the use of heated aqueous fluid as antisolvent, this process is limited to solvents such as dichloromethane that can be easily evaporated and because precipitation occurs at elevated temperature, it may not be suitable for the stability of ASD. Modification of EPAS process, CP involves in-line removal of solvent by vacuum distillation. The CP process also uses low boiling point solvents such as methanol as the preferred solvent. Alternatively, use of organic solvents as antisolvents has also been examined in technologies such as Nanomorph but robust development into a commercially viable product needs to be demonstrated (Keck and Muller 2006). Along with the selection of appropriate solvent/antisolvent pair and the processing conditions, these systems may be preferred to produce nanocrystals rather than amorphous dispersions.

3.5 Supercritical Fluid Processing

Over the past two decades, utility of supercritical fluids (SCF) has gained substantial momentum in the pharmaceutical industry. Although customarily used in the food industry for extraction (caffeine, essential oils, etc.) or in separation science for purification, the SCF offer promising opportunities in the development of specialized drug delivery systems such as particle design, nanoparticles, and amorphous dispersions. The key advantage of using supercritical fluids lies in their liquid- and gas-like properties that provide excellent media for solubilization with very low solvent burden. Due to the flexibility in designing the system, SCF can be used either as a solvent or antisolvent depending on the solubility of API and the stabilizing polymer. Its applications to ASD development is as diverse as the technology itself, e.g.:

- **HME:** As a processing aid in HME, SCF can serve multiple purposes ranging from lowering the melt viscosity, lowering processing temperature, modifying solubility of the drug in the molten polymer, and increasing the porosity of the extrudates that can improve dissolution and compaction.

- **Spray drying:** As an extraction solvent, SCF can be used to extract residual solvents from spray-dried material.
- **Microprecipitation:** As a stand-alone system, depending on the solubility, SCF may be used as a solvent or an antisolvent for microprecipitation technology that is akin to rapid expansion of supercritical solvent (RESS) or SCF as an antisolvent for precipitation (SAS).

Depending on how SCF is used, several techniques have evolved over the years especially in the particle engineering area. The commonly used variations of different processes are delineated below:

- Rapid expansion of supercritical solutions (RESS)
- Gas antisolvent precipitation (GAS)
- Supercritical antisolvent precipitation (SAS)
- Precipitation with compressed fluid antisolvent (PCA)
- Solution-enhanced dispersion by supercritical fluids (SEDS)
- Precipitation from gas-saturated solutions (PGSS)

Although there are very few case studies where SCF has been evaluated for production of ASD, the literature is rich with its application in particle engineering areas such as nanoparticles, and applications requiring low-temperature processing. Few examples showing the utility of RESS in producing amorphous particles include cefuroxime axetil (Varshosaz et al. 2009), ibuprofen, and indomethacin (Pathak et al. 2004). Similarly, there are few examples demonstrating the potential of using SAS techniques to produce ASD, e.g., itraconazole (Lee et al. 2005), rifampicin (Reverchon et al. 2002) and amoxicillin (Kalogiannis et al. 2005). While some formulation and processing factors may be similar for SAS or RESS system, it is critical to optimize the temperature and pressure in the SCF chamber to ensure that solubility conditions are fine-tuned to induce rapid supersaturation to ensure the precipitation of amorphous system.

The formulation and processing factors that can be tailored to customize the product attributes include:

- Use of cosolvents
- Nozzle dimension, spray rate, temperature, and pressure
- Conditions in the extraction chamber
  - Temperature
  - Pressure
  - Volume
  - Precipitation in aqueous phase with stabilizers (surfactants and polymers)

The selection of SCF technology to produce ASD depends primarily on the solubility of API and polymer in the most commonly used SCF, supercritical CO_2. Further formulation modification may be necessary to achieve desired particle morphology, e.g., polymers and surfactants are widely used to deagglomerate the particles and improve dissolution. Application of SCF in the development of ASD is still in its infancy, however, based on the flexibility in designing the process and properties of the SCF, it offers great potential for future advancement. For instance:

- Supercritical fluids could potentially enable the fastest rate of quenching and hence may open new possibilities in the solubilization space especially for challenging compounds.
- Differential solubility of API and polymer in the SCF may provide novel means of stabilizing the amorphous form.
- Processing temperatures may be suitable for thermo-labile compounds.
- By process design, the true particle size can be controlled in the submicron to nano range, thus offering dual advantage in improving the dissolution rate.

Once a suitable amorphous system has been produced, the downstream processing considerations will need to be addressed. Based on the nanoparticles work that has been conducted in this field, it is apparent that the amorphous product produced by the SCF will generally be of low density and high porosity and further densification will be required to make final dosage form.

3.6 KinetiSol

Poor aqueous solubility is a growing challenge in the pharmaceutical industry. Although several technologies have been successfully developed to produce commercially viable products, there is still a need for newer technologies that can be applied to challenging compounds and/or provide additional benefit of simplifying the process or increasing drug load. KinetiSol® is a promising new technology that has specific advantage for compounds that cannot be processed with more established processes such as ASD and HME. Similar to microprecipitation technology, KinetiSol is developed to address the processing needs of difficult compounds that are limited by either high melting point and/or low solubility in volatile organic solvents (DiNunzio et al. 2010; Hughey et al. 2010).

The core aspect of the technology is a specific type of equipment that has been used in the plastic industry to mix high-melting, high-viscosity products. The primary mechanism of making amorphous form is a variation of the fusion method. Similar to HME, it utilizes the frictional and shear energy to melt the drug and polymer blend. However, its distinguishing features are the intensity of mixing that causes material to melt within few seconds as opposed to HME where total residence time can vary from 30 s to few minutes. Faster heat transfer and melting result in shorter exposure time to high temperature that is specifically useful for high-melting and thermo-labile compounds. Due to the short exposure times, chemically labile compounds can be processed by KinetiSol® (Miller et al. 2012). Although this technology is in the early stages of development, prototype equipment have already been designed to provide insights into scale-up and production. Laboratory-scale equipment is generally run in batch mode to conserve the API, however, the pilot- and production-scale equipment are being designed to run in semicontinuous mode with relatively high-throughput rate

In addition to being suitable for thermo-labile compounds, the short exposure to high temperature also expands the range of polymers that are generally not stable for high-temperature HME. From a downstream processing perspective, the material appears to be similar to HME and requires particle size reduction prior to processing into the final dosage form. Additives such as plasticizer and wetting agents may also be included to improve product performance.

3.7 Ultrasonic-Assisted Compaction

To harness the full potential of amorphous systems for all types of chemical compounds, alternate technologies are constantly being added to the toolbox. Ultrasonic-assisted compaction is a modified tabletting process that can provide heat, pressure, and shear due to ultrasonic energy to the powder mixture during compaction. The application of ultrasound to solubility enhancement is based on the fusion method and in some ways mimics the extrusion process (Fini et al. 1997; Sancin et al. 1999). The ultrasonic frequency vibration is applied at the same time as compaction force. The key features of the technology include:

- Need small amount of material to conduct feasibility.
- Eliminates need for downstream processing since the manufacturing process delivers the final product.
- Current tablet presses may be retrofitted with the needed components.
- Product may show some inhomogeneity due to lack of distributive mixing with ultrasonic energy.
- The low porosity of compressed tablet may require use of hydrophilic fillers to improve the dissolution rate that may be at the expense of drug/polymer interactions.

A schematic of the process is shown in Fig. 3.10 with a representative tablet sample showing amorphous glass. Although research in this area is still limited, if successful, this may be a useful tool for early screening and for minimization of downstream processing.

3.8 Cryogenic Processing

Bottom-up particle engineering technologies based on cryogenic processing such as SFD, spray freezing into liquid, and TFF can produce amorphous nanostructured aggregates (Yang 2010). Cryogenic technologies involve use of cryogens such as liquid nitrogen to introduce a change in the temperature of the solubilized system that causes supersaturation, nucleation, and precipitation. Use of cryogens combined with a particular mechanism of addition can produce very high cooling rates thus resulting in rapid quenching of the amorphous form. These technologies are further classified based on the differences in the type of injection devices (capillary, rotary, pneumatic, and ultrasonic), location of nozzle (spray into the liquid or applying

Fig. 3.10 Amorphous compacts generated using ultrasound-assisted compaction unit

the solution onto cryogenic substrate), and the composition of the cryogenic liquid (hydrofluoralkanes, liquid nitrogen, liquid argon, compressed CO_2). Generally, these technologies involve rapid freezing of the solvent that can then be removed by sublimation, thus producing a powder. These techniques are particularly useful for temperature-sensitive materials such as proteins and peptides. Key considerations in applying these technologies are:

- Formation of feed solution: For amorphous processing, a solution formulation is preferred over suspension or emulsion. The total solid content may affect particulate properties.
- Ease of lyophilization of solvents: Solvents with high vapor pressure, melting point close to room temperature, high viscosity, and low toxicity. Commonly used solvents include acetonitirile, dioxane, and t-butanol.
- Due to the nature of the process, it may be possible to obtain amorphous materials at relatively high drug loading; however, stability during storage and dissolution may still limit the drug loading.
- Downstream considerations will be similar to spray-dried material.

3.9 Electrospinning and Rotating Jet Spinning

Analogous to HME, electrospinning is also a widely used technique in the polymer industry. A schematic of electrospinning process is shown in Fig. 3.11. A polymer solution is drawn through a capillary tube that is subjected to an electric field. As the electric field increases, the feed solution forms a Taylor cone at the tip of the capillary. Once the electric field overcomes the surface tension of the solution, the polymer solution is ejected as an electrically charged jet. Due to the increase in surface area, the solvent evaporates leaving thin filaments of material (50 nm to 5 microns). These fibers are then collected on collector screens for further processing. This technique has been applied for pharmaceutical systems by several researchers

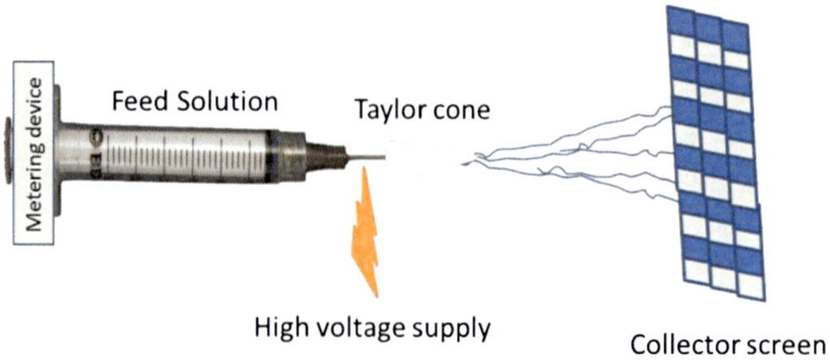

Fig. 3.11 Schematic of the electrospinning process

(Verreck et al. 2003; Nagy 2010). For amorphous processing, drug and polymer are generally dissolved in a common solvent similar to spray drying. Key factors in the processing include:

- Selection of the common solvent (generally ethanol is used).
- Electric potential from 16 to 24 kV has been used in some case studies.
- Downstream processing of fibers may be performed by milling.

Although this technique relies on solvent-based processing, the ability to form nanofibers can provide further advantage compared to other processing techniques. As the research in this area grows, there will be an opportunity to better understand the properties of pharmaceutical materials under high electric voltage. For most application in the literature, solvent-based processing has been evaluated, but nonsolvent-based processing using polymer melt is also feasible.

Rotating jet-spinning process is an evolution of the "cotton-candy" manufacturing equipment and uses centrifugal force of the rotor to create thin fibers that are deposited on the receiving chamber. Instead of a sugar solution, the drug:polymer solution in a suitable solvent is sprayed through the rotating jet. As shown in Fig. 3.12, the apparatus consists of a perforated reservoir containing polymer solutions attached to a motor. When the reservoir is spun about its axis of symmetry at a rate that exceeds the capillary and centrifugal forces, a viscous jet is ejected from a small orifice (Badrossamay et al. 2014). This jet is thrown outwards along a spiral trajectory as the solvent evaporates due to the creation of a high surface area. While moving, it is extended by centrifugal forces and solvent evaporates at a rate dependent on the diffusion coefficient of solvent through the polymer (Mellado et al. 2011). Compared to spray drying, the key limitations of this process may be the ability to remove the residual solvents to a satisfactory level, batch mode processing and downstream processing.

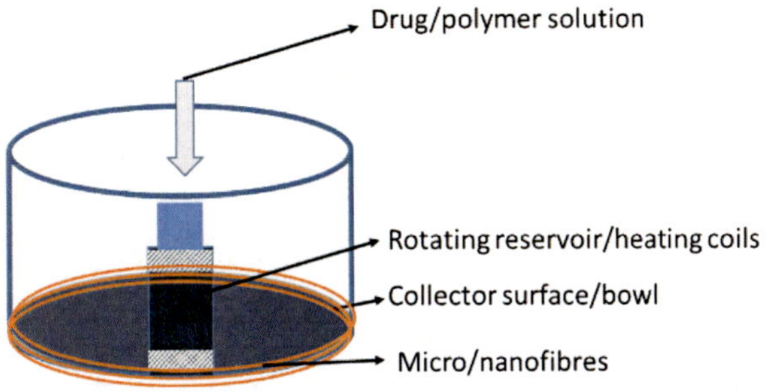

Fig. 3.12 Overview of the rotating jet-spinning process

3.10 Milling and Cryogrinding

Particle size reduction has been known to reduce crystallinity and induce amorphous characteristics for a long time (Mura et al. 2002). Since the naked amorphous API does not have adequate physical or chemical stability, co-grinding with polymers or stabilizers has also been used. Because milling is a standard unit operation in solid dosage form processing, this appears to be the most convenient means to produce the amorphous form, however, this simplicity comes with much higher risks. Due to the fact that milling is a top-down approach, there is always a risk that some material may exist in a nanocrystalline state that could act as seeds to induce nucleation and cause reversion of amorphous form to crystalline state. Several studies have been conducted to evaluate different milling mechanisms as well as stabilizers albeit with limited success. Media milling such as ball mill or cryo-milling with wide range of excipients such as Neusilin (magnesium aluminometasilicate), crospovidone, sodium chloride, or sugar (Gupta et al. 2002, 2003) have met with limited success. Although not claimed as one hundred percent amorphous, an anti-inflammatory product has been successfully manufactured using SoluMatrix® technology that involves dry milling the crystalline drug with a hydrophilic carrier (iCeutica 2014). Similarly, another milling technology that involves media milling in the dry state with crospovidone has been employed in a commercially available drug product (Perret 2014) by Aptalis. Considering that dry milling may have challenges for compounds with a high tendency to convert, it may be suitable for compounds that are inherently amorphous or have low tendency to crystallize. The products where drug could exist as a mixture of amorphous and nanocrystalline forms present much higher development risk and require stringent controls to ensure product consistency.

3.11 Hot-Melt Coating/Granulation

In an effort to extend the concept of lipid solubilization to produce solid dosage forms, a solution of drug substance in molten lipid is either coated or dispersed on an inert carrier (Faham et al. 2000; Holm et al. 2007). Several technologies have been

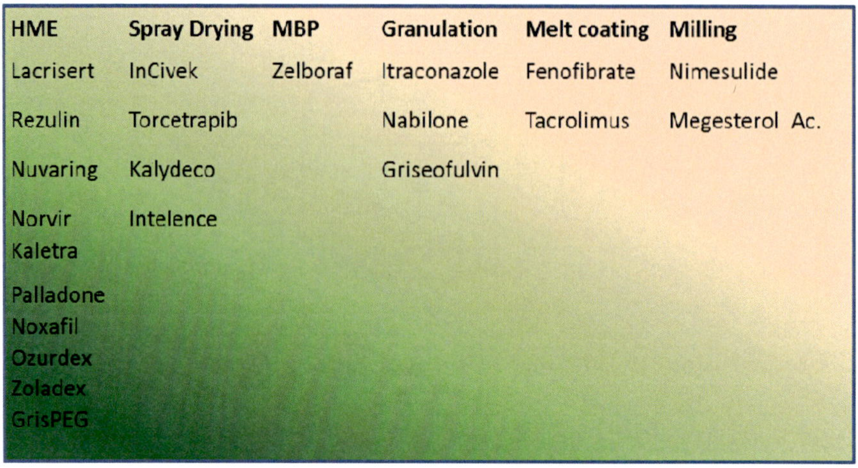
HME
Spray Drying
MBP
Granulation
Melt coating
Milling
Lacrisert
InCivek
Zelboraf
Itraconazole
Fenofibrate
Nimesulide
Rezulin
Torcetrapib
Nabilone
Tacrolimus
Megesterol Ac.
Nuvaring
Kalydeco
Griseofulvin
Norvir
Intelence
Kaletra
Palladone
Noxafil
Ozurdex
Zoladex
GrisPEG

Fig. 3.13 Distribution of compounds and technologies based on late-stage experience (all formulations may not be amorphous solid dispersion (ASD))

developed where amorphous drug can be trapped in the molten lipid which is cooled during processing. The fluid bed processing used for this purpose is retrofitted with a temperature-control setup to ensure that the product can be maintained in the molten state. Depending on the carrier, the processing conditions and the properties of the drug, the amorphous form of the drug may be obtained by these processes. However, it is critical to ensure that molten feed material is stable for the duration of the process and the quality of amorphous material is consistent and reproducible. Since drug and polymer melt requires spraying, these technologies are generally limited to polymers that melt at relatively low temperature and have relatively low melt viscosity such as poloxamers and/or high HLB gelucires. Generally, these carriers are not highly regarded as suitable stabilizers for amorphous form.

3.12 Process Selection Guide

The path to making an amorphous form requires two basic types of processes, i.e., either dissolve the crystalline form in a suitable solvent or melt the crystalline form with the stabilizing polymer. Numerous variations have been developed in each of these two categories to match the compound's properties, product needs, and organizational preference. Several compounds are in development using one of the many ASD technologies; however, melt extrusion and spray drying are leading the way with regard to the number of commercially successful products (see Fig. 3.13). The chart also depicts the degree of difficulty in assuring the conversion to complete amorphous form with some technologies. For example, technologies such as milling and spray coating perform similar to nanocrystalline formulations rather than the true

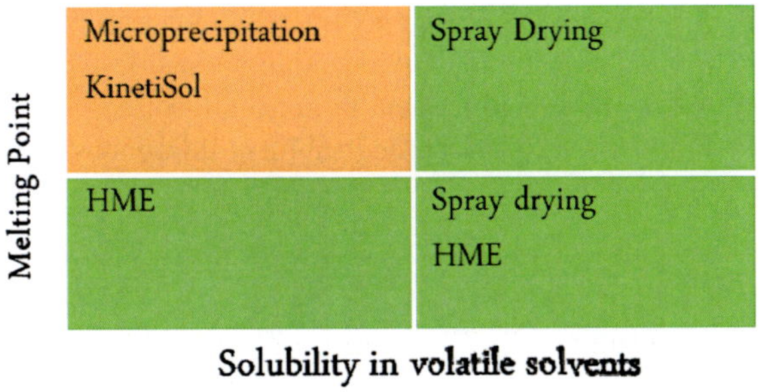

Fig. 3.14 Empirical guide to select a solid dispersion technology based on physicochemical properties of the active pharmaceutical ingredient (API)

amorphous form. Some of the newer technologies on the horizon have yet to meet the rigors of full-scale development as well as regulatory challenges to demonstrate their utility.

Usually the solubility in volatile organic solvents and melting point serve as the first level screen. The selection paradigm based on these two attributes is shown in Fig. 3.14. Compounds with melting point below 200 °C are generally suitable for melt extrusion and compounds with solubility of 10 mg/mL or greater in low boiling point solvents such as ethanol and acetone may be suitable for spray drying. Microprecipitation and KinetiSol provide alternate options for compounds that are not suitable for melt extrusion or spray drying due to processing difficulties.

3.13 Summary

As a first principle, it may be possible to estimate the solubility advantage that can be gained by completely destroying the crystalline lattice of a compound; however, it does not necessarily predict the impact on dissolution and bioavailability. Despite having totally similar X-ray amorphous structure and no apparent melting endotherm, the material produced by one process could have a widely different PK behavior than the material produced by another method. In some cases, the differences are attributed to certain physical properties of the amorphous material such as porosity but in other cases they are truly due to the type of interactions that may occur in solvent-based systems versus nonaqueous melts resulting in different product performance (Dong et al. 2008; Huang et al. 2011; Tominaga 2013). Therefore, the challenge to select the right processing method goes beyond the ability to make the amorphous material. In cases wherever multiple methods are possible, the selection criteria should take into consideration bioavailability followed by other factors such as stability, robustness, downstream processing, organizational capability, and cost. Important considerations in the selection of the processing technologies include:

- Physicochemical properties of the compound, e.g., solubility in aqueous, volatile, and other organic solvents
- Thermal stability of the compound and the polymer
- Extent of improvement in bioavailability

- Selection of stabilizing polymer and other processing aids
- Formulation complexity and ability to achieve highest drug loading
- Availability of equipment train from laboratory scale to commercial scale
- Product robustness (processability, amorphous stability, and dissolution performance)

References

Abeysena I, Darrington E (2013) Understanding evaporation and concentration technologies labmate. http://www.labmate-online.com/articles/laboratoryroducts/3/dr_induka_abeysena_application_specialist_rob_darrington_product_manager_genevac_ltd_ipswich _uk.this_paper _explains_the_basic_principles_of_evaporation_and_concentration_and_outlines _some_of_the_commonly_used_technologies._it_also_reviews_the_wid/understanding_evaporation_and _concentration_technologies/1442/. Accessed 12 Dec 2013

Appel L (2009) Amorphous solid dispersion. http://www.pharmatek.com/pdf/PTEKU/Jul302009.pdf. Accessed Dec 2013

Badrossamay MR, Balachandran K, Capulli AK, Golecki HM, Agarwal A, Goss JA, Kim H, Shin K, Parker KK (2014) Engineering hybrid polymer-protein super-aligned nanofibers via rotary jet spinning. Biomater 35(10):3188–3197

Conine JW (1980) Of ethanol insoluble pharmaceutically acceptable excipients, US 4195078 A. Mar 25, 1980

Desai D (2007) Solid-state plasticizers for melt extrusion. Ph.D., University of Rhode Island

DiNunzio JC, Miller DA (2013) Formulation development of amorphous solid dispersions prepared by melt extrusion. In: Repka MA, Langley N, DiNunzio J (eds) Melt extrusion. AAPS advances in the pharmaceutical sciences series, vol 9. Springer, New York, pp 161–204

DiNunzio JC, Brough C, Miller DA, Williams RO III, McGinity JW (2010) Applications of KinetiSol dispersing for the production of plasticizer free amorphous solid dispersions. Eur J Pharm Sci 40(3):179–187

Dong Z, Chatterji A, Sandhu H, Choi DS, Chokshi H, Shah N (2008) Evaluation of solid state properties of solid dispersions prepared by hot-melt extrusion and solvent co-precipitation. Int J Pharm 355(1–2):141–149

Douroumis D (2012) Hot melt extrusion: pharmaceutical applications. Wiley, Chichester

Faham A, Prinderre P, Piccerelle P, Farah N, Joachim J (2000) Hot melt coating technology: influence of Compritol 888 Ato and granule size on chloroquine release. Pharmazie 55(6):444–448

Fini A, Fernandez-Hervas MJ, Holgado MA, Rodriguez L, Cavallari C, Passerini N, Caputo O (1997) Fractal analysis of beta-cyclodextrin-indomethacin particles compacted by ultrasound. J Pharm Sci 86(11):1303–1309

Friesen DT, Shanker R, Crew M, Smithey DT, Curatolo WJ, Nightingale JA (2008) Hydroxypropyl methylcellulose acetate succinate-based spray-dried dispersions: an overview. Mol Pharm 5(6):1003–1019

Ghebre-Sellassie I, Martin CE, Zhang F, Dinunzio J (2003) Pharmaceutical extrusion technology, drugs and the pharmaceutical sciences (Series editor Swarbrick J)

Gupta MK, Tseng YC, Goldman D, Bogner RH (2002) Hydrogen bonding with adsorbent during storage governs drug dissolution from solid-dispersion granules. Pharm Res 19(11):1663–1672

Gupta MK, Vanwert A, Bogner RH (2003) Formation of physically stable amorphous drugs by milling with Neusilin. J Pharm Sci 92(3):536–551

Holm P, Buur A, Elema MO, Møllgaard B, Holm JE, Schultz K (2007) Controlled agglomeration, Google Patents

Huang J, Li Y, Wigent RJ, Malick WA, Sandhu HK, Singhal D, Shah NH (2011) Interplay of formulation and process methodology on the extent of nifedipine molecular dispersion in polymers. Int J Pharm 420(1):59–67

Hughey JR, DiNunzio JC, Bennett RC, Brough C, Miller DA, Ma H, Williams RO III, McGinity JW (2010) Dissolution enhancement of a drug exhibiting thermal and acidic decomposition characteristics by fusion processing: a comparative study of hot melt extrusion and KinetiSol dispersing. AAPS PharmSciTech 11(2):760–774

iCeutica's SoluMatrix Technology (2014) http://www.iceutica.com/solumatrix. Accessed 12 Dec 2013

Iyer R, Hegde S, Zhang YE, Dinunzio J, Singhal D, Malick A, Amidon G (2013) The impact of hot melt extrusion and spray drying on mechanical properties and tableting indices of materials used in pharmaceutical development. J Pharm Sci 102(10):3604–3613

Joe JH, Lee WM, Park YJ, Joe KH, Oh DH, Seo YG, Woo JS, Yong CS, Choi HG (2010) Effect of the solid-dispersion method on the solubility and crystalline property of tacrolimus. Int J Pharm 395(1–2):161–166

Kadir MF, Sayeed MSB (2012) Drug Delivery strategy: coprecipitation method of solid dispersion: a formulaion stratgey for poorly soluble drugs. LAP LAMBERT Academic, Saarbrücken

Kalogiannis CG, Pavlidou E, Panayiotou CG (2005) Production of amoxicillin microparticles by supercritical antisolvent precipitation. Ind Eng Chem Res 44(24):9339–9346

Keck CM, Muller RH (2006) Drug nanocrystals of poorly soluble drugs produced by high pressure homogenisation. Eur J Pharm Biopharm 62(1):3–16

Lee S, Nam K, Kim MS, Jun SW, Park JS, Woo JS, Hwang SJ (2005) Preparation and characterization of solid dispersions of itraconazole by using aerosol solvent extraction system for improvement in drug solubility and bioavailability. Arch Pharm Res 28(7):866–874

Mellado P, McIlwee HA, Badrossamay MR, Goss JA, Parker K, Mahadevan L (2011) A simple model for nanofiber formation by rotary jet-spinning. ACEEE

Miller D, Gil M (2010) Spray-drying technology. In: Williams RO III, Watts AB, Miller DA (eds) Formulating poorly soluble drugs. AAPS advances in the pharmaceutical sciences series, vol 9. Springer, New York, pp 363–442

Miller D, DiNunzio JC, Hughey JR, Williams RO III, McGinity JW (2012) KinetiSol®: a new processing paradigm for amorphous solid dispersion systems. Drug Dev Deliv 11(2011):22–31

Mosquera-Giraldo LI, Trasi NS, Taylor LS (2014) Impact of surfactants on the crystal growth of amorphous celecoxib. Int J Pharm 461(1–2):251–257

Mura P, Cirri M, Faucci MT, Gines-Dorado JM, Bettinetti GP (2002) Investigation of the effects of grinding and co-grinding on physicochemical properties of glisentide. J Pharm Biomed Anal 30(2):227–237

Murugesan S, Sharma P, Tabora JE (2011) Design of filtration and drying operations. In: Am Ende DJ (ed) Chemical engineering in the pharmaceutical industry: R & D to manufacturing. Wiley, Hoboken, pp 315–346

Nagy ZK, Nyul K, Wagner I, Molnár K, Marosi G (2010) Electrospun water soluble polymer mat for ultrafast release of Donepezil HCl. eXPRESS Polym Lett 4(12):763–772

Pathak P, Meziani MJ, Desai T, Sun YP (2004) Nanosizing drug particles in supercritical fluid processing. J Am Chem Soc 126(35):10842–10843

Perret S, Venkatesh G (2014) Enhancing the bioavailability of insoluble drug compounds: innovations in pharmaceutical technology. http://www.iptonline.com/articles/public/IPT. Accessed 20 Jan 2014

ProCept Information Brochure (2014) Spray dryer. http://www.procept.be/spray-dryer-chiller. Accessed 20 Jan 2014

Repka MA, Langley NA, DiNunzio J (2013) Melt extrusion materials, technology and drug product design. AAPS advances in the pharmaceutical sciences series, vol 9. Springer, New York

Reverchon E, De Marco I, Della Porta G (2002) Rifampicin microparticles production by supercritical antisolvent precipitation. Int J Pharm 243(1–2):83–91

Rosenberg J, Reinhold U, Liepold B, Berndl G, Alani L, Ghosh S, Brietenbach J (2005) A solid pharmaceutical dosage formulation comprising lopinavir, European Patent, Abbott Laboratories. EP 2283844 A1

Sancin P, Caputo O, Cavallari C, Passerini N, Rodriguez L, Cini M, Fini A (1999) Effects of ultrasound-assisted compaction on Ketoprofen/Eudragit S100 mixtures. Eur J Pharm Sci 7(3):207–213

Schenck L, Troup GM, Lowinger M, Li L, McKelvey C (2011) Achieving a hot melt extrusion design space for the production of solid solutions. Chemical engineering in the pharmaceutical industry: R & D to manufacturing. Wiley, Hoboken, pp 819–836

Sekiguchi K, Obi N, Ueda Y (1964) Studies on absorption of eutectic mixture. II. Absorption of fused conglomerates of chloramphenicol and urea in rabbits. Chem Pharm Bull (Tokyo) 12:134–144

Shah N, Sandhu H, Phuapradit W, Pinal R, Iyer R, Albano A, Chatterji A, Anand S, Choi DS, Tang K, Tian H, Chokshi H, Singhal D, Malick W (2012) Development of novel microprecipitated bulk powder (MBP) technology for manufacturing stable amorphous formulations of poorly soluble drugs. Int J Pharm 438(1–2):53–60

Shah N, Iyer RM, Mair HJ, Choi DS, Tian H, Diodone R, Fahnrich K, Pabst-Ravot A, Tang K, Scheubel E, Grippo JF, Moreira SA, Go Z, Mouskountakis J, Louie T, Ibrahim PN, Sandhu H, Rubia L, Chokshi H, Singhal D, Malick W (2013) Improved human bioavailability of vemurafenib, a practically insoluble drug, using an amorphous polymer-stabilized solid dispersion prepared by a solvent-controlled coprecipitation process. J Pharm Sci 102(3):967–981

Tominaga K (2013) Effect of manufacturing methods used in the stability or amorphous solid solutions and predictions to test them. Ph.D., University of Rhode Island Dissertations and Master's Theses

Varshosaz J, Hassanzedh F, Mahmoudzadeh M, Sadeghi A (2009) Preparation of cefuroxime axetil nanoparticles by rapid expansion of supercritical fluid technology. Powder Technol 189(1): 97–102

Vaughn JM, Gao X, Yacaman MJ, Johnston KP, Williams RO III (2005) Comparison of powder produced by evaporative precipitation into aqueous solution (EPAS) and spray freezing into liquid (SFL) technologies using novel Z-contrast STEM and complimentary techniques. Eur J Pharm Biopharm 60(1):81–89

Verreck G, Chun I, Peeters J, Rosenblatt J, Brewster ME (2003) Preparation and characterization of nanofibers containing amorphous drug dispersions generated by electrostatic spinning. Pharm Res 20(5):810–817

Verreck G, Decorte A, Heymans K, Adriaensen J, Cleeren D, Jacobs A, Liu D, Tomasko D, Arien A, Peeters J, Rombaut P, Van den Mooter G, Brewster ME (2005) The effect of pressurized carbon dioxide as a temporary plasticizer and foaming agent on the hot stage extrusion process and extrudate properties of solid dispersions of itraconazole with PVP-VA 64. Eur J Pharm Sci 26(3–4):349–358

Williams RO III, Watts AB, Miller DA (2010) Formulating poorly soluble drugs. Springer, New York

Williams HD, Trevaskis NL, Charman SA, Shanker RM, Charman WN, Pouton CW, Porter CJ (2013) Strategies to address low drug solubility in discovery and development. Pharmacol Rev 65(1):315–499

Yang G, Ran Y, Yalkowsky SH (2002) Prediction of the aqueous solubility: comparison of the general solubility equation and the method using an amended solvation energy relationship. J Pharm Sci 91(2):517–533

Yang W, Owens DE III, Williams RO III (2010) Pharmaceutical cryogenic technologies. In: Williams RO III, Watts AB, Miller DA (eds) Formulating poorly soluble drugs. AAPS advances in the pharmaceutical sciences series, vol 9. Springer, New York, pp 443–500

Zhu Y, Shah NH, Malick AW, Infeld MH, McGinity JW (2002) Solid-state plasticization of an acrylic polymer with chlorpheniramine maleate and triethyl citrate. Int J Pharm 241(2):301–310

Chapter 4
Excipients for Amorphous Solid Dispersions

Siva Ram Kiran Vaka, Murali Mohan Bommana, Dipen Desai, Jelena Djordjevic, Wantanee Phuapradit and Navnit Shah

4.1 Introduction

Amorphous solid dispersions (ASD) have been a topic of interest in recent years for the pharmaceutical community due to their potential in improving the oral bioavailability of poorly water-soluble drugs (Craig 2002; Leuner and Dressman 2000). Amorphous forms, which have high free energy and greater chemical and thermodynamic activity as compared to crystalline polymorphs, provide faster dissolution rates and higher apparent solubility. However, the most common concerns of ASD are the lack of thermodynamic stability. One of the approaches typically used to overcome the stability problems with amorphous active pharmaceutical ingredients (APIs) is to formulate them with pharmaceutically acceptable polymers to form ASD. The ability to form intermolecular interactions (hydrogen bonding, ionic interaction, or dipole–dipole interactions) is one of the most important criteria in the formation of amorphous molecular dispersions. In case of miscible ASD, improved physical stability can be attributed to the reduction of molecular mobility of the API molecules and/or by inhibition of nucleation and crystal growth through preferential API–polymer interactions (Ivanisevic 2010).

The polymer serves as a carrier in which API is dispersed in an ASD. Polymer selection is very important as it influences manufacturing, bioavailability, and stability of the ASD. Initial assessment of potentially "useful" excipients should be based on basic physicochemical properties of the polymers such as glass transition temperature (T_g), hygroscopicity, solid solution capacity and solubilization capacity to name a few.

S. R. K. Vaka (✉) · M. M. Bommana · D. Desai · J. Djordjevic · W. Phuapradit · N. Shah
Kashiv Pharma LLC, Bridgewater, NJ, USA
e-mail: sivav@kashivpharma.com

N. Shah et al. (eds.), *Amorphous Solid Dispersions*,
Advances in Delivery Science and Technology, DOI 10.1007/978-1-4939-1598-9_4

4.2 Challenges of Amorphous Solid Dispersions

ASD are certainly a valuable formulation strategy to enhance bioavailability of poorly soluble drugs by improving their solubility and dissolution rate. However, inherent thermodynamic instability leading to relaxation, nucleation, and crystallization during storage is one of the biggest challenges associated with the development of ASD. Critical parameters, namely temperature, moisture, and pressure, which are generally encountered during manufacturing, could adversely affect physical stability of amorphous solids. Some amorphous solids easily get plasticized with water resulting in low T_g. Typically, plasticization enhances the molecular mobility, leading to gelling or crystallization of the amorphous solid. Temperature naturally enhances molecular mobility and crystallization rate of an amorphous drug. As a rule of thumb, the storage temperature for an amorphous solid should be at least 50 °C below its T_g, irrespective of other factors such as the impact of water and pressure. Similarly, pressure may initiate nucleation of the drug, which could act as seeds and adversely impact long-term physical stability of the amorphous formulation. Crystallization of amorphous solids could also occur during the transit through gastrointestinal (GI) tract.

Pure amorphous drugs are not commonly developed as commercial dosage forms but are manufactured in combination with excipients to stabilize the amorphous state during storage as well as to prevent crystallization of API during *in vivo* dissolution in the GI tract. The high kinetic solubility of the amorphous form can drop to the equilibrium solubility of the crystalline form if devitrification is induced by the dissolution medium. Therefore, appropriate carriers that can serve as stabilizers of the amorphous state of the API are needed in the formulation. The dissolved carrier can also influence the supersaturated drug solution that is formed following dissolution. Some carriers solubilize the released drug, whereas others stabilize the supersaturated drug solution. Ideally, it is preferred to have an ASD with improved extent and rate of dissolution and one which maintains supersaturation of the drug in the GI fluids to maximize drug absorption. It should be noted that in the case of amorphous systems, kinetic solubility carries a thermodynamic representation of a high-energy form, and quantifies the degree of metastability of the amorphous phase relative to the crystalline form. Therefore, a supersaturation kinetic study is typically performed for initial screening of ASD.

4.3 Role of Excipients in Amorphous Solid Dispersions

In many instances, amorphous drug by itself cannot withstand the processing conditions involved in manufacturing. Polymers impact shelf-life stability of amorphous solid dosage forms by immobilizing and isolating amorphous drug in a rigid glass, preventing drug crystallization. ASD stabilized by polymers can be categorized into solid solutions and solid suspension. However, in general, ASD refers to solid solution.

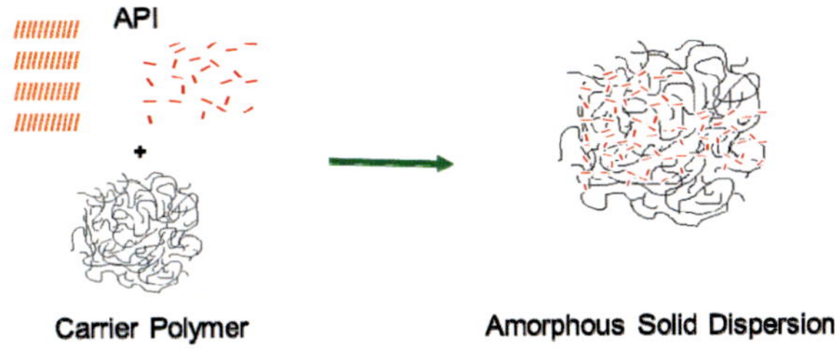

Fig. 4.1 Schematic representation of the amorphous solid dispersion stabilization by embedding the drug in the polymer matrix

A. Solid Solution If an amorphous drug is miscible with the polymer, the system is known as an amorphous solid solution or molecular dispersion distinguished by one T_g value. The physical stability of these systems is expected to be API concentration dependent. The major determining factors for designing solid solutions include solubility parameters, drug loading, and other properties of drug and polymer. Van Krevelen and Fedor group contribution methods are useful for solubility parameter calculation as a first screening tool in selecting appropriate polymers. The differences in solubility parameters of less than 7.0 $MPa^{1/2}$ between materials predict miscibility or a one-phase system (Greenhalgh et al. 1999). For an amorphous solid solution, T_g of the drug/polymer can be predicted by using the Gordon–Taylor (GT) equation:

$$T_{gmix} = \frac{w_1.T_{g1} + K.w_2.T_{g2}}{w_1 + K.w_2},$$

where T_g is the glass transition temperature, w_1 and w_2 are the weight fractions of components, and K is calculated from the densities ρ and T_g of amorphous components. One-phase system is preferred only when the system has sufficiently high T_g.

B. Solid Suspension If the amorphous drug is dispersed in the polymer matrix at the particle level, it is referred to as an amorphous solid suspension, distinguished by two separate T_g values of the drug and the polymer. The physical stability relies on immobilization and isolation of the amorphous particles in a rigid polymer matrix.

To maximize the stabilization effect, it is critical to ensure that the amorphous drug is molecularly embedded in the polymer matrix as solid solution (Fig. 4.1).

Polymers play a pivotal role in (a) attaining and maintaining supersaturation, (b) preventing API from nucleation and crystallization, and (c) modulating the hygroscopicity of the amorphous API. In addition, drug polymer interactions can also impart stability to ASD by providing mechanical rigidity due to increase in the glass transition temperature of a given matrix.

It is essential to understand the molecular and thermodynamic properties that contribute to the solubility and stability of an ASD. The properties include glass transition temperature, fragility, molecular mobility, devitrification kinetics, and chemical interactions. A thorough understanding of all of these aspects is imperative for a rational formulation strategy.

4.3.1 Glass Transition Temperature (T_g)

Glass transition temperature is defined as a temperature at which the material is converted from a "rubbery" to a "glassy" state. Generally speaking, polymers with a "hard" monomer and high molecular weight have high glass transition temperature (e.g., cellulose ethers: hydroxypropyl methylcellulose, HPMC, and hypromellose acetate succinate, HPMCAS). The glass transition temperature of the polymer can be lowered by using a plasticizer that should be perfectly mixed with the polymer at the molecular level. When choosing a polymer for ASD, one has to be careful to select one that has a high enough glass transition temperature to reduce molecular mobility and hence decrease crystallization tendency of the API while still having acceptable attributes from the processing point of view.

4.3.2 Molecular Mobility

Molecular mobility in amorphous materials is related to the macromolecular properties like viscosity; it is generally quantified in terms of mean relaxation time and it determines physical stability and reactivity. The relaxation time is defined as the time necessary for a molecule or chain segment to diffuse across the distance of one molecule or chain segment. The relaxation time varies with temperature and the typical relaxation times at T_g are estimated to be 100–200 s (Ediger et al. 1996). Molecular relaxation times can be characterized by the change of several bulk properties like enthalpy or volume or spectroscopic properties. The extent of relaxation is described empirically by the Kohlrausch–Williams–Watts equation (Hodge 1994):

$$\phi(t) = \exp\left[-\left(\frac{t}{\tau}\right)^{\beta}\right],$$

where $\Phi(t)$ is the extent of relaxation at time t, τ is mean relaxation time constant, and β is relaxation time distribution parameter. Molecular mobility is also viewed as antiplasticizing of the drug by the polymer. Miscible ASD will typically have a higher glass transition temperature (T_g) compared to the amorphous API due to the antiplasticizing effect of a high T_g polymer in the formulation. In addition, certain specific chemical interactions between the drug and polymer can also limit the molecular motion of the drug in the amorphous state resulting in stabilization of the system. Polymer content and its molecular weight have been found to be a major contributory factor in restricting the molecular mobility of amorphous drugs (Kaushal et al. 2004; Albano et al. 2002).

4.3.3 Polymer Molecular Weight

Selection of an ideal polymer with the desired structural features is very important to the performance of ASD. Molecular weight of the polymer is directly related to its T_g; hence, a polymer with high molecular weight has a high T_g, thereby favoring its use as a stabilizing carrier. In addition, the molecular weight of the polymer is also directly related to its intrinsic viscosity which in turn affects drug dissolution. A high molecular weight polymer will form a high viscosity diffusion boundary layer around the ASD particles, resulting in diffusion-controlled release of drug, whereas a low molecular weight polymer will dissolve rapidly, resulting in release of drug as a single entity (Kaushal et al. 2004; Omelczuk and McGinity 1992).

4.3.4 Drug–Polymer Ratio

The drug–polymer ratio in ASD is based on the influence of the polymer on the drug's physical form. Moreover, the maximum amount of the polymer that can be employed is governed by its ability to formulate the ASD into a dosage form of administrable size. High drug loading may lead to crystallization within the dispersion. On the other hand, a high polymer amount in the formulation may ensure absence of drug crystallinity, but at a cost. The low drug loading would potentially result in a higher pill burden. The polymer also prevents fusion/nucleation of amorphous API particles under compaction. Hence, drug–polymer ratio needs to be optimized to formulate an ASD into a stable dosage form (Kaushal et al. 2004).

4.3.5 Solubility Parameters

Solubility parameters are used to predict drug–polymer miscibility. Systems with similar solubility parameter values are likely to be miscible because the energetics of interactions within one component are similar to those in other component. As a result, the overall energy needed to facilitate the mixing of components will be small because the energy required to break the interactions within like molecules will be equally compensated for the energy released by interactions between unlike molecules. Hilderbrand solubility parameters have been used to predict drug–polymer miscibility and were found useful in selecting suitable polymers. However, the limitation of Hilderbrand solubility parameters is that it does not consider the various types of forces such as hydrogen bonding/polar/dispersion operating in the system. Hence, Hansen solubility parameters were developed to take into account all of these forces, in addition to molar volumes and molar attraction constants. As it has been reported that drug's crystallinity affects the solubility parameter values, due care must be taken while calculating these parameters (Kaushal et al. 2004).

4.3.6 Solid Solution Capacity

Solid solution capacity is the maximum concentration of an active ingredient which can be completely dissolved in a polymer. Typically, solid solution capacity is influenced by drug lipophilicity, solubility parameters, presence of hydrogen bonds, as well as presence of amide structures that can act as hydrogen bond acceptors. For example, polymers with amide structure such as polyvinyllactam polymers (e.g., Kollidon® VA 64) have better solid solution capacity than the polymers with other structures.

4.3.7 Solubilization Capacity

Solubilization capacity is defined as solubilization effect of polymers on active ingredients in an aqueous solution. Needless to say, if the polymer can retain the drug in supersaturation state in the GI tract, it will significantly enhance bioavailability. Of all polymers used in ASD, amphiphilic polymers such as Soluplus® have better solubilization capacity due to their ability to create micellar structures. On the other hand, most of ionic polymers such as methacrylate copolymers can create complexes with the drug and thus increase its solubility.

4.3.8 Hygroscopicity

Moisture is known to have a profound effect on the T_g of amorphous solids, acting as a plasticizer by increasing the free volume of the material, enhancing structural mobility, and thereby decreasing T_g. At any particular temperature, the amorphous system may change from the glassy to the rubbery state if water uptake takes place. Apart from plasticization, moisture can accelerate chemical degradation and crystallization. Therefore, water vapor sorption analysis is very useful for the early evaluation of amorphous solids. Also, storage at high relative humidity is also an important factor influencing solid-state properties of the amorphous system. As shown previously by other researchers, the enteric polymers, e.g., HPMCAS, and acrylate polymers such as Eudragits® are somewhat less hygroscopic than the water-soluble polymers like povidone or copovidone, thereby imparting better stability to the amorphous form (Rumondor and Taylor 2010). In addition, due to the hydrophobicity and somewhat low hygroscopicity, the ionic polymers also offer distinct advantage with respect to water immiscibility (pH-dependent solubility). As these polymers are water insoluble, they can absorb water without dissolving and hence the polymer: API interactions may be preserved to ensure stability of the amorphous form. As discussed previously, the amorphous state has greater free volume, molecular mobility, and enthalpy relative to the crystalline state, resulting in higher dissolution rates.

4.3.9 *Chemical Reactivity*

Amorphous forms tend to degrade at a higher rate than the crystalline forms. This may be due to increased specific surface area and enhanced level of molecular mobility which reduces the activation energy for solid-state chemical reactions. Moreover, higher hygroscopicity of amorphous forms may enhance the rate of degradation by plasticization. Increasing molecular mobility consequently mediates degradation reactions. It has been reported that in presence of moisture, thermal degradation rates of β-lactam antibiotics were a magnitude higher in amorphous form than in the crystalline form (Pikal et al. 1977). The chemical reactivity in case of amorphous systems could be reduced by formulation interventions such as the addition of high molecular weight polymers to the amorphous formulation; this results in a high level of positional specificity between reacting components. In addition, the enhanced chemical reactivity in case of amorphous systems can be overcome through the use of appropriate packaging and storage conditions (Kaushal et al. 2004).

4.4 Classification of Excipients

The excipients used in solid dispersions can be broadly classified as (a) polymeric and (b) non-polymeric excipients. Polymeric excipients are the primary excipients, whereas the non-polymeric ones are the auxiliary excipients. Polymeric excipients are further classified based on their charge into the following categories: (a) nonionic or non-pH-dependent and (b) ionic or pH-dependent polymers. Further, nonionic polymers are classified as polyvinyllactam polymers and cellulose ethers. The ionic polymers are further classified as cationic and anionic polymers (Fig. 4.2).

A summary in Table 4.1 lists important properties of commonly used polymers in solid dispersions, including glass transition temperature (T_g), hygroscopicity, solubility parameters, and degradation temperature, based on which initial assessment of potentially "useful" excipients can be made. In general, polymeric materials having higher glass transition temperatures will result in solid dispersions with higher glass transition temperatures and lower molecular mobility. However, stability will also be influenced by intermolecular drug polymer interactions and moisture absorption during storage. Therefore, when selecting excipient, one should consider polymer chemistry together with the properties of the API and manufacturing aspects.

4.4.1 *Nonionic/Non-pH-Dependent Polymers*

4.4.1.1 Polyvinyllactam Polymers

This family of excipients is typically synthesized using vinylpyrrolidone as a monomer. This monomer is polymerized to the homopolymer polyvinylpyrrolidone

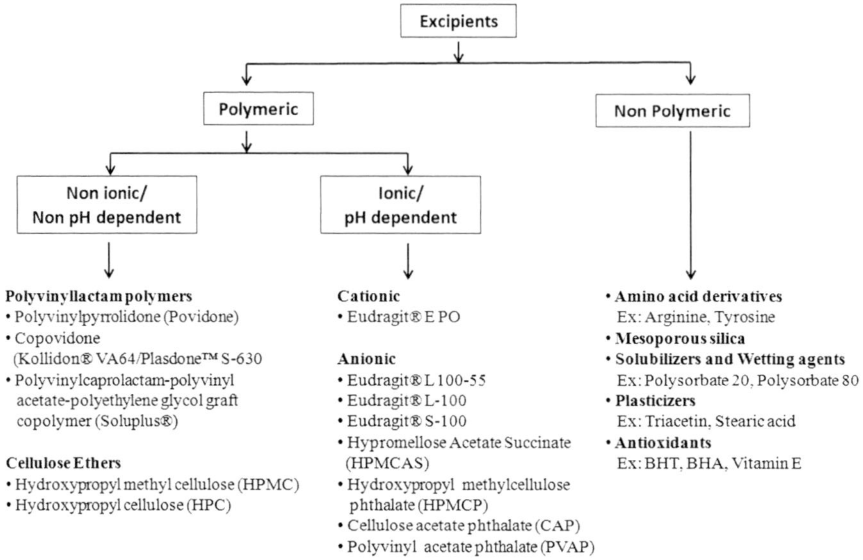

Fig. 4.2 Classification of excipients used in solid dispersions

(povidone; PVP) or copolymerized with vinyl acetate to copovidone. In recent years, a new addition to this family of excipient is vinylcaprolactam (Soluplus®).

4.4.1.1.1 Polyvinylpyrrolidone (Povidone)

The different grades of povidone are predominantly based on their molecular weight (Reintjes 2011). They have good solubility in water and organic solvents with medium lipophilicity (Table 4.1) and most importantly have the ability to interact with both hydrophilic and lipophilic active ingredients. Due to their hydrophilic nature, they have been the most commonly used as precipitation inhibitors. PVP-based hydrophilic matrices prevent drug crystallization by arresting reorientation and forming stronger drug–polymer interactions. Furthermore, once the matrix is in the GI tract, it maintains the supersaturation state by inhibiting drug crystallization by preventing aggregation of nuclei formed due to increased mobility of the matrix. The crystal inhibition is very drug specific and may be dependent on the type of the polymeric excipient. In one of the classic examples, Lindfors et al. used PVP as a crystal inhibitor for the bicalutamide; in this case, PVP gets adsorbed on the fresh nuclei of the drug. However, PVP did not control the formation of nuclei; it rather inhibited the addition of solute onto the nuclei, preventing crystal growth. Hydrogen bonding between the drug and the polymer excipients in the aqueous solution also led to the crystal inhibition (Lindfors et al. 2008).

Table 4.1 Physicochemical properties of the polymers commonly used in ASD

Polymer	Chemical category and classification	Glass transition temperature (T_g)	Molecular weight (MW)	Solubility	Hygroscopicity	Solubility parameter[a]	Degradation temperature	Technologies
Polyvinylpyrrolidone (Povidone)	Polyvinyllactam polymers; Nonionic	150–180 °C	50,000 g/mol	Water soluble	High	19.4	175 °C	Hot melt extrusion
Copovidone (Kollidon® VA64/Plasdone™ S-630)	Polyvinyllactam polymers; Nonionic	101 °C	45,000–70,000 g/mol	Water soluble	High	21.2	230 °C	Hot melt extrusion Spray drying Fluid bed spraying
Polyvinyl caprolactam-polyvinyl acetate-polyethylene glycol graft copolymer (SOLUPLUS®)	Polyvinyllactam polymers; Nonionic	~ 70 °C	118,000 g/mol	Water soluble	High	19.4	250 °C	Hot melt extrusion
Hypromellose (HPMC)	Cellulose ethers; nonionic	~ 190 °C	85,000–150,000 g/mol	Water soluble	High	23.7	190 °C	Hot melt extrusion Spray drying Fluid bed spraying
Hydroxypropyl cellulose (HPC)	Cellulose ethers; nonionic	0 °C	40,000–115,000	Water soluble	High	26.4	220 °C	Hot melt extrusion Spray drying

Table 4.1 (continued)

Polymer	Chemical category and classification	Glass transition temperature (T_g)	Molecular weight (MW)	Solubility	Hygroscopicity	Solubility parameter[a]	Degradation temperature	Technologies
Poly(butyl methacrylate-co-(2-dimethylaminoethyl) methacrylate-co-methyl methacrylate) 1:2:1 (Eudragit® EPO)	Methacrylate copolymers; cationic	~ 48 °C	~ 47,000 g/mol	Soluble in gastric fluid up to pH 5.0	Low	19.6	200 °C	Hot melt extrusion Spray drying Solvent-controlled precipitation
Poly(methacrylic acid-co-ethyl acrylate) 1:1 (Eudragit® L100-55)	Methacrylate copolymers; anionic	~ 110 °C	~ 320,000 g/mol	Above pH 5.5	Low	22.5	300 °C	Hot melt extrusion Solvent-controlled precipitation
Eudragit® L100	Methacrylic acid copolymer; Anionic	> 150 °C	~ 125,000 g/mol	Above pH 6.0	Low	22.9	380 °C	Hot melt extrusion Solvent-controlled precipitation
Eudragit® S100	Methacrylic acid copolymer; anionic	> 130 °C	~ 125,000 g/mol	Above pH 7.0	Low	–	380 °C	Hot melt extrusion Solvent-controlled precipitation Spray drying
Hypromellose acetate succinate (HPMCAS)	Cellulose ethers; anionic	~ 130 °C	~ 50,000 g/mol	Above pH 5.0	Low	29.1	200 °C	Hot melt extrusion Spray drying Solvent-controlled precipitation Fluid bed spraying

Table 4.1 (continued)

Polymer	Chemical category and classification	Glass transition temperature (T_g)	Molecular weight (MW)	Solubility	Hygroscopicity	Solubility parameter[a]	Degradation temperature	Technologies
Hydroxypropyl methylcellulose phthalate (HPMCP)	Cellulose ethers; anionic	137 °C (for HP-50) 133 °C (for HP-55)	80,000–130,000	Above pH 5.0	Hygroscopic	28	185 °C	Hot melt extrusion Spray drying
Cellulose acetate phthalate (CAP) or Cellacefate	Cellulose esters; anionic	175 °C	2534.12	In acid form, soluble in organic solvents and insoluble in water	Hygroscopic	27	200 °C	Ultra rapid freezing
Polyvinyl acetate phthalate (PVAP) (Opaseal®, Sureteric®)	Polyvinyllactam polymers; anionic	42.5 °C	47,000–60,700	Soluble in methanol and ethanol; < 1 % soluble in isopropanol and acetone and insoluble in chloroform and methylene chloride	Nonhygroscopic	25	150 °C	Spray drying

[a]Solubility parameter calculated by Molecular Modeling Pro Plus™ from ChemSW Inc. Polymeric excipients are categorized into two main categories: (1) nonionic/non-pH dependent polymers and (2) ionic/pH dependent polymers

4.4.1.1.2 Copovidone (Kollidon® VA64/ Plasdone™ S-630)

Kollidon® VA64 is a vinylpyrrolidone–vinyl acetate copolymer, soluble in water and alcohols. This polymer is amorphous in nature, with a T_g value of 101 °C, with a degradation temperature around 230 °C (Table 4.1). It has good processability and is commonly used for manufacturing of solid dispersions using either hot-melt extrusion (HME) or spray drying (SD). Kaletra® is the most successful example of copovidone-based solid dispersions on the market. Kaletra®, a combination of lopinavir and ritonavir, is used for the treatment of HIV-1 infected individuals. This tablet formulation is practically a solid solution, which serves two purposes: (a) increases the dissolution rate of the APIs and (b) stabilizes the amorphous drug as a solid solution in the solid glassy hydrophilic polymer. In case of Kaletra®, Meltrex® technology using melt extrusion with copovidone was employed to manufacture the extrudates; these were further subjected to downstream processing to produce a stable product with an acceptable shelf life.

4.4.1.1.3 Polyvinylcaprolactam–Polyvinyl Acetate–Polyethyelne Glycol Graft Copolymer (Soluplus®)

This polymer was designed to be amphiphilic in nature, soluble in organic solvents, having a high molecular weight, low glass transition temperature, and a high degradation temperature (Table 4.1). All of these characteristics indicate that this polymer is an excellent candidate for the manufacture of solid dispersions using either HME or SD. Also, due to its amphiphilic nature, Soluplus® provides high solid solution and solubilization capacity.

Soluplus®-based solid dispersions showed promising results when used with model drugs like itranconazole, fenofibrate, and carbamazepine, to name a few. For example, solid solution of itranconazole in Soluplus® showed significantly enhanced absorption by 26-fold, while the absorption of fine crystals of API was enhanced approximately twofold for the marketed product Sempera®. Overall, due to its amphiphilic properties, Soluplus® can serve as an excellent solubilizer and matrix former in solid dispersions.

4.4.1.2 Cellulose Ethers

Several research groups have successfully used cellulose ethers in solid dispersions. These polymers are hydrophilic in nature and have high molecular weight and good thermal and mechanical properties, which makes them good candidates for HME (Table 4.1). The most commonly used polymers in this family of excipients are HPMC and hydroxypropyl cellulose (HPC).

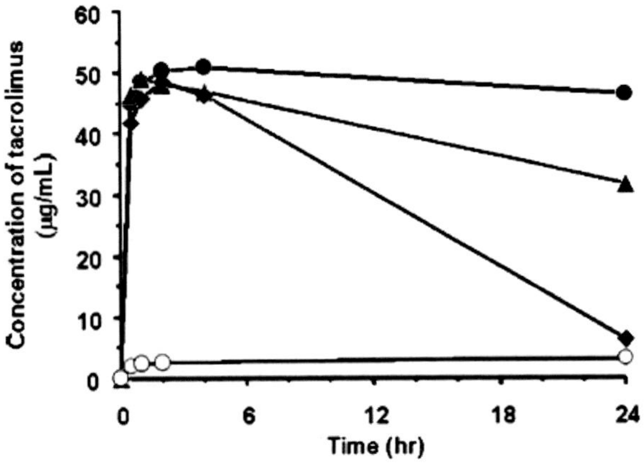

Fig. 4.3 Dissolution profiles of tacrolimus from solid dispersion formulations, (•) with HPMC, (▲) with PVP, (◇) with PEG 6000 and (○) tacrolimus crystalline powder. JP 14 Paddle method, 200 RPM, medium- JP 14 first fluid (pH 1.2)

4.4.1.2.1 Hydroxypropyl Methyl Cellulose

HPMC, a cellulosic derivative with a melting point of approximately 190 °C is a water-soluble polymer used extensively in the pharmaceutical solid dispersions. HPMC has been shown to be a good stabilizer for amorphous tacrolimus by maintaining a kinetic supersaturation over a much more prolonged period of time as compared to PVP and PEG (Yamashita et al. 2003; Fig. 4.3). The PVP and PEG polymers are well known for solubility enhancement, but they lack stabilization effect for ASD.

To date, there are numerous commercial products in the market such as Certican®, Nivadil®, Crestor®, Prograf®, and Sporanox® in which HPMC is used as a carrier for the solid dispersions. Sporanox® utilized spray-dried layering of the HPMC and the drug onto non-pareil beads. The amorphous API was shown to be stabilized by the hydrophilic matrix.

4.4.1.2.2 Hydroxypropyl Cellulose

HPC has excellent thermoplastic properties, low-melt viscosity, fast melt-flow properties, and low glass transition temperature (−4 °C) which makes it a good candidate for HME (Table 4.2). Low molecular weight grades such as Klucel EF or ELF are typically processed at lower temperatures (120 °C) and are commonly used in immediate release applications to enhance solubility of low solubility drugs. In general, Klucel acts as a matrix in which APIs are immobilized and dispersed in either nanocrystalline or amorphous state. On the other hand, high molecular weight grade (HF) is processed at high temperature (200 °C) and is used for controlled release applications.

Due to the unique mechanical properties of HPC, it is widely used in many applications such as extruded films, solid dispersions, and hot-melt extruded tablet formulations to name a few.

Table 4.2 Effect of HPC grade on thermoplastic and mechanical properties

Type	MW	Processing Temperature (° C)	Melt viscosity	Tensile strength	Solubility rate
HF	1,150,000	205	High	High	Slow
MF	850,000	190			
GF	370,000	176			
JF	140,000	160			
LF	95,00	150			
EF	80,000	137			
ELF	40,000	120	Low	Low	Fast

4.4.2 *Ionic/pH-Dependent Polymers*

Ionic/pH-dependent polymers are further categorized into two categories: (1) cationic polymers and (2) anionic polymers.

4.4.2.1 Cationic Polymers

4.4.2.1.1 Eudragit® EPO

Eudragit® EPO is a cationic copolymer composed of dimethylaminoethyl methacrylate, butyl methacrylate, and methyl methacrylate. The polymer gets ionized and solubilized at pH below 5.5. It is swellable and permeable at higher alkaline pH conditions. The molecular weight of the polymer is approximately 47,000 g/mol with a glass transition temperature of 48 °C (Table 4.1). Although glass transition temperature is low, due to its cationic nature, Eudragit® EPO has the capability to form a complex with anionic drugs, thus stabilizing the amorphous drug in the matrix due to strong intermolecular drug polymer interactions. Due to these unique properties, Eudragit® EPO is commonly used as an excipient in solid dispersions as well as in taste-masking applications using the melt extrusion and SD process. For example, Eudragit® EPO with cationic tertiary amine groups was shown to form a complex with anionic drugs like ibuprofen and masking the taste of the bitter API. High loading of drug > 35 % with 10 % talc using an extrusion process gave good results in terms of taste masking (Gryczke et al. 2011). In another study, bioavailability of fenofibrate solid dispersions manufactured using HME with Eudragit® EPO was significantly enhanced compared to conventional formulations (He et al. 2010).

4.4.2.2 Anionic Polymers

4.4.2.2.1 Eudragit® L 100-55

Eudragit® L100-55 is an anionic polymer based on methacrylic acid and ethyl acrylate, which starts dissolving at pH 5.5. It is a high molecular weight polymer, with a T_g value of 110 °C, and capable of strong molecular interactions which results in improved supersaturation of amorphous drugs. Since it is an anionic polymer, it has a strong intermolecular interaction with the cationic drugs. For example, Maniruzzaman et al. have demonstrated that Eudragit® L100-55 interacted strongly with propranolol and diphenhydramine hydrochloride salts. The drugs were shown to be stable and maintained the amorphous state in the Eudragit® L100-55 polymer matrices (Maniruzzaman et al. 2013). Further, this polymer has ideal attributes for the formation of solid solution using the microprecipitated bulk precipitation (MBP) technique. The great potential of this polymer for application to MBP technology due to its anionic nature can be tapped to great advantage. Shah et al. successfully demonstrated the utility of the polymer in the preparation of solid dispersions using the MBP technique (Shah et al. 2012). Eudragit® L100-55 shows onset of significant degradation at 160 °C.

4.4.2.2.1 Eudragit® L-100

Eudragit® L100, a pH-dependent anionic polymer that is fully ionized at pH 6.80, is extensively used as an excipient for controlled MBP, HME, and fluid bed layering to stabilize the amorphous dispersions. This polymer is commonly used for enteric functional coating as well as for controlled release delivery applications. Fan et al. 2009 have studied the effect of anionic (Eudragit® L100) and nonionic (Kollidon® K30) polymers on the dissolution profile of an amorphous gellable drug with low glass transition temperature (60 °C). The API was coated on cellet beads in a fluid bed with the help of either Eudragit® L100 or Kollidon® K30. The authors have successfully demonstrated that the anionic polymer (Eudragit® L100) protected API by preventing its gelling and clumping in situ, while the nonionic polymer (Kollidon® K30) promoted gelling. The observed phenomena can be explained by the fact that API molecules were dissolved in Eudragit® L100 matrix; in this manner, intermolecular interaction of drug molecules with water was minimized during the dissolution process and the surface area of interaction of the water molecules with that of the drug was increased before the drug molecules could be clumped into small particles. Eudragit® L100, being an ionic polymer, dissolves by exchanging ions with the alkaline phosphate buffer ions; hence, surface erosion is mainly the mechanism of dissolution. Ion-exchange ability results in fast erosion of the film. Additionally, the steady and fast hydration of the polymer is accelerated by the ion water-absorbing capacity of the API (Fig. 4.4; Fan et al. 2009).

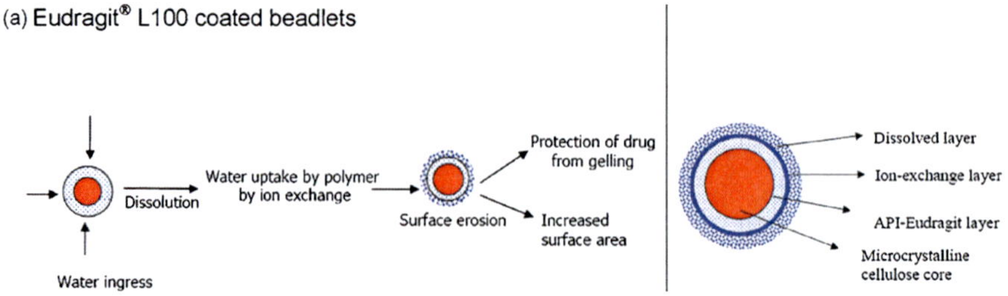

Fig. 4.4 API release mechanism from Eudragit® L100 coated beads

4.4.2.2.3 Eudragit® S100

Eudragit® S100, an anionic methacrylate copolymer which ionizes at pH 7.0, is primarily used as an excipient for colonic drug delivery. However, this polymer has recently gained importance in the amorphous formulation development. Chauhan et al. showed that Eudragit® S100 inhibited the precipitation kinetics of the dipyridamole primarily due to drug–polymer interaction and increase in glass transition temperature. Also, Eudragit® S100 was shown to be superior compared to polymers with similar glass transition temperature such as Eudragit® E100, HPMC, PVP K90, and Eudragit® L100 (Chauhan et al. 2013). In another case, solid dispersions of piroxicam with Eudragit® S100 were prepared using spherical crystallization technique. The dissolution rate of piroxicam increased in vitro, and the amorphous state of the drug was stabilized over its shelf life (Maghsoodi and Sadeghpoor 2010).

4.4.2.2.4 Hypromellose Acetate Succinate

HPMCAS is a commonly used excipient in solid dispersions due to its desirable melt viscosity, high glass transition temperature, good thermal stability, and low hygroscopicity. It is soluble in organic solvents and insoluble in water and acidic media ($pH < 5.5$), but it dissolves at pH higher than 5.5. It is important to note that selection of the appropriate grade of HPMCAS polymer plays a significant role in terms of solubilization and crystallization inhibition. There are three available chemical grades (MF, AF, and LF) based on the succinyl to acetate ratios, each of which has two physical grades with different particle size (Table 4.1). The LF and LG grades are soluble at $pH \geq 5.5$, MF and MG at $pH \geq 6.0$, and HF and HG at $pH \geq 6.80$. Higher succinyl to acetate ratio leads to higher hydrophilicity compared to lower ratios, which are more hydrophobic in nature. In case of drugs having higher melting temperature, lower succinyl to acetate ratio HPMCAS produce greater crystallization inhibition. In contrast, higher succinyl to acetate ratio HPMCAS produce better solubilization of the lipophilic drugs.

HPMCAS was extensively researched in the field of amorphous spray-dried dispersions, HME, and controlled precipitation and was proven to significantly enhance the solubilization of APIs, as well as physical stability and manufacturing reproducibility (Morgen et al. 2013; Friesen et al. 2008). Miller et al. reported that HPMCAS ASD enhanced the solubility and permeability of progesterone over other solubility enhancement techniques such as use of surfactant (SLS)/cyclodextrin (HPβCD) cosolvent (PEG-400). HPMCAS is indeed the excellent candidate for solid dispersions technology due to its high T_g in the un-ionized state, high solubility in organic solvents, and low hygroscopicity. In addition to its amphiphilic nature, it has the capability to interact with the hydrophobic and hydrophilic pockets of the drug molecules. Moreover, its low adsorption of water molecules enhances the physical stability of the ASD (Miller et al. 2012).

4.4.2.2.5 Hydroxypropyl Methylcellulose Phthalate

Hydroxypropyl methylcellulose phthalate (HPMCP) is a phthalic half ester of HPMC. Two types of HPMCP with different solubility (HP-55 and HP-50) are available. In addition, there is HP-55S, a special type of HP-55 which has higher molecular weight, higher film strength, and higher resistance to simulated gastric fluid compared to the regular grades. It has been reported that the ASD of griseofulvin prepared by coevaporation and of a new triazol antifungal drug candidate by SD using an enteric cellulosic ester HPMCP showed drastic increase in the dissolution rate compared to the pure drugs (Hasegawa et al. 1985; Kai et al.1996). Engers et al. reported that the amorphous SDD of itraconazole with HPMCP displayed the best homogeneity (the narrowest T_g width) and the highest physical stability among the different stabilizers tested (Engers et al. 2010).

4.4.2.2.6 Cellulose Acetate Phthalate

Cellulose acetate phthalate (CAP) is a partial acetate ester of cellulose. One carboxyl group of the phthalic acid is esterified with the cellulose acetate. The finished product contains about 20 % acetyl groups and about 35 % phthalyl groups. In the acid form, it is soluble in organic solvents and insoluble in water. The salt formed is readily soluble in water. DiNunzio et al. investigated the effect of CAP on the bioavailability of itraconozole (ITZ) solid dispersions prepared by ultra-rapid freezing. The results indicated that ITZ to CAP ratio formulations provided the greatest degree and extent of supersaturation in neutral media. Although not fully investigated, it has been reported that the stabilization mechanism was due to interactions between the drug and polymer, primarily attributed to steric hindrance resulting from the molecular weight of the polymer chain and chemical composition of the polymer backbone relative to position of hydrogen-bonding sites. In addition, in vivo testing conducted in Sprague-Dawley rats ($n = 6$) demonstrated a significant improvement in oral bioavailability from the 1:2 ITZ:CAP (AUC = 4516 ± 1949 ng*h/mL) compared to the Sporanox pellets (AUC = 2132 ± 1273 ng*h/mL; $p \leq 0.05$). From the results, it was concluded that amorphous compositions of ITZ and CAP provided improved bioavailability due to enhanced intestinal targeting and increased durations of supersaturation (DiNunzio et al. 2008).

4.4.2.2.7 Polyvinyl Acetate Phthalate

Polyvinyl acetate phthalate (PVAP) is a vinyl acetate polymer that is partially hydrolyzed and then esterified with phthalic acid. It has been reported to have a promising ability as a solid dispersion polymer for low solubility APIs due to a high T_g and its propensity for hydrogen bond donating and accepting ability. Minikis et al. reported that PVAP spray-dried dispersions of fenofibrate, carbamazepine, and dipyridamole are found to be amorphous by powder X-ray diffraction (PXRD) and exhibited high T_g values relative to the crystalline drug. Non-sink dissolution performance of the solid dispersions formulated with PVAP also showed increased solubility in vitro compared to the respective native crystalline drug. In addition, it has also been reported that the stability studies with PVAP as a dispersion polymer indicated no change in performance under accelerated storage conditions (Minikis et al. 2013).

4.4.3 Non-Polymeric Excipients

4.4.3.1 Amino Acid Derivatives

The high T_g of the polymer does not always lead to protection of the amorphous drug from crystallization. Hence, very few drugs are commercially available on the market due to the physical instability of the drug in solid solutions. The smart concept of "co-amorphous drugs" utilizes low molecular weight polymers together with the amorphous drug. They protect the amorphous drugs by strong specific molecular interactions, which are better than the higher T_g effect of the solid solutions. Löbmann et al. used the concept with the low molecular weight amino acids (e.g., phenylalanine, arginine, tyrosine, and tryptophan) as the polymer excipients for the co-amorphous drug formulations. The low molecular weight of the amino acids results in lower fraction of the excipient in the formulation. These materials are generally regarded as safe (GRAS) materials (Löbmann 2013).

4.4.3.2 Mesoporous Silica

Mesoporous silica was recently investigated as an excipient for formulations of molecules with low water solubility. These materials have very high specific surface area and small pore size. The customized template synthesis produces highly porous silica materials which can enhance the drug dissolution of hydrophobic molecules. Due to the porous nature and the controlled pore size volume of these materials, surface adsorption of the molecules to the mesoporous silica not only enhances the dissolution but also prevents the recrystallization of the amorphous materials. Due to the relatively finite space available to the amorphous molecules, the probability to align with their crystalline counterparts is low to negligible, resulting in amorphous stabilization of the drug. Van Speybroeck et al. used fenofibrate as the model drug for evaluating the SBA-15 (mesoporous silica) solid dispersion formulations. The DSC study showed the glassy nature of fenofibrate at a 40 % drug load, compared to 20 % load in their previous formulations. The amorphous nature could be attributed to the decrease in the availability of pore space, decreased surface adsorption of the fenofibrate molecule, as well as no molecular interaction with silanoyl groups. The formulations are stable over 6 months, thus the mesoporous silica could be a viable option for those drugs which are less miscible with the established polymers (Van Speybroeck et al. 2010).

4.4.3.3 Solubilizers and Wetting Agents

Surfactants are most commonly used as solubilizers or emulsifying agents in ASD. Their primary objective is to increase the apparent aqueous solubility and bioavailability of the drug. As with polymers, solubility in organic solvents is an important consideration when preparing ASD from solutions in solvents. In the case of HME, surfactants can have a plasticizing effect, which allows processing at lower temperatures. Some of the commonly used surfactants include Polysorbate 20, Polysorbate 80, Vitamin E polyethylene glycol succinate, Polyoxyl 40 hydrogenated castor oil, etc. (Padden et al. 2011).

4.4.3.4 Plasticizers

The use of polymeric carriers in the case of HME processes generally requires the incorporation of a plasticizer into the formulation in order to improve the processing conditions of certain high molecular weight polymers or to improve the physical and mechanical properties of the final product. According to the free volume theory, with the inclusion of plasticizers (usually small molecules) in the polymers, the free volume between the polymer chains is increased, resulting in increased molecular motion, which is referred to as the plasticization effect. The choice of the suitable plasticizers depends on factors such as plasticizer–polymer compatibility and plasticizer stability (McGinity et al. 2000). Plasticizers help in lowering the processing temperatures necessary for production and improving the stability profile of the active compound and/or of the polymeric carrier (Repka and McGinity 2000). Plasticizers also lower the shear forces needed to extrude a polymer, thereby improving the processing of certain high molecular weight polymers (Zhang and McGinity 1999; Follonier et al. 1994).

Although researchers have investigated triacetin, citrate ester, and lower molecular weight polyethylene glycols as plasticizers in hot-melt extruded systems, most of them are in liquid state (Zhang and McGinity 1999; Follonier et al. 1994, 1995). It is difficult to get a homogeneous blend of ingredients prior to extrusion in case of liquid plasticizers. An incomplete mixing of a polymer powder with a liquid additive has been shown to result in unstable mass flow when feeding the mixture into the extruder

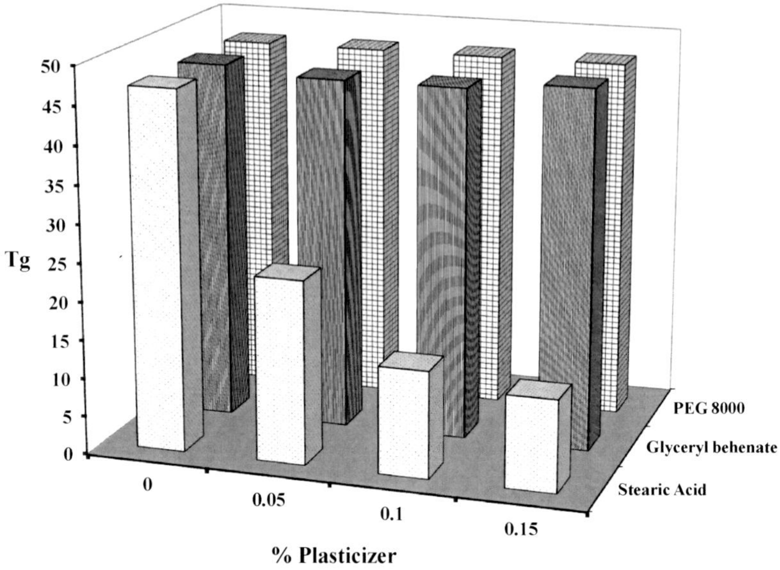

Fig. 4.5 Effect of solid-state plasticizers on T_g of Eudragit® EPO

(Tate et al. 1996). Studies have shown that the evaporation and loss of plasticizer, during a high-temperature process, may lead to stability problems in the finished dosage forms (Frohoff-Hulsmann et al. 1999; Gutierrez-Rocca and McGinity 1993). To overcome the shortcomings of liquid plasticizers, it may be useful to evaluate solid-state pharmaceutical excipients with plasticizing properties.

Desai investigated the effect of three plasticizers, stearic acid (284.48 g/mol), glyceryl behenate (414.66 g/mol), and PEG 8000 (8000 g/mol), on Eudragit® EPO during HME processing with respect to T_g and percentage motor load. From the thermal analysis results, it was reported that with increasing concentration of the plasticizer, the T_g of the polymer was found to decrease in case of stearic acid and had no effect on glyceryl behenate and PEG 8000, indicating that stearic acid is miscible with the polymer, whereas glyceryl behenate and PEG 8000 are immiscible (Desai 2007; Fig. 4.5).

In case of the HME process, the motor load is generally considered as a dependent parameter and mainly depends on feed rate, screw speed, as well as molecular and rheological properties of polymers and overall formulation. In another study, feed rate and screw speed were kept constant and the motor load was used as a response variable to determine the effect of solid-state plasticizers on the HME process. The results indicated that all the three plasticizers were successful in lowering percentage motor load with increase in concentration of plasticizers. This is attributed to increase in the free volume of the polymer which permits greater freedom of movement, thereby reducing the viscosity resulting in lowering the motor load (Desai 2007; Fig. 4.6).

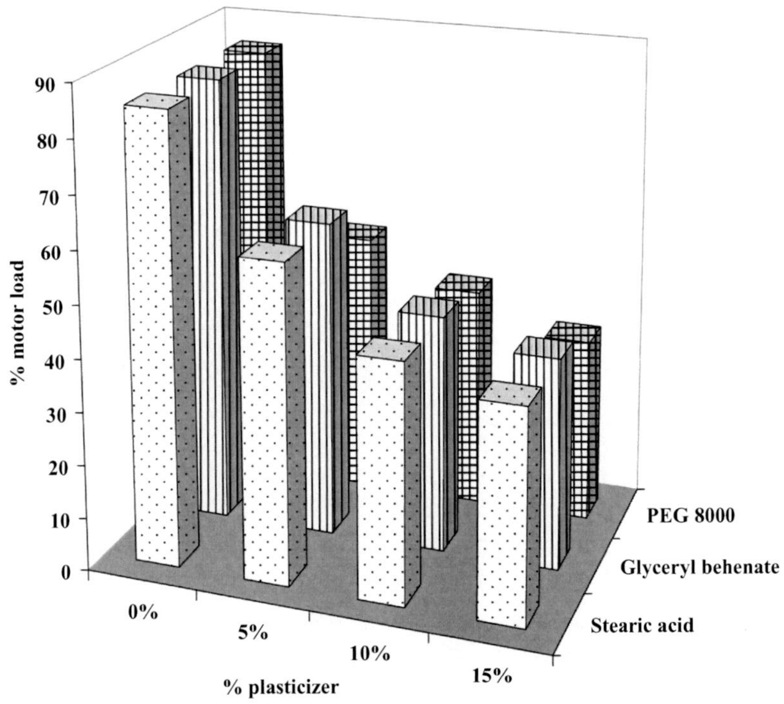

Fig. 4.6 Effect of solid-state plasticizers on the percentage motor load

However, the impact of plasticizer on long-term stability of ASD and maintenance of supersaturation kinetics of the amorphous drug needs to be carefully assessed.

Repka et al. prepared films with HPC and polyethylene oxide by HME with and without Vitamin E TPGS. It was shown that Vitamin E TPGS reduces glass transition temperature by almost 11 °C compared to the films without Vitamin E TPGS (Fig. 4.7). In addition, films containing 3 % Vitamin E TPGS had similar mechanical properties to the films plasticized with PEG 400 and showed improved processing conditions by decreasing barrel pressure and torque during extrusion (Fig. 4.8; Repka et al. 2007).

4.4.3.5 Antioxidants

Antioxidants are most effective in stabilizing oxidation-prone drug formulations. They have the ability to inhibit or slow down chain reaction oxidative processes at relatively low concentrations. This property of the antioxidant substances is of considerable importance with respect to formulations because of the large number of chemically diverse medicinal agents known to undergo oxidative decomposition. Antioxidants are classified as preventive antioxidants or chain-breaking antioxidants based upon their mechanism. Preventive antioxidants include materials that act to

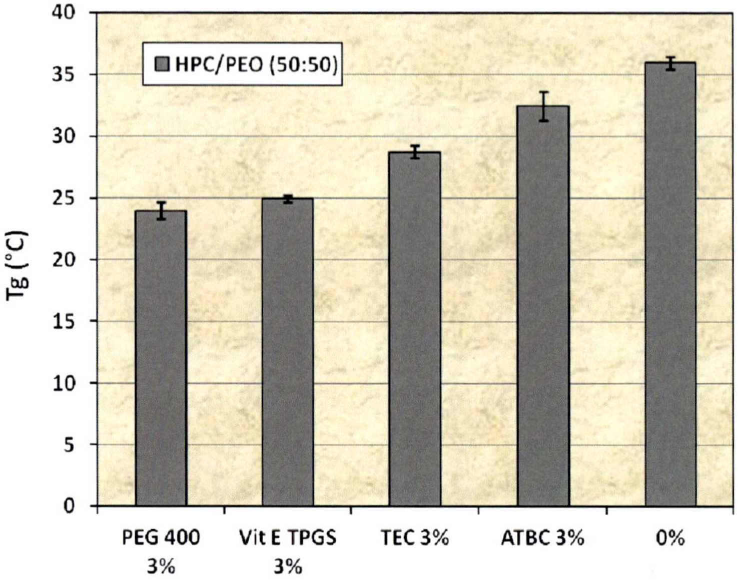

Fig. 4.7 Effect of plasticizer on glass transition temperature of HPC/PEO (50:50) films [Triethyl citrate (TEC), Acetyl tributyl citrate (ATBC)]

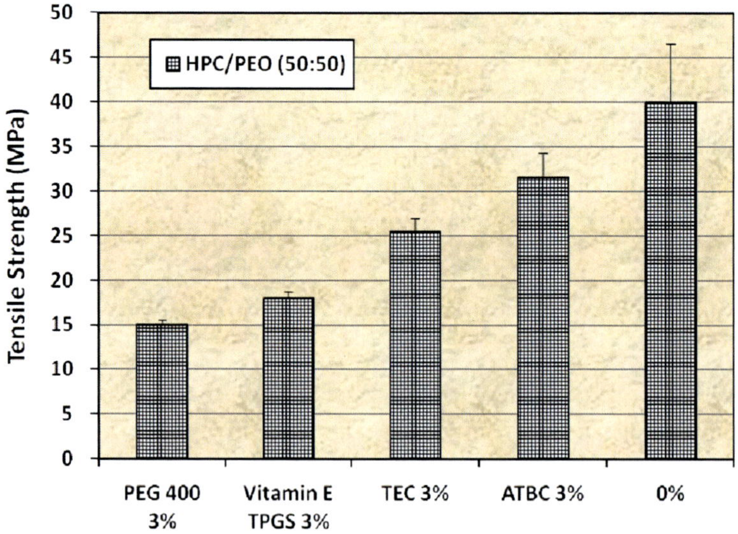

Fig. 4.8 Effect of plasticizer on tensile strength of HPC/PEO (50:50) films

prevent initiation of free radical chain reactions. Reducing agents, such as ascorbic acid, are able to interfere with autoxidation in a preventive manner since they preferentially undergo oxidation. The preferential oxidation of reducing agents protects drugs, polymers, and other excipients from attack by oxygen molecules. Chelating agents such as edetate disodium (EDTA) and citric acid are another type of preventive antioxidant that decrease the rate of free radical formation by forming a stable complex with metal ions that catalyze these reduction reactions.

Hindered phenols and aromatic amines are the two major groups of chain-breaking antioxidants that inhibit free radical chain reactions. Commonly used antioxidants such as butylated hydroxyanisole (BHA), butylated hydroxytoluene (BHT), and vitamin E are hindered phenols. Because the O–H bonds of phenols and the N–H bonds of aromatic amines are very weak, the rate of oxidation is generally higher with the antioxidant than with the polymer (Crowley et al. 2007).

4.4.4 Selection and Optimization of Excipients

As discussed in previous sections, there are various types of polymers, solubilizers, plasticizers, antioxidants, and other suitable fillers that can be used in the formulation of ASD. However, based on the physicochemical properties of the API, the type and level of excipients need to be carefully selected, as these would significantly impact the overall stability of ASD and eventually the bioavailability. Hence, it is strongly recommended to perform a proper study design based on quality by design (QbD) by varying ratios within the specifications of the excipients to better understand their effect on the selected API.

For example, there are three commercial grades of HPMCAS with fixed succinyl and acetyl content (wt %). In order to better understand the effect of the succinyl and acetyl content on solubility enhancement, Dow Pharma & Food Solutions in collaboration with Bend Research carried out the QbD studies within the United States Pharmacopoeia (USP) specifications of succinyl and acetyl content. The results of the studies of ASD of various drugs with varying physicochemical properties prepared by SD indicated that there is a big difference in solubility enhancement with respect to substitution of succinyl and acetyl content. However, some compounds are insensitive to changes in substitution level of succinyl and acetyl content. Hence, selection of right grade of HPMCAS is very crucial to maximize the solubility enhancement or find an area within the substitution space that will give an overall robust formulation and be less sensitive to change. In addition, the effect of molecular weight (succinate/acetate, wt %/wt %) on solubility enhancement has been studied and was found that in case of spray-dried dispersions, the solubility enhancement would depend on the type of API. In order to overcome the effect of molecular weight (high/low) during the SD process, Dow Pharma & Food Solutions developed Affinisol™ High Productivity HPMCAS (HP-HPMCAS) which falls within the USP monograph. It is a low molecular weight, low viscosity grade which allows increased solid loading as compared to commercial HPMCAS.

4.5 Impact of Excipients on Amorphous Solid Dispersion Processes

ASD are generally prepared by melting (fusion) or solvent methods (Chiou and Riegelman 1971). Polymers are the critical components in the manufacturing of ASD as they act as carriers for the drug and inhibit crystallization in both the dosage form and *in vivo*. The most important properties of the polymers such as the glass transition temperature (T_g), solubility in organic solvents, and hygroscopicity need to be considered to make ASD that are stable and manufacturable so as to achieve target pharmacokinetic profiles for bioavailability enhancement.

4.5.1 Melting (Fusion) Methods

The critical polymer attributes that need to be considered for the manufacture of ASD by melt extrusion includes the melt viscosity, melting point/T_g, and miscibility/solubility.

Melt Viscosity of the Polymer Melt viscosity determines the extent of miscibility of the drug and polymer as well as the efficiency of the process. Polymers with low-melt viscosities and high thermal conductivity exhibit a more efficient melting process (Crowley et al. 2007). In contrast, if the melt viscosity of the polymer is too high, it may limit miscibility of the API and polymer (Forster et al. 2001). The melt viscosity of Kollidon® VA-64 is shown to be much lower than that of HPMCAS as shown in Fig. 4.9. Melt viscosity regulates motor load and diffusivity during processing. With respect to melt viscosity and solid solution capacity, Kollidon® VA-64 is a good candidate for HME, as it enables lower processing temperature with lower motor load and faster melt-dissolution rates compared to HPMCAS.

Melting point/T_g To facilitate easy material transfer during the melt extrusion, the processing is performed at temperatures at least 20 °C above the melting point of a semi-crystalline polymer (or drug) or the T_g of an amorphous polymer (Chokshi et al. 2005). Other material variables such as molecular weight and molecular weight distribution of the polymer, hygroscopicity, and presence of monomeric impurities can affect the melting point/T_g and should be taken into consideration. As a rule of thumb, the processing temperature should be lower than T_M of the crystalline drug substance but greater than T_M or T_g of the polymer.

Miscibility/Solubility In order to form a one-phase system, the two molten components (drug and polymer) have to be miscible. It has been reported that the changes in melting point/T_g as a function of polymer concentration provide a phase diagram to establish the boundary of solid-state miscibility and helps in selecting the processing temperature (Chokshi et al. 2005).

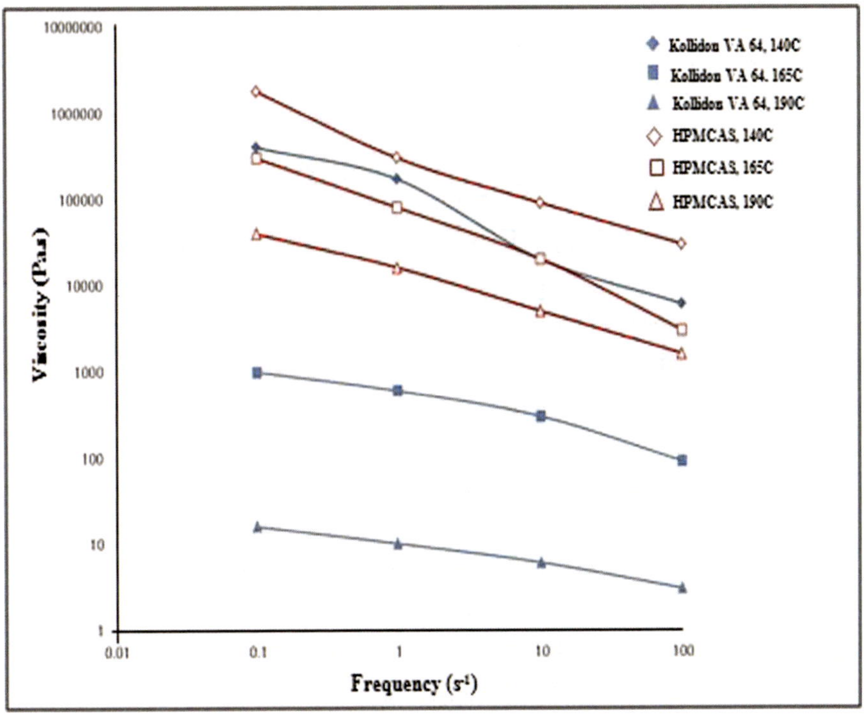

Fig. 4.9 Melt viscosity of Kollidon® VA-64 and HPMCAS at different temperatures

In addition, the selection of optimal melt-extrusion conditions depends on the chemical stability of the drug and polymer and the physical properties of the polymer. The processing parameters for melt extrusion and the impact of solid-state intermolecular drug–polymer interactions on supersaturation have been investigated by several research groups. HME performed on physical mixtures of poorly water-soluble drugs (Indomethacin, Itraconazole, and Griseofulvin) and hydrophilic polymers (Eudragit® EPO, Eudragit® L100-55, Eudragit® L100, HPMCAS-LF, HPMCAS-MF, Pharmacoat® 603, Kollidon® VA-64) at different drug to polymer ratios (30:70, 50:50, 70:30) indicated that higher supersaturation could be achieved for indomethacin, itraconazole, and griseofulvin using Eudragit® EPO, HPMCAS-LF, and Eudragit® L100-55, respectively. Transparent glassy extrudates were produced from most of the physical mixtures of indomethacin, itraconazole, and griseofulvin at the temperatures within $\pm 20\,°C$ of their softening temperatures and speeds of 100, 150, and 200 rpm, respectively. It was reported that when the temperatures for HME were reduced significantly below their softening temperatures in order to compensate for low zero-rate viscosity of physical mixtures, nontransparent extrudates were produced as a dispersion of crystalline drug into the polymer matrix, whereas when the temperatures for HME were significantly increased above the softening temperatures, charring was reported due to degradation of the polymers (Sarode et al. 2013).

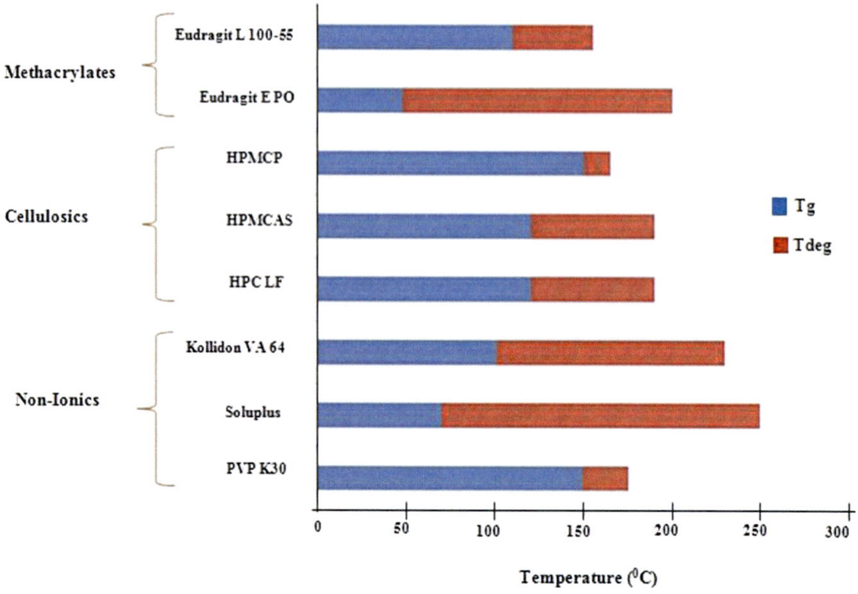

Fig. 4.10 Overview of various polymers with respect to their T_g and degradation temperatures (T_{deg})

Based on rheological properties of the materials, the HME conditions such as the lowest processing temperature and speed can be determined to prevent degradation of the drug and the polymer. An overview of various polymers with respect to their T_g and degradation temperature (T_{deg}) is summarized in Fig. 4.10. The polymer utility for melt extrusion is strongly related to the T_g toT_{deg} ratio. The processing conditions must be below the degradation temperature of the polymer, as thermal stability of the polymer can become an operating constraint in HME.

HPMCAS exhibits trend of increasing polymer degradation with temperature and rate of shear; the LF grade appears to be the most stable as shown in Fig. 4.11 (Sarode et al. 2014).

The stability of HPMCAS at higher temperatures for shorter periods of time has been examined to understand its behavior while considering processes such as HME (Shin-Etsu Chemicals Co., Ltd). In this study, the stability was tested at 150–180 °C, for 15–30 min. A powder sample was stored in an oven, and the tests were carried out according to *Japanese Pharmaceutical Excipients* (JPE), with the exception of the yellowness index, which was measured using a color computer. The observed changes were a decrease in viscosity (due to a decrease in the molecular weight of the polymer), an increase in free acid, and discoloration. From the results, it is suggested that the polymer itself will be fairly stable up to about 150 °C when subjected to this temperature for a short period of 15 min. The results are shown in Table 4.3.

Drug dissolution in molten polymer can be accelerated to achieve a solid solution in the same way dissolution in aqueous media is improved:

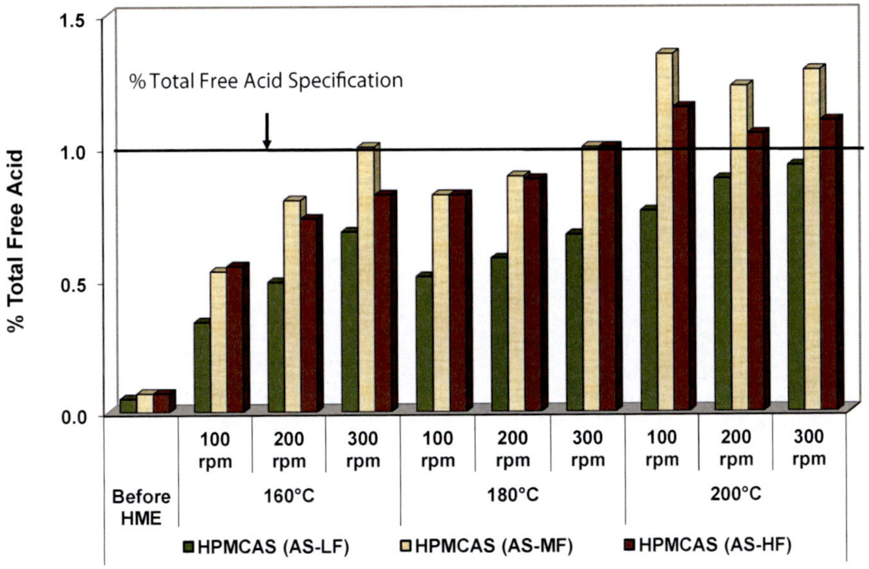

Fig. 4.11 Degradation product indicated by % total free acid presented in various grades of HPMCAS when exposed to different temperatures and shear rates

Table 4.3 The effect of processing temperatures at different time intervals on the viscosity and yellowness index of different grades of HPMCAS

		HPMCAS-LF		HPMCAS-MF		HPMCAS-HF	
Temperature (°C)	Minutes	Viscosity (mm^2/s)	Yellowness index	Viscosity (mm^2/s)	Yellowness index	Viscosity (mm^2/s)	Yellowness index
150	0	2.61	4.4	2.65	4.0	2.65	4.4
	30	2.57	15.0	2.47	11.4	2.59	12.6
165	0	2.61	4.4	2.65	4.0	2.65	4.4
	15	2.43	4.3	2.62	4.4	2.63	6.5
180	0	2.61	4.4	2.65	4.0	2.65	4.4
	15	2.13	60.9	2.18	51.6	2.37	24.2

- Increase A (drug surface area): pre-micronization $\frac{dM}{dt} = \frac{DA(C_S - C)}{h}$
- Increase C_S: Increase T, choice of polymer, cosolvents
- Increase D: reduce viscosity, addition of plasticizers
- Decrease h: screw design, screw speed, reduce viscosity
- Increase Δt: HME residence time

In addition to the processing conditions, the miscibility of drug and polymer relies on their solubility parameter and interactions, hydrophobicity, and interfacial tension. An overview of the key points to consider for the commonly used polymers during HME process is summarized in Table 4.4. To reduce the processing temperature for

Table 4.4 Key points to consider for the commonly used polymers during HME process

Polymer	Grade	T_g (°C)	Points to consider
Nonionic polymer			
Hydroxypropyl methycellulose	Methocel® E5	170–180	1. Non-thermoplastic 2. API must plasticize 3. Excellent nucleation inhibition 4. Difficult to mill
Vinylpyrrolidone	Povidone® K30	163	1. API must plasticize 2. Hygroscopic 3. Residual peroxides 4. Easily milled
Vinylpyrrolidone–vinylacetate copolymer	Kollidon® VA 64	163	1. Easily processed by melt extrusion 2. No API plasticization required 3. More hydrophobic than PVP 4. Processed around 130 °C
Polyethylene glycol, vinyl acetate, vinyl caprolactam graft copolymer	Soluplus®	70	1. Newest excipient for HME 2. Low T_g can limit stability 3. Not of compendial status 4. Stable up to 180 °C
Ionic polymer			
Amino methacrylate copolymer	Eudragit® EPO	56	1. Processing at 100 °C 2. Degradation onset is > 200 °C 3. Low T_g can limit stability
Polymethacrylates	Eudragit® L100-55 Eudragit® L100 Eudragit® S100	130	1. Not easily extruded without plasticizer 2. Degradation onset is 155 °C
Hypromellose acetate succinate	AQOAT®-L AQOAT®-M AQOAT®-H	120–135	1. Easily extruded without plasticizer 2. Process temperatures > 140 °C 3. Stable up to 190 °C depending on processing conditions

high T_g polymers, such as HPMC and PVP, it is a prerequisite that API is plasticized in the polymer.

4.5.2 Solvent-Based Methods

An important prerequisite for the manufacture of amorphous formulation using this process is that both the drug and the carrier polymer are sufficiently soluble in a low boiling point solvent (practically less than 75 °C). The solvent can be removed

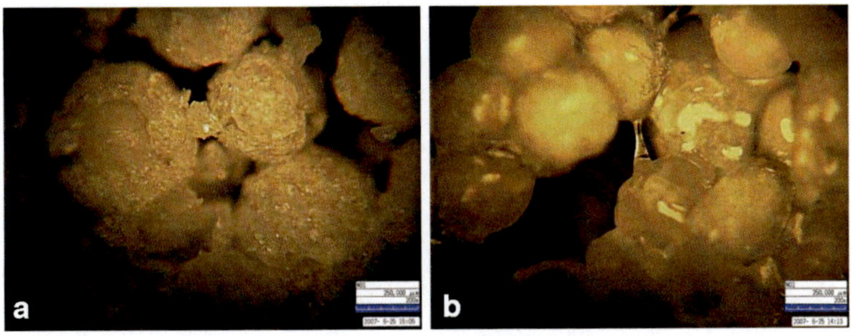

Fig. 4.12 Crushed API beadlets in the pH 7.4 phosphate buffer solution, (**a**) Eudragit® L100 beadlets with exposed cores and (**b**) PVP® K30 beadlets exhibiting the characteristic gelling of the amorphous API

commonly by SD. SD is a common method to produce amorphous pharmaceuticals. In case of ASD prepared by solvent method like SD, the drug is dissolved in a solution of the hydrophilic polymer in an organic solvent. The critical material attributes that need to be considered include the solubility of the polymer in the solvent or solvent mixture and the viscosity of the feed solution. In general, 10 % or higher solubility of the polymer and the drug is desired to achieve a sufficient SD efficiency. The affinity of the solvent to the polymer and the drug as well as the drying conditions will determine the amount of residual solvent, which would impact the stability of the ASD. The viscosity of the feed solution should be kept below 250 cps for the pressure nozzles and centrifugal atomizers to assure adequate atomization (Gibson 2001).

Some amorphous drugs may be easily plasticized by water, resulting in gelling and incomplete dissolution recovery. Solid dosage form development of such amorphous drugs is considered quite challenging. Fan et al. (2009) have successfully shown that by understanding the drug and polymer properties together with appropriate selection of a manufacturing process, it is possible to develop an ASD with a low glass transition temperature of 60 °C and aqueous solubility of 0.8 mg/mL by overcoming gelling issues of the amorphous drug during dissolution. The drug and polymer (e.g., Eudragit® L100 versus PVP K30) are first dissolved in a solvent. This solution mixture is sprayed through a nozzle onto the surface of microcrystalline cellulose spheres in a fluid-bed coater. No drug–polymer interactions were reported when examined using FTIR, implying that this factor did not play a role in the differences observed in the release profiles. The anionic polymer protected the drug by preventing its gelling and clumping in situ, while the nonionic polymer promoted gelling (Fig. 4.12). On the other hand, gelling, clumping, and agglomeration were observed on the surface of the particles coated with PVP K30 which resulted in slow and incomplete release of the drug. From the anionic polymer coating, greater than 90 % drug was dissolved in 50 min, whereas the nonionic polymer coating released 60 % drug in 5 h (Fig. 4.13). As the drug gels at a critical moisture level and at a critical time interval, any delivery system that can protect the drug from reaching the critical moisture level can control the drug release. The drug is released via surface

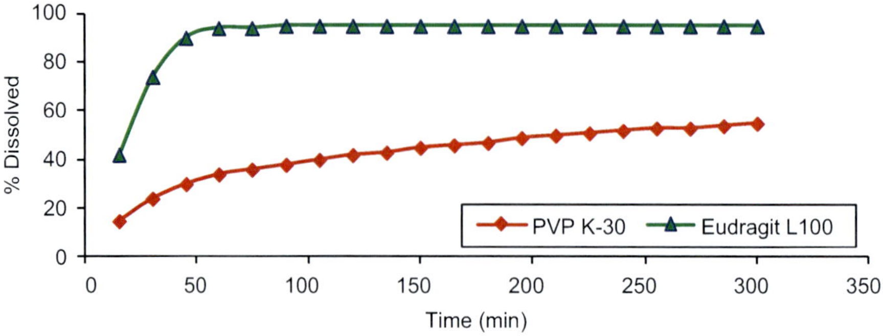

Fig. 4.13 Dissolution profiles of API from PVP K30 and Eudragit® L100 surface coatings in pH 7.4 phosphate buffer, respectively (T = 37 °C). USP basket method, 100 RPM

erosion from the Eudragit® L100 coating, whereas PVP K30, the nonionic polymer, releases the drug via a diffusion process. The results indicate that polymer properties can play a critical role in the release mechanism and kinetics of gellable drugs. An understanding of mechanisms involved in drug–polymer interactions will be useful to screen the polymers that are useful in engineering suitable delivery systems for such drugs.

4.5.3 Solvent-Controlled Precipitation/Microprecipitated Bulk Powder

This method is useful for the manufacture of ASD of poorly soluble compounds that do not have adequate solubility in low boiling point solvents and those that have very high melting points, rendering them less attractive for SD or melting (fusion) methods. Due to the nature of precipitation process employed in this technology, it is applicable only to ionic polymers that have pH-dependent solubility. Shah et al. (2012) investigated the efficiency of this technology in the manufacturing of ASD of two oncology compounds with different physicochemical properties using the solvent-controlled precipitation method. The polymers which were evaluated in this study included anionic polymers like Eudragit® L100, Eudragit® L100-55, and HPMCAS. The MBP formulation showed approximately 20-fold higher bioavailability compared to the micronized crystalline drug, suggesting that the amorphous form of API produced using MBP process was able to maintain the desired stability that resulted in complete dissolution and absorption. Based on the dog PK results, it was observed that the MBP process provided consistent pharmacokinetic profiles at different batch scales. The stabilization of amorphous dispersion was attributed to the high T_g, ionic nature of the polymer that helped to stabilize the amorphous form by possible ionic interactions, and/or due to insolubility of polymer in water. As these polymers are water insoluble, they can absorb water without dissolving and hence the polymer: API interactions may be preserved to ensure stability of the amorphous form. Commonly used solid dispersions where water-soluble polymers such as povidone or copovidone are used as hydrophilic carrier can produce amorphous form due to high drug concentration: polymer miscibility, hydrogen bonding, and/or high T_g. These, however, fail to maintain amorphous solid state because the carrier has high affinity for water, resulting in a drop in T_g followed by disruption of the stabilizing hydrogen bonds. The enteric-polymers provide added advantage of ionic interactions that can better withstand the heat and moisture stress. This has been observed with other processes as well as HPMCAS in SD or melt extrusion (Rumondor and Taylor 2010; Dong et al. 2008). In addition to being an alternate technology to SD or HME, MBP technology provides advantages with respect to stability, density, and downstream processing.

An intrinsic primary particle size of the drug in the two-phase ASD system is one of the key factors that are critically important for bioavailability enhancement and amorphous stabilization. Shah et al. have shown that in an investigational oncology drug present in a two-phase ASD system, the particle size can be determined by stripping the polymer in an appropriate medium (i.e., Eudragit® L100 in phosphate buffer, pH 10, in which the drug is practically insoluble). The intrinsic primary particle size of the drug present in the ASD called MBP is much finer with narrower distribution than that produced by SD process. MBP is produced by a solvent-controlled precipitation (CP) method. Microscopic examination with high magnification, such as a Hirox digital camera, revealed that phase separation between drug and polymer was observed in the ASD produced by SD process due to differences in precipitation rate between drug and polymer (Fig. 4.14).

The bioavailability of the drug from ASD produced by SD was substantially reduced after downstream densification processing by roller compaction (Table 4.5). The densification processing of ASD via roller compaction may not be robust for handling the segregated amorphous API in the spray-dried formulation; hence, appropriate processing methods need to be established based on the solid-state properties of the ASD. In contrast, bioavailability of the drug from ASD produced by MBP process was maintained after downstream processing by roller compaction. Micro-embedding amorphous drug in the polymer matrix enhanced wettability of the ASD; therefore, the intrinsic dissolution of the ASD produced by MBP process was superior to the SD process. Wettability and intrinsic primary particle size of the drug present in the ASD are of critical importance to ensure bioavailability of poorly soluble compounds.

In addition, physical stability of the ASD produced by MBP process was maintained after 6-month storage at 40 °C/75 %RH, while crystal formation was observed in the ASD produced by the SD process.

Availability and utilization of various analytical techniques are essential to ensure the quality of the drug product throughout the development of ASD including process selection to effectively micro-embed the amorphous drug into the polymer matrix, downstream processing, and physical stability upon storage.

Intrinsic dissolution rate has been shown by Dong et al. to differentiate the quality attributes of an identical HPMCAS- based ASD composition of a poorly soluble drug,

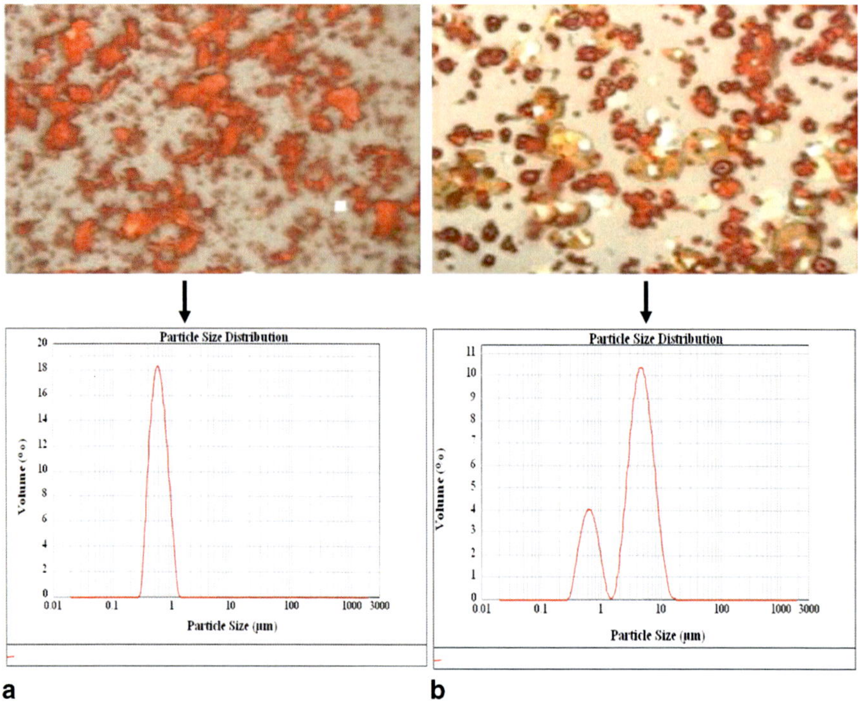

Fig. 4.14 Comparison of intrinsic primary particle size of the drug present in the ASD produced by (**a**) MBP and (**b**) spray-dried (SD) process

Table 4.5 Pharmacokinetic parameters of a poorly soluble drug from the ASD-produced MBP and SD process

Drug product	Cmax/Dose (ng/mL)/(mg/kg)	AUC/Dose (ng.h/mL)/(mg/kg)	% Relative bioavailability[a]
As Is			
MBP	113 ± 39	630 ± 221	100
SD	96 ± 32	509 ± 214	81
After densification by roller compaction			
MBP	109 ± 44	653 ± 310	104
SD	61 ± 24	329 ± 162	52

[a]Compared to the MBP (as is) administered orally at 10 mg/kg in beagle dogs ($N = 6$) under fasting condition

prepared by two different methods: (1) CP and (2) HME. The CP product was more porous and had a larger specific surface area than the HME product, as indicated by the BET results and SEM micrographs. Dissolution study using USP apparatus 2 showed that the CP product had a faster dissolution profile but slower intrinsic

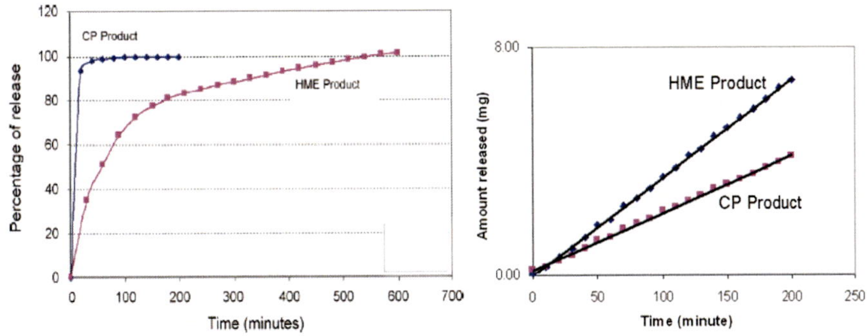

Fig. 4.15 Dissolution results of the HME and CP product using USP paddle method (on the *left side*) in comparison with intrinsic dissolution rate of the HME and CP products using USP method (on the *right side*)

Fig. 4.16 Extent of drug absorption from the HME and CP products in dogs when administered orally at various doses

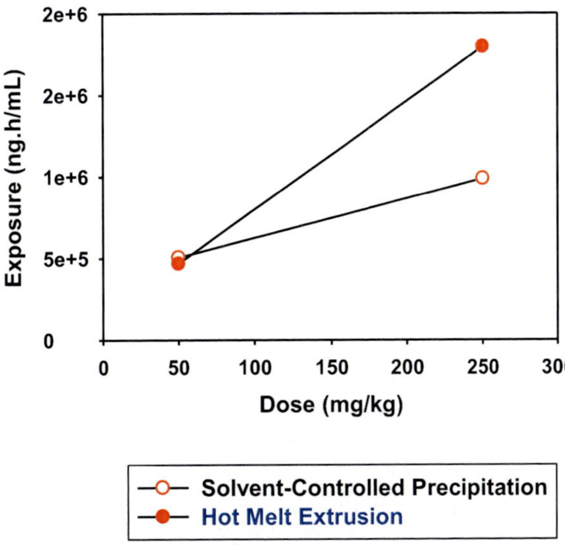

dissolution rate than the HME product (Fig. 4.15). The intrinsic dissolution rate of the HME product was shown to be higher than the CP product and it seems to have correlated well with the extent of drug absorption observed in dogs, particularly when given at a higher dose level (Fig. 4.16; Dong et al. 2008).

4.6 Marketed Products Using Amorphous Solid Dispersions

The selection of the polymer and manufacturing process are key factors in the success of the ASD development. An overview of marketed products using ASD is summarized in Table 4.6. Itraconazole is an interesting example of a drug product that was

Table 4.6 Marketed products using amorphous solid dispersion

Trade name	Manufacturer	Drug	Processing technology	Polymer	Dosage form	FDA approval
Isoptin® ER-E	Abbott	Verapamil	HME	HPC/HPMC	Tablet	
Cesamet®	Valeant	Nabilone	Solvent evaporation	PVP	Tablet	1985
Sporanox®	Janssen	Itraconazole	Fluid-bed bead layering	HPMC	Capsule	1992
Nivadil®	Fujisawa	Nivaldipine		HPMC	Tablet	1989
Prograf®	Fujisawa	Tacrolimus	Solvent evaporation	HPMC	Capsule	1994
Kaletra®	Abbott	Ritonavir/ Lopinavir	HME	PVP-VA64	Tablet	2007
Intelence®	Janssen	Etravirine	Spray drying	HPMC	Tablet	2008
Zortress®	Novartis	Everolimus	Spray drying	HPMC	Tablet	2010
Norvir®	Abott	Ritonavir	HME	PVP-VA64	Tablet	2010
Onmel®	Stiefel	Itraconazole	HME	HPMC	Tablet	2010
Zelboraf®	Roche	Vemurafenib	Solvent-controlled precipitation	HPMCAS	Tablet	2011
Incivek®	Vertex	Telaprevir	Spray drying	HPMCAS	Tablet	2011
Kalydeco®	Vertex	Ivacaftor	Spray drying	HPMCAS	Tablet	2012

commercialized using an ASD technology and is among the first marketed solid amorphous dispersion products. The compound is a potent broad-spectrum triazole antifungal drug and is practically insoluble in water (solubility 4 ng/ml). Itraconazole is so insoluble in intestinal fluids that drug therapy with the compound could not be achieved without substantial solubility enhancement by formulation intervention. The original solid oral formulation, Sporanox® Capsule, was produced by a fluid-bed bead layering process that used a cosolvent system of dichloromethane and methanol to dissolve itraconazole and HPMC which was then sprayed on inert sugar spheres (Verreck et al. 2003). The resultant product provided a significant enhancement of itraconazole bioavailability with approximately 55 % of the administered dose absorbed (Lee et al. 2005). Itraconazole has recently been reformulated into a tablet composition that contains an amorphous dispersion in HPMC by HME utilizing MeltRx Technology®. The trade name is Onmel®; it is available in 200 mg strength for once-daily administration and was approved by the Food and Drug Administration (FDA) in April 2010 for the treatment of onychomycosis. The HME formulation not only eliminated the use of organic solvents in manufacturing but also reduced dosing frequency from twice daily to once daily (Six et al. 2004). Zelboraf™ is a

Table 4.7 Summary of safety information of commonly used excipients in ASD

Polymer	IID limits	LD50
Poly(butyl methacrylate-co-(2- dimethylaminoethyl) methacrylate-co-methyl methacrylate) 1:2:1 (Eudragit® EPO)	10 mg	N/A
Poly(methacylic acid-co-methyl methacrylate) 1:1 (Eudragit® L100)	93.36 mg	LD_{50} rat >15,900 mg/kg LD_{50} mouse >10,000 mg/kg LD_{50} dog >10,000 mg/kg
Poly(methacylic acid-co-ethyl acrylate) 1:1 (Eudragit® L100–55)	99.99 mg	N/A
Hypromellose acetate succinate (HPMCAS)	560 mg	> 2.5 g/kg
Hypromellose (HPMC)	480 mg	> 4000 mg/kg/day
Copovidone (Kollidon® VA64)	853.8 mg	> 10,000 mg/kg (BASF test)
Polyvinyl caprolactam–polyvinyl acetate–polyethylene glycol graft copolymer (Soluplus®)	N/A	> 5000 mg/kg (BASF test)
Polyvinylpyrrolidine (PVP)	80 mg	> 100 g/kg

tablet dosage form containing an amorphous dispersion of vemurafenib in HPMCAS-LF produced by a solvent/anti-solvent precipitation method called MBP technology (Shah et al. 2012). In initial phase I clinical studies with a conventional formulation of vemurafenib, patients did not respond, i.e., no tumor regression, to doses as high as 1600 mg (Harmon 2010b). The issue was identified as low oral bioavailability stemming from poor solubility, which caused halting of the clinical study until it could be reformulated into a more bioavailable form. Due to melting point and organic solubility limitations, traditional ASD processes could not be applied, therefore necessitating the application of the MBP technology. When clinical trials resumed with the new MBP-based formulation, substantial tumor regression was achieved in majority of patients as a result of the enhanced formulation (Harmon 2010a). The application of the MBP technology to vemurafenib is a compelling case study for the application of ASD technology because formulation intervention was directly responsible for enabling the drug therapy and prolonging the lives of patients suffering from metastatic melanoma.

4.7 Safety and Regulatory Consideration of Excipients

Most of the pharmaceutical polymers used in ASD have already been approved for oral applications by major regulatory agencies (e.g., FDA, EMA) and have been published in the pharmacopeias (USP; European Pharmacopoeia, Ph. Eur.; Japanese Pharmacopoeia, JP). When evaluating safety of excipients used in solid dispersions, several factors have to be considered such as maximum allowable limit (IID) and LD50 (Table 4.7).

4.8 Summary

A thorough understanding of excipients and processes is crucial for achieving stable amorphous formulations with maximum bioavailability, as excipients and processes play a vital role in stabilization of the amorphous drug throughout its shelf life and in maintaining supersaturation of drug in solution *in vivo*. While selecting polymers, desirable attributes such as high T_g, moisture scavenger capability, high molecular weight, and nucleation inhibition properties need to be evaluated. Micro-embedding amorphous drug in nano or micron sizes in the polymer matrix tremendously improves the wettability and physical stability of amorphous drugs. In addition, downstream processing needs to be selected appropriately based on the physicochemical and particulate properties of the ASD.

References

Albano AA, Phuapradit W, Sandhu HK, Shah N (2002) Stable complexes of poorly soluble compounds in ionic polymers. U.S. Patent 6.350,786 B1

Chauhan H, Hui-Gu C, Atef E (2013) Correlating the behavior of polymers in solution as precipitation inhibitor to its amorphous stabilization ability in solid dispersions. J Pharm Sci 102(6):1924–1935

Chiou WL, Riegelman S (1971) Pharmaceutical applications of solid dispersion systems. J Pharm Sci 60:1281–1302

Chokshi RJ, Sandhu HK, Iyer RM, Shah NH, Malick AW, Zia H (2005) Characterization of physico-mechanical properties of indomethacin and polymers to assess their suitability for hot-melt extrusion process as a means to manufacture solid dispersion/solution. J Pharm Sci 94: 2463–2474

Craig DQM (2002) The mechanism of drug release from solid dispersions in water soluble polymers. Int J Pharm 231:131–144

Crowley MM, Zhang F, Repka MA, Thumma S, Upadhye SB, Battu SK, McGinity JW, Martin C (2007) Pharmaceutical applications of hot-melt extrusion: part I. Drug Dev Ind Pharm 33: 909–926

Desai D (2007) Solid state plasticizers for melt extrusion. Dissertations and Master's Theses Paper AAI3276980

DiNunzio JC, Miller DA, Yang W, McGinity JW, Williams RO III (2008) Amorphous compositions using concentration enhancing polymers for improved bioavailability of itraconazole. Mol Pharm 5(6):968–980

Dong Z, Chatterji A, Sandhu H, Choi D, Chokshi H, Shah N (2008) Evaluation of solid state properties of solid dispersions prepared by hot–melt extrusion an solvent co-precipitation. Int J Pharm 355:141–149

Ediger MD, Angell CA, Nagel SR (1996) Super cooled liquids and glasses. J Phys Chem 100: 13200–13212

Engers D, Teng J, Jimenez-Novoa J, Gent P, Hossack S, Campbell C, Thomson J, Ivanisevic I, Templeton A, Byrn S, Newman A (2010) A solid-state approach to enable early development compounds: selection and animal bioavailability studies of an itraconazole amorphous solid dispersion. J Pharm Sci 99:3901–3922

Fan C, Pai-Thakur R, Phuapradit W, Zhang L, Tian H, Malick W, Shah N, Kislalioglu MS (2009) Impact of polymers on dissolution performance of an amorphous gelleable drug from surface-coated beads. Eur J Pharm Sci 37(1):1–10

Follonier N, Doelker E, Cole ET (1994) Evaluation of hot-melt extrusion as a new technique for the production of polymer-based pellets for sustained release capsules containing high loadings of freely soluble drugs. Drug Dev Ind Pharm 20:1323–1339

Follonier N, Doelker E, Cole ET (1995) Various ways of modulating the release of diltiazem hydrochloride from hot-melt extruded sustained-release pellets prepared using polymeric material. J Control Release 36:342–250

Forster A, Hempenstall J, Tucker I, Rades T (2001) Selection of excipients for melt extrusion with two poorly water-soluble drugs by solubility parameter calculation and thermal analysis. Int J Pharm 226:147–161

Friesen DT, Shanker R, Crew M, Smithey DT, Curatolo WJ, Nightingale JAS (2008) Hydroxypropyl methylcellulose acetate succinate-based spray-dried dispersions: an overview. Mol Pharm 5(6):1003–1019

Frohoff-Hulsmann MA, Schmitz A, Lippold BC (1999) Aqueous ethyl cellulose dispersions containing plasticizers of different water solubility and hydroxypropyl ethyl cellulose as coating material for diffusion pellets, I. Drug release rates from coated pellets. Int J Pharm 177:69–82.

Gibson SG (2001) How to optimize your spray dryer's performance. Powder Bulk Eng 15:31–41

Greenhalgh DJ, Williams AC, Timmins P, York P (1999) Solubility parameters as predictors of miscibility in solid dispersions. J Pharm Sci 88(11):1182–1190

Gryczke A, Schminke S, Maniruzzaman M, Beck J, Douroumis D (2011) Development and evaluation of orally disintegrating tablets (ODTs) containing Ibuprofen granules prepared by hot melt extrusion. Colloids Surf B Biointerfaces 86(2):275–84

Gutierrez-Rocca JC, McGinity JW (1993) Influence of aging on the physical-mechanical properties of acrylic resin films cast from aqueous dispersions and organic solutions. Drug Dev Ind Pharm 19:315–332

Harmon A (2010a) After long fight. Drug gives sudden reprieve. The New York Times, February 23, Page A1

Harmon A (2010b) A roller coaster chase for a cure. The New York Times, New York, February 22, Page A1

Hasegawa A, Kawamura R, Nakagawa H, Sugimoto I (1985) Physical propertiesof solid dispersions of poorly water-soluble drugs with enteric coating agents. Chem Pharm Bull (Tokyo) 33: 3429–3435

He H, Yang R, Tang X (2010) In vitro and in vivo evaluation of fenofibrate solid dispersion prepared by hot-melt extrusion. Drug Dev Ind Pharm 36(6):681–687

Hodge IM (1994) Enthalpy relaxation and recovery in amorphous materials. J Non-Cryst Solids 169:211–266

Ivanisevic I (2010) Physical stability studies of miscible amorphous solid dispersions. J Pharm Sci 99(9):4005–4012

Kai T, Akiyama Y, Nomura S, Sato M (1996) Oral absorption improvement of poorly soluble drug using solid dispersion technique. Chem Pharm Bull (Tokyo) 44:568–571

Kaushal AM, Gupta P, Bansal AK (2004) Amorphous drug delivery systems: molecular aspects, design and performance. Crit Rev Ther Drug Carrier Syst 21(3):133–193

Lee SL, Nam K, Kim MS, Jun SW, Park J-S, Woo JS, Hwang S-J (2005) Preparation and characterization of solid dispersions of itraconazole by using aerosol solvent extraction system for improvement in drug solubility and bioavailability. Arch Pharm Res 28:866–874

Leuner C, Dressman J (2000) Improving drug solubility for oral delivery using solid dispersions. Eur J Pharm Biopharm 50:47–60

Lindfors L, Forssén S, Westergren J, Olsson U (2008) Nucleation and crystal growth in supersaturated solutions of a model drug. J Colloid Interface Sci 325(2):404–413

Löbmann K (2013) Co-amorphous drug delivery systems. Pharm Solid State Res Cluster 18:19

Maghsoodi M, Sadeghpoor F (2010) Preparation and evaluation of solid dispersions of piroxicam and Eudragit S100 by spherical crystallization technique. Drug Dev Ind Pharm 36(8):917–925

Maniruzzaman M et al (2013) A review on the taste masking of bitter APIs: hot-melt extrusion (HME) evaluation. Drug Dev Ind Pharm:1–12 [Ahead of Print]

McGinity JW, Koleng JJ, Repka MA, Zhang F (2000) Hot-melt extrusion technology. In: Swarbrick J, Boylan JC (eds) Encyclopedia of pharmaceutical technology, 19th edn. Marcel Dekker, New York, pp 203–225

Miller JM, Beig A, Carr RA, Spence JK, Dahan A (9 Jun 2009–2016; 2012) A win win solution in oral delivery of lipophilic drugs: supersaturation via amorphous solid dispersions increases apparent solubility without sacrifice of intestinal membrane permeability. Mol Pharm 9(7):2009–2016

Minikis R, Konagurthu S, Freauff A, McVey A, Wilmoth J, House B, Kerkmann M, Pickens C (2013) Polyvinyl Acetate Phthalate (PVAP) as a solid dispersion polymer for improving bioavailabilityof poorly soluble compounds. AAPS J. San Antonio, Texas

Morgen M, Lyon D, Schmitt R, Brackhagen M, Petermann O (2013) New excipients for solubilizing APIS: tablets and capsules 2013-BASF

Omelczuk MO, McGinity JW (1992) The influence of polymer glass transition temperature and molecular weight on drug release from tablets containing poly (DL-lactic acid). Pharm Res 9:26–32

Padden BE, Miller JM, Robbins T, Zocharski PD, Prasad L, Spence JK, LaFountaine J (1 January 2011) Amorphous solid dispersions as enabling formulations for discovery and early development. Am Pharm Rev 14(1): pp 66

Pikal MJ, Lukes AL, Lang JE (1977) Thermal decomposition of amorphous beta-lactam antibacterials. J Pharm Sci 66:1312–1316

Reintjes T (2011) Solubility enhancement with BASF polymers: solubility compendium. October 2011

Repka MA, McGinity JW (2000) Influence of vitamin E TPGS on the properties of hydrophilic films produced by hot melt extrusion. Int J Pharm 202:63–70

Repka M, Battu S, Upadhye S, Thumma S, Crowley M, Zhang F, Martin C, McGinity J (2007) Pharmaceutical applications of hot-melt extrusion: part II. Drug Dev Ind Pharm 33:1043–1057

Rumondor AC, Taylor LS (2010) Effect of polymer hygroscopicity on the phase behavior of amorphous solid dispersions in the presence of moisture. Mol Pharm 7:477–490

Sarode AL, Sandhu H, Shah N, Malick W, Zia H (2013) Hot melt extrusion for amorphous solid dispersions: predictive tools for processing and impact of drug-polymer interactions on supersaturation. Eur J Pharm Sci 48:371–384

Sarode AL, Obara S, Tanno FK, Sandhu H, Iyer R, Shah N (2014) Stability assessment of hypromellose acetate succinate (HPMCAS) NF for application in hot melt extrusion (HME). Carbohydr Polym 101:146–53

Shah N, Sandhu H, Phuapradit W, Pinal R, Iyer R, Albano A, Chatterji A, Anand S, Choi DS, Tang K, Tian H, Chokshi H, Singhal D, Malick W (2012) Development of novel microprecipitated bulk powder technology for manufacturing stable amorphous formulations of poorly soluble drugs. Int J Pharm 438:53–60

Six K, Verreck G, Peeters J et al (2004) Increased physical stability and improved dissolution properties of itraconazole, a class II drug, by solid dispersions that combine fast- and slow-dissolving polymers. J Pharm Sci 93:124–131

Tate S, Chiba S, Tani K (1996) Melt viscosity reduction of poly (ethylene terephthalate) by solvent impregnation. Polymer 37 (19):4421–4424

Van Speybroeck M, Mellaerts R, Mols R, Thi TD, Martens JD, Van Humbeeck J, Annaert P, Van den Mooter G, Augustijn P (2010) Enhanced absorption of the poorly soluble drug fenofibrate by tuning its release rate from ordered mesoporous silica. Eur J Pharm Sci 41(20105) 623–630

Verreck G, Six K, Van den Mooter G et al (2003) Characterization of solid dispersions of itraconazole and hydroxypropylmethylcellulose prepared by melt extrusion—part 1. Int J Pharm 251: 165–174

Yamashita K, Nakate T, Okimoto K, Ohike A, Tokunaga Y, Ibuki R, Higaki K, Kimura T (2003) Establishment of new preparation method for solid dispersion formulation of tacrolimus. Int J Pharm 267(1–2):79–91

Zhang F, McGinity JW (1999) Properties of sustained-release tablets prepared by hot-melt extrusion. Pharm Dev Tech 4:241–250

Part II
Technologies

Chapter 5
Miniaturized Screening Tools for Polymer and Process Evaluation

Qingyan Hu, Nicole Wyttenbach, Koji Shiraki and Duk Soon Choi

5.1 Introduction

Amorphous solid dispersion (ASD) is one of the most remarkable formulation technologies in the delivery of poorly water-soluble compounds in recent years (Hancock and Zografi 1997). Earlier chapters in this book described the chemical and physical characteristics of amorphous API and ASD, and various processing technologies in detail. As discussed in Chaps. 1 and 2, an ASD is thermodynamically unstable and has a tendency to revert back to the crystalline state. The key factors in the development of successful ASD are designing and preparing complete amorphous dispersion systems, maintaining the amorphous state over the product shelf life, and maintaining supersaturation during dissolution and absorption. These attributes can be achieved by employing the right polymer at the right drug concentration, i.e., designing a one-phase ASD system in which the drug molecule and the matrix polymer are intermixed at the molecular level using the right processing technology. The selection of the right polymer is the most important determinant in the development of ASD. The best matching polymer(s) to the API has to be determined at the onset of ASD development, as it is impractical to change the polymer in the middle of development. It should be clearly understood that the polymer plays a major role in ASD performance and dictates the overall ASD properties such as glass transition temperature, molecular mobility, hygroscopicity, hardness, microstructure, etc. Numerous approaches have been reported in literature to identify the right polymer for

Q. Hu (✉)
Formulation Development, Regeneron Pharmaceuticals, Inc., Tarrytown, NY, USA
e-mail: qingyan.hu@regeneron.com

N. Wyttenbach
pRED, Roche Innovation Center, F. Hoffmann-La Roche Ltd., Basel, Switzerland

K. Shiraki
Research division, Chugai Pharmaceutical Co., Ltd., Gotemba, Shizuoka, Japan

D. S. Choi
Kashiv Pharma LLC, Bridgewater, NJ, USA

N. Shah et al. (eds.), *Amorphous Solid Dispersions*,
Advances in Delivery Science and Technology, DOI 10.1007/978-1-4939-1598-9_5

ASD including the use of in silico solubility parameter (δ) calculation (Ghebremeskel et al. 2007), Flory–Huggins (F–H) interaction parameter calculation (Marsac et al. 2006b; Zhao et al. 2011), drug–polymer thermodynamic phase diagrams prediction (Tian et al. 2013), crystallization inhibition with molecular dynamic calculation (Pajula et al. 2012), etc. However, in spite of their use and popularity, these theoretical methods have limitations and lack predictability, reliability, and thereby have limited utility.

Several experimental-based miniaturized high throughput (HT) ASD screening systems have been published to facilitate the selection of polymer, drug loading, and ASD performance evaluation (Chiang et al. 2012; Hu et al. 2013; Wyttenbach et al. 2013). These screening systems rely on amorphous preparation at miniaturized scale with no presumptions or a priori theories. Typically, HT miniaturized ASD screening is conducted in a 96-well format using minimal quantity of API (less than 10 mg of compound per test condition). The general scheme of miniaturized ASD screening can be broken down into two major components: (1) The generation of ASD at miniaturized scale exploring large experimental design space with minimum consumption of API; (2) fast and reliable characterization scheme to evaluate the quality and in vitro performance of ASD. The output of miniaturized screening can be used to guide amorphous formulation development in a systematic way.

Authors believe that the practice of miniaturized ASD screening is extremely valuable and should be employed, particularly in the early formulation development phase when availability of API is limited. Miniaturized ASD screening not only provides early readout whether a drug molecule is a viable candidate for ASD, but also provides a pool of polymer candidates and drug loadings that can be evaluated for the scale up, using milligram quantity of a drug in a few days.

5.2 ASD Screening Assessment

In general, ASD processing technologies have two main classes, solvent based and melt based. Solvent-based technology includes spray drying, spray granulation, freeze drying, fluid bed layering, coprecipitation, supercritical fluid, electrospinning, etc. Melt-based technology includes melt extrusion, melt granulation, cogrinding, Kinetisol®, etc. Among these, spray drying, fluid bed layering, hot melting extrusion, and MBP (Microprecipitated Bulk Powder) technologies have been successfully adopted for the manufacture of commercial ASD products. Selection of a processing technology depends on the API properties. Understanding the physicochemical properties of API, therefore, will streamline the amorphous formulation screening design. Table 5.1 lists the key physicochemical properties that should be evaluated in advance related to screening system.

Table 5.1 Physicochemical properties of drug/polymer related to ASD screening systems

Attributes	Applicable screening system[a]
Solubility and stability in organic solvent (acetone, methanol, ethanol, THF, DMA, DMSO, DMF, etc.)	Solvent casting Solvent shift Spin coating Freeze drying Coprecipitation
pH solubility and stability (pH 1–12)	Coprecipitation Solvent shift
Thermal properties (melting point, T_g, thermal stability)	Melt fusion
Melt rheology	Melt fusion

[a]Screening systems are reviewed in Sect. 5.3

THF tetrahydrofuran, *DMA* dimethylacetamide, *DMSO* dimethyl sulfoxide, *DMF* dimethylformamide

5.2.1 API Physicochemical Properties Evaluation for ASD Feasibility

As discussed in this book, certain drugs are readily amenable to ASD whereas other drugs are not suitable for ASD owing to their unique molecular properties. Therefore, assessment of ASD feasibility is the first step in the development of successful ASD.

Glass-forming property of an API is suggested to be one of the key factors in ASD formation (Baird et al. 2010; Pajula et al. 2010). The glass-forming property can be extracted from the physicochemical properties of API as summarized in Table 5.2. Among these, glass transition temperature has been most frequently used for the prediction of glass-forming properties, although the use of T_g alone has limited predictability (Marsac et al. 2006a). When thermal parameters along with chemical structure information were treated by statistical analysis, glass-forming prediction is improved (Baird et al. 2010; Mahlin et al. 2011; Mahlin and Bergström 2013). Molecular weight seemed to be one of the major indicators of the glass-forming property with the higher the molecular weight, the better the glass-forming ability.

DSC has been extensively used to study thermal properties of drugs and polymers. To study glass-forming properties, a number of drugs were subjected to DSC heating and cooling cycles, and from the DSC results, compounds were categorized into three classes depending on melt and crystallization behaviors (Van Eerdenbrugh et al. 2010; Mahlin et al. 2011):

Class 1: Highly crystallizing compounds crystallized during the cooling segment.

Class 2: Moderately crystallizing compounds crystallized during the second heating segment.

Class 3: Noncrystallizing compounds that did not crystallize at all during cooling and heating cycles.

Table 5.2 Physicochemical properties used for prediction of good or poor glass formers

Literature	Attribute	Drug property	Glass-forming property[a]
Baird et al. (2010); Mahlin et al. (2011); Mahlin and Bergström (2013)	Chemical structure	MW	↑
		Rotational bonds	↑
		Benzene rings	↓
		Molecular symmetry	↓
		Branches	↑
		H-bond acceptors/H-bond donors	↑
Hancock et al. (1995); Marsac et al. (2006a); Friesen et al. (2008); Graeser et al. (2009); Baird et al. (2010); Mahlin and Bergström (2013)	Thermal properties	Glass transition temp, T_g	↑
		Crystallization temp, T_c	↑
		Configurational entropy, Sc (above T_g)	↑
		Free energy ΔG_v (negative)	↑
		Melting point/glass transition temp, T_m/T_g	↓

[a] ↑Drug property correlates with glass-forming property positively, ↓ drug property correlates with glass-forming property negatively

Although not explicitly discussed, it implies that class 1 compounds may not be amenable for ASD formation, while class 3 compounds are the ideal candidates for ASD formation. This DSC-based classification system provides an early readout from a simple experiment to assess ASD feasibility (with exception of thermally labile compounds).

Microprecipitated bulk powder (MBP) is a unique ASD technology based on coprecipitation of drugs and ionic polymers using acidic water as an antisolvent. MBP technology has been used to manufacture Zelboraf®, and frequently used for toxicology-enabling formulations of poorly soluble compounds. Realizing that not all poorly soluble compounds are viable for MBP processing, Shiraki (unpublished report) studied relationship between physicochemical properties of clinical candidate compounds and MBP formation. By studying more than 20 model compounds, Shiraki developed an MBP classification system (MCS) shown in Fig. 5.1.

In Fig. 5.1, API was classified as "easy" when the resulting MBP was XRPD amorphous and the drug recovery was greater than 98 %. Otherwise, API was classified as "difficult" for making MBP. Among all the parameters investigated, log D value, hydrogen-bonding acceptors and MW of API were identified as the critical physicochemical parameters to classify the drug into "easy" or "difficult". Shiraki rationalized that hydrophobic interaction between a drug and a polymer is important in MBP formation, as it could be the major attractive force between nonpolar molecules in aqueous solutions (Jancsó et al. 1994). The MCS predicts "easy" compounds with the following parameters: Log $D > 2$ and total number of hydrogen bond acceptors > 8–10. In addition, molecular weight alone is also an important indicator to predict ASD propensity. "Easy" compounds usually have molecular weight over 500.

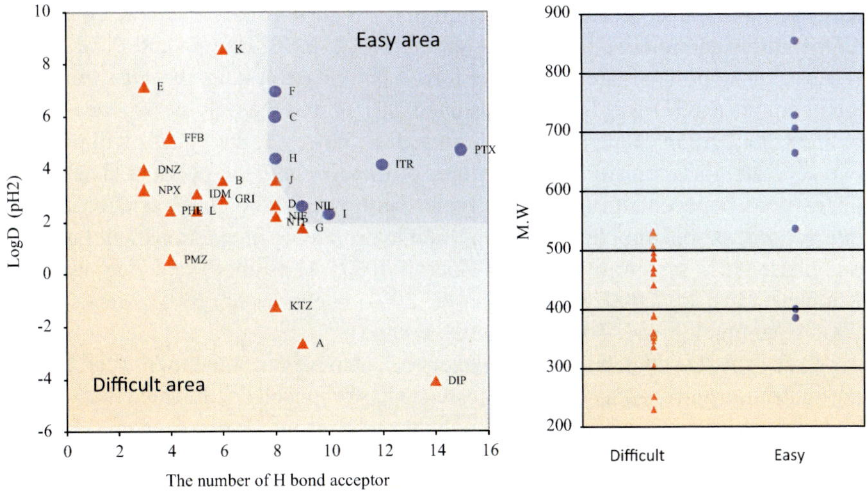

Fig. 5.1 MBP classification system (*MCS*). *Circle* stands for "Easy" drug and *triangle* stands for "Difficult" drug. *DIP* dipyridamole, *DNZ* danazol, *FFB* fenofibrate, *GRI* grisoefulvin, *IDM* indomethacin, *ITR* itraconazole, *NPX* naproxen, *NIF* nifedipine, *NIL* nilvadipine, *NTP* nitrendipine, *PHE* phenytoin, *PMZ* pimozide, *PTX* paclitaxel, *KTZ* ketoconazole. A-L, Roche in-house drugs. MBP preparation condition: polymer; HPMCAS; pH 2; drug/polymer ratio = 3/7

Interestingly, the two independently developed classification systems (DSC-based glass forming classification and MCS) generate comparable results. All drugs in MCS categorized as "easy" were class 3 drugs (although not necessarily vice versa), and no class 1 drugs were categorized as "easy" drugs by MCS.

5.2.2 Drug–Polymer Interaction for ASD Feasibility

ASDs should have good physical and chemical stability and provide enhanced API dissolution and oral bioavailability. Optimal performance can be delivered by single-phase amorphous mixtures of the API and a polymer, i.e., glass solutions with molecularly dispersed drug molecules in the polymer (Janssens and Van den Mooter 2009; Padden et al. 2011). Molecular interactions play a key role in such systems. They (1) ensure the long-term physical stability of the amorphous drug and (2) prevent drug precipitation from the supersaturated solution formed upon in vivo dissolution. Miniaturized screening methods probe both of these aspects. The relevant drug–polymer interactions are briefly discussed below.

5.2.2.1 Drug–Polymer Interactions in the Solid State

The ability of polymers to inhibit crystallization in glass solutions is primarily related to the overall increase in the glass transition temperature (T_g) of the dispersion, which reduces the molecular mobility of the API at the normally encountered storage temperatures and relative humidity (Shamblin et al. 1998; Bhugra and Pikal 2008). If a drug and a polymer are miscible to form a one-phase system, the glass transition temperature of ASD will lie in between the API T_g and the polymer T_g, depending on the composition of the system. If immiscible, however, the system will produce two separate glass transition temperatures of the drug and the polymer. The strong interactions between a drug and a polymer via hydrogen bonding, acid–base ionic interactions, dipole–dipole interactions, and hydrophobic interactions will favor the one-phase ASD system (Taylor and Zografi 1997; Matsumoto and Zografi 1999; Khougaz and Clas 2000; Miyazaki et al. 2004; Weuts et al. 2005; Marsac et al. 2006b; Rumondor et al. 2009a; Yoo et al. 2009).

Mode of ASD stabilization by polymers can also be attributed to interfering with the nucleation and crystal growth processes. It has been well documented that specific hydrogen bonding between drug and polymer inhibits nucleation process (Hancock et al. 1995; Taylor and Zografi 1997; Matsumoto and Zografi 1999). In fact, other types of molecular interactions between a drug and a polymer can also interfere nucleation process of API (Van den Mooter et al. 2001; Weuts et al. 2005).

Moisture can disrupt drug and polymer interactions, promoting demixing of drug and polymer with eventual crystallization of the drug (Konno and Taylor 2006; Rumondor et al. 2009b). Recently published work showed that the physical stability of solid dispersions depends on the hygroscopicity of the ASD and the strength of drug–polymer interactions (Rumondor et al. 2009b; Rumondor et al. 2009c, Rumondor and Taylor 2010). Hygroscopic polymer tends to pose stability challenges and should be well protected from moisture.

Recently, Van Eerdenbrugh and Taylor (Van Eerdenbrugh and Taylor 2011) applied crystal engineering principles to arrive at an ab initio rank order of polymers for the ability of crystallization inhibition. The working hypothesis of this study was that polymers inhibit the crystallization of a drug better if the hydrogen-bonding interactions between the drug and the polymer are more favorable than those present in the crystalline drug. Relative strength of hydrogen bonding of drugs and polymers was calculated to assess the ability of a given drug–polymer mixture to prevent crystallization. The predicted rank order was in good agreement with the observation from an extensive experimental dataset. The crystallization inhibition was strongly dependent on the functional groups of the drug and polymer. As summarized in Table 5.3, the results of this study facilitate the rational selection of polymers for the development of stable ASDs and even guide the design of novel polymers.

From a practical point of view, various analytical methods are commonly applied to study the solid phase behavior of ASDs (Table 5.4). Their use in combination is generally recommended as each has its own advantages and limitations. FTIR, Raman, and solid-state nuclear magnetic resonance (SS-NMR) spectroscopy are generally used to investigate molecular drug–polymer interactions. Changes in the FTIR and/or Raman spectra (new bands, disappearing bands, widening and intensity changes of existing bands, or band shifts) can indicate such interactions. With SS-NMR changes of chemical shifts, relaxation times and/or cross-signals in HETCOR spectra (Pham et al. 2010) may be induced by closer spatial proximity of the

Table 5.3 Predicted and observed best API–polymer combinations with favorable drug–polymer hydrogen bond interactions for optimal crystallization inhibition. (Adapted from data by Van Eerdenbrugh and Talyor 2011)

API functional group	Best polymers	Remarks
Carboxylic acids	E100[a] PVP[b] PVPVA[c]	Molecular API self-association will be most disrupted by polymers bearing strong acceptor groups that can effectively compete with the drug acceptor groups. Indeed, the best results were obtained with polymers bearing strong acceptors and no donors
Acidic NH groups	E100 PVP PVPVA	As for the carboxylic acids, the best crystallization inhibiting performance was obtained for polymers containing strong acceptor groups that provide a competitive hydrogen bond alternative for the acidic N–H group
Alcohols	E100 PVP PVPVA	As the OH group acts as a strong donor and medium acceptor, polymers with strong acceptors would be expected to compete successfully for these donors. Indeed, the polymers having strong acceptors showed the best results in terms of crystallization inhibition
Amides and bases	PSSA[d] PAA[e] PVP PVPVA	Competitive hydrogen bond formation would be expected in the presence of polymers with strong donor and/or extremely strong acceptor groups. Accordingly, the polymers containing strong donors (PSSA and PAA) were the best crystallization inhibitors for this category of APIs, and polymers containing the very strong pyrrolidone acceptor group (PVP/PVPVA) also performed well. The lower performance observed for HPMC[f] and HPMCAS[g] is explained by the lower strength of its donors (compared to PAA/PSSA) and acceptors (compared to PVP/PVPVA)

[a]Eudragit® E100
[b]Poly(vinylpyrrolidone) (PVP, K 12, Ph. Eur., USP)
[c]Poly(vinylpyrrolidone-vinyl acetate) (PVPVA, K 28, Ph. Eur.)
[d]Poly(styrene sulfonic acid)
[e]Poly(acrylic acid) (M_v 450.000)
[f]Hypromellose USP, substitution type 2910, viscosity 6 mPa s
[g]Hydroxypropylmethylcellulose acetate succinate, grade AS-MF

components in the drug–polymer system. However, SS-NMR cannot reliably distinguish between force-like interactions and spatial proximity. Detailed characterization principle and application should be referred to Chap. 14.

5.2.2.2 Drug–polymer Interactions in Aqueous Media for ASD Feasibility

An amorphous formulation gives rise to the higher apparent solubility than crystalline drug and become supersaturated (Brouwers et al. 2009). The supersaturated solution

Table 5.4 Commonly used analytical instruments to study ASD

Detection of:	AFM[a]	DSC[b]	XRPD[c]	PLM[d]	FTIR[e]	Raman[f]	SS-NMR[g]
Single-phase amorphous mixture[h]	x	x					x[i]
Crystallinity	x	x	x	x	x	x	x
Molecular drug–polymer interactions					x	x	x[i]

[a]Atomic force microscopy
[b]Differential scanning calorimetry
[c]X-ray powder diffraction
[d]Polarized light microscopy
[e]Fourier transform infrared spectroscopy
[f]Raman spectroscopy
[g]Solid state nuclear magnetic resonance spectroscopy
[h]Phase separation detectable at a nanometer scale (Newman et al. 2008; Lauer et al. 2011)
[i]Amorphous mixtures and strong drug–polymer interactions require the components in the system to be in close spatial proximity, which can be detected by SS-NMR

has the tendency to return to equilibrium by drug precipitation. Supersaturated condition can be maintained by incorporating the right polymers to the ASD. Polymer can modulate the dissolution rates and extents of precipitation by various modes as described below, and these effects should be considered in the selection of polymers for ASD. The most important factors that generally influence drug precipitation and, more specifically, the interaction between polymers and drug molecules are:

1. Polymer as a antinucleation agent: A polymer can inhibit nucleation and crystal growth of a drug by specific interaction with functional groups of the drug (Curatolo et al. 2009; Alonzo et al. 2010). Polymers such as PVP, HPMC and HPMC-AS have been extensively studied for the amorphous stabilization effect in aqueous solution (Lindfors et al. 2008; Miller et al. 2008; Alonzo et al. 2010).
2. Ionic interactions: Polymers with opposite charge to the drug can form ion pair complexes and stabilize drug solution (Warren et al. 2013).
3. Hydrogen bonding: Increasing the number of hydrogen bonding sites increase the interaction with drug. Itraconazole interacts with HPMC stronger than with PVP (Miller et al. 2008).
4. Viscosity of solution: High viscosity decreases the rate of molecular diffusion and molecular collision, retarding nucleation and crystal growth (Wyttenbach et al. 2013).
5. Molecular weight of polymer: High MW polymers interact with drug molecules strongly. This effect has been observed for large MW PVP and HPMC, which have been shown to maintain the supersaturation of itraconazole longer for a long period of time (Miller et al. 2008). It can be attributed to either an increase in viscosity or large number of functional groups in the polymer chain that can interact with API.

6. Temperature: Interaction between drug and polymer is weaker at higher temperatures because of thermal motions of the molecules. The concentration of felodipine during dissolution of amorphous felodipine was higher at 25 °C than at 37 °C, suggesting that weaker drug–polymer interaction at higher temperature can be the cause of faster crystallization (Alonzo et al. 2010; Wyttenbach et al. 2011).
7. pH shifts: The dissolution of ionic polymers or drugs will shift the pH of dissolution media, which can influence the solubility and precipitation behaviors of drugs and polymers (Wyttenbach et al. 2013).
8. Interfacial tension: Polymers can reduce the interfacial tension and prevent the aggregation of fine drug particles upon dissolution of ASDs. On the other hand, a decreased interfacial tension also can increase the nucleation and induce drug crystallization (Lindfors et al. 2008).
9. Co-solvent effect of dissolved polymers: Polymers in solution can act as solubilizers and increase the solubility of drugs, thus reducing the degree of supersaturation and the risk of drug precipitation (Rodríguez-Hornedo and Murphy 1999; Warren et al. 2010).

In addition, amorphous formulations often contain surfactants such as Tween 80, Span 80, TPGS, or Cremophor for a variety of reasons as processing aids. One should be mindful as the inclusion of surfactants to the ASD formulation can either prevent crystallization (Pouton 2006) or promote crystallization (Rodríguez-Hornedo and Murphy 2004). Bile salt micelles and other lipids (e.g., digestion products) presented in the GI tract may help to maintain high levels of supersaturation of drugs.

In silico prediction of the in vivo dissolution behavior of amorphous systems is currently almost impossible. Thus, in vitro screening for the identification of suitable drug-excipient combinations with appropriate dissolution and/or high supersaturation potential has become a vital step in the development of ASDs. Today, supersaturation screening is commonly carried out by solvent-shift approaches (e.g., the co-solvent quench method) or by dissolution testing (e.g., amorphous film dissolution in solvent-casting approaches). These topics are discussed further in Sect. 5.3.

5.3 ASD Miniaturized Screening

The aims of ASD screening study are: (1) to explore large experimental design space while consuming small quantity of API; (2) to handle large number of experiments in parallel preferably using an automated system; (3) to complete the screening in a short period of time. The outcome of miniaturized ASD screening should be applicable to the quality by design (QbD) and to large production scale of ASD development.

In this section, several miniaturized ASD screening methods are discussed.

5.3.1 *Solvent-Casting Method*

Solvent-casting method is one of the most commonly used miniaturized ASD screening method in the literature, which can be operated either in manual mode or automated mode (Table 5.5). The schematic representation of experimental procedure of solvent-casting method is shown in Fig. 5.2. Drug and polymer(s) are dissolved in volatile organic solvents or solvent mixtures, respectively. Drug and polymer solution can then be combined at the desired drug/polymer ratio. The prepared solutions are dispensed into a 96-well plate or an appropriate vessel using suitable liquid-handling device. Organic solvent is allowed to evaporate from the plate using a stream of inert gas or vacuum with heat, leaving a film of drug and polymer residue in the bottom of the vessel.

The residue films are then analyzed for crystallinity, solubility, dissolution, supersaturation behavior, and solid-state stability. Two recent publications on solvent casting-based screening approaches are described in more detail in the following subsections.

5.3.1.1 96-Well Plate Vacuum Dry System for ASD Screening

Chiang et al. reported a 96-well plate-based high-throughput miniaturized screening method for the identification of the polymer and optimum drug loading for ASD (Chiang et al. 2012). Screening samples were prepared in special 96-well plates with a glass substrate (Freeslate S120464) by solvent casting. A 40 μL of API stock solution (10 mg/ml in ethanol or acetone) and 80 μL of polymer stock solution (2.1–45 mg/mL in ethanol or acetone) were transferred to the 96-well plate. The solvent was evaporated using an EZ-2 Plus centrifuge vacuum dry system (SP Scientific, Stone Ridge, NY) with the maximum temperature set at 80 °C and a vacuum of 6–8 mbar or lower. Duplicate plates were made for each sample set. One plate was used for XRPD analysis, and the second plate for the solubility measurements. After drying, plate 1 was disassembled and the glass plate (substrate) containing residue films was examined by XRPD in reflection mode for the crystallinity. After initial examination, the glass plate was placed in a 50 °C and 75 % RH oven for 7 and 14 days to assess physical stability. The plate 2 was directly placed in a 50 °C and 75 % RH oven without analysis. On day 14, the plate 2 was removed from the oven and (kinetic) solubility was measured. To each well, added was 500 μL of 50 mM pH 6.5 sodium phosphate buffer with 0.1 % Tween 80 (preheated to 37 °C). The plate was then placed on a shaker heated to 37 °C for 1 h. After 1 h of shaking, samples were vacuum filtered for HPLC analysis. The best set of polymer(s) and drug loadings were selected from the solid-state stability and solubility data for the large-scale ASD development.

Table 5.5 Examples of solvent-casting approaches

References	Screening format	Drug/polymer solvent	Evaporation method	Characterization	Analysis method	Drug(s) tested
Barillaro et al. (2008)	10 mL glass vials	Acetone	Heating at 40 °C overnight	Dissolution	HPLC	Phenytoin
Shanbhag et al. (2008)	96-well microplate	Acetone/ethanol (1:1)	Vacuum centrifuge	Dissolution	UV[c]	JNJ-25894934
Dai et al. (2008)	96-well microplate	Acetone/methanol (85/15, v/v)	Vacuum centrifuge	Dissolution	HPLC	Not specified
Weuts et al. (2011)	Teflon plate (15 cm × 15 cm)	DCM[a]-ethanol (80:20) and DMF[b]-methanol	Vacuum oven	SS[d] d–p[e] miscibility crystallinity and stability	m-DSC[f] XRPD	Etravirine
Chiang et al. (2012)	96-well microplate	Acetone or ethanol	Vacuum centrifuge	Solubility SS crystallinity and stability	HPLC XRPD	Acetaminophen, indomethacin, celebrex, griseofulvin, compound A
Wyttenbach et al. (2013)	96-well microplate and film fracture surfaces on glass slides	Acetone or acetone/ethanol (1:1)	Heated nitrogen stream (Ultravap™)	Dissolution/supersaturation, SS d–p interactions SS d–p miscibility and stability	UPLC FTIR AFM	CETP(2) inhibitor

[a]Dichloromethane [b]Dimethylformamide
[c]UV spectroscopy
[d]Solid state
[e]Drug-polymer
[f]Modulated differential scanning calorimetry

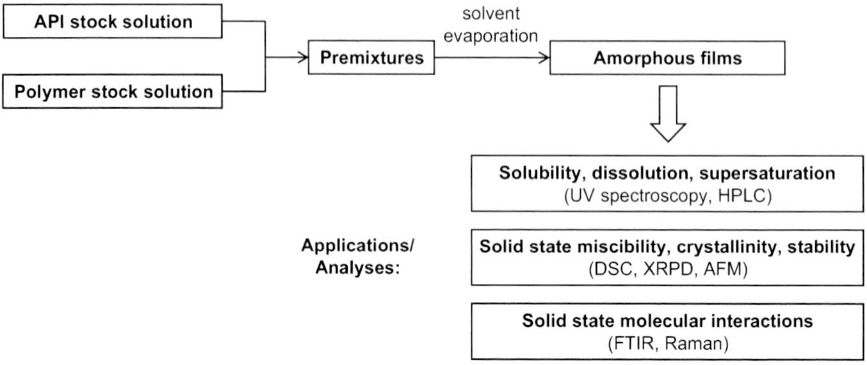

Fig. 5.2 Schematic representation of the experimental procedure used in solvent casting approaches

5.3.1.2 SPADS Approach

Wyttenbach et al. published a characterization scheme of amorphous film generated by solvent casting method, "Screening of Polymers for Amorphous Drug Stabilization (SPADS)" (Wyttenbach et al. 2013). It focuses on the evaluation of supersaturation potential, drug and polymer miscibility, and stability. In the SPADS, three assays are performed in a 2-step process.

Step 1: SPADS dissolution assay. Dissolution profile of an amorphous film is measured to assess the spring and parachute effect (Guzman et al. 2007). Selected polymer and drug loading set of ASD that meets the preset dissolution criteria moves to step 2 characterization.

Step 2: SPADS interaction and SPADS imaging assays. FTIR microspectroscopy is used to study intermolecular interactions. Atomic force microscopy (AFM) is used to examine the molecular homogeneity (uniformity) of ASD and the effect of physical stress on uniformity.

5.3.1.2.1 SPADS Dissolution Assay

The SPADS dissolution assay is performed on amorphous drug–polymer films prepared in 96-well plates by solvent casting using 0.2 mg of sample per well. Specific drug–polymer mixtures are prepared from polymers and drug stock solutions (typically 10 mg/ml) in volatile solvents (e.g., acetone, ethanol, methanol etc.) and dispensed to a 96-well polypropylene microtiter plate according to a predetermined filling scheme. Organic solvent is evaporated by a heated nitrogen flow at 60 °C delivered through 96 stainless-steel needles (Ultravap^{TM} device) to leave residue films on the bottom of the well plates.

After solvent evaporation, stirring elements and 200 μL of dissolution medium (e.g., simulated fasted state intestinal fluid, FaSSIF, pH 6.5) are added to each well. The 96 microtiter plate is sealed with a cover plate and agitated by head-over-head rotation at 37 °C. After 1 h and 3 h of agitation (one microtiter plate per time point), 100 μL of the solution is transferred to a 96 microtiter filter pate. Filtration is performed by centrifugation. The filtrates are collected in a new receiver 96-well plate. Drug contents in the filtrates are determined by UPLC after appropriate dilution. To evaluate the extent of supersaturation for each drug–polymer system, the equilibrium solubility of the crystalline drug is determined in an additional experiment in the presence of the corresponding polymers dissolved in FaSSIF (24 h equilibration at 37 °C by head-over-head rotation).

Figure 5.3 shows the results of a SPADS dissolution screening with felodipine and seven polymers at two drug loading (50 and 20 % w/w). In this example, the most promising polymers at both drug loadings tested were PVP VA 64 and HPC LF. Outstanding supersaturation was observed for Eudragit E PO with high drug concentrations for at least 3 h, but only at a drug loading of 20 %. High supersaturation was also observed with HPMCAS-MF at 50 % drug loading. The result demonstrates that the ability of amorphous drug–polymer mixtures to generate and maintain supersaturation is strongly dependent on both the type of carrier polymers and the drug loadings. In this experiment, the polymers (PVP VA 64, HPC LF, HPMCAS-MF, and Eudragit E PO in the felodipine example) with the best dissolution and supersaturation behavior are progressed to step 2 for SPADS interaction and SPADS imaging assays.

5.3.1.2.2 SPADS Interaction Assay

The SPADS interaction assay is used to examine molecular interactions between the drug and polymer set selected from the SPADS dissolution assays. Amorphous films are prepared in DSC aluminum pan by dispensing 100 μL of selected drug–polymer solution and allowing solvent to evaporate. After the solvent is evaporated, amorphous films are examined using FTIR spectroscopy in reflection mode using a standard FTIR microscope. The FTIR spectra of the amorphous films are compared to the spectra of the pure amorphous drug and the pure polymers to detect if any change in IR spectra as indicative of molecular interactions.

5.3.1.2.3 SPADS Imaging Assay

The AFM method developed by Lauer et al. (Lauer et al. 2011) is applied to analyze the molecular homogeneity and stability of promising amorphous API–polymer combinations on the nanometer scale. For these investigations, glassy film fracture surfaces are generated on glass slides. Subsequent examination in the AFM allows homogenous (one phase) and heterogeneous (two phase) drug–polymer films to be distinguished. Homogenous films are analyzed further to assess their physical stability after exposure to stress conditions (increased temperature and humidity, e.g., 40 °C/75 % RH) for a few hours. Physical instability is generally initiated

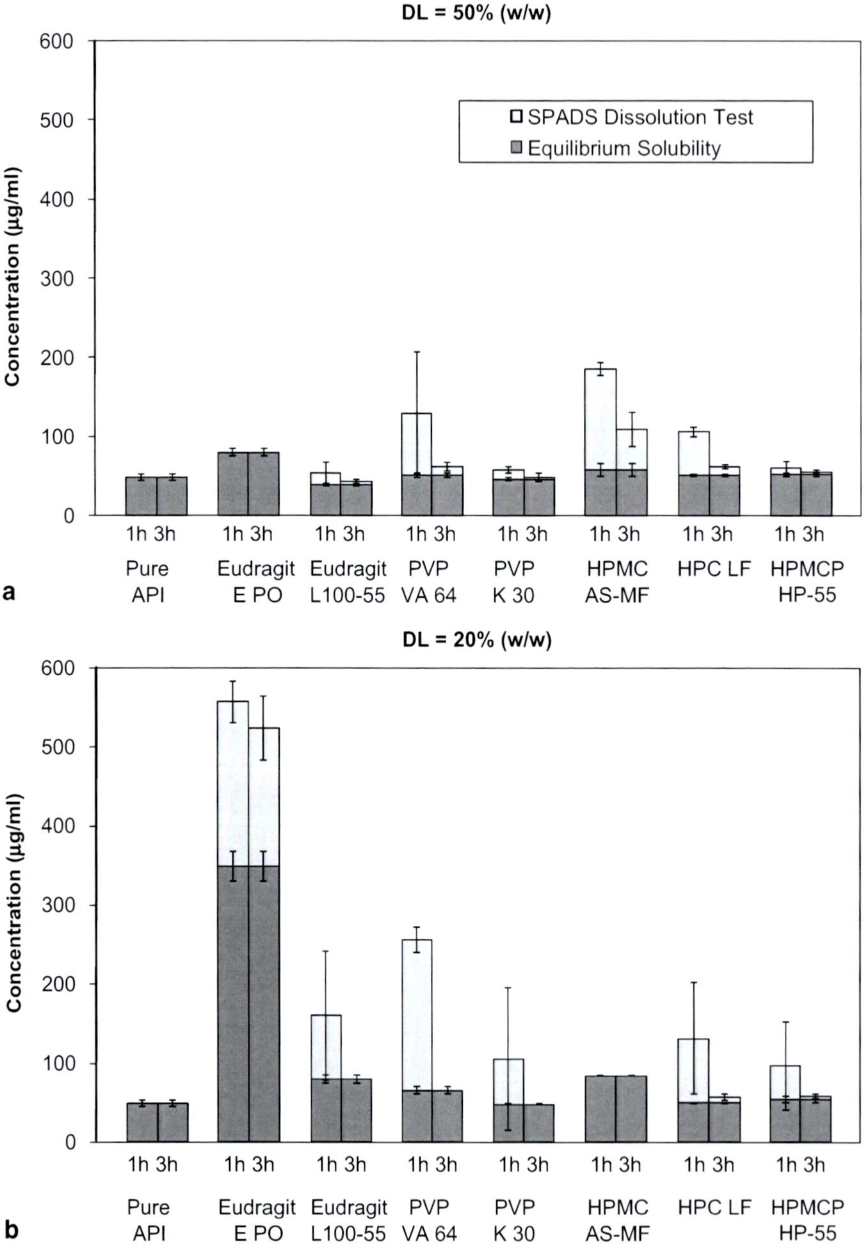

Fig. 5.3 Results of SPADS dissolution experiments with felodipine in FaSSIF at pH 6.5. The SPADS dissolution results of the amorphous drug–polymer films (*light gray*) are depicted on the same graph as the equilibrium solubility data of the crystalline drug (*dark gray*) to allow the degree of supersaturation to be roughly estimated. Data are reported as the mean of three replicates, and the error bars represent the standard deviation

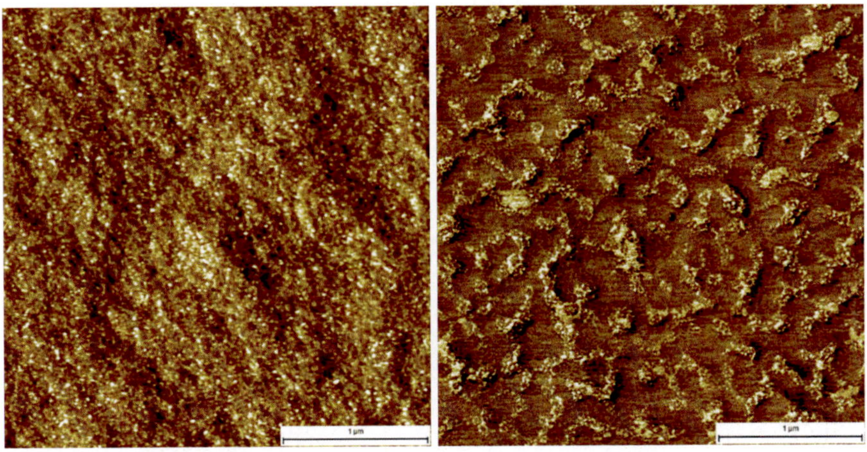

Fig. 5.4 Phase maps recorded from fracture surfaces of a 50 % drug loading (w/w) test compound A—PVP VA 64 combination by tapping-mode AFM. The maps indicate the surface homogeneity of the mixture on the nanometer scale. *Left*, homogeneous surface before stress treatment. *Right*, heterogeneous surface after sample exposure to stress conditions (40 °C/75 % RH for 2 h) indicating phase separation. The scale bars correspond to a length of 1 μm

by amorphous–amorphous phase separation, and crystallization starts in the drug-enriched amorphous phase. These changes are easy to detect by AFM and can be studied directly on the molecular level and, importantly, over the corresponding short-time scale during which they occur in the glassy films (Fig. 5.4).

Predicting the long-term stability of amorphous systems is challenging. Traditionally, amorphous formulations are analyzed for recrystallization by various techniques, e.g., PLM, DSC, or XRPD. Since crystallization kinetics might be very slow, time-consuming stress tests (weeks or months) have to be performed to identify appropriate formulations with sufficient shelf life. In contrast to these classical analytical methods, the above AFM approach allows API-polymer systems with a low long-term phase separation potential to be identified within hours (Lauer et al. 2013). Together, the SPADS interaction and SPADS imaging assays provide a comprehensive view of drug–polymer miscibility and, in combination with the enhanced stress test (e.g., 2 h at 40 °C/75 % RH), allow long-term physical stability to be predicted.

The hierarchic SPADS approach ensures that only systems with promising dissolution performance proceed to the second step where the objective is to identify ASDs with long-term physical stability (long shelf life). Prediction of the latter is greatly facilitated by the incorporated imaging assay. Combinations that successfully absolve all three assays are promising formulation candidates from which it is reasonable to prepare tailored prototype ASDs for examination by classical processes, such as spray drying, hot-melt extrusion or fluid bead layering.

Although solvent-casts cannot be considered as true formulations, their solid state physical stability and dissolution behavior can provide useful information for solid dispersion development. Solvent casting allows components to be accurately dispensed and mixed, and permits the development of rapid, miniaturized, parallel, and automated screening assays with minimum compound consumption. However, the level of residual solvent after evaporation can be a concern, and the screening compounds need to be soluble and stable in the same volatile solvents (e.g., ethanol, acetone, etc.) for the polymers and/or other excipients. Further, drug crystallization during the evaporation process can be an issue for compounds with a high crystallization tendency (class 1 compounds). The impact of particle size and morphology of processed ASDs (e.g., prepared by spray drying) on the dissolution cannot, of course, be predicted by amorphous film dissolution studies.

5.3.2 *Solvent-Shift Method for the Selection of Polymers*

Solvent-shift method has been extensively used to identify polymers that can maintain supersaturation of poorly soluble compounds in an aqueous media. In this procedure, drug is dissolved in a solvent(s) at a high concentration and then slowly added to the aqueous media containing polymer(s) to study polymer's ability to inhibit drug precipitation. An excellent paper reviewed various screening assays for the polymer's ability to maintain supersaturation and inhibit precipitation with focus on solvent-shift method (Bevernage et al. 2013). Generally, two solvent-shift methods are described in the literature: (1) the co-solvent quench method and; (2) the pH-shift method. Examples of both are given in Table 5.6.

In the co-solvent quench method, a drug is dissolved in a water-miscible organic solvent at a high concentration. The drug solution is then slowly added to an aqueous media (typically simulated gastric fluid (SGF) or simulated intestinal fluid (SIF)) that contains the dissolved polymer(s) till the drug precipitation is apparent. The pH-shift method is an alternative way to assess the supersaturation potential of ionizable drugs. In the pH-shift method, the drug is dissolved in acidic or basic media at high concentration, and then the pH of the media is changed by adding acid or base to induce precipitation. The concentration of drug in aqueous media can be monitored by UV or HPLC, or alternatively by turbidimetric or nephelometric methods (Fig. 5.5).

Solvent-shift method is simple and fast in the selection of polymers that can be used in ASD development, and does not require any special laboratory equipment (standard laboratory equipment can be used). This method can be readily automated in a 96-well format or can run in any size of laboratory scale. The co-solvent quench method is broadly applicable to most poorly water-soluble drug candidates while pH-shift method is limited to ionic compounds.

Although solvent-shift method enjoys simplicity and versatility in selection of the polymer, the outcome of this study is somewhat questionable as the organic solvent may interfere with precipitation kinetics and thermodynamics, and may not reflect amorphous solid system. In addition, supersaturation or precipitation inhibition potential of specific polymers is highly concentration dependent. Since local in vivo drug–polymer concentrations are difficult to predict, it is thus challenging to define suitable biorelevant polymer concentrations for in vitro experiments.

Table 5.6 Examples of solvent-shift approaches for supersaturation/precipitation inhibition screening

References	Screening format	Co-solvent/solvent	Polymer aqueous phase	Polymer conc.	Co-solvent conc.	Analysis method	Drug(s) tested
Co-solvent quench method							
Vandecruys et al. (2007)	10 mL	DMF[a], DMA[b]	0.01N HCl, pH 4.5 buffer USP, pH 6.8 buffer USP, water	2.5 % (w/v)	1–5 % (v/v)	UV[J]	25 different (not specified) drug candidates
Brewster et al. (2008)	10 mL	DMF	0.01N HCl	2.5 % (w/v)	< 1 %	UV	Itraconazole
Janssens et al. (2008)	10 mL	DMF	SGF[e]$_{sp}$ USP 24	0.01, 0.05, 0.1, 1, 4, 8 % (w/v)	4.8 % (v/v)	UV	Itraconazole
Warren et al. (2010)	96-well microplate	PG[c]	Phosphate buffer pH 6.5	0.001 %, 0.1 % (w/v)	9.4 % (w/w)	Nephelometry	Danazol
Bevernage et al. (2011)	20 mL and 2 mL	DMSO[d]	Aqueous buffer, FaSSIF[f], FeSSIF[g], HIF[h]	0.05 % (w/v)	≤ 2 %	HPLC	Etravirine, ritonavir, loviride, danazol, fenofibrate
Yamashita et al. (2011)	96-well microplate	DMSO	FaSSIF	0.015 % (w/v)	2 % (v/v)	HPLC	Itraconazole
Petruševska et al. (2013)	96-well microplate	DMSO	McIlvaine's buffer (pH 6.8)	0.001 %, 0.01 %, 0.1 %(w/v)	1 % (v/v)	UPLC	Fenofibrate carbamazepine

Table 5.6 (continued)

References	Screening format	Co-solvent/solvent	Polymer aqueous phase	Polymer conc.	Co-solvent conc.	Analysis method	Drug(s) tested
Warren et al. (2013)	96-wellmicroplate	PG	Phosphate buffer pH 6.5	0.001 %, 0.1 % (w/v)	9.4 % (w/w)	Nephelometry	Carbamazepine, danazol, ethinylestradiol, amiodarone, halofantrine, itraconazole, meclofenamic acid, mefenamic acid, tolfenamic acid
pH-shift method							
Guzman et al. (2007)	96-well microplate	1N NaOH	SGF	2 mg/mL	N.a.[i]	Nephelometry	Celecoxib
Yamashita et al. (2010)	96-well microplate	SGF	FaSSIF	0.036, 0.36 % (w/w)	N.a.	HPLC	Model compound X

[a]Dimethylformamide
[b]Dimethylacetamide
[c]Propylene glycol
[d]Dimethyl sulfoxide
[e]Simulated gastric fluid
[f]Fasted state simulated intestinal fluid
[g]Fed state simulated intestinal fluid
[h]Human intestinal fluid
[i]Not applicable
[j]UV spectroscopy

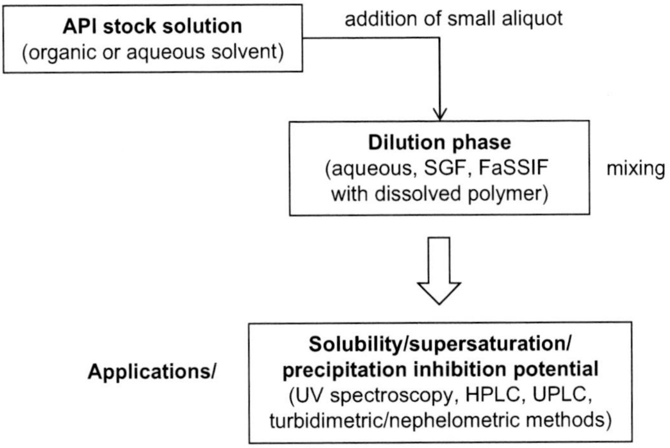

Fig. 5.5 Schematic representation of the experimental procedure of solvent-shift approaches

5.3.3 Coprecipitation Method

Coprecipitation method uses combination of solvent and antisolvent to generate ASD (Karnachi et al. 1995; Huang et al. 2006; Dong et al. 2008; Shah et al. 2012). With the coprecipitation method, a drug and a polymer are dissolved in a common organic solvent and then coprecipitated by a common antisolvent. MBP technology is a subset of the coprecipitation technique with demonstrated superiority in generation of ASD of poorly water-soluble compound. MBP technology is unique as it uses water as the antisolvent to generate ASD.

A high throughput miniaturized coprecipitation screening (MiCoS) based on a 96-well plate format was reported by Hu et al. (2013). The screening consists of five steps, and the schematic representation of the method is shown in Fig. 5.6. In the first step, the polymer and the drug stock solutions are prepared in water miscible solvents such as DMA. The drug stock and polymer stock solutions are mixed to make the desired drug/polymer ratio typically in the range of 10–60 % of drug, with around 10 % total solid content. Polymers in MBP preparation are limited to the ionic polymers such as HPMC-AS, Eudragit® L100, and Eudragit® L100–55. In the second step, coprecipitation is carried out in 1-mL glass vials with V-shaped magnetic stir bars in a 96-well insert. Each glass vial is pre–filled with acidic water chilled to 5 °C. The drug–polymer solutions are added dropwise using a multi-channel pipet to the acidic water to induce coprecipitation. Recommended settings of the MiCoS screening, such as the pH of the acidic water, stirring speed, antisolvent/solvent ratio, and coprecipitation temperature are shown in Fig. 5.6. In the third step, filtration and washing are carried out using two 96-well filter plates. One plate is used for solid-state characterization of ASD, and the second plate is used for kinetic solubility test. The amount of the drugs in resulting MBP is calculated by measuring the drug concentration in the filtrate. In the fourth step, collected solids

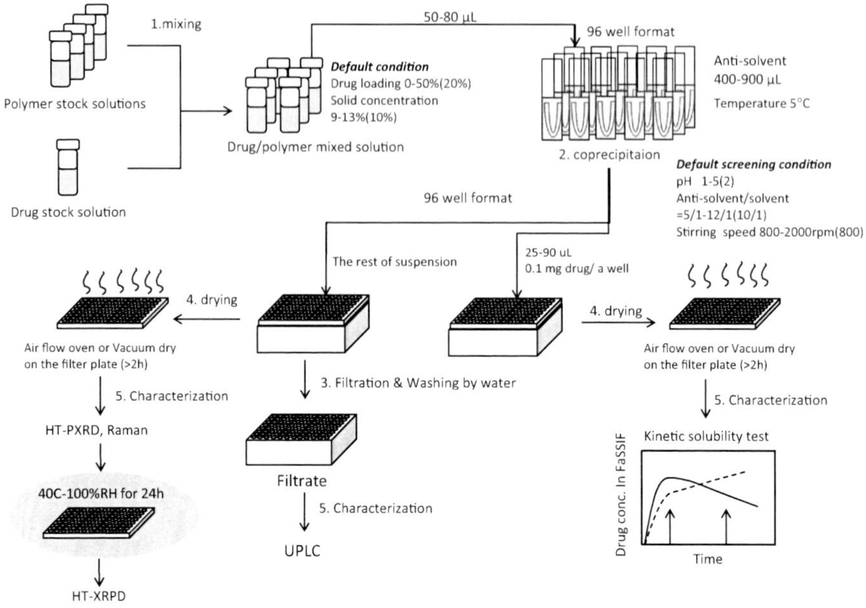

Fig. 5.6 Schematic presentation of coprecipitation screening method

are dried on the 96-well filter plates with an air-forced drier or vacuum drier. In the final step, the dried solids are characterized by HT-XRPD and Raman spectroscopy to assess if amorphous ASD is formed. Optionally, a polarized microscope can be used. The same plate is stored at 40 °C at 100 %RH for 24 h to determine stability of ASD. The second plate is used to determine kinetic solubility profile using FaSSIF at two time points (1 and 3 h).

With the above method, MiCoS has been applied to glybenclamide ASD screening with four enteric polymers at five drug loadings (Hu et al. 2013). Based on both the solid state properties and kinetic solubility, HPMCAS LF was recommended to prepare glybenclamide ASD at up to 40 % of drug loading.

MiCoS shares similar pros and cons to solvent-casting methods. With parallel preparation, it is highly efficient and effective in evaluating polymer types, drug loadings and antisolvent/solvent ratio comprehensively. However, the residue solvent and antisolvent content, which are critical for amorphous stability, cannot be determined due to low amount of solid products. The kinetic solubility results can only be interpreted qualitatively rather than quantitatively, as the particle size of the miniaturized products are not tightly controlled.

5.3.4 Melt-fusion Method

Melt extrusion is a solvent-free ASD process, in which ASD is generated by co-melting of a drug and a polymer or dissolution of a drug in a molten polymer. Because of high viscosity, aggressive physical mixing is needed to ensure complete mixing of viscous liquids. Well mixed molten solution is solidified to form amorphous solid dispersion upon cooling. Clear glass often indicates solid solution whereas turbid or translucent glass often indicates incomplete conversion to amorphous drug.

HME is a complex engineering process composed of blending, heating, mixing, kneading, and extrusion. The concept of melt-fusion is simple. The development of a miniaturized melt-fusion screening system is, however, technically challenging owing to the aggressive mixing and blending requirements at elevated temperature. Additional issues in HT miniaturization may include delivery of miniscule quantity of drug and polymer to small reaction vessel, control of temperature and heating time, and mechanical mixing of viscous molten polymer. Many attempts have been made for automated small scale melt extrusion, but currently there is no report of high-throughput miniaturized melt-fusion screening of polymers for ASD development.

A few small scale low throughput melt-fusion methods were reported at milligram to gram scale to assess the feasibility for HME process. Table 5.7 summarizes the small-scale fusion methods to assess drug and polymer miscibility and HME feasibility. These small scale fusion methods utilize a heating apparatus such as hot plate, oil bath, oven, DSC thermal cycle, capillary melting and hot-stage microscope to prepare small quantities of ASD. Usually, in these methods, aggressive mixing of viscous liquid cannot be accomplished. The ASDs were characterized by a DSC, XRPD, FTIR, dissolution and solubility as discussed in previous sections.

In a specific example, Forster et al. conducted melt-fusion study using DSC and hot plate heating with five drugs, indomethacin, lacidipine, nifedipine, piroxicam, tolbutamide and polyvinylpyrrolidone (PVP) at 1:1 mass ratio (Forster et al. 2001). Desired quantity of the drug and the polymer were weighed into a DSC pan, respectively, and the DSC pan was heated to the melting point of drug. After cooling to room temperature, the melt-fusion mixture was recovered from the DSC pan for analysis. Except piroxicam which degraded, all fused dispersions showed high solubility at 60 min compared to crystalline drugs and physical mix of drugs and polymers. The fused solid dispersion was also characterized by XRPD and DSC before and after stress condition. Below 60 % RH, all solid dispersions remained amorphous at 30 °C after 5 weeks. At 75 % RH, all solid dispersions, except for indomethacin, recrystallized after 5 weeks of storage. The results were in good agreement with the solid stability and dissolution of HME extrudates.

Chokshi et al. utilized a torque rheometer to study the physical and viscoelastic properties of binary mixtures of indomethacin and polymers considered for HME processing (Chokshi et al. 2005). Selected polymers were Eudragit EPO (EPO), polyvinylpyrrolidone/vinyl acetate copolymer (PVP-VA), polyvinylpyrrolidone K30 (PVPK30), and poloxamer 188 (P188). The zero rate viscosity (η_o) for binary mixtures of indomethacin and EPO, PVP-VA and PVPK30 were lower than the pure polymers, whereas η_o for indomethacin/P188 mixture was higher than the pure

Table 5.7 Small-scale ASD screening by melt-fusion methods

Representative reference	Drug/polymer	Solid dispersion preparation		Assessment	
		Heat	Quench	Solid state properties	Biopharmaceutical properties
Betageri and Makarla (1995)	Glyburide/PEG	On hot plate at 120 °C with stirring	Cool at room temperature	NA	Dissolution at 37 °C in phosphate buffer (pH 7.4 at initial and after aging
Forster et al. (2001)	Indomethacin etc/PVP K30	DSC heating	Placing a stainless steel holder filled with liquid nitrogen over the DSC cell	DSC	Dissolution in pH 6.8 phosphate buffer or 1 % SDS at 10 and 60 min
Nepal et al. (2010)	CoQ10/poloxamer, PEG	In the oven set at 70 °C	Cool at room temperature for 15 min	DSC, XRPD, SEM[a]	Solubility in distilled water at 37 °C (initial and after storage) and dissolution at 37 °C in 6.8 phosphate buffer
Moore and Wildfong (2011)	Felodipine etc/PVP VA64	Samples in a crucible heated in oil bath at 10 °C above fusion temperature	Ice water bath	DSC, XRPD, Pair distribution function	Not specified
Djuris et al. (2013)	Carbamazepine/Soluplus	150–180 °C using a microscope hot-stage	Not specified	FTIR, DSC	Not specified

polymer. The activation energy (E_a) of indomethacin/EPO increased with the increasing drug concentration, whereas E_a of indomethacin/PVP-VA decreased with the increasing drug concentration. Using thermal and rheological properties of various drug/polymer mixtures and the actual hot melt processing, Chokshi et al. developed a model correlating viscoelastic properties to the HME processes.

Although these examples demonstrated polymer selection and performance of melt fused ASD, the screening throughput was low. The learnings from small scale melt fusion, though valuable in providing guidance, showed limited applicability to HME manufacturing.

5.3.5 Freeze-Drying method

Freeze-drying is a solvent removal process utilizing the sublimation of frozen liquid. It is a mature technology in biotech industry for manufacturing lyophilized drug products. The technique, however, is poorly exploited to study ASDs. For compounds with poor aqueous solubility, organic solvents are usually required for drug solubilization. Removing organic solvents by freeze-drying is challenging as most organic solvents have low freezing point or low vapor pressure.

Although there is no marketed ASD drug product manufactured by freeze-drying, the technology has been reported to prepare lab-scale ASDs which showed superior bioavailability over crystalline drugs. Betageri and Makarla (Betageri and Makarla 1995) reported glyburide ASD with PEG by freeze-drying at − 75 °C using cyclohexanol as the solvent. The ASD prepared by freeze-drying showed significantly higher dissolution rate than glyburide/PEG physical mix and the corresponding ASD by melt-fusion. This is likely due to the porous structure, increased surface area and intimately-mixing of the freeze-dried ASD. Later, Engers et al. (Engers et al. 2010) reported polymer screening for itraconazole (ITZ) ASD by freeze-drying method. ASD were prepared with seven polymers at four drug loadings by removing *p*-dioxane using freeze-drying. Based on physical characterization and solid stability, 1:2 (w/w) ITZ/HPMC-P dispersion was selected for further characterization, testing, and scale-up. In a recent study (Moes et al. 2011), anticancer drug docetaxel ASDs were prepared with seven polymers and SLS at various drug loadings by removing water/t-butanol using freeze-drying. Based on stability and dissolution studies, formulation of docetaxel/PVP-K30/SLS (sodium lauryl sulfate) in weight ratio of 1/9/1 was selected for the clinical study.

One variation of freeze-drying method is ultra-rapid freezing (also known as thin film freezing), during which droplets of API/excipient(s) solution are rapidly frozen onto a cryogenically-cooled substrate to form an ASD (Overhoff et al. 2007). The process produces amorphous high surface area powders with submicron primary particles to enhance dissolution. Zhang et al. (2012) prepared and characterized solid dispersions with fenofibrate/polymer blends at ratios of 1:4, 1:6, and 1:8 by ultra-rapid freezing. The resulting solid dispersion showed trace amount of crystallinity at low polymer content (1:4), while the solid dispersion was amorphous at high polymer content (1:6 and 1:8). Dissolution testing showed the HPMC E5 provided higher degree of supersaturation than Soluplus®.

Although freeze drying is not widely used in preparing ASDs presumably owing to technical challenges and costs, it may provide special advantages particularly for thermally labile drug. The low operational temperatures of freeze-drying minimize the risk of thermal degradation of drug candidates. In addition, the low temperature lowers molecular mobility, and thus reduces the recrystallization potential of the amorphous phase during processing.

5.3.6 Spin-coating Method

Spin-coating technique has been widely used to study solid-state drug–polymer miscibility in academia as well as pharmaceutical labs. Drug and polymer are dissolved in a suitable volatile solvent or solvent mixture and 1–2 drops of this solution are placed in the center of a substrate rotating at typically 2000–8000 rpm. The solution is spread across the substrate and the excess is spun away by the centrifugal force. The solvent evaporates by forced convection near the rotating surface, leaving a thin film of dried solid dispersion across the glass substrate. This process is very quick and requires approximately 3–20 s. Spin-coating is usually performed at room temperature and in some cases the thin film obtained is additionally heated (e.g., up to 90 °C) for several seconds to remove residual solvent if nonvolatile solvents are used. Different substrates can be used in this technique, depending on the analytical methods used to characterize the films. Lee and Lee (2003) used 2 × 2 cm silicon wafer chips, and other authors used glass microscope cover slips (Konno and Taylor 2006; Van Eerdenbrugh and Taylor 2010) in combination with polarized light microscopy to investigate the ability of different polymers and drug–polymer ratios to inhibit the crystallization of amorphous drugs. Spin-coated glassy drug–polymer films have also been prepared on ZnS discs to evaluate drug–polymer molecular interactions by FTIR measurements (Marsac et al. 2006a; Konno et al. 2008).

Spin coating is a small-scale rapid method for ASD screening that requires minimum amounts of API. Sample preparation is simple and the technique is compatible with a number of analytical characterization methods. Automation of the process is possible but parallel processing obviously has certain limitations. Similar to solvent casting approaches, the level of residual solvent after evaporation can be an issue, and the compounds need to dissolve and remain stable in the same volatile solvents (e.g., ethanol, acetone, etc.) as the polymers and/or other excipients used.

5.4 Miniaturization Screening Strategy and Decision Making

The primary objective of ASD screening is to assess ASD feasibility and to identify the suitable polymer and drug loading when ASD is feasible. The selected combination of polymer and drug loading can be further optimized during ASD scale up. As the ASD generation is a highly engineered process, evaluation of manufacturing parameters is beyond the scope of ASD screening.

With parallel ASD preparation and multi-dimensional characterization in miniaturized screening, large quantity of data is collected. It is important to interpret the data scientifically and make rational decision on polymer and process evaluation.

5.4.1 Polymer Evaluation in Miniaturized Screening

At the beginning of miniaturized screening study, the polymer candidates are selected based on their physicochemical properties and in-silico assessment of drug–polymer interaction. The selected polymers are then evaluated in miniaturized screening experiments using the methods described in Sect. 5.3 for the ASD formation.

The prepared ASDs are generally characterized in two steps: physical stability and degree of supersaturation in dissolution medium. The analyses can be performed in parallel or in a stepwise approach. Firstly, amorphous physical stability of ASDs is evaluated by XRPD, DSC, polarized microscopy, IR and/or Raman spectroscopy. XRPD is commonly used as the primary characterization tool for the physical stability evaluation of the ASDs. XRPD, however, is not sensitive for the detection of trace level of crystallinity, particularly when total amount of sample analyzed is miniscule as commonly encountered in the miniaturized screening study. As XRPD suffers from poor detection limits of crystallinity, caution must be exercised and any sign of crystallinity should be considered as incomplete conversion to amorphous ASD. Alternative techniques such as PLM, IR, or Raman spectroscopy should be employed whenever secondary confirmation is needed. Secondly, the degree of supersaturation as a surrogate of biopharmaceutical performance is measured by solubility time profile or kinetic solubility. Obviously, polymer should provide high degree of supersaturation as long as possible.

The physical stability and kinetic solubility of ASD do not necessarily go together side by side. For example, in the miniaturized coprecipitation screening for glyburide (also known as glybenclamide), physical stability evaluation indicated that Eudragit L100–55 was the best stabilizing polymer for glyburide (Hu et al. 2013) as its ASDs remained amorphous even at 50 % drug loading after 100 % RH storage (24 h at 40 °C). However, subsequent dissolution test indicated that Eudragit L100–55 was not the best polymer in maintaining glyburide supersaturation (glyburide precipitated quickly after initial supersaturation). The final polymer selection has to balance both aspects.

The chemical stability of ASDs is equally important if not more. Since the amorphous state of drug generally has higher chemical reactivity as compared to the crystalline drug, the compatibility between drug and polymer(s) should be evaluated thoroughly early in the ASD development. It is critically important that the selected polymers or any additives do not chemically interact with the drug substance.

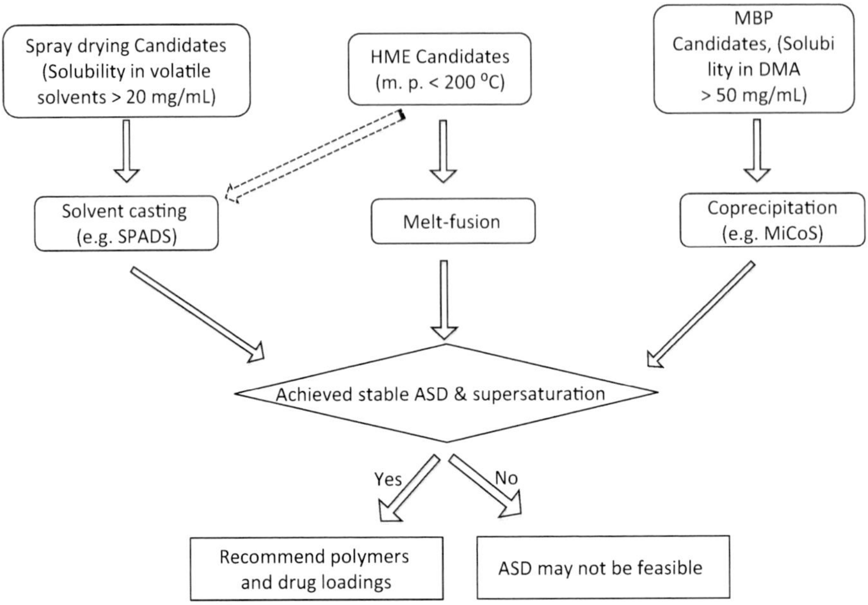

Fig. 5.7 Flow chart of miniaturized screening to evaluate amorphous solid dispersion preparation and processing methods

5.4.2 Process Evaluation in Miniaturized Screening

Figure 5.7 is a flow chart of miniaturized screening to evaluate ASD preparation and processing methods. Generally, at early ASD screening stage when API supply is limited, a step-by-step strategy is recommended. Drug candidates with good solubility in volatile solvents (spray-drying candidates) are first screened with solvent casting methods for selection of stabilizing polymers. HME candidates with low melting point can be evaluated by small scale melt-fusion methods. For high throughput screening, HT solvent casting methods can also be used for HME candidates if they have good solubility. Coprecipitation candidates (with poor solubility in volatile organic solvents and high melting point) with good solubility in non-volatile water-miscible solvents are screened by MiCoS for ASD feasibility.

Alternatively, a parallel screening approach with different screening methods (solvent-casting, melt-fusion and coprecipitation) can also be adopted in case: (1) ASD with high solubility and stability is required; (2) drug candidates are amendable for different ASD processing (3) API supply is not the limiting factor.

Once the best performing polymers, drug loading and processes are identified in miniaturized screening, the ASD can be prepared in lab-scale for further in vitro and in vivo assessment (animal PK studies).

5.5 Summary

Miniaturized high throughput screening can add tremendous value to ASD development, particularly during early drug development. It allows exploring vast experimental design space for the selection of polymer and optimal drug loading as well as guidance to processing conditions. High throughput mode of screening uses a 96 well plate format and consumes only a few mg of API per well. Solvent-casting methods, (e.g., 96-well plate vacuum dry system; SPADS) and the MiCoS approach are well suited for high throughput experimentation. Initial readout of the experiment usually comes out within a few days although stability assessment may take longer.

ASD screening relies on experimental results without any presumption or a priori theory; thereby allowing practical selection of polymer and drug loading. Selected combination of polymer and drug can be further scaled up for the prototype ASD generation and provide the basis for further ASD development.

Acknowledgments The authors would like to thank Dr. Michael Brandl for his comments on this manuscript and Dr. Matthias Eckhard Lauer for the AFM pictures.

References

Alonzo DE, Zhang GG, Zhou D, Gao Y, Taylor LS (2010) Understanding the behavior of amorphous pharmaceutical systems during dissolution. Pharm Res 27:608–618.

Baird JA, Van Eerdenbrugh B, Taylor LS (2010) A classification system to assess the crystallization tendency of organic molecules from undercooled melts. J Pharm Sci 99:3787–3806

Barillaro V, Pescarmona PP, Van Speybroeck M, Thi TD, Van Humbeeck J, Vermant J et al (2008) High-throughput study of phenytoin solid dispersions: formulation using an automated solvent casting method, dissolution testing, and scaling-up. J Comb Chem 10:637–643

Betageri G, Makarla K (1995) Enhancement of dissolution of glyburide by solid dispersion and lyophilization techniques. Int J Pharm 126:155–160

Bevernage J, Forier T, Brouwers J, Tack J, Annaert P, Augustijns P (2011) Excipient-mediated supersaturation stabilization in human intestinal fluids. Mol Pharm 8:564–570

Bevernage J, Brouwers J, Brewster ME, Augustijns P (2013) Evaluation of gastrointestinal drug supersaturation and precipitation: strategies and issues. Int J Pharm 453:25–35

Bhugra C, Pikal MJ (2008) Role of thermodynamic, molecular, and kinetic factors in crystallization from the amorphous state. J Pharm Sci 97:1329–1349

Brewster ME, Vandecruys R, Peeters J, Neeskens P, Verreck G, Loftsson T (2008) Comparative interaction of 2-hydroxypropyl-β-cyclodextrin and sulfobutylether-β-cyclodextrin with itraconazole: phase-solubility behavior and stabilization of supersaturated drug solutions. Eur J Pharm Sci 34:94–103

Brouwers J, Brewster ME, Augustijns P (2009) Supersaturating drug delivery systems: the answer to solubilityâŁłlimited oral bioavailability? J Pharm Sci 98:2549–2572

Chiang P-C, Ran Y, Chou K-J, Cui Y, Sambrone A, Chan C et al (2012) Evaluation of drug load and polymer by using a 96-well plate vacuum dry system for amorphous solid dispersion drug delivery. AAPS PharmSciTech 13:713–722

Chokshi RJ, Sandhu HK, Iyer RM, Shah NH, Malick AW, Zia H (2005) Characterization of physico mechanical properties of indomethacin and polymers to assess their suitability for hot melt

extrusion processs as a means to manufacture solid dispersion/solution. J Pharm Sci 94:2463–2474
Curatolo W, Nightingale JA, Herbig SM (2009) Utility of hydroxypropylmethylcellulose acetate succinate (HPMCAS) for initiation and maintenance of drug supersaturation in the GI milieu. Pharm Res 26:1419–1431
Dai WG, Pollock-Dove C, Dong LC, Li S (2008) Advanced screening assays to rapidly identify solubility-enhancing formulations: high-throughput, miniaturization and automation. Adv Drug Deliv Rev 60:657–672
Djuris J, Nikolakakis I, Ibric S, Djuric Z, Kachrimanis K (2013) Preparation of carbamazepine-Soluplus solid dispersions by hot-melt extrusion, and prediction of drug–polymer miscibility by thermodynamic model fitting. Eur J Pharm Biopharm 84:228–237
Dong Z, Chatterji A, Sandhu H, Choi DS, Chokshi H, Shah N (2008) Evaluation of solid state properties of solid dispersions prepared by hot-melt extrusion and solvent co-precipitation. Int J Pharm 355:141–149
Engers D, Teng J, JimenezâŁłNovoa J, Gent P, Hossack S, Campbell C et al (2010) A solid state approach to enable early development compounds: selection and animal bioavailability studies of an itraconazole amorphous solid dispersion. J Pharm Sci 99:3901–3922
Forster A, Hempenstall J, Tucker I, Rades T (2001) The potential of small-scale fusion experiments and the Gordon-Taylor equation to predict the suitability of drug/polymer blends for melt extrusion. Drug Dev Ind Pharm 27:549–560
Friesen DT, Shanker R, Crew M, Smithey DT, Curatolo WJ, Nightingale JA (2008) Hydroxypropyl methylcellulose acetate succinate-based spray-dried dispersions: an overview. Mol Pharm 5:1003–1019
Ghebremeskel AN, Vemavarapu C, Lodaya M (2007) Use of surfactants as plasticizers in preparing solid dispersions of poorly soluble API: selection of polymer–surfactant combinations using solubility parameters and testing the processability. Int J Pharm 328:119–129
Graeser KA, Patterson JE, Zeitler JA, Gordon KC, Rades T (2009) Correlating thermodynamic and kinetic parameters with amorphous stability. Eur J Pharm Sci 37:492–498
Guzman HR, Tawa M, Zhang Z, Ratanabanangkoon P, Shaw P, Gardner CR et al (2007) Combined use of crystalline salt forms and precipitation inhibitors to improve oral absorption of celecoxib from solid oral formulations. J Pharm Sci 96:2686–2702
Hancock BC, Zografi G (1997) Characteristics and significance of the amorphous state in pharmaceutical systems. J Pharm Sci 86:1–12
Hancock BC, Shamblin SL, Zografi G (1995) Molecular mobility of amorphous pharmaceutical solids below their glass transition temperatures. Pharm Res 12:799–806
Hu Q, Choi DS, Chokshi H, Shah N, Sandhu H (2013) Highly efficient miniaturized coprecipitation screening (MiCoS) for amorphous solid dispersion formulation development. Int J Pharm 450:53–62
Huang J, Wigent RJ, Bentzley CM, Schwartz JB (2006) Nifedipine solid dispersion in microparticles of ammonio methacrylate copolymer and ethylcellulose binary blend for controlled drug delivery: effect of drug loading on release kinetics. Int J Pharm 319:44–54
Jancsó G, Cser L, Grosz T, Ostanevich YM (1994) Hydrophobic interactions and small-angle neutron scattering in aqueous solutions. Pure Appl Chem 66:515–520
Janssens S, Van den Mooter G (2009) Review: physical chemistry of solid dispersions. J Pharm Pharmacol 61:1571–1586
Janssens S, Nagels S, Armas HNd, D'Autry W, Van Schepdael A, Van den Mooter G (2008) Formulation and characterization of ternary solid dispersions made up of Itraconazole and two excipients, TPGS 1000 and PVPVA 64, that were selected based on a supersaturation screening study. Eur J Pharm Biopharm 69:158–166
Karnachi AA, De Hon RA, Khan MA (1995) Compression of indomethacin coprecipitates with polymer mixtures: effect of preparation methodology. Drug Dev Ind Pharm 21:1473–1483
Khougaz K, Clas SD (2000) Crystallization inhibition in solid dispersions of MKâŁł0591 and poly (vinylpyrrolidone) polymers. J Pharm Sci 89:1325–1334

Konno H, Taylor LS (2006) Influence of different polymers on the crystallization tendency of molecularly dispersed amorphous felodipine. J Pharm Sci 95:2692–2705

Konno H, Handa T, Alonzo DE, Taylor LS (2008) Effect of polymer type on the dissolution profile of amorphous solid dispersions containing felodipine. Eur J Pharm Biopharm 70:493–499

Lauer ME, Grassmann O, Siam M, Tardio J, Jacob L, Page S et al (2011) Atomic force microscopy-based screening of drug-excipient miscibility and stability of solid dispersions. Pharm Res 28:572–584

Lauer ME, Siam M, Tardio J, Page S, Kindt JH, Grassmann O (2013) Rapid assessment of homogeneity and stability of amorphous solid dispersions by atomic force microscopy–from bench to batch. Pharm Res 30:2010–2022

Lee T, Lee J (2003) Drug-carrier screening on a chip. Pharm Tech 27:40–49

Lindfors L, Forssen S, Westergren J, Olsson U (2008) Nucleation and crystal growth in supersaturated solutions of a model drug. J Colloid Interface Sci 325:404–413

Mahlin D, Bergström CA (2013) Early drug development predictions of glass-forming ability and physical stability of drugs. Eur J Pharm Sci 49:323–332

Mahlin D, Ponnambalam S, Heidarian Ho⊂ckerfelt Höckerfelt M, Bergstro⊂m Bergström CA (2011) Toward in silico prediction of glass-forming ability from molecular structure alone: a screening tool in early drug development. Mol Pharm 8:498–506

Marsac PJ, Konno H, Taylor LS (2006a) A comparison of the physical stability of amorphous felodipine and nifedipine systems. Pharm Res 23:2306–2316

Marsac PJ, Shamblin SL, Taylor LS (2006b) Theoretical and practical approaches for prediction of drug–polymer miscibility and solubility. Pharm Res 23:2417–2426.

Matsumoto T, Zografi G (1999) Physical properties of solid molecular dispersions of indomethacin with poly(vinylpyrrolidone) and poly(vinylpyrrolidone-co-vinyl-acetate) in relation to indomethacin crystallization. Pharm Res 16:1722–1728

Miller DA, DiNunzio JC, Yang W, McGinity JW, Williams RO 3rd (2008) Enhanced in vivo absorption of itraconazole via stabilization of supersaturation following acidic-to-neutral pH transition. Drug Dev Ind Pharm 34:890–902

Miyazaki T, Yoshioka S, Aso Y, Kojima S (2004) Ability of polyvinylpyrrolidone and polyacrylic acid to inhibit the crystallization of amorphous acetaminophen. J Pharm Sci 93:2710–2717

Moes J, Koolen S, Huitema A, Schellens J, Beijnen J, Nuijen B (2011) Pharmaceutical development and preliminary clinical testing of an oral solid dispersion formulation of docetaxel (ModraDoc001). J Pharm Sci 420:244–250

Moore MD, Wildfong PL (2011) Informatics calibration of a molecular descriptors database to predict solid dispersion potential of small molecule organic solids. Int J Pharm 418:217–226

Nepal PR, Han H-K, Choi H-K (2010) Enhancement of solubility and dissolution of Coenzyme Q_{10} using solid dispersion formulation. Int J Pharm 383:147–153

Newman A, Engers D, Bates S, Ivanisevic I, Kelly RC, Zografi G (2008) Characterization of amorphous API: polymer mixtures using XâŁłray powder diffraction. J Pharm Sci 97:4840–4856

Overhoff KA, Engstrom JD, Chen B, Scherzer BD, Milner TE, Johnston KP et al (2007) Novel ultra-rapid freezing particle engineering process for enhancement of dissolution rates of poorly water-soluble drugs. Eur J Pharm Biopharm 65:57–67

Padden BE, Miller JM, Robbins T, Zocharski PD, Prasad L, Spence JK et al (2011) Amorphous solid dispersions as enabling formulations for discovery and early development. Am Pharm Rev 14:66–73

Pajula K, Taskinen M, Lehto VP, Ketolainen J, Korhonen O (2010) Predicting the formation and stability of amorphous small molecule binary mixtures from computationally determined Flory-Huggins interaction parameter and phase diagram. Mol Pharm 7:795–804

Pajula K, Lehto VP, Ketolainen J, Korhonen O (2012) Computational approach for fast screening of small molecular candidates to inhibit crystallization in amorphous drugs. Mol Pharm 9:2844–2855

Petruševska M, Urleb U, Peternel L (2013) Evaluation of a high-throughput screening method for the detection of the excipient-mediated precipitation inhibition of poorly soluble drugs. Assay Drug Dev Technol 11:117–129

Pham TN, Watson SA, Edwards AJ, Chavda M, Clawson JS, Strohmeier M et al (2010) Analysis of amorphous solid dispersions using 2D solid-state NMR and 1H T 1 relaxation measurements. Mol Pharm 7:1667–1691

Pouton CW (2006) Formulation of poorly water-soluble drugs for oral administration: physicochemical and physiological issues and the lipid formulation classification system. Eur J Pharm Sci 29:278–287

Rodríguez-Hornedo N, Murphy D (1999) Significance of controlling crystallization mechanisms and kinetics in pharmaceutical systems. J Pharm Sci 88:651–660

Rodríguez-Hornedo N, Murphy D (2004) Surfactant-facilitated crystallization of dihydrate carbamazepine during dissolution of anhydrous polymorph. J Pharm Sci 93:449–460

Rumondor AC, Taylor LS (2010) Effect of polymer hygroscopicity on the phase behavior of amorphous solid dispersions in the presence of moisture. Mol Pharm 7:477–490

Rumondor AC, Ivanisevic I, Bates S, Alonzo DE, Taylor LS (2009a) Evaluation of drug–polymer miscibility in amorphous solid dispersion systems. Pharm Res 26:2523–2534

Rumondor AC, Jackson MJ, Taylor LS (2009b) Effects of moisture on the growth rate of felodipine crystals in the presence and absence of polymers. Cryst Growth Des 10:747–753

Rumondor AC, Stanford LA, Taylor LS (2009c) Effects of polymer type and storage relative humidity on the kinetics of felodipine crystallization from amorphous solid dispersions. Pharm Res 26:2599–2606

Shah N, Sandhu H, Phuapradit W, Pinal R, Iyer R, Albano A et al. (2012) Development of novel microprecipitated bulk powder (MBP) technology for manufacturing stable amorphous formulations of poorly soluble drugs. Int J Pharm 438:53–60

Shamblin SL, Taylor LS, Zografi G (1998) Mixing behavior of colyophilized binary systems. J Pharm Sci 87:694–701

Shanbhag A, Rabel S, Casadevall G, Shivanand P, Eichenbaum G, Mansky P (2008) Method for screening of solid dispersion formulations of low-solubility compounds—miniaturization and automation of solvent casting and dissolution testing. Int J Pharm 351:209–218

Taylor LS, Zografi G (1997) Spectroscopic characterization of interactions between PVP and indomethacin in amorphous molecular dispersions. Pharm Res 14:1691–1698

Tian Y, Booth J, Meehan E, Jones DS, Li S, Andrews GP (2013) Construction of drug–polymer thermodynamic phase diagrams using Flory–Huggins interaction theory: identifying the relevance of temperature and drug weight fraction to phase separation within solid dispersions. Mol Pharm 10:236–248

Van den Mooter G, Wuyts M, Blaton N, Busson R, Grobet P, Augustijns P et al (2001) Physical stabilisation of amorphous ketoconazole in solid dispersions with polyvinylpyrrolidone K25. Eur J Pharm Sci 12:261–269

Van Eerdenbrugh B, Taylor LS (2010) Small scale screening to determine the ability of different polymers to inhibit drug crystallization upon rapid solvent evaporation. Mol Pharm 7:1328–1337

Van Eerdenbrugh B, Taylor LS (2011) An ab initio polymer selection methodology to prevent crystallization in amorphous solid dispersions by application of crystal engineering principles. Cryst Eng Comm 13:6171–6178

Van Eerdenbrugh B, Baird JA, Taylor LS (2010) Crystallization tendency of active pharmaceutical ingredients following rapid solvent evaporation—classification and comparison with crystallization tendency from undercooled melts. J Pharm Sci 99:3826–3838

Vandecruys R, Peeters J, Verreck G, Brewster ME (2007) Use of a screening method to determine excipients which optimize the extent and stability of supersaturated drug solutions and application of this system to solid formulation design. Int J Pharm 342:168–175

Warren DB, Benameur H, Porter CJ, Pouton CW (2010) Using polymeric precipitation inhibitors to improve the absorption of poorly water-soluble drugs: a mechanistic basis for utility. J Drug Target 18:704–731

Warren DB, Bergstrom CA, Benameur H, Porter CJ, Pouton CW (2013) Evaluation of the structural determinants of polymeric precipitation inhibitors using solvent shift methods and principle component analysis. Mol Pharm 10:2823–2848

Weuts I, Kempen D, Verreck G, Peeters J, Brewster M, Blaton N et al (2005) Salt formation in solid dispersions consisting of polyacrylic acid as a carrier and three basic model compounds resulting in very high glass transition temperatures and constant dissolution properties upon storage. Eur J Pharm Sci 25:387–393

Weuts I, Van Dycke F, Voorspoels J, De Cort S, Stokbroekx S, Leemans R et al (2011) Physicochemical properties of the amorphous drug, cast films, and spray dried powders to predict formulation probability of success for solid dispersions: etravirine. J Pharm Sci 100:260–274

Wyttenbach N, Achtziger C, Page S (2011) Development of a miniaturized dissolution method for solid dispersion screening (SPADS): effect of temperature and different polymers on supersaturation. Abstract of 2011 AAPS Annual Meeting and Exposition, vol 13, p S2

Wyttenbach N, Janas C, Siam M, Lauer ME, Jacob L, Scheubel E et al (2013) Miniaturized screening of polymers for amorphous drug stabilization (SPADS): rapid assessment of solid dispersion systems. Eur J Pharm Biopharm 84:583–598

Yamashita T, Kokubo T, Zhao C, Ohki Y (2010) Antiprecipitant screening system for basic model compounds using bio-relevant media. J Assoc Lab Autom 15:306–312

Yamashita T, Ozaki S, Kushida I (2011) Solvent shift method for anti-precipitant screening of poorly soluble drugs using biorelevant medium and dimethyl sulfoxide. Int J Pharm 419:170–174

Yoo Su, Krill SL, Wang Z, Telang C (2009) Miscibility/stability considerations in binary solid dispersion systems composed of functional excipients towards the design of multiâŁłcomponent amorphous systems. Int J Pharm 98:4711–4723

Zhang M, Li H, Lang B, O'Donnell K, Zhang H, Wang Z et al (2012) Formulation and delivery of improved amorphous fenofibrate solid dispersions prepared by thin film freezing. Eur J Pharm Biopharm 82:534–544

Zhao Y, Inbar P, Chokshi HP, Malick AW, Choi DS (2011) Prediction of the thermal phase diagram of amorphous solid dispersions by Flory-Huggins theory. J Pharm Sci 100:3196–3207

Chapter 6
Hot-Melt Extrusion for Solid Dispersions: Composition and Design Considerations

Chad Brown, James DiNunzio, Michael Eglesia, Seth Forster, Matthew Lamm, Michael Lowinger, Patrick Marsac, Craig McKelvey, Robert Meyer, Luke Schenck, Graciela Terife, Gregory Troup, Brandye Smith-Goettler and Cindy Starbuck

6.1 Introduction

Pharmaceutical hot-melt extrusion (HME) has been an area of great interest in academia and pharmaceutical industry alike since the 1980s (Crowley et al. 2007), with numerous patents and research papers having been published since then. However, extrusion technology is a very mature platform widely used in the polymer and food industries. Some examples of plastic products manufactured through extrusion include medical tubing, electric cables, pipes, and plastic bags, among others. In the food industry, the extrusion process, often referred to as extrusion cooking, is used to manufacture numerous products such as cereals, snacks, pet food, flours, and precooked mixtures for infant feeding (Singh et al. 2007).

Through the HME process, one or more active pharmaceutical ingredients (API) are blended with at least one molten excipient in an extruder. The API in the extrudate (or HME product) may exist in its crystalline or amorphous state. Some of the applications of pharmaceutical HME include products designed to promote oral absorption, sustained release (either for oral delivery or implants; Follonier et al. 1995), targeted release (Doelker 1993; Follonier et al. 1995; Andrews et al. 2008), and prevention of substance abuse (Oshlack et al. 2001; Arkenau-Maric and Bartholomaus 2008). Some of these applications are listed in Table 6.1, where the commercial status and the purpose of the HME process are summarized for several drug products.

HME is a continuous melt manufacturing process consisting of the elementary steps of solids conveying, melting, mixing, devolatilization, pumping, and pressurization for shaping (Tadmor and Gogos 2006; Todd 1998). The API, the polymer

J. DiNunzio (✉)
Pharmaceutical Sciences & Clinical Supplies, Merck & Co., Inc.,
556 Morris Ave, Summit, NJ 07901, USA Tel.: 908-473-7329
e-mail: james.dinunzio@merck.com

C. Brown · M. Eglesia · S. Forster · M. Lamm · M. Lowinger · P. Marsac · C. McKelvey · R. Meyer · L. Schenck · G. Terife · G. Troup · B. Smith-Goettler · C. Starbuck
Merck & Co., Inc., Whitehouse Station, NJ, USA

N. Shah et al. (eds.), *Amorphous Solid Dispersions*,
Advances in Delivery Science and Technology, DOI 10.1007/978-1-4939-1598-9_6

Table 6.1 Examples of drug products manufactured by the HME process

Name	API	Polymer excipient	Delivery form	Indication	Status[b]	HME purpose
Dapivirine–maraviroc	Dapivirine + maraviroc[a]	EVA[a]	Implant	Antiviral (HIV)	UD[a]	Shape
Lacrisert®	None	HPMC	Implant	Dry eye syndrome	M	Shape
NuvaRing	Etonogestrel + ethinyl estradiol	EVA[a]	Implant	Contraceptive	M	Shape
Zoladex	Goserelin acetate	PLGA	Implant	Prostate cancer	M	Shape
Implanon	Etonogestrel	EVA	Implant	Contraceptive	M	Shape
Ozurdex®	Dexamethasone	PLGA	Implant	Macular edema	M	Shape
Kaletra®	Lopinavir + ritonavir	PVP-VA	Tablet	Antiviral (HIV)	M	Amorphous dispersion
Norvir®	Ritonavir	PVP-VA	Tablet	Antiviral (HIV)	M	Amorphous dispersion
Eucreas®	Vildagliptin + metformin hydrochloride	HPMC	Tablet	Diabetes	M	Melt granulation
Zithromax®	Azithromycin	HPMC	Tablet	Antibiotic	M	Taste masking
Gris-PEG®	Griseofulvin	PEG	Tablet	Antifungal	M	Crystalline dispersion
Rezulin®	Troglitazone	PVP	Tablet	Diabetes	W	Amorphous dispersion
Palladone™	Hydrophone	EC + ERS	Tablet	Pain	W	Controlled release
Posaconazole	Posaconazole	HPMCAS	Tablet	Antifungal	M	Amorphous dispersion
Anacetrapib	Anacetrapib	–	–	Cardiovascular disease	UD	Amorphous dispersion

API active pharmaceutical ingredient, *EVA* ethyl vinyl acetate, *HPMC* hydroxypropyl methylcellulose, *HPMCAS* hydroxypropylmethylcellulose acetate succinate, *PLGA* poly(lactic-co-glycolic acid), *PVP-VA* polyvinyl pyrrolidone-co-vinyl acetate, *PEG* polyethylene glycol, *PVP* polyvinyl pyrrolidone, *EC* ethyl cellulose, *ERS* Eudragit® RS, *UD* under development, *M* marketed product, *W* withdrawn from the market

[a]Loxley 2010

[b]DiNunzio 2011

carrier, and other excipients are fed as solid particulates, either as a preblend or independently, through the hopper. Additional solids and/or liquids can be independently fed downstream. Solids are conveyed by one or more screws down the length of the extruder barrel, followed by melting of the polymer carrier. In the case of miscible systems, the API is progressively dissolved in the molten polymer. In the case of immiscible systems, a crystalline API is homogenously dispersed in the process stream. Devolatilization may be required to remove entrapped air, moisture, and/or residual solvent. Finally, pressure is generated and the molten blend is forced through the die with the desired shape. After the material exits the die, the process stream is then cooled and subjected to secondary processing steps, such as milling, pelletization, or direct shaping.

Although both single-screw and twin-screw extruders are widely used for polymer processing and have been utilized in pharmaceutical research, the following discussion is centered on the latter. Fully intermeshing corotating twin-screw extruders are of the greatest interest for pharmaceutical applications since they provide more efficient mixing, tight residence time distributions (RTD), and minimal material stagnation (McCrum et al. 1997; Tadmor and Gogos 2006).

As schematically shown in Fig. 6.1, the properties of the HME product, or extrudate, are a function of three groups of variables: (1) design variables, (2) process variables, and (3) material variables. It is important to point out that these three groups are not fully independent but strongly interrelated.

The design variables can be further subdivided into three groups: extruder, screw, and die design. A twin-screw extruder is schematically depicted in Fig. 6.2, and it consists of a heated barrel that encloses the screws, which convey the material forward and force it through the die. Extruders are primarily defined by the diameter of their screws and their length to diameter ratio (*L*/*D*). The barrel can be modular or fixed, and is independently heated and cooled by means of a control system.

Modular screws are often used since they provide an additional degree of freedom to the design space. Furthermore, screw configuration should be defined based on the formulation and process objectives. Screw configurations are built by combining the three basic types of screw elements: conveying, kneading blocks, and special mixing elements. Conveying elements are employed for material transport and pressure buildup. They also provide some degree of mixing through shearing and linkage or backflow. The main geometrical characteristics of conveying elements are pitch, flight angle, length, and number of flights. A comprehensive geometrical description of these elements and fully intermeshing twin-screw extruders has been published (Booy 1978). Kneading blocks consist of a stack of paddles, of a given thickness and offset angle. Depending on their design, kneading block sections may be conveying, neutral, or reversing and cause varying extents of polymer melting and mixing. Mixing is predominantly due to elongational flows (i.e., dispersive mixing) and the multiple divisions and recombinations (i.e., distributive mixing) of the process stream. Specialized mixing elements are sometimes used to promote dispersive or distributive mixing. Detailed description of the flow patterns and mixing mechanisms in diverse screw elements is out of scope for this discussion but can be found elsewhere (Brouwer et al. 2002; Ishikawa et al. 2002; Tadmor and Gogos 2006; Kohlgrüber and Bierdel 2008).

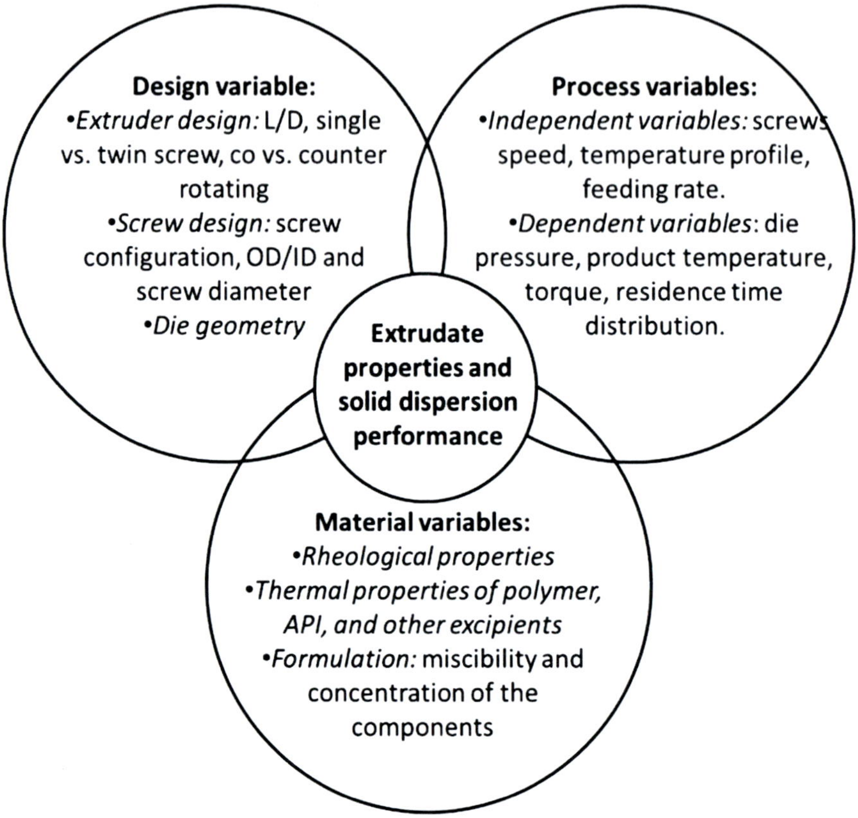
Design variable:
•Extruder design: L/D, single vs. twin screw, co vs. counter rotating
•Screw design: screw configuration, OD/ID and screw diameter
•Die geometry
Process variables:
•Independent variables: screws speed, temperature profile, feeding rate.
•Dependent variables: die pressure, product temperature, torque, residence time distribution.
Extrudate properties and solid dispersion performance
Material variables:
•Rheological properties
•Thermal properties of polymer, API, and other excipients
•Formulation: miscibility and concentration of the components

Fig. 6.1 Summary of the variables that affect the properties and performance of the extrudate or HME product

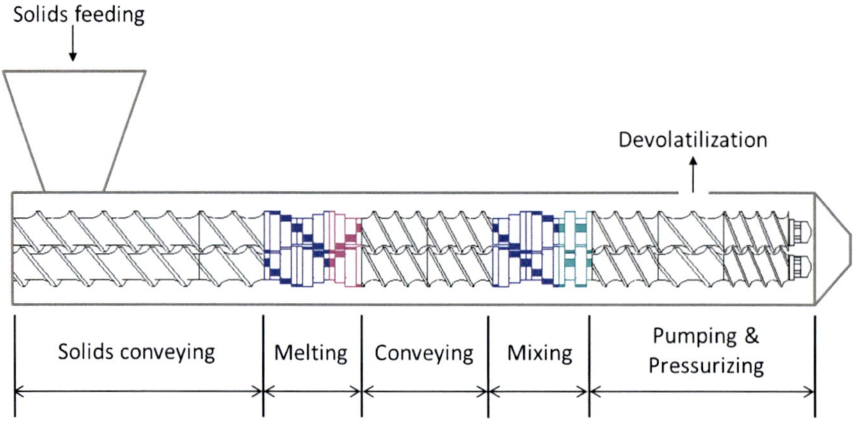
Solids feeding
Devolatilization
Solids conveying
Melting
Conveying
Mixing
Pumping & Pressurizing

Fig. 6.2 Schematic representation of a twin-screw extruder and elementary steps

Process variables can be subdivided into two groups: (1) independent variables such as screw's rotating speed, temperature profile of the barrel, and feeding rates and (2) dependent variables such as product temperature or actual temperature of the process stream, RTD, pressure, and torque.

Typically, the barrel temperature profile is set at least 30 °C above the glass transition temperature (T_g) of the polymer or above its melting point, in the case of a semicrystalline polymer excipient. Furthermore, these temperatures are generally below the melting point of the API, although process temperatures can be above the API's melting point, if the components do not degrade. It is important to bear in mind that although the extruder barrel is heated, much of the energy utilized for melting is provided by the rotation of the screws—particularly at larger scales. As the solids are conveyed, heat is generated through frictional energy dissipation, followed by a combination of plastic and viscous energy dissipation in fully filled kneading blocks (Todd 1998; Tadmor and Gogos 2006).

Feeding rate in twin-screw extrusion is very important as it defines the manufacturing throughput. Twin-screw extruders are starve-fed, where the amount of material fed to the extruder does not completely fill its free volume. In general, conveying elements tend to be partially filled, while kneading blocks tend to be fully filled. The residence time of the melt in partially filled elements is solely dependent on the screw speed and screw element pitch; while in the fully filled sections, it is independent of the screw speed, i.e., only depends on throughput (Todd 1998). However, the length of the fully filled sections is a function of the screw speed. A practical implication of this is that the residence time of the material—for a fixed-screw configuration—is predominantly controlled by the feeding rate.

Finally, material variables will have a direct impact on the design and process variables. Both the properties of the individual components and those of phases formed during processing are important in the design of extrusion processes. For example, the melt viscosity of a polymer can be lowered by the addition of plasticizer or increased by the addition of an immiscible dispersed phase. This behavior was clearly shown (Yang et al. 2011) for an API–polymer binary system.

As such, it is clear that the design of extruded amorphous dispersions will be dependent on formulation and process considerations. The flexibility provided by the extruder yields unique opportunities to address many of the challenges faced during development. The subsequent sections detail the considerations for selection of the extrusion platform, classification of dispersion systems, formulation design, characterization, commercialization, integration within the supply chain, scale-up, inline monitoring through process analytical technologies (PAT), and implementation of extrusion operations within a quality-by-design (QbD) framework.

6.2 Enabled Technology Platform Selection

At its core, an amorphous solid dispersion formulation is simply a single-phase mixture of drug with other components. However, multiple paths exist for achieving that single-phase mixture, including mechanical activation, spray drying, and HME, among others. At a high level, all process routes to manufacture an amorphous solid dispersion follow the same generalized set of activities: mixing of individual components followed by a quench step. In the case of spray drying, the mixing of individual components is achieved through dissolution of the components in a common solvent system, whereas the quench step is the actual spray drying process where atomized droplets are rapidly dried. In contrast to spray drying where mixing is relatively simple and achieved through dissolution in a solvent system, the melt extrusion process itself is where mixing takes place. For melt extrusion, quenching occurs following extrusion, where the extrudate is rapidly cooled by forced air, dry ice, chilled rolls, or other techniques.

Much has been written in the literature on the impact of amorphous solid dispersion *composition* on the stability or performance of the drug product, including studies covering the stability of polyethylene glycol (Zhu et al. 2013), the stability of PVP (Taylor and Zografi 1997), the stability of PVP-VA64 (Wang et al. 2005), the performance of HPMC (Suzuki and Sunada 1998), and the performance of HPMCAS (Friesen et al. **?**). Some published studies have examined the impact of *process route* on the stability or performance of amorphous solid dispersions with the same composition, including publications covering melt/quench methods relative to ball milling (Patterson et al. 2005), comparing HME to spray drying and ball milling (Patterson et al. 2007), evaluating HME and solvent coprecipitation (Dong et al. 2008), and examining spray drying relative to HME (Patterson et al. 2008). However, very few studies have examined the impact of *process parameters* on the stability or performance of amorphous solid dispersions having the same composition.

One seeking to determine the best process technology to leverage for a given amorphous solid dispersion often attempts to determine whether spray drying or HME is a more appropriate route. In some ways, the choice is straightforward. All else equal, an HME process occupies a smaller facility footprint, requires comparatively lower-cost capital equipment, enables higher throughput, and fits into many existing pharmaceutical processing suites (Breitenbach 2002). However, all else is generally not equal: both the process route and the process parameters themselves can substantially impact the stability or performance of the drug product. Any given process could result in a product of poor quality, and in some cases, several different processes can produce an amorphous solid dispersion with similar quality attributes.

Each process presents different challenges toward achieving the desired product. Amorphous solid dispersions manufactured by HME are often challenged by the ability to achieve a single-phase mixture, whereas this goal is rather easily achieved during spray drying by dissolving all components in a solvent system. Those formulations manufactured by spray drying may be challenged by the capability to maintain a single-phase mixture throughout the manufacturing process, given relative drying rates and the presence of residual solvent in the spray-dried product that may substantially plasticize the material. The HME process largely decouples the phase state of the amorphous solid dispersion from the physical properties of the particles generated through the use of a separate milling step. In contrast, the spray drying process largely links the phase state of the formulation to the physical properties of the resultant particles. Changing the size and density of the spray-dried particles requires changes in heat and mass transfer in the spray dryer, which may impact homogeneity and phase state.

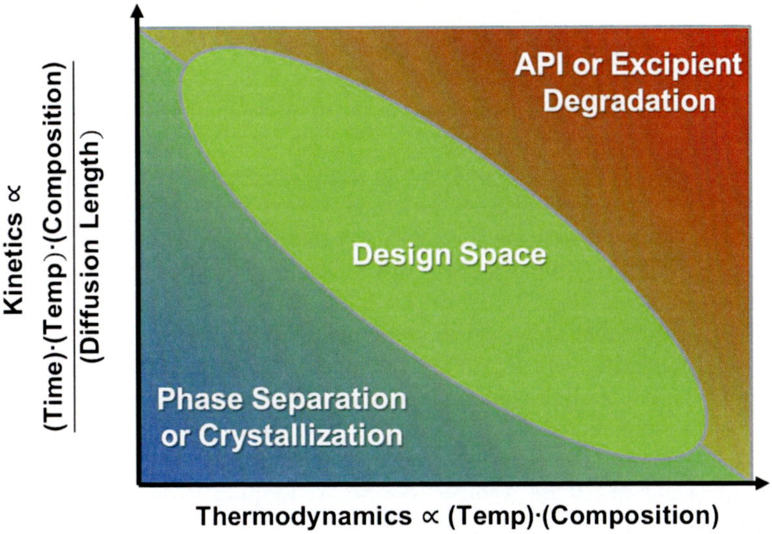

Fig. 6.3 Conceptual design space for an amorphous solid dispersion manufactured by hot-melt extrusion, where failure modes are depicted with respect to thermodynamic and kinetic variables

Achieving a single-phase mixture by HME requires a balancing of thermodynamic and kinetic driving forces, as shown by Fig. 6.3. From a thermodynamic perspective, both the temperature and composition of the formulation may impact its risk of crystallization or phase separation. From a kinetic perspective, single-phase mixtures require sufficient temperature, time, and surface area for diffusion to occur. These principles are made more complex by the reality that there is not a single temperature, time, composition, or surface area within the extruder. Instead, there is a distribution of temperatures and compositions, owing to a physical mixture of particles flowing through a barrel with axially varying screw profile, axially changing barrel temperature profile, and even radially different temperature and shear profiles within a given screw segment (Griffith 1962).

Too much energy input is not necessarily a good thing. Excessive time, temperature, or stress may result in degradation of API, polymer, or components. Limits of API degradants in drug products have been well established through International Conference on Harmonisation International Conference on Harmonization(ICH) guidelines (ICH Q3B(R2): Impurities in New Drug Products, 2006); however, degradation of excipients may play a more critical role for amorphous solid dispersions than for other drug products. Although nearly all pharmaceutical excipients are functional, the excipients present in the amorphous solid dispersion often link directly to the stability and performance of the formulation. Consequently, excipient degradation may result in loss of functionality, which could translate to a change in the stability or dissolution behavior of the drug product.

High melting point drugs usually present considerable challenges toward achieving a single-phase mixture while avoiding degradation of any components. Diffusion of the individual components into a single phase over the timescales relevant to a continuous process is facilitated by mixing all components in the liquid state, so drugs with melting points in considerable excess of 200 °C require that the components in the extruder experience very high temperatures or be formulated with excipients that melt soluble materials far below the melting point of the pure API. Some polymers commonly utilized for amorphous solid dispersion formulations have been reported to degrade above 220 °C (Schenck et al. 2011), with some materials even showing instability as low as 150 °C. Thermally labile drugs also present a significant challenge to the extrusion of a single-phase formulation without the onset of degradation (Verreck et al. 2006a). In particular, the gap between the melting point of the crystalline drug and its onset of decomposition needs to be wide enough to ensure a sufficient operating window exists.

While high melting point and thermally labile drugs add complexity to HME process development, there are still opportunities to develop a process that ensures a single-phase mixture while avoiding degradation. Depending on specific interactions, polymers may depress the melting point of drugs considerably, such that these drug compounds will dissolve into the polymer at temperatures well below degradation onset (Marsac et al. 2006). Another opportunity to mitigate degradation risk is to incorporate components into the amorphous solid dispersion formulation whose sole function is to depress the melting point of the drug and enable lower processing temperatures (Ghebremeskel 2007). A liability with this approach is that the very components which facilitate processing may plasticize the resultant amorphous solid dispersion, potentially increasing the physical stability risk of the drug product during shelf life. A compelling response to this risk is the injection of supercritical fluid into the extruder barrel, which can dissolve into the formulation, temporarily depress the melting point, and subsequently evaporate from the extrudate (Nalawade 2006). This technique has the advantage of having no impact on the glass transition temperature of the amorphous solid dispersion on storage, thereby avoiding additional physical stability risk.

Another consideration in the selection of technology platform is the polymer chosen for the amorphous solid dispersion formulation. The fact that many pharmaceutical polymers degrade, crosslink, or lose functionality at high temperatures has already been discussed. However, the melt viscosity of a polymer is critical to the ability to extrude the amorphous solid dispersion within the capabilities of the extrusion equipment. The melt viscosity as a function of temperature and shear rate varies considerably across pharmaceutical polymers (Chokshi et al. 2005). Formulation melt viscosities in the range of 10–100,000 Pa s are generally acceptable for HME, although the range depends heavily on the torque limit capability of the particular extruder.

Process technology selection for the manufacture of amorphous solid dispersions requires consideration of the particular complexities of the drug and excipients. HME offers the possibility to manufacture drug products in a continuous, cost-effective manner, yet it presents unique challenges that must be tackled. Noting the significant interplay between formulation and process, a risk-based classification system has been developed to aid in the early assessment of dispersion success using melt extrusion.

6.3 Drug: Polymer Systems for Extrusion

The complexity of compounding an API and a polymer into an amorphous dispersion is dependent on the physical and mechanical properties of the constituent ingredients, and the processing conditions. Both thermodynamic and kinetic mixing considerations are at play during the formation of a solid dispersion in a hot-melt extruder. HME compounding classification schemes have been reported previously (DiNunzio et al. 2012; Liu et al. 2012). Categorization of formulations into two types of systems may provide insight into ultimate process development. The first system is characterized by the dissolution of solid API particles into a "liquid-like" polymer melt. The second system is described by the mixing of miscible liquids of differing viscosities. These two systems can be further subdivided based on the following system attributes: the melting point of the API, the extent of API melting point depression observed in the presence of the polymer, and the melt viscosities of the API and polymer. In this section, an expanded classification system for binary API/polymer amorphous dispersion compounding problems is presented based on the above attributes (Troup classification system; TCS) and summarized in Table 6.2. The main features of solid/liquid and liquid/liquid systems and details on each class are explained in the following subsections.

6.3.1 *Classes I and II: Solid/Liquid Systems*

Solid/liquid systems are categorized by high-melting-point APIs that exhibit negligible melting point depression in the presence of polymer. In class I systems, the polymer is highly viscous, while in class II systems, the polymer is inviscid. The system behavior of these two classes can be described as a solid drug dissolution problem (Liu et al. 2010), and can be understood in terms of the Noyes–Whitney equation (Noyes and Whitney 1897), given as Eq. 6.1:

$$\frac{dC(t)}{dt} = D_{api,polymer} \times A_{surface} \times \frac{[C_{sat} - C(t)]}{h \times V_{melt}} \tag{6.1}$$

where $D_{\mathrm{api,polymer}}$ is the diffusion coefficient of the API in the polymer melt at the processing temperature, A_{surface} is the total surface area of the API in contact with the polymer melt, C_{sat} is the saturation solubility of the API in the polymer melt at the processing temperature, C is the concentration of API in the bulk polymer melt, h is the diffusion boundary layer thickness, and V_{melt} is the volume of the polymer melt. Analysis of Eq. 6.1 suggests that increasing API surface area, reducing the boundary layer thickness, and increasing convective mixing are required to drive homogeneity of solid/liquid systems. General processing guidelines for classes I and II systems are briefly described in the following subsections.

Table 6.2 Troup classification system characterizing the risk of dispersion production

Class	Melting temperature of API	Extent of melting point depression	Polymer system	Complexity	Phase attributes
I	High	Negligible	Viscous	Mixing degradation	Solid/viscous liquid
II	High	Negligible	Inviscid	Mixing degradation	Solid/inviscid liquid
III	High	Significant	Viscous	Mixing	Liquid/liquid
IV	High	Significant	Inviscid	Mixing	Liquid/liquid
V	Low	NA	Viscous	Mixing for extreme viscosity ratios	Liquid/liquid
VI	Low	NA	Inviscid	Mixing for extreme viscosity ratios	Liquid/liquid

API active pharmaceutical ingredient, *NA* not available

6.3.2 Class I: High T_{melt}^{API}, Negligible Melting Point Depression, and Viscous Polymer System

Class I systems require high processing temperatures and long residence times to fully compound the API and polymer into an amorphous dispersion. At these processing conditions, thermal degradation of the polymer and/or the API is often an issue. From Eq. 6.1, increasing the total surface area by jet-milling the API should improve processing performance by reducing the required residence time. Preblending the feedstock prior to melt extrusion may also improve processing performance by maximizing the initial amount of API in contact with bulk polymer. Distributive mixing sections in the extruder will promote drug dissolution into the bulk. Higher viscosity polymers are anticipated to be more challenging in these cases because it is more difficult to refresh the boundary layer during mixing in the extruder. The addition of low levels of melt-solubilizing polymers and/or plasticizers should be considered for this class.

6.3.3 Class II: High T_{melt}^{API}, Negligible Melting Point Depression, and Inviscid Polymer System

Similar to class I, class II systems also typically require high processing temperatures and long residence times to fully melt and disperse the API. Thermal degradation of the polymer and/or the API is again an issue. The inviscid polymers may possibly be prone to thermal degradation at these temperatures, but the lower viscosity should lead to improved mixing performance due to rapid surface renewal of bulk polymer at the boundary layer during mixing. Both jet-milling the API and preblending the feedstock should improve processing performance. Distributive mixing sections are also recommended for this class. If the API dissolution rate is sufficiently high, lower temperature processing may be possible.

6.3.4 *Classes III and VI: Liquid/Liquid Systems*

In contrast to classes I and II, where the drug dissolution into the polymer dominates system behavior, classes III and VI are better characterized by liquid/liquid mixing phenomenon. In these cases, the API rapidly melts, forming discrete fluid pockets enriched with API in a continuous matrix of pure polymer melt. A disparity in viscosity ratio will transiently exist between the discrete API-enriched phase and the continuous polymer-enriched phase as the two components are mixed and as the API diffuses and dissolves. For a rigorous theoretical treatment of laminar mixing of homogeneous fluids, interested readers should refer to (Tadmor and Gogos 2006). Laminar mixing theory reveals that mixing in liquid/liquid systems is dependent on the total strain, the volume fraction of the minor component, and the initial striation thickness, in this case the droplet diameter. The final striation thickness, which is a measure of mixedness, as a function of these parameters is given in Eq. 6.2 (Tadmor and Gogos 2006), which is derived for an arbitrarily oriented surface element in a homogeneous simple shear flow field

$$r = \frac{2L}{3X_{vol}\gamma} = \frac{2r_0}{\gamma} \tag{6.2}$$

where r is the final striation thickness, L is the characteristic length, X_{vol} is the volume fraction of the minor component, γ is the total strain, and r_0 is the initial striation thickness. This simplified model shows that the key variable in liquid/liquid systems is the total strain allowed by the screw design and process conditions. Additionally, from inspection of Eq. 6.2, it is evident that the problem can be amplified if a low volume fraction of API is being incorporated. For nonhomogeneous liquid mixing, it is generally regarded that mixing a low-viscosity minor component into a viscous matrix or mixing high-viscosity minor component into a low viscosity matrix are the two most challenging scenarios (Rauwendaal 1998, 2002, Tadmor and Gogos 2006). The former case is the common situation for most pharmaceutical compounding problems. In liquid/liquid systems, the droplet breakup theory developed for immiscible systems (Grace 1982) could also partially apply, as there will be a transient surface tension difference between the discrete and continuous phases. In particular, glass-forming APIs in inviscid polymer systems may exhibit droplet breakup behavior. General processing guidelines for classes III and VI are briefly described in the following subsections.

6.3.5 Class III: High T_{melt}^{API}, Significant Melting Point Depression, and Viscous Polymer System

The melting point depression of the API/polymer systems exhibited in class III systems should result in more moderate processing temperatures compared to classes I and II. The complexity in this system arises from the potential for large differences in the viscosities of the API-enriched phase and the bulk polymer phase. Distributive mixing sections will be beneficial to reduce the length scale of discrete-phase API-enriched fluid droplets, and dispersive mixing may aid in API-enriched droplet deformation and breakage. In these systems, it may be useful to have a distributive mixing section, followed by a dispersive mixing section, then followed by a second distributive mixing section to homogenize the dissolving API into the polymer matrix.

6.3.6 Class IV: High T_{melt}^{API}, Significant Melting Point Depression, and Inviscid Polymer System

Class IV systems are simpler to process than class III systems due to the lower viscosity polymer, leading to a lower viscosity ratio and the potential for improved mixing efficiency with low-viscosity APIs. Less dispersive mixing should be required in this class compared to class III systems, as length-scale reduction should proceed more readily. However, APIs that can transform into a viscous glass requiring mixing of a viscous minor component into a lower viscosity matrix can complicate processing.

6.3.7 Class V: Low T_{melt}^{API} and Viscous Polymer System

Class V systems are analogous to class III systems in that they result in a liquid/liquid mixing problem. Class V systems should result in lower complexity due to the higher degree of freedom afforded by low melting point APIs to increase process temperatures above the melting point of the pure drug substance. Also, in the cases where the API plasticizes the polymer, further processing benefits may be realized, for example, lower absolute processing temperature and improve mixing efficiency. Again both classes III and V will benefit from both dispersive and distributive mixing sections. Complexity in this class may arise if a low-viscosity minor component needs to be compounded.

6.3.8 Class VI: Low T_{melt}^{API} and Inviscid Polymer System

This class is expected to be the least complex system to compound since it requires slower processing temperatures and should have improved mixing efficiency by virtue of its lower viscosity polymer system. The low melting point APIs in classes V and VI should result in processing temperatures that are more a function of API properties than viscosity reduction of the polymer. These systems are less prone to thermal degradation issues and should be reasonably robust to changes in extrusion operating conditions.

6.4 Formulation Design

Formulation and process design for the production of solid solutions must be considered simultaneously. As discussed previously, the properties of the API have a significant influence on the way in which the dispersion is formed and also influence the thermodynamic end point for the process. In general, the TCS can be used to describe the relative risk for producing amorphous dispersions. While many examples exist in the literature covering classes IV–VI systems (Verreck et al. 2003, Keen et al. 2013, DiNunzio et al. 2010, Chokshi et al. 2005), only a few have been described for classes I–III (Hughey et al. 2010). Likely, the absence of examples for classes I–III is tied to the basic challenges of appropriately identifying and manufacturing these systems. However, even with these challenges, extrusion remains a preferred manufacturing technology for a number of solid dispersion products.

For extrusion, viewed as a mixing process at elevated temperature and subsequent quenching, it becomes possible to describe the phase behavior of the dispersion. Shown in Fig. 6.4, this diagram describes the melting temperature of the API as a function of composition as well as the glass transition temperature of the dispersion. Phase envelopes can also be described in this space, leading to a comprehensive understanding of dispersion behavior at relevant temperatures. Serving as a guide for design, additional kinetic factors must also be accounted for, which contribute to the final dispersion properties.

The concentration of API in a solid solution formulation is typically evaluated to understand the effect of drug loading on solid solution properties such as propensity for phase instability. The addition of some APIs directly influences properties critical to melt extrusion process design and development. For example, APIs influence melt rheology as plasticizers, anti-plasticizers, or fillers.

Compatibilizers, excipients that help promote miscibility or interactions between one component (often the API) and other components (e.g., the polymer; Work et al. 2004), may be incorporated into solid solution compositions. Compatibilizers may be added to manipulate solid-state properties and/or the properties of solid solutions upon dissolution. Surfactants often serve the role of compatibilizers in solid solution formulations, influencing dissolution behavior, and, ultimately, bioavailability.

The dependence of formulation properties (e.g., supersaturation maintenance upon dissolution) on both formulation and production process complicates aspects of early formulation screening. Specific formulation compositions may be erroneously disregarded because of the way in which they are prepared during screening. The use of heated ovens and thermogravimetric analysis (TGA) to simulate extrusion

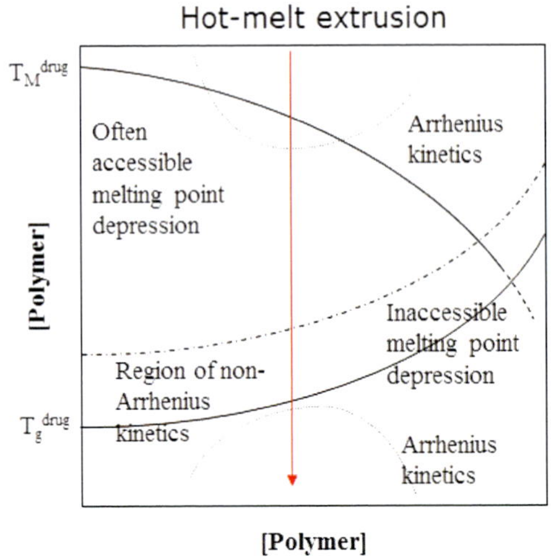

Fig. 6.4 Phase diagram of amorphous dispersion in temperature and composition space

can lead to relatively long exposures of formulations to heat compared to typical extrusion residence times. Extended heating times can lead to polymer and/or API degradation (DiNunzio et al. 2010). Polymers like HPMCAS do not appreciably mix during differential scanning calorimetry (DSC), making cyclical DSC experiments suboptimal for screening many formulations based on this polymer. Attempts have been made to experimentally improve miscibility assessment via thermal methods for polymers like HPMCAS by particle size reduction, cryomilling, and systematically varying heating rates (Sun et al. 2009, 2010; Tian et al. 2012; Mahieu et al. 2013a, 2013b). Solvent casting (Verreck et al. 2003), a technique directly amenable to high-throughput screening (Chiang et al. 2012), may require multiple solvents and/or relatively slow quench kinetics, both of which have the potential to lead to phase separation during preparation which could undesirably bias formulation definition.

Conversely, process constraints may lead to changes in formulations. Plasticizers, including surfactants (Ghebremeskel et al. 2007) and dissolved gases (Verreck et al. 2006a, 2007), may be employed in extrusion processing in order to reduce the temperature and/or stress required to form a homogeneous melt.

Given the complexity of amorphous dispersions, both in terms of criteria related to production as well as stability and bioperformance, it is necessary to develop such systems using a structured design approach. In this type of approach, outlined generally in Fig. 6.5, feedback between process performance, stability, and bioperformance are all necessary to define the optimum system (DiNunzio et al. 2012). This requires strong communication between multiple functions within an organization and also necessitates the appropriate characterization tools for performance assessment. When developing amorphous dispersion formulations, one can consider several paradigms based on the stage for which the technology is utilized. In general, many limitations exist that are prohibitive for implementation of extrusion in the early development space. Specifically, restrictions on equipment scale and small

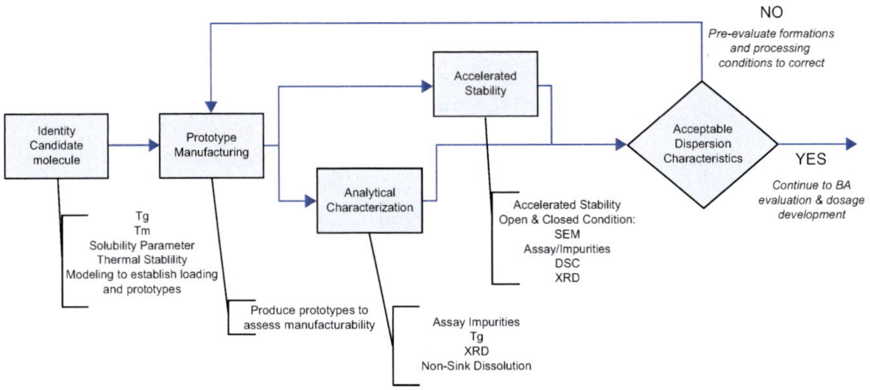

Fig. 6.5 Pathway for prototype dispersion development

batch size can make implementation logistically challenging. Additionally, restrictions based on API/polymer systems that were previously discussed can also limit the utility. As such, many organizations will adopt a strategy of developing dispersions using another processing technology, such as milling, coprecipitation, or spray drying, and then transitioning to extrusion to leverage process advantages for larger production runs. Alternatively, by nature of the properties of the compound and/or organizational philosophy, an end-to-end development of extrusion may be utilized. This section outlines the general approaches for designing melt-extruded solid dispersions under each of these paradigms, with a focus on compositional design to optimize manufacturability, bioavailability, and stability.

6.4.1 Early Formulation Development Considerations

Amorphous solid dispersions are leveraged at varying stages in development for a number of reasons. For extruded dispersions, a limited number of polymer systems summarized in Table 6.3 form the backbone of the compositional definition. In early development, they are most commonly used to support elevated exposures necessary for preclinical assessment and/or assure phase stability when a crystalline form is not readily isolated. At this stage of development, the amount of material available for development will be restricted. As discussed previously, this constraint can challenge the utility of extruded systems where minimum batch sizes are significantly larger than for development of spray-dried dispersions or coprecipitated material.

The small-scale characterization approaches are often conducted in an automated format where the dispersion is produced using solvent casting and then exposed to thermal cycling to simultaneously devolatilize and anneal the system (DiNunzio and Miller 2013). While an effective approach is to regulate the thermal history of the product, these types of approaches do not accurately reflect the quench rate kinetics

Table 6.3 Properties of common excipients used in solid dispersions

Polymer	T_g or T_m (°C)	Grades	Notes
Hypromellose	170–180 (T_g)	Methocel® E5	Non-thermoplastic API must plasticize Excellent nucleation inhibition Difficult to mill
Vinylpyrrolidone	168 (T_g)	Povidone® K30	API must plasticize Potential for H-bonding Hygroscopic Residual peroxides Easily milled
Vinylpyrrolidone–vinylacetate copolymer	106 (T_g)	Kollidon® VA 64	Easily processed by melt extrusion No API plasticization required More hydrophobic than vinylpyrrolidone Processed around 130 °C
Polyethylene glycol, vinyl acetate, vinyl caprolactam graft copolymer	70 (T_g)	Soluplus®	Newest excipient for melt-extruded dispersions Easily processed by melt extrusion Low T_g can limit stability Not of compendial status Stable up to 180 °C
Polymethacrylates	130 (T_g)	Eudragit® L100-55 Eudragit® L100	Not easily extruded without plasticizer Degradation onset is 155 °C Ionic polymer soluble above pH 5.5
Hypromellose acetate succinate	120–135 (T_g)	AQOAT®-L AQOAT®-M AQOAT®-H	Easily extruded without plasticizer Process temperatures 140 °C Ionic polymer soluble above pH 5.5 depending on grade Excellent concentration-enhancing polymer Stable up to 190 °C depending on processing conditions
Amino methacrylate copolymer	56 (T_g)	Eudragit® E PO	Processing at 100 °C Degradation onset is 200 °C Low T_g can limit stability
Methacrylic acid ester	65–70 (T_g)	Eudragit® RS Eudragit® RL	Extrudable at moderate temperatures (> 100°C) Excellent CR polymer
Poly(ethylene vinylacetate)	35–205 (T_m)	Elvax®	Extrudable at low temperatures (60 °C) Excellent controlled-release polymer but nonbiodegradable

Table 6.3 (continued)

Poly(ethylene oxide)	< 25–80 (T_m)	Polyox®	Mechanical properties ideal for abuse-deterrent applications and CR Process temperatures 70 °C Excellent CR polymer
Poly(lactic-co-glycolic acid)	40–60 (T_m)	RESOMER®	Low-melt viscosity for certain grades is challenging to process Biodegradation rate controlled by polymer chemistry Excellent for implantable systems

API active pharmaceutical ingredient, *CR* controlled release

associated with typical spray drying processes or the impact of mechanical energy associated with typical extrusion processes on the critical product attributes. Some researchers have utilized rheometers as surrogates to the extrusion process to assess material performance under a stress field (Yang et al. 2011). While able to simulate shear stresses in extrusion, they do not provide the distributive mixing experienced in extrusion operations. The maximum shear rate in an extruder can be estimated from the clearance between the screw and barrel (overflight; C), screw speed (N), and the outer diameter of the screw (D):

$$\frac{\pi DN}{C} = \dot{\gamma}.$$

The maximum shear rate is on the order of $1000\,\mathrm{s}^{-1}$ for a typical intermeshing 16–18-mm corotating extruder ($D = 16$ mm; $N = 200$ revolutions/min; $C = 0.1$ mm). Alternative methods leveraging DSC to identify the solubility of drug in molten polymer (DiNunzio et al. 2010) have also been advocated as an approach to support selection of the optimum dispersion compositions; however, viscosity limitations associated with several pharmaceutical polymers may inhibit sufficient diffusion during the timescale of the experiment. Forming a homogeneous composition during prototype screening is a critical first step in designing amorphous dispersions. Assessing the stability and bioperformance of these compositions is needed to define the compositional design space that will result in successful products. While methods like TGA and stressed stability can provide insight into the performance (Hughey et al. 2010), there is not an effective way to conduct all of these tests in a truly representative fashion without direct manufacturing on an extrusion platform.

Supporting formulation identification at this stage can be facilitated by small-scale characterization and manufacturing of prototype batches using customized low-volume extruders. To address the scale limitations, a number of small-scale extrusion options are available, ranging in size from 3 to 16 mm that are capable of producing batch sizes as low as 5 g. At this size, geometric similarity to pilot and production scale units may not be preserved as designs are engineered to maximize yield and minimize batch size. However, these systems do serve an important role by providing a representative platform for assessing formulation performance using melt extrusion. Within these systems, a design approach can be implemented through stepwise manufacturing of probe formulations. Supported by early characterization that identifies the optimum formulation for dissolution performance and stability, manufacturability attributes can be assessed in the early development space and used to identify compositions of interest for extrusion development. During the extrusion process, the performance is evaluated based on the operating temperatures and motor loads that will be predictive of larger batch production. Additionally, samples of the dispersion are analyzed for attributes covering the physical and chemical stability of components in the formulation. The scope of this characterization may also be more limited at this stage, focusing primarily on API stability during production and initial assessment of amorphous form generation. Homogeneous and stable dispersions can be further evaluated at this point via dissolution behavior and preclinical pharmacokinetic studies. In this manner, critical attributes related to bioperformance and stability can be optimized, in addition to setting a basis for development of manufacturability attributes in later development.

6.4.2 Pilot-Scale Development Considerations

Optimized formulations developed for clinical trials will typically be produced on larger-scale extrusion equipment than the equipment used during early screening. As extruders transition between the lab and pilot scale, a number of geometric differences can drive changes in performance. These differences, including changes in screw type (i.e., conical to parallel), element design difference (for example, outer to inner diameter ratio), and feed method (manual vs. volumetric vs. gravimetric), can all influence the energy input to the system and the approach for scaling. It is not generally possible to quantitatively map the operating space from these early-screening extruders to pilot-scale extruders because of these differences. Experience and empirical correlations more typically guide process development as programs transition between these disparate pieces of equipment.

Maintaining a constant maximum shear stress between extruders is often not possible because of the exceptionally small clearances that would be required to do this for screening scale extruders. These smaller extruders are often shorter to minimize their free volume (maximizing yield), which necessarily means they will be more limited in the amount of distributive mixing that can be incorporated into the extrusion process. As a result, small-scale equipment may provide misleading results with respect to the mixing that can be achieved readily using larger-scale extruders. Reducing the feed rate can compensate for the reduced mixing in these small-scale extruders. However, many benchtop systems are manually fed or fed with poor control due to the relatively low feed rates required. Variable feed rates inherently cause variation in both the residence time and specific energy input the processed material experiences. The ability to effectively cool or heat the product via the barrel wall with the smaller-scale extruders often used in early pharmaceutical development (e.g., < 16 mm) can lead to challenges with process scale-up.

In order to address these limitations, development at the pilot scale may be necessary prior to current good manufacturing practice (cGMP)—particularly for products requiring a narrow processing window (e.g., where significant degradation is observed near temperatures required to ensure a homogeneous glass is formed). Development and optimization are designed around addressing issues associated with energy input and residence time through formulation and process modification, where uniformity of the dispersion and thermal stability of the formulation are paramount. As such, characterization techniques designed to determine the physical and chemical stability of the drug and polymer are routinely utilized. Specific chemical approaches include gel permeation chromatography, infrared (IR)-coupled gel permeation chromatography (DiNunzio et al. 2010), and chemometric titration (DiNunzio et al. 2010), all of which are intended to determine backbone and side-group changes. Advanced characterization of solid-state properties which are discussed in more detail in the following section are also used at this stage to provide a more comprehensive understanding of molecule distribution and molecular interactions that govern performance of the system.

The extrusion RTD will often be characterized at this stage to provide a baseline for the process and assess the impact of process changes that are conducted to yield a final optimized system. In one approach, the effect of a bolus tracer is measured at the discharge of the extruder, by (i) visual determination with colored tracer, (ii) offline analysis using a chemical tracer, and/or (iii) inline analysis using a chemical tracer. With any of these techniques, it is possible to extract mean and moment information about the distribution that can be related back to the performance. In a recent example of this, Keen et al. (2013) characterized RTDs of corotating and counter-rotating extruders, highlighting performance differences between the units as shown in Fig. 6.6. In general, process modifications (as opposed to compositional modifications) are the most preferred approach to address manufacturing challenges facing prototype formulations. The section below discusses options for addressing two of the most common challenges observed at the early stage of extrusion development: (i) formulation modifications to expand processing windows and (ii) compositional changes to reduce chemical impurity formation during extrusion.

Plasticizers have been well documented for their ability to reduce processing temperatures during extrusion, which translates into a wider operational space (DiNunzio et al. 2010). Although many plasticizers are liquids at room temperature, addition can easily be facilitated by pregranulation or direct injection during extrusion. Solid-state plasticizers, such as citric acid, can also provide a convenient way to reduce processing temperature while facilitating addition via gravimetric feeders (Schilling et al. 2007). However, careful consideration must be given when adding plasticizers into a formulation as the material will also reduce the glass transition temperature of the dispersion, thereby potentially negatively impacting stability and/or dissolution of the dispersion. One method for addressing this is with the use of transient plasticizers, such as supercritical carbon dioxide, where the gas is injected into the processing section under supercritical conditions. Within this environment, the injected material behaves as a supercritical fluid, facilitating molten flow of the melt while functioning as a molecular lubricant. On discharge from the die, the material

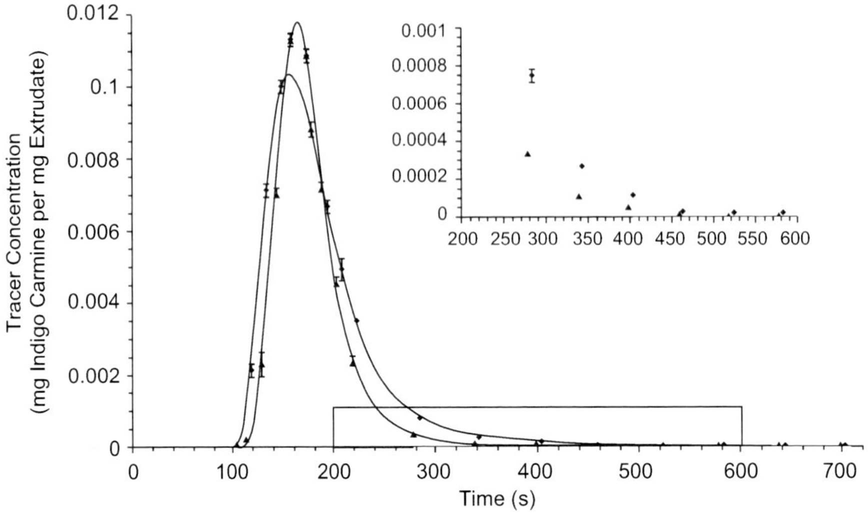

Fig. 6.6 Residence time distribution of corotating and counter-rotating extruders

experiences a dramatic pressure drop that drives a rapid expansion of gas within the melt and creates a foam structure as the additive leaves the system. By this mechanism, supercritical fluids or subcritical gases added directly to the extruder or incorporated separately provide reductions in melt viscosity through transient plasticization (Verreck et al. 2007).

Beyond plasticizers, it is also well known that polymer selection influences molten solubility of drug substances in a system. Careful selection of polymer type has been shown to improve solubilization and reduce impurity formation during thermal processing (DiNunzio et al. 2010). Adapting this approach, researchers have recently illustrated the utility of polymer blends for enhancing the processing characteristics of solid dispersion formulations by incorporating low levels of melt-solubilizing polymer into the dispersion (Albano et al. 2012). This addition allows for greater levels of drug substance to be dissolved in the molten polymer, which provides a viable approach for expanding the operational space of extrusion when dealing with group classes I and II systems. Importantly, because many polymers have high glass transition temperatures, often > 100°C, the polymeric additive will generally have less impact on physical stability when compared with plasticizers or surfactants that are typically characterized by low T_g's. However, care must be given to maintain levels of the solubilizing polymer below which they would impact bioperformance aspects associated with the primary polymer system.

Addressing the chemical impurity formation during extrusion is also generally achieved through process modification by altering the mechanical energy input and residence time of the process. However, it may not be possible to adjust these attributes independently of the composition while balancing requirements for amorphous material formation and minimization of impurity formation. Formulations may also be modified to incorporate additives, such as pH modifiers and antioxidants, that help to reduce degradation during the process (Crowley et al. 2007). These materials provide a means to reduce impurity formation by altering the local environment or scavenging free radicals that would drive decomposition. Another important point to take note of is the purity of the starting materials since many pharmaceutical polymers, such as polyvinylpyrrolidone and polyethylene oxide, have high levels of peroxides, which can be detrimental at elevated temperatures. Although antioxidants can improve performance of extruded dispersions, many vendors now supply high purity grades of these excipients that can also aid in performance. Careful selection of the composition, therefore, begins with the identification of appropriate raw materials and continues on to additives that facilitate manufacturing.

6.5 Solid-State Characterization of Melt-Extruded Amorphous Dispersions

Solid-state characterization of amorphous solid dispersion systems prepared by HME is essential to understand their physical behavior. Several tools and techniques to detect physical failure modes such as crystallization or amorphous–amorphous phase separation will be outlined with an emphasis on the strengths and weaknesses of each approach. Characterization tools such as thermal methods may help to inform process development, specifically the phase diagram, inherent restrictions in processing space, and potential thermal liabilities. Finally, approaches aimed at understanding the fundamentals of amorphous solid dispersions will be discussed. In particular, tools and techniques which offer insight into the thermodynamics and molecular mobility of amorphous systems will be emphasized.

6.6 Detection of Crystallization and Amorphous–Amorphous Phase Separation

Demonstrating the absence of physical failure most often requires the application of multiple characterization tools and techniques. Detectability of relevant failure modes must be demonstrated along with the absence of failure at all relevant processing and storage conditions. The two most common modes of failure are crystallization and amorphous–amorphous phase separation. Crystallization is most often detected using X-ray powder diffraction (XRPD). The exceptional discriminating power of XRPD is largely the result of amorphous materials, lacking in long-range order, giving no constructive interference of incident X-rays. Relevant crystalline forms of the API most often display peaks which are resolved from those peaks associated with excipients used in the formulated product. The absence of API peaks, therefore, provides strong evidence for the stability of the amorphous solid dispersion.

Other spectroscopic techniques may also be used to detect crystalline API in the amorphous matrix. Solid-state nuclear magnetic resonance (ssNMR) spectroscopy can be used to push limits of detection exceptionally low when the API contains an atom which is of high natural abundance and is present exclusively in the API. For instance, ^{19}F and ^{31}P sometimes present exclusively in the API and data can be acquired with reasonable speed. Raman spectroscopy most often provides discriminating power as a result of the API having unique chemical moieties and, therefore, unique vibrational bands as compared to most excipients. Differences between the crystalline and amorphous forms in some instances allow for excellent limits of detection (Sinclair et al. 2011). Advances in nonlinear spectroscopy may provide yet another tool in the solid-state pharmaceutical scientist's toolbox (Strachan et al. 2011). Second-harmonic generation (SHG) operates under the principle that crystalline materials possessing a chiral space group will double the frequency of the incident radiation. Amorphous materials show no second-harmonic signal, and therefore, the discriminating power of SHG can be exceptional (Wanapun et al. 2010, 2011; Kestur et al. 2012). In the presence of finished dosage forms, excipient interferences have been observed. Coupling SHG with two-photon fluorescence may provide additional discriminating power (Toth et al. 2012).

Although the above approaches may all be amenable to detection of crystallization in finished products, they can also be used to characterize the HME (i.e., prior to downstream processing). Further, many other techniques are often applied exclusively to the HME intermediate. For instance, optical microscopy offers excellent detectability of crystalline material in transparent extrudates. Dielectric analysis (DEA; Alie et al. 2004; Bhugra et al. 2007, 2008) and thermally stimulated current IR spectroscopy (Shah et al. 2006; Rumondor and Taylor 2010), atomic force microscopy (AFM; Lauer et al. 2013; Marsac et al. 2012; Price and Young 2004), and calorimetric methods have also been used to detect crystallization from an amorphous matrix (Baird and Taylor 2012; Pikal and Dellerman 1989; Avella et al. 1991).

Phase separation into two amorphous phases may also be of concern. Most generally, a property which discriminates between the amorphous dispersion and a physical mixture of each of the component amorphous materials can be leveraged to detect amorphous phase separation. In practice, detecting amorphous–amorphous phase separation can be very difficult. This is because amorphous materials inherently present analytical signatures which are less well defined as compared to the crystalline counterpart. Further, amorphous phase separation will not often present as well-defined phases of discrete composition as is the case for crystallization. Instead, it is likely that a distribution of compositions may be observed, making detectability very difficult. Most often, DSC is used to detect the presence of multiple amorphous phases (Lu and Zografi 1988). Specifically, if the glass transition temperature (T_g) of each component of the dispersion is unique and if the T_g shows a strong functional dependence with composition, phase separation may be detected with DSC. Alternatively, if the components have similar T_g's or if the compositional dependence of T_g is subtle near the target composition, detection may be difficult. Further, the measurement itself may homogenize the sample, the samples may have an inherently small change in heat capacity across the T_g, or the distribution of molecular environments may be very broad as a result of the distribution of molecular weight in the components which make up the solid dispersion. These and other difficulties may present challenges in measuring a well-defined T_g and are not necessarily unique to DSC. Nevertheless, there is a clear need to consider orthogonal approaches to measure the phase separation. Mathematically transformed X-ray data may be used to understand the phase behavior of amorphous solid dispersions. Specifically, the X-ray signal from amorphous materials may be used to produce a pair distribution function (PDF) via Fourier transformation with the results describing the probability of finding two atoms separated by a specific interatomic distance. Mixing, of course, influences the result and provides useful information about the miscibility of a system or lack thereof (Newman et al. 2008). Vibrational spectroscopy can also be used to detect amorphous phase separation. When interactions between species within the mixture manifest as a change in the frequency and distribution of vibrational modes, this may be detected using approaches such as IR spectroscopy (Rumondor et al. 2009; Marsac et al. 2010; Rumondor and Taylor 2010; Rumondor et al. 2011). Raman spectroscopy has also shown sensitivity to detect amorphous phase separation. In one example, two solid dispersions prepared at different HME processing conditions showed differences in physical stability despite both displaying a single T_g (Qian et al. 2010). Confocal Raman spectroscopy was used to explain the varying degrees of compositional homogeneity between the samples. Although Raman mapping is quite time consuming, nonlinear approaches such as broadband coherent anti-Stokes Raman scattering may expedite the collection process significantly (Hartshorn et al. 2013). In addition to detecting crystalline material within an amorphous matrix, ssNMR may also be used to demonstrate compositional heterogeneity. For instance, two-dimensional correlation techniques and 1H T1 relaxation methods are showing utility in understanding amorphous systems (Pham et al. 2010). Dynamic mechanical analysis (DMA) has found great utility in the study of polymer processing and may be extended to pharmaceutical systems (Karabanova et al. 2008; Carpenter et al. 2009; Szczepanski et al. 2012). DEA and thermally stimulated current have also been shown to provide sensitivity in understanding the homogeneity of the amorphous phase (Power et al. 2007; Shmeis et al. 2004a). Yet another approach, AFM, has been shown to detect amorphous phase separation in samples presented as thin films. For instance, felodipine and polyvinylpyrrolidone were shown to phase separate as indicated by changes in surface roughness and phase shifts after exposure to high relative humidity (Marsac et al. 2010). In another study, differences in HME processing conditions were shown with AFM. Specifically, preparation of a solid dispersion at two processing conditions showed differences in the homogeneity as measured by AFM. The material produced at the higher temperature showed a signal more similar to a control sample with the lower temperature signal showing signs of heterogeneity (Lauer et al. 2013).

Regardless of the method used and the mode of failure detected, all approaches share the common issues of having to discriminate the API from the excipients. Further, desired and undesired phases must show reasonable discrimination and therefore the limit of detection is inherently a function of the system. Most often, several techniques must be explored in parallel during development before a commercial quality control approach that balances the ease of measurement and discriminating power is selected.

Characterization Informs Process Development and Provides Insight into the Fundamentals of Amorphous Solid Dispersions

Characterization of HME-based solid dispersions is not solely motivated by the need to directly measure physical failure. By extension, characterization tools and techniques provide insight into the fundamental properties which facilitate physical failure. For instance, measures of the thermodynamic properties and modes of motion associated with amorphous systems serve to better assess risk of physicochemical failure. Also, as was noted above, in several instances, although the material may be rendered amorphous, the differences in length scale of mixing may manifest as differences in performance, and, thus, characterization tools also inform process development.

Process space can be better understood through construction of the phase diagram and definition of temperature boundaries where failure modes occur. Various characterization approaches may be used to define phase boundaries and robust processing space. For instance, consider a binary API–polymer system. The liquidus line defining equilibrium between the crystalline API and the molten binary API–polymer phase represents the lowest temperature at which the extruder can be operated while still achieving a homogeneous single-phase amorphous solid dispersion system. This line can be generated via approaches such as melting point depression experiments and variations thereof (Marsac et al. 2006,. 2009; Marsac 2006, 2009; Mahieu et al.; Tao et al. 2009; Tian et al. 2009; Sun et al. 2010). However, it is often difficult to access the thermodynamic end state for highly viscous samples and so the kinetics of mixing should always be considered when interpreting DSC results. In the extruder, the combination of dispersive mixing, distributive mixing, and thermal homogenization expedites the formation of a single-phase amorphous system. Viscosity measurements as a function of shear rate and temperature may inform screw design but may also provide insight into the location of the liquidus line for materials of high viscosity. Once a homogeneous system is achieved, cooling below the liquidus line creates a thermodynamic driving force for crystallization. DSC may be used to understand the tendency for a material to crystallize on cooling below the liquidus line (Baird and Taylor 2012). The kinetics of crystallization is system dependent and inhibited by increased viscosity of the material below the liquidus line. Further, the temperature dependence of viscosity as T_g is approached can vary significantly across materials and so the risk of crystallization is case dependent. Yet another important limitation in extrusion is the temperature at which thermal liabilities become relevant over the timescale of the extrusion run. Samples from the DSC experiments may be analyzed by appropriate chemical assays to determine risk. Further, thermogravimetric experiments may be conducted with various time–temperature profiles and assays conducted with the same end in mind. A view of the liquidus line, the glass transition temperature as a function of composition, and the temperature at which thermal degradation exists provides a baseline understanding of processing limitations.

Applying thermodynamic principles to unstable amorphous systems allows for a more complete understanding of the driving forces associated with various failure modes. The tools used to measure these failure modes were discussed above, and here the focus is on how these tools can be used to measure fundamental thermodynamic properties of the amorphous systems. Melting point depression experiments not only provide definition of the phase boundary as discussed above, they also provide insight into the thermodynamic changes which occur as a result of mixing. Specifically, the extent of melting point depression reflects the change in chemical potential of the API, as shown in Fig. 6.7. The greater the melting point depression, the greater the reduction in chemical potential of the drug in the molten phase (Marsac et al. 2006, 2009). More recently, a method for measuring the chemical potential of an API in the presence of a polymer at room temperature was developed. Specifically, solution calorimetry provides a direct method to measure the heat of mixing drugs and polymers and can be used to calculate the solubility of an API in a polymer matrix at room temperature (Marsac 2012). Thermodynamics provides insight into the driving force for crystallization, but mobility facilitates crystallization. Although physical stability risk is most often considered negligible below the T_g, examples of sub-T_g crystallization exist (Vyazovkin and Dranca 2007), nucleation may occur during production of amorphous materials (Baird et al. 2010), and growth rates of crystalline materials may exceed those expected based on diffusion control by an order of magnitude at temperatures below the T_g (Hikima et al. 1995; Ishida et al. 2007; Sun et al. 2008a, 2008b; Yu 2006). These results, among others, have motivated the research toward linking molecular mobility with crystallization tendency. Many of the tools outlined above offer access to various modes of molecular mobility and may offer fundamental insight into the motions linked to physicochemical changes. Most generally, if the activation energy associated with a particular molecular motion matches the activation energy associated with a failure mode, this provides strong evidence that the two are linked. Many of the tools and techniques outlined above can also be used to access various timescales of motion. Most notably, DEA offers access to motions spanning the range of about 10^{-11}–10^4 s, the complimentary TSC approach offers access to motions which occur over timescales of roughly 20–300 s, and ssNMR provides insight into motions on the order of 10^{-11}–10^3 s (Ediger et al. 1996; Correia et al. 2001). In some instances, these approaches have shown some success in linking sub-T_g motions with crystallization tendency, but this remains an area of active research (Alie et al. 2004; Shmeis et al. 2004a, 2004b; Bhugra et al. 2007, 2008; Bhattacharya and Suryanarayanan 2009; Dantuluri et al. 2011; Bhardwaj and Suryanarayanan 2012a; Bhardwaj et al. 2013).

Given the versatility of current amorphous characterization techniques, it is clear that a range of resolutions and data can be generated on amorphous dispersions. As discussed previously, within a risk-based development approach, it becomes possible to triage testing to yield the appropriate balance of resolution and resource utilization so that the extruded product can be successfully positioned for commercialization.

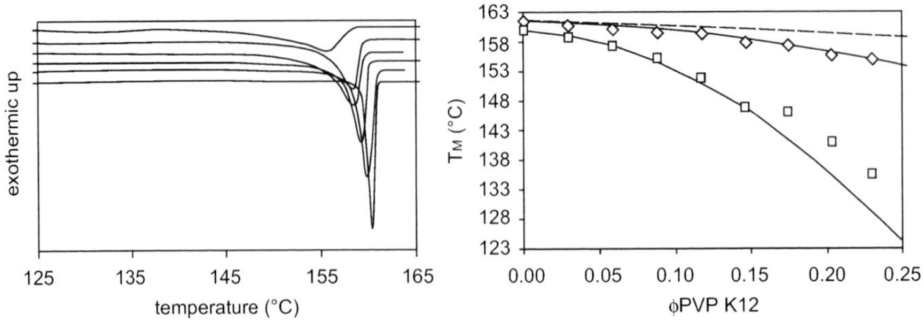

Fig. 6.7 Melting point depression as a function of drug loading for amorphous dispersions for indomethacin using onset (*square*) and offset (*diamond*) data. (Reproduced with permission from Marsac et al. 2009)

6.7 Mechanical Properties of Melt-Extruded Amorphous Dispersions

The mechanical properties of amorphous solid dispersions prepared by HME are an important yet often overlooked feature of these materials. The impact of an extrudate's mechanical properties is realized in further downstream processes such as particle size reduction and compaction.

The particle size distribution resulting from a milling operation is primarily determined by both the method of particle size reduction as well as the mechanical properties of the material such as fracture toughness, elastic modulus, and hardness. Thus, two extrudate samples with different mechanical properties milled under the same conditions will yield different particle size distributions. Beyond the intrinsic properties of the system, the mechanical behavior of extruded material is also affected by features of the bulk extrudate itself such as air bubbles, particle inclusions, or other defects that can increase the apparent brittleness of the material. Foamed extrudate, for example, could have different milling behavior as compared to a nonfoamed extrudate of the same composition.

The milled extrudate's particle size is often a critical quality attribute for the drug product performance for many reasons. It is well known that the dissolution rate of a particle is determined in part by the particle's size and surface area. For polymer-based materials such as extrudate, particle size can influence phenomena such as swelling and gelling, which may or may not be desirable for the product performance. Particle size may also affect powder flow in feeders and hoppers and can result in segregation risks that impact content uniformity in the final drug product.

In addition to particle size reduction, roller compaction and tableting are other downstream processes that will likely to be impacted by the mechanical properties of the milled extrudate. In the case of tableting, the extrudate may be subjected to localized high stresses which can induce particle breakage, elastic deformation, and/or plastic flow that affect compactability and tablet hardness.

Amorphous solid dispersions are prepared primarily with amorphous and/or semicrystalline materials, and therefore, the mechanical behavior of the extrudate is generally viscoelastic in nature. The materials' viscoelasticity implies a strain-rate dependence of the mechanical response and time-dependent mechanical behavior such as creep and stress relaxation. For example, in cases of high strain rates, these materials tend to be more brittle than under slower strain rates where viscous flow and other molecular relaxations can dissipate the energy without fracture. Thus, high strain rates are beneficial for particle size reduction operations.

Some of the important mechanical property descriptors of polymeric materials such as hot-melt extrudate are as follows:

- Elastic modulus: Stiffness, resistance to deformation, analogous to the spring constant in Hooke's law.
- Yield strength: The stress at which the behavior deviates from the linear elastic region and permanent plastic deformation is achieved.
- Ductility: The amount of plastic deformation that occurs before fracture.
- Fracture toughness: The resistance to fracture in the presence of a crack.
- Hardness: Resistance to localized plastic deformation.
- Creep modulus: A measure of the continued, time-dependent strain for a constant applied stress.

Mechanical testing of hot-melt extrudate can be performed on a variety of equipment typically used to test other types of materials. Loading configurations such as tensile, three-point bend, and cantilever deflection can assess different mechanical properties of the material under different stress states. If quasi-static methods are used, tests may be performed under different strain rates to assess viscoelastic effects of mechanical properties as discussed above. From a practical standpoint, the specimen tested should have uniform dimensions devoid of defects and ideally be of regular cross-sectional shape such as a circle or rectangle, enabling accurate determination of the cross-sectional area and the stress state for a given applied load. Perhaps the most ubiquitous device for testing polymers and therefore, hot-melt extrudate is the DMA. In addition to the ability to perform quasi-static tests in multiple loading configurations, a DMA can also test materials with oscillatory loading with varying frequency, temperatures, and even relative humidity for some models. The complex modulus obtained from a dynamic test can be separated into its elastic (storage modulus) and viscous (loss modulus) components. With the ability to ramp temperature during the test, changes in the mechanical properties can be assessed as a function of temperature and frequency, thus enabling not only temperature-dependent mechanical properties but also other sub-T_g relaxations that other techniques such as DSC may not be sensitive enough to detect. When amorphous polymers are heated through their glass transition, the elastic modulus can drop by a few orders of magnitude, and since the glass transition is related to changes in molecular dynamics, the transition temperature itself as measured with DMA will be a function of the applied strain rate with increases in T_g observed with increasing strain rates.

In addition to temperature, other environmental factors such as relative humidity can have a strong impact on the mechanical properties of the extrudate as many of

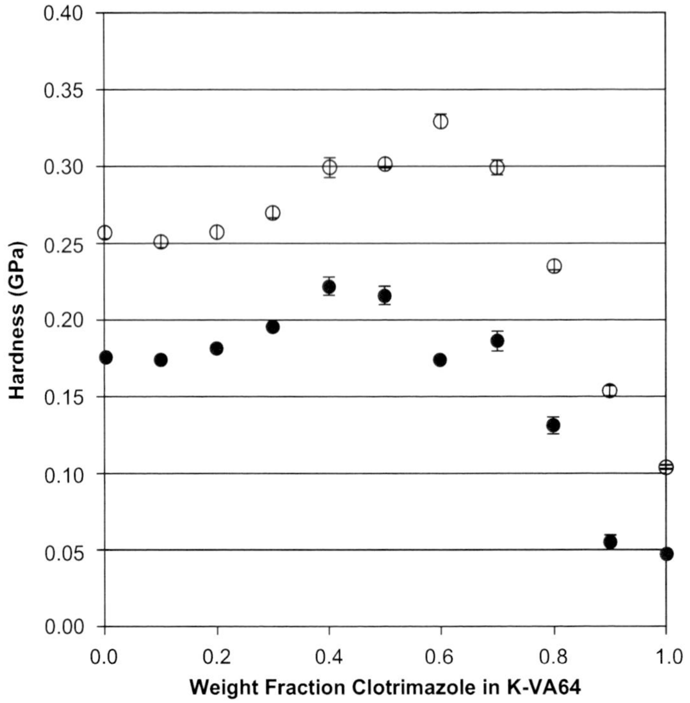

Fig. 6.8 Hardness versus weight fraction of clotrimazole in Kollidon VA 64 measured by nanoindentation at 18 % (o) and 49 % (o) relative humidity

the polymers used in the process tend to be hygroscopic. With some exceptions, water generally acts as a plasticizer for these polymers, lowering the glass transition temperature and reducing mechanical properties, such as modulus and hardness. It is important to control the relative humidity during mechanical tests performed on extrudates and also to be aware of the storage conditions the materials were exposed to prior to testing. For example, if extrudate is stored under desiccated conditions but tested at ambient laboratory conditions, the mechanical properties could change over time as the materials slowly absorb moisture. For accurate measurement and comparison between samples, it is recommended to equilibrate the materials at desired environmental conditions and then test at the same. Equilibration times will depend on the thickness of the samples and rates of moisture diffusion into and out of the sample.

A final consideration with respect to mechanical properties of hot-melt extrudate is the composition of the amorphous solid dispersion itself. Just as water content described above can impact the mechanical properties of the material, so can the other components such as plasticizers, surfactants, and the API itself. In one study, the effect of both API loading and humidity on the mechanical properties of amorphous solid dispersions was determined using nanoindentation and nanoDMA (Lamm et al 2012). This is illustrated in Fig. 6.8, where dispersions of clotrimazole and copovidone were tested, and it was found that adding the drug to the polymer actually increased the hardness and modulus of the materials up to approximately 50 % drug loading despite the fact that glass transition temperature decreased with increasing drug load. This phenomenon, known as anti-plasticization, can have significant impact on the materials performance in downstream processes as discussed above. For all the extrudate compositions tested, increasing humidity lowered the hardness and modulus of the dispersions, thus highlighting water's plasticizing effects on the dispersions.

Beyond this, there are also a limited number of examples describing mechanical properties that can be used to indirectly relate mechanical properties of extrudates to milling and compression performance. In a recent study, using three-point bend analysis of extruded parts, the modulus, yield strength, and toughness of materials were characterized with particular properties ascribed to the downstream processability of these materials (DiNunzio et al. 2012). For HPMCAS, a material known to be particularly challenging to mill, both brittle and ductile behavior was observed. As drug loading increased, the yield strength and toughness decreased; however, the modulus remained largely unchanged, explaining why drug loading may favorably influence milling performance of these systems. When compared to other extruded polymers, specifically copovidone and amino methacrylate copolymer, a significantly greater toughness is observed for HPMCAS that falls in line with the millability of these systems. Additional characterization of polyethylene glycol showed no brittle failure of the sample, only a continuous deformation. This behavior, unique to polyethylene oxide among the systems studied, illustrated why this material exhibits challenges during milling operations and can provide significant benefits for abuse-deterrent formulations.

6.8 Summary

Among process options for commercial amorphous solid dispersion generation, HME is often preferred due to its continuous nature, small manufacturing footprint, and lack of solvents. Preclinical development may require alternative processes, but these can often be transitioned to HME. Designing HME amorphous solid dispersions requires a thorough understanding of polymer, plasticizer, and surfactant selection, extrusion equipment design, and process parameters, guided by increasingly effective characterization tools to assure drug particle dispersal and dissolution into the matrix and stabilization throughout the shelf life of the product to finally deliver an effective dose to the patient. Furthermore, the limitations of the use of HME are quickly being overcome by application of formulation understanding and the use of supercritical fluids to allow processing of high melting point APIs with additional understanding of mechanical properties, leading to improved milling efficiency and compaction performance.

By building on the extensive product and process design experience of the polymer and food industries, pharmaceutical development is now able to add its unique considerations to incorporate HME as a core capability for commercial manufacturing. As will be described in a following chapter, HME can be rapidly scaled up based on product and process design space understanding after successfully demonstrating drug product quality in early development.

References

Albano A, Desai D et al (2012) Pharmaceutical composition. United States Patent & Trademark Office US20130172375 A1

Alie J, Menegotto J et al (2004) Dielectric study of the molecular mobility and the isothermal crystallization kinetics of an amorphous pharmaceutical drug substance. J Pharm Sciences 93:218–233.

Andrews GP, Jones DS et al (2008) The manufacture and characterisation of hot-melt extruded enteric tablets. Eur J Pharm Biopharm 69(1):264–273

Arkenau-Maric E, Bartholomaus J (2008) Process for the production of an abuse proofed dosage form, US 2008031197 A1, USPTO, pp 11.

Avella M, Martuscelli E et al (1991) Crystallization behaviour of poly (ethylene oxide) from poly (3-hydroxybutyrate)/poly (ethylene oxide) blends: phase structuring, morphology and thermal behaviour. Elsevier 32:1647–1653

Baird JA, Taylor LS (2012) Evaluation of amorphous solid dispersion properties using thermal analysis techniques. Adv Drug Deliv Rev 64:396–421

Baird JA, Van Eerdenbrugh B, et al (2010). A classification system to assess the crystallization tendency of organic molecules from undercooled melts. J Pharm Sci 99(9):3787–3806

Bhardwaj SP, Suryanarayanan R (2012a) Molecular mobility as an effective predictor of the physical stability of amorphous trehalose. Mol Pharm 9(11):3209–3217

Bhardwaj SP, Suryanarayanan R (2012b) Use of dielectric spectroscopy to monitor molecular mobility in glassy and supercooled trehalose. J Phys Chem B 116(38):11728–11736

Bhardwaj SP, Arora KK et al (2013) Correlation between molecular mobility and physical stability of amorphous itraconazole. Mol Pharm 10(2):694–700

Bhattacharya S, Suryanarayanan R (2009) Local mobility in amorphous pharmaceuticals—characterization and implications on stability. J Pharm Sci 98:2935–2953

Bhugra C, Pikal MJ (2008) Role of thermodynamic, molecular, and kinetic factors in crystallization from the amorphous state. J Pharm Sci 97(4):1329–1349

Bhugra C, Rambhatla S et al (2007) Prediction of the onset of crystallization of amorphous sucrose below the calorimetric glass transition temperature from correlations with mobility. J Pharm Sci 96(5):1258–1269

Bhugra C, Shmeis R et al (2008) Prediction of onset of crystallization from experimental relaxation times. II. Comparison between predicted and experimental onset times. J Pharm Sci 97:455–472

Booy M L (1978) Geometry of fully wiped twin-screw equipment. Polym Eng Sci 18(12):973–984

Breitenbach J (2002) Melt extrusion: from process to drug delivery technology. Eur J Pharm Biopharm 54(2):107–117

Brouwer T, Todd DB et al (2002) Flow characteristics of screws and special mixing enhancers in a co-rotating twin screw extruder. Int Polym Process 17(1):26–32

Carpenter J, Katayama D et al (2009) Measurement of T_g in lyophilized protein and protein excipient mixtures by dynamic mechanical analysis. J Therm Anal Calorim 95:881–884

Chiang P-C, Ran Y et al (2012) Evaluation of drug load and polymer by using a 96-well plate vacuum dry system for amorphous solid dispersion drug delivery. AAPS Pharm Sci Tech 13(2):713–722

Chokshi RJ, Sandhu HK et al (2005) Characterization of physico-mechanical properties of indomethacin and polymers to assess their suitability for hot-melt extrusion process as a means to manufacture solid dispersion/solution. J Pharm Sci 94(11):2463–2474

Correia NT, Alvarez C et al (2001) The beta–alpha branching in D-sorbitol as studied by thermally stimulated depolarization currents (TSDC). J Phys Chem B 105(24):5663–5669

Crowley MM, Zhang F et al (2007) Pharmaceutical applications of hot-melt extrusion: part I. Drug Dev Ind Pharm 33(9):909–926

Dantuluri AKR, Amin A et al (2011) Role of alpha-relaxation on crystallization of amorphous celecoxib above T_g probed by dielectric spectroscopy. Mol Pharm 8(3):814–822

DiNunzio JC, Miller DA (2013) Formulation development of amorphous solid dispersions prepared by melt extrusion. In: Repka MA, Langley N, DiNunzio J (eds) Melt extrusion: materials, technology and drug product design. Springer, New York, pp 161–204

DiNunzio JC, Brough C et al (2010) Fusion production of solid dispersions containing a heat-sensitive active ingredient by hot melt extrusion and Kinetisol® dispersing. Eur J Pharm Biopharm 74(2):340–351

DiNunzio JC, Zhang F et al (2012) Melt extrusion. In: Williams III RO, Watts AB, Miller DA (eds) Formulating poorly water soluble drugs. Springer, New York, pp 311–362

Doelker E (1993) Cellulose derivatives. In: Langer IR, Peppas N (eds) Biopolymers, vol 107. Springer, Berlin, pp 199–265

Dong, Zedong, Chatterji, Ashish, Sandhu, Harpreet, Choi, Duk Soon, Chokshi, Hitesh, Shah, Navnit (2008) Evaluation of solid state properties of solid dispersions prepared by hot-melt extrusion and solvent co-precipitation Int J Pharm 355(1–2):141–149

Ediger MD, Angell CA et al (1996) Supercooled liquids and glasses. J Phys Chem 100(31):13200–13212

Follonier N, Doelker E et al (1995) Various ways of modulating the release of diltiazem hydrochloride from hot-melt extruded sustained release pellets prepared using polymeric materials. J Control Release 36(3):243–250

Ghebremeskel A, Vemavarapu C et al (2007) Use of surfactants as plasticizers in preparing solid dispersions of poorly soluble API: selection of polymer–surfactant combinations using solubility parameters and testing the processability. Int J Pharm 328(2):119–129

Gogos C, Liu H (2012) Laminar dispersive and distributive mixing with dissolution and applications to hot-melt extrusion. In: Douroumis D (ed) Hot-melt extrusion: pharmaceutical applications. Wiley, New York, pp 261–284

Grace HP (1982) Dispersion phenomena in high viscosity immiscible fluid systems and application of static mixers as dispersion devices in such systems. Chem Eng Comm 14:225–277

Griffith RM (1962) Fully developed flow in screw extruders, industrial & engineering chemistry fundamentals. Ind Eng Chem Fundam 1(3):180–187

Hartshorn CM, Lee YJ et al (2013) Multicomponent chemical imaging of pharmaceutical solid dosage forms with broadband CARS microscopy. Anal Chem 85(17):8102–8111

Hikima T, Adachi Y et al (1995) Determination of potentially homogeneous-nucleation-based crystallization in O-terphernyl and an interpretation of the nucleation-enhancement mechanism. Phys Rev B 52(6):3900–3908

Hughey JR, DiNunzio JC et al (2010) Dissolution enhancement of a drug exhibiting thermal and acidic decomposition characteristics by fusion processing: a comparative study of hot melt extrusion and KinetiSol® dispersing. AAPS Pharm Sci Tech 11(2):760–774

Ishida H, Wu T et al (2007) Sudden rise of crystal growth rate of nifedipine near T_g without and with polyvinylpyrrolidone. J Pharm Sci 96(5):1131–1138

Ishikawa T, Amano T et al (2002) Flow patterns and mixing mechanisms in the screw mixing element of a co-rotating twin-screw extruder. Polym Eng Sci 42(5):925–939

Karabanova LV, Boiteux G et al (2008) Phase separation in the polyurethane/poly (2-hydroxyethyl methacrylate) semi-interpenetrating polymer networks synthesized by different ways. Polym Eng Sci 48(3):588–597

Karwe MV, Godavarti S (1997) Accurate measurement of extrudate temperature and heat loss on a twin-screw extruder. J Food Sci 62(2):367–372

Keen JM, Martin C et al (2013) Investigation of process temperature and screw speed on properties of a pharmaceutical solid dispersion using corotating and counter-rotating twin-screw extruders. J Pharm Pharmacol 66(2):204–217
Kestur US, Wanapun D et al (2012) Nonlinear optical imaging for sensitive detection of crystals in bulk amorphous powders. J Pharm Sci 101:4201–4213
Kohlgrüber K, Bierdel M (2008) Co-rotating twin-screw extruders: fundamentals, technology, and applications. Carl Hanser Publishers, Cincinnati
Lamm MS, Simpson A, McNevin M, Frankenfeld C, Nay R, Variankaval N (2012) Probing the effect of drug loading and humidity on the mechanical properties of solid dispersions with nanoindentation: antiplasticization of a polymer by a drug molecule. Mol Pharm 9(11):3396–3402
Lauer ME, Grassmann O et al (2011) Atomic force microscopy-based screening of drug-excipient miscibility and stability of solid dispersions. Pharm Res 28:572–584
Lauer ME, Siam M et al (2013) Rapid assessment of homogeneity and stability of amorphous solid dispersions by atomic force microscopy—from bench to batch. Pharm Res 30(8):2012–2022
Liu H, Wang P et al (2010) Effects of extrusion process parameters on the dissolution behavior of indomethacin in Eudragit® E PO solid dispersions. Int J Pharm 383(1–2):161–169
Liu H, Zhang X et al (2012) Miscibility studies of indomethacin and Eudragit® E PO by thermal, rheological, and spectroscopic analysis. J Pharm Sci 101:2204–2212
Lu Q, Zografi G (1998) Phase behavior of binary and ternary amorphous mixtures containing indomethacin, citric acid, and PVP, vol 15. Springer, pp 1202–1206
Mahieu A, Willart J-F et al (2013a) A new protocol to determine the solubility of drugs into polymer matrixes. Mol Pharm 10(2):560–566
Mahieu A, Willart J-F et al (2013b) On the polymorphism of griseofulvin: identification of two additional polymorphs. J Pharm Sci 102:462–468
Marsac PJ, Li T et al (2009) Estimation of drug-polymer miscibility and solubility in amorphous solid dispersions using experimentally determined interaction parameters, vol 26. Springer, pp 139–151
Marsac P J, Rumondor ACF et al (2010) Effect of temperature and moisture on the miscibility of amorphous dispersions of felodipine and poly (vinyl pyrrolidine). J Pharm Sci 99(1):169–185
Marsac PJ, Shamblin SL et al (2006) Theoretical and practical approaches for prediction of drug: polymer miscibility and solubility. Pharm Res 23:2417–2426
Marsac PJ, Taylor LS, Hanmi X, Lisa B, Zhen L, Hang L (2012) A novel method for accessing the enthalpy of mixing active pharmaceutical ingredients with polymers. American Association for Pharmaceutical Scientists National Meeting, Chicago
McCrum NG, Buckley CP et al (1997) Principles of polymer engineering. Oxford University Press, New York
Nalawade SP, Picchioni F et al (2006) Supercritical carbon dioxide as a green solvent for processing polymer melts: processing aspects and applications. Prog Polym Sci 31(1):19–43
Newman A, Engers D et al (2008) Characterization of amorphous API: polymer mixtures using X-ray powder diffraction. J Pharm Sci 97:4840–4856
Noyes AA, Whitney WR (1897) The rate of solution of solid substances in their own solutions. J Am Chem Soc 19:930–934
Oshlack B, Wright C et al (2001) Tamper-resistant oral opioid agonist formulations. United States 36
Patterson JE, James MB et al (2005) The influence of thermal and mechanical preparative techniques on the amorphous state of four poorly soluble compounds. J Pharm Sci 94(9):1998–2012
Patterson JE, James MB et al (2007). Preparation of glass solutions of three poorly water soluble drugs by spray drying, melt extrusion and ball milling. Int J Pharm 336(1):22–34
Patterson JE, James MB et al (2008) Melt extrusion and spray drying of carbamazepine and dipyridamole with polyvinylpyrrolidone/vinyl acetate copolymers. Drug Dev Ind Pharm 34(1):95–106

Pham TN, Watson SA et al (2010) Analysis of amorphous solid dispersions using 2D solid-state NMR and 1H T 1 relaxation measurements, vol 7. ACS Publications, New York, pp 1667–1691
Pikal MJ, Dellerman KM (1989) Stability testing of pharmaceuticals by high-sensitivity isothermal calorimetry at 25 C: cephalosporins in the solid and aqueous solution states. Int J Pharm 50:233–252
Power G, Vij JK et al (2007) Dielectric relaxation and crystallization of nanophase separated 1-propanol-isoamylbromide mixture. J Chem Phys 127:094507
Price R, Young PM (2004) Visualization of the crystallization of lactose from the amorphous state. J Pharm Sci 93:155–164
Qian F, Huang J et al (2010) Is a distinctive single Tg a reliable indicator for the homogeneity of amorphous solid dispersion? Int J Pharm 395:232–235
Rauwendaal C (1998) Polymer mixing: a self study guide. Hanser Gardner Publications, Cincinnati
Rauwendaal C (ed) (2002) Polymer extrusion, 4th edn. Hanser Gardner Publications, Cincinnati
Rumondor ACF, Marsac PJ et al (2009) Phase behavior of poly (vinylpyrrolidone) containing amorphous solid dispersions in the presence of moisture. Mol Pharm 6:1492–1505
Rumondor ACF, Taylor LS (2010) Application of partial least-squares (PLS) modeling in quantifying drug crystallinity in amorphous solid dispersions. Int J Pharm 398:155–160
Rumondor ACF, Wikström HK et al (2011) Understanding the tendency of amorphous solid dispersions to undergo amorphous-amorphous phase separation in the presence of absorbed moisture, vol 12. Springer, Berlin, pp 1209–1219
Schenck L, Troup GM et al (2011) Achieving a hot melt extrusion design space for the production of solid solutions. Chem Eng Pharm Ind 14(3):1034–1044
Schilling S, Shah N et al (2007) Citric acid as a solid-state plasticizer for Eudragit RS PO. J Pharm Pharmacol 59(11):1493–1500
Shah B, Kakumanu VK et al (2006) Analytical techniques for quantification of amorphous/crystalline phases in pharmaceutical solids. J Pharm Sci 95:1641–1665
Shamblin SL, Hancock BC et al (2006) Coupling between chemical reactivity and structural relaxation in pharmaceutical glasses, vol 23. Springer, Berlin, pp 2254–2268
Shmeis RA, Wang Z et al (2004a) A mechanistic investigation of an amorphous pharmaceutical and its solid dispersions, part I: a comparative analysis by thermally stimulated depolarization current and differential scanning calorimetry, vol 21. Springer, Berlin, pp 2025–2030
Shmeis RA, Wang Z et al (2004b) A mechanistic investigation of an amorphous pharmaceutical and its solid dispersions, part II: molecular mobility and activation thermodynamic parameters, vol 21. Springer, Berlin, pp 2031–2039
Sinclair W, Leane M et al (2011) Physical stability and recrystallization kinetics of amorphous ibipinabant drug product by Fourier transform Raman spectroscopy. J Pharm Sci 100:4687–4699
Singh S, Gamlath S et al (2007) Nutritional aspects of food extrusion: a review. Int J Food Sci Technol 42(8):916–929
Strachan CJ, Windbergs M et al (2011) Pharmaceutical applications of non-linear imaging. Int J Pharm 417:163–172
Sun Y, Xi H et al (2008a) Crystallization near glass transition: transition from diffusion-controlled to diffusionless crystal growth studied with seven polymorphs. J Phys Chem B 112(18):5594–5601
Sun Y, Xi H et al (2008b) Diffusionless crystal growth from glass has precursor in equilibrium liquid. J Phys Chem B 112(3):661–664
Sun Y, Xi H, Ediger MD, Richert R, Yu L (2009) Diffusion-Controlled and "Diffusionless" Crystal Growth near the Glass Transition Temperature: Relation between Liquid Dynamics and Growth Kinetics of Seven ROY Polymorphs. J Chem Phys, pp 131
Sun Y, Tao J et al (2010) Solubilities of crystalline drugs in polymers: an improved analytical method and comparison of solubilities of indomethacin and nifedipine in PVP, PVP/VA, and PVAc. J Pharm Sci 99(9):4023–4031
Suzuki H, Sunada H (1998) Influence of water-soluble polymers on the dissolution of nifedipine solid dispersions with combined carriers. Int J Pharm 303(1–2):54–61

Szczepanski CR, Pfeifer CS et al (2012) A new approach to network heterogeneity: polymerization induced phase separation in photo-initiated, free-radical methacrylic systems. Polymer 53:4694–4701

Tadmor Z, Gogos CG (2006) Principles of polymer processing. Wiley-Interscience, Hoboken

Tao J, Sun Y et al (2009) Solubility of small-molecule crystals in polymers: d-Mannitol in PVP, indomethacin in PVP/VA, and nifedipine in PVP/VA. Pharm Res 26(4):855–864

Taylor LS, Zografi G (1997) Spectroscopic characterization of interactions between PVP and indomethacin in amorphous molecular dispersions. Pharm Res 14(12):1691–1698

Tian Y, Booth J et al (2009) Construction of drug-polymer thermodynamic phase diagrams using Flory–Huggins interaction theory: identifying the relevance of temperature and drug weight fraction to phase separation within solid dispersions. Mol Pharm 10(1):236–248

Todd DB (1998) Introduction to compounding. Polymer Processing Institute Books from Hanser Publishers, D. B. Todd. Hanser/ Gardner Publications, Inc., Cincinnati, pp 1–12

Toth SJ, Madden JT, Taylor LS, Marsac P, Simpson GJ (2012) Selective Imaging of Active Pharmaceutical Ingredients in Powdered Blends with Common Excipients Utilizing Two-Photon Excited Ultraviolet-Fluorescence and Ultraviolet-Second Order Nonlinear Optical Imaging of Chiral Crystals Anal Chem 84(14):5869–5875

Verreck G, Decorte A et al (2006a) The effect of pressurized carbon dioxide as a plasticizer and foaming agent on the hot melt extrusion process and extrudate properties of pharmaceutical polymers. J Supercrit Fluids 38(3):383–391

Verreck G, Decorte A et al (2006b) Hot stage extrusion of p-amino salicylic acid with EC using CO_2 as a temporary plasticizer. Int J Pharm 327(1–2):45–50

Verreck G, Decorte A, Heymans K, Adriaensen J, Liu D, Tomasko DL, Arien A, Peeters J, Rombaut P, Van den Mooter G, Brewster ME (2007) The effect of supercritical CO_2 as a reversible plasticizer and foaming agent on the hot stage extrusion of itraconazole with EC 20 cps. J Supercrit Fluids 40(1):153–162

Verreck G, Six K et al (2003) Characterization of solid dispersions of itraconazole and hydroxypropylmethylcellulose prepared by melt extrusion—part I. Int J Pharm 251(1–2):165–174

Vyazovkin S, Dranca I (2007) Effect of physical aging on nucleation of amorphous indomethacin. J Phys Chem 111:7283–7287

Work WJ, Hess KHM et al (2004) Definitions of terms related to polymer blends, composites, and multiphase polymeric materials. Pure Appl Chem 76(11):1985–2007

Wanapun D, Kestur US et al (2010) Selective detection and quantitation of organic molecule crystallization by second harmonic generation microscopy. Anal Chem 82:5425–5432

Wanapun D, Kestur US et al (2011) Single particle nonlinear optical imaging of trace crystallinity in an organic powder. Anal Chem 83:4745–4751

Wang X, Michoel A et al (2005) Solid state characteristics of ternary solid dispersions composed of PVP VA64, Myrj 52 and itraconazole. Int J Pharm 303(1–2):54–61

Wu T, Yu L (2006) Origin of enhanced crystal growth kinetics near T_g probed with indomethacin polymorphs. J Phys Chem B 110(32):15694–15699

Yang M, Wang P et al (2011) Determination of acetaminophen's solubility in poly(ethylene oxide) by rheological, thermal and microscopic methods. Int J Pharm 403(1–2):83–89

Zhu Q, Harris MT et al (2013) Modification of crystallization behavior in drug/polyethylene glycol solid dispersions. Mol Pharm 9:546–553

Chapter 7
HME for Solid Dispersions: Scale-Up and Late-Stage Development

Chad Brown, James DiNunzio, Michael Eglesia, Seth Forster, Matthew Lamm, Michael Lowinger, Patrick Marsac, Craig McKelvey, Robert Meyer, Luke Schenck, Graciela Terife, Gregory Troup, Brandye Smith-Goettler and Cindy Starbuck

7.1 Introduction to Commercialization of Extruded Dispersions

Solid dispersions are typically produced using nonextrusion-based approaches to support early screening. Small, often customized extruders can be used to produce more representative materials with compositions identified from these early screening experiments. The throughput and yields obtained from small twin-screw extruders (e.g., < 20 mm D_o) are well suited to supply early clinical studies. Scale-up from these small pilot-scale extruders is often facilitated by the geometric similarity of extruders in this size range (although differences in design from one manufacturer to another can make this more challenging).

Several key design principles are critical including product degradation temperature, feed configurations, screw/barrel layout, and process control elements. Product temperature can be utilized as a scale-independent attribute and is an important parameter in design space definition because it is generally related to amorphous conversion and thermal degradation. Several methods are described in Table 7.1 for measuring product temperature. Immersion probes and infrared (IR) methods provide the best translation across scales and line of sight for implementation in commercial extrusion lines (Godavarti and Karwe 1997; Karwe and Godavarti 1997; Baugh et al. 2003). While product temperature can be utilized throughout development as a key scale-independent attribute, one must exercise caution when trying to relate measured process temperatures to thermal stability. At small scale, users may utilize relatively low product flow rates in an effort to conserve material during process

J. DiNunzio (✉)
Pharmaceutical Sciences & Clinical Supplies, Merck & Co., Inc.,
556 Morris Ave, Summit, NJ 07901, USA Tel.: 908-473-7329
e-mail: james.dinunzio@merck.com

C. Brown · M. Eglesia · S. Forster · M. Lamm · M. Lowinger · P. Marsac · C. McKelvey · R. Meyer · L. Schenck · G. Terife · G. Troup · B. Smith-Goettler · C. Starbuck
Merck & Co., Inc., Whitehouse Station, NJ, USA

N. Shah et al. (eds.), *Amorphous Solid Dispersions*,
Advances in Delivery Science and Technology, DOI 10.1007/978-1-4939-1598-9_7

Table 7.1 Product temperature measurement methods for extruded dispersions

	Advantages	Disadvantages
Handheld	Accurate Inexpensive	No continuous monitoring Not applicable to manufacturing Invasive User bias
Flush mount	Continuous monitoring Applicable to manufacturing Noninvasive No user bias	Biased by conduction from barrel
Immersion	May be accurate Continuous monitoring Applicable to manufacturing Relatively inexpensive No user bias	May disrupt flow stream Shear effects
IR	Accurate Continuous monitoring Enhanced spatial resolution across flight Applicable to manufacturing Noninvasive No user bias	Requires method development Environmental factors
Thermal imaging	May be accurate Continuous monitoring Noninvasive	Expensive User bias Environmental factors

development and/or prototype production. These low-flow rates result in longer residence times which can lead to thermal degradation. Challenges with thermal stability can be improved upon scale-up as shorter residence times are achievable.

One-dimensional modeling software such as WinTXS™ and Ludovic® (Carneiro et al. 2000) can be utilized to further probe the interaction between residence time, product temperature, and thermal degradation across scales.

There are several design and geometry considerations to keep in mind as one progresses through the product development life cycle toward commercialization of a hot-melt extrusion (HME) product. Early in development at the small/pilot scale, it may be necessary to preblend feed components to ensure homogenous and consistent material supply to the extruder. This is particularly relevant while active pharmaceutical ingredient (API) attributes are still under investigation and prone to change from lot to lot. A variety of feeders can be selected to accurately feed materials with a wide range of flow characteristics. In addition, as a last resort, the cost of preblending feed materials for a low- to mid-volume product is likely economically beneficial over constraining the API design space to produce material required for independent feed streams. Once the final API attributes are identified, thoughtful feeder studies can be conducted for appropriate feeder selection. Laboratory scale tests of physical properties such as particle size distribution, Flodex, bulk density, and Carr's Index can be utilized to help guide appropriate feeder selection.

Once a feed strategy is identified, appropriate screw configuration and barrel design for commercial extrusion must be identified (Schenck et al. 2011). As a process is scaled from pilot to commercial scale, often a need for more efficient moisture removal is identified. This requires optimal placement and design of melt seal sections near the feed zone to prevent premature plasticization of the feedstock by water vapor. In addition, vents must be placed accordingly along the extruder barrel with

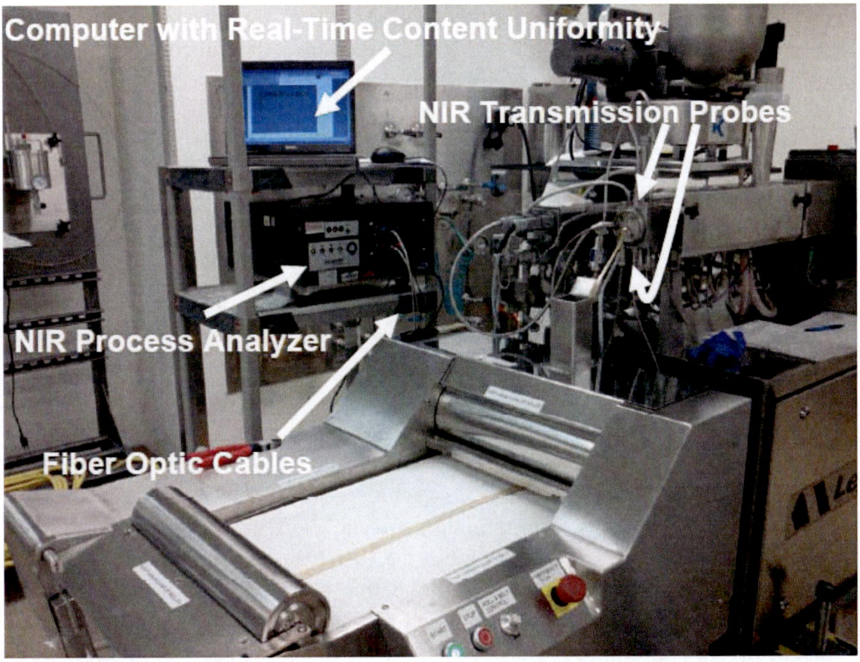

Fig. 7.1 Integration of in-line monitoring into melt extrusion production line. (Bigert and Smith-Goettler 2011)

material experiencing necessary residence time under the vents to efficiently remove residual moisture in the system. Finally, upon scale-up to larger extruders, there is less reliance on thermal energy from the barrel to melt the material as mechanical energy imparted by the screw becomes a more dominant source of energy input. At pilot scale, a particular screw profile may be adequate to provide mixing of components once melted via thermal energy input (e.g., via the barrel walls) for a wide range of formulations. However, at commercial scale, greater concern over the aggressiveness of the downstream melting/mixing section is needed to ensure the process provides a quality solid solution over an adequate design space.

Process analytical technology (PAT) can be developed early in development with a line of sight to commercial process control (ICH Q11 2011a). Easily integrated into the process equipment, as shown in Fig. 7.1, the application of in-line probes can provide unique opportunities to monitor processes in line. Spectroscopic techniques such as ultraviolet/visible (UV/VIS), near-infrared (NIR), and Raman can be utilized in line at the extruder die to ensure critical quality attributes (CQAs) such as composition and amorphous conversion of the API. Validation of these techniques at the pilot scale and transfer to the commercial scale allow for efficient manufacture of commercial material in a continuous manner and the implementation of a process control strategy that enables material outside the design space to be diverted to waste in real time. The successful development and validation of appropriate process analytical tools within the control strategy framework allows the full commercial benefit of utilizing extrusion to produce solid dispersions to be realized.

7.2 Integrating Melt Extrusion into Drug Substance Manufacturing

Beyond dispersion compounding, melt extrusion provides a platform for completing a number of unit operations, including devolatilization. Noting that one of the last steps in API manufacturing is drying, this offers an enticing opportunity to improve manufacturing efficiency by concurrently conducting compounding and drying operations. An efficient process for removing solvent from hard to dry APIs can be developed by integrating the drug substance and drug product processes with HME devolatilization. Drying of stoichiometric solvates with bound solvent, nonstoichiometric solvents with unbound solvent, or APIs with elevated levels of unbound, high-boiling residual solvent to the International Conference on Harmonization (ICH) levels can take an extended period of time in traditional API drying equipment such as pressure filters, agitated filter dryers, and Summix dryers (ICH Q11 2011a). In addition to higher manufacturing costs at larger production scales, these long-drying cycle times can lead to greater thermal product degradation, reduced API particle size, and poor API flowability; all of which may negatively impact final drug product CQAs. For drug products that already use HME to form amorphous solids, HME devolatilization offers a potential optimal solution to these challenging drug substance drying issues.

7.2.1 HME Devolatilization

HME devolatilization is an established process within other nonpharmaceutical industries such as the plastics industry, where it is performed to remove or recover solvent and monomers, to improve the properties of the polymer, and to increase the extent of polymerization (Albalak 1996). In the pharmaceutical field, the use of devolatilization and concurrent amorphous dispersion production can help to realize significant manufacturing efficiencies using the HME platform. HME can overcome the mass transfer limited, falling rate drying regime which handicaps the traditional API drying process by increasing the drying rate through the use of high temperatures, melt surface renewal, vacuum, and short residence times. APIs, even with thermal degradation concerns, can be processed via HME at temperatures in excess of 150 °C because materials are only exposed to these temperatures for a short period of time.

Utilizing HME to devolatilize APIs may allow for the delivery of API solvates or APIs with high-boiling residual solvents as the amount of residual solvent in these

APIs is well within the devolatilization capabilities of hot-melt extruders. Delivering API solvates to the drug product process can also result in improved product quality as the isolation of API solvates can yield improved purity profiles, better flow properties, and compressibility (Fachaux et al. 1993; Wirth and Stephenson 1997; Che et al. 2010). Additionally, drug substance process cycle time improvements may be seen when delivering APIs with low levels of high-boiling solvents to the HME drug product process as lengthy solvent distillations are typically performed to remove these solvents prior to the final API isolation step in traditional API drying equipment.

The amount of solvent that can be removed during HME is dependent on a number of extruder design and extrusion process variables. While some of these variables, such as whether a single-screw or twin-screw extruder is used, are fixed by the equipment available to the user, the HME process can be configured and optimized to remove solvents with a wide range of boiling points down to ICH levels. Mathematically described in Eq. 7.1, devolatilization efficiencies greater than 90 % of the solvent removed can be seen through the manipulation of the number of extruder vents used, proper screw profile design, and the optimization of extruder process parameters such as mass throughput, screw speed, temperature, and pressure (Biesenberger and Kessidis 1982):

$$E_F = \frac{1 - \frac{w_f}{w_o}}{1 - \frac{w_e}{w_o}} \tag{7.1}$$

where E_F, the devolatilization efficiency, can be calculated by analytically measuring the volatile content in the extruder feed, w_o, and in the extrudate, w_f, and through the assumption that the equilibrium volatile mass fraction, w_e, is zero at the high temperatures used during HME.

7.2.2 *Extrusion Process Parameters and Raw Material Properties for Devolatilization*

The drying rate (d*M*/d*t*) and devolatilization efficiency is dependent upon the HME operating conditions used to form the amorphous API solid dispersion. This can be seen in Eqs. 7.2–7.4 which describe the mass transfer limited drying seen in HME devolatilization. More surface area (*A*) for mass transfer will be generated when the melt is exposed to vacuum which results in foam/bubble generation (Albalak et al. 1990). More solvent will be removed when extruding at higher temperatures as this lowers the viscosity of the melt (μ), raises the diffusion coefficient (*D*), and increases the vapor pressure of the volatile material. Placing the vent(s) under vacuum will increase the pressure difference (d*P*) between the solvent vapor pressure and the operating pressure thereby increasing devolatilization efficiency. Running with higher screw speeds will increase melt surface renewal ensuring that the melt surface is continuously enriched with volatiles in a very thin melt surface layer (d*x*):

$$\frac{dM}{dt} = -\frac{M_w}{RT} DA \frac{dP}{dx} \tag{7.2}$$

$$D = \frac{kT}{6\pi \ r_o \eta} \tag{7.3}$$

$$\eta = \eta_0 \exp\left(\frac{E}{RT}\right) \tag{7.4}$$

where A is the heat or mass transfer surface area (m^2), D is the diffusion coefficient (m^2/s), D_o is the maximum diffusion coefficient at infinite temperature (m^2/s), E is the activation energy (J/mol), h_y is the heat transfer coefficient (W/m^2K), k is Boltzmann's constant (J/K), M is the mass of the volatile component (g), M_w is the molecular weight of the volatile component, P is pressure (N/m^2), r_o is radius of the volatile (m), R is the universal gas constant (J/mol K), t is time (s), T is the temperature (K), T_b is the boiling point (K) of the volatile component, x is the mass transfer distance or melt thickness (m), λ_i is the latent heat of vaporization (J/g), η is viscosity (kg/s m), and η_0 is the Arrhenius viscosity coefficient (kg/s m).

Raw material properties are also an important factor to consider for HME devolatilization operations. Using lower molecular weight polymers or adding plasticizers like surfactants will lower the melt viscosity and improve the rate of mass transfer. Meeting the targeted end of drying residual solvent levels will be more difficult when dealing with higher boiling solvents or when trying to remove ICH class I (benzene, carbon tetrachloride, etc.) or class II (acetonitrile, toluene, dichloromethane, etc.) solvents which have much lower permitted daily exposure limits than ICH class III solvents such as acetone, ethanol, and heptane.

7.2.3 Extruder Design for Devolatilization

The efficiency of the devolatilization step will be impacted by the type of extruder used and the design of the venting system. Higher devolatilization efficiencies will be seen in twin-screw extruders (TSEs) compared to single-screw extruders (SSEs) as TSEs provide greater mixing and melt surface renewal capabilities. Greater melt surface renewal will lead to higher mass transfer rates and better drying of the melt. Extruders can also be designed as single venting stage or multiple venting stage systems. Increasing the length of the vented screw sections will increase the area (A in Eq. 7.2) for devolatilization. The use of a multi-vent design also allows for staged venting (i.e., different operating pressures in each vented section) or the addition of stripping agents such as water, nitrogen, or supercritical carbon dioxide (CO_2), which can enhance devolatilization efficiency through the generation of additional mass transfer surface area.

Extruder screw profile design is critical to HME devolatilization. For optimal devolatilization performance, the screw must completely melt the material fed into the extruder in a mixing zone of the screw, prior to the first vent zone, and this mixing zone must be filled with molten material. This melt seal upstream of the first vent zone will ensure that any vacuum applied to the first vent zone will not leak past the first mixing zone and pull powders from the feed zone up into the vent. The melt seal downstream of the first vent zone can be formed by adding another mixing

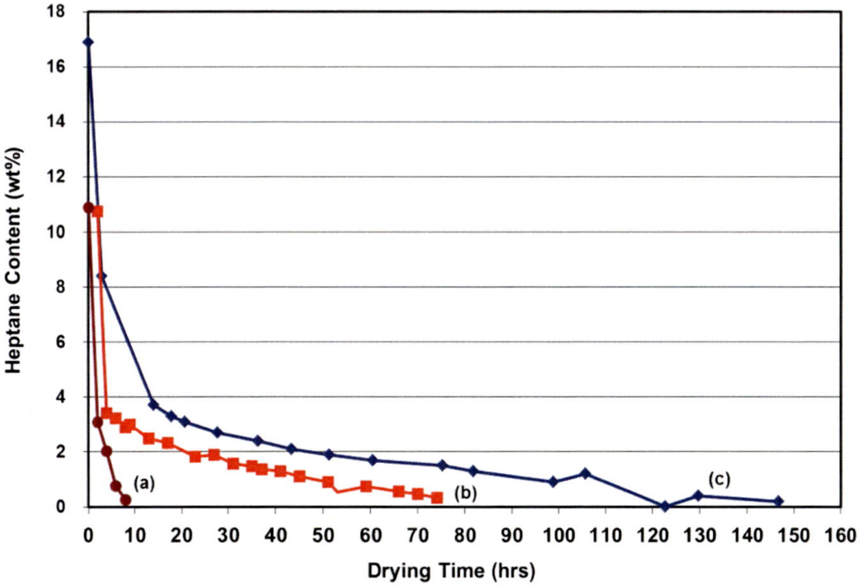

Fig. 7.2 Drying of a nonstoichiometric heptane solvate in a Summix dryer with **a** continuous agitation throughout drying **b** continuous agitation to 3 wt% residual heptane and then intermittent agitation afterwards and **c** intermittent agitation throughout drying

section or through the die pressure build section at the end of the screw. The material residence time within the mixing sections of the screw must be optimized for the devolatilization process. Short mixing sections with minimal residence time will not impart enough shear to completely melt the feed, preventing melt seal formation. Mixing sections that are too lengthy will cause material to backflow into the feed zone or vent(s) as mixing elements have very little (if any) forward conveying capability. The conveying screw elements chosen for the vented sections of the screw should have low degrees of fill when the HME process is operating at targeted flow rates. This will help prevent molten material from flowing into the vent(s) and will allow for proper expansion of the surface of the molten material when it is exposed to the lower operating pressures seen in these vented screw sections.

7.2.4 Integrated Drug Substance and Drug Product Processing

Figure 7.2 shows the drying of three batches of a nonstoichiometric API heptane solvate in a Summix dryer with varying amounts of agitation used during the drying cycle. These batches were approximately of the same size and similar temperatures and vacuum levels were used during drying. Going from intermittent to continuous agitation increased the mass transfer rate and decreased the amount of time needed to

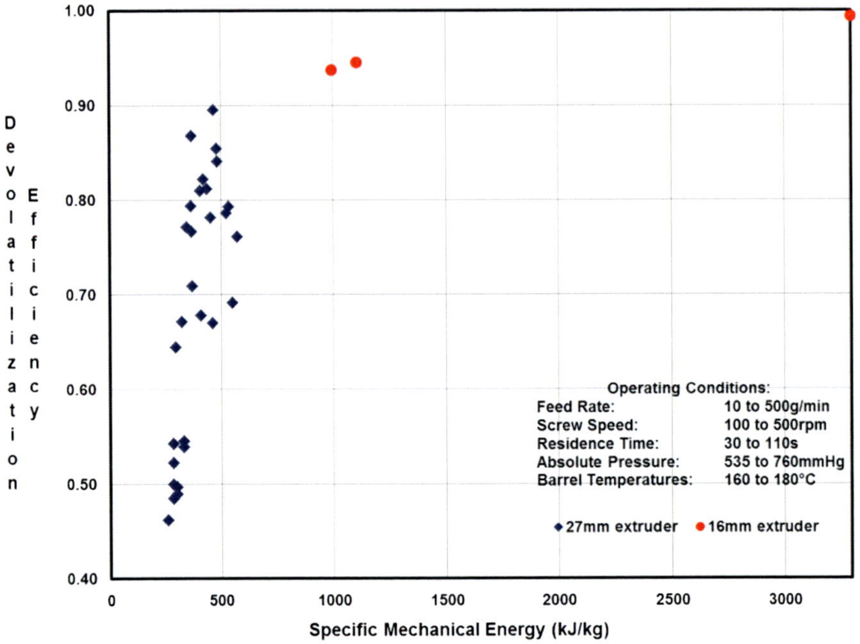

Fig. 7.3 Hot-melt extrusion devolatilization efficiencies for drying of nonstoichiometric heptane solvate with various HME operating conditions

reach the target residual solvent level (< 0.5 wt%) from approximately 150 h to 10 h. When the continuous agitation drying process was scaled up, however, a cohesive, poorly flowing API powder was produced which could not be easily fed into the downstream HME step.

An improved production process was developed by integrating the API drying and HME steps. The heptane solvate was dried down to approximately 3 wt% residual heptane in the Summix dryer as this part of the drying process was robust and scalable and took slightly less than 5 h to complete. The 3 wt% heptane solvate was then mixed with a polymer and a surfactant and desolvated in a corotating TSE. Figure 7.3 shows the HME devolatilization efficiencies seen at the 16-mm and 27-mm corotating TSE scales. Devolatilization efficiencies are directly correlated with the specific mechanical energy (SME) input from the extruder to the material fed into the extruder. Higher SME input levels are seen with the smaller TSE and this leads to 100 % solvent removal at the small scale. Devolatilization efficiencies of 0.45–0.90 were seen at the 27-mm scale with SME levels of 260–575 kJ/kg. Higher devolatilization efficiencies can be achieved at a given SME input level through any combination of raising the product temperature in the vent zones, lowering the vent zone pressures, or increasing screw speeds. Assuming geometrically similar screw profiles are used, maintaining a constant specific feed rate when scaling up will give a similar degree of fill within the conveying elements of the vented sections of the screw and help prevent material vent flow upon scale-up.

Table 7.2 Production rate and equipment sizes at various scales

Product stage	Scale	Screw outer diameter (mm)	Production rate (kg/hr)	Typical limitations	Manufacturers
Phase I	Micro	3–12	< 0.1–1	Torque, material feeding, lack of geometric similarity, short L/D, long residence time	Brabender, Steer, Thermo, ThreeTec, custom manufacturers
Phase II and biocomparison studies	Small	16–20	1–10	Torque, long residence time at low-flow rates	Thermo, Leistritz, Coperion Thermo, Leistritz, Coperion
Phase III and launch	Intermediate	24–30	10–50	Material handling such as refilling, feeding	Thermo, Leistritz, Coperion
Supply	Large	40–50	50–250	Heat transfer, venting, product cooling	Leistritz, Coperion

This integrated process resulted in a more robust and scalable process compared to the traditional continuous agitation API drying process and this will help ensure that all final drug product CQAs are met. Additionally, no increase in HME cycle time was needed to achieve the target residual heptane level (<0.5 wt% heptane relative to the weight of API in the amorphous solid dispersion) resulting in a reduction of at least 5 h of traditional API drying cycle time as shown in Fig. 7.2 (5 h needed to go from 3 wt% residual heptane to <0.5 wt% with continuous agitation throughout drying). This is possible, even with the shortened material residence times within the 27-mm extruder (30–110 s as shown in Fig. 7.3), due to the optimization of the extruder temperature, pressure, and screw speeds.

7.3 Scale-Up of Extrusion Operations

When considering the scale-up of the extrusion process, it is instructive to first consider the scale of production which might be needed in the course of drug development, and the scale of equipment needed to reach that level of production. Because extrusion is an inherently continuous process, the level of production can be classified in terms of mass of extrudate produced per unit time, whereas the equipment used to achieve a given production rate is primarily classified by the outer diameter of the screw cross section. Table 7.2 lists typical production rates and equipment sizes for various product stages encountered in pharmaceutical development.

7.3.1 Production Considerations at Different Scales

Because HME is a continuous process with wide flexibility in terms of production rate and time, scale-up does not need to happen very often. Generally, only two equipment scales would be required, with additional scales being useful to minimize raw material usage during the development process.

For early phase formulation screening experiments around the time of phase I, a typical flow rate might be 1–10 g/min, which can be achieved on machines ranging from 3 to12 mm in diameter. Although numerous manufacturers advertise equipment on the larger end of the size range, for screw diameters of 10 mm or less, there are few "off-the-shelf" options commercially available. At this scale of production, it is difficult to produce a product in a manner that is completely analogous to larger scales. Because of the size of powders relative to the size of equipment, it can be difficult to even deliver the material into the TSE without overfilling the feed screws, a condition known as flood feeding. Furthermore, at very low rates, loss-in-weight feeders begin to lose accuracy because scale fluctuations are large relative to the weight loss being measured. The torque limit of the extruder screws is also quickly reached, further limiting mass throughput. Finally, due to the small size (e.g., length:diameter ratio/L:D), there are fewer barrel segments, and thus barrel temperatures cannot be as finely controlled compared to larger scales. For these reasons, formulation scientists will often move directly into small-scale manufacturing if enough material is available.

Small-scale manufacturing using 16–20 mm diameter screws is often the easiest, because the 1–10 kg/h production rate is easy to handle, and the equipment size is on par with the size of those operating the equipment. Operating conditions can be quickly tested, but care must be taken that the conditions chosen for production have analogues at larger scales, a topic that will be covered later in the chapter. As with the micro scale, machine torque or flood-feeding limits are often what restrict overall production.

At the intermediate scale encountered during phase III and product launch, production rates grow to around 10–50 kg/h, and a screw diameter in the 24–30 mm range is typical. The systems required to operate the TSE process become more complex, with operations often switching from a single feed of preblended materials to multiple raw material feed streams entering the extruder at multiple feed ports. Adding to the complexity, raw material refill systems must be operated simultaneously without disrupting the extruder feeding process, and the quenching of molten extrudate after exiting the extruder becomes an important consideration. Due to the lower barrel surface area available compared to smaller scales, venting, devolatilization, and heat transfer become more difficult, thus requiring careful consideration during process development at the small scale. Because production times grow longer, phenomena like product buildup in feeders, vents, and around the die can cause intermittent disruptions or even product degradation. Product temperature rise generally occurs above what is seen at small scale, because the shear rate is typically higher due to the faster rotational speed. Ultimately, production is limited by whatever process occurs at the slowest rate, from refilling, feeding, extrusion, or product quenching.

During large-scale commercialization and supply, a simple analysis can provide insight into the maximum production rate which might be needed. An extrudate drug loading of 20 % and a dose of 100 mg would lead to 500 mg of extrudate per dose, and a high volume product may sell 2 billion doses per year, equivalent to 1 million kg of extrudate. Because extruders are normally operated around the clock, with periodic shutdowns for cleaning and maintenance, the maximum operating time per year would be about 6700 h, resulting in a required average production rate of 150 kg/h. Because this figure is well within the capacity of a 40–50 mm extruder, we need not consider larger equipment except in exceptional circumstances. For smaller volume products, on the order of 10–100 million doses per year, intermediate scale equipment can readily meet production demand, and parallel intermediate scale machines are often considered in lieu of a single large-scale machine if there is uncertainty in the product demand or when multiple products must share extrusion equipment. Although the same scale-up principles apply broadly, it is expected that in most circumstances, pharmaceutical products will not require equipment larger than the intermediate scale.

7.3.2 Tools for Assessing Scale-Up

Before considering the different methods used to scale-up, the objective of the extrusion process deserves to be revisited. In asking the question "What should our extruder do?" it is obvious that the intent is to mix API, polymer, and surfactant, if present. Oftentimes, the desired result is a single phase, amorphous solid solution, though other mixtures containing multiple phases are occasionally targeted. This is achieved through the addition of heat and mechanical shear (i.e., mixing) to the product ingredients without degradation.

So how does one measure scale-up success? The answer depends on the product being considered for scale-up. For some products, the degree of mixing is of utmost importance, and a quality product is produced by ensuring a stable single-phase mixture as the result. For products where a single phase is not achievable, ensuring that the domain size of the mixture's phases reaches the smallest dimensions possible is desirable. Other products may be sensitive to residual moisture or solvents from the API production process, and, thus, devolatilization is a key concern. Finally, degradation of the API or excipients is often a limitation. In some cases, these attributes might be considered CQAs, depending on the definition that is applied. In practice, more than one quality attribute will govern which path towards scale-up is selected.

Looking first at the online analyses, extrudate API concentration ($< c_{\mathrm{API}} >$) is generally considered a CQA, and the API standard deviation (σ_{API}) is an indicator of process control. When PAT enables online measurement of API concentration, both measurements can be quickly captured as a function of operating conditions and equipment scales. PAT also enables straightforward measurement of the residence time distribution (RTD; Bigert and Smith-Goettler 2011). If the RTD and product exit temperatures are maintained, then this can indicate that the product's time-at-temperature is maintained across scales. Although product temperature and residence time vary in a complex way throughout the extruder, the simple measures of mean residence time, t_{mean}, and product exit temperature, T_{melt}, indicate that the amount of time and energy allowed for mixing, diffusion, and devolatilization are approximately equal across scales. Further, it indicates that the amount of any degradation products will remain similar across scales. Thus, these measures are recommended for use in any scale-up study.

In addition to the online analyses described, offline tests are also required. Specifically, standard release tests such as a product's dissolution profile and any degradation products must be assessed. If not obtained online via PAT, the assay and content uniformity of extrudate must be measured via off-line lab tests. Any residual solvents (including water) should be measured as they can have a plasticizing effect potentially influencing both extrudate physical and chemical stability. Finally, tests such as differential scaling calorimetry (DSC), X-ray diffraction (XRD), or others must be conducted before and after holding the product at controlled environmental conditions, to ensure that the phase state of the extrudate is understood and controlled over time.

7.3.3 Types of Scale-Up

In a typical scenario, a lot (i.e., batch) of material will be manufactured successfully using a certain equipment scale, screw and barrel configuration, and operating conditions like screw speed, mass throughput, and barrel temperature profile. Subsequently, the need for additional material mandates that the process be scaled up. For a conventional batch process such as bin blending, scaling the number of lots produced requires repeating the same process over and over again, or changing the size of the equipment. However, to scale-up a continuous process like melt extrusion, there are four primary options:

1. Scaling in parallel—duplicating unit operations so that identical equipment and production conditions can be used
2. Scaling with time—increasing quantity produced on same equipment at same production conditions
3. Scaling with operating conditions—increasing production rate on same equipment using different production conditions
4. Scaling with equipment size—changing equipment size and matching smaller scale production conditions

For each of the above options, there is a unique set of risks and benefits which are discussed in the following paragraphs.

7.3.4 Scaling in Parallel

Ostensibly the easiest, this option is tempting for risk-averse organizations because a known equipment design and operating conditions can be copied, thereby reducing risk. In general, scaling with time or with operating conditions should be attempted first, and only after maximizing production with these options should scaling in parallel be considered. Although it is tempting to assume that the previously mentioned analyses for assessing scale-up are not needed when using this option, it is wise to check $< c_{API} >$, T_{melt}, and the RTD, as subtle differences between machine dimensions, calibrations, wear, etc. must be understood.

7.3.5 Scaling with Time

Scaling with time involves simply operating at known conditions for longer periods of time. In many industries outside of pharmaceuticals, extruders are operated around the clock with limited shutdown periods. Within the pharmaceutical industry, cleaning and maintenance schedules can restrict TSEs from operating on the order of months, but it is common to find around the clock production for periods lasting 1 week or more. Thus for a dedicated production line, equipment uptime of 5000–7000 h can be achieved. When extended periods of uptime are being considered, equipment and product must be analyzed for signs of material buildup, hang-up and/or degradation, as this can often limit how long equipment can be run between cleaning and maintenance cycles. For long production time periods, steady-state analyses looking at $< c_{API} >$, T_{melt}, and other parameters should be conducted based on concepts from statistical process control, such as the Western Electric rules (Montgomery et al. 1994). If PAT is available, advanced analyses can be conducted to infer the RTD based on the system's dynamic response to upsets in API feed rate or other process disturbances (Bigert and Smith-Goettler 2011), and these RTDs can then be compared across time to identify any drift in process performance.

For products with a filed design space or proven acceptable range, scaling production by changing operating conditions is a straightforward endeavor. In this case, scaling up production simply means increasing the mass throughput of the extruder, and this should be done while keeping response parameters (i.e., both noncritical and CQAs) within the range of what is known to be acceptable.

In many situations, an extruder will be operated with fixed screw configuration, barrel configuration, and barrel temperature set points, while screw speed and mass throughput can be adjusted in response to production demands. Typically, screw and barrel configuration are set early on in product development based on the product characteristics such as number and type of feed streams, degree of mixing needed, and venting requirements. Once these are set, the barrel temperature profile is adjusted to minimize heat flow into and out of the screw chamber, a condition known as adiabatic operation. When this is done correctly, all energy input to the product will come from the screw, and thus changes to the material residence time will not result in additional

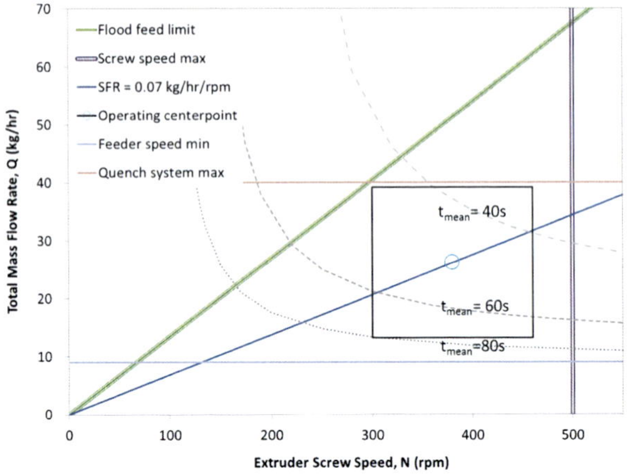

Fig. 7.4 *N*-*Q* operating diagram for a 27-mm twin-screw extruder (TSE) with adiabatic barrel temperature settings

heat addition or removal via the barrel. Since only two operating parameters remain, the space can be conveniently visualized in a screw speed (N) and mass flow rate (Q) operating diagram, as shown in Fig. 7.4.

When attempting to move about in the operating space, it is nearly impossible to change the production rate without impacting the system's RTD. However, when the mass flow rate is changed simultaneously in proportion to the screw speed, the system's specific feed rate, $SFR = Q/N$, is held constant (Gao et al.1999). Because the numerator and denominator both have units of time, the units of SFR, kg/hr/rpm, can be reduced to a more convenient form, g/rev. Thus, if SFR is held constant, the average degree of fill in the extruder screw will be held constant. Even though t_{mean} changes, the RTD at different SFRs will take on a common form, and will overlie one another when normalized by t_{mean} (Zhang et al. 2008). Furthermore, the average mixing rotations $= t_{\text{mean}} \times N$ will remain constant, which implies that, to a first approximation, the degree of mixing will remain constant. And finally, the SME for the product will remain constant, which will result in an approximately equal T_{melt} if adiabatic barrel temperature settings are maintained. For the example system shown in Fig. 7.4, the operating space is based on $N = 380 \pm 80$ rpm and $Q = 26 \pm 13$ kg/hr, the bounds being defined primarily by equipment limitations. For high-viscosity polymers, the TSE's torque limit might also restrict production. Other limitations related to system residence time, venting, heat transfer, or shear rate could be dominant when considering different products. For the product described in Fig. 7.4, moving to opposite ends of the operating space along the line $SFR =$ 0.07 kg/hr/rpm results in a change in Q from 21 kg/hr to 32 kg/hr, and a change in t_{mean} from 60 s to 40 s. For most products, a change of this magnitude would not be substantial enough to impact product quality, though that determination must be made case-by-case based on the individual product characteristics. For example, time-dependent reactive extrusion processes would be one exception to this.

In summary, moving about the operating space at a given scale along lines of constant SFR has an empirical backing to justify that the product manufactured

Table 7.3 Analyses for assessing scale-up of melt extrusion

Online steady-state analyses	Online dynamic analyses	Off-line analyses
T_{melt} (a.k.a. product or exit temperature)	**Overall residence time distribution (RTD, a.k.a. exit age distribution = *E(t)*)**	**Standard release tests (e.g., assay, content uniformity, dissolution profile, degradation products)**
$<c_{API}>$ *and* σ_{API} (mean and standard deviation of API concentration)	**t_{mean}, σ^2 (mean and variance of RTD)**	**Standard solid dispersion tests (e.g., DSC, XRD)**
$T_{post-quench}$	t_{lag}, τ (lag time and time constant of distribution, found when modeling RTD)	**Residual moisture/solvents stability studies**
Specific mechanical energy (SME = screw energy input/kg of product)	Extruder average percentage fill, mixing mass, and mixing rotations	Mixedness (e.g., NMR, AFM, or ΔCp via DSC if more than one phase)
ΔP_{die}		
Barrel heat flow		

DSC differential scaling calorimetry, *XRD* X-ray diffraction, *NMR* nuclear magnetic resonance,*AFM* atomic force microscopy

at different operating conditions are nearly equivalent. When changing conditions results in SFR not being held constant, the analyses listed in Table 7.3 can be used to assess the impact of the changes on the resultant product properties.

7.3.6 Scaling with Equipment Size

Scaling between different equipment sizes would be expected only one or two times for most products. When it does occur, it is usually because scaling with time and operating conditions has already been conducted, and still additional production is required. When this is the case, the main scale-up strategy recommended here is based upon the geometry of the extruder, as shown in Fig. 7.5.

For a perfectly self-wiping corotating bilobal TSE, only two parameters define the entire cross-sectional geometry (Booy 1963), namely the screw inner and outer diameters, D_i and D_o. Together, these parameters set the screw centerline distance, $C_L = (D_i + D_o)/2$. In practice, screws are not perfectly self-wiping with each other or the barrel, but instead a small clearance is maintained that also must be defined. Thus, the diameter of the barrel, D_b, is slightly greater than D_o, so typically $0.1\text{ mm} < D_b - D_o < 0.2\text{ mm}$. The interstitial space between the screws and the barrel is referred to as the free volume, V_{free}, and it is this volume that is of most interest during scale-up calculations.

Fig. 7.5 Critical element dimensions for extruder components

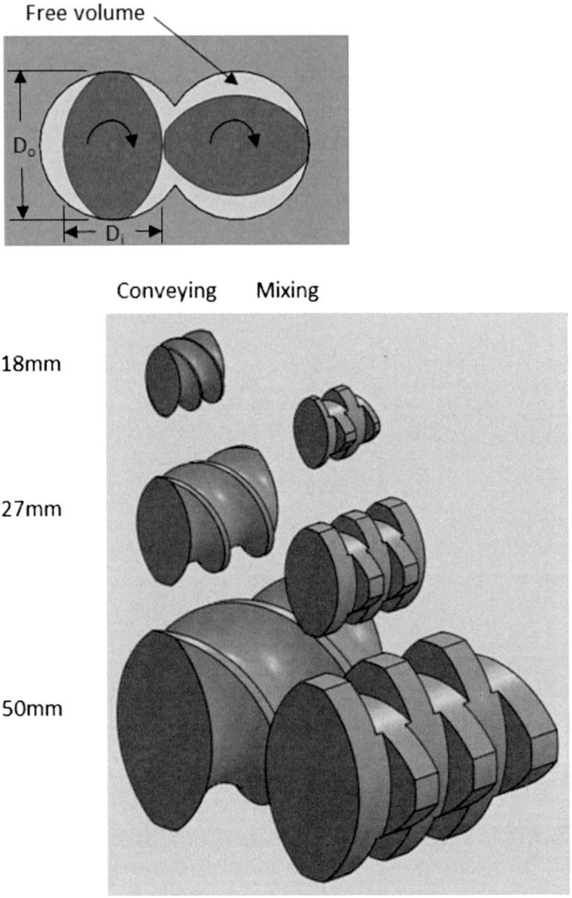

7.3.7 Volumetric Scale-Up Strategy

Although D_o/D_i may vary from one manufacturer to another, this parameter is often held constant as a function of scale within equipment made by the same manufacturer (these machines are referred to as having geometrically similar cross sections). This cross section of the screw is the same for nearly all element types, be they conveying or mixing segments. Figure 7.5 shows conveying and mixing elements for three scales in the Leistritz ZSE HP family of TSE, where $D_o/D_i = 1.5$.

For machines with geometrically similar cross sections, the free area of the cross section can be determined from a single parameter, D_o. As the length of the TSE is also specified in terms of its length to diameter ratio, L/D_o, V_{free} of a TSE can be determined by the definition of D_o, and consequently $V_{free}\ \alpha\ D_o^3$.

The volumetric scale-up strategy, also referred to as scaling by free volume, requires that the mass throughput to free volume ratio, Q/V_{free}, be held constant when moving between extruder scales. As a result, the mass throughput between different scales for geometrically similar machines is simply $Q_2 = Q_1(D_{o,2}/D_{o,1})^3$. If a change in extruder manufacturers results in dissimilar geometry, then the mass throughput relationship changes to $Q_2 = Q_1(V_{\text{free},2}/V_{\text{free},1})^3$.

Overall, the volumetric scale-up strategy, described extensively by a number of authors (Tadmor and Gogos 2006) can be summarized by the following rules which describe what to do when changing scales:

1. The length to diameter ratio of the extruder screws is maintained
2. The extruder screw profile (i.e., mixing zone and conveying zone types and positions) is matched to the extent possible
3. The extruder barrel configuration (i.e., feed position, vent port position, etc.) is maintained
4. The extruder barrel temperature profile remains the same
5. The extruder barrel temperature is set for adiabatic operation
6. The extruder screw speed is kept constant
7. The mass flow rate through the extruder is scaled in proportion to the extruder screw free volume
8. In the die, the pressure drop is maintained, which is achieved when the diameter of bores is held constant and the cross-sectional area is increased proportional to the increase in mass flow rate

When these principles are used, the following process responses will be similar between different extruder scales:

a. Specific mechanical energy, SME
b. Product temperature at the extruder exit, T_{melt}
c. Fill fraction along the length of the extruder
d. Overall RTD, $E(t)$
e. Average mixing rotations $= N \times t_{\text{mean}}$

By way of example of how the volumetric scale-up strategy can work, Fig. 7.6 shows RTD obtained from two Leistritz ZSE series extruders. Using this strategy, a Leistritz ZSE18-HP extruder with $D_o = 17.8$ mm and operating at $N = 380$ rpm and $Q = 6.8$ kg/hr is expected to scale to a Leistritz ZSE 27-HP with $D_o = 27.0$ mm and operated at a throughput of $Q = (27\text{ mm}/17.8\text{ mm})^3 \times 6.8$ kg/hr $= 23.7$ kg/hr.

The RTD is only one measure of the similarity between scales, but because it plays such a dominant role in determining product characteristics, it is the focus here, and other analyses have been omitted for sake of brevity. It is recommended that anyone conducting scale-up operations for pharmaceuticals manufactured with TSEs conduct all of the analyses listed in bold in Table 7.3 to ensure that products are equivalent across scales. Additionally, verification of steady-state process performance using the testing strategy outlined in Table 7.4 to probe various aspects of the operation is recommended to confirm manufacturing robustness.

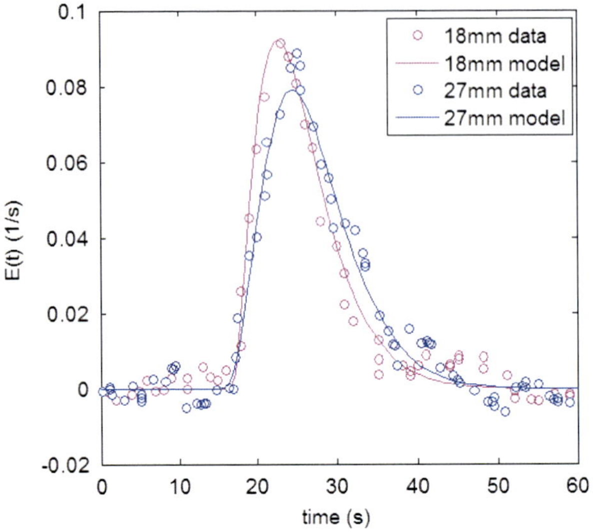

Scale (mm)	18	27
Mass flow rate (kg/hr)	7.0	23.7
Screw speed (rpm)	380	380
Mean residence time (s)	25.7	26.7
StDev residence time (s)	5.4	3.6
Lag time (s)	16.7	15.0
"Mixers in series"	3.1	4.6
Time constant (s)	8.8	12.0

Fig. 7.6 Residence time distribution (RTD) monitoring of extrusion operations

Table 7.4 Critical quality attributes (CQAs) and formulation/process impact during extrusion

Critical quality attribute	Excipient feed	Melt/mixing	Die outlet/shaping	Quench	Downstream: dry granulation, compression, and packaging
Content uniformity					......
Assay					
Purity/degredates					
API form					
Physical properties—density, particle size					......

Suspected or confirmed impact of the processing step on the CQA

7.3.8 Heat Transfer Limited Scale-Up Strategy

Although it is recommended that extruder barrel temperatures be set for adiabatic operation, this is not always achievable, especially when the desired product temperature is well below the temperature at which mixing takes place. In this case, the barrel might be set higher around the more restrictive mixing elements, and then set to cooler temperatures downstream, such that the product temperature decreases as it approaches the die. Because heat transfer occurs across the barrel surface, and the barrel surface area changes proportional to D_o^2 across scales, a first approximation would be that $Q_2 = Q_1(D_{o,2}/D_{o,1})^2$. But because scaling throughput in this way changes the RTD, more time will be allotted for heat transfer with the barrel at longer residence times, and the flow rate Q_2 that results in equal product temperature T_{melt} will be somewhere between $Q_1(D_{o,2}/D_{o,1})^2 < Q_2 < Q_1(D_{o,2}/D_{o,1})^3$. Because average shear rate also increases with increasing D_o, SME can increase even when N is held constant, and this effect becomes more prominent at low product temperatures where the viscosity is higher. As can be seen, scaling based on heat transfer limitations is a complex process with many interacting parameters. To effectively scale these types of processes, additional experimentation and modeling may be necessary to yield an optimum process at scale.

7.4 Process Analytical Technology for Melt Extrusion

Extrusion and process analytical technologies (PAT) have been used for nearly a century in the plastics and food industries, yet both are relatively new to the pharmaceutical industry (Akdogan 1999; Workman et al. 2005; Crowley et al. 2007). In 2002 the Food and Drug Administration (FDA) began an initiative, "Pharmaceutical CGMPs for the twenty-first century: A Risk-Based Approach," that yielded the 2004 guidance "Process Analytical Technology (PAT)— A Framework for Innovative Pharmaceutical Manufacturing and Quality Assurance" (Guidance for Industry 2004a, b). The scope of PAT as defined by the FDA guidance encompasses on-line, in-line, or at-line measurements for enhanced process understanding to enable optimization, control and product quality by design via the following mechanisms:

- Multivariate tools for design, data acquisition, and analysis
- Process analyzers
- Process control tools
- Continuous improvement and knowledge management tools

Process analyzer measurements, e.g., spectra or chemical images, typically require a mathematical transformation, e.g., multivariate data analysis, to correlate the process analytical data to a more relevant critical product attribute for design space definition. For brevity, throughout this section the measurement system that yields process analytical data is noted as a PAT method and the subsequent mathematical transformation is described as a model. To forego a debate regarding what constitutes a process analyzer, i.e., temperature sensor versus a Raman fiber optic probe, herein focuses on process analyzers that yield multivariate data.

Particular advantages of HME cannot be fully realized without the use of PAT. For example, the utilization of a polymer matrix is advantageous from a drug delivery perspective; however, it complicates drug substance extraction from extrudate for off-line analysis such as high-performance liquid chromatography (HPLC; Tumuluri et al. 2008). Additionally, continuous processing, or the potential thereof, necessitates real-time quality assurance. PAT better enables continuous manufacturing as any quality attribute deviations can be identified in real time as opposed to having to discard the entire batch.

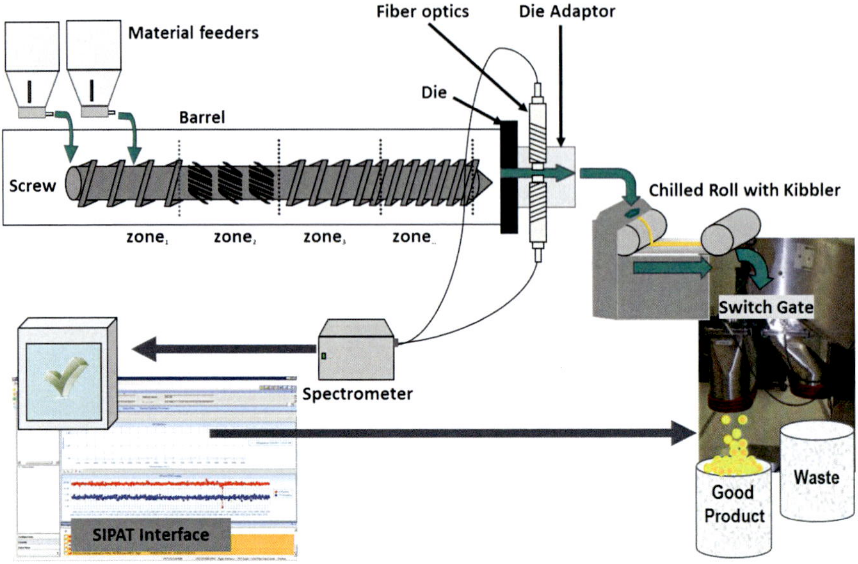

Fig. 7.7 Schematic of process analytical technology (PAT) integration into melt extrusion

In 2008, Tumuluri et al. used an on-line Raman method and model to quantitate drug substances in two respective formulations, confirmed transferability of the model from lab scale to pilot scale, and inadvertently demonstrated form change detection (Tumuluri et al. 2008). Additionally in 2008, in-line UV/VIS spectroscopy for thermal degradation detection was reported by Wang et al. (2008). Since then, multiple PAT applications of NIR and Raman spectroscopy during pharmaceutical HME have been reported for real-time drug substance quantitation (Saerens et al. 2011, 2012; Krier et al. 2013; Wahl et al. 2013), solid-state characterization (Saerens et al. 2011, 2012; Almeida et al. 2012; Kelly et al. 2012), and material residence time determination (Markl et al. 2013).

The implementation of an in-line, transmission mode, Fourier transform (FT)-NIRS method and a partial least squares (PLS) model to support the HME manufacturing platform is detailed herein to demonstrate how PAT can yield process understanding, process fault detection, and real-time quality assurance of intermediate drug product.

The NIR spectrometer was integrated to the extruder via a custom in-line temperature-controlled die adapter with a fixed optical path length, as shown in Fig. 7.7. The single-fiber transmission probes have standard 1/2–20 UNF- mounting threads and are screwed into the die adapter such that they are perpendicular to the plane of the material. The probes have been designed to be seated flush within the die adaptor to minimize optical path length fluctuations, to be resistant to the exterior operating temperatures, and to withstand pressures up to 1500 psi.

PLS is a commonly used multivariate linear regression technique and was applied to spectral data for real-time drug substance and surfactant concentration predictions. For such a model to be utilized, an upfront investment needs to be made as PAT calibration samples are typically manufactured rather than prepared in a laboratory. Calibration samples were purposely extruded with varying levels of drug substance and surfactant while varying material throughput rates. Samples were thieved upon exit from the die adaptor, allowed to cool, labeled with the NIRS time stamp at the time of collection and sent for off-line reference measurements, i.e., HPLC assay. The potential risk regarding inaccurate alignment of spectra to the off-line reference measurement was mitigated by ensuring a steady state was established during sample collection.

An accurate and robust PAT model requires a calibration set that contains chemical or physical insight to the attribute of interest and includes anticipated sources of variation. Anticipated variation includes but is not limited to, different lots of raw material, different scales of production, and different spectrometers and processing equipment. It is of utmost importance that an independent validation set that spans the operational space defined by the calibration set be used for model validation. Achieving all these model requirements in a single campaign is improbable. As such, models are typically a work in progress until they are transitioned to support routine commercial production. Thus PAT model development should be viewed as a lifecycle and ideally would progress in parallel with process development. Concurrent development is an efficient approach to yield a model that is robust to parameters varied during processing scoping experiments, e.g., screw speed and extrudate T_{melt}. A model update approach has been with each manufacturing campaign to add calibration samples into the existing calibration set and to regenerate the model. Once routine, commercial production is achieved, model updates should be solely dependent upon manufacturing process changes.

A validated model is required for real-time quality assurance of intermediate drug product. Once a model development lifecycle has been initiated, the model can be deployed for process understanding and process fault detection. It is recommended that validation elements such as specificity, linearity, precision, and accuracy be assessed prior to deploying the model for development purposes and then the model be formally validated using a representative and independent validation set prior to deployment for the purpose of real-time quality assurance.

7.4.1 Process Understanding

Process understanding applications of the given NIRS HME method has been used to monitor the transition from placebo extrusion to active extrusion during active batch start-up and purging of the extruder during shutdown and cleaning. Additionally, the NIRS HME method has been used to study the RTD of drug substance in the extruder and the extent of mixing. Significant process understanding has been achieved via application of the given PAT method for the determination of drug substance RTDs as

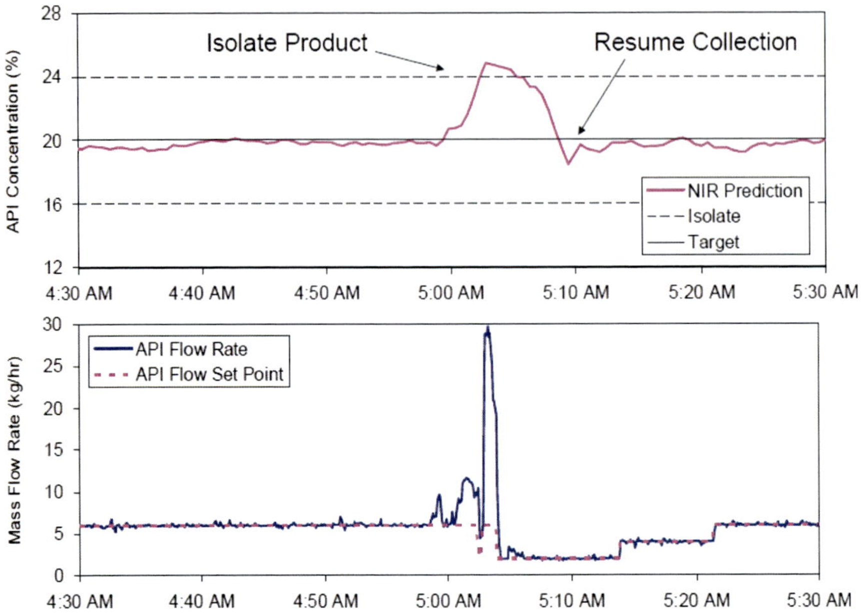

Fig. 7.8 Real-time monitoring of extrudate composition using in-line monitoring. (Bigert and Smith-Goettler 2011)

a function of varying process conditions (e.g., material throughputs, varying barrel temperatures, and different screw profiles). Illustrated in Fig. 7.8, a multicomponent analysis allows for the assessment of other material levels in the formulation as well which can be integrated to support rejection in the event of system perturbations. Ultimately, this can be extended to provide a comprehensive understanding of extrusion performance capable of supporting real-time release.

7.4.2 *Process Fault Detection*

A balance between process monitoring and manufacturing support has to be maintained by the manufacturing scientists and operators. In regard to the process monitoring aspect, it is easier to simplify the multivariate nature of such a complicated process into a reduced data stream that is representative of product quality, i.e., NIRS prediction of drug substance and surfactant concentration. During HME, process faults have been immediately and easily identified from alarms triggered by concentration predictions exceeding specifications. When this happens, drug product intermediate collection is diverted to waste. Then various process parameters are checked to identify the root cause of the compositional deviation.

The importance of PAT instead of simply relying on mass feed rates and mass balance to determine concentration is best described by an example. During a representative development example of such a process fault, drug substance concentration

(%wt/wt) drifted upward, ultimately exceeded the alarm limit. The extrudate stream was immediately diverted to waste and while being diverted to waste, process parameters associated with the material feeders were investigated. In this circumstance, a change in drug substance lots with different bulk densities, led to material being fed at a faster rate than the set point. The material continued to be diverted until the feeder reached steady state around an adjusted set point and the NIR response was closely watched to deem the result of this action. In a matter of minutes, the problem was identified, the suspect extrudate was isolated from quality extrudate, the fault was corrected and quality product collection was resumed. Without the real-time concentration predictions provided by PAT, the drum of extrudate would have needed to be quarantined for further testing. The worst-case scenario is that the entire batch would have been lost. As earlier noted, one could argue that monitoring the material flow would have ultimately identified the process upset if PAT monitoring was not in place. However, the issue is much more complex in that if the upset had been detected by a process parameter, i.e., flow rate, it would not have been known if the upset was of enough significance to put product quality at risk, and, if so, for how long.

7.4.3 Real-Time Quality Assurance

For the specific example described, the calibration model was tested for linearity, accuracy, and bias with independent validation samples, meaning that they were not included in the calibration set. The respective drug substance and surfactant concentration prediction models yielded correlation coefficients > 0.990 and a root-mean-squared error of prediction (RMSEP) values $< 1\,\%$ (wt/wt) which correspond to accuracy and linearity expectations defined in the United States Pharmacopoeia (USP) and ICH for lab-based testing methodologies. There was no significant bias between the estimates from the PAT model and those from the reference off-line HPLC method. Noteworthy is that scale-up from development to commercialization for continuous manufacturing processes, if even applicable, is not as significant relative to batch processes. This is important in regard to ease of PAT model transfer. In the given example, as well as in other hot-melt extruded drug products, a formal PAT model transfer was mitigated given the PAT model was built across development and commercialization sites as the same model spectrometer and same scale extruder were used from development to commercialization.

A risk mitigation strategy in multivariate regression models, such as the model described here for concentration prediction, is employing outlier detection methods. Outlier detection methods are statistical tests which are conducted to determine if the analysis of a multivariate response using a calibration model represents a result outside the calibration space. Hotelling's T^2 values and residual variance values can be reported in real time. Both metrics are representative of model distances, where Hotelling's T^2 values represent the distance of an observation (spectrum) to the model origin and residual variance values represent the distance of an observation (spectrum) to the model plane. Thus, the residual variance metric would be sensitive to an outlier that was at the XY origin, but displaced from it along the Z-axis. Yet, the Hotelling's T^2 metric is more sensitive to other types of outliers; for example, where the Z coordinate is described by the model but the X or Y coordinate is quite different. Hence it is advisable to look at both types of outlier detection plots. Thus, on a routine basis for every spectral measurement, these diagnostics have been calculated to reduce the risk of making concentration predictions with spectra that are outside the validated range or calibration space. There are multiple reasons a spectrum may be identified as an outlier, including: there is no extrudate at the spectroscopic interface, there is probe fouling, the concentration is outside of the validated prediction range, or there is a difference such as chemical degradation or a phase transformation.

7.4.4 PAT Operationalization

PAT operationalization is achieved through integrated hardware and software which enables instrument control, data analysis, visualization and archiving, and integration of the instrument to an automation system. PAT methods/models are intended for implementation into routine supply when the drug product is released to market. The next step in the previously described example was to use the drug substance and surfactant concentration prediction values to automate diversion to waste whenever there is an out of specification alarm. Automation logic uses packages such as SIPAT inputs to control the extruder switch gate (quality product collection versus waste). An enabled component which is outside of the limits will generate an alarm and a nonenabled component which is outside of the limits will generate a warning. Alarm acknowledgement is required. Alarming is enabled once the drug product status has been established as "good," to allow the process to come to a normal, steady-state condition. Examples of enabled components for switch gate control have been material constituent concentrations (drug substance and surfactant), outlier diagnostics (Hotelling's T2 and spectral residuals), and process parameters.

The benefits of PAT span development, commercialization, and supply. PAT integrated into the HME platform early in process development has facilitated a model robust to process variations and thus will yield process flexibility post-drug product regulatory approval. In multiple HME drug product applications, PAT monitoring has identified and facilitated the correction of process faults while yielding quality assurance via waste gate switch feed-forward control.

7.5 Quality by Design for Melt Extrusion

The quality by design (QbD) initiative arose to encourage innovation and implementation of new manufacturing technology along with enhancing science and risk-based regulatory approaches. As defined in ICH Q8, (2009a), QbD is viewed as a systematic approach to development beginning with predefined objectives and emphasizing product and process understanding and process control, based on sound science and quality risk management.

The "carrot" for QbD initially resided in regulatory agencies proposing that effective demonstration of enhanced product understanding in the registration application would create regulatory flexibility, perhaps in the form of real-time release or reduced end-product testing, reduced post approval change notifications, manufacturing scale flexibility, and opportunities for continued process optimization once in commercial space. The degree of regulatory flexibility actually achieved with QbD can certainly be debated. What seems to have consensus is that structured, QbD-focused development yields improved process understanding that results in more robust processes at manufacturing scale. These processes in general should have less risk at validation, higher assurance of supply, less atypical events, and more consistent performance in commercial space even in the event that the full range of regulatory flexibility is not realized.

Expectations for QbD are that development activities progress by first considering the patient experience via the quality target product profile. This should be followed by a definition of product CQAs. Risk assessment (RA) activities occur next to identify how these CQAs are influenced. The RA activities also ensure that development efforts are proportional to the degree of risk to the patient. The natural end to development activities should be when enhanced understanding exists to assure the consistent achievement of CQAs through a rigorous control strategy. Upon tech transfer and transition into routine commercial production, system lifecycle elements come into play, assuring the process is monitored, process changes are managed, and the process is maintained in a state of control. Each of these stages of the quality risk management process that achieves QbD will be discussed in turn in greater detail within the context of HME.

7.5.1 Quality Target Product Profile, CQA definition, and Initial RA

A guiding principle within QbD is driving the enhanced understanding with a focus on the product quality defined in the context of the intended patient experience. This effort begins with the quality target product profile (QTPP), where elements including intended dosage form, strength, release profile, purity, and appearance (while not directly linked to the safety and efficacy, this is important for dose recognition and adherence) are outlined. The QTPP allows for an easier identification of those CQAs that must be controlled to ensure product quality.

Focusing on these CQAs, activities progress by considering where within the process the CQAs might be influenced, for instance by parameter settings or incoming material attributes. Here, structured RA tools come into play, ensuring the process in its entirety is considered, with all sources of potential impact on CQAs vetted.

Table 7.5 Release attributes and control strategy for melt extrusion

CQA	Material control	In process control	Control by process design	Drug product release criteria
Final API form in tablet	Polymer and surfactant meet CoA specifications	*Melt/Mix:* Actual barrel temperature values in zones upstream from the die	*Melt/Mix:* fixed screw profile. RPM range, barrel temperature, and feed rate w/in design space parameters. Screw speed designated CPP	NA (no crystalline API at HME outlet)
		Die outlet: Monitor pressure and die melt temperature. Online spectroscopy for form	*Die outlet/shaping:* barrel temperature	
			Quench: chilled rolls gap, and temperature design space ranges	
			Downstream dry granulation, compression and packaging: specified max humidity range	

COA certificate of analysis, *CQA* critical quality attribute, *API* active pharmaceutical ingredient, *HME* hot-melt extrusion, *RPM* revolutions per minute, *CPP* controllable pitch propeller

Prudent RA exercises ensure appropriate effort is deployed in accordance with the degree of risk. An example of such an RA for final API form in the drug product is shown in Table 7.5.

The ICH guidance does indicate that "it is neither always appropriate nor always necessary to use a formal risk management process" (ICH Q9 2006). However, RA activities are most value added when they are attempted early enough within a development program to help prioritize activities. This also helps ensure quality is designed into the product, rather than relying solely on product specifications and release testing as the means for process control. Early RA activities can also help ensure elements of scale are considered and are factored into the development strategy. Considering the material sensitivities, for instance shear, time at temperature, or minimum temperature to achieve conversion to the amorphous state, can help prioritize where development activities might progress, and do so in a way mindful of production scale equipment. For instance, if the polymer is known to be shear sensitive, perhaps focusing on understanding the impact of screw tip speed on performance and maintaining this across scales would trump attempts to achieve identical max temperature and residence time. Effectively addressing these concerns can reduce risk around verifying design space or proven acceptable ranges at scale. This is also in-line with regulatory guidance specifying the "... need to consider the effects of scale. However, it is not typically necessary to explore the entire operating range at commercial scale if assurance can be provided by process design data."

Early experiments may include exploratory one-factor at-a-time trials to understand process sensitivity, or highly fractionated Plackett–Burman type designs to screen for main effects, or high-level second-order interactions. Once again with a line of sight to commercial manufacture, evaluations could start to consider design space versus proven acceptable range designations. The expectation for design space ranges would be to provide the technical details to support that operating within the defined ranges assures the achievement of the CQAs. When this process understanding is multivariate in nature, that is the interdependence between process variables is understood to allow multiple parameters to be modified and understanding of their additive influence still achieves quality attributes, this is said to be a design space. When the influence of individual parameters is not understood in the context of other process parameters, these ranges are said to be proven acceptable ranges.

7.5.2 Late-Stage Risk Assessment and Control Strategy

As knowledge about the process increases, so should the rigor of the RA activities, requiring revisiting early stage activities. The expectation for the RA activities is that they should achieve decision making that is transparent and reproducible, and can entail adding elements including probability of occurrence, detectability, and the likelihood of occurrence to the early stage RAs.

It is also helpful to consider where the iterations on the RA end. The natural end occurs when the enhanced knowledge gained through focused design of experiments (DOE's) or targeted one-factor at-a-time experiments is sufficient to demonstrate understanding of the sources of variability. Here, process parameters can be adjusted, and the impact on CQAs effectively predicted. This suggests RA activities were prudent, with the significant sources of variability identified and controlled. An example of the opposite would be if parameters were changed to achieve CQA target only to find upon implementation at pilot or manufacturing scale that the actual CQA value does not match the target. This would suggest there were parameters that are influencing quality attributes that were missed during the RA activities, or the impact of process parameters on CQAs is poorly understood. Beyond the regulatory benefits of QbD, this is the type of residual risk that could put validation or launch at risk but can be successfully managed through QbD activities.

With the sources of variability better understood, the route to control and assure CQAs can be effectively mapped. The routes for control should span (1) control on raw materials, (2) in process control, and (3) control by process design. The following example shows how detailed RA could evolve into this control strategy. The example below has built off of the early RA grid, once again considering the various stages across the HME process that were identified to impact the specific

CQA, and how each of these is controlled. There could be many other ways to map the control strategy, provided each of the three modes of control is explicitly defined.

The elements of in-process control and control by process design should weigh whether it is sufficient to control the variability solely via process design via parameter range definition. A rigorous quality risk management process can also enable multiple control strategies for the same process if processing activities occur at different sites. Perhaps one site opts to leverage rigorous in-process testing via at-line testing or PAT, whereas another site chooses to utilize automation systems and tight batch sheet controls with parameter settings further constrained within defined design space or proven acceptable ranges.

7.5.3 Product Lifecycle Management

At commercial scale, additional RA activities are prudent, though now shifting to focus on risks to the control strategy specific to the commercial process train. For example, if using an in-process test to augment control by process design to assure extrudate assay, what happens if PAT goes down, or the spectroscopy window is occluded, or what would happen if the die melt temperature is considered a critical process control and the thermocouple fails.

These elements of quality risk management recognize that once approved, processing can remain dynamic to allow ongoing optimization, accommodating the fact that the "the full spectrum of input variability typical of commercial production" is generally not known following development activities (Industry Guidance 2011b). This provides assurance that the process will continue to be monitored, controlled, and periodically reassessed. Having the assumptions captured in earlier RA activities can prove useful as process changes are evaluated and considered against the composite and relevant process knowledge accumulated to data.

7.6 Summary

Using a structured design approach, a drug product formulation and process can be developed that has been appropriately engineered to support commercialization. Within the late-stage space, there are significant opportunities to couple with drug substance manufacturing to leverage novel capabilities with devolatilization. Scale-up across a range of potential commercial volumes is also possible with extrusion, using well-established scaling methodologies and computer simulation to ensure success while effectively probing the operational space. In-line measurements can also be facilitated with internally mounted probes to support NIR and Raman measurements, as well as melt viscosity, temperature, and pressure leading to a comprehensive PAT approach to support real-time release and out of specification material isolation. Ultimately, these attributes make the extrusion platform uniquely suitable for QbD and highly effective for commercialization of both small and large volume pharmaceutical dispersions.

References

Akdogan HI (1999) High moisture food extrusion. Int J Food Sci Tech 34:195–207

Albalak RJ (1996) An introduction to devolatilization in polymer devolatilization. 1–12

Albalak RJ, Tadmor Z et al (1990) Polymer melt devolatilization mechanisms. Wiley Online Libr 36:1313–1320

Almeida A, Saerens L et al (2012) Upscaling and in-line process monitoring via spectroscopic techniques of ethylene vinyl acetate hot-melt extruded formulations. Int J Pharm 439:223–229

Baugh DAB, Koppi K, Spalding M, Buzanowski W (2003) Temperature gradients in the channel of a single-screw extruder. ANTEC Pap 2003:202–206

Biesenberger JA Kessidis G (1982) Devolatilization of polymer melts in single-screw extruders. Polym Eng Sci 22:832–835

Bigert M, Smith-Goettler B (2011) PAT to support the hot melt extrusion platform. Leistritz Extrusion Seminar, Clinton, NJ, June 2011

Booy ML (1963) Influence of Channel Curvature on Flow, Pressure Distribution and Power Requirements of Screw Pumps and Melt Extruders, Soc Plastics Engrs Trans., 3 pp 176.

Carneiro OS, Covas JA et al (2000) Experimental and theoretical study of twin-screw extrusion of polypropylene. J Appl Polym Sci 78(7):1419–1430

Che D, Kotipalli U et al (2010) Atorvastatin calcium propylene glycol solvates, USPTO

Crowley MM, Zhang F et al (2007) Pharmaceutical applications of hot-melt extrusion: part I. Drug Dev Ind Pharm 33(9):909–926

Fachaux JM, Guyot Hermann AM et al (1993). Compression ability improvement by solvation/desolvation process: application to paracetamol for direct compression. Int J Pharm 99:99–107

Gao J, Walsh GC, Bigio D, Briber RM, Wetzel MD (1999) Residence-Time Distribution Model for Twin-Screw Extruders, AIChE Journal 45(12) pp 2541

Godavarti S, Karwe MV (1997) Determination of specific mechanical energy distribution on a twin-screw extruder. Elsevier 67:277–287

Guidance for Industry (2004a) PAT—A framework for innovative pharmaceutical manufacturing and quality assurance (2004)

Guidance for Industry (2004b) Pharmaceutical cGMPs for the 21st Century: a risk-based approach (2004)

Industry Guidance (2006) ICH Q9 quality risk management

Industry Guidance (2009a) ICH Q8 (R2) pharmaceutical development

Industry Guidance (2011a) Q11 Development and manufacture of drug substances

Industry Guidance (2011b) Process validation: general principles and practices

Karwe MV. Godavarti S (1997) Accurate measurement of extrudate temperature and heat loss on a twin screw extruder. J Food Sci 62:367–372

Kelly AL, Gough T et al (2012) Monitoring ibuprofen-nicotinamide cocrystal formation during solvent free continuous cocrystallization (SFCC) using near infrared spectroscopy as a PAT tool. Int J Pharm 426:15–20

Krier F, Mantanus J et al (2013) PAT tools for the control of co-extrusion implants manufacturing process. Int J Pharm 458:15–24

Markl D, Wahl PR et al (2013) Supervisory control system for monitoring a pharmaceutical hot melt extrusion process. AAPS PharmSciTech 14:1034–1044

Montgomery DC, Keats JB et al (1994) Integrating statistical process control and engineering process control, J Qual Technol 26:79–87

Saerens L, Dierickx L et al (2011). Raman spectroscopy for the in-line polymer-drug quantification and solid state characterization during a pharmaceutical hot-melt extrusion process. Eur J Pharm Biopharm 77:158–163

Saerens L, Dierickx L et al (2012) In-line NIR spectroscopy for the understanding of polymer-drug interaction during pharmaceutical hot-melt extrusion. Eur J Pharm Biopharm 81:230–237

Schenck et al (2011) Achieving a hot melt extrusion design space for the production of solid solutions. In: Chemical engineering in the pharmaceutical industry

Tadmor Z, Gogos CG (2006) Principles of polymer processing, 2nd edn. Wiley, Hoboken

Tumuluri VS, Kemper MS et al (2008) Off-line and on-line measurements of drug-loaded hot-melt extruded films using Raman spectroscopy. Int J Pharm 357:77–84

Wahl PR, Treffer D et al (2013). Inline monitoring and a PAT strategy for pharmaceutical hot melt extrusion. Int J Pharm 455:159–168

Wang Y, Steinhoff B et al (2008) In-line monitoring of the thermal degradation of poly (l-lactic acid) during melt extrusion by UV-vis spectroscopy. Polymer 49:1257–1265

Wirth DD, Stephenson GA (1997) Purification of dirithromycin. Impurity reduction and polymorph manipulation. ACS Publi 1:55–60

Workman J, Koch M et al (2005) Process analytical chemistry. Anal Sci 77:3789–3806

Zhang XM, Feng LF et al (2008) Local residence time, residence revolution, and residence volume distributions in twin screw extruders. Polym Eng Sci 48:19–28

Chapter 8
Spray Drying: Scale-Up and Manufacturing

Filipe Gaspar, Joao Vicente, Filipe Neves and Jean-Rene Authelin

List of Abbreviations

A	Area of the spray dryer chamber
API	Active pharmaceutical ingredient
BD	Bulk density of the product
CCF	Central composite face-centered design
C_{feed}	Solids content in the feed
Cp	Heat capacity coefficient
d_D	Droplet size
D_{noz}	Diameter of the nozzle orifice
DoE	Design of experiments
D_v	Diffusion coefficient in the gas phase
Dv50	Volumetric median particle size
F_{feed}	Flow rate of feed fed to the spray dryer
F_{drying}	Flow rate of drying nitrogen
m	Mass flow rate
MFP	Maximum free passage
MW	Molecular weight
P_{feed}	Atomization pressure of the feed
P_{out}	Vapor pressure at outlet temperature
P^{sat}	Saturation pressure of the solvent
PSD	Pharmaceutical spray dryer
P_{wb}	Vapor pressure at wet bulb temperature
QbD	Quality by design

F. Gaspar (✉)
Particle Engineering Services, Hovione, Lisbon, Portugal
e-mail: fgaspar@hovione.com

J. Vicente · F. Neves
R&D Drug Product Development, Hovione, Loures, Portugal

J.-R. Authelin
Pharmaceutical Sciences Operation, Sanofi R&D, Vitry Sur Seine, France

N. Shah et al. (eds.), *Amorphous Solid Dispersions,*
Advances in Delivery Science and Technology, DOI 10.1007/978-1-4939-1598-9_8

Q_{loss}	Heat lost from the drying chamber walls
RS_{out}	Relative saturation at the outlet of the drying chamber
T_b	Boiling temperature
T_{cond}	Drying gas temperature at the exit of the condenser
T_{dew}	Dew point temperature at the outlet of the drying chamber
T_{feed}	Temperature of the feed solution
T_g	Glass transition temperature
T_{in}	Drying gas temperature at the inlet of the drying chamber
T_{out}	Drying gas temperature at the outlet of the drying chamber
T_{room}	Room temperature
T_{wb}	Web bulb temperature
U	Overall heat transmission coefficient
V_m	Molecular volume
x	Molar fraction in the liquid phase
y	Molar fraction in the gas phase
σ	Surface tension
μ	Viscosity
ρ_l	Liquid density
ρ_g	Gas density
ρ_P	Particle density
ΔH_{vap}	Vaporization heat
γ	Activity coefficient

8.1 Technology Transfer and Scale-Up to Commercial Units

In the pharmaceutical industry, the production of spray-dried powders is still widely based on the batch concept. The quantities required in the early stages of the development are typically small but may increase by many orders of magnitude as the drug candidate advances through the clinical phases and reaches the market. This requires the scale-up to different units along the process. Pharmaceutical spray dryers are available in a wide range of scales: from lab units where milligrams of material can be produced to large commercial units capable of handling multiple tons of powder per day. If necessary, a spray drying process can be scaled up directly from the laboratory to a final production scale. However, the quantities required for clinical trials are more efficiently produced in pilot or small production scales, where product losses and scale-up risk are considerably lower. Therefore, these intermediate production scales are commonly used during the development of the process. Nevertheless, it is important to mention that during scale-up, some quality attributes of the product can change, and there is need to understand whether these changes are acceptable, and if so, desirable. For example, powder properties such as flowability and compressibility can be improved significantly when moving from lab units to larger ones. Changes in the particle size distribution, level of residual solvents, friability, density, and compressibility of the powder may strongly influence the properties of the final tablet, viz. hardness, friability, disintegration time, and dissolution rate. A careless scale-up strategy may lead to considerable losses of expensive materials and ultimately jeopardize the timelines of a clinical program.

Despite its criticality, the scale-up of the spray drying process is still vastly empirical and based on costly experiments, and their statistical interpretation. To minimize the experimental burden of such an approach, recent efforts have focused on applying mechanistic models and simulation tools to describe the process of spray drying. In fact, mechanistic modeling and process simulation tools have been successfully used in chemical and oil industries for more than half a century. This rational approach has gained wide recognition, and pharmaceutical scientists are now making use of it during development, scale-up, and manufacturing (Koulouris and Lagonikos 2002). Nevertheless, pharmaceutical process development will require some sort of simulation and experimental testing at small scale, and at least some level of verification in a production environment.

The following sections highlight some important considerations regarding the most critical decisions related to the spray drying process, viz. selection of scale, atomizer, and key process parameters. Common challenges associated with the operation of the process will also be addressed. At the end of the section, a scale-up methodology based on scientific first principles, simulation models, and process characterization techniques are presented.

8.1.1 Manufacturing Scales

The way droplets are dried within the drying chamber dictates the characteristics of the final product. Several aspects like evaporation rate, particles trajectories, residence time, and wall deposition are governed by the factors like atomization device and conditions, design and positioning of the gas disperser, the dimensions of the chamber, and the location of the atomizer and exhaust gas duct. Spray dryers are available in multiple configurations, including cocurrent, countercurrent, mixed flow, fountain, or fluidized spray drying mode. The pharmaceutical industry predominantly uses the cocurrent mode since it minimizes the exposure of the product to high temperatures which may be crucial when processing thermally labile products or materials with a low glass transition temperature. The cocurrent mode is therefore considered the most suitable for the majority of the pharmaceutical applications.

Spray drying is a continuous process capable of full automation, and can be designed to meet any capacity required in the pharmaceutical industry. A process can be run in a large size unit for as short as 30 min or 1 h, or can be run continuously for many days. The selection of the right scale involves several considerations, but ultimately, it is primarily driven by the targeted process throughput and batch size requirements.

a Lab unit (Buchi B-290 Advance), **b** pilot unit (Niro Mobile Minor) **c** commercial unit (Niro PSD4)

Fig. 8.1 Lab-, pilot-, and commercial-scale spray dryers at Hovione

8.1.1.1 Lab to Production Equipment

Laboratory scale spray dryers are particularly useful for producing small quantities of prototype formulations in early stages of development. They can process small quantities of solution (as low as 2 ml) with relatively high yield. On the other extreme, the process can run continuously for hours or days providing the flexibility of producing hundreds of grams or even a few kilograms of material. It is not surprising that lab-scale units have been used to produce commercial quantities of very-low-volume products. A typical feature of these small-scale systems is that the drying chamber and cyclone are constructed in glass (Miller and Gil 2012), enabling the privileged visualization of the drying and separation processes (see Fig. 8.1a). The main limitation of the lab units is the powder properties of the resulting materials, namely particle size. The small dimensions of the drying chambers limit the residence time, and therefore, the droplets need to be small in order to be dried completely before leaving the drying chamber, or they will collide with the walls. Therefore, most small units produce powders with mean particle sizes below 10–20 μm, more often between 3 and 10 μm. There are, however, lab-scale units (e.g., ProCepT R&D Spray Dryer) that operate under laminar flow, allowing the drying of much larger droplets (100 μm and larger). These units represent an excellent platform to mimic the size and morphology of the industrial-scale spray-dried particles and can be used at early stages of development to assess the criticality of particle size and other powder attributes in the quality of the product.

Pilot- and commercial-scale spray dryers (see Fig. 8.1b, 8.1c) are suitable for a wide range of batch sizes, ranging from less than 1 kg to several metric tons. They share many similarities regarding the configuration, materials of construction (typically stainless steel), ability to handle most organic solvents, and level of automation. Additional features such as cleaning-in-place or recirculation of the drying gas (close-loop units) may be included. Some units can operate under vacuum (to minimize risk of powder exposure of highly potent drugs) but commonly operate under slight

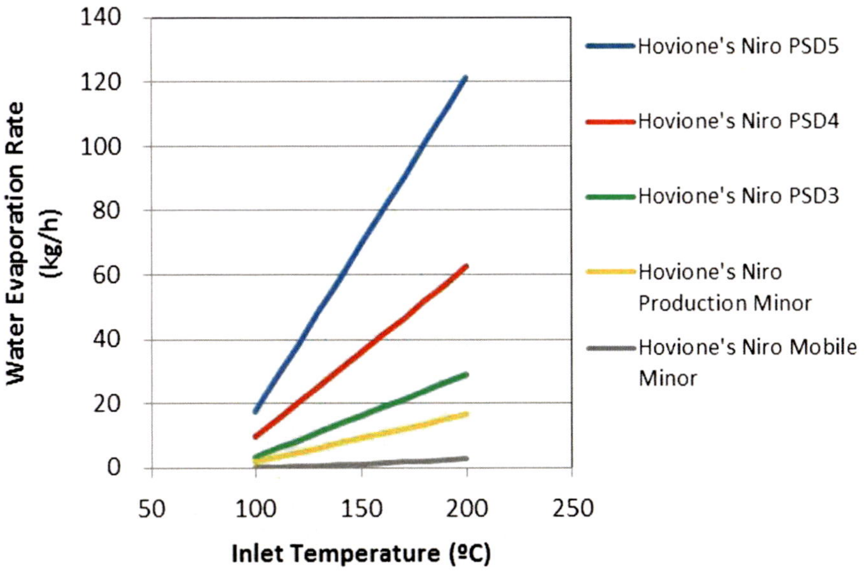

Fig. 8.2 Water evaporation capacity for different spray dryer units

positive pressure. Despite these many options, the main difference between scales is the evaporation capacity and the throughput.

The evaporation capacity depends mainly on the drying gas flow rate, solvent, temperature profile (inlet and outlet temperature), and heat loss of the spray drying unit. Fig. 8.2 shows the water evaporation rate at different production scales and nominal drying gas flow rates.

8.1.1.2 Considerations for the Selection of Scale

Scaling up a spray drying process offers in most cases a more energy-efficient process, lower manpower input per kilogram produced, and greater flexibility in adjusting and optimizing product attributes. The latter is particularly advantageous if the purpose is to obtain large and denser particles for solid oral dosage forms. For inhalation, however, the challenge is to maintain particle properties unchanged throughout the scale-up processes. The selection of ideal scale for a given product is dictated by the commercial demand projections and the real throughput of the process. To calculate real throughput, one needs to know not only the throughput in the spray drying step (and how it changes at different scales—see example in Fig. 8.2) but also which step (e.g., solution preparation, spray drying, or secondary drying) is the bottleneck of the process. Other variables with significant impact on the real throughput of the process are batch size, cleaning regime, and the duration of the cleaning process.

8.1.2 Nozzle Selection

Once the scale of spray dryer is established, it is important to select an atomization nozzle appropriate to the scale. The purpose of the atomization stage is to produce a fine mist (spray) from a liquid feed to substantially increase the liquid surface area and improve the efficiency of heat and mass transfer. For example, 50 ml of a solvent atomized in 800 million droplets of 50 μm creates a surface area about 6 m^2. By the generation of such high surface area, droplets dry fast, in the order of seconds or fraction of a second depending on the drying conditions. Moreover, the control of the atomization process dictates droplet size and consequently the particle size.

Sprays may be produced in various ways, but essentially, all that is needed is a high relative velocity between the liquid to be atomized and the surrounding gas. Some atomizers accomplish this by discharging the liquid at high velocity into a nearly stagnant gas. Notable examples include pressure nozzles and rotary atomizers. An alternative approach is to expose the relatively slow-moving liquid to a high-velocity gas stream. The latter method is generally known as two-fluid atomization. Independent of the device, atomization is a complex phenomenon of inertial, shearing, and surface tension forces, the balance of which determines the angle and penetration of the spray as well as the density number, droplet velocity, and size distribution. All these characteristics are markedly affected by the internal geometry of the atomizer, the properties of the gaseous medium, and the physical properties of the liquid itself, particularly its surface tension and viscosity.

Atomizers are generally classified according to the type of energy used. Rotary atomizers (centrifugal energy) use high velocity discharge of liquid from the edge of a wheel or disk. Two-fluid nozzles (kinetic energy) rely on the breakup of liquid on impact with high-speed gas at the orifice. Pressure nozzles (pressure energy) feature the discharge of liquid under pressure through an orifice, and ultrasonic nozzles (acoustic energy) breakup of liquid is promoted through sonic excitation. For each class of atomizer, there are several configurations and designs available to handle the diversity of feed materials and to meet the specific spray-dried product characteristics (Masters 2002). The liquid feed properties (viscosity, surface tension, solids concentration) impact the atomization performance in all types of atomizers. However, their sensitivity to each property depends on the particular type of nozzle.

8.1.2.1 Rotary Nozzles

For rotary nozzles, atomization is achieved by centrifugal energy transmitted to the liquid stream by a disk or wheel rotating at high speed (from 10,000 to 50,000 rpm). The liquid is fed into the center of a rotating wheel, moves to the edge of the wheel under the centrifugal force, and is disintegrated at the wheel edge into droplets. A spray angle of about 180° is best accommodated in large-diameter chambers (Mujumdar 2006). Rotary nozzles can be used to atomize slurries, suspensions, or solutions of high viscosity. Besides the feed properties, the operating variables that influence droplet size are feed flow, rotational velocity, wheel diameter, and design. Rotary nozzles typically produce droplets of a wide range of sizes: from 20 to 200 μm (Masters 2002).

8.1.2.2 Two-Fluid Nozzles

Two-fluid nozzles, also known as pneumatic nozzles, use a compressed gas to atomize the liquid feed. There are different designs of nozzles on the market. The two major groups of two-fluid nozzles are known as an external mixing nozzle and an internal mixing nozzle.

External mixing nozzle is operated with low liquid pressure. The liquid feed is provided through an inner duct while an atomization gas is fed by an external annular opening around the liquid orifice. On the other hand, internal mixing nozzles take advantage of gas expansion at the nozzle outlet. Part of the pressure energy applied is used to scatter the liquid fragments within and beyond the nozzle orifice by the sudden gas expansion (Walzel 2011).

Although an external mixing two-fluid nozzle is the most common in the lab- and the pilot-scale spray dryers, an internal mixing nozzle is far more efficient in regards to the gas to liquid ratio and, therefore, preferred for larger-scale spray dryers, especially when small particle sizes (less than 10 μm) are required (Miller and Gil 2012). The main disadvantage is the air/nitrogen pressure required to overcome the high pressure drop of this type of nozzle. Nevertheless, both internal and external mixing nozzles produce droplet sizes within the range of 5–75 μm (Masters 2002) and offer, probably, the best control over droplet size since feed and gas flow rates can be controlled independently. However, the gas consumption and pressure required may limit their use at industrial scales mainly when drying organic solvents. Therefore, pneumatic nozzles are most suitable for small scales or when very small particle sizes are required.

8.1.2.3 Pressure Nozzles

In pressure nozzles, atomization is achieved by converting pressure energy into kinetic energy. Often the design of pressure nozzles includes the inlet slots to impart a swirling motion to the liquid at the swirl chamber entry and a convergent section to accelerate the flow as it enters the orifice. The swirl motion of the liquid pushes it to the wall and, consequently, the liquid is ejected from the orifice as a conical sheet that spreads outwards due to centrifugal forces. These nozzles require the use of high-pressure pumps as pressures can go up to 450 bar. Droplet size can be manipulated with the operating pressure, but feed flow is dependent on that pressure. With increasing atomization pressure, the droplet size decreases and the feed flow increases. This dependent manipulation of the droplet size is one of the major drawbacks of this type of nozzles, i.e., in order to change droplet size at constant feed flow, it is necessary to change the nozzle dimensions or design. Further, as they involve the acceleration of the liquid feed, they are not suitable for high viscous feeds.

In large-scale spray dryers, these nozzles are used for production of medium to large particles (30–200 μm). They also produce more uniform powders with a narrower particle size distribution than pneumatic or rotary nozzles and, therefore, are preferred for the production of powders for oral dosage forms.

8.1.2.4 Ultrasonic Nozzles

The principle of the ultrasonic nozzles is based on the usage of high-frequency sound waves to atomize the feed and produce very narrow droplet size distributions with low velocities. For ultrasonic nozzles, the feed delivery pump controls the liquid flow. The ultrasonic nozzle generates a uniform droplet size distribution ranging approximately from 20 to 100 μm. However, the frequency of vibration is specific for a given nozzle, and so does the droplet size produced. In practice, in order to change the droplet size, it is required to use a different nozzle. Higher-frequency nozzles produce smaller droplets. The major drawback of these devices is that the throughput is limited (typically up to 50 ml/min) which limits their applicability to laboratorial- and pilot-scale units.

8.1.2.5 Considerations for the Selection of Nozzle

In the selection of the atomizer and atomization parameters, two general requirements should be considered: one is to provide the throughput that meets the required powder production and the second is to generate a droplet size that provides for the target particle size distribution. In the pharmaceutical industry, the most commonly used nozzles are two-fluid and pressure nozzles, owing to their simplicity of use, easy of cleaning, ability to handle wide variety of feeds and the reduced tendency (when compared to rotary nozzles) to generate wall deposits. In most applications, a pressure nozzle is preferred than a two-fluid nozzle, primarily because pressure nozzle provides powders with a narrower particle size distribution. Exceptions include very fine powders or when feeds have very high viscosities or large suspended particles which may block or damage the pressure nozzle. In the former case, the greater flexibility to manipulate and control particle size in the fine range favors two-fluid nozzles. Powders with higher densities are generally obtained from pressure nozzles compared to two-fluid nozzles. This is associated with the degree of aeration of feed during the atomization process. Low particle densities, if required, can be obtained by optimizing the aeration effect (gas ejection or pressurization of the feed).

Table 8.1 below summarizes the main characteristics of pressure and two-fluid nozzles.

Table 8.1 Guidelines for nozzle performance

Pressure Nozzles	Two-fluid nozzles	
Pressure swirl	External Mixing	Internal Mixing
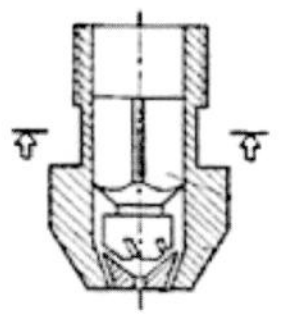	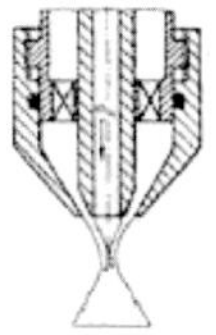	
Feed flow > 5 kg/h (not suitable for lab units)	High atomization gas consumption restricts the production of small particles at larger scales	Efficient atomization at low gas-to-liquid ratios
Flexible angles: 40-70°	Narrow angles: ~20°	Flexible angles: 20-60°
Eroded by suspensions	Suited for suspensions	Eroded by suspensions
Not suitable for viscous feeds	Insensitive to viscosity	Sensitive to viscosity
Weak manipulation of droplet size (for the same throughput)	High degree of control over droplet size	
Favors flowability Particle size – 20 to 500 μm	Fine powder with low densities Particle size – 3 to 200 □m	
Narrow distribution span ~ 1.4-1.8	Wide distribution; span ~2.0-2.4	

8.1.3 *Typical Challenges*

Most of the causes of unplanned shutdowns and/or limited run times are related either to (1) excessive buildup of material or the equipment walls, (2) improper atomization of the feed, and/or (3) chemical stability constraints (feed solution and/or powder). These issues and the ways to overcome (or account for) them are addressed in this section.

8.1.3.1 Product Accumulation

Product accumulation on the walls of the equipment is one of the most common occurrences when developing or during scale-up of a spray drying process. Product may accumulate on the walls of the drying chamber, cyclones, conveying ducts, or at the nozzle tip. Wall deposits are more commonly observed in the small scales since radial distances from the atomizer and residence time are shorter. Product buildup can be caused by several factors:

- Product stickiness; materials with a low glass transition temperature exhibit sticky properties and tend to build up on the walls of the equipment. The level of buildup is related to the content of solvent in the powder, the glass transition temperature of wet product, and the drying temperature at which the product is exposed. In the production of amorphous forms, this is particularly important since the deposits

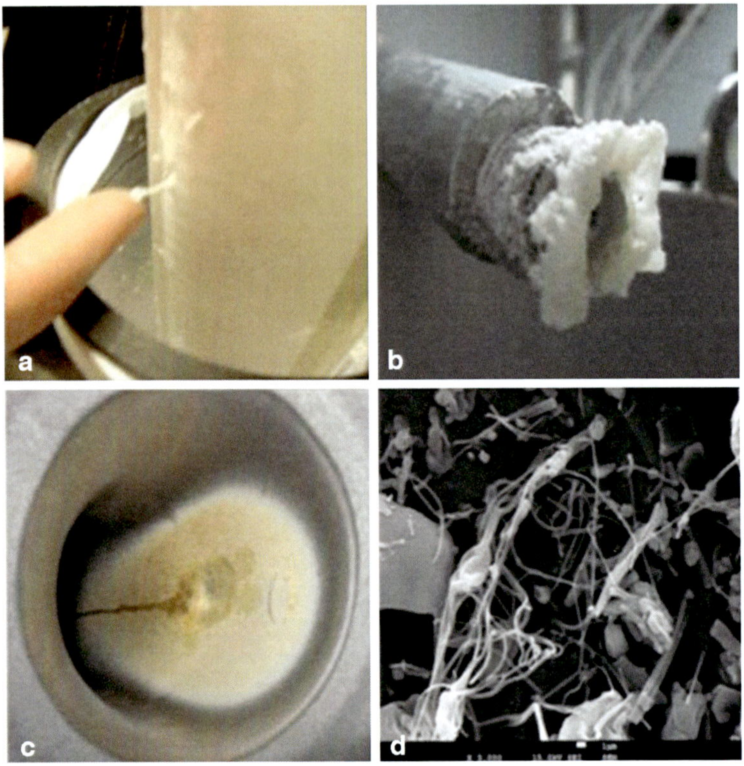

Fig. 8.3 Typical problems associated with spray drying operations: **a** condensation in the bag filter, **b** nozzle bearding, **c** dripping, and **d** stringing

on the equipment surfaces will occur when drying at temperatures close or above the glass transition temperature. The problem can be mitigated by reducing the outlet temperature (T_{out}) or by operating at lower relative saturation (RS_{out}) at the same outlet temperature. The latter is achieved by reducing the condenser temperature (T_{cond}) and/or the feed flow rate (F_{feed}). The stickiness tendency can be predicted offline, for example, using a hot stage and exposing the product to different temperatures.

- Solvent condensation; the gas inside a spray drying process equipment is partially saturated with solvent and, therefore, is prone to condense if exposed to cooler surfaces. If the surface is allowed to cool down below the dew point of the gas stream, then condensation will occur and the powder will accumulate on those wet surfaces (Fig. 8.3a illustrates an example where condensation occurred in the filter bag). To prevent this, the spray dryer is typically insulated or heat jacketed and the heating of the chamber prior to start of the atomization process is done in a gradual manner and over an extended period of time (typically between 20 and 60 min).

- Bearding; during spray drying, some droplets entrapped in the eddies around the nozzle may collide with the tip of the nozzle, dry on its surface and start building up around it. (see Fig. 8.3b). This process is commonly referred as bearding. If left unattended, the buildup can interfere with the spray formation and/or fall down into the drying chamber, promoting further powder deposition and ultimately clogging of the equipment. There are several approaches to overcome bearding. The simplest way is the repositioning of the nozzle (by changing the depth of the nozzle tip inside the chamber) to reduce or eliminate the droplet entrapment into the eddies. A change of atomization conditions as well as the drying gas flow may also be beneficial, though often there are narrow margins to manipulate these parameters. An engineering alternative is to include an additional gas stream concentric to the nozzle tip to prevent the collision of the droplets at the nozzle level. Another option is to use a nozzle with an anti-bearding cap which provides less deposition area for buildup to occur. More commonly, a combination of the above approaches is needed to successfully overcome bearding.
- Too large droplets; if droplets are too large for the drying chamber, they will touch the chamber wall before drying is complete. The solution is to decrease the droplet size or increase the drying temperature. Production scale may need to be changed if the target particle size cannot be achieved. This should be well thought out early in the development since all units have their own requirements regarding the maximum droplet size allowed.

8.1.3.2 Improper Atomization

Atomization is probably the most critical step involved in the spray drying process. The problems that can be associated with poor atomization include:

1. Inadequate atomization settings; this is the most common cause of poor atomization. Often in an attempt to increase particle size, atomization is set to maximize droplet size either by using low operating pressures (for pressure nozzles) or by using low atomization ratios (for two-fluid nozzles). However, there is a fine line between the conditions that allow maximum droplet size and those that lead to not fully developed spray.
2. Dripping; in pressure nozzles, the spray is fully developed only above a critical pressure. Below the critical pressure, often between 10 and 30 bar, large droplets may form which will fall into the drying chamber, which in turn may cause the formation of a wet layer of product in the discharging pipe of the drying chamber (see Fig. 8.3c). Dripping may also occur when restarting the operation after an unexpected shutdown. To prevent dripping, pressure nozzle with check valves that only allow the passage of liquid above a preset pressure can be employed. Some suppliers offer integrated check valves that can be adjusted for different pressures. Other option is the use of a compressed gas to purge the feed line once the high-pressure pump is switched off.
3. Stringing; stringing is the term used to refer to a bad atomization that happens in high evaporative systems. If a fine spray with high specific area is exposed to high

temperatures, string like particle formation may occur before the spray pattern is completely developed. String-like particles are obtained and agglomeration occurs, see Fig. 8.3d. In order to avoid stringing, it is required to delay the onset of particle formation by operating at lower temperatures or by reducing the concentration of the feed. Both approaches will reduce the throughput. If not resolved, solvent system or even the type of atomization system may have to be replaced.

4. Poor assembly of the nozzle; most nozzles require careful assembly of the internal and external components. Incorrect installation will result in poor atomization and often results in the shutdown of the process and loss of product. Nozzle assembly procedure should be carefully followed as per instruction and thoroughly tested before use.

8.1.3.3 Chemical Stability

The product purity profile is a very important attribute for any product. During spray drying operation, chemical degradation may occur:

- When the product is dissolved in the solvent prior to spray drying. This is particularly true at commercial scale, as the holding time may be long. It is therefore critical to define the solution storage conditions, especially the temperature and duration.
- When the freshly spray-dried product is still hot and not fully dried. In particular, some deposits of product may remain for a long time in the spray drying chamber, exposed to hot temperature (the walls of the chamber typically exhibit temperature close to the exit temperature of the drying gas).

Aforementioned chemical stability risks will be discussed below using a solid dispersion (API to excipient ratio of 1:4), where the solvent system used was a mixture of dichloromethane to ethanol ratio (97.5:2.5, % w/w), as an illustration.

8.1.3.3.1 Degradation of the Feed Solution

The degradation kinetics of a feed solution, at three different temperatures, is shown in Fig. 8.4. In order to determine accurate degradation rates, the timescale in this study was much longer than typical holding time. Although the formation of multiple impurities was observed, depicted in Fig. 8.4 is only the main (and most critical) impurity, for which an upper limit of 0.1 % was set.

As shown in Fig. 8.4, the kinetics show a linear increase over time, more pronounced at higher temperature. Using a simple Arrhenius equation, the activation energy (Ea) can be calculated, from which a mathematical expression capable of describing the growth of the impurity can be obtained. Equation can be readily used to construct a two-dimensional plot of temperature versus hold time (see Fig. 8.5).

Obtaining this type of representations enables the selection of the most appropriate feed solution temperature, in agreement with the target run time; for example, for a

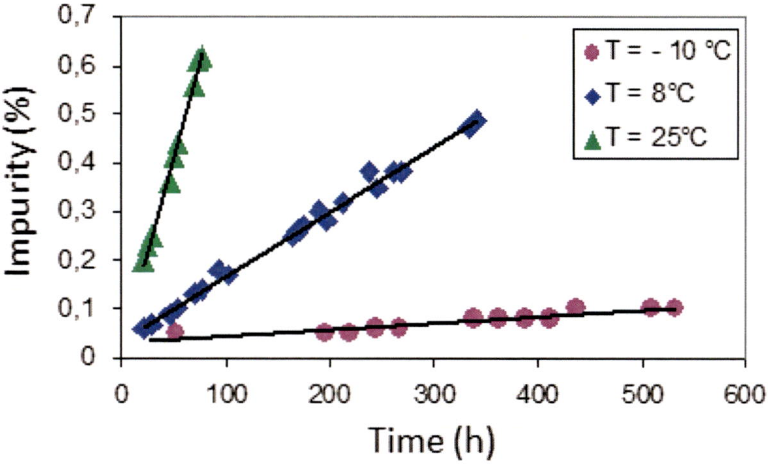

Fig. 8.4 Main impurity formation over time, as a function of the feed solution temperature

Fig. 8.5 Maximum (hold time vs. temperature), considering the main impurity formation

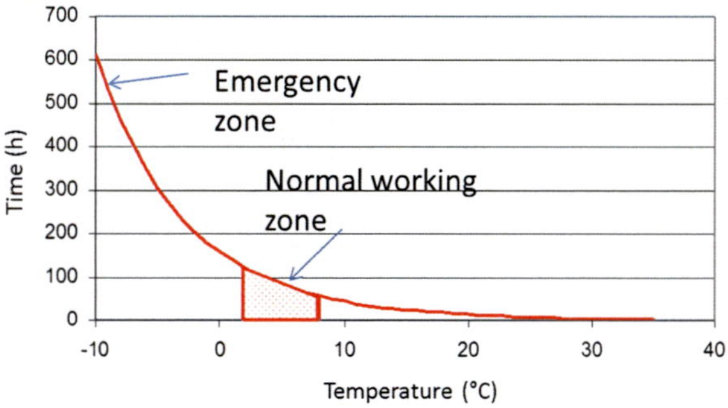

target spray drying time of about 24 h, a hold temperature range of 2–8 °C keeps the degradation level well below the 0.1 % limit. Additionally, in the case of emergency (e.g., equipment failure, requiring full process stop in order to allow for maintenance), Fig. 8.5 plot also shows that feed solution could be kept for more than 1 week (168 h) in case hold temperature is lowered to − 10 °C.

8.1.3.3.2 Degradation Inside the Spray Dryer Chamber

The spray-dried product accumulated in the chamber is exposed to harsh conditions (solvent vapor and high temperature) for an extended period of time and, therefore, it is expected to show a less favorable purity profile. Considering a worst-case scenario, where the entire amount of powder deposited on the walls would suddenly fall into the main product container, the entire batch could be jeopardized; therefore, it is important to evaluate the degradation of the deposited material as a function of its age (e.g., to define the optimal cleaning frequency of the spray dryer: too frequent would

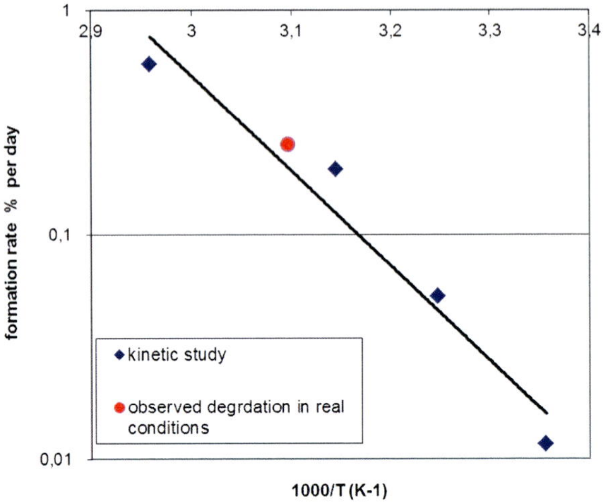

Fig. 8.6 Arrhenius plot for main impurity formation of the powder (lab kinetic results vs. spray drying data)

impact cycle time, whereas too rare would imply a risk of product contamination). These types of studies can be conducted using a two-step approach (the same example previously used to address the feed solution chemical stability is being considered):

- Acquisition of the degradation kinetics using a sample of freshly spray-dried product. In this illustrative case, a product with 3 % of residual solvents (the typical value obtained at normal operating conditions) was enclosed in containers and kept at four different temperatures (25, 35, 45, and 65 °C).
- Verification of the degradation kinetics at the real spray drying scale. In this illustrative case, a spray drying run (using a PSD3 unit, with an inlet and outlet drying gas temperatures of 95 and 50 °C, respectively) was performed for about 1 h (with a feed rate of 50 kg/h of feed solution), in order to obtain some product deposited on the walls; afterwards, the spray drying of pure solvent (at similar flow rate and thermal profile) was maintained for 24 h, in order to keep the product on the wall under normal degradation conditions. At the end of the trial, the powder deposited on the walls of the chamber was manually recovered (scrapped) and analyzed.

The results obtained for the current illustrative example are shown in Fig. 8.6.

As shown in Fig. 8.6, the rate of impurity growth observed in the laboratory is typically comparable to the degradation observed in the powder deposited on the spray dryer; considering the good alignment of the data, the degradation in real manufacturing conditions can be estimated from lab data (in this case 0.25 %/day). Therefore, by knowing the overall amount of deposited solids during a typical run (in this case known to be less than 1 % of the batch size) and considering that the drug product is typically homogenized and sieved, it is possible to anticipate the maximum impact for a given time window (e.g., 0.01 % of degradation during 4 days) and define the cleaning frequency (e.g., every 4 days of production would be, in this case, a good compromise).

8.1.4 *Modeling Tools and Mechanistic Interpretation*

The scale-up of spray drying processes has been primarily conducted based on actual experimental data and experience mainly because the process, characterized by rapid and simultaneous heat and mass transfer between the droplets and the drying gas, is difficult to describe mathematically, and some of the parameters are often not readily measured. Furthermore, the whole process is extremely dependent on the feed properties and equipment scale and design. Despite this, fundamental modeling approaches for process characterization have been proposed in the literature. Among these are thermodynamic models to estimate the humidity of the exhaust air (Berman et al. 1994), atomization models to predict droplet size (Lefebvre 1989; Senecal et al. 1999), or drying kinetics studies to anticipate the morphology of the particles (Larhrib et al. 2003; Littringer et al. 2013).

In this section, several tools, which will serve as the basis for the scale-up method, are presented: (1) a thermodynamic model to predict the outlet conditions of the spray drying, viz. relative saturation and temperature of the outlet gas; (2) an atomization model to predict the droplet size; and finally (3) a simplified model for droplet drying to define the general trends of particle size and morphology.

8.1.4.1 Thermodynamic Modeling

The thermodynamic model allows the characterization of the process in terms of drying gas flow (F_{drying}); relative saturation at the outlet of the drying chamber (RS_{out}); inlet, outlet, and condenser temperatures (T_{in}, T_{out}, T_{cond}); dew point of wet drying gas (T_{dew}); and feed flow (F_{feed}). A multivariate relationship of such variables can be used to define the process operation range that respects the desired outlet conditions (T_{out}, T_{dew}, and RS_{out}). The determination of such variables plays a role in the development of the spray drying process in several distinct ways. T_{out} is one of the most important process parameters since it has an impact on particle and powder properties such as density, surface area, mechanical strength, and physical stability. RS_{out} is the main driving force in the drying process and, therefore, has direct impact on the level of residual solvents in the final powder. The latter is of particular importance when producing amorphous materials since the level of residual solvent in the solids strongly affects its glass transition temperature (T_g). Dew point temperature (T_{dew}) is also an important parameter to have in mind in order to prevent solvent condensation in the equipment.

Essentially, the thermodynamic modeling consists in a set of equations that relate process parameters through mass and heat balances and liquid–vapor equilibrium equations. Below are the critical equations:

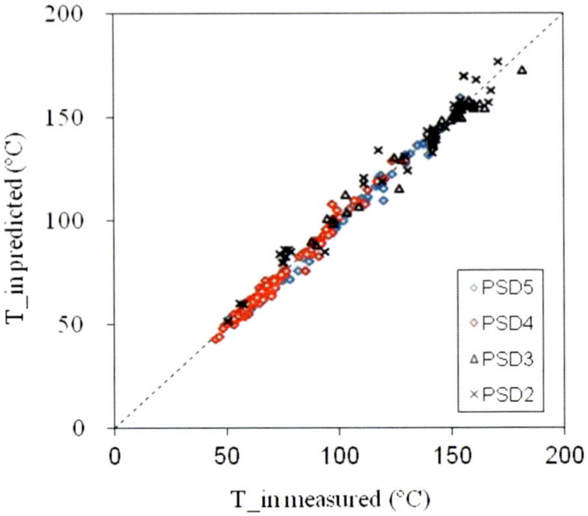

Fig. 8.7 Accuracy of the thermodynamic model. (Data from 386 batches, using multiple solvent systems and four scales of Hovione's Niro spray dryers)

$$F_{\text{drying}} \cdot \overline{Cp_{\text{drying}}} \cdot (T_{\text{in}} - T_{\text{out}}) = \sum_{i=A}^{B} \left[\overline{Cp_i} \cdot (T_{\text{out}} - T_{\text{feed}}) \cdot \dot{m}_i^1 \right] + \sum_{i=A}^{B} \left[\overline{\Delta H_{\text{vap}\_i}} \cdot \dot{m}_i^2 \right] + Q_{\text{loss}} \tag{8.1}$$

$$Q_{\text{loss}} = U \cdot A \cdot (T_{\text{out}} - T_{\text{room}}) \tag{8.2}$$

$$y_i \cdot P = x_i \cdot \gamma_i \cdot P_i^{\text{sat}}. \tag{8.3}$$

The accuracy of the thermodynamic model can be significantly improved by measuring experimentally the heat loss of a particular unit. Figure 8.7 depicts the accuracy of the model by comparing experiment with model inlet temperatures.

The development of a thermodynamic model is of utmost importance for modeling and scale-up purposes and provides an expeditious way to anticipate process conditions at any scale. Figure 8.8 below shows the typical information that is obtained through this model.

8.1.4.2 Droplet Size Estimation

The second stage of process modeling consists in applying a reliable correlation between atomization parameters, liquid properties, and droplet size. Numerous experimental studies have been carried out, and several equations have been proposed to relate droplet size to nozzle design, atomization energy, and physical and flow properties of the gas and liquids employed. Although some published models have proved

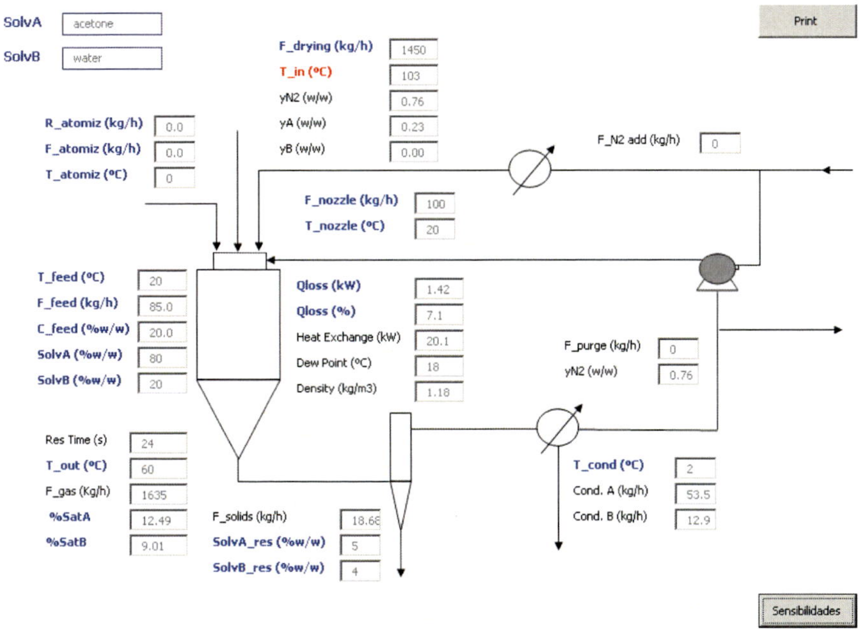

Fig. 8.8 Output of thermodynamic modeling for a process using pressure nozzle

good predictive capabilities, it is worth noting that these models were developed for specific nozzle geometry, and extrapolation to other nozzles needs to be done with care. An example of such correlations is shown in Eq. 8.4 below. The correlation was developed based on data from 12 nozzles from Spraying Systems (Maximum Free Passage, SK Series SprayDry®) with different geometric dimensions.

$$d_D = 2.41\sigma^{0.25}\mu^{0.25}F_{feed}^{0.25}P_{feed}^{-0.450}\rho_g^{-0.25}. \tag{8.4}$$

Figure 8.9 shows that despite many influences of process and formulation variables on the particle formation process, droplet size is still the major factor controlling the particle size, and it can be estimated by Eq. 8.4. The atomization model is a valuable tool for scale-up since it can be used to select the nozzle that best suits the targets for process throughput and particle size.

8.1.4.3 Particle Formation

To describe the particle formation process, several authors (Vehring 2008) emphasize the usefulness of the Peclet number (ratio between droplet evaporation rate and diffusional motion of the solutes) as a mean to predict the morphology of spray-dried powders. For low Peclet numbers, the diffusion motion of the solutes is fast compared to the velocity of the receding droplet surface and the droplet is allowed

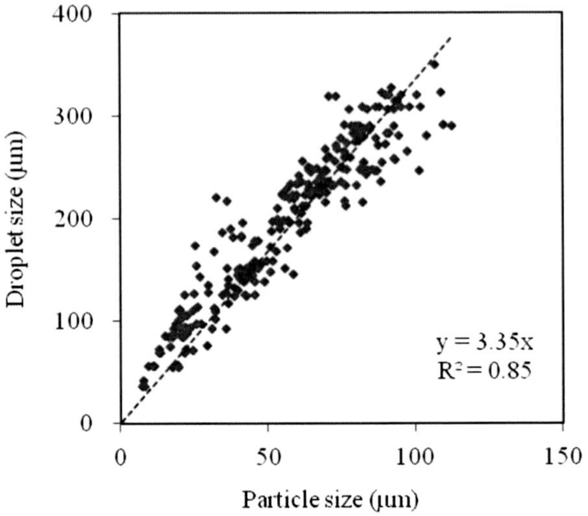

Fig. 8.9 Dv50 is highly correlated with the droplet size produced during the atomization process. Trendline was forced to pass through the origin

to shrink while solutes migrate to the droplet center. At a critical supersaturation level, dense and solid particles are produced. For high Peclet numbers, on the other hand, the evaporation predominates over diffusion and the surface becomes rapidly enriched in solutes that precipitate. In these cases, an outer layer is formed almost instantaneously at the droplet surface leading to hollow, light, and porous particles.

Often in pharmaceutical applications, viscous feeds are obtained as a result of the formulations used (e.g., polymers, proteins, carbohydrates among other large molecules) and particles tend to be hollow. However, the plasticity of the pharmaceutical materials during drying has resulted in greater morphologies diversity (Walton 2000). Due to the low diffusivity, most pharmaceutical particles form a shell earlier during drying and the rate of evaporation decays gradually as the shell becomes thicker. At this point, the shell mobility is determined not only by the diffusion of the dissolved solids but also by their solubility and, more importantly, by the mechanical properties of the formed shell (Vehring et al. 2007). If the drying rate is high, the critical thickness, i.e., the thickness that assures the mechanical stability of the shell, is reached very early in the drying process and the resulting particles sustain the spherical form of the droplet. On the other extreme, if the drying is slow, the thin shell formed in the early stages of drying will recede until its thickness is stable enough to sustain the particle structure. So by controlling the evaporation rate and drying time, one can exert control over particle morphology (see Fig. 8.10). The mechanism of evaporation in a still gas, based on boundary-layer theory, can be justifiably applied to many spray drying conditions (Masters 2002) and used to estimate the drying time of the droplets. While the constant evaporation rate is applied, the drying time can be expressed by the following equations:

$$t = \frac{\rho_L}{8D_v\left(P_{wb} - P_{out}\right)}\left(d_D^2 - d_P^2\right) \tag{8.5}$$

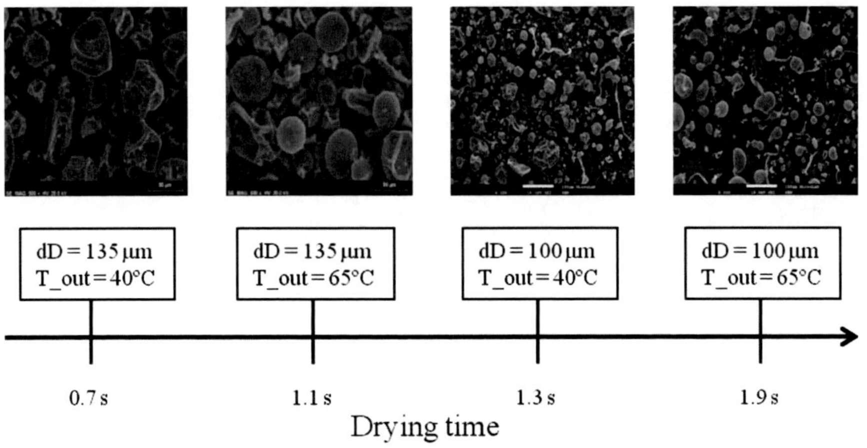

Fig. 8.10 Illustrative relationship between drying time and the sphericity of the spray-dried particles

$$Dv = \frac{0.00143T^{1.75}}{MW_{AB}^{1/2}\left[(\Sigma_v)_A^{1/3} + (\Sigma_v)_B^{1/3}\right]^2} \tag{8.6}$$

$$T_{wb} = 137\left(\frac{T_b}{373.15}\right)^{0.68} \log(T_{out}) - 45 \tag{8.7}$$

In Eq. 8.5, the vapor pressure at wet bulb temperature can be calculated from the Antoine equation, while the solvents concentration (P_{out}) in the drying gas can be determined in the thermodynamic step described previously. For example, fast evaporation and low drying times can be imposed by manipulating T_{out} and droplet size (d_{D}) and used to promote the production of smooth spherical particles.

The other feature throughout the formation of particles is the creation of internal pressure when droplets are dried at temperatures close or above the boiling point of the solvent. In this case, the vapor pressure inside the particles is higher than the outer surface and particles can, depending on the shell properties, inflate or break apart as the result of the pressure gradient. When particles expand, then particle size becomes a function of droplet size and outlet temperature (Fig. 8.11a). On the other hand, if the material is friable, the particles tend to break apart and the degree of breakage is typically more pronounced at higher temperatures. In those cases, particle size is still dependent on the droplet size, but the outlet temperature or relative saturation has a negative effect on particle size (Fig. 8.11b).

The tools presented throughout this section are intended to be used in all the steps of the scale-up to assure the production of powders with the desired quality, viz. particle size, particle morphology, and level of residual solvents. This enhanced understanding and mechanistic thinking, in line with the quality by design (QbD) initiative, will support the establishment of ample design spaces that are both unit and scale independent.

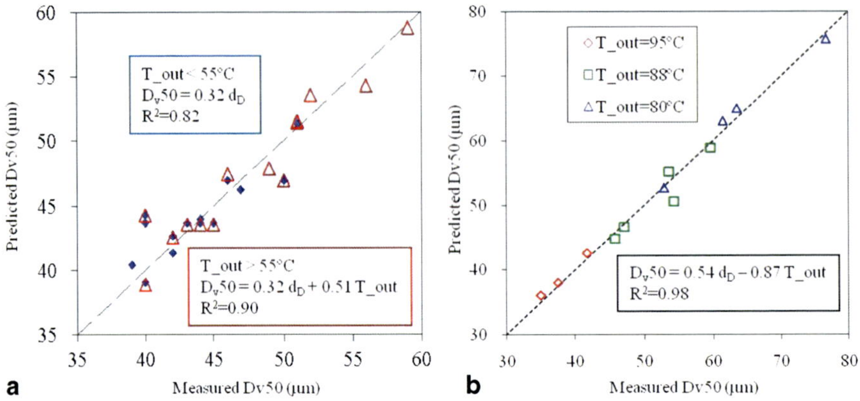

Fig. 8.11 **a** Particle inflation observed above boiling point; **b** typical behavior of a friable material

8.1.5 Scale-Up Methodology

As discussed in the Sect. 8.1.1, the spray dryer scale may have a significant impact on the properties of the spray-dried material and general scale-up procedures have to be established to assure an uneventful transfer to larger scales. Before embarking on the scale-up of any process, it is highly recommended to attain a stable and robust process at lab scale. Only then one can clearly understand how key process parameters such as temperature profile in the drying chamber (T_{in} and T_{out}), condenser temperature (T_{cond}), drying gas and feed flow (F_{drying} and F_{feed}), and atomization conditions should be set at the larger-scale unit. The present methodology is based on the mechanistic understanding described in the previous section and comprises three modeling steps: thermodynamic, atomization, and particle formation.

8.1.5.1 Thermodynamic Step

Through thermodynamic modeling (see Sect. 8.1.4), one can calculate the relative saturation of the drying gas. This combined with a small set of experiments at lab scale provides the relationship between the level of residual solvents in the product and relative saturation of the drying gas. This is particularly important when producing amorphous materials since their glass transition temperature (T_g) is affected by the residual solvent since solvent acts as a plasticizing agent. Figure 8.12a illustrates how these relationships can be used for scale-up purposes: If the target is to manufacture a spray-dried (wet) powder at a $T_g > 60\,°C$, then the level of residual solvent in the spray-dried material should not exceed 9 % w/w. At lab scale, this residual solvent level was obtained when operating with a relative saturation of 8 %. Therefore, a possible scale-up condition, which can be seen as a conservative or safe approach, is to maintain at the larger-scale equipment a relative saturation at the exit of the drying chamber (RS_{out}) at a similar level. If droplet size is maintained somewhat

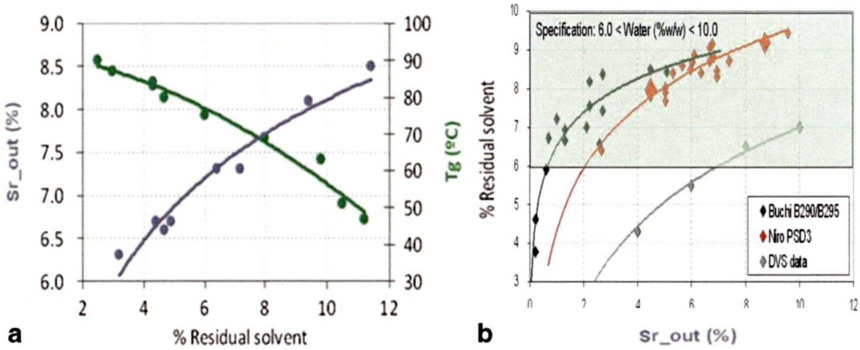

Fig. 8.12 **a** Effect of the relative saturation of the drying gas on the level of residual solvents and T_g and **b** effect of the scale on the desorption curves of a spray-dried product

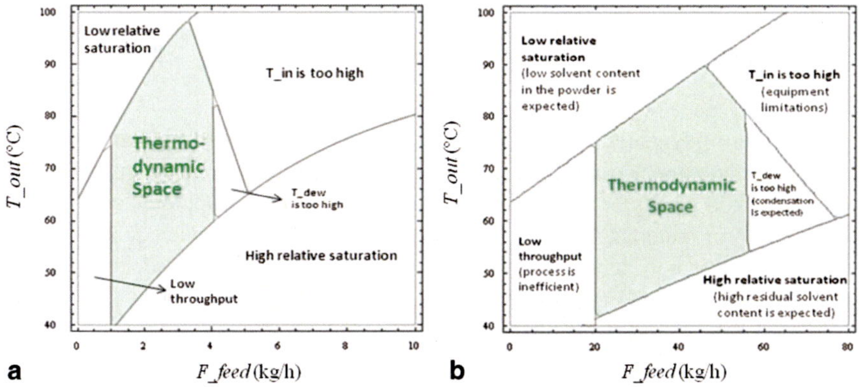

Fig. 8.13 **a** Thermodynamic space at pilot scale and **b** projection to commercial scale

unchanged, the extended residence time in larger scales will provide a safety margin to the assumed relationship between RS_{out} and level of residual solvents. As can be seen in Fig. 8.12b, the sorption curve approaches the equilibrium when increasing the scale (equilibrium data is obtained with dynamic vapor sorption studies).

The multivariate thermodynamic relationship of the process parameters with RS_{out} can also be used to define the process operation range that respects the imposed constraints. Apart from the limits imposed on the RS_{out}, the constraints usually include equipment limitations (e.g., maximum inlet temperature or minimum condenser temperature), product requirements (e.g., outlet temperature limited by the product degradation profile, physical stability, or stickiness behavior), and other process constraints (e.g., F_{drying} limited by the gas disperser or flow requirements, F_{feed} limited to avoid high dew points). These theoretical relationships provide a bridge between processes at different scales, as depicted in Fig. 8.13 below.

Fig. 8.14 **a** stable spray drying process, **b** product degradation in the cyclone, and **c** bad atomization leading to product accumulation in the bottom of the drying chamber

This thermodynamic analysis is a very powerful tool often replacing the need for experimentation at different scales. However, as mentioned before, product constraints should ideally be studied at lab scale to minimize development costs. Visual degradation of material, bearding, heavy accumulation due to inadequate temperature profiles, or poor atomization among other are all easier to understand and to solve at lab scale (see Fig. 8.14).

Some of the most common studies performed at lab scale include:

- Stability of the drug in the feed. This can be critical to the success of the scale-up. In larger scales, product may be held in solution for many hours or days, and therefore, the kinetics of degradation at different feed temperatures should be known. In some cases, the feed temperature needs to be reduced to prevent impurity growth.
- Process yield (expect $> 80\,\%$ for a sample of more than 5 g). The reasons for a low yield are very diverse (e.g., product stickiness or bad atomization) and should be solved at small scale before scale-up.
- Ability/ease of the secondary drying step. At lab scale, drying profiles and physical/chemical stability of the product can and should be evaluated.
- Process edge of failure by spray drying, for example, at elevated relative saturation (RS_{out}) can be very useful to understand the limits of the process and provide extra information for the scale-up process.
- When stable process conditions are found, it is recommended to run the process for an extended period to monitor the robustness of the process. Note, for example, that some processing issues (e.g., nozzle bearding or heavy product accumulation in the equipment walls) may not be obvious in very short tests.

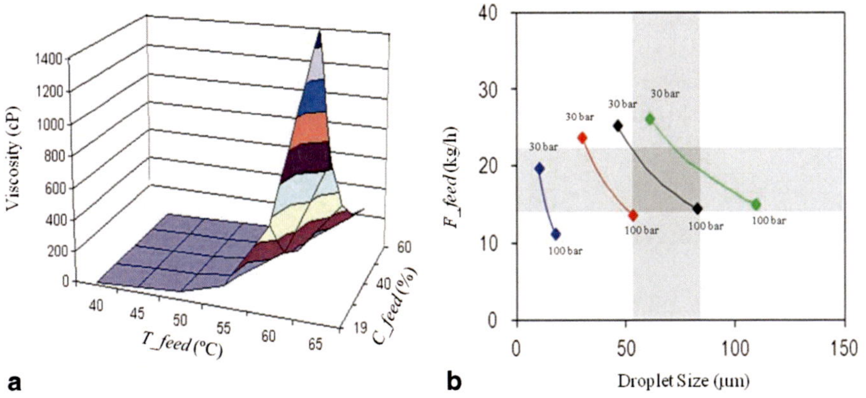

Fig. 8.15 **a** Feed solution characterization regarding the impact of C_{feed} and T_{feed} on feed viscosity and **b** simulation of four nozzles MFP (maximum free passage) SK series SprayDry nozzles from Spraying Systems; the core/orifice codes are –80/16, –70/20, –65/21, and –65/17

8.1.5.2 Atomization Step

Once the thermodynamic conditions have been established, there is a need to select the nozzle that best suits the targets of droplet size and process throughput. The most common nozzles used at lab scale are external two-fluid nozzles. However, during scale-up, there is typically the opportunity to improve powder properties by switching to a pressure nozzle. This results from the greater ability to produce and dry larger droplets in the larger drying chambers of the commercial units.

Feed properties like viscosity, density, and surface tension are well known to affect droplet size (and hence particle size). Therefore, for an accurate estimation of droplet size, it is recommended to characterize the solution regarding those properties. Frequently, all of these properties are dependent on the feed temperature (T_{feed}) and solids content (C_{feed}).

C_{feed} is a critical parameter which impacts the process viability and product quality in several manners, viz. process throughput, particle/powder density, or feed viscosity. Economic considerations of the process favor high C_{feed}, but concentrations close to the saturation point should be further studied to minimize the risk of product precipitation. Further, the use of high concentrations may lead to highly viscous feeds which may be difficult to atomize (Fig. 8.15a illustrates a typical relationship between feed concentration and temperature with feed viscosity). T_{feed} is rarely manipulated to obtain target powder properties. Nevertheless, it may affect solution stability and also influences viscosity and solubility. A strict control of T_{feed} is required, namely when operating close to solubility limits.

After determination of the feed properties, correlations like the one shown in the Sect. 8.1.4 (Eq. 8.4) are commonly used to predict the droplet size.

In the example given in Fig. 8.15b, the ranges of interest were 52–80 μm for droplet size and 14–21 kg/h for the feed flow rate. The most adequate nozzle, according to the feed properties measured and simulations performed, was the 65/21. The 65/21 nozzle fulfills both criteria within the typical operating conditions of pressure 30–100 bar.

8.1.5.3 Particle Formation Step

In order to establish a target for droplet size, it is required to study the effect of the drying condition on droplet drying and consequent particle formation. In the particle formation step, which is by far the most complex physical mechanism to describe, two approaches may be considered to estimate the final particle size.

The first approach assumes a constant shrinking ratio, i.e., a characteristic ratio between droplet size and particle size. When there is no information available about the product drying behavior, a pragmatic approach can consider a general shrinking ratio of about 3.3 (as seen in Fig. 8.9, most spray drying products follow the general trend of droplet size, independently of their nature and drying conditions).

A second approach is to use experimentally found shrinking ratio. This requires some prior knowledge of product/drying behavior. For example, the shrinking ratio observed at a smaller scale (and using similar drying conditions) can be used to model particle size at larger scale. In this approach, there is a need to measure or estimate the apparent density of the particles (measured, for example, by mercury intrusion porosimetry—Fig. 8.16), which can then be used to estimate particle size using a mass balance to the solids in the droplet and particle (Eq. 8.8).

$$\frac{d_{droplet}}{d_{particle}} \sqrt[3]{\frac{\rho_{particle}}{\rho_{droplet} \times C_{feed}}} \tag{8.8}$$

Besides particle size, the control of the particle morphology may also be critical for some applications. Surface area, particle density, and roughness or porosity are all known to affect the performance of spray-dried powders. Therefore, it is common during the development of a spray-dried product to explore the process to produce powders with distinct characteristics. This step of powder optimization can be conducted at pilot scale where a good compromise is achieved in terms of the range of particle/powders that can be obtained and the ability to target these ranges throughout the remaining of the scale-up process.

Particle properties are related to the drying kinetics of the droplets inside the drying chamber and are primarily dependent on the mechanical and chemical properties of the spray-dried material. These interactions are very complex to model, and therefore, process development is typically ruled by some general trends, as described below (see Fig. 8.17). The parameters that most influence particle morphology are T_{out} (or RS_{out}) and C_{feed}. Note that by adjusting the droplet size produced during the atomization process, one can control the particle size in such a way that size and morphology become almost independent of each other.

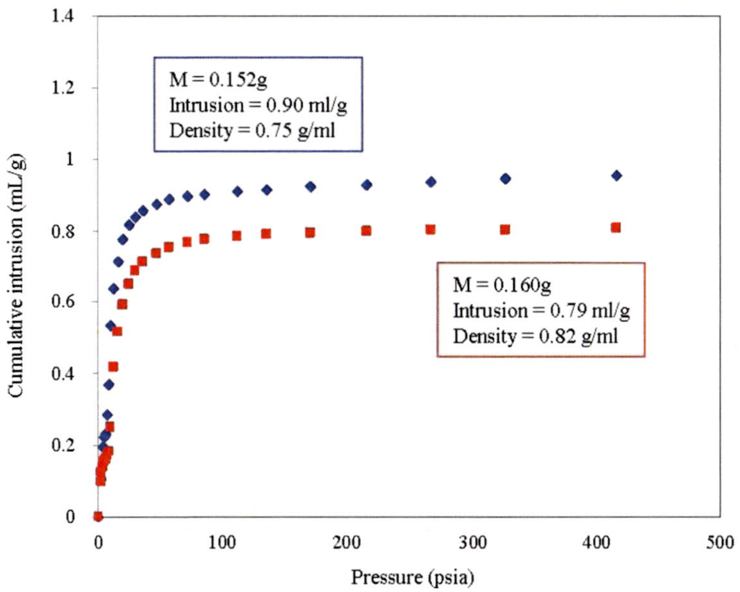

Fig. 8.16 Mercury intrusion in HPMCAS spray-dried particles manufactured at a Niro PSD4

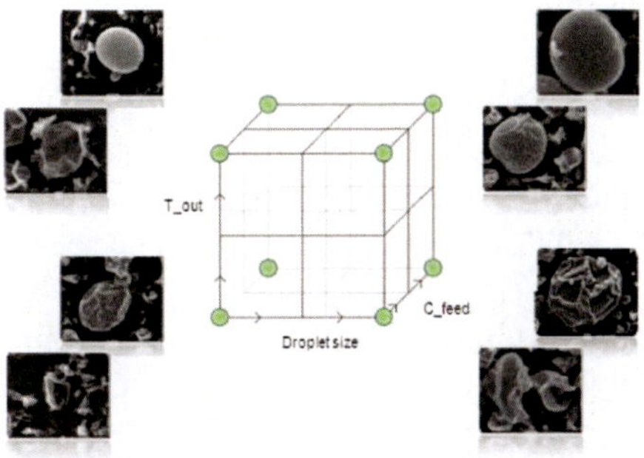

- Increase *T_out* to promote the production of smooth spherical particles
- Increase *C_Feed* to promote the production of dense particles
- Decrease *C_feed* to promote the production of shriveled particles (increased surface area)
- Size and morphology are roughly independent

Fig. 8.17 General guidelines to produce particles of different morphologies

8.1.6 Process Intensification

When moving to a full commercial scale, one of the main goals is to increase the process throughput while maintaining or improving the attributes of the powder. In other words, the goal is to increase F_{feed} and keep constant the RS_{out} and droplet size, already optimized at a smaller scale. Droplet size and throughput can be controlled by nozzle selection and atomization conditions, while RS_{out} is controlled by T_{out}, F_{feed}, and T_{cond}. This is achieved by using the thermodynamic and atomization models

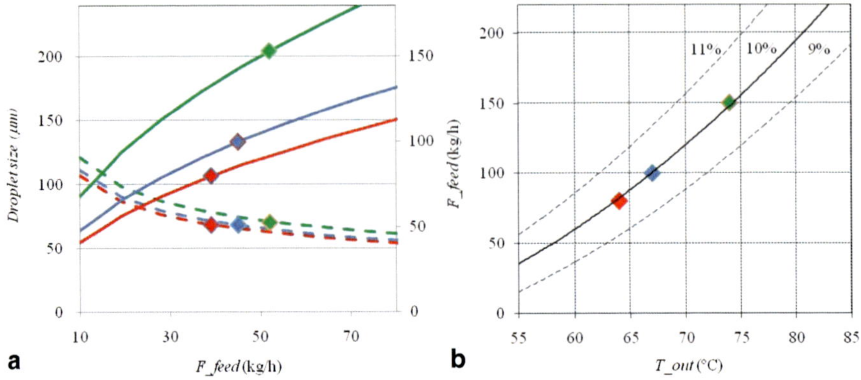

nozzle	A	B	C
P_feed (bar)	39	45	50
d_D (μm)	66	66	69
F_feed (kg/h)	80	100	150
RS_out (%)	10	10	10

Fig. 8.18 Simulation of **a** nozzle performance and **b** relative saturation for three possible scenarios

described before. Figure 8.18 describes an illustrative example of how this can be simulated.

As depicted in the simulations of Fig. 8.18a, several nozzles can produce droplets within approximately the same size range but at very different feed flow rates. By changing from nozzle A to nozzle B or C, droplet size can be maintained while the process throughput is increased. With the feed flow defined, then, as can be seen in Fig. 8.18b, the heat requirements can be determined by the thermodynamic modeling in such a way that outlet temperature is defined in order to have approximately similar levels of relative saturation of those optimized in previous stages. It is schematized in Fig. 8.18; three different scenarios that, according to the developed models, have similar conditions of droplet size and relative saturation. It is expected that the three scenarios result in the production of powders with equivalent quality.

8.1.7 Conclusions

A scale-up method was proposed for spray drying processes. The method encompasses three steps: thermodynamics, atomization, and particle formation. A scale-up approach based on keeping the relative saturation of the drying gas at constant level can be derived using only lab-scale data to define the thermodynamic design space. The atomization condition and nozzle selection are decided based on target process throughput and droplet size considerations. The methodology presented here reduces the risk of failure during the initial scale-up of the process. The process can then be

Table 8.2 Summary of the scale-up methodology

Feed properties	
T_{feed}	Use the T_{feed} defined at lab scale. Keep a strict control over T_{feed}, mainly if operating close to the solubility limit
C_{feed}	Preferably defined at lab scale. May need adjustments for adjusting powder properties or for process intensification
F_{feed}	Work within the thermodynamic space defined by the target throughput and drying capacity of the equipment
Drying gas variables	
F_{drying}	Use the nominal flow of the equipment and adjust for drying gas density to keep pressure drop through gas disperser at nominal level
T_{out} and T_{cond}	Work within the thermodynamic space. Use initially a conservative scale-up approach; keep RS_{out} similar to the lab scale. Adjust at larger scale based on desorption/T_g data
Atomization variables	
Nozzle/P_{feed}	Pressure nozzle preferred for most applications. Very small particles (e.g., for inhalation powders), feeds with large suspended particles and very viscous feeds, may require other atomization systems, namely two-fluid nozzle. Use droplet size correlations to select the most suited nozzle and atomization conditions

T_{in} is a dependent variable that is limited by equipment constraints

gradually intensified to improve throughput while keeping material attributes roughly unchanged. Table 8.2 summarizes the methodology suggested.

8.2 Development of a Manufacturing Process of a Spray-Dried Dispersion Under a Quality by Design Approach

8.2.1 Methodology Overview

Pharmaceutical QbD is a systematic scientific risk-based holistic and proactive approach to pharmaceutical development that begins with predefined objectives that address product and process understanding. Successful product development relies on consistent application of a proven methodology. The key steps are the same, irrespective of the product or formulation being developed. One proven methodology is described within this chapter. The framework is shown in Fig. 8.19, while a short description of the main steps is given below.

Target Product Profile and Critical Quality Attribute The target drug profile consists of prospective and dynamic summary of the characteristics of a drug that should be achieved in order to reproducibly deliver the therapeutic benefit; the target product profile (TPP) sets an important number of performance parameters that will be the

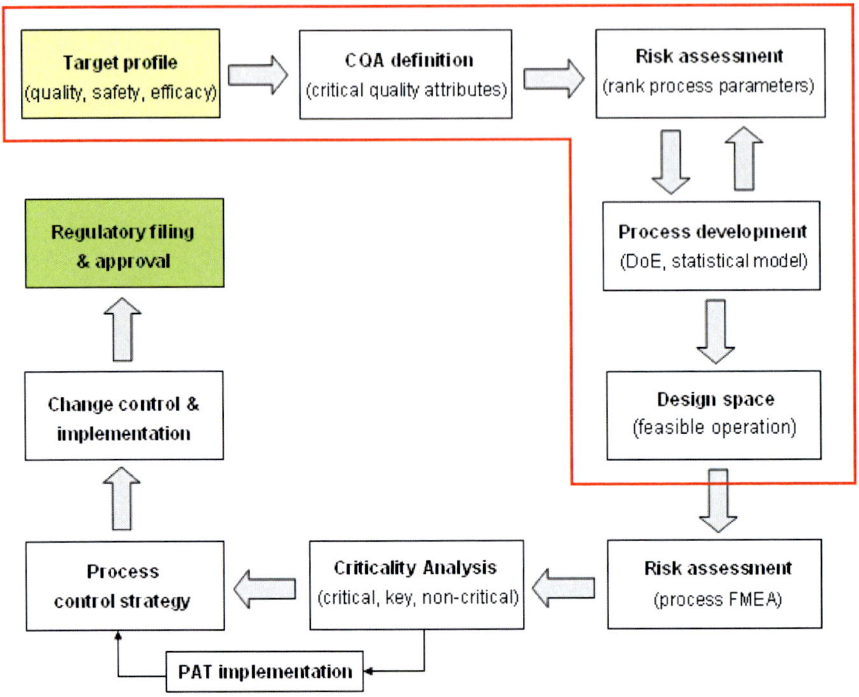

Fig. 8.19 Quality by design framework (main stages)

basis of the critical quality attribute's (CQA's) definition, that is, the attributes of the drug that must be kept within appropriate limits in order to ensure the desired product quality.

Risk Assessment (Development Phase) For each CQA, an analysis of the potential critical process parameters (pCPPs) and potential critical material attributes (pCMAs) is conducted. The aim is to evaluate, in each process step, which operating parameters or raw materials have the potential to impact a CQA, within the known ranges, and therefore should be monitored or controlled, in order to ensure the desired quality. Since the number of parameters is usually high, a risk assessment, based on prior knowledge of product/process, is used to rank the parameters in terms of perceived criticality; the ultimate goal is to keep the development process as lean as possible, by focusing the studies on those parameters and material attributes with a higher likelihood of having a critical impact.

Process Development The output of the previous risk assessment is a qualitative match between CQAs and pCPPs/pCMAs. To confirm the dependences and quantify the effects, a process development stage is conducted. If a statistical approach is followed, a sequence of design of experiments (DoE) is usually performed with different objectives: screening, optimization, and robustness studies. This development stage constitutes the core of the QbD methodology since most of the specific process knowledge is generated during this stage. Although not mandatory, a model, either statistical and/or mechanistic, is a usual outcome of this stage. Process analytical tools can also to be considered at this stage; based on the need to improve, the CQA's monitoring as the process is scaled up.

Design Space and Normal Operating Range Once the impacts of the pCPPs/pCMAs are quantified on the CQAs, a feasible operating space can be defined. This space, known as the design space, will consider all the interactions between operating parameters and material attributes and will often be multidimensional. The normal operating range (NOR) is established within the design space, and can be thought of as the ranges where the process typically operates.

Risk Assessment (Manufacturing) After defining the design space and NOR, an exhaustive analysis of the process is conducted at the manufacturing scale. In this study, a failure mode effect analysis (FMEA) of all manufacturing aspects are reviewed, challenging the equipment operating ranges and procedures against the process knowledge gathered in the previous steps. The purpose of this study is to understand and quantify the risk of failure and to define actions to minimize it.

Criticality Analysis By knowing the feasible operating regions, the design space, and after evaluating the equipment/procedures at the manufacturing scale, directly evaluating the practical NORs, a final criticality analysis will take place in order to identify parameters and/or material attributes that will require tight monitoring or control; for example, all those for which the corresponding NORs are close to the boundaries of the design space.

Process Control Strategy Once the criticality around a process parameter and/or raw material attribute is confirmed, adequate control strategies will be set in place. The ultimate goal is to assure that operation is always taking place within the design space, therefore assuring the quality of the final product. For this purpose, and considering the dependence of a control strategy on a given monitoring capability, the final implementation of process analytical tools may be carried out at this stage.

The subsequent steps of this methodology are mainly focused on documentation aspects associated with the filing process and, given the purpose of this current article, will not be further discussed. This work focuses on the steps highlighted in Fig. 8.19, which will be discussed in detail in the sections below.

8.2.2 Case Study Overview

Two particular difficulties are generally recognized during the formulation of solid dispersions by spray drying: the need to dissolve both the drug and the polymer in a common solvent system and the need to prevent phase separation during the removal of the solvent. The selection/optimization of formulations will not be addressed in this chapter as the corresponding approaches/methodologies are extensively discussed throughout this book by other authors; the formulation presented in this case study, optimized in previous stages of development, consisted of (1) a binary solvent system (methylene chloride and ethanol, 95/5 % w/w), (2) hypromellose phthalate (HPMCP) mixed with the drug at a ratio of 4:1, and (3) a solids concentration of 9 % w/w.

The solution is prepared in a reactor with a mechanical stirrer and a thermal circuit for temperature control. After complete dissolution, the solution is fed to the spray dryer. Droplet size is controlled by the liquid feed flow and by the type of atomizer and atomization conditions (pressure nozzles were used during this work). T_{out} is used to define the morphology of the particles and assure an efficient drying. The particles obtained were separated from the drying gas through a cyclone. The unit is operated in closed-loop mode, i.e., with recirculation of the drying gas. The solvent was removed by a condenser temperature within the gas recycling unit. Finally, the spray-dried material is collected and transferred to a double cone dryer for a secondary drying operation to fulfill the applicable limits for residual organic solvents.

8.2.3 *Target Product Profile and Critical Quality Attributes*

As introduced in Sect. 8.2.1, the roadmap of any QbD approach starts with the Target Product Profile (TPP) definition; this summary of drug characteristics (e.g., pharmacokinetic properties and stability) will serve as the basis for a set of performance parameters (e.g., immediate release drug: 80 % in $\leq$ 30 min, 36-month shelf life at room temperature, respectively) that, in turn, will be linked to a set of Critical Quality Attributes (CQAs; e.g., shelf life will depend on the amount of *residual solvents* due to its impact on chemical stability; release profile will depend on *particle size* for some drugs due to its impact on dissolution).

As one may expect, not all attributes of a drug will be classified as critical; Q6A (Conference and Harmonisation 1999) offers guidance on this matter, by distinguishing groups of attributes that should always be classified as critical regardless of the drug's end use (e.g., identification, assay, purity) from others whose classification (critical, noncritical) will depend on the final product nature (see decision trees in Q6A Conference and Harmonisation 1999). In order to conduct this exercise, it is therefore important to have the full picture of all manufacturing processes that are associated with a given product since, as depicted in Fig. 8.20, some of the attributes of an intermediate manufacturing step (e.g., density of the bulk powder) are established as critical due to their importance for subsequent manufacturing steps (e.g., by affecting tablet hardness during downstream operations).

For the sake of simplicity, the current case study will only focus in one of the manufacturing steps (the spray drying of the bulk powder, as previously introduced in Sect. 8.2.2) and, within this one, only show the application of the QbD methodology for two of the CQAs: particle size (Dv50) and bulk density (BD). The considered specifications for these CQAs (31–57 μm for particle size and 0.100–0.200 g/ml for BD) were set based on the interaction/interdependence with the downstream process

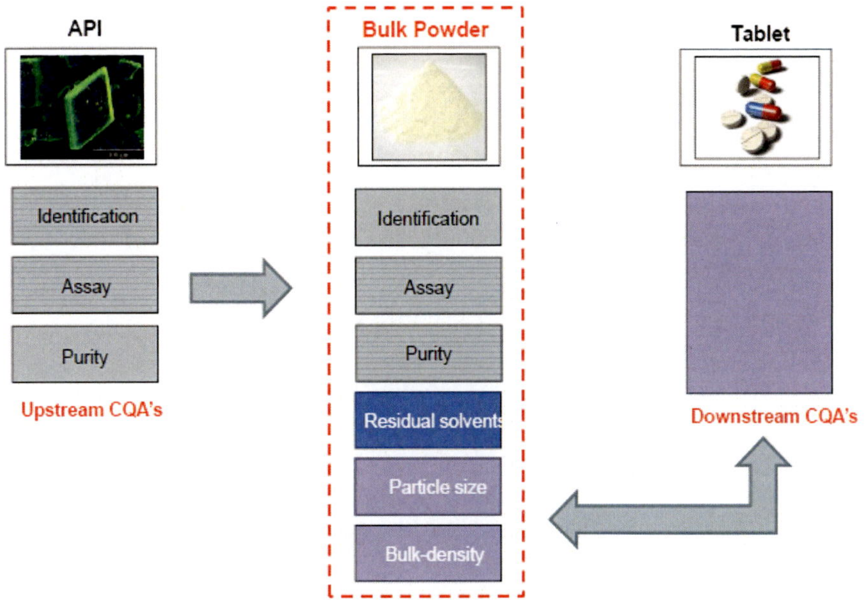

Fig. 8.20 Interactions between upstream and downstream processes during CQA definition

where, as introduced before, these attributes reveal a critical importance during manufacturing of the final oral dosage form.

8.2.4 Risk Assessment

The number of parameters involved in any spray drying process is relatively large, and evaluating the impact of each one, in all CQAs, would be difficult to manage (both from a cost and time perspective); therefore, one of the main goals of the risk assessment, as previously introduced in Sect. 8.2.1, is to reduce the number of pCPPs that will be studied in subsequent stages of process development. As shown Fig. 8.21, the procedure adopted considers the ranking, for each CQA, of all process parameters, according to the perception of criticality (that relates to the mechanistic understanding and relevant manufacturing experience with similar products).

Once a ranking of perception of criticalilty is obtained, the process parameters can then be divided into three groups: a first group (T_{out}, P_{feed}, D_{noz}, T_{cond}), that is considered to have a potentially relevant impact on the CQAs and, therefore, will be studied in detail in order to establish the design space; a second group (F_{drying}, C_{feed}), considered to have a lower potential of criticality—these parameters will not be optimized (they will be fixed at preestablished set point values based on prior knowledge of the process or similar processes) but will be used during the final evaluation of the process robustness (in order to confirm the assessment made); a

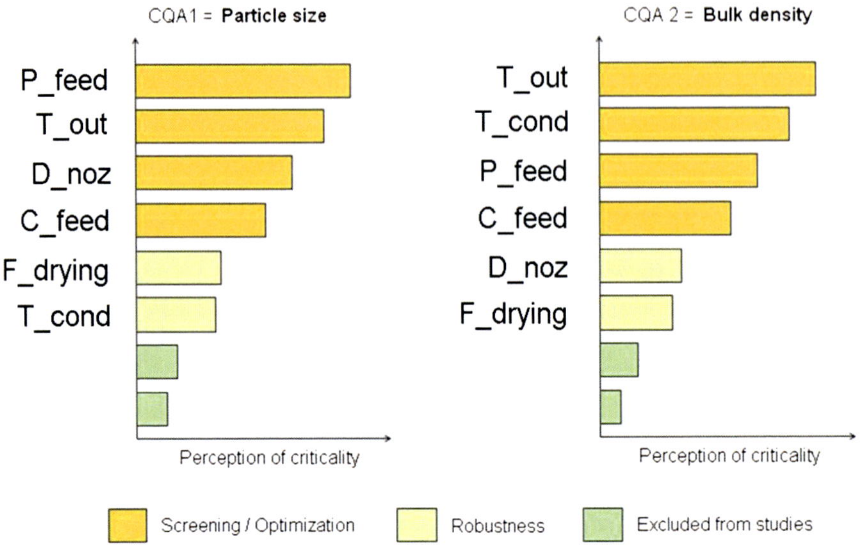

Fig. 8.21 Ranking of pCPP per each CQA

third group (the least ranked parameters), that are expected to have no impact on the process and that will be excluded from further studies. The rational for exclusion of these parameters needs to be appropriately justified in a QbD application.

8.2.5 *Process Modeling*

The outcome of the previous risk assessment is a qualitative match, based on perceptions, which needs to be confirmed and quantified during a subsequent process development stage. A mechanistic description of the process, as introduced in Sect. 8.1.5, can be used as a powerful tool to establish the design space with minimum need for experimentation at final scale. However, this type of description is not always readily available, and the use of a statistical approach constitutes a pragmatic alternative. The limitation of the statistical approach is its reliance on scale-dependent experimentation and the difficulty in extrapolating relationships to other scales or equipment. Selection of the most adequate approach (mechanistic or statistical) depends, therefore, on the (1) existence/reliability of the mechanistic understanding, (2) availability of material for experimentation, and (3) flexibility required on the design space. In this section, a statistical approach will be illustrated in order to quantify relationships between process parameters and product attributes; the three involved steps are described in Fig. 8.22.

The objectives of a screening stage, used to confirm the most significant factors, are to determine the ranges of process parameters to be investigated and to reduce the

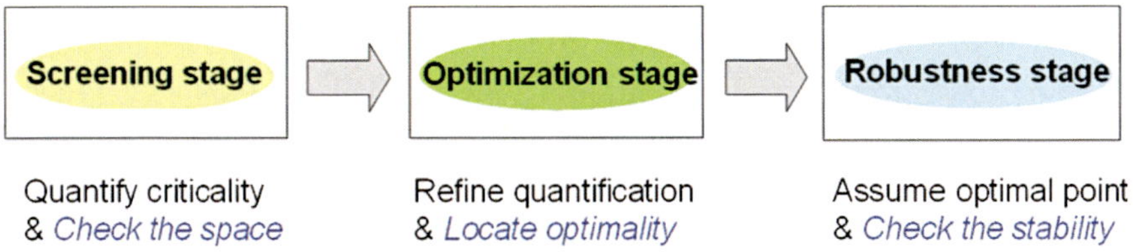

Fig. 8.22 Statistical modeling approach: structured sequence of design of experiments

number of parameters to be further studied. Since an accurate quantification of the effects is not crucial at this stage, low-resolution experimental designs can be considered. Hence, few experiments per studied factor are typically required. The second level of the experimental plan consists of the optimization of the process conditions over the most promising subregion found during the screening stage (considering only the parameters that have previously shown statistical significance). Since the ability of capturing interactions between process parameters may greatly influence the outcome of this stage, high-resolution experimental designs are advised. Finally, the robustness evaluation is conducted in order to determine the sensitivity of a CQAs toward small changes in some process parameters. Typically, the robustness evaluation is centered at the target operating point (defined during the optimization stage), by considering variations on parameters that have not been studied in detail.

8.2.5.1 Screening Stage

Among the pCPPs of the spray drying step, the risk assessment identified feed concentration (C_{feed}), feed pressure (P_{feed}), outlet temperature (T_{out}), condenser temperature (T_{cond}), and nozzle orifice diameter (D_{noz}) as the highest-ranked ones.

For the screening stage, a DoE with resolution ≥ 4 is recommended in order to retrieve unconfounded relationships between first-order terms. The structure of the DoE is shown in Fig. 8.23. In order to study a broader range of process throughputs, six star points were added to a 2^{4-1} fractional factorial.

The results obtained, shown in Fig. 8.24, reveal that the response of BD seems to be well described by a linear function of T_{out}. For particle size, although a reasonable model has been obtained, a linear structure seems to be insufficient to explain the observed variance (as denoted by the relatively low R^2 and Q^2).

Additionally, this screening phase shows that two factors (T_{cond} and D_{noz}) were found to have no statistical significance (within the tested ranges) and, therefore, should be subsequently excluded from the optimization stage. Finally, the screening model also showed that a great share of the evaluated operational ranges is feasible (considering the target ranges for Dv50 and BD).

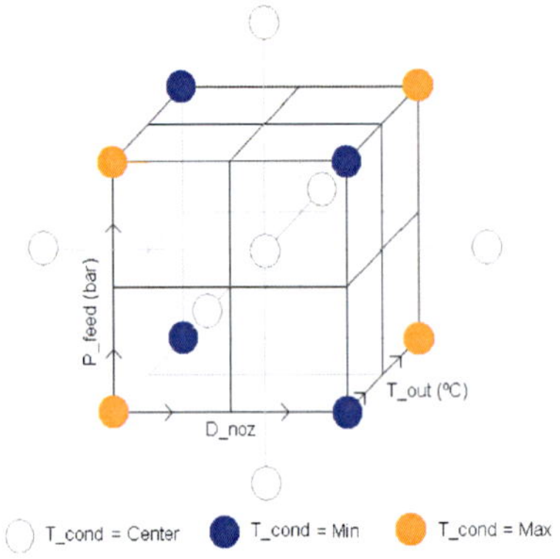

Input variable	Min	Center	Max
P_feed (bar)	50	85	120
T_out (ºC)	25	55	85
T_cond (ºC)	-10	-5	0
D_noz (mm)	1.01	0.91	0.78

Fig. 8.23 Screening study: DoE structure (*left side*) and ranges of variation studied (*right side*)

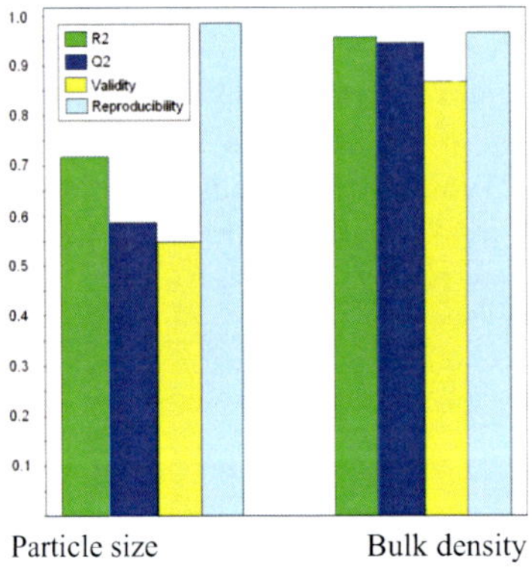

	p-value	
	D50	BD
P_feed	$6.72x10_{-3}$	0.280
T_out	$8.93x10_{-4}$	5.78x10-10
D_nozzle	0.846	0.381
T_cond	0.742	0.322

Fig. 8.24 Screening study: model adequacy (*left side*) and statistical significance data (*right side*)

8.2.5.2 Optimization Stage

During the optimization stage, the operating ranges were narrowed down and an experimental design (central composite face-centered design) was considered in order to support nonlinear interaction models. Only P_{feed} and T_{out} were selected as input variables at this stage, in agreement with the conclusions taken before. The center point assumed during the screening study was maintained since product attributes obtained at 85 bar and 55 °C were close to the target values. The DoE selected to support the optimization stage, as well as the considered ranges for the input variables, are shown in Fig. 8.25.

As shown in Fig. 8.26, the quadratic interactions improved the accuracy of the model for particle size prediction. For powder BD, the optimization study confirmed the suitability of a linear model (only function of T_{out}).

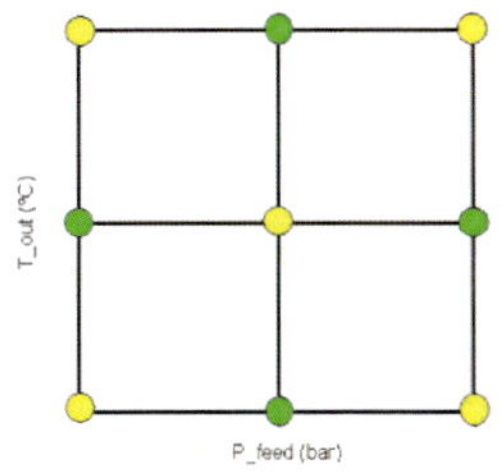

Input variable	Min	Center	Max
P_feed (bar)	70	85	100
T_out (°C)	40	55	70

D_ nozzle = 0.91mm; *T_cond* = Cte = -5°C

Fig. 8.25 Optimization study: DoE structure (*left side*) and ranges of variation studied (*right side*)

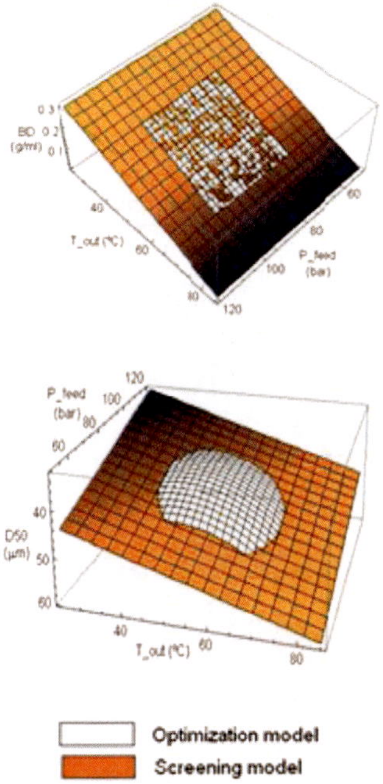

		D50 (□ m)	BD (g/ml)
Constant		141.2	0.449
P_feed	(bar)	-1.46	---
T_out	(°C)	-1.35	-5.44 x 10^{-3}
P_feed · *T_out*	(bar °C)	---	---
$(P\_feed)^2$	$(bar)^2$	7.64 x 10^{-3}	---
$(T\_out)^2$	$(°C)^2$	1.47 x 10^{-2}	---
R^2	(-)	0.87	0.97
Q^2	(-)	0.79	0.96

Fig. 8.26 Model predictions: Dv50 (*left bottom*) and BD (*left top*), coefficients, and adequacy data

8.2.5.3 Robustness Stage

The DoE considered in this stage is shown in Fig. 8.27. The selected ranges for the input variables translate possible deviations around the parameters settings. The optimal values for P_{feed} and T_{out} were determined through the optimization model, considering the target values for product attributes.

No reliable model was obtained for particle size prediction (a desired outcome, considering the nature of this study), which is in agreement with the small response of Dv50 observed in Fig. 8.28 (between 44 and 48 µm). For BD, however, it was still

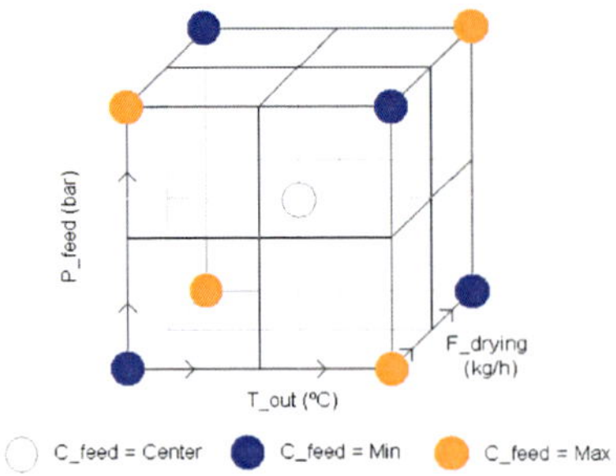

Input variable	Min	Center	Max
P_feed (bar)	P_opt -5	P_opt	P_opt +5
T_out (°C)	T_opt -2	T_opt	T_opt +2
F_drying (kg/h)	1300	1400	1500
C_feed (% w/w)	8.5	9.0	9.5

D_noz = 0.96 mm; *T_cond* = -5°C

Fig. 8.27 Robustness study: DoE structure (*left side*) and ranges of variation studied (*right side*)

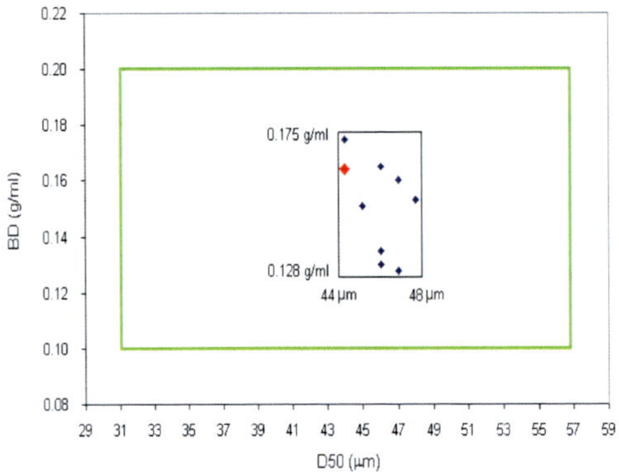

	p-value	
	D50	BD
P_feed	0.118	0.220
T_out	0.793	0.019
F_drying	0.433	0.395
C_feed	0.514	0.608

Fig. 8.28 Robustness study: results against targets in place (*left side*) and statistical significance (*right side*)

possible to obtain a model as a function of T_{out} (denoting an undesirably high sensitivity), confirming the strong dependence predicted by the screening/optimization models. Based on the results of Fig. 8.28, it can be concluded that (1) the process is robust towards the parameters F_{drying} and C_{feed} (thus validating the risk assessment) and that (2) although all obtained results are within the specification limits, T_{out} will need to be carefully monitored and controlled during the control strategy definition (due to its high potential of criticality via a high process sensitivity).

8.2.6 Design Space

The design space was defined as the multidimensional combination of the pCPPs where, according to the optimization model, all CQAs' values are obtained within the applicable target ranges. Although considering the intersection of the feasible operating ranges for each CQA, the final design space was mainly constrained by the BD target range, as shown in Fig. 8.29.

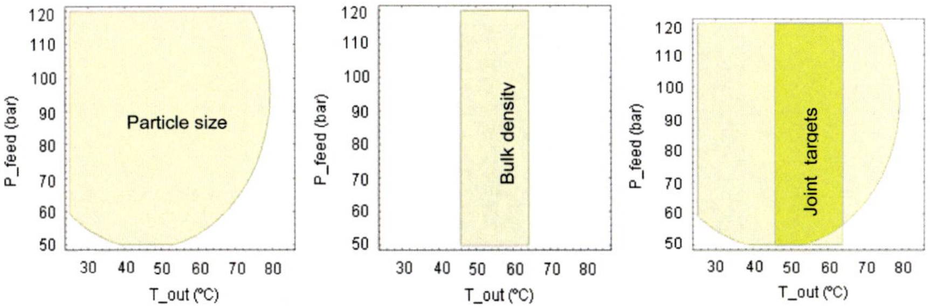

Fig. 8.29 Design spaces for Dv50, BD, and joint Dv50 and BD targets

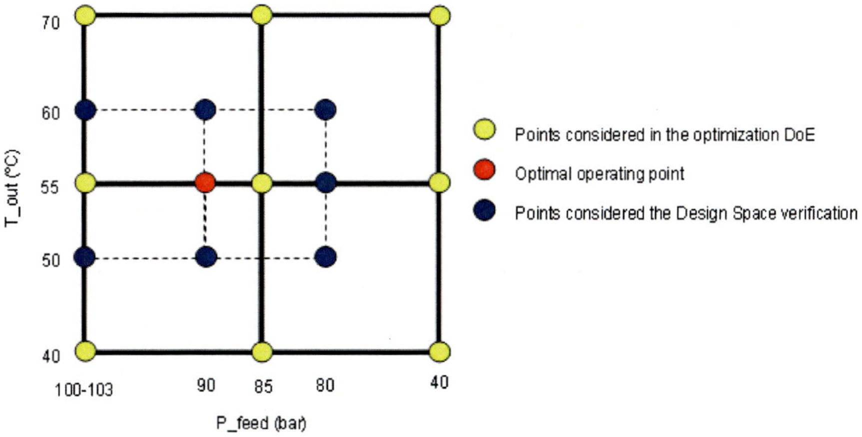

Fig. 8.30 DoE structure for design space validation

Regarding the validation of the design space, there is no universally recommended procedure for verification of a model, and there is limited literature or regulatory guidance that addresses the extent of verification required to justify a design space (Hallow et al. 2010). In the present work, eight additional runs were performed inside the design space and confronted with the predictions of the optimization model. A 3^2 DoE was centered in the optimized operation point (see Fig. 8.30). The results obtained showed very small errors (the root mean squared error was 0.06 μm for particle size and 0.003 g/ml for BD).

Nevertheless, this type of assessment can be misleading, as the magnitude of the model prediction errors (small or large, acceptable or unacceptable) can only be properly judged when confronted with the target specification ranges, during an uncertainty evaluation exercise (García-Muñoz et al. 2010; Peterson 2008). In fact, all models will have a corresponding error distribution, causing the boarders of the design space to be a less safe operating region. These error distributions are the sum of all the variability that cannot be explained or controlled (e.g., sampling errors, analytical method variability, influence of unknown factors, etc.) and should

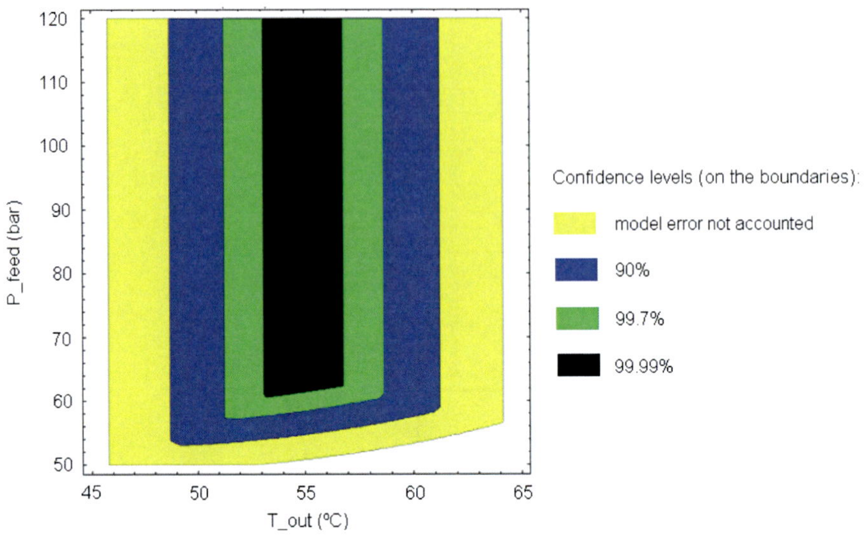

Fig. 8.31 Design space boundaries as a function of the imposed confidence level

be accounted during definition of the design space limits. In this work, the error distribution was calculated using the residuals of the optimization model and the eight runs performed for the design space validation. A normal distribution was fitted to the data (17 points in total) through a nonlinear parametric regression. The error distribution was then used to estimate error intervals at different confidence levels. By accounting the errors intervals on the limits of the CQAs' target ranges, it was possible to establish the design space in a true risk-based approach where broader operating ranges for the pCPPs imply greater risks of excursions outside the CQAs' target ranges. As shown in Fig. 8.31, the 90 % confidence boundaries lead to a significant reduction of the original design space. However, if the operating point is properly selected ($P_{feed} = 90$ bar and $T_{out} = 55\,°C$), a temperature variation of $\pm 3\,°C$ and a pressure variation $\pm$ 30 bar will have a very low likelihood of threatening the target CQAs' ranges (as the 99.7 % confidence boundary is not crossed), an observation that is in agreement with the outcome of the robustness study.

8.2.7 *Mechanistic Understanding*

The design space reported in the previous section was built based on statistical models; however, the ultimate goal of any QbD framework should be the enhanced understanding of the process, and this is ultimately achieved when a mechanistic interpretation of the underlying phenomena is derived. Theoretical description of the process of particle formation comprises two sequential steps: the atomization

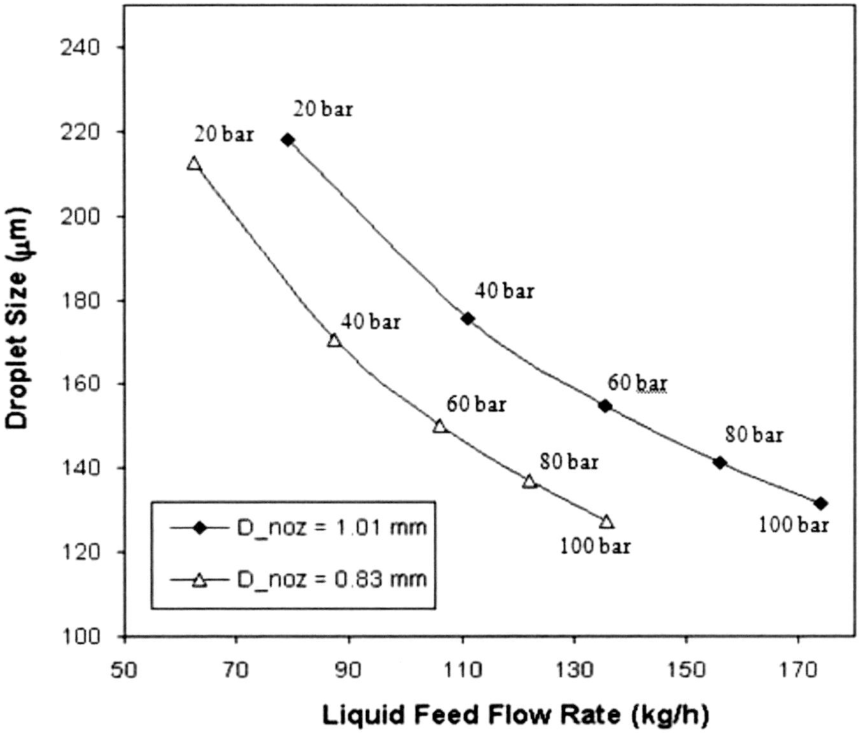

Fig. 8.32 Atomization model predictions: Influence of nozzle orifice diameter on droplet size and throughput, P_{feed}, ranged from 30 to 100 bar in both nozzles

process to describe the liquid breakup and drying kinetics to explain the final particle morphology.

In the present work, droplet size was calculated using the atomization model introduced in Sect. 8.1.5, and the obtained simulations (Fig. 8.32) were used to support the final nozzle selection, considering all relevant production goals (i.e., process throughput and target particle size). As forecasted by the theoretical simulations (and confirmed by the screening statistical studies of Sect. 8.2.5), the nozzle orifice diameter has little impact on droplet size. However, it is one of the most important factors that determine the process throughput; changing the diameter of the orifice is actually the most expeditious way to control the process throughput and keep the droplet size roughly unchanged.

Regarding the particle morphology, the maximum structural stability of the spray-dried particle is in a spherical form; however, the interactions between several process parameters, which are related to the mass and heat transfer of the droplet, can result in different particle morphologies. From the theoretical point of view, the evaporation begins as soon as the droplets are ejected from the nozzle. Consequently, the droplet

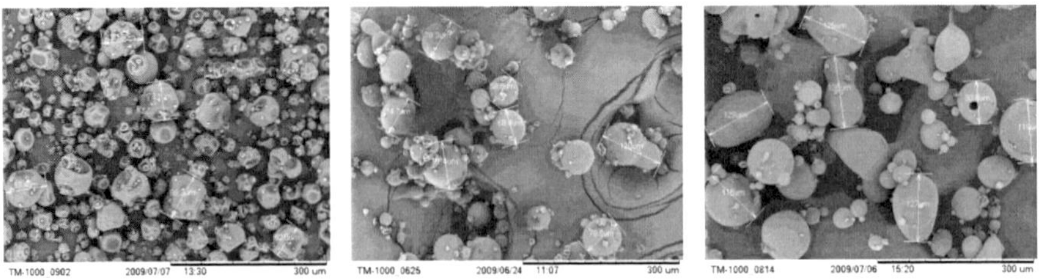

T_out = 25ºC → Dv50 = 43□ m   T_out = 55ºC → Dv50 = 44 □ m   T_out = 85ºC → Dv50 = 59□ m

Fig. 8.33 SEM pictographs: Influence of T_{out} on particle size, at the same conditions ($P_{feed} = 85$ bar; $D_{noz} = 0.96$ mm; $T_{cond} = -5\,°C$)

surface rapidly becomes enriched in solutes. At a certain supersaturation level, precipitation will occur and a rigid particle will be formed with a given particle diameter. From this standpoint, it is expectable that particle size closely follows droplet size. However, at low temperatures, the mechanisms of solvent evaporation are slower, allowing more time for particle surface to deform, shrink, or collapse. On the other hand, at temperatures close or above the boiling point of the solvents, the increased internal vapor pressure may lead the particles to inflate. Both phenomena could be observed during the current work, being more notorious when the experimental data is grouped in two sets: drying temperatures below and above boiling point. Two models for particle size were built with different sensitivities towards T_{out}. Below the boiling point, particle size is only dependent on the droplet size; above boiling point, particle size is dependent on both droplet size and T_{out}. Both models of the particle size showed good correlation coefficients (see Fig. 8.11a), and, together with the different morphologies shown in Fig. 8.33, it is therefore possible to consolidate and bridge the current mechanistic interpretation with the statistical model derived for particle size in Sect. 8.2.5.

Since the atomization process is not affected by the drying condition (T_{out}, T_{cond}, and F_{drying}), the droplet size distribution produced by a given nozzle is mainly dependent on P_{feed}. Hence, since the same applies to the mass distribution of the droplets (at constant C_{feed}), particles produced at high temperatures have lower densities. Thus, powders dried at low temperatures have relatively high BD, because the individual particles are shriveled and not very porous; on the other hand, powders dried at high temperatures have relatively low BD, because the individual particles preserve their inflated state and are, consequently, very porous (Langrish et al. 2006). Both interpretations are well aligned with the statistical model derived in Sect. 8.2.5 for BD, where T_{out} shows up as the most important operating parameter.

Therefore, through the development of mechanistic analyses, it is therefore possible not only to predict and control particle size but also to understand the main physical phenomena that rule particle morphology and powder BD.

8.3 Conclusions

In the current work, a QbD methodology was applied during the development of a spray drying process for the manufacture of a pharmaceutical solid dispersion. The risk assessment stage (where mechanistic knowledge and past experience play an important role) was found to be of major importance in order to keep development as lean as possible (by focusing on the most important parameters). Although not mandatory, the development of predictive models is strongly advisable to establish a reliable design space where model uncertainty should be addressed. Statistical approaches are a pragmatic way of establishing these relationships; however, a fundamental mechanistic understanding will always be advantageous as this one portrays the physical principles that rule the process and, therefore, enables more general considerations. For the current spray drying process, the combination of statistical and mechanistic information enabled to conclude that (1) particle size depends mainly on the droplet size and outlet drying temperature, (2) nozzle diameter dictates the throughput and does not affect particle size significantly, and (3) BD can be controlled just by manipulation of the outlet drying temperature due to its strong influence on the porosity of the particles.

References

Berman J, Pierce P, Page PE (1994) Scale-up of a spray dry tablet granulation process: thermodynamic considerations. Drug Dev Ind Pharm 20(5):731–755

Conference I, Harmonisation (1999) Specifications: test procedures and a cceptance criteria for new drug substances and new drug products: Q6A, (October)

García-Muñoz S, Dolph S, Ward HW (2010) Handling uncertainty in the establishment of a design space for the manufacture of a pharmaceutical product. Comput Chem Eng 34(7):1098–1107

Hallow DM, Mudryk BM, Braem AD, Tabora JE (2010) An example of utilizing mechanistic and empirical modeling in quality by design. J Pharm Innov 5(4):193–203

Koulouris A, Lagonikos PT (2002) The role of process simulation in pharmaceutical process development and product commercialization. Pharm Eng 22(1):1–8

Langrish TAG, Marquez N, Kota K (2006) An investigation and quantitative assessment of particle shape in milk powders from a laboratory-scale spray dryer. Drying Technol 24(12):1619–1630

Larhrib H, Peter G, Marriott C, Prime D (2003) The influence of carrier and drug morphology on drug delivery from dry powder formulations. Int J Pharm 257(1–2):283–296

Lefebvre AH (1989) Properties of sprays. Part Part Syst Char 6:176–186

Littringer EM, Paus R, Mescher A, Schroettner H, Walzel P, Urbanetz NA (2013) The morphology of spray dried mannitol particles—the vital importanceof droplet size. Powder Technol 239:162–174

Masters K (2002) Spray drying in practice. In: SprayDry consult, International ApS, Denmark

Miller DA, Gil M (2012) Spray drying technology. In: Robert O. Williams III, Alan B. Watts and Dave A. Miller Formulating poorly water soluble drugs. Springer, New York pp 363–442

Mujumdar AS (2006) Handbook of industrial drying, 3rd edn. CRC Press, Taylor Group, Boca Raton, USA

Peterson JJ (2008) A Bayesian approach to the ICH Q8 definition of design space. J Biopharm Stat 18(5):959–975

Senecal PK, Schmidt DP, Nouar I, Rutland CJ, Reitz RD, Corradini ML (1999) Modeling high-speed viscous liquid sheet atomization, Int. J. Multiphase Flow. 25

Vehring R (2008) Expert review pharmaceutical particle engineering via spray drying. Pharm Res 25(5):999–1022

Vehring R, Foss WR, Lechuga-ballesteros D (2007) Particle formation in spray drying. J Aerosol Sci 38(7):728–746

Walton DE (2000) The morphology of spray-dried particles a qualitative view. Drying Technol 18(9):1943–1986

Walzel P (2011) Influence of the spray method on product quality and morphology in spray drying. Chemical Engineering & Technology 34(7):1039–1048

Chapter 9
Design and Development of HPMCAS-Based Spray-Dried Dispersions

David T. Vodak and Michael Morgen

9.1 Introduction

Poor oral bioavailability due to the low aqueous solubility of potential drug candidates is an increasingly common challenge facing the pharmaceutical industry (Friesen et al. 2008). Nearly one third of compounds in early development have poor bioavailability due to low solubility, representing a significant loss in economic and therapeutic opportunity (Government Accounting Office (GAO) 2006). Although they may not fit Lipinski's "rule of five," many of these low-solubility compounds, which fall into classes II and IV of the biopharmaceutics classification system (BCS), have the potential to be safe and efficacious, so it is critical that their development is not halted by solubility limitations (Amidon et al. 1995). To address low active pharmaceutical ingredient (API) solubility, multiple drug delivery technologies have been advanced in an attempt to solubilize these molecules and enhance their oral bioavailability. Solubilization technologies can improve oral absorption of BCS class II compounds by:

1. Increasing solubilized drug levels (i.e., increasing the concentration of dissolved drug above the equilibrium concentration of the solubility of bulk crystalline drug)
2. Increasing dissolution rate
3. Sustaining the enhanced dissolved drug concentration in the intestinal milieu for a physiologically relevant time.

This chapter presents an overview of amorphous spray-dried dispersions (SDDs), which have been successfully used as a platform technology to enhance the oral bioavailability of hundreds of compounds with low aqueous solubility. SDDs can be prepared with several nonionic polymers, such as polyvinylpyrrolidone (PVP) and cellulosic polymers, as well as with ionic polymers, such as hydroxypropyl

D. T. Vodak (✉) · M. Morgen
Bend Research, Inc., Bend, OR, USA Tel: 541 382 4100
e-mail: david.vodak@BendResearch.com

N. Shah et al. (eds.), *Amorphous Solid Dispersions*,
Advances in Delivery Science and Technology, DOI 10.1007/978-1-4939-1598-9_9

methylcellulose acetate succinate (HPMCAS-based and methacrylic-acid-, methyl–methacrylate-, and ethyl-acrylate-based copolymers. However, SDDs based on HPMCAS are highlighted, since this polymer has been found to have widespread utility for low-solubility compounds.[1]

We provide background information on past solubilization technologies and describe the attributes of HPMCAS that make it ideal for use as a dispersion polymer for SDD platform technology. Speciation theory, formulation and process selection methodology, and performance of amorphous HPMCAS-based SDDs are then described.

9.2 Background: Efforts to Enhance the Solubility of Pharmaceutical Compounds

Typically, solubilization technologies are used to achieve rapid dissolution and enhance drug concentrations in two ways: (1) By formulating the drug as a solution in which the drug is predissolved (e.g., lipid systems or self-emulsifying drug delivery systems, SEDDS) or (2) by formulating the drug as a high energy solid form (e.g., crystals formed by attrition, crystals formed by bottom-up nucleation and controlled growth, or amorphous forms formed by melting or solvent removal).

To improve the dissolution rate and solubility of a compound relative to its lowest energy crystal form, one general approach is the generation of an amorphous form, usually stabilized as an amorphous dispersion of the drug in a polymeric material. The major challenge for this approach is selecting the appropriate formulation and process to develop a high energy amorphous form that has adequate physical stability, and achieves and maintains an in vivo drug concentration that is well above the crystalline solubility.

In the 1960s and 1970s, a variety of reports described the use of solid solutions and dispersions of drugs with polymers and with small molecules to improve drug dissolution rate and bioavailability. In an early report, Sekiguchi and Obi (1961) presented data for a single human subject indicating that a eutectic mixture of sulfathiazole and urea resulted in higher blood levels than sulfathiazole alone. Goldberg et al. (1965) described the use of solid solutions of sulfathiazole with urea and chloramphenicol with urea that offered improved dissolution rate. In another study, Goldberg et al. reported the use of eutectic mixtures for this purpose (1966). Stoll et al. (1969, 1973) reported dissolution and bioavailability improvements using coprecipitates with bile acids.

Early reports on dispersions and coprecipitates with polymers were focused on the use of PVP (Chiou and Riegelman 1969, 1971; Simonelli et al. 1976). However, the mechanism of solubility enhancement with PVP was somewhat unclear, and in

[1] HPMCAS is also known as hypromellose acetate succinate and is commercially available from Shin-Etsu Chemical Company and The Dow Chemical Company.

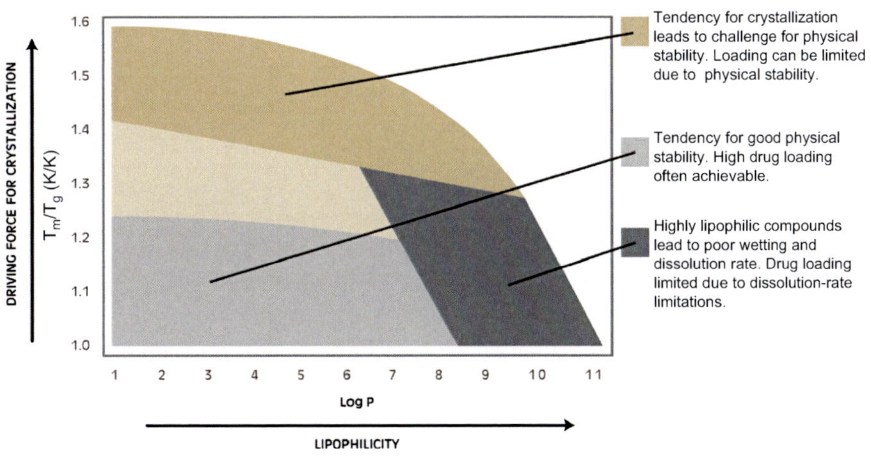

Fig. 9.1 SDD formulation guidance plot, showing the ratio of melting temperature (T_m) to glass-transition temperature (T_g) as a function of log P

fact, other reports described specific drug/PVP complexes designed to slow drug release in solution (Higuchi and Kuramoto 1954; Horn and Ditter 1982).

In other work, Chiou and Riegelman (1970) demonstrated enhanced canine oral absorption of griseofulvin in drug/polymer dispersions prepared by melting using polyethylene glycol (PEG) 6000. Many subsequent reports of drug/polymer dispersions have been published, and some have been summarized in excellent reviews by Serajuddin (1999), and Leuner and Dressman (2000).

9.3 SDD Formulation Selection and Manufacture

The selection of SDD formulations and manufacturing conditions can be conducted based on a rational methodology that relies on extensive experience with a wide variety of low-solubility compounds. The process for selecting the type of SDD formulation, polymer, and active loading is based on the product concept (dose and type, size, and number of dosage forms) and the properties of the compound. Guidance maps based on historical experience, such as the plot shown in Fig. 9.1, can be leveraged to formulate compounds of interest (Friesen et al. 2008). For example, experience has shown that compounds with a high tendency to crystallize (i.e., those with a T_m/T_g ratio > 1.4) will likely require higher dilution (lower active loading) in the polymer dispersion to achieve appropriate physical stability. Based on this information, formulation efforts should focus on lower active loadings, e.g., SDD containing 10 wt% active compound and 90 wt% polymer.

Once a formulation or formulations have been selected, manufacturing of homogeneous dispersions becomes the critical factor. Spray drying is a well-established and widely used industrial process for transforming solutions, emulsions, and suspensions of materials into dry powdered forms (Morgen et al. 2013). A general process configuration is shown in Fig. 9.2.

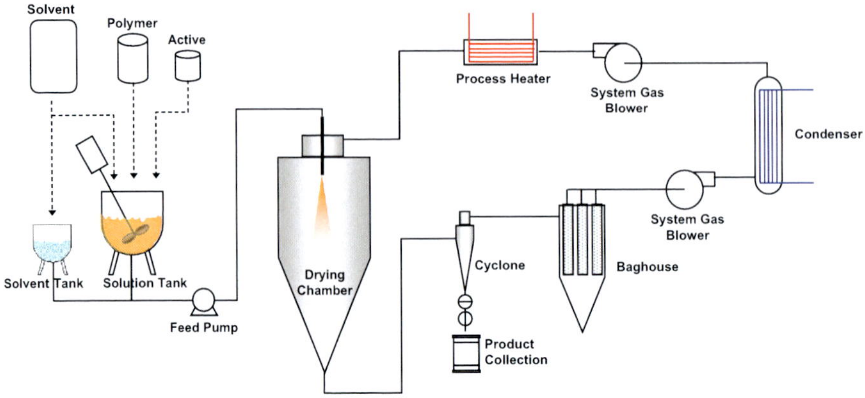

Fig. 9.2 General spray-drying process configuration

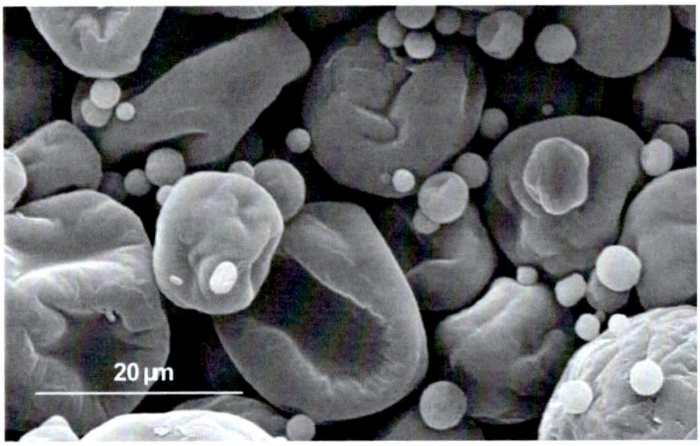

Fig. 9.3 SEM image of an HPMCAS-M SDD at 1500-fold magnification (Friesen et al. 2008)

In this process, a feed solution is prepared by dissolving drug and polymer (e.g., HPMCAS) in a volatile solvent and then pumping the solution to an atomizer inside a drying chamber. The atomizer breaks the solution into a plume of small droplets (typically, less than 100 μm in diameter). In the drying chamber, the droplets are mixed with a hot drying gas stream (typically, nitrogen for organic solvents). Heat is transferred from the hot drying gas to the droplets to provide the latent heat of vaporization required for rapid evaporation of the solvent from the droplets. As the solvent is removed from a droplet containing film-forming ingredients, a high-viscosity gel or "skin" forms on the outside of the droplet. Typically, at this stage of drying, the skin is sufficiently plasticized (due to the high solvent-to-solids ratio) that the particle skin collapses on itself as the solvent evaporates from the core, yielding particles with the "shriveled raisin" morphology shown in the scanning electron micrography (SEM) image in Fig. 9.3.

By controlling the temperature at the inlet and outlet of the spray dryer, along with the rate at which spray solution and drying gas are introduced to the spray dryer, the morphology, particle size, and density of the resulting SDD powder can be controlled. The solid powder is typically collected from the gas stream using a cyclone or filter system.

Based on an evaluation of the physicochemical properties of the active compound, several initial formulations (generally, four to six) are selected and screened in this step (Dobry et al. 2009). A small-scale spray dryer designed for maximizing yields from SDD batches of less than 100 mg is used. This dryer is not designed to replicate optimized bulk powder properties (e.g., particle size, density) of larger scale spray dryers, but rather is used to guide formulation decisions based on physicochemical properties and fast, efficient formulation-screening studies.

Process and formulation selection flowcharts, which refer predictive physical stability models, rapid chemical stability screens, and biorelevant in vitro performance tests, are used to select a lead SDD formulation (including the drug/polymer ratio) and process parameters (Dobry et al. 2009).

Additional formulation information is gathered during this stage of product development, including preferred spray solvents and spray solution solids content. At the end of this step, a robust formulation has been selected based on fundamental physicochemical properties. Typically, the entire formulation-screening step can be completed with 200–400 mg, and sometimes as little as 100 mg of active compound. The formulation and process development flowchart methodology uses time and resources similar to those required for conventional immediate-release crystalline formulations. The methodology, which is based on fundamental engineering models and state-of-the-art process characterization tools, is an alternative to traditional empirical spray-drying process development methods, and results in streamlined and robust process development.

Using a quality-by-design (QbD) approach, formulation and process are linked through identification of critical quality attributes (CQAs) and key quality attributes (KQAs), which are related to critical process parameters (CPPs) and key process parameters (KPPs). CQAs, KQAs, and CPPs are defined in criticality and risk assessment (Babcock et al. 2009). Using this methodology, process development is focused on the selection of spray-drying process parameters that result in the desired KQAs (e.g., particle size and density) and process performance (e.g., yield) with minimal impact on the CQAs of bioperformance and stability. This model-based process development represents a QbD approach that lays the groundwork for continuous improvement and eventual design space process regulatory filings. This approach is in alignment with the FDA's current guidance on pharmaceutical development (US Food and Drug 2008; Pharmaceutical Development Q8, Revision 1).

9.4 HPMCAS Attributes for Use in SDD Platform Technology

HPMCAS has been identified as a particularly effective polymer for preparing SDDs of low-solubility drugs. HPMCAS-based SDDs have proven broadly applicable at improving the oral exposure of low-solubility compounds by (1) enhancing aqueous solubility compared with bulk crystalline drug, (2) enhancing the dissolution rate relative to bulk crystalline drug, and (3) sustaining the enhanced solubility in the intestinal milieu for a physiologically relevant time.

SDDs are often formed using HPMCAS in its unionized (protonated) form which is quite soluble in volatile organic solvents, such as methanol and acetone. Since many drug candidates are soluble in these solvents, they can be processed into HPMCAS-based SDDs readily and economically using spray drying.

Curatolo et al. (2009) described a large study in which HPMCAS was compared to other common dispersion polymers using in vitro solution performance. This work showed that among the dispersion polymers studied, HPMCAS was the most effective in achieving and maintaining drug supersaturation. HPMCAS-based SDDs achieved and maintained drug supersaturation in vitro more consistently and effectively than SDDs prepared with other polymers. In addition, dispersions prepared by spray drying (i.e., SDDs) had better homogeneity and better performance than dispersions prepared by a rotary evaporation process, presumably due to the significantly faster drying kinetics in the spray dryer.

HPMCAS has unique attributes that make it ideal for use in SDDs, as described by Friesen et al. (2008). These attributes include the following:

(1) A high T_g in its unionized state. This high T_g results in low drug mobility, which is responsible for the excellent physical stability of HPMCAS SDDs. The T_g also remains relatively high at elevated relative humidity (RH).
(2) Solubility in volatile organic solvents, such as acetone and methanol, allowing for economical and controllable processes for preparation of SDDs.
(3) When the polymer is at least partially ionized (as it is at any pH above approximately 5), the charge on it minimizes the formation of large polymer aggregates, stabilizing drug/polymer colloids (e.g., amorphous nanostructures).
(4) The amphiphilic nature of HPMCAS allows insoluble drug molecules to interact with the hydrophobic regions of the polymer, whereas the hydrophilic regions of the polymer ensure these structures will remain as stable colloids in aqueous solution.

HPMCAS is a cellulosic polymer with four types of substituents semi-randomly substituted at the saccharide hydroxyls:

- Methoxy, with a mass content of 12–28 wt%
- Hydroxypropoxy, with a mass content of 4–23 wt%
- Acetate, with a mass content of 2–16 wt%
- Succinate, with a mass content of 4 –28 wt% (National Formulary (NF) 2006).

The succinate groups of HPMCAS have a logarithmic acid dissociation constant (pKa) of about 5, so the polymer is less than 10 % ionized at pH values below approximately 4 and is at least 50 % ionized at pH values of approximately 5. Due to the presence of relatively hydrophobic methoxy and acetate substituents, HPMCAS is insoluble in water when unionized (i.e., at pH values < approximately 5) and remains predominantly colloidal at intestinal pH (i.e., at pH values of 6.0–7.5).

Traditionally, three grades of HPMCAS have been sold commercially, designated –L, -M, and –H, as illustrated in Fig. 9.4. The approximate pH values above which

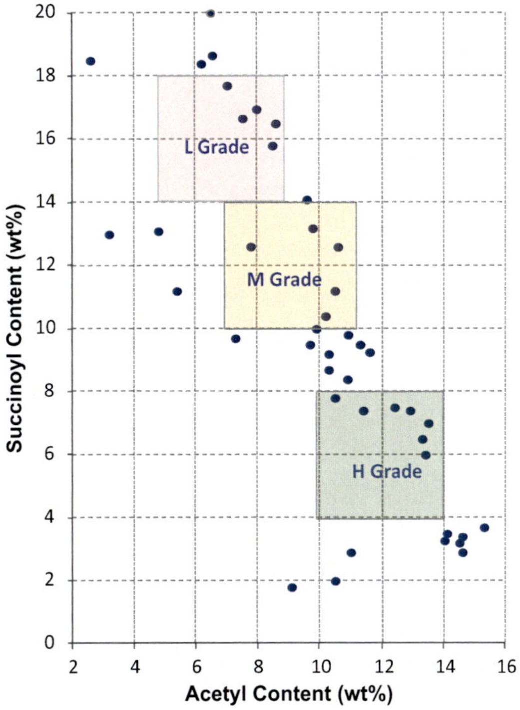

Fig. 9.4 Degree of substitution map for HPMCAS, showing the three commercially available grades of HPMCAS and other substitutions sampled in an R & D setting that fall within the US Pharmacopeia (USP) specification for HPMCAS, but outside the supplier specifications for the individual grades

each grade becomes aqueous dispersible or soluble are 6.8 (-H grade), 6 (-M grade), and 5.5 (-L grade).

HPMCAS contains several hydrophobic substituents. As a result, even when HPMCAS is ionized, as it is at intestinal pH, the polymer is only sparingly soluble, and exists predominantly as colloidal polymer aggregates in aqueous solutions. The negative charge of the ionized succinate groups ensures the colloids will remain stable, avoiding large hydrophobic aggregates of the polymer in aqueous solution.

This colloidal nature of HPMCAS when ionized, combined with the hydrophobic nature of the substituents on the polymer, allows insoluble drug molecules to interact with the polymer to form amorphous drug/polymer nanostructures in solution. These drug/polymer nanostructures constitute a high energy ("high solubility") form of amorphous drug that is quite stable for hours or days and, in selected cases, for weeks in aqueous suspensions. In vitro measurements have shown that drug in these nanostructures can rapidly dissolve to provide a high free drug concentration that is supersaturated relative to bulk crystalline drug.

In vivo, drug partitions into bile salt micelles and is absorbed from the intestine into systemic circulation. Additional drug can subsequently be rapidly sourced from these nanostructures to maintain a supersaturated free drug concentration. These properties ultimately lead to the enhanced absorption observed when HPMCAS-based SDDs are dosed orally.

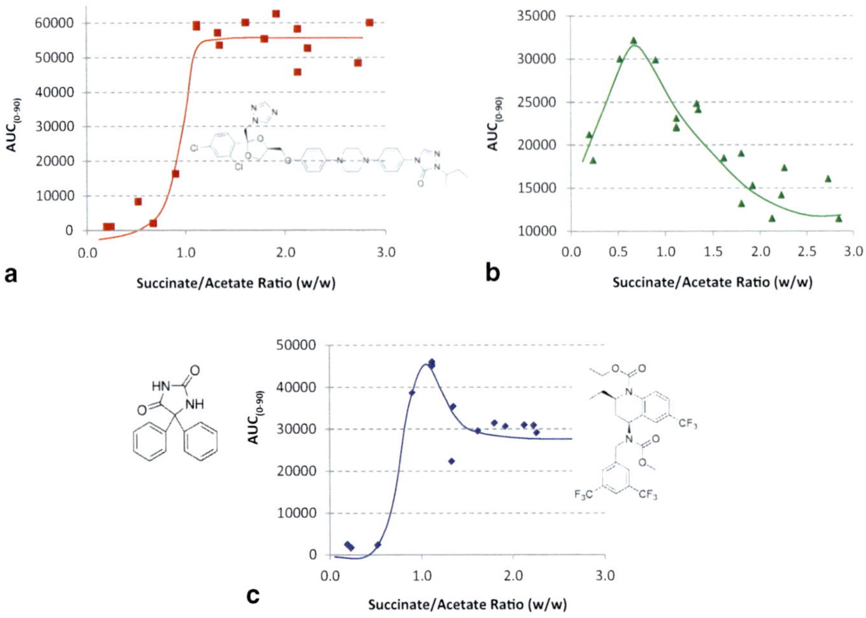

Fig. 9.5 Effect of succinate/acetate ratio on the HPMC backbone on the in vitro performance of SDDs prepared for three model compounds: itraconazole (**a**), phenytoin (**b**), and torcetrapib (**c**)[2]

Recent work has shown that the in vitro performance of SDDs can be optimized by altering the succinate/acetate ratio on the hydroxypropoxy methylcellulose (HPMC) backbone (Morgen et al. 2013). Figure 9.5 shows how small changes in the hydrophilic to hydrophobic substitution profiles, i.e., the succinate/acetate ratio, can be used to maximize in vitro performance for three low-solubility-model compounds, and illustrates that a specific optimal ratio can be identified for an individual active compound. This work illustrates the rich opportunity that exists to develop new functional excipients that are optimized for the performance of specific classes of molecules (Vodak 2013).

9.5 Speciation of HPMCAS-Based SDDs

When added to an aqueous solution simulating the environment of the small intestine, SDDs rapidly dissolve and/or disperse to produce a wide variety of species that facilitate absorption. To enhance the bioavailability of poorly soluble drugs, fundamental understanding of the drug species formed and the mechanism of action of SDDs is essential. Two general routes of HPMCAS SDD dissolution and drug speciation have been observed, which seem to bracket the behavior for most SDDs that have been studied. The two mechanisms of action—referred to as nanoparticle formation and erosion, respectively—are illustrated in Fig. 9.6a and 9.6b, respectively, and

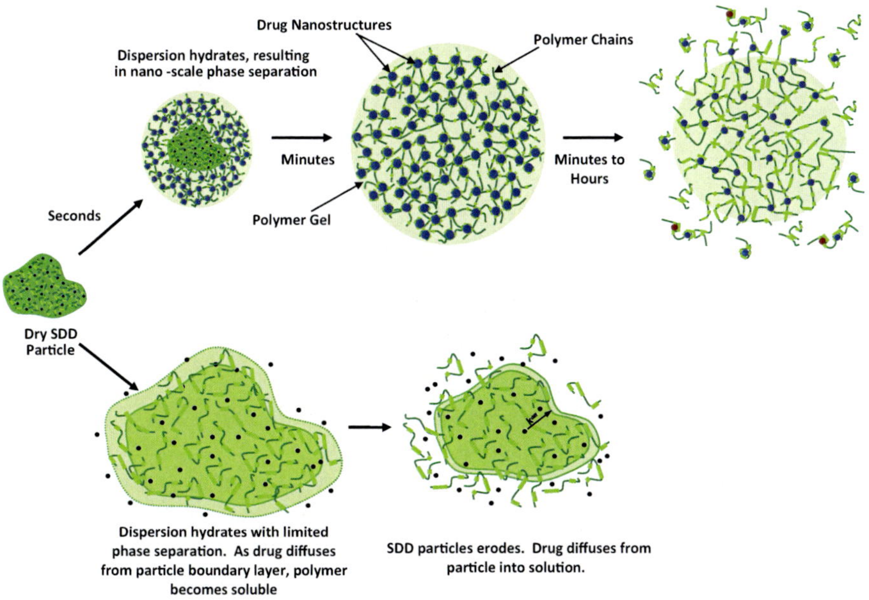

Fig. 9.6 Illustration of two SDD dissolution mechanisms and formation of drug–containing species critical to in vivo performance: nanoparticle formation (**a**) and erosion (**b**)

are described below. We also describe the species present during these dissolution mechanisms and test methods used to determine their presence.

9.5.1 Nanoparticle Formation Mechanism

In the first dissolution mechanism, the drug has limited solubility in the polymer and the solubility decreases upon absorption of water in biorelevant media. As the water enters the SDD particle, two factors, decreased drug solubility in the polymer and increased overall mobility of the components in the dispersion, lead to spinodal phase separation. The drug phase separates into drug-rich nanodomains that break off from the larger SDD particle and produce high energy amorphous nanoparticles. HPMCAS in its ionized state can then act as a surface stabilizer to the drug-rich nanoparticles. The same effect can be achieved using nonionic polymers with the addition of a surfactant in the formulation. Due to their small size (20–300 nm), these nanoparticles can rapidly source free drug that crosses the intestinal epithelial wall or partitions into bile salt micelles. It is believed that these nanoparticles also have a stabilizing influence in inhibiting rapid precipitation of drug from the supersaturated state in the intestine. This mechanism is illustrated in Fig. 9.7 with images that show the different stages of dissolution.

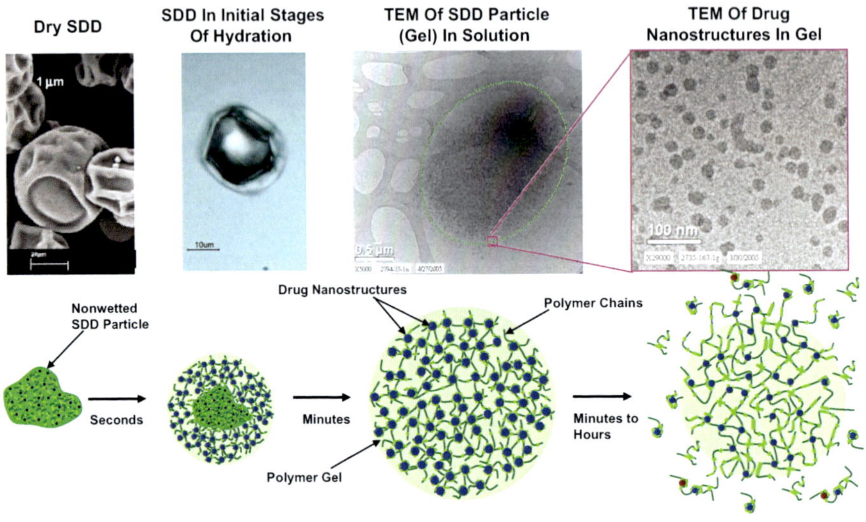

Fig. 9.7 SDD dissolution via the nanoparticle formation and dissolution mechanism

9.5.2 *Erosion Dissolution Mechanism*

In the erosion mechanism, the SDD particle does not disintegrate, but rather erodes from the surface to generate supersaturated free drug species and dissolved polymer chains. Typically, no nanoparticles are formed when this mechanism occurs, and performance usually is tied to the size and surface area of the particles, since the mechanism is a surface phenomenon.

Generally, this type of dissolution mechanism is observed when (1) the formulation has a high drug loading ($> 35\,\%$ active), (2) the drug has high solubility in the polymer, or (3) the drug solubility in the polymer increases as the water content of the SDD increases. This mechanism is illustrated in Fig. 9.8, with images that show the different stages of dissolution. Note in the SEM image, which was taken after the dissolution test, the "shriveled raisin" morphology of the original SDD particles remains.

9.5.3 *Dissolution Species*

For convenience in characterizing and comparing the species formed by SDDs under various conditions, we have divided these species, based on their size and composition, into the following seven classes: (1) free or solvated drug, (2) drug in bile-salt micelles, (3) free or solvated polymer, (4) polymer colloids, (5) amorphous drug/polymer nanoparticles, (6) large amorphous particles (i.e., "precipitate"), and (7) drug crystals, can be observed when things are improperly formulated.

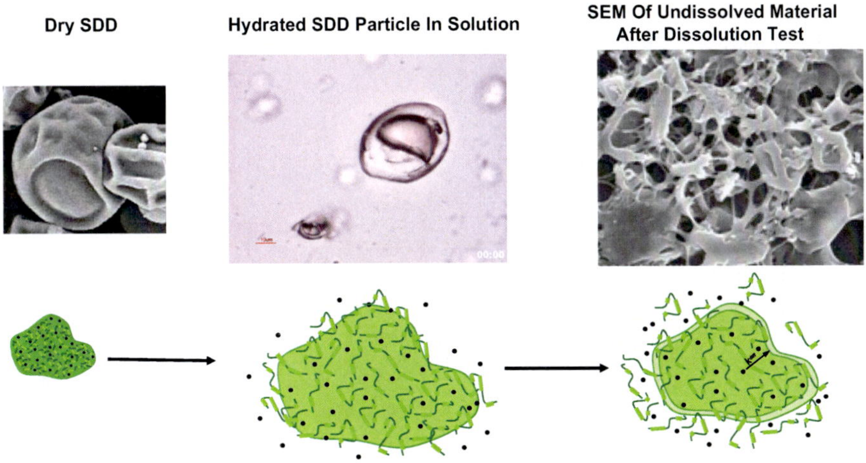

Fig. 9.8 SDD dissolution via the erosion mechanism

9.6 Testing Methods

To understand the performance of each SDD, a number of characterization tests have proven useful to measure and quantify individually the drug species that are present. In early development, bulk sparing methods are critical, due to the cost and limited quantities of drug compound available.

Two types of bulk sparing in vitro methods are described: (1) the centrifugal dissolution tests and (2) the membrane permeation test. These tests are used to identify the critical performance attributes of the system that are important to improve absorption and to rank the relative performance of SDD formulations.

9.6.1 Centrifugal Dissolution Tests

Centrifugal dissolution tests are used to measure the capability of SDDs to increase dissolution rate and levels of solubilized drug relative to crystalline drug. One key measure is the ability of SDDs to supply and sustain high energy, neutrally buoyant drug/polymer nanoparticles (Friesen et al. 2008; Curatolo et al. 2009). These nanoparticles are important because they can rapidly and continually source free drug during absorption.

The microcentrifuge dissolution test measures total drug arising from several species in solution, separated based on size and density: free drug ($[D_{free}]$), drug in bile-salt micelles ($[D_{micelles}]$), and drug in drug/polymer nanoparticles ([DPN]). The total drug ($[D_{total}]$) measured is:

$$[D_{total}] = \left[D_{free}\right] + [D_{micelles}] + [DPN]. \tag{9.1}$$

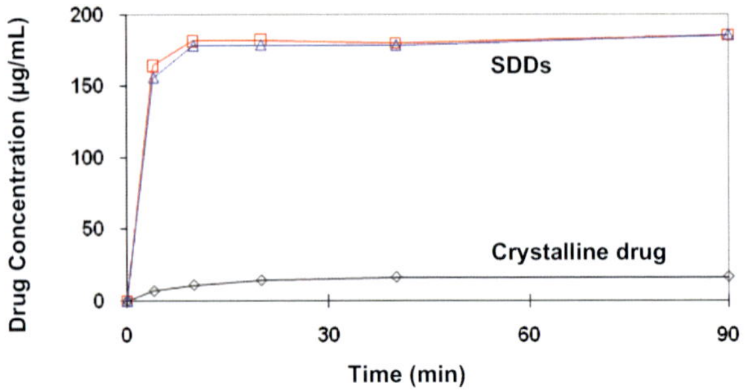

Fig. 9.9 Representative microcentrifuge dissolution results for a model compound comparing two HPMCAS-based SDDs and bulk crystalline drug

In this test, samples are dosed with suspension vehicle. Sample is weighed into a centrifuge tube, suspension vehicle is added, and the tube is vortexed to mix the sample with suspension vehicle. At each time point, the tubes are centrifuged at 13,000 g for 1 min. This step pellets any undissolved solids that are too dense to remain buoyant in the aqueous medium, predominantly undissolved SDD and API that precipitates or crystallizes. High-performance liquid chromatography (HPLC) is used to analyze aliquots of the supernatant for $[D_{free}]$, $[D_{micelles}]$, and [DPN].

These species often form rapidly when an SDD is added to simulated intestinal media. As illustrated in Fig. 9.9 for a model compound, the dissolution rate of SDD particles is at least two orders of magnitude faster than that of bulk crystalline drug. As the figure shows, the SDDs dissolve completely within 3 min, whereas the crystalline drug requires approximately 60 min to reach its equilibrium solubility.

The microcentrifuge test may also be used in conjunction with an ultracentrifuge test to generate additional size separation data, allowing separation of drug in nanoparticles from free drug and drug partitioned into bile salt micelles.

As Fig. 9.10 shows, the microcentrifuge test is also used (1) to quantify precipitation inhibition for compounds that rapidly crystallize and (2) to compare dissolution rates for SDDs of more lipophilic compounds, which tend to dissolve more slowly as particle size increases during process scale-up. A simulated gastric exposure step before dissolution in simulated intestinal media can also be added. This option is useful when evaluating weakly basic compounds that have pH-dependent solubility (Mathias et al. 2013).

9.6.2 *Membrane Permeation Test*

The membrane permeation test is another bulk sparing in vitro dissolution technique. It was developed at Bend Research and has been used for more than a decade (Babcock et al. 2009). This biphasic dissolution test is designed to assess the ability of a formulation to rapidly establish a high free drug concentration and then sustain that concentration for a physiologically relevant time period.

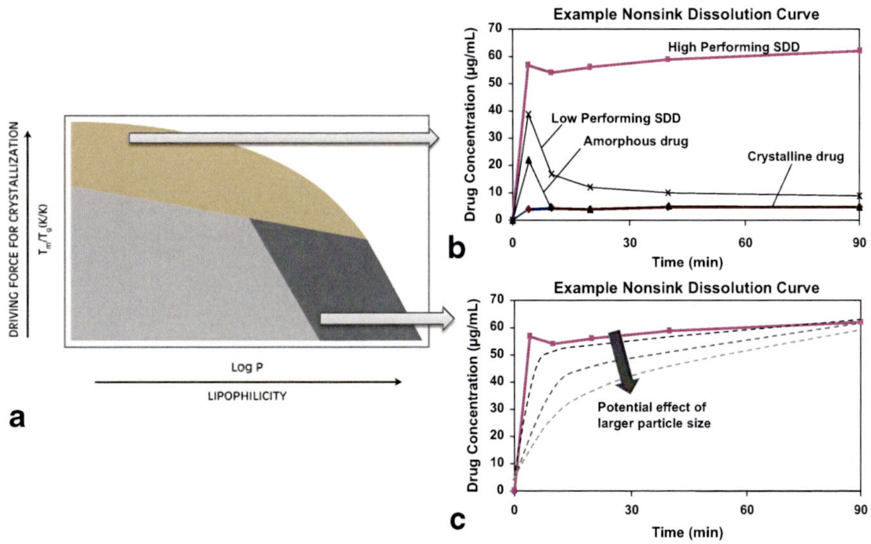

Fig. 9.10 Representative drug properties and data for a wide range of SDD formulations from a solubilization technology map (**a**), showing how the microcentrifuge test can be used to quantify precipitation inhibition (**b**), and to show negative impacts on dissolution rate for properties such as increased particle size (**c**)

The membrane permeation test measures the flux of drug across a synthetic membrane into an organic sink (permeate). For the test, a synthetic semipermeable membrane is used to separate the feed solution (i.e., simulated intestinal medium) and permeate (sink) solution (e.g., 80 % decanol and 20 % decane, by weight). Aliquots of permeate are taken at specific time points and the concentration of drug is measured by HPLC. High flux indicates a formulation's ability to rapidly dissolve and source a high concentration of free drug.

In the membrane permeation test, only free drug molecules from the feed solution can diffuse into the sink permeate. The test is intended to simulate the in vivo situation in which rapid passive diffusion of lipophilic molecules across the intestinal membrane occurs. In this situation, the ability of a formulation to establish a high level of free drug and its ability to maintain that level of free drug are critical formulation attributes for improved absorption. While the membrane permeation test does not enable the correlation of in vitro/in vivo performance, this test is useful in ranking the relative performance of SDDs. Representative results for the membrane permeation test is shown in Fig. 9.11, which compares results for an HPMCAS-based SDD formulation to that of bulk crystalline drug for a model compound. When combined with data from other in vitro dissolution tests (e.g., the microcentrifuge dissolution test), the results give mechanistic insight into relative formulation performance.

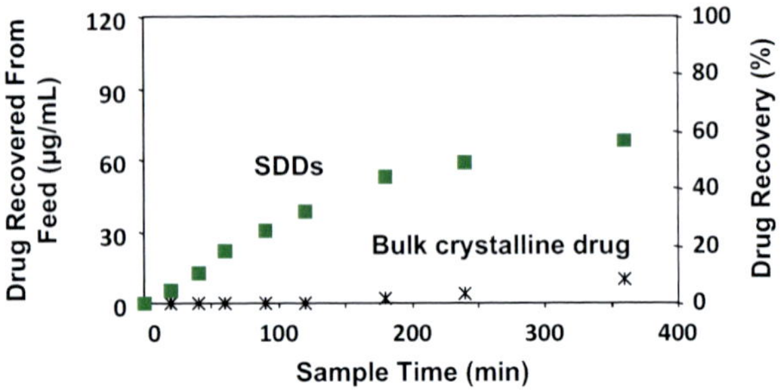

Fig. 9.11 Representative membrane permeation test results for a model compound comparing an HPMCAS-based SDD and bulk crystalline drug

9.7 Performance of the SDD Platform

SDDs have a proven track record for improving bioavailability for BCS II or IV compounds. Many in vivo studies have been performed in preclinical animal models and in human clinical studies demonstrating the enhancement. The following section describes some of these in vivo results, as well as an approach to understand the physical stability of these high energy formulations.

9.7.1 SDD Performance In Vivo

More than 500 different drugs have been formulated as SDDs at Bend Research and tested in various animal models.[2] Absorption enhancement relative to crystalline drug ranges from around 1.5-fold to nearly 100–fold, but varies widely based on the dose and drug properties. Figure 9.12 shows representative preclinical in vivo data for BCS class II compounds.

In addition, SDDs of 65 different drugs have been successfully tested in humans.[4] In all cases, the fraction of dose absorbed was at least twofold higher for the SDD than for the poorly absorbed control formulation. Figure 9.13 shows representative results from human clinical studies for BCS class II compounds.

As the data in Figs. 9.12 and 9.13 show, in cases where the crystalline drug (or comparison formulation) is poorly absorbed, the average AUC enhancement is approximately tenfold higher for SDDs dosed orally.[3]

---

[2] Numbers are much higher for global testing experience.

[3] The enhancement over bulk crystalline drug is lower in cases where the crystalline drug control is moderately well absorbed.

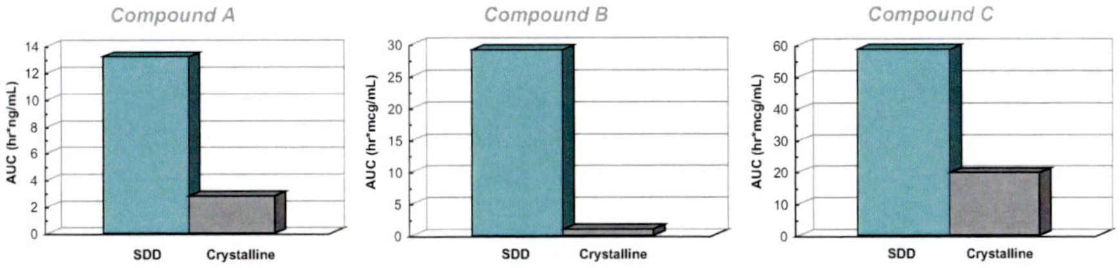

Fig. 9.12 Representative preclinical in vivo data, illustrating enhanced bioavailability of BCS class II compounds for SDDs relative to bulk crystalline drug

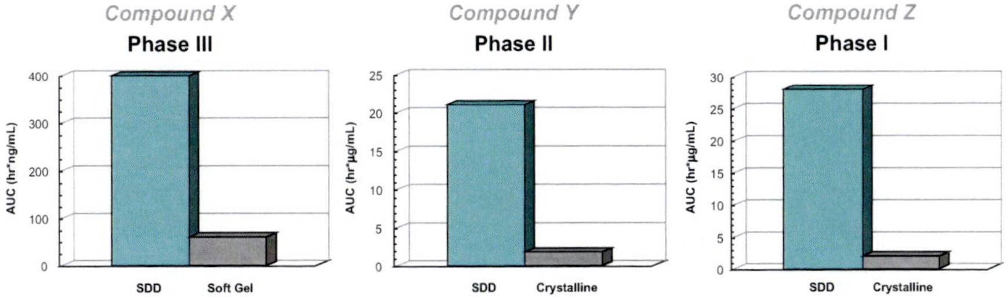

Fig. 9.13 Representative clinical data illustrating enhanced bioavailability of BCS class II compounds for SDDs relative to bulk crystalline drug or soft-gel formulations

9.7.2 Stability

HPMCAS SDDs have demonstrated long-term kinetic physical stability, routinely demonstrating shelf lives of more than 2 years under standard storage conditions. This is due, in part, to the high T_g of the polymer, and the resulting high T_g of the HPMCAS-based SDDs. As described below, T_g is a primary indicator of SDD physical stability.

In its unionized state (as it is in the solid SDD before dissolution), HPMCAS has a high T_g, even when exposed to high RH. Figure 9.14a shows the T_g for three commercially available grades of HPMCAS that had been equilibrated with air having varying RH.[6,7] Under dry conditions, the T_g is on the order of 120 °C. Like all amorphous materials, when exposed to humid air HPMCAS absorbs water, which plasticizes the polymer, increasing its mobility. This is reflected in the decrease in its T_g. However, the relative hydrophobicity of HPMCAS results in absorption of much less water than for typical water soluble polymers. Figure 9.14b shows dynamic vapor sorption (DVS) data taken at 25 °C for selected polymers. At 75 % RH, PVP and HPMC absorbed approximately 23 wt% and 10 wt% water, respectively, whereas HPMCAS absorbed only about 6 wt% water. As a result, the T_g value of HPMCAS remained above about 70 °C, even when equilibrated with 75 % RH air (Friesen et al. 2008). The low mobility of drug molecules dispersed in such high T_g glassy polymers leads to the excellent physical stability observed for HPMCAS-based SDDs.

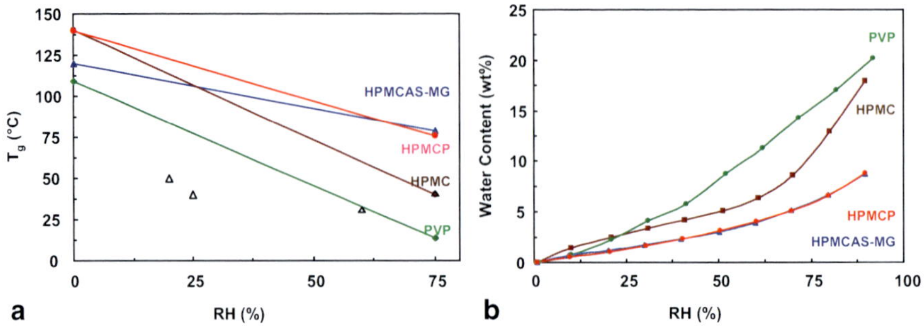

Fig. 9.14 Effect of RH on HPMCAS, PVP, HPMC, and hydroxypropyl methylcellulose phthalate (HPMCP): T_g versus the RH to which samples were equilibrated at ambient temperature for representative polymers (**a**), and equilibrium water absorption versus RH measured at 25 °C (**b**)

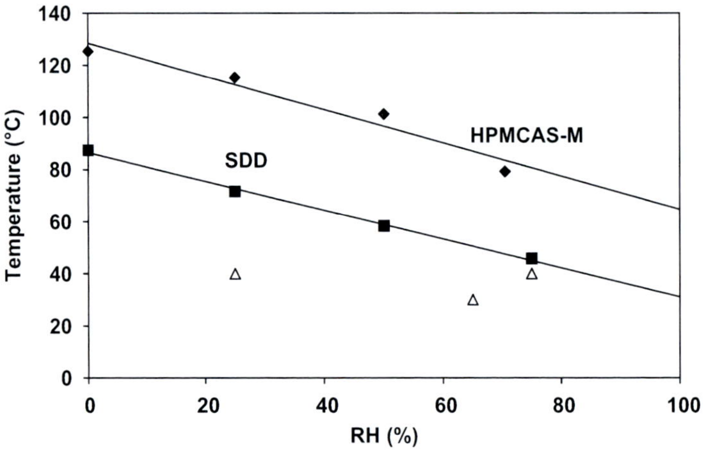

Fig. 9.15 Tg of a 25wt%-HPMCAS-based SDD and HPMCAS alone as a function of RH at 25 °C. Key: lines are least-square fits to the data and triangle data points show typical storage conditions

Figure 9.15 shows the T_g for a 25-wt%-drug-loaded HPMCAS-M SDD as a function of the RH of air to which it was equilibrated (at ambient temperature, about 22 °C). As the figure shows, the T_g of the SDD is high, well above the typical storage temperatures for RH values, up to about 60 % RH. As a result, drug mobility within the SDD (that is, the diffusion coefficient of drug in the SDD) is low even at temperatures of 40 °C and at water contents associated with RH values up to 60 %. This low rate of diffusion of drug in an SDD at or below the T_g of the SDD results in the diffusion of drug being the rate-limiting step for drug to phase separate and crystallize. For such homogeneous fluids near their T_g, the diffusion coefficient of a solute with a size of about 1 nm decreases by about tenfold for every 10°C decrease in temperature (Friesen et al. 2008; Angell 1985; Wang et al. 2002).

Using the approach introduced by Angell (1985), the temperature dependence of the viscosity of glasses can be presented in a T_g scaled Arrhenius plot. The minimum slope of the log10 viscosity versus T_g/T occurs for so-called strong liquids. For all organic glass forming materials, the slope at temperatures near T_g (0.9 to 1.1 with the T_g/T measured in Kelvin) is at least two- to threefold this minimum value (Wang et al. 2002). This slope is a measure of the "fragility" of the amorphous material. Taking a conservative estimate of fragility to be two- to threefold that of the strong fluid limit, for a 10 °C decrease in temperature from a T_g value of 60 °C (333 K), viscosity increases between 10-fold and 20-fold. Assuming that the diffusion coefficient of drug in the SDD decreases in inverse proportion to the viscosity, the diffusion coefficient of drug in an HPMCAS dispersion with a T_g near 60 °C would be expected to decrease 10-fold to 20-fold for every 10 °C decrease in temperature.

This suggests that a drug molecule dispersed in a polymer matrix is essentially immobilized and unable to migrate in the powder, in order to find other drug molecules and crystallize. This is defined as kinetic stabilization of the high energy form. Kinetic stabilization in tandem with the rapid quenching kinetics of the spray drying process enables higher drug loadings in SDD formulations, and also allows for predictive models to be developed based on the mobility (T_g) of the dispersion.

As a result, for the regime where (1) the temperature is between 30 °C below and 20 °C above the T_g, (2) there is a homogeneous dispersion, and (3) the drug concentration is above its solubility in HPMCAS but below about 70 wt%, the diffusion of drug is sufficiently slow that it is the rate-limiting step for crystallization.

The time to 5 % phase separation for an SDD increases by about tenfold for every 10 °C increase in the value of T_g–$T_{storage}$ (T_g is the T_g of the SDD at the storage conditions and $T_{storage}$ is the storage temperature). As a result, as long as the value of T_g–$T_{storage}$ is greater than about 5 °C to 30 °C and the SDD is initially homogeneous, less than 5 % phase separation is expected over a period of 2 years.[4]

The theory described above can be assessed by analysis of data from SDDs of more than 500 compounds that have been evaluated for physical stability. The data are summarized in the histogram presented in Fig. 9.16, which shows the fraction of stable SDDs (i.e., no phase separation) after storage for 6–13 weeks in stability challenges. Based on these data, physical stability estimates can be made using only the SDD T_g versus RH data. For example, as shown in this Fig. 9.16, 95 % of SDDs having a T_g more than 20 °C above $T_{storage}$ show no phase separation during the stability challenge, suggesting the rule of thumb that SDDs stored at 20 °C or more below their T_g are very likely to be physically stable for an extended period of time.

The physical stability of SDDs is further illustrated by the data in Table 9.1, which show that SDDs can be stored for long periods of time with no change in the

[4] The actual rate of phase separation and the corresponding time to 5 % phase separation has been measured for 17 different SDDs over a wide range of storage temperatures—both above and below the T_g of the SDD. Based on linear extrapolation of the data for temperatures near or above the T_g (plotted as the $\log_{10}$ of the time to 5 % phase separation versus $T_g/T_{storage}$), SDDs should be stable for at least 2 years if stored at temperatures from 5 °C to 33 °C below the T_g of the SDD. This prediction is based on data from HPMCAS SDDs for seven different active compounds.

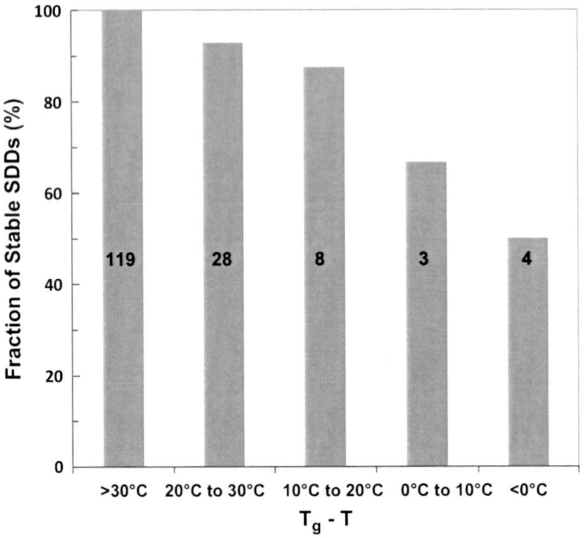

Fig. 9.16 Histogram summarizing selected bend research experience in formulating stable SDDs after 6- to 13-week stability challenges for given T_g–$T_{storage}$. Key: bold numerals on the bars represent the number of formulations used to generate the percentages

Table 9.1 Physical stability of HPMCAS-based SDDs aged at ambient conditions

SDD formulation	Aging time (year)	Observations
25-wt% compound 1:HPMCAS-M	3.0	No change in T_g, appearance, or dissolution performance
66-wt% compound 2:HPMCAS-M	2.2	No change in appearance or dissolution performance
10-wt% compound 7:HPMCAS-H	2.0	No change in appearance or dissolution performance
33-wt% compound 8:HPMCAS-L	0.7	No change in appearance or dissolution performance
33-wt% compound 9:HPMCAS-M	0.7	No change in appearance or dissolution performance

amorphous nature of the drug in the SDD. This is corroborated by the similarity of the SDD appearance in SEM images taken before and after storage. SEM images have been shown to be a sensitive measure of crystallinity, down to 1 wt% or less, allowing more sensitive detection than by powder X-ray diffraction (PXRD). Even more importantly, the dissolution properties of the SDDs, as reflected in their AUC values, show no significant changes over the time of storage. Based on the model above, these HPMCAS-based SDDs are expected to remain physically stable for even longer storage times than those used in this study.

9.8 Conclusions

HPMCAS SDDs are a particularly effective platform for enhancing the oral bioavailability of poorly aqueous-soluble pharmaceutical compounds, and have been successfully used for drug candidates having a wide range of physicochemical properties.

These SDDs provide significant enhancements in oral absorption of compounds with low aqueous solubility by (1) rapidly providing a free drug concentration well in excess of their crystalline solubilities and (2) maintaining these enhanced concentrations for long times. The composition and resulting physicochemical properties of HPMCAS are responsible for the formation of bioavailability-enhancing colloidal structures. Furthermore, the high T_g of the HPMCAS-based SDDs, combined with the homogeneous, single phase amorphous nature of the SDD, the result of the spray-drying process used to form the SDDs, produces physically stable formulations that have shelf lives of more than 2 years under standard storage conditions.

Spray drying has proven to be a robust and scalable method to manufacture SDDs from early formulation screening through commercial manufacture. Spray drying from an organic solution enables rapid drying kinetics, which is critical for preparing homogeneous amorphous dispersions of drug and HPMCAS.

The advantageous features of SDDs described above make them an attractive and broadly applicable platform technology for formulating poorly aqueous-soluble (BCS classes II and IV) compounds in a robust, scalable manner from very early development through commercial manufacture.

References

Amidon GL, Lennernas H, Shah VP, Crison JR (1995) A theoretical basis for a biopharmaceutic drug classification: the correlation of in vitro drug product dissolution and in vivo bioavailability. Pharm Res 12:413–420

Angell CA (1985) Strong and fragile liquids in relaxations in complex systems. In: Nagai K, Wright GB (eds) National Technical Information Service, US Department of Commerce, p 1

Babcock WC, Friesen DT, McCray SB (2009) Method and device for evaluation of pharmaceutical compositions. U.S. Patent No.7,611,630 B2

Chiou WL, Riegelman S (1969) Preparation and dissolution characteristics of several fast-release solid dispersions of griseofulvin. J Pharm Sci 58:1505–1510

Chiou WL, Riegelman S (1970) Oral absorption of griseofulvin in dogs: increased absorption via solid dispersion in polyethyleneglycol 6000. J Pharm Sci 59:937–942

Chiou WL, Riegelman S (1971) Pharmaceutical applications of solid dispersion systems. J Pharm Sci 60:1281–1302

Curatolo W, Nightingale J, Herbig S (2009) Utility of hydroxypropyl methylcellulose acetate succinate (HPMCAS) for initiation and maintenance of drug supersaturation in the GI milieu. Pharm Res 26(6):1419–1431

Dobry DE, Settell DM, Baumann JM, Ray RJ, Graham LJ, Beyerinck RA (2009) A model-based methodology for spray-drying process development. J Pharm Innov 4(3):133–142

Friesen DT, Shanker R, Crew MD, Smithey DT, Curatolo WJ, Nightingale JAS (2008) Hydroxypropyl methylcellulose acetate succinate-based spray-dried dispersions: an overview. Mol Pharm 5(6):1003–1019

Goldberg AH, Gibaldi M, Kanig JL (1965) Increasing dissolution rates and gastrointestinal absorption of drugs via solid solutions and eutectic mixtures I–theoretical considerations and discussions of the literature. J Pharm Sci 54:1145–1148

Goldberg AH, Gibaldi M, Kanig JL (1966) Increasing dissolution rates and gastrointestinal absorption of drugs via solid solutions and eutectic mixtures II– experimental evaluation of a eutectic mixture: urea-acetaminophen system. J Pharm Sci 55:482–487

Government Accounting Office (GAO) (2006) Report to congress, new drug development, GAO-07-49

Higuchi T, Kuramoto R (1954) Study of possible complex formation between macromolecules and certain pharmaceuticals. J Amer Pharm Assn Sci Ed 47:393–397

Horn D, Ditter W (1982) Chromatographic study of interactions between polyvinylpyrrolidone and drugs. J Pharm Sci 71:1021–1026

Leuner C, Dressman J (2000) Improving drug solubility for oral delivery using solid dispersions. Eur J Pharm Biopharm 50:47–60

Morgen M, Lyon D, Schmitt R, Brackhagen M, Petermann O (2013) New excipients for solubilizing APIs: expansion of the HPMCAS chemistry space. Tablets Capsul 11(5):10–16

Mathias NR, Xu Y, Patel D, Grass M, Caldwell B, Jager C, Mullin J, Hansen L, Crison J, Saari A, Gesenberg C, Morrison J, Vig B, Raghavan K (2013) Assessing the risk of pH-Dependent absorption for new molecular entities: a novel in vitro dissolution test, physicochemical analysis, and risk assessment strategy. Mol Pharm 10(11):4063–4073. doi: 10.1021/mp400426f

National Formulary (NF) (2006) Official monograph for hypromellose acetate succinate, NF 25, pp 1136–1138

Sekiguchi K, Obi N (1961) Studies on absorption of eutectic mixtures. I. A comparison of the behavior of eutectic mixtures of sulphathiazole and that of ordinary sulphathiazole in man. Clin Pharm Bull 9:866–872

Serajuddin A (1999) Solid dispersion of poorly water-soluble drugs: early promises, subsequent problems, and recent breakthroughs. J Pharm Sci 88:1058–1066

Simonelli AP, Mehta SC, Higuchi WI (1976). Dissolution rates of high energy sulfathiazole-povidone coprecipitates II–characterization of form of drug controlling its dissolution rate via solubility studies. J Pharm Sci 65:355–361

Stoll RT, Bates TR, Nieforth KA, Swarbrick J (1969) Some physical factors affecting the enhanced blepharoptotic activity of orally administered reserpine-cholanic acid coprecipitates. J Pharm Sci 58:1457–1459

Stoll R, Bates T, Swarbrick J (1973) In vitro dissolution and in vivo absorption of nitrofurantoin from deoxycholic acid coprecipitates. J Pharm Sci 62:65–68

US Food and Drug Administration (2008) International conference on harmonisation (ICH)–draft guidance: Q8(R1) pharmaceutical development revision 1. US FDA; Rockville, Maryland

Vodak DT (2013) Development of amorphous solid dispersions using the spray-drying process. Presentation at the 15th international workshop on physical characterization of pharmaceutical solids (IWPCPS®-15), Philadelphia, Pennsylvania, 24–27 June 2013

Wang L, Velikov V, Angell CA (2002) Direct determination of kinetic fragility indices of glass-forming liquids by differential scanning calorimetry. J Chem Phys 117(22):10184–10192

Chapter 10
MBP Technology: Composition and Design Considerations

Navnit Shah, Harpreet Sandhu, Duk Soon Choi, Hitesh Chokshi, Raman Iyer and A. Waseem Malick

10.1 Introduction

Over the past several decades, amorphous solid dispersion (ASD) technology has been increasingly utilized to address the challenges of poorly soluble compounds, which are becoming more prevalent in the current drug discovery environment. From the first practical application of solid dispersions in improving the solubility and bioavailability in the pharmaceuticals, the science and practice of amorphous technology have advanced considerably (Sekiguchi and Obi 1961; Chiou and Riegelman 1970; Leuner and Dressman 2000; Williams et al. 2010). Since then, various techniques have evolved and handful of compounds have made their way to the market using amorphous technology.

Despite extensive research and advancement in this area, application of ASD technology has not gained widespread use in the pharmaceutical industry, and still remains a niche technology applicable only to small number of compounds. The primary reasons for this reluctant adoption can be attributed to fear over the inherent physical instability of the amorphous material during manufacturing, storage, and dissolution, as well as lack of accessibility of robust and commercially viable manufacturing facilities.

N. Shah (✉) · D. S. Choi
Kashiv Pharma LLC, Bridgewater, NJ, USA
e-mail: navnits@kashivpharma.com

H. Sandhu
Merck & Co., Inc., Summit, NJ, USA

H. Chokshi
Roche Pharma Research & Early Development, Roche Innovation Center, New York, NY, USA

R. Iyer
Novartis, East Hanover, NJ, USA

A. W. Malick
Pharmaceutical and Analytical R&D, Hoffmann-La Roche Ltd., Nutley, NJ, USA

N. Shah et al. (eds.), *Amorphous Solid Dispersions*,
Advances in Delivery Science and Technology, DOI 10.1007/978-1-4939-1598-9_10

With regard to manufacturing technologies, considerable progress was made with the introduction of spray drying and melt extrusion processes, enabling successful commercialization of several challenging molecules (Williams et al. 2010; Repka et al. 2013). Recently, advancement in supercritical fluid and cryogenic freezing technologieshas shown promise in demonstrating the production of fine powders of ASDs (Yang et al. 2010). However, despite extensive research, the use of these technologies is limited to thermally stable low melting drug molecules or compounds soluble in volatile organic solvents. Effective technologies for the so-called brickdust-like molecule remain elusive. Moreover, as drug discovery becomes more sophisticated with respect to maximizing receptor binding, the percentage of such difficult compounds will be ever growing in modern pharmaceutical drug discovery.

To address the needs of such challenging compounds, a solvent-controlled coprecipitation technology, also known as microprecipitated bulk powder (MBP) technology, was developed in the late 1990s and is covered in detail in this chapter (Albano et al. 2002; Shah et al. 2012). The MBP technology has been applied to numerous development compounds in preclinical and clinical stage including a recently marketed product Zelboraf®.

10.2 Precipitation and Coprecipitation

The MBP technology is based on solvent–antisolvent precipitation. Precipitation occurs when the concentration of a compound in solution exceeds its saturation solubility. Solvent-controlled crystallization is a well-established process in the chemical industry and is briefly reviewed here before starting the discussion about solvent-controlled amorphous precipitation (McKeown et al. 2011). It should be noticed that the driving force for both processes, crystallization and amorphous coprecipitation, is supersaturation but the key difference lies in the rate at which the supersaturation conditions are created for the binary system comprised of drug and polymer. For crystallization, it is generally understood that formation of precipitate in supersaturated solution starts with the onset of nuclei formation. The onset of nuclei formation starts, hypothetically, from the formation of an interface between the solid and the solution (De Yoreo and Vekilov 2003). When molecules adhere to the nuclei in an orderly fashion with a specific motif, crystalline solid emerges. This self-assembly of molecules into crystal is governed by such factors as the degree of supersaturation, purity of solute, and diffusion rate, and it occurs usually after a certain induction period. If molecules adhere to the nuclei in a disorderly manner due to the lack of induction period, an amorphous solid emerges.

During crystallization, the degree of supersaturation is carefully controlled to obtain a desired polymorphic form and crystal habit. In general, crystallization occurs in the crystallization zone, which lies in between solubility line and precipitation line of a solubility phase diagram (Fig. 10.1).

If the system is shocked and forwarded to the precipitation zone with minimum time in crystallization zone, owing to huge solubility differentials, molecules seek

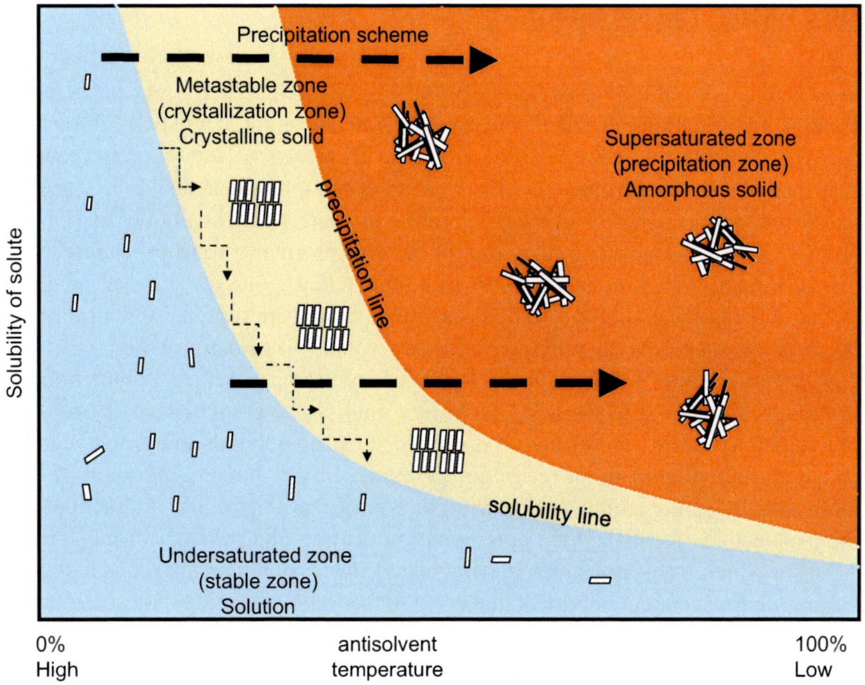

Fig. 10.1 Schematics of solubility phase diagram. Crystallization can occur in crystallization zone which lies in between solubility line and precipitation line, whereas amorphous precipitation occurs by jumping into precipitation zone by passing crystallization zone

immediate release of the excess energy (supersaturation) by disorderly stochastic precipitation in the form of amorphous solid. This extreme supersaturation condition can be created by adding drug solution to the larger volume of antisolvent together with reduced temperature. Upon contact with the chilled antisolvent, the solubility of compound rapidly falls below the saturation solubility, resulting in the precipitation as amorphous solid.

Coprecipitation occurs when the two compounds exceed their saturation solubility simultaneously. Coprecipitate usually forms in such a way that a minor component is incorporated in the matrix of a major component where polymer further inhibits nucleation. Although precipitation and coprecipitation are commonly used in the chemical industry, its application in pharmaceutical industry has been quite limited. The term "coprecipitation" in pharmaceutical literature was first used to produce ASDs by precipitating drug and polymer together by changing the solubility conditions (Simonelli et al. 1969). Coprecipitation was induced by either addition of organic antisolvent to drug solution or by evaporation of the solvent but not the aqueous solvent. In rare attempts when aqueous phase was used as antisolvent to induce precipitation, the resulting material was partially crystalline, suggesting the use of aqueous antisolvent was not suitable for producing amorphous form (Kislalioglu et al. 1991).

10.3 MBP Development

In principle, both orderly process of crystallization and disorderly process of amorphous solid formation can occur from supersaturated solution. It will be a competition between the orderly assembly of molecules with greater reduction of free energy (thermodynamically favored) and the stochastic assembly of molecule to noncrystalline amorphous solid for immediate release of energy (kinetically favored). MBP process is designed to maximize amorphous solid formation and to minimize the crystal formation by kinetically prompting rapid energy release. This was achieved by introducing drug solution to cold antisolvent with appropriate agitation in the presence of an amorphous polymer to further reduce molecular mobility.

Amorphous ionic polymer plays a key role in MBP manufacture. Polymer almost always precipitates stochastically, providing multiple heterogeneous sites for drug precipitation. Precipitate formation is governed by the polymer and growth can occur in all directions without restriction. Polymer further hinders diffusion of drug molecules by acting as physical barrier, as well as by nonspecific interaction with drug molecules. Result is the coprecipitation of drug and polymer in the form of ASD, in which drug molecules are imbedded in the polymer matrix. Among the various pharmaceutical polymers that provide favorable conditions for precipitation in aqueous antisolvent are ionic polymers, such as hypromellose acetate succinate (HPMCAS) and polymethacrylates (Eudragit L100, Eudragit L100-55, and Eudragit S100).

Thus, the MBP technology takes advantage of the solvent-controlled coprecipitation of a drug and ionic polymer under controlled conditions to produce stable ASD. By virtue of coprecipitation, the amorphous drug is molecularly dispersed in the polymer to provide a stable ASD, referred to as MBP. The fast quenching that occurs during precipitation also helps in retaining the intermolecular interactions between drug and polymer. The rapid coprecipitation is achieved by maintaining the solvent–antisolvent ratio, temperature, and shear rate. Additionally, the rate of addition of the drug and polymer solution into antisolvent and subsequent dispersion are also critical for the rapid extraction of solvent. Insufficient agitation and inappropriate solvent to antisolvent ratio may affect the simultaneous precipitation of drug and polymer resulting in drug- or polymer-rich domains that may result in phase separation and/or crystallization.

The MBP process is particularly useful for compounds that have low solubility in volatile solvents such as acetone, ethanol, or compounds that have high melting point or are thermally labile. By design, MBP process can be carried out either in acidic or in basic conditions, depending on the ionic nature of the molecule and polymer used. Although the concept of MBP appears seemingly simple and straightforward, the design and execution of MBP could be tricky because generating amorphous form in aqueous phase is counterintuitive. The successful implementation requires a thorough understanding of active pharmaceutical ingredient (API) properties, polymer properties, and processing parameters including the solvent, antisolvent, temperature, shear force, and drug loading in the solvent.

The following section provides a detailed guidance with regard to the key formulation and processing parameters that need to be understood to obtain a viable product.

10.3.1 Process Overview

The MBP manufacturing process starts from dissolution of drug and an ionic polymer in polar nonvolatile (super) solvents such as dimethylacetamide (DMA), dimethylformamide (DMF), dimethylsulfoxide (DMSO), or n-methylpyrrolidone (NMP). These organic solvents have higher solubilization power than polar volatile organic solvents such as acetone, methanol, or tetrahydrofuran (THF). The drug polymer solution is then delivered to the "pH-controlled" and "temperature-controlled" aqueous media (antisolvent) to cause instant coprecipitation of drug and polymer upon contact with aqueous media. Organic solvent diffuses out from the initial precipitate and water diffuses in until chemical potential of inside and outside of the precipitate becomes equal. Constant agitation during coprecipitation expedites the solvent exchange and the time to reach the chemical potential equilibrium. The formation of coprecipitates, solvent exchange within the coprecipitates, and maturation of solid mass (coprecipitates) continue until all drug solution has been added and the chemical potential equilibrium is reached. After completion of coprecipitation, the solid precipitates are isolated via filtration or centrifugation. These solid precipitates contain relatively high amount of organic solvent, which is equal to that of the solvent composition in the reaction vessel viz around 10 %. This residual solvent must be removed from the precipitate to below the acceptable level (ICHQ3 Guidance 2009). The organic solvent removal is achieved by washing the coprecipitates with aqueous medium until residual organic solvent falls below the set value, viz 0.1 %. Once organic solvent level falls below the set value, the precipitate is further processed for drying. Typically, the precipitate contains high amount of water in the range of 60–90 %, and the efficient removal of water is a key process in MBP manufacture. After drying, based on the particulate properties, the final dried powder can be further processed (milling and densification) to be ready for the final formulation.

The key process aspects in MBP include:

- Total solid content (API and polymer) in organic solvent is in the range of 10–40 % by weight
- Ratio of solvent to antisolvent can be from 1:5 to 1:20 depending on the solubility profile of the drug
- Temperature of aqueous phase (antisolvent) is typically 5 ± 3 °C
- The precipitate is collected as wet cake by vacuum filtration or by a centrifugal filtration device
- Final MBP can be dried in a forced air oven or fluid bed dryer typically at 45 ± 5 °C

The schematic of the MBP process is shown in Fig. 10.2.

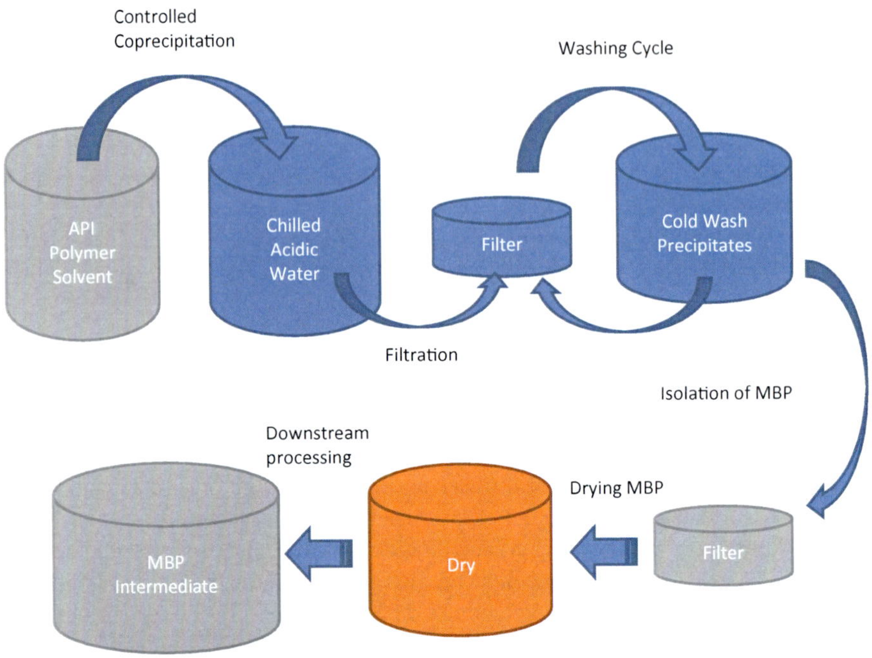

Fig. 10.2 Schematic of MBP process. MBP process is comprised of several unit operations including stock solution preparation, controlled coprecipitation, filtration, washing cycles, isolation of MBP, drying, and downstream processing

10.3.2 API Properties

During the early drug discovery phase, basic physicochemical properties of molecule are determined. Ordinarily, these sets of data are useful to support clinical candidate selection, early chemical process development, and formulation development. Even promising clinical candidates face dire consequences of termination (No Go decision) when conventional formulations fail to provide adequate bioavailability to enable efficacy and toxicology evaluation. Various formulation strategies can be employed to rescue molecules with poor "drug-like" biopharmaceutical properties. Among these strategies, amorphous formulation strategy is one of the most remarkable formulation approaches. Furthermore, within the various ASD technologies, MBP provides the best alternative because of its versatility, excellent API recovery rate ($> 90\,\%$), ability to handle small quantity of material (less than gram), and an excellent track record. The process can be readily scaled up from few milligrams to kilograms with minimal investment in equipment and facilities. Ability to handle a small quantity of API with good recovery is a particularly attractive feature in early drug development where the availability of drug substance is oftentimes limited.

However, not all drug substances can be candidates for MBP technology. Therefore, it is logical to critically examine API properties to (a) assess if the molecule is feasible for MBP, (b) guide selection of suitable polymer and drug loading for

Table 10.1 Physicochemical properties of API relevant for MBP process

Criticalproperties	Criteria	Comments
Ionization constant	pKa of molecule should not be higher than the pH of aqueous buffer (antisolvent)	Acidity or alkalinity dictates the pH solubility profile of the drug and degree of ionization
Functional group	Potential of interaction and reactivity with polymer estimated	Polymer selection for best interaction
Solubility profile	Solubility $>3\,\%$ in polar solvent such as DMA, DMSO, DMF or NMP Solubility <10 ppm in $10\,\%$ organic solvent in aqueous buffer	Drug must have good solubility in organic solvent and insolubility in aqueous media to be viable for the MBP process
Chemical stability	Chemically stable in extreme pH ranges where coprecipitation occurs Chemically stable in organic solvent for at least 24 h	Drug must be stable in acidic or basic aqueous media, or organic solvent
Crystallinity	API should not form solvate(s) with the solvent(s) Low crystallization tendency	Solvates may form during manufacturing process. Process at below saturation solubility to avoid solvate formation
Molecular Weight	MW > 500	Larger molecule tends to crystallize slowly
H bonding acceptors and donors	H acceptor > 7	H bonding donors and acceptors increase interaction with polymer in MBP
Lipophilicity	$\log P > 3$	Primary bonding between drug and polymer is likely to be hydrophobic interaction. High log P compounds have tendency to form stable MBP
Miscibility estimate	Negative free energy of mixing estimated by solubility parameter	To guide the selection of polymer

a feasible candidate, and (c) identify the optimal process conditions. The key physicochemical properties are listed in Table 10.1.

10.3.3 Solubility of API

Understanding solubility behavior of API is probably the most critical factor in MBP development. API must be highly soluble in organic solvent and must be insoluble in acidic water. Commonly used solvents for MBP are DMA, DMF, DMSO, or NMP. Practicality dictates that the API must have greater solubility than 3 % in the solvent. Acidified water is the most commonly used antisolvent. The amorphous drug must be insoluble in this aqueous media, and in practice, this translates into less than 0.01 mg/mL.

When the coprecipitation process is complete, the solvent composition in the reaction vessel is around 10 % if the solvent to antisolvent ratio is targeted to be 1:10, for example. API must have minimal solubility at this composition to ensure maximum recovery of the API. One should be mindful that API solubility in this solvent mixture is that of amorphous form (not of crystalline form!) which is higher than crystalline API. The API concentration in the supernatant after initial coprecipitation should always be monitored. Needless to say, if drug has good solubility in this solvent composition, then only a partial amount of drug will be precipitated as MBP. Not only will recovery be low but the unprecipitated drug in solution may also precipitate out during subsequent rinsing and washing cycles without the protection of polymer. If that happens, crystalline API seeds adhering to MBP particles can promote crystallization of amorphous drug, jeopardizing the entire MBP batch.

If API is a basic molecule, attention should be paid to the pH of antisolvent and the p*K*a of the molecule. The pH of aqueous buffer should be controlled at a level not to exceed the p*K*a of API to minimize API solubility. Ideally, the p*K*a of the drug should not be higher than 5 to ensure complete precipitation during initial coprecipitation and to minimize dissolution during subsequent rinsing and washing cycles. Like other precipitation processes, there may be sensitivity to the type of counterion used for pH control, and in some cases, it may be critical to select optimal counterion that provides best mode for the production of ASD.

10.3.4 Chemical Stability of API

Total solid mass, i.e., total amount of API and polymer in organic solvent, can range from 10 to 40 % w/w. Since many polymers dissolve slowly, heat may be applied to aid dissolution and to reduce the solution viscosity. Temperature as high as 80 °C has been used if the API has acceptable stability under these conditions. MBP is formed by coprecipitation of drug and polymer in acidic or basic media (mostly in acidic media), and it may stay exposed to this aqueous media from several hours to several days to weeks during precipitation, rinsing, washing, and storage. This is an important consideration since acid- or base-labile compound may degrade during MBP manufacture. Therefore, it is important to conduct risk assessment using the stability profile of the API and use this as a guide to select the pH and buffer.

10.3.5 Assessment of MBP Feasibility

MBP is an excellent ASD technique, particularly during early phase of drug development and for toxicology-enabling studies. Because of the excellent recovery rate and flexibility in scale, even milligram quantity drug can be processed to enable animal pharmacokinetic (PK) studies. This makes MBP technique particularly attractive for early drug assessment studies.

When due consideration is given to the physicochemical properties of the API and aforementioned precautions, the MBP technique works well in most instances. However, the authors did encounter some molecules that were very difficult to convert to amorphous MBP. The following are our findings from many years of experience.

10.3.5.1 Molecular Weight

The authors have applied MBP technology to hundreds of newly discovered compounds to enable toxicology and animal PK studies. In general, drugs with molecular weight greater than 500 had a higher propensity for conversion to amorphous MBP, whereas drugs with molecular weight less than 500 needed greater effort and closer attention. This observation is in good agreement with other researchers in the ASD field (Zhou et al. 2002). It is suggested that large molecules may assume more complex conformations and molecular configurations which retard nucleation and crystal growth, thereby making them more prone to conversion to amorphous state, and vice versa.

10.3.5.2 API Hydrophobicity

Hydrophobicity is measured by evaluating the partitioning behavior of a drug between n-octanol and an aqueous buffer. High $\log P$ suggests that the molecule favors van der Waals type nonpolar interactions, whereas low P indicates that molecule favors polar interactions such as H bonding or dipole–dipole interaction. Experimentally, it has been observed that molecules with high $\log P$ have a better chance of forming stable MBP than molecules with low $\log P$. A low $\log P$ suggests that the molecule is likely to have a high affinity toward polar solvents like water. It can be postulated that when water is used as an antisolvent, the low $\log P$ molecule may show high affinity toward water, resulting in phase separation during the MBP process (Qian et al. 2010).

10.3.5.3 H-bonding Donor and Acceptors

The MBP process uses mostly polar aprotic organic solvents and water as the antisolvent. One can imagine various interactions taking place in the reaction vessel including ionic, H bonding, dipole–dipole, and van der Waals interaction. At the end of the MBP process, when all solvents including water and organic solvent are removed, leaving the drug molecule imbedded in a polymer matrix, depending on functional group of drug and polymer, only a few interactions are possible between drug–drug, drug–polymer, and polymer–polymer molecules. A system that maximizes drug–polymer interaction while minimizing drug–drug and polymer–polymer interaction would be the best in stabilization of ASD. Over the many years of experience, molecules with high H-bonding acceptors were found to form more stable MBP, and one can speculate this may help in drug–polymer interaction.

It is almost ironic to examine physicochemical properties of an API against Lipinsky's rule of 5 (Lipinski 2000). The molecules that would be rejected based on Lipinsky's rule are good candidates for MBP suggestive of the saying that the stone the builders rejected has become the cornerstone.

10.3.6 Polymer Properties

Earlier chapters in this book have provided a detailed treatise on the selection of polymers in the development of stable ASDs of poorly soluble compounds. The importance of the polymer cannot be over emphasized as it is the polymer that helps maintain the amorphous state during processing, storage, and dissolution, leading to a viable ASD product. This is true for any ASD that is manufactured and stabilized by the help of a polymer. An important consideration for MBP is the exclusive use of an ionic polymer. Due to the nature of the MBP process, the use of an ionic polymers is a requirement. In addition to enabling the processing, ionic polymer can add an additional stabilization effect through ionic interaction with the drug molecule (Rumondor et al. 2009). Cellulosic polymers, specifically HPMCAS, have been shown to be superior in maintaining supersaturation during dissolution, presumably due to the formation of colloidal aggregates in solution (Curatolo et al. 2009).

The most commonly used polymers for the MBP process are HPMCAS (L, M, and H grades) and polymethacrylate-based polymers, e.g., Eudragit L100, L100-55, and S100. Other ionic polymers such as cellulose acetate phthalate (CAP), hypromellose phthalate (HP), polyvinyl acetate phthalate (PVAP), and cationic polymer Eudragit E100 can also be used. The polymer use levels are determined based on the drug loading and amorphous form stability.

The selection and use of the polymer should be adequately supported by safety data and appropriate toxicological assessment. For example, some polymers have residual synthetic materials that may include monomers, low molecular weight impurities, and processing aids such as surfactants and stabilizers. All of these may have a negative impact on safety. It may be necessary to establish appropriate controls around the additives to ensure safety and stability. For example, presence of surfactants such as polysorbate 80 and sodium dodecyl sulfate in the Eudragit L100-55 may limit the levels that can be safely used based on toxicological assessments(Evonik 2014). Furthermore, these additives can affect the performance of the MBP either negatively by lowering the T_g or positively by micellization. Finally, for compounds with narrow window of absorption, polymer selection could be critical in translating the in vitro results to in vivo performance, and polymer selection can play an important role. Commonly used polymers for MBP are listed in Table 10.2.

Table 10.2 Commonly used ionic polymers for MBP

Polymer	Common name	Soluble pH
Polymethacrylic	Eudragit L100	pH > 6.0
	Eudragit L100-55	pH > 5.5
	Eudragit S100	pH > 7.0
	Eudragit E100	pH < 5.0
Cellulosic	HPMC-AS LF	pH > 5.5
	HPMC-AS MF	pH > 6.0
	HPMC-AS HF	pH > 6.5
	Cellulose acetate phthalate	pH > 6.0
	Hypromellose phthalate 5	pH > 5.0
	Hypromellose phthalate 55	pH > 5.5
	Polyvinylphthalate (PVAP)	pH > 5.0

10.3.7 Drug Loading

For the sake of visualization, ASD can be viewed analogous to a solid solution where drug (solute) is dissolved in a polymer (solvent). Solubility is a thermodynamic parameter which is a function of temperature, pressure, and composition. If solubility of drug is high in a polymer, more drug can be incorporated in the polymer matrix, thus achieving a higher drug loading. However, the solubility of drug in polymer at room temperature is typically quite low, making it impractical to achieve the solubility and miscibility limit. Generally, ASDs are supersaturated systems, i.e., higher drug levels are incorporated in the polymer than allowed by its solubility limit. The most important aspect in developing supersaturated amorphous systems is the assurance of physical stability over the period of product shelf life. It is generally understood that the degree of supersaturation may influence the kinetics of physical instability, that is, the higher the supersaturation, the faster the phase separation. Since polymer provides the framework for stabilizing the amorphous drug, selecting the polymer and defining maximum drug loading while maintaining long-term physical stability are two critical goals in ASD development.

Based on the aforementioned primary factors, the drug loading can be initially selected at the small production scale and then further refined as the process is scaledup. Due to the uncertainty around amorphous form stability during long-term storage, a conservative approach is to operate at 5–10 % below the maximum drug loading determined during the small scale study. As more experience is gained, it may be possible to further increase the drug loading without compromising the stability and quality of the ASD.

10.3.8 MBP Process Design

MBP is a complex engineering process with multiple unit operations. Every step is important because any mismatched operational parameters can negatively impact the formation of amorphous MBP and the maintenance of amorphous state during subsequent processing and storage. In this section, process requirements are discussed in detail.

10.3.8.1 Solvent

The selection of organic solvent is primarily based on solubility and stability of the API and the polymer. Other factors that affect the selection of organic solvent include:

- Miscibility of solvent with the antisolvent: Solvent must be miscible with the antisolvent. This is an important factor as the rapid precipitation is primarily afforded by rapid mixing of solvent and antisolvent. If organic solvent is partially miscible, precipitation inefficiency may result owing to liquid-phase separation. In addition, the extraction of solvent out of the precipitates will not be efficient in subsequent washing and rinsing cycles.
- Permissible residual solvents: Level of residual solvents must be controlled based on the International Conference on Harmonization (ICH) solvent classification guidance permissible daily limit (ICHQ3 Guidance 2009). In addition, the impact on the T_g of the amorphous product and stability of the amorphous form must be considered.
- Stability of API in the solvent to support the manufacturing at the desired scale.
- Viscosity of the solution to maintain uninterrupted smooth flow rates during precipitation.

The most commonly used solvents for MBP are DMA, DMF, DMSO, and NMP. Other solvents such as alcohols and acetone may also be used, but they are less favored for MBP because if there is adequate solubility in these solvent, other means of ASD manufacture such as spray drying or fluid bed granulation/layering may be feasible. List of solvents with their relevant properties is summarized in Table 10.3.

10.3.8.2 Antisolvent

Selection of antisolvent depends on the properties of API, polymer, and the selected solvent. Both polymer and API should be insoluble in antisolvent system even when mixed with 10–20 % of the solvent. There can be many choices of antisolvents; however, water is almost exclusively used in the MBP process. In fact, the use of water makes MBP unique compared to other coprecipitation techniques, where other organic antisolvents may be used. MBP, in this regard, represents a special subset of coprecipitation techniques. Use of aqueous media along with ionic polymer and dispersing mechanism to produce finely dispersed homogeneous ASD particles

Table 10.3 Some relevant properties for the commonly used solvents are summarized

Solvent	Mol wt (g/mol)	BP (°C)	MP (°C)	Density (g/mL)	Water solubility (g/mL)	Dielectric constant	Flash point (°C)	ICH/PDE[a] (mg/day)
Acetic acid	60.1	118	16.6	1.049	Miscible	6.2	39	Class 3
Acetone	58.1	56.2	−94.3	0.786	Miscible	20.7	−18	Class 3
Acetonitrile	41.1	81.6	−46	0.786	Miscible	37.5	6	Class 2/4.1
DMA	87.1	165	−20	0.937	Miscible	37.8	63	Class 2/10.9
DMF	73.1	153	−61	0.944	Miscible	36.7	58	Class 2/8.8
DMSO	78.1	189	18.4	1.092	Miscible	47.0	95	Class 3
Dioxane	88.1	101.1	11.8	1.033	Miscible	2.2	12	Class 2/3.8
Ethanol	46.1	78.5	−114.1	0.789	Miscible	24.6	13	Class 3
Methanol	32.0	64.6	−98	0.791	Miscible	32.6	12	Class 2/30
NMP	99.1	202	−24	1.033	Miscible	32.0	91	Class 2/5.3
2-propanol	88.2	82.4	−88.5	0.785	Miscible	18.3	12	Class 3

[a]Residual solvents' limit based on permissible daily exposure (ICH Guidance)

forms the core of MBP technology. Additionally, the aqueous precipitation provides material with superior particulate and wetting properties. This is attributed to the removal of solvent during precipitation process, i.e., as the solvent diffuses out of the droplet, the material that is left is of a sponge-like porous nature filled with aqueous fluid. The aqueous fluid in the pores is eventually removed during drying, thus leaving material with superior wetting, compaction, and dissolution properties.

Acidic cold water is typically used in MBP when an anionic polymer is used. The pH and temperature are controlled to maximize the precipitation of drug and polymer. Acidic water at pH around 2 and temperature around 2–8 °C usually provides adequate conditions for most drug molecules unless the API is a strong basic molecule. Weakly basic molecules of p*K*a up to 5 have been processed successfully using aqueous buffer systems of pH up to 4.

10.3.8.3 Operation

A schematic representation of MBP process is shown in Fig. 10.3. The different steps (left to right) are (a) crystalline drug, (b) dissolution in DMA, (c) coprecipitation in acidic water, (d) cold acidic water rinse, and (e) final MBP.

10.3.8.4 Description and Details of Unit Operations

- Stock solution preparation: Predetermined amounts of drug and polymer are dissolved in the solvent. In a small laboratory-scale operation, both drug and polymer

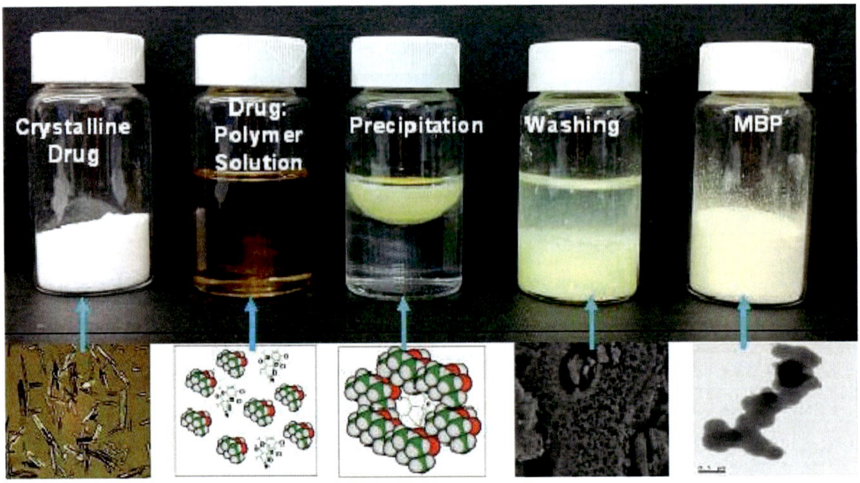

Fig. 10.3 Pictorial view of MBP processes (Shah et al. 2012)

can be added together to a vessel containing solvent and dissolved using standard laboratory mixer. As the scale increases, stock solution preparation becomes non-trivial. Usually, the polymer takes a long time to dissolve and often heating is needed. The homogenizer mixer can be used to improve the process efficiency. Chemical stability of drug and possible formation of solvates have to be investigated during this process. Total solid content of up to 40 % has been demonstrated (Shah et al. 2013). However, if solid content is too high, viscosity of stock solution can cause transfer and diffusion problems, impacting droplets/particle formation during coprecipitation and solvent exchange during washing and rinsing steps. Therefore, total solid contents should be decided based on the solution transfer efficiency, coprecipitation mechanism, and resulting particulate properties.

- Coprecipitation: Coprecipitation is the most critical step in the entire MBP processes. There are two viable manufacturing modes of coprecipitation—batch mode and continuous mode (see Fig. 10.4). In batch mode, stock solution containing drug and polymer is delivered to a vessel containing a large volume of aqueous media. At the small laboratory scale, stock solution can be added carefully to the reactor containing aqueous media, while contents in the reaction vessel are being stirred by an overhead propeller or homogenizer. At larger scale, the stock solution can be sprayed over the antisolvent while stirring, or sprayed into the antisolvent in a manner that allows the solution to be rapidly taken up in cavitational zone to break the precipitate into finer particles. Coprecipitation occurs instantly when drug polymer solution contacts the cold acidic water. In continuous mode, streams of stock solution and acidic water are continuously pumped into a homogenizer chamber where two streams of liquid are mixed and sheared by high-speed rotors to produce fine droplets of the coprecipitate. The precipitate along with the solution is then pumped into a holding tank or to the filtration unit. Regardless

Fig. 10.4 MBP manufacture reactors showing batch mode reactor (left) and continuous mode reactor (right)

of the precipitation mode, reaction temperature, shear rate, and mixing time, the solvent to antisolvent ratio has significant impact on initial MBP formation (see Fig. 10.5).

- Coprecipitation parameters: Critical coprecipitation parameters that should be under tight control are reaction temperature, shear rate, mixing time, and solvent to antisolvent ratio.

Temperature control during coprecipitation plays a significant role in producing a quality ASD. When the stock solution is added to the aqueous media, substantial amount of heat is generated due to heat of mixing. Additional energy input from high shear mixing can add additional heat to the system. Unless dissipated, the heat can cause undesired impact on the quality of MBP, resulting in poor drug recovery and inadequate conversion to amorphous state (Fig. 10.5). Incomplete conversion may leave dreaded crystalline seeds in the MBP. As discussed, any crystalline seeds in the system may propagate crystallization; therefore, should be avoided at any cost.

Extensive research has shown that the best temperature range is $5 \pm 3\,°C$. The positive temperature deviation coupled with high shear may result in the traces of crystalline API in the MBP cake and the final MBP powder. The impact of temperature, shear, and mixing time is shown in Fig. 10.5. These parameters should be carefully controlled to achieve high quality and consistency of amorphous material.

The effects of shear rate (2500, 4000, 5000 rpm), mixing time (30, 60, 180 min), and reaction temperature (2, 10, 15 °C) were examined during laboratory-scale production of MBP using Eudragit L100-55 polymer and investigational drug ROX35. It was found that the high-speed mixing during coprecipitation was detrimental in amorphous MBP formation, presumably due to localized energy input to the precipitate at the point of contact. Temperature of the vessel also played a key role in MBP integrity. At higher temperature, more undesirable crystalline API was found in MBP. Duration of coprecipitation and subsequent mixing time was crucial in MBP stability; longer duration was found to be detrimental. This observation suggests

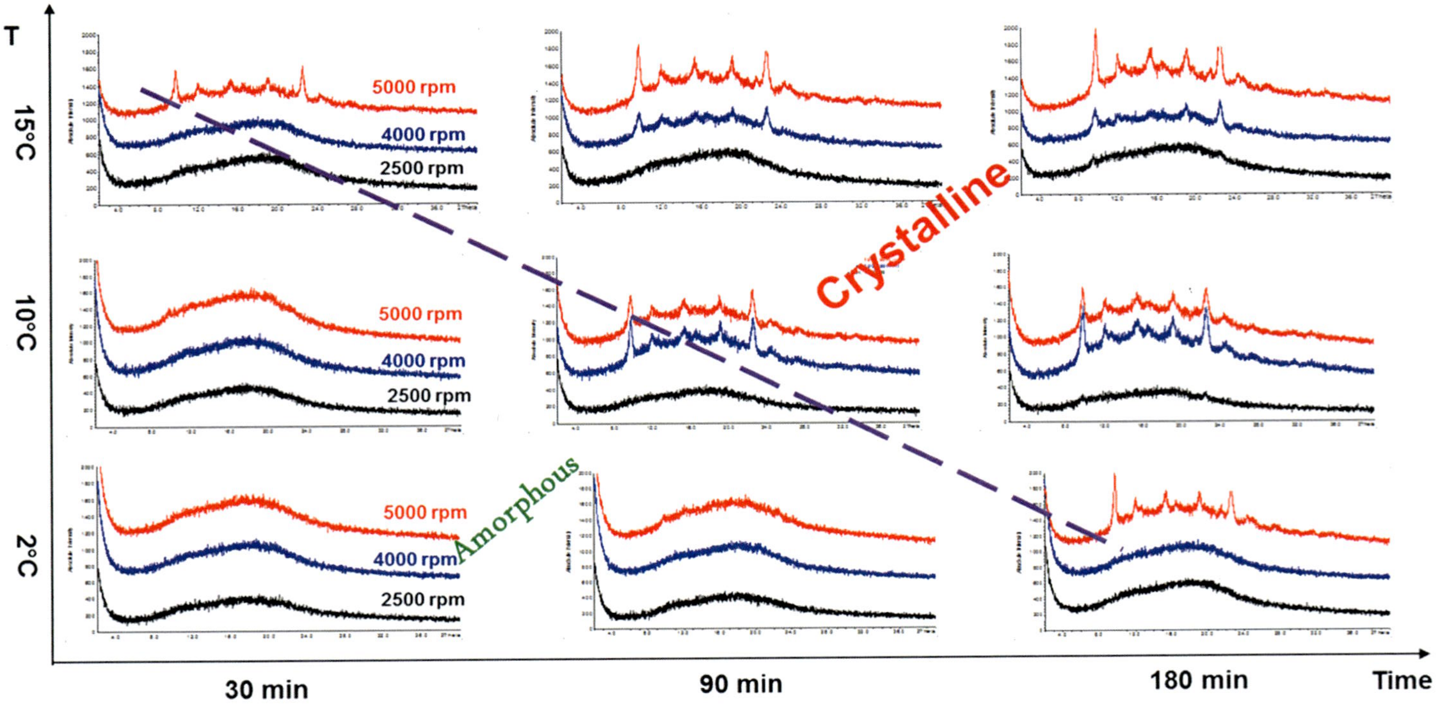

Fig. 10.5 Effect of shear rate, reactor temperature, and mixing time in MBP formation. The risk of crystallization during MBP manufacture increases with the increase of temperature, mixing time, and shear rate

Fig. 10.6 Microstructure of MBP particle reproduced using 3D printing (courtesy of Dr. Siegfried Krimmer)

that as soon as coprecipitation is complete, unnecessary additional mixing should be avoided.

The overall impact of the various parameters was found to be in the following order: shear rate > mixing time > temperature. The combination of slower shear, shorter mixing time, and lower reaction temperature consistently produced quality amorphous MBP. A similar observation was made in the experiment with continuous mode procedure.

In summary, it was observed that, from process perspective, shear force is the most sensitive parameter in the formation of amorphous coprecipitate. Shorter mixing and churning time was considered more favorable in production of MBP.

- Solvent removal, washing, and isolation: At the end of coprecipitation, the precipitates contain substantial amount of organic solvent depending on solvent to antisolvent ratio used. During subsequent washing and rinsing cycles, the residual organic solvent must be removed from the precipitate. Washings are typically conducted with acidic water maintained at 2–8 °C and can be performed either on the filter media or by resuspending the material in acidic water. Depending on the process selected, care should be taken to ensure that the washing is complete. In addition to the number of washings, washing time, temperature, and pH of the washing solution need to be controlled. Although all washings are performed with the cold-acidified water, the last wash is generally carried out with purified water only to minimize residual acidic component in the final MBP. As mentioned before, the solvent exchange during precipitation and washing steps produces sponge-like material with high porosity that in turn offers rapid dissolution and high compressibility (Fig. 10.6)

- Isolation: The wet MBP can be isolated by vacuum-assisted filtration, filter press, or a centrifugal filter. Filtration efficiency mainly depends on the particle size of the precipitate; however, other factors such as type of polymer as well as drug loading can also influence the filtration efficiency. Depending on the filtration and washing mechanism used to remove the residual organic solvent, the wet solids can be isolated by using impeller-driven scrapper or scoops. Care should be taken to ensure removal of most of the free liquid during filtration step. The resulting wet solid can be dried immediately or can be held in proper storage conditions for later drying. Usually, refrigerated conditions are preferred to minimize risk of crystallinity and/or microbial growth. In batch mode, several sub-batches of coprecipitates can be collected for drying.

- Drying: Drying of wet cake is the second most critical step in the MBP process as the product is subjected to elevated temperature for an extended period of time to drive off the excess water. After final isolation, typical wet MBP cake contains about 60–90 % of its mass as water. Obviously, removing this amount of water out of MBP cake is not a trivial matter. It is well documented in literature that combination of heat and moisture is the number one culprit in amorphous material destabilization. In this drying step, both heat and moisture are present, the same elements that should be avoided. The key success factor for overcoming these destabilizing forces is an efficient and rapid drying process. Typically, a forced-air oven, filter dryer, agitated conical or spherical dryer, drum dryer, or fluid bed dryer has been used. For compounds with a high tendency to crystallize, fluid bed dryer provides the best mode of drying. As a rough rule of thumb, the product temperature should be maintained 50 °C below the T_g of ASD during drying to have the least impact on the product quality (Taylor and Zografi 1997).
- Milling: On a microscopic level, the MBP particle is homogenous material which contains the amorphous drug either molecularly dispersed or nanoscale dispersed in the polymer matrix. On a macroscopic level, the bulk powder properties of MBP can vary substantially. To normalize material properties for downstream processing, the MBP material can be milled using standard milling technologies such as impact milling or air-jet milling. Bulk density of the powder ranges between 0.1 and 0.3 g/cc and may require further densification prior to the final dosage form manufacture.

The key formulation and process factors are summarized in Table 10.4.

10.4 In-Process Characterization

As discussed in the previous section, MBP is a highly complex industrial process comprising of multiple unit operations including (a) stock solution preparation in which API and polymer are mixed in a common solvent, (b)dissolution in which API and polymer are heated and agitated to ensure complete dissolution without any residual crystals, (c) coprecipitation in which stock solution is brought in contact with antisolvent such as chilled acidic water in a controlled manner to induce well-dispersed amorphous coprecipitate, (d) isolation in which the coprecipitated solid is separated from the suspension by means of filtration or centrifugation, (e) washing and rinsing in which the coprecipitated solid is further rinsed with water to reduce residual organic solvent, (f) drying in which the washed cake is dried to remove excess water, and finally (g) milling in which dried material is delumped and milled

Table 10.4 Overview of MBP processing steps

Unit operations	Key considerations
API and polymer dissolution	Solubility and stability in solvent Temperature and time Solid content and viscosity Mixer design and speed
Coprecipitation	Batch mode vs. continuous mode Flow rate Droplet size distribution Mixer design (shear and energy, e.g., stir bar, vortex, propeller, homogenizer, rotor–stator) Solvent to antisolvent ratio Temperature Processing times (scale dependent)
Extraction/washing/filtration	Volume of antisolvent Filter media (filter paper, filter press, centrifugal filter) Compressibility may cause blinding Channeling may reduce extraction efficiency Stability of wet cake Wet cake stability (physical, chemical, microbiological) Residual solvents, moisture content Cycle time
Isolation/discharge	Moisture content, particulate properties
Drying	Tray dryer Agitated dryer (rotary tumble and conical) Filter dryer Fluid bed dryer Cycle time and stability Particulate properties
Milling	Delumping (Conical mill) Impact milling (hammer or pin) Air-jet milling Media and ball milling
Densification	Dry granulation (roller compaction), fluid bed granulation, or wet granulation Blending/compatibility with other excipients Compression or encapsulation

to a target particle size. As part of in-process evaluation, it is important to ensure that amorphous state is maintained throughout the manufacturing process, as it not only helps in establishing the controls but also to troubleshoot the process in case of failure. The whole operation can take a few days to several weeks that may be performed in multiple locations/sites depending on the batch sizes and the logistics.

It is important to establish the critical quality attributes for each operational step. Above all, keeping the product as non-crystalline amorphous state during each operation is the utmost critical element in the MBP process. During the initial dissolution operation, API should be completely dissolved and no crystalline seeds should be found anywhere in the vessel. Any residual crystalline seeds can potentially act as further nucleation sites in subsequent steps. Heating may be applied to aid dissolution of drug and polymer and to reduce the viscosity. If need be, the drug–polymer solution can be filtered prior to precipitation to ensure no undissolved material is introduced into precipitation step. Needless to say, controlled coprecipitation is at the heart of the MBP process. First, drug and polymer must be precipitated together in such a way that drug molecules are incorporated uniformly in the polymer matrix at a molecular level. Second, coprecipitate should disperse in aqueous media as uniform fine particles without forming large solid aggregates. Well-dispersed MBP suspension maximizes solvent–antisolvent exchange and ensures optimal downstream processing. Any mishandling of the wet cake in subsequent operations may cause phase separation leading to crystallization. As such, drying is another key operation in the MBP process where material can convert to crystalline state if proper care is not taken.

Although crystallinity is the most critical parameters in MBP manufacture, residual solvent level and water content are equally important. ICH guideline dictates the allowable organic solvent level in the final pharmaceutical product. It has been observed that residual organic solvent, if not removed, can negatively impact stability of ASD either by lowering the T_g or by dissolving drug in micro-domains, resulting in recrystallization during storage. Water content in any ASD is critical since it can adversely impact stability of the ASD by lowering its T_g. Although MBP is best prepared in aqueous media using water as the antisolvent and the penultimate wet cake is more than 70 % water, however, establishing appropriate moisture control is critical for long-term storage. Because of the presence of water, one should be mindful of bioburden as the wet cake can support mold or fungi growth. Particle size of the precipitate and its microstructure have significant impact on downstream processing and performance of drug product.

The following subsections address the key in-process control parameters.

10.4.1 Crystallinity

Ensuring complete amorphous state is a key element in any ASD manufacturing process. Analytically, amorphousness is only assumed by lack of crystallinity. Various techniques can be employed, but two techniques are most useful. The first technique, X-ray diffraction (XRD) is the primary tool in assessing crystallinity and is useful in assessing successful MBP production. Secondly, birefringence examination using a cross-polarized light microscope (PLM) is utilized. Occasionally, birefringence can be used for fast feedback. Only after assurance of the lack of crystallinity, the process can move to the next step of the manufacturing chain.

Wet MBP cake is typically 60–90 % water causing X-ray signal substantially attenuated. Any small suspect crystalline peak can become quite large when material is dried. Any suspect peaks in the wet cake, therefore, should be examined thoroughly to investigate potential incomplete conversion. XRD is not a sensitive tool in the determination of trace levels of crystallinity. Secondary techniques such as Raman spectroscopy should be employed where applicable.

10.4.2 Residual Solvent

Typically, nonvolatile solvents such as DMA, DMSO, DMF, or NMP are used in manufacture of MBP and these are only removed by the rinsing and washing operation. Higher level of organic solvent not only poses health concerns but can also raise long-term MBP stability concerns. High level of organic solvent has been shown to induce crystallization in the MBP. One can postulate that residual organic solvent can lower the glass transition temperature, and thereby increase molecular mobility. In addition, it can dissolve the drug which can crystallize over a period of time. Usually, three to five washing cycles are sufficient to reduce the organic solvent level to below 0.1 %. This is dependent on the particle size, type of polymer, and design of the vessel. GC methods are commonly used for the detection of organic solvents. Organic solvent level of less than 0.1 % has been found to be acceptable without any negative impact on the quality of the MBP.

10.4.3 Moisture

The wet MBP cake after filtration generally contains about 60–90 % water by weight, and this must be removed to an acceptable level. Thus, the drying step is one of the most critical steps in MBP manufacturing and should be closely monitored. Water level in the cake can be monitored by water activity of the outlet air and further confirmed by more sensitive moisture measurements such as potentiometric titration.

10.4.4 Bioburden

Drying is a unit operation where multiple batches of coprecipitate can be combined to gain production efficiencies. In such cases, wet MBP can be stored in a refrigerator for an extended period, but only if the amorphousness of wet MBP is ensured. It has been observed that, if not properly stored, mold and fungi can grow in the wet cake. Bioburden tests should be performed prior to drying in cases where the wet cake needs to be stored for extended period of time before drying.

Table 10.5 Tiered MBP IPC testing

Tier	Tests	Targets	Comments
Tier 1	DMA level Water level Granule size	<0.1 % <2.0 % Monitor	In-process control to minimize residual solvents and granule size
Tier 2	XRD DSC PLM	No crystalline peak Monitor T_g No crystalline drug particles in MBP	First-hand testing to evaluate the success of initial amorphous conversion
Tier 3	Suspension stability in aqueous vehicle	No crystallization within at least 8 h, preferably longer periods	Initial testing to evaluate stability aspects of the amorphous product
Tier 4	Comprehensive solid-state characterization Discriminating dissolution testing	Satisfactory	Comprehensive testing for long-term stability and dosage-form development

10.4.5 Tiered Testing

MBP is a complex multiunit operational process. Identifying and assessing quality of MBP at each step expedites the successful development of MBP. The following tiered approach can be used to assess MBP during early screening (Table 10.5).

10.5 Characterization of MBP

The MBP is a special type of ASD obtained from a controlled coprecipitation process and differs from other ASDs in physicomechanical properties such as porosity, surface area, bulk density, microstructure, flow properties, wettability, etc. Nevertheless, the overall characterization scheme of MBP is the same as any other ASD, which will be described elsewhere in this book. This may include crystallinity by XRD, glass transition temperature evaluation by differential scanning calorimetry (DSC), molecular structure by IR and Raman, solubility assessment and dissolution profile, and micromeritics such as bulk density, particle size, porosity, flowability, etc. Physicomechanical properties of MBP are addressed in Examples of Bioavailability Enhancement section of this chapter.

10.6 Formulation for Preclinical Toxicology Studies

As noted earlier, the efficiency and versatility of the MBP technology makes it an ideal option for preparing formulations for preclinical studies. Owing to the flexibility of dosing, liquid formulations are preferred for preclinical PK and toxicological studies. However, the development of liquid formulations of ASD such as MBP requires a careful consideration of stability, wettability, and dispersibility, as well as storage and holding time. High concentrations of ASD suspensions tend to exhibit gelling and difficulty in dosing, especially upon storage over time. Further, given the hydrophobic nature of the drug, poor wettability and dispersibility can present problems in reconstitution of the powder.

In developing liquid formulations for toxicological dosing, the liquid intake volume in animals, especially rodents, can be quite limiting. For example, at 5 mL/kg, the volume intake in a rat is around 1 mL for a single dose which limits the total amount of dispersion that can be administered. At high dose levels, the solids' concentration achievable in such small volumes can be a major challenge.

10.6.1 Toxicological Vehicle Selection

The vehicle used to prepare amorphous MBP suspension should be able to maintain the physical stability of the ASD for at least 4 h and preferably 24 h to support the typical time period of constitution, mixing, and dosing in toxicology studies. Vehicle pH is often critical for amorphous formulations containing enteric polymers. The vehicle pH should be on the acidic side to minimize the API and polymer dissolution.

The inclusion of nucleation inhibitors such as silicon dioxide can modulate nucleation process, thus prolonging the suspension stability. Particle size control of amorphous formulations is essential for homogeneity and withdrawability for dosing accuracy.

10.6.2 Evaluation of MBP Toxicology Formulation

A standard approach is to prepare various concentrations of MBP suspension in a vehicle with different additives. Stability of MBP formulation should be monitored both chemically and physically. The amorphous nature of MBP can be investigated by XRD or other techniques. To be viable, MBP suspension should demonstrate at least 4 h, preferably 24 h stability in the toxicology vehicle.

10.6.2.1 Effect of MBP Concentration

The effect of MBP concentration in aqueous vehicles has been studied by the authors. Physical stability of various concentrations of MBP suspension (up to 100 mg/mL) was examined in aqueous vehicles containing Klucel LF as a wetting and suspending agent. Solid residue was analyzed by XRD after 4 h of mixing. The relative level of crystallinity by XRD suggested that the higher concentration MBP suspension (e.g., 100 mg/mL) was more stable than the lower concentration MBP (e.g., 5 mg/mL).

10.6.2.2 Effect of Additives

Commonly used additives in the toxicology vehicle are preservatives and stabilizers. Use of antimicrobial preservatives in aqueous vehicles is needed if toxicology vehicle is not prepared fresh but prepared in bulk for later use. Unfortunately, certain preservatives such as methyl and propyl parabens may influence the physical stability of the MBP negatively. For example, in the study described above, the relative stability of MBP suspensions prepared without parabens was better than that of the suspension with the parabens. Therefore, the use of preservatives in these preparations should be carefully evaluated.

Based on extensive research and years of experience, it has been shown that dispersing agents can be quite useful. The inclusion of small amounts of dispersing agents in the suspension was found to prolong the MBP suspension stability over extended periods of time during the conduct of toxicology studies.

Although pH of the vehicle prevents dissolution of API, the ingress of water into polymeric system cannot be avoided and that can have a negative impact on the stability. In some cases, use of hydrophilic fumed silica was found to prolong the stability of MBP formulation during toxicology enabling studies. It is postulated that the nano-sized silica particles adhered to the MBP particles. This results in amorphous MBP particles being "coated" with silica agglomerates, thus minimizing fusion, nucleation, and crystallization (Planinsek et al. 2011).

10.7 Design of Final Dosage Form

The powder obtained after drying needs to be processed into final dosage form. From practical perspective, the particulate properties of MBP such as bulk density, porosity, particle size, and size distribution are similar or slightly superior to spray-dried material. The two key features of the MBP particulate properties that standout quite favorably compared to other amorphous technologies are high porosity and superior wetting. They are attributed to the nature of the process, i.e., solvent extraction versus surface drying and the use of aqueous medium as antisolvent. These characteristics can have direct impact on the downstream processing (densification) and dissolution. In contrast to melt-extruded products where the compaction to final dosage form is limited due to porosity of the extrudate, the high porosity of MBP provides superior compaction properties without loss of dissolution. The particulate properties of the MBP depend on the MBP processing and factors ranging from preparation of solvent (solid content) to the final step of drying can influence the particulate properties. To produce granules suitable for high-speed tablet machines, the intermediate powder of MBP is densified using preferably dry granulation method. The densified material with additional excipients such as disintegrant and lubricant can be converted into capsules or tablet. More details about the downstream processing are presented in the other chapters in this book.

Table 10.6 Bioavailability Comparison of ROX45J

Formulation	AUCO-∞/dose (ng.h/mL)/(mg/kg)	Tmax (hr)	Cmax (ng/ml)	Percent bioavailability
Micronized drug suspension	29.5 ± 8.3	1.0 ± 0.0	55 ± 17	3.9
Nanosized drug suspension	86.1 ± 13.7	1.5 ± 0.6	142 ± 53	11.2
Pluronic F68 dispersion (10% drug loading)	532 ± 152	0.8 ± 0.4	2044 ± 374	69.0
Microprecipitated Bulk Powder (MBP)(50% drug loading)	686 ± 237	2.5 ± 0.9	1212 ± 358	89.0
IV formulation	766 ± 8.3	n/a	n/a	100.0

[1] N = 4.2 males and 2 females with a parallel design

10.8 Examples of Bioavailability Enhancement

Experimental drug ROX45J is a potent kinase inhibitor demonstrating an excellent efficacy and toxicology profile during early preclinical studies. Unfortunately, an unacceptable bioavailability (3.9 % in dog) put this candidate in danger of premature termination. Subsequent effort to enhance bioavailability using nanomilling proved to be insufficient (11 % in dog).This molecule has no measurable p*K*a in physiological pH ranges and showed aqueous solubility of 0.0001 mg/mL. The compound decomposed upon melting at 230 °C. Solubility in acetone was less than 0.5 % and the compound also showed tendency to form solvate. These physicochemical properties made it difficult to make an ASD using either hot-melt extrusion or spray-drying processes.

Employing MBP technology, ASD of ROX45J was developed at a 50 % drug loading using Eudragit L100 as a stabilizing polymer. MBP of this compound was tested in dogs, resulting in an outstanding bioavailability improvement (89 %) as seen in Table 10.6. This molecule progressed to clinical studies for a full evaluation, which was possible only because of MBP technology (Dupont et al. 2004).

10.9 Challenges and Future Innovation in MBP Technology

Since the introduction of MBP technology in the late 1990s, it has made tremendous progress, enabling advancements of hundreds of compounds into preclinical and clinical studies. Zelboraf® is the culmination of these efforts and technical advancements (Chapman et al. 2011; Heakal et al. 2011). The development and marketing of this important oncology medicine were only possible because of the application of the MBP technology:

Despite its successes and advancements, MBP is still in a stage of infancy with much room for improvement. One of the major challenges that the authors observed was that certain molecules are very difficult to convert to stable amorphous MBP. Although they identified the anecdotal cause and effect relationships over many years of experience, no robust theoretical relationship has been established. As discussed in the API section, the relationship of API physicochemical properties to successful conversion to a stable MBP must be further explored.

Polymers are limited to several ionic polymers due to the design of the coprecipitation process as well as the available safety data of polymers. Excipient manufacturers are striving to design additional polymers with different functional groups exhibiting different solubility and interaction potential. This certainly will broaden MBP applications. As discussed earlier, the authors observed the drug–polymer interaction as one of the key elements in the success of MBP. Increased pool of polymers may expand the usefulness of MBP and move the technology beyond the current limitations.

MBP is a complex process with several unit operations, and appropriate in-process controls are absolutely essential. The scale-up from laboratory scale to production scale is challenging and complex. This technology has been developed with the capacity to produce several tons supply; however, additional process efficiency may be feasible by further investing in understanding and optimizing the key unit operations. Continuous manufacturing must evolve in order to reduce production cost and increase productivity (Mascia et al. 2013).

10.10 Summary

A solvent-controlled coprecipitation technology, known as MBP technology, can be used for the manufacture of ASDs of poorly soluble drugs. In this technology, a solution containing drug and polymer is carefully delivered to an antisolvent in order to induce fine droplets of coprecipitates, which are then isolated, rinsed, washed, dried, and milled. MBP produced by this technique is an ASD with unique physicochemical and physicomechanical properties that provides enhanced bioavailability not seen from products manufactured using other ASD techniques. MBP is a complex engineering process comprising multiple unit operations. The critical operational parameters related to this process have been fully described in this chapter.

MBP technology is particularly useful for highly insoluble so-called brick dust-like compounds where other conventional amorphous techniques fail. Compounds that can benefit by application of MBP technologies are molecules with the following properties:

- Poor solubility in water, typically less than 0.001 mg/mL
- Poor solubility in volatile organic solvent, typically less than 1 %
- High melting point, typically higher than 200 °C
- High lipophilicity with log P greater than 3
- High molecular weight greater than 500
- High H-bonding acceptors greater than 7

It is believed that in MBP, the drug molecule is dispersed in an inert polymer carrier at a molecular or nanoscale level. The drug molecules in MBP are immobilized by the polymer preventing drug molecules to migrate, resulting in inhibition of nucleation and crystallization. Furthermore, the polymer protects the amorphous drug from moisture enabling maintenance of physical stability.

The solid dispersion produced by the MBP process can achieve high degree of supersaturation during dissolution in the GI tract resulting in enhanced absorption with minimum food effect. If desired, MBP formulations can even be engineered to provide sustained release profiles. The MBP technology provides a viable alternative for ASD technology when other technologies such as spray drying and hot-melt extrusion are not suitable.

References

Albano A, Phuapradit W, Sandhu H, Shah N (2002) Amorphous form of cell cycle inhibitor having improved solubility and bioavailability. U.S. Patent. 6,350,786

Chapman PB, Hauschild A, Robert C, Haanen JB, Ascierto P, Larkin J, Dummer R, Garbe C, Testori A, Maio M, Hogg D, Lorigan P, Lebbe C, Jouary T, Schadendorf D, Ribas A, O'Day SJ, Sosman JA, Kirkwood JM, Eggermont AM, Dreno B, Nolop K, Li J, Nelson B, Hou J, Lee RJ, Flaherty KT, McArthur GA, Group B-S (2011) Improved survival with vemurafenib in melanoma with BRAF V600E mutation. N Engl J Med 364(26):2507–2516

Chiou WL, Riegelman S (1970) Oral absorption of griseofulvin in dogs: increased absorption via solid dispersion in polyethylene glycol 6000. J Pharm Sci 59(7):937–942

Curatolo W, Nightingale J, Herbig S (2009) Utility of hydroxypropylmethylcellulose acetate succinate (HPMCAS) for initiation and maintenance of drug supersaturation in the GI milieu. Pharm Res 26(6):1419–1431

De Yoreo JJ, Vekilov PG (2003) Principles of crystal nucleation and growth. Rev Mineral Geochem 54(1):57–93

Dupont J, Bienvenu B, Aghajanian C, Pezzulli S, Sabbatini P, Vongphrachanh P, Chang C, Perkell C, Ng K, Passe S, Breimer L, Zhi J, DeMario M, Spriggs D, Soignet SL (2004) Phase I and pharmacokinetic study of the novel oral cell-cycle inhibitor Ro 31–7453 in patients with advanced solid tumors. J Clin Oncol 22(16):3366–3374

Evonik (2014) Eudragit L100–55product profile. Evonik, Germany

Heakal Y, Kester M, Savage S (2011) Vemurafenib (PLX4032): an orally available inhibitor of mutated BRAF for the treatment of metastatic melanoma. Ann Pharmacother 45(11):1399–1405

ICHQ3 Guidance (2009) International Conference on Harmonisation of Technical Requirements for Registration of Pharmaceuticals for Human Use; Impurities: Guideline for Residual Solvents Q3C(R4). http://westpalmbeachanalytic.homestead.com/Q3C_R4__Guide_Res_Solv.pdf

Kislalioglu MS, Khan MA, Blount C, Goettsch RW, Bolton S (1991) Physical characterization and dissolution properties of ibuprofen: eudragit coprecipitates. J Pharm Sci 80(8):799–804

Leuner C, Dressman J (2000) Improving drug solubility for oral delivery using solid dispersions. Eur J Pharm Biopharm 50(1):47–60

Lipinski CA (2000) Drug-like properties and the causes of poor solubility and poor permeability. J Pharmacol Toxicol Methods 44(1):235–249

Mascia S, Heider PL, Zhang H, Lakerveld R, Benyahia B, Barton PI, Braatz RD, Cooney CL, Evans JM, Jamison TF, Jensen KF, Myerson AS, Trout BL (2013) End-to-end continuous manufacturing of pharmaceuticals: integrated synthesis, purification, and final dosage formation. Angew Chem Int Ed Engl 52(47):12359–12363

McKeown RR, Wertman JTA, Dell'Orco PC (2011) Crystallization design and scale-up. In: am Ende DJ Chemical enginnering in the pharmaceutical industry R & D to manufacturing. John Wiley & Sons, New Jersey, pp 213–248

Planinsek O, Kovacic B, Vrecer F (2011) Carvedilol dissolution improvement by preparation of solid dispersions with porous silica. Int J Pharm 406(1–2):41–48

Qian F, Huang J, Hussain MA (2010) Drug-polymer solubility and miscibility: stability consideration and practical challenges in amorphous solid dispersion development. J Pharm Sci 99(7):2941–2947

Repka MA, Langley NA, DiNunzio J (2013). Melt extrusion materials, technology and drug product design. AAPS press and Springer, New York

Rumondor AC, Stanford LA, Taylor LS (2009) Effects of polymer type and storage relative humidity on the kinetics of felodipine crystallization from amorphous solid dispersions. Pharm Res 26(12):2599–2606

Sekiguchi K, Obi N (1961) Studies on absorption of eutectic mixtures I. A comparison of the behavior of eutectic mixtures of sulfathaiazole and that of ordinary sulfathaiazole in man. Chem Phar Bull 9(11):866–872

Shah N, Sandhu H, Phuapradit W, Pinal R, Iyer R, Albano A, Chatterji A, Anand S, Choi DS, Tang K, Tian H, Chokshi H, Singhal D, Malick W (2012) Development of novel microprecipitated bulk powder (MBP) technology for manufacturing stable amorphous formulations of poorly soluble drugs. Int J Pharm 438(1–2):53–60

Shah N, Iyer RM, Mair HJ, Choi DS, Tian H, Diodone R, Fahnrich K, Pabst-Ravot A, Tang K, Scheubel E, Grippo JF, Moreira SA, Go Z, Mouskountakis J, Louie V, Ibrahim PN, Sandhu H, Rubia L, Chokshi H, Singhal D, Malick W (2013) Improved human bioavailability of vemurafenib, a practically insoluble drug, using an amorphous polymer-stabilized solid dispersion prepared by a solvent-controlled coprecipitation process. J Pharm Sci 102(3):967–981

Simonelli AP, Mehta SC, Higuchi WI (1969) Dissolution Rates of High Energy Polyvinylpyrrolidone (PVP)-Sulfathiazole Coprecipitates. J Pharm Sci 58(5):538–549

Taylor LS, Zografi G (1997) Spectroscopic characterization of interactions between PVP and indomethacin in amorphous molecular dispersions. Pharm Res 14(12):1691–1698

Williams IRO, Watts AB, Miller DA (2010) Formulating poorly soluble drugs. AAPS press and Springer, New York

Yang W, Williams RO, Donald EO (2010) Pharmaceutical cryogenic technologies. In: Williams III RO, Watts AB, Miller DA (eds) Formulating poorly soluble drugs. AAPS press and Springer, New York, pp 443–500

Zhou D, Zhang GGZ, Law D, Grant DJW, Schmitt EA (2002) Physical stability of amorphous pharmaceuticals: Importance of configurational thermodynamic quantities and molecular mobility. J Pharm Sci 91(8):1863–1872

Chapter 11
MBP Technology: Process Development and Scale-Up

Ralph Diodone, Hans J. Mair, Harpreet Sandhu and Navnit Shah

11.1 Introduction

The successful development of a drug product encompasses the overarching goal of providing consistent material during the development and ensuring robust supply chain throughout the life cycle of the product. Because of the stage–gate approach used in the drug development, the activities in the drug product tracks closely with the clinical phase. Therefore, the quantity and demand of active pharmaceutical ingredient (API) and drug product build over time. In order to streamline the supply requirements and minimize surprises, it is important that the product provides comparable performance as the process is scaled up from small scale to pilot scale and eventually to commercial scale. To conserve the API during development, the primary requirements in the early stage is to provide the material with desired attributes while ensuring that when scaled-up material with similar characteristics (downstream processing, stability, and performance) can be produced while maximizing the process efficiency and yield (Hu et al. 2013).

Because amorphous solid dispersion (ASD) are thermodynamically metastable and generally supersaturated systems, the scale-up of process had been one of the major reason for the slow uptake of ASD in the pharmaceutical industry (Sekiguchi et al. 1964). Over the past 2 decades, the development of fluid-bed processing, spray drying, and melt extrusion processes have greatly changed the landscape leading to the successful development of several products (Williams et al. 2010; Repka et al. 2013). For this reason, any new technology that is being developed for ASD

H. J. Mair (✉) · R. Diodone
Pharmaceutical Technical Development Chemical Actives,
F. Hoffmann-La Roche Ltd., Basel, Switzerland
e-mail: hans-juergen.mair@roche.com

H. Sandhu
Merck & Co., Inc., Summit, NJ, USA

N. Shah
Kashiv Pharma LLC, Bridgewater, NJ, USA

N. Shah et al. (eds.), *Amorphous Solid Dispersions*,
Advances in Delivery Science and Technology, DOI 10.1007/978-1-4939-1598-9_11

Fig. 11.1 Key operations in the MBP process and relevant processing considerations

manufacture must ensure that desired product characteristics can be achieved as the product goes through different development stages. The novel microprecipitated bulk powder (MBP) technology that was developed to address the solubility needs of challenging compounds had to fulfill this requirement (Albano et al. 2002; Dong et al. 2008; Shah et al. 2012, 2013). The key process attributes, product characteristics, and the in-process controls that are required to obtain reproducible material using MBP process, are discussed in this chapter.

From scale-up perspective, the MBP process can be divided into two main parts: preparation of solution and antisolvent and the actual manufacturing step. Key considerations for each step are summarized below and depicted in Fig. 11.1:

1. Solvent/Antisolvent
   a. Selection of solvent
   b. Physicochemical properties of API and polymer (solubility and ionic nature)
   c. Concentration of API and polymer in solution
   d. Stability of API/polymer solution
   e. Solution preparation.
   f. Selection and preparation of antisolvent.
2. MBP Manufacturing
   a. General considerations
   b. Precipitation techniques
   c. Processing conditions for high-shear precipitation
      – High-shear mixing-based processes
      – Continuous process
      – Semi-Batch process

d. Mixing tools
e. Ratio solution/ antisolvent
f. Isolation and drying of MBP

Each of these aspects is elaborated individually in the subsequent sections. Since the output of one unit operation is an input for the next step and may affect the properties of the final material, it is important to establish meaningful controls as we go through the process. The relevant in-process tests and the preferred controls are also summarized in each section.

11.2 Selection of Solvent and Antisolvent

11.2.1 Solvent

Selection of solvent plays an important role in the successful execution of MBP manufacturing. The solvent characteristics that are considered important for MBP processing include:

- Provide sufficient solubility for API and polymer
- Miscible with the antisolvent over a sufficient broad range
- Does not form solvate or promote crystallization of API
- Chemically compatible with API and polymer
- Can be removed below a certain threshold limit (ICH 2009)

Compounds with low solubility in commonly used volatile solvents and high melting point are likely candidates for MBP process. Therefore, for the preparation of the API/polymer solution, solvents that provide sufficient solubility of API and polymer are preferred. Typically aprotic, polar solvents with high boiling points generally provide good solubility for most organic compounds. Commonly used solvents that meet the desired characteristics and provide sufficient solubility (ideally $> 10\,\%$, m/m) are limited. Some of these solvents that may be suitable for preparing API and polymer solution are listed in Table 11.1 along with the key properties.

The goal is to select a solvent that provides highest solubility of API and polymer. Since the solubility of each component can be affected by the presence of other constituents, it is obligatory to determine the solubility of API with and without the polymer. In addition, other factors that can affect solubility such as temperature, API properties (particle size, solid state, and purity profile), and compatibility with container and transfer tubing should also be evaluated. Increase in temperature is frequently used to increase the solid content for efficiency reasons; solubility studies should include relevant temperature cycling.

Table 11.1 Examples for organic solvents which can be used in the MBP process to make a solution of API and polymer

Solvent	Abbreviation	mp (°C)	bp (°C)	ICH class	ICH limit (ppm)	Miscibility with water
Dimethyl sulfoxide	DMSO	19	189	3	5000	Complete
N, N-Dimethylformamide	DMF	– 61	153	2	880	Complete
N, N-Dimethylacetamide	DMA	– 20	165	2	1090	Complete
N-Methylpyrrolidone	NMP	– 24	203	2	530	Complete
Tetrahydrofuran	THF	– 108	66	3	720	Complete
Acetone	–	– 95	56	3	5000	Complete
Ethanol	EtOH	– 114	78	3	5000	Complete

11.2.2 API and Polymer Physicochemical Properties

The impact of API and polymer properties on the suitability of MBP process is described in the previous chapter. However, briefly API with high log *P* (greater than 4), high molecular weight, high melting point (> 200 °C), and hydrogen bond-accepting group greater than 7 are considered most suitable for MBP process. Success of MBP process relies on the coprecipitation of API and polymer; the selection of polymer that has low solubility in the antisolvent is preferred. The current experience in the amorphous coprecipitation area is primarily based on using aqueous phase as antisolvent, therefore ionic polymers that have pH-dependent solubility provide the best opportunity to achieve the appropriate solubility conditions during processing as well as dissolution. In addition to enabling the processing and dissolution, the ionic polymers may also help to achieve higher miscibility of drug in the polymer due to electrostatic interactions translating better product performance. In literature, this effect is attributed to the hydrophobicity of the polymer as well as stronger ionic interactions between API and polymer (Friesen et al. 2008; Rumondor and Taylor 2010). As described in the previous chapters, the most commonly used polymers are hypromellose acetate succinate (HPMCAS), and polymethacrylate (Eudragit L100 and Eudragit L100-55).

11.2.3 Concentration in Solution

The higher the concentration of API and polymer in solution the higher is the productivity of the subsequent process. By the use of polymers, the viscosity of the solution increases with increasing concentration of the polymer. Typically, the concentration of API and polymer ranges between 10 and 40 % in the solution (ratio of API to polymer is typically in a range from 2:8 to 6:4). Therefore, the selection of final concentration of API and polymer in the solvent depends on three factors, i.e., solubility, stability, and viscosity. All these characteristics can be affected by changing the temperature. Temperature as high as 80°C have been used to maximize the concentration of API and polymer.

11.2.4 Stability of API/Polymer Solution

Sufficient stability of API/polymer solutions at the applied dissolution and/or process temperatures need to be performed to minimize the degradation and/or formation of undesired byproducts. Depending on their solubility in solvent or antisolvent, these degradation products may precipitate at different rates that can affect the amorphous nature of the final material or simply gets trapped in the MBP. Therefore, it is critical to perform stability studies of the API/polymer solution at different temperatures to identify and confirm a safe operating range.

11.2.5 Solution Preparation

For the precipitation process a clear solution of API and polymer in a suitable solvent needs to be prepared at the target ratio. The preparation of the solution is generally performed sequentially, i.e., dissolution of API followed by addition of polymer. The API is dissolved with appropriate mixing device and if necessary supported by heating. After complete dissolution, the solution may be filtered to ensure no solid particles remain. Subsequently, the polymer is added and dissolved in the API solution. To ensure a complete dissolution of the polymer, vigorous stirring and heating may be applied. Complete dissolution of API and polymer is important to maintain the final API/polymer ratio as well as drug content in the final MBP. If high-shear mixing is used, it is important to evaluate the effect of shear on the product properties, especially the viscosity and stability. The shear impact should be carefully evaluated based on API and polymer properties. API with surface-active properties or presence of surfactant in polymer such as Eudragit L100-55 should be handled with moderate shear to avoid foaming.

11.2.6 Key Characteristics of the Final Solution

During process development, as the product is scaled up, the dissolution time of API and polymer may increase, therefore it is critical to ensure that solution integrity is maintained with respect to solubility, stability, and viscosity. The key in-process tests should consider solution appearance (color and clarity), assay, density, and viscosity to ensure solution consistency. Appropriate acceptance controls can be established based on the process/product need.

11.2.7 Antisolvent

An antisolvent performs two key functions in the MBP processing, i.e., induce precipitation and extract residual solvent. While doing so, the antisolvent should also maintain product stability and be removed from the product to produce final material. Therefore, key criteria for selecting an antisolvent are the immiscibility and stability of API and polymer in the antisolvent. Any solubility of API or polymer under these conditions may cause phase separation, crystallization, or simply loss of potency. To further reduce API:polymer solubility in the antisolvent, the process is preferably carried out at low temperatures. Typically, the solubility of API and polymer in the solvent/antisolvent mixture at the precipitation conditions should be less than 1 μg/mL. The second most important function of the antisolvent is the removal of solvent from the wet precipitate. The antisolvent that is miscible with the solvent provides the best extraction efficiency. Aqueous systems are generally preferred for MBP processing as it performs all functionality effectively, i.e., providing suitable conditions for precipitation, removing solvents from the precipitate, and being able to be removed from the product. In addition, it has been observed that the resulting MBP has better handling and downstream processing due to porosity and wetting characteristics. Caution must be used to ensure that amorphous from of the product is protected during processing (precipitation and washing), storage, and handling. Although the risk is low, the microbial purity of the material should also be monitored. As the processing may complete in relatively short time, it may become critical if the wet cake needs to be stored prior to drying.

Commonly used ionic polymers in the MBP manufacturing are anionic and hence the ideal precipitation conditions are two units below the pKa (generally between 4 and 5). Based on the current practice, the pH of the antisolvent is controlled with dilute hydrochloric acid. Use of buffers is avoided as some salts may interfere with the precipitation process and compromise the amorphous integrity of MBP. However, if the drug is a weak base and/or requires rapid release in the upper part of gastrointestinal tract, cationic polymer such as Eudragit EPO can also be used. In that case, the precipitation is carried out at a pH above the pKa of API and polymer, usually 7.

11.2.8 Key Characteristics of the Antisolvent

From a process control perspective, the antisolvent should have optimal pH, and be maintained at desired temperature prior to use. Appropriate acceptance controls can be established based on the process/product need.

11.3 MBP Manufacturing

11.3.1 General Considerations

The glass transition temperature (T_g) of amorphous MBP, ideally in a molecular dispersion of API and polymer, is a function of the T_g of both, the amorphous API and the polymer. Due to the amount of residual solvents from the precipitation process, a further reduction of the theoretical T_g is possible. If the processing temperature is close to the T_g of the MBP, it may result in the phase separation and/or crystallization of the API in the precipitate (Taylor and Zografi 1997). Therefore, the manufacturing process needs to be performed under controlled conditions. Appropriate temperature range and acceptable exposure time for different unit operations of MBP manufacturing need to be established. The goal is to achieve the highest energy difference and the shortest time that could be achieved by maintaining high temperature difference between solvent and antisolvent during precipitation. For instance, given the duration of the precipitation process and the relative high quantity of solvent/antisolvent at this stage, a stable manufacturing process can only be established using a suitable low-temperature range, e.g., 1–5 °C in aqueous systems. The same considerations apply for establishing the holding time for MBP suspension and the isolated wet MBP cake.

11.3.2 Precipitation Technique

In the precipitation process, the API/polymer solution and the antisolvent need to be mixed efficiently to ensure an instantaneous supersaturation/precipitation of MBP maintaining the homogeneous distribution of API and polymer from the solution. The mixing of the two phases can be performed by several techniques. The key features of different mixing techniques are summarized in Table 11.2. Techniques other than those listed here, i.e., continuous in-line mixer (Sulzer Mixer), could also be applied but are considered outside the scope of this discussion.

The high-shear mixer-based precipitation technique is most suited for MBP process compared to drop-in or nozzle spraying because precipitation occurs instantaneously in a scale-independent manner, and it can process highly viscous phases. The principle setup for drop-in-, nozzle spraying-, and high-shear mixing-precipitation is depicted in Fig. 11.2. In a simplified sectional drawing (Fig. 11.3), the functional principle of high-shear mixing precipitation is illustrated. High-shear mixing approach is described in more detail in the subsequent sections as it has been successfully applied to produce MBP in a large scale.

Table 11.2 Key characteristics of commonly used precipitation techniques for the manufacturing of MBP

Technique	Advantage	Disadvantage
Drop-in Drop-in of API/polymer solution into antisolvent	Easy to perform Suited only for small scale No special designated equipment needed	Not suited for solutions with high viscosities (disintegration of drops not warranted) Ideal drop-in area is dependent on stirrer speed, mixing efficiency, and filling level of the equipment Local concentration gradient may cause crystallization Difficult to scale-up
Nozzle spraying Spraying of the API/polymer solution through a nozzle on top of the surface of the antisolvent	Only solutions with low viscosity leads to sufficiently small droplets Easy equipment setup	Not suited for solutions with high viscosities Pumping pressure influence product Foam emerging on top of the antisolvent due to heavy mixing can lead to partially crystalline material Difficult to scale-up
High-shear mixing High-shear mixing of API/polymer solution with the antisolvent	Highly efficient mixing of solvent/antisolvent Mixing is done in a scale-independent manner in a small compartment of the high-shear mixer High viscosity solutions can be handled	Designated equipment setup needed

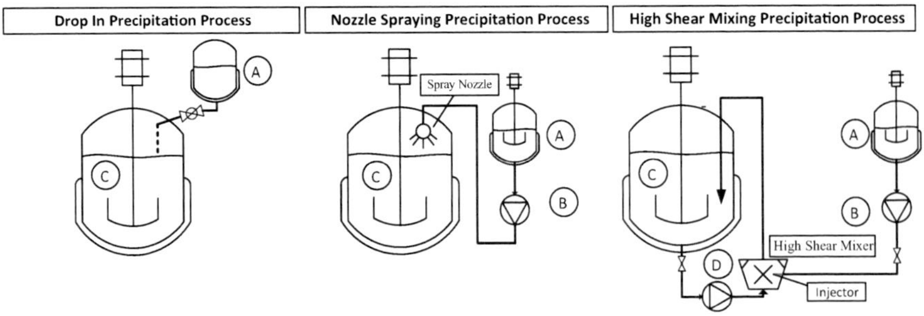

Fig. 11.2 Simplified setup used for drop-in-, nozzle spraying-, and high-shear mixing-precipitation process. **A** Container for API/polymer solution. **B** Dosing pump API/polymer solution. **C** Reactor for MBP suspension. **D** Pump for circulation of antisolvent respective MBP suspension

Fig. 11.3 Sectional drawing of a high-shear mixer (simplified). Antisolvent is pumped into the high-shear mixer (*blue arrow*). The API/polymer solution (*yellow*) is dosed via injector into the mixing chamber close to the rotor. The MBP suspension (MBP particles as yellow spheres) is leaving the mixer after precipitation

11.3.3 Processing Conditions for High-Shear Precipitation

Several factors need to be considered and elaborated to establish a robust and scalable manufacturing process for the manufacturing of MBP by high-shear mixing. For the evaluation of different processing parameters, the critical quality attributes are recognized to be amorphicity, purity, and assay. Purity and assay of the API are tested typically by high-performance liquid chromatography (HPLC) and amorphicity is tested by X-ray powder diffraction (XRPD). It is generally regarded that if amorphicity is maintained the product performance will also be comparable. However, in some cases it is important to extend the analysis using in vitro dissolution and short-term stability studies to further corroborate the findings.

It should be noted that any small crystalline impurity that may not be detectable in the initial condition could potentially impact the long-term stability of the product. Therefore, during process optimization, it is important to evaluate the stability of the product with different processing conditions. In the absence of reliable predictive tools to assess the stability of the amorphous form, the accelerated stress conditions can be used to estimate the effect of process variables and help establish the operating ranges. Further confirmation can be made based on the real-time data. As shown in Fig. 11.4, clear differences in the stability of the product were observed when comparing two process conditions. Process A on the left showed traces of crystallinity after 8 days while an alternate process (on right) showed crystallinity after 12 days when the sample was stored at accelerated condition (50 °C/90 % RH).

In a separate study shown in Fig. 11.5, two different modes of addition of the API/polymer solution were compared. As shown by the scanning electron microphotographs, the initial samples were consistent with the amorphous nature of the drug and appear to be comparable. However, discrimination was observed when these samples were stored at accelerated conditions; the rods observed on the surface of high-pressure homogenization sample after accelerated storage corresponded to crystalline API.

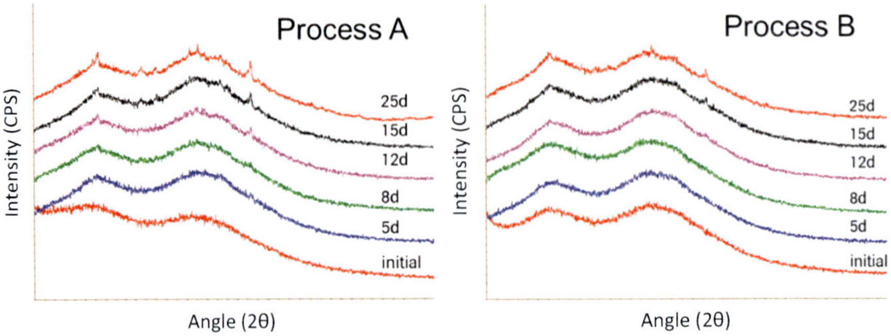

Fig. 11.4 Overlay of XRD pattern showing accelerated stability of MBP produced by two processing conditions (50 °C/90 % RH, open storage; after 0 h (initial), 5 days, 8 days, 12 days, 15 days and 25 days). In the *left* overlay, crystallization was detectable after 8 days of sample storage while in the *right* overlay crystallization was detectable after 12 days

These studies show the importance of judicious analytical tools and strategies to discriminate between various processing conditions and help to select the optimal processes.

11.3.4 High-Shear Mixing-Based Processes

Along the development path, mixing conditions used during precipitation can range from simple stirring to high-shear homogenization depending on the availability of API and equipment. However, for large-scale manufacturing, two principles considered relevant are high-shear mixing performed in continuous and semi-batch mode. Figure 11.6 shows the process flow schemes for both processes.

11.3.4.1 Continuous Process

In the continuous process, the precipitation step is performed as one passage through the high-shear mixer. The API/polymer solution (by injector) and antisolvent are pumped at the desired flow rate into the mixing chamber of the high-shear mixer (Fig. 11.6). The continuous precipitation process is favored for shear and temperature sensitive products; however, the throughput of the continuous process is low due to considerable high-volume factor. Continuous precipitation process adds more value if washing and isolation can also be performed in a continuous fashion because efficient extraction of solvent/antisolvent mixture is critical to ensure the amorphous stability during processing.

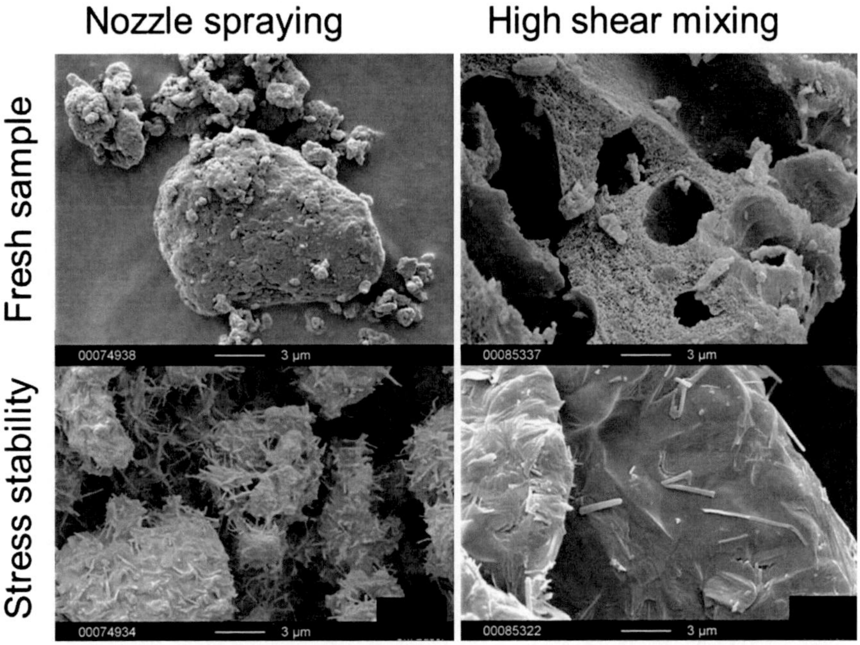

Fig. 11.5 Electron microphotographs showing amorphous MBP obtained by nozzle spraying and high-shear mixing at initial and after storage at stress conditions; nozzle spraying for 24 days at 50 °C/90 % RH and high-shear mixing for 18 days at 60 °C/80 % RH. The rods on the surface of the particles after storage at stress conditions are attributed to crystalline API

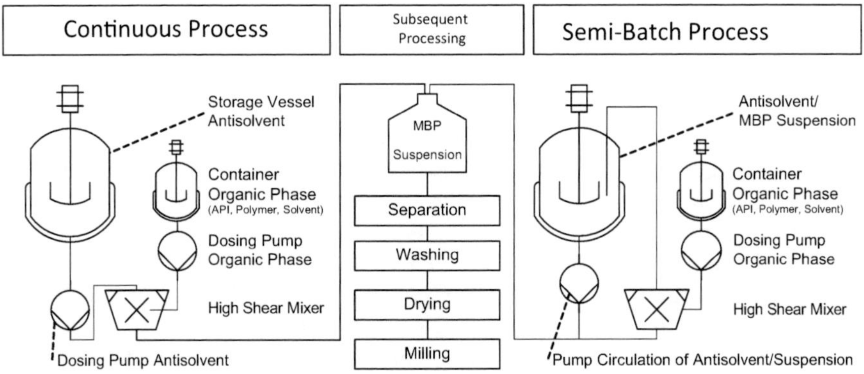

Fig. 11.6 Process flow diagram depicting continuous- (*left side*) and semi-batch (*right side*) processes for the manufacturing of MBP. Further processing of the MBP suspension (shown in the *middle*) is similar for both processes

11.3.4.2 Semi-Batch Process

In a semi-batch manufacturing process, the antisolvent is circulated through the high-shear mixer and back into the storage vessel. The API/polymer solution is dosed via injector into the mixing chamber of the high-shear mixer. The resulting suspension is circulated until the end of dosing operation (Fig. 11.6). The final suspension may be circulated for additional time through the high-shear mixer until a homogeneous and/or specific particle size range is reached. The continuous circulation also provides more time for extraction of solvent in to aqueous phase, thus improving the stability of MBP particles that further aids in efficient filtration and washing process. The batch process may not be suitable for shear-sensitive products but offers the advantage of high productivity due to significantly lower-volume factor compared to the continuous process.

11.3.5 Mixing Tools

The precipitation conditions afforded by the mixer design, speed of mixing, and the fluid addition rates form the core of MBP manufacturing. Based on the high dissipation energy, high-shear mixers are preferred for MBP production. High-shear mixers, which can be used during the precipitation process typically consists of tooth-disc rotor–stator system. The rotor and stator tools can be arranged either as single-stage- or multiple-stage tools in axial and/or radial orientation with variable gap between rotor and stator. A schematic of the high-shear mixer is shown in Fig. 11.3 (mixer head). The shear forces can be adjusted by using the tools with different geometry and by changing the tip speed of the rotor, typically in a range between 8 and 25 m/s. A number of possible options are shown in Fig. 11.7 for comparison. The precipitation process can be controlled by variation of the shear force, the dosing speed of the API/polymer solution, and antisolvent into the mixing chamber of the high-shear mixer. In a semi-batch process, the number of circulations of the MBP suspension through the high-shear mixer for homogenization also needs to be considered. By varying different process parameters, the solid state properties of the MBP can be modified. Future work in this area using more advanced tools such as discrete element analysis and computational fluid modeling can help elucidate the effect of different shear conditions on the product characteristics (Paul et al. 2004; Halla et al. 2012). The scale-up of high-shear mixing process is achieved by increasing the volume of the mixing chamber. For reliable scale-up, it is important to link the energy dissipation to the process by controlling the design (size and geometry) and process variables (rotor speed and flow rate) (Cooke et al. 2012).

Figure 11.8 shows the effect of shear generated by using different shear head designs and its impact on the process. Depending on the properties of the solvent-rich precipitate as it is being produced under high-shear conditions, the processing can be problematic. As shown in Fig. 11.8, by reducing the number of shear pins in the rotor, the precipitates were easily removed from the mixer without causing any material buildup.

Fig. 11.7 Possible rotor–stator design to vary the shear and flow conditions in the mixing chamber

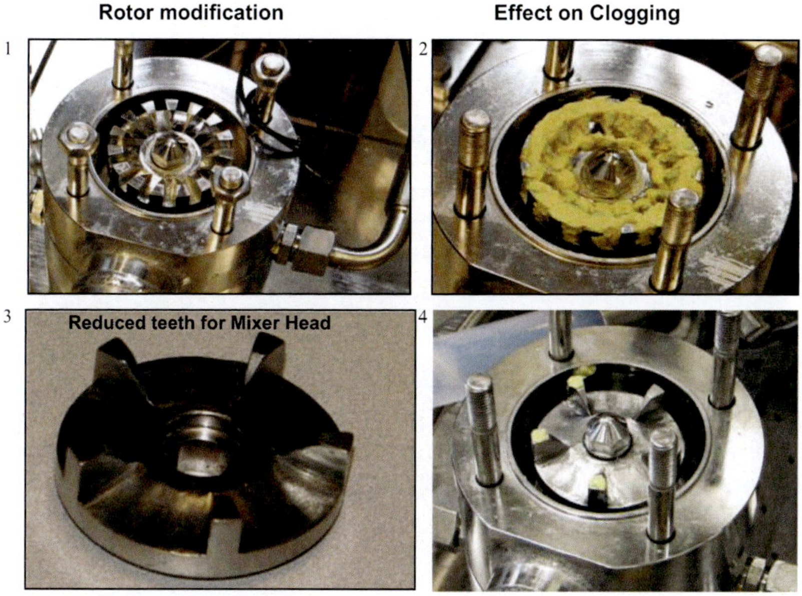

Fig. 11.8 Design of rotor head and the clogging issues. Photographs *1* and *2* show high-intensity shear head before processing and after 1 h of processing. Photographs *3* and *4* show low-intensity shear head before and after processing for 3 h

A common problem during the scale-up of batch mode process is the processing times and the batch turnover in order to achieve similarity at different scales. Figure 11.9 shows the effect of mixing speed and time on the crystallinity of the dried MBP.

11.3.6 Ratio of API/Polymer Solution (Solvent)/Antisolvent

The ratio of API/polymer solution (solvent) to antisolvent is one of the most important factors for the success of the precipitation process. For continuous process, it is controlled by the feed rates of the API/polymer solution and the antisolvent into the high-shear mixing chamber and is generally constant for the process. For semi-batch

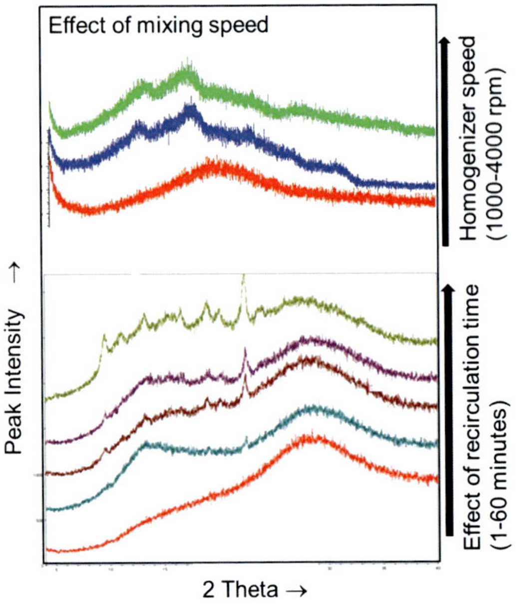

Fig. 11.9 Effect of mixing speed and process time on the crystallinity of the final MBP

process, this ratio is constantly changing until all the API/polymer solution is added to the fixed amount of antisolvent. Therefore, for semi-batch process two ratios need to be considered. The first one is the ratio of solvent/antisolvent in the precipitation zone (mixing chamber) and the second ratio is the final ratio of solvent/antisolvent in the MBP suspension. In continuous processing, the ratio of solvent/antisolvent during precipitation is equivalent to the ratio in the final MBP suspension. In the semi-batch process, the ratio of solvent to antisolvent varies from start of the precipitation to the end of precipitation. Depending on the selected processing conditions this may vary from 1:1000 in the beginning to 1:1 towards the end of precipitation. This is achieved by varying the antisolvent flow rate while keeping the solvent dosage constant or vice versa. Consequently, an intensive mixing of the API/polymer solution with an excess amount of antisolvent resulting in an instantaneous precipitation of MBP can be achieved. As mentioned earlier in the case of semi-batch process, the ratio of solvent to antisolvent in the final MBP suspension also needs to be taken into consideration. Regardless of the mixing process used, it is critical to ensure that the solubility of both components of MBP (API and Polymer) in the final composition of solvent/antisolvent is sufficiently low to prevent the risk of crystallization and/or a potential change in the API/polymer ratio in the MBP. The antisolvent conditions can be further optimized with respect to pH in particular for aqueous systems, and temperature to minimize the solubility of API and polymer.

11.3.7 Isolation and Washing of MBP

Efficient and rapid removal of solvent and antisolvent is a vital component of the successful MBP process. Once the precipitation has been completed (concurrently for continuous process and at the end of the solvent addition for the semi-batch process), the MBP is filtered from the suspension. This is generally performed by any of the standard filtration processes such as vacuum filtration ranging from Buchner funnel to Nutsche filter or centrifugal filter. Centrifugal filters are preferred for filtration of MBP suspension because of the particle size, the hydrogel nature of the polymer, and the effectiveness in solid/liquid separation

A washing media that is miscible with the solvent, preferably the antisolvent, provides the most efficient extraction of the residual solvent (EndeJam 2011). The success of the washing is ideally monitored via process analytical technology (PAT) analytics. Washing is continued until the acceptable threshold is reached to indicate that residual solvent in the final product has reached the target value. If high-boiling solvents like dimethylamine (DMA) or N-methylpyrrolidone (NMP) are used as solvents for the API/polymer solution, an efficient washing is necessary as removal of those solvents in the subsequent drying step is limited. Because MBP at this stage is in fully hydrated state, care should be taken to ensure that filtration and washing processes are performed at conditions ensuring acceptable stability.

To help with material handling and drying, efforts should be made whenever water is used as the antisolvent and/or washing solvent to remove as much water as possible. Water retained during filtration depends on the cake resistance which in turn depends on the material properties such as particle shape and size of the precipitate, process temperature, cake thickness, porosity, density, and the filtration system. Filtration can be further influenced by the hydrogel nature of the polymer and the drug to polymer ratio that can affect the compressibility of the cake. Due to the buildup of cake resistance, this can become the rate-limiting step in the entire process during scale-up (Murugesan et al. 2011). Depending on the material properties, filtration system, and the scale of the material, the residual water in the final wet cake may range from 60 to 95 % by weight. If the drying does not occur immediately after filtration, it is judicious to store the material at refrigerated conditions to minimize risk of crystallization and control microbial growth.

11.3.8 Drying

The potential conversion of amorphous form to crystalline form is always a risk when working with ASD, however, this risk is the highest during the drying of MBP as the wet material is subjected to relatively high-temperature and high-humidity conditions. Prudent selection of the drying process and conditions are necessary to avoid the risk of crystallization. Commonly used dryers include tray, filter, tumble, conical, freeze and spray, and fluid bed dyers (Murugesan et al. 2011). More frequently used dryers are shown in Fig. 11.10.

Fig. 11.10 Types of dryers commonly used in the MBP manufacturing

Between all the drying techniques evaluated, fluid bed drying was found to be the most efficient drying process (Leuenberger et al. 2011). In addition to providing the best heat and mass transfer, fluidized bed also offers the most flexibility to fine-tune the process as the sample is drying. Unlike conventional materials, the ASD often have high heat sensitivity and hence may benefit from the customized drying profiles, i.e., increase temperature as the drying proceeds. There needs to be a balance between time and temperature as low temperature is preferred during initial drying to avoid rapid surface drying. The drying temperature can then be progressively increased to achieve higher drying rates and reduce processing times. To aid the drying process, the wet mass can be milled through low-energy impact mills before charging into the dryer (similar to wet granulation) and the fluid bed dryers can also be custom-fitted with agitators to further help with the drying.

In addition to having an obvious effect on the amorphous nature of the API, the drying process can also impact the particulate properties such as granule size, density, and surface area. As shown in Fig. 11.11, surface area of granules decreased with increase in the drying air humidity because higher humidity results in a longer drying time. The slow and prolonged drying tends to fuse amorphous particles forming hard granules that are difficult to mill to produce granules suitable for downstream processing. An example of such hard-to-mill granules, resulting from fusion of amorphous MBP, is shown in Fig. 11.12. The particulate fusion due to formation of liquid bridges can also occur during stability assessment that can have adverse effect on disintegration and/or dissolution. Therefore, appropriate moisture protection is critical for the dried MBP as well as finished dosage form.

In contrast to fusion of particles, the increase in fluidization air volume during drying can cause attrition of the particles resulting material with smaller particle size and low density.

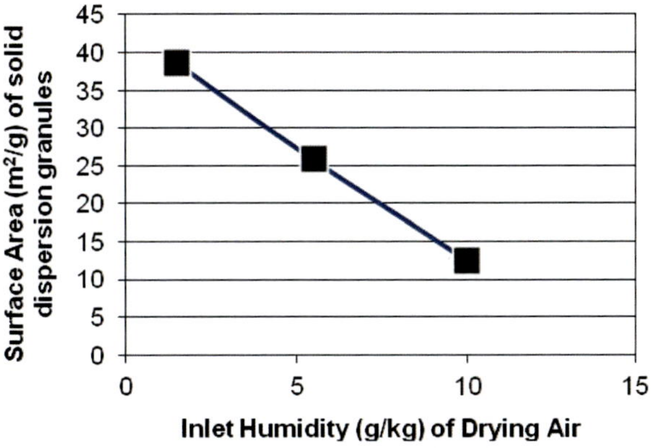

Fig. 11.11 Impact of inlet drying air humidity on drying of MBP granules

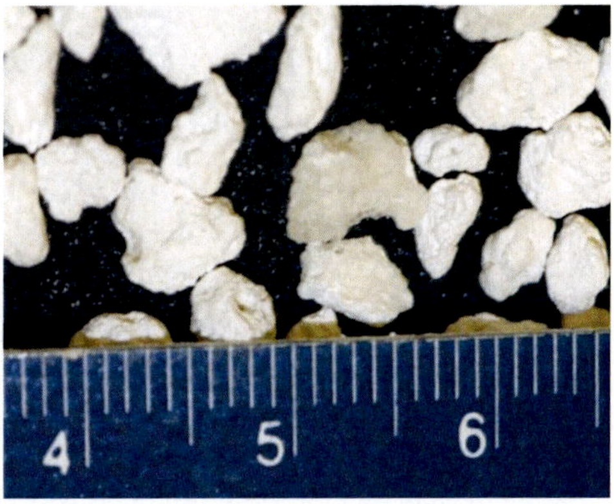

Fig. 11.12 Photograph of fused, hard aggregates of amorphous MBP

11.3.9 Milling and Size Reduction

Akin to the wet granulation process, the dried material is milled to control the particle size for downstream processing such as flow, uniformity, and compaction (Jones 2011). Air jet or impact mills such as hammer or conical mills can be used for this purpose. For pharmaceutical operations, hammer milling is sufficient to reduce the particle size, however due to the risk of heat generation, for some sensitive material air jet milling may be necessary. As alluded earlier, any process involving temperature increase requires careful assessment with respect to the quality attributes.

Based on the experience with powder processing, the preferred range for bulk density of the material is approximately 0.2–0.4 g/cm^3 with average particle size in the range of 200–600 μm to achieve flow and compaction. Bulk density less than 0.2 g/cm^3 may generally require densification step prior to final dosage form processing. Particle size of MBP (dried granules) or roller compacted granules could have significant effect on the disintegration and dissolution properties of the final product, thus requiring an assessment of the milling technique, scalability before establishing final specification. In addition to dissolution, the particulate properties such as granule size and bulk density also need to be evaluated. The MBP being ASD with polymers, viscoelastic properties may be impacted by particle size and density and it should be evaluated with respect to compaction and dissolution. The material after milling needs to be appropriately characterized before further processing.

11.4 Common Issues and Troubleshooting

Given the detailed background of MBP process and other considerations discussed previously, this section aims to provide additional tools for troubleshooting in case a problem arises. The key components of developing MBP depend on three components: API, polymer, and process. Troubleshooting needs to be evaluated with respect to its impact on the critical quality attributes as well as the process. Summary of various processes and the key considerations is provided in Table 11.3.

11.5 Summary and Conclusions

MBP is a unique process for production of amorphous solid dispersion, when other technologies such as hot-melt extrusion, spray drying, and solvent layering are not feasible. MBP technology is preferably used for molecules with high melting point, and low solubility in low boiling point solvents. With suitable controls in place, the MBP technology has been successfully scaled up from 10 mg to up to 200 kg/day. In addition to selecting the right polymer, solvent and antisolvent, the unit operations such as solution preparation, precipitation, filtration, washing, drying, and milling are critical steps to ensure consistency of the product.

Table 11.3 A summary of various processes and the key considerations

Observation	Unit operation						
	API/polymer solution	Antisolvent	Precipitation	Filtration	Washing	Drying	Milling
Crystallinity	Impurity Solvate Incomplete dissolution	pH Temp Amount Miscibility w/solvent	Mixer type(high-shear in-line versus suspended mixer homogenizer) Mixer design: Rotor/stator design and gap Tip speed S/AS flow rates and ratio Batch size Processing time Temperature	Too slow	Incomplete removal of solvent	Temp. Time Air flow Humidity	Shear/temp Milling mechanism
Residual solvent	–	Miscibility w/solvent	Flow rate Shear PS/PSD Morphology (S/AS ratio)	Incomplete	Incomplete	–	–
Assay	Purity Solubility Temp Stability	API/polymer solubility	Stability due to shear/temperature	–	API vs polymer solubility in the wash	Stability	Stability
Particulate properties	Viscosity of the solution and precipitation method	–	Precipitation conditions	Compaction	–	Compaction	Mill and conditions
Dissolution	Crystallinity + assay + residual solvent + particle size + particle size distribution + milling						
Long-term stability	Trace amount of crystallinity (see factors related to crystallinity + residual solvent)						

Acknowledgments Authors wish to acknowledge the contributions from Dr. M. Matchett and Dr. S. Anand for Figs. 11.8–11.10 and their valuable experience.

References

Albano A, Phuapradit W, Sandhu H, Shah N (2002) Amorphous form of cell cycle inhibitor having improved solubility and bioavailability. U.S. Patent 6486329

Am Ende DJ (2011) Chemical engineering in the pharmaceutical industry: R & D to manufacturing, John Wiley and Sons, NJ

Cooke M, Rodgers TL, Kowalski AJ (2012) Power consumption characteristics of an in-line silverson high shear mixer. AIChE J 58(6):1683–1692

Dong Z, Chatterji A, Sandhu H, Choi DS, Chokshi H, Shah N (2008) Evaluation of solid state properties of solid dispersions prepared by hot-melt extrusion and solvent co-precipitation. Int J Pharm 355(1–2):141–149

Friesen DT, Shanker R, Crew M, Smithey DT, Curatolo WJ, Nightingale JA (2008) Hydroxypropyl methylcellulose acetate succinate-based spray-dried dispersions: an overview. Mol Pharm 5(6):1003–1019

Halla S, Paceka AW, Kowalskib AJ, Cookec M and, Rothmand D (2012) The effect of scale of liquid-liquid dispersion in line Silverson Rotor Stator Mixers. 14th European conference on mixing, Warszawa

Hu Q, Choi DS, Chokshi H, Shah N, Sandhu H (2013) Highly efficient miniaturized coprecipitation screening (MiCoS) for ASD formulation development. Int J Pharm 450(1–2):53–62

ICH QG (2009) International conference on harmonisation of technical requirements for registration of pharmaceuticals for human use; impurities: guideline for residual solvents q3c(r4). http://westpalmbeachanalytic.homestead.com/Q3C_R4__Guide_Res_Solv.pdf. Accessed March 10, 2014

Leueneberger H, Betz G, Allen M, Donsmark S, and Jones DH (2011) Scale-up in the field of granulation and drying, in Pharmaceutical Process Scale-up edited by M. Levin, Informa Healthcare, NY

Murugesan S, Sharma P, Tabora JE (2011) Design of filtration and drying operations. In: Am Ende DJ (ed) Chemical engineering in the pharmaceutical industry R & D to manufacturing. Wiley, NJ, pp 315–346

Paul EL, Atiemo-Obeng VA, Kresta SM (2004) Handbook of industrial mixing science and practice. John Wiley and Sons, NJ

Repka MA, Langley Na, DiNunzio J (2013) Melt extrusion materials, technology and drug product design. AAPS and Springer, NY

Rumondor AC, Taylor LS (2010) Effect of polymer hygroscopicity on the phase behavior of amorphous solid dispersions in the presence of moisture. Mol Pharm 7(2):477–490

Sekiguchi K, Obi N, Ueda Y (1964) Studies on absorption of eutectic mixture. Ii. Absorption of fused conglomerates of chloramphenicol and urea in rabbits. Chem Pharm Bull (Tokyo) 12:134–144

Shah N, Sandhu H, Phuapradit W, Pinal R, Iyer R, Albano A, Chatterji A, Anand S, Choi DS, Tang K, Tian H, Chokshi H, Singhal D, Malick W (2012) Development of novel microprecipitated bulk powder (MBP) technology for manufacturing stable amorphous formulations of poorly soluble drugs. Int J Pharm 438(1–2): 53–60

Shah N, Iyer RM, Mair HJ, Choi DS, Tian H, Diodone R, Fahnrich K, Pabst-Ravot A, Tang K, Scheubel E, Grippo JF, Moreira SA, Go Z, Mouskountakis J, Louie T, Ibrahim PN, Sandhu H, Rubia L, Chokshi H, Singhal D, Malick W (2013) Improved human bioavailability of vemurafenib, a practically insoluble drug, using an amorphous polymer-stabilized solid dispersion prepared by a solvent-controlled coprecipitation process. J Pharm Sci 102(3): 967–981

Taylor LS, Zografi G (1997) Spectroscopic characterization of interactions between PVP and indomethacin in amorphous molecular dispersions. Pharm Res 14(12): 1691–1698

Williams I, RO, Watts AB and Miller DA (2010) Formulating poorly soluble drugs. AAPS Press and Springer, NY

Chapter 12
Pharmaceutical Development of MBP Solid Dispersions: Case Studies

Raman Iyer, Navnit Shah, Harpreet Sandhu, Duk Soon Choi, Hitesh Chokshi and A. Waseem Malick

12.1 Introduction

The biopharmaceutical classification system (BCS) was first introduced in 1995 to facilitate the drug development, and it is based on two independent variables that influence bioavailability, viz., aqueous solubility and intestinal permeability (Amidon et al. 1995; FDA 2000). Compounds that belong to BCS class II or IV are of primary interests from a formulation perspective and, therefore, solubility enhancement using formulation intervention is the key driver for improving the bioavailability of poorly soluble drugs. From a conceptual perspective, the dissolution rate can be expressed by the following equation (Noyes and Whitney 1897):

$$\frac{M}{t} = KAC_s \tag{12.1}$$

where *M/t,* the amount of drug dissolved at a given time *t,* is a function of the permeability coefficient (K), saturation solubility C_s, and surface area A of the dissolving particles. The saturation solubility refers to the thermodynamic or equilibrium solubility which is attained quickly for highly soluble drugs. In case of poorly soluble

R. Iyer (✉)
Novartis, East Hanover, NJ, USA
e-mail: raman.iyer@novartis.com

N. Shah · D. S. Choi
Kashiv Pharma LLC, Bridgewater, NJ, USA

H. Sandhu
Merck & Co., Inc., Summit, NJ, USA

H. Chokshi
Roche Pharma Research & Early Development, Roche Innovation Center, New York, NY, USA

A. W. Malick
Pharmaceutical and Analytical R&D, Hoffmann-La Roche Ltd., Nutley, NJ, USA

N. Shah et al. (eds.), *Amorphous Solid Dispersions,*
Advances in Delivery Science and Technology, DOI 10.1007/978-1-4939-1598-9_12

drugs, it may also refer to kinetic solubility $C_S f(t)$, which changes over time during the course of dissolution to the thermodynamic value.

A major thrust of the formulation intervention effort involves not only maximizing the kinetic solubility C_S but also modulating the rate of change of kinetic solubility to the thermodynamic value. The solubility C_S can be enhanced by using conventional solubilizers such as surfactants, micellar systems such as self-emulsifying drug delivery systems, and complexing agents such as cyclodextrins (Loftsson and Brewster 1996; Liu 2008; Williams et al. 2010, 2013). An approach to increase rate of dissolution is to increase the surface area by milling or micronization and, where feasible, to develop stabilized nanoparticulate systems using nanomilling or nanocrystallization techniques (Rabinow 2004; Keck and Muller 2006).

Over the past two decades, amorphous solid dispersion systems (ASD) where the drug is embedded in polymeric matrices as crystalline or amorphous form (solid dispersions) and/or drug–polymer solutions (solid solutions) have been studied extensively as a means of improving the bioavailability of poorly soluble drugs (Kai et al. 1996; Okimoto et al. 1997; Shin and Cho 1997; Kohri et al. 1999; Serajuddin 1999; Leuner and Dressman 2000). Despite their successes in improving bioavailability, the major concern is the reduced physical stability of these systems. Substantial efforts are required to achieve an optimal balance between the solubility gain and the risk of reversion to more stable form. Approaches to improve the stability of these high-energy amorphous systems rely on the use of polymers that help to stabilize the system by means of physical as well as chemical interactions. The physical stabilization is ascribed to the restricted molecular mobility and diffusional barriers that physically limit the motion of molecules, and the chemical stabilization is ascribed to Van der Waal's forces, hydrogen bonding, and electrostatic interactions between the drug and the polymer (Hancock et al. 1995; Rolfes et al. 2001; Faure et al. 2013). Due to the nature of processing, these interactions are maximized in microprecipitated bulk powder (MBP) process in contrast to other processing methods such as:

- Melt extrusion: solution-state interactions in MBP process may be more favorable than molten state
- Spray drying: solvent extraction with aqueous fluid in MBP process renders the particles more hydrophilic and porous, thereby providing superior compaction and wetting

These interactions increase the kinetic solubility of the drug, $C_S f(t)$, and help maintain the extent and duration of supersaturation which leads to enhanced bioavailability. Since only disolved drug can be absorbed, enhanced absorption can only occur in the "supersaturation maintenance window," the time during which the kinetic solubility $C_S(ft)$ is maintained at a high level. Beyond this window, the solubility reverts to the thermodynamic value via precipitation or crystallization resulting in lower solubility, and therefore loss of bioavailability. The stabilizing polymer in ASD is thus a critical component of the ASD that governs the drug's solubility and bioavailability. Several techniques are available to create a stabilized drug-carrier solid dispersion where the drug may exist in partial states of crystallinity or in an amorphous state (Williams et al. 2010). Non-polymer-based amorphous conversion such as co-milling/co-grinding with inorganic silicates has also been used for some drugs (Bahl et al. 2008).

Polymer-based techniques of solid dispersion can be simple, moderately difficult, or complex. Co-melting and melt quenching are simple approaches, while examples of moderate ones are solvent evaporation under vacuum, fluid bed granulation or layering, spray drying, and lyophilization (El-Badry and Fathy 2006; Kim et al. 2006; Moser et al. 2008; Bley et al. 2010). Solvent–anti-solvent precipitation and hot-melt extrusion are examples of more complex techniques where solubility characteristics and thermal stability have to be considered in preparation of solid dispersion (Sertsou et al. 2002a, b; Wu et al. 2009; Evonik 2014). This chapter presents the case studies of ASD manufacture of highly crystalline compounds using MBP technology, the experiences gained during ASD development, galenical processing, and dosage form development. The details of MBP technology can be found in prior literature (Albano et al. 2002; Shah et al. 2012).

12.2 Factors to Consider in MBP Development

In a typical ASD, the drug (solute phase) is dispersed in an inert carrier (e.g., a polymeric continuous phase) with molecular level distribution being the most desirable. Depending on the interactions between drug and polymer and the method of preparation, the ASD may exist as a one-phase, two-phase, or mixed-phase system. In the one-phase system, the drug is immobilized within the polymer matrix at a molecular level such that it

- Prevents nucleation (and crystallization)
- Protects from moisture initiated mobility
- Maintains supersaturation (higher kinetic solubility)

Polymer-based amorphous dispersions attain their stability when the polymer molecules disrupt the self-assembly of drug molecules via positive drug–polymer interaction, for example, hydrogen bonding to form a stable matrix at the molecular level, akin to the concept of crystallization poisoning. Therefore, the selection of polymer and technologies of processing are critical in the development of ASDs with long-term stability. A polymer with a high glass transition temperature and several hydrogen-bonding sites is preferred. On the other hand, a polymer with high hygroscopicity and degradation potential is undesirable. Table 12.1 shows the factors that need to be considered in selection of polymer and technology.

The suitability of MBP technology depends on the physicochemical properties of drug or API active pharmaceutical ingredients (API) and polymer. As mentioned earlier, the stabilization of amorphous form in the ASD is attained by the physical and chemical interactions between drug and polymer. The strength of various interactions is ranked as electrostatic interactions > hydrogen bonding > Van der Waal's dispersion forces. The fact that MBP process uses ionic polymers helps to

Table 12.1 Factors in the selection of polymer and technologies for ASD

Factors to consider	Technologies
Glass transition temperatures (T_g) and melting points of API and polymer	Microprecipitation
Degree of lowering of T_g by water or residual solvent(s)	Hot-melt extrusion
MW and viscosity of the polymers	Fluid bed granulation/drying
Compatibility of API and the polymer (solubility parameters)	Spray drying/spray coating
Solubility of drug and polymer in solvents	Supercritical fluid extraction

API active pharmaceutical ingredients, *MW* molecular weight

maximize these interactions. Based on the assessment of more than 20 development compounds, the following characteristics of API are considered to be preferred for the MBP process (Hu et al. 2013):

1. *Non-covalent interaction:* Since the process involves water as anti-solvent, based on hydrophobicity of the compound, drug–polymer interactions are favored over the drug–water or polymer–water interactions. Therefore, compounds with log *P* greater than 3 and polymers having hydrophobic functional groups provide the best prospect for interactions.
2. *High molecular weight*: APIs with molecular weight greater than 500 tend to perform better. Although scientific literature in this regard is not definitive, the heuristic knowledge suggests that it may be more difficult for high molecular weight compounds to achieve the desired orientation for nucleation irrespective of the ASD-manufacturing technique.
3. *Hydrogen bond accepting group* > 8 preferred to enhance hydrogen bond interactions.
4. *Miscibility with polymer*: estimated by Flory Huggins interaction parameter and negative Gibbs free energy.

The crystallization inhibitory effect of polymer seems to play an important role in the stabilization of ASD (Miller et al. 2012). Additional criteria such as crystallization tendency determined by thermal cycling and polymer miscibility may also contribute to the overall performance. In addition to the API properties and the ASD stability considerations, the manufacturing process dictates that the drug and polymer should have good solubility in the solvent. The most suitable solvents include dimethylacetamide (DMA), dimethylformamide (DMF), dimethylsulfoxide (DMSO), terahydrofuran (THF), and N-methyl pyrrolidone (NMP). Also, the anti-solvent should have least solubility for drug and polymer to ensure maximum recovery and yield. By selecting appropriate solvent, polymer, and processing conditions, a high-quality ASD can be produced by MBP technology.

Fig. 12.1 Flowchart of MBP microprecipitation process (Shah et al. 2012)

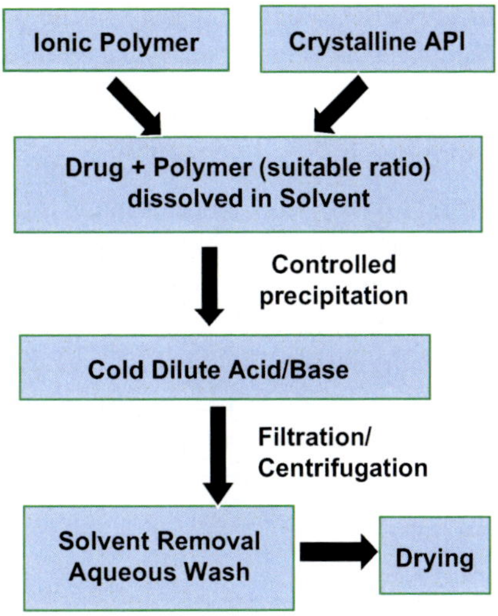

12.3 MBP Preparation and Characterization

The flowchart of MBP microprecipitation process is shown in Fig. 12.1. In a typical run, the drug and an ionic polymer are dissolved in a suitable solvent. The solution is introduced into an anti-solvent under conditions that prevent nucleation and crystallization. Since solvent and anti-solvent are miscible, under the precipitation conditions, the solvent exchange occurs rapidly preserving the drug–polymer interactions. These interactions are further maintained during subsequent processing by maintaining a low temperature as well as avoiding conditions where either drug or polymer could dissolve. The resulting precipitate is washed, filtered, and dried at a relatively low temperature. Under appropriate processing conditions, the amorphous drug is precipitated as uniformly embedded in the polymer. Although precipitation conditions can also be generated by using organic anti-solvent, aqueous fluids are commonly used as anti-solvents. The MBP process is applicable to ionic polymers that have pH-dependent solubility and favor the use of aqueous acidic or basic anti-solvents. This helps in coprecipitation and also provides the added advantage of ionic interactions between the polymer and the API. Typically, enteric polymers that dissolve at physiological pH are more suitable for MBP process as they facilitate release of drug in an enteric environment resulting in a larger window of absorption.

Specific application of MBP lies in improving the bioavailability of poorly soluble crystalline drugs with high melting point or thermal liability that are not amenable to melt extrusion or spray drying (Shah et al. 2012). Further, it is more suitable for drugs that are prone to recrystallization since the MBP process is a relatively low energy process requiring relatively low temperature for processing as compared to extrusion or spray drying.

Since the biopharmaceutical performance of MBP is directly linked to the integrity of the amorphous form, suitable analytical methods are needed during development of the MBP to ensure quality of the product. Characterization of an MBP product with respect to the physical state of the drug, dissolution, and stability follows similar protocols as other ASDs. Because water is used in processing, control of water content in ASD throughout the processing is critical to maintain stability of MBP.

Tools most commonly used for ASD characterization include powder x-ray diffraction (pXRD), thermal analysis (differential scanning calorimetry—DSC, thermogravimetric analysis—TGA), microscopy (atomic force microscopy, scanning electron microscopy, transmission electron microscopy), and hygroscopicity. In addition, spectroscopic tools can be employed to gain deeper insights into the nature of molecular interactions such as near-IR imaging, Raman mapping, solid-state NMR spectroscopy, dielectric spectroscopy, thermally simulated current, etc. The particulate properties including bulk and mechanical properties are relevant from a downstream processing perspective as well as dissolution performance of the final dosage form.

12.4 Pharmaceutical Development of MBP

In order to realize the benefits of MBP technology, the process needs to be scaled up beyond the laboratory into the manufacturing plant where large quantities can be prepared using a robust process. Further, the MBP ASD needs to be formulated into a dosage form that can be mass produced, packaged, stored, and transported to the distribution centers (pharmacy, hospitals, etc.). It is critical that the MBP ASD retains its integrity, stability, and bioavailability characteristics all through these phases of development, commercial manufacturing, packaging, and distribution.

The formulation of MBP ASD into a solid dosage form requires an integrated understanding of its stability profile, mechanical properties, interactions with the environment during storage (moisture, heat, light), and patient needs (dosage form size, convenience of administration, patient compliance). The factors that can affect the processability of MBP product during scale-up and its impact on the product quality are outlined in Fig. 12.2. The general prerequisites for amorphous stability of MBP material at the commercial manufacturing stage are: (a) that the aqueous MBP suspension remains amorphous at 5 °C for more than 48 h and (b) the final dried MBP powder remains amorphous for more than 2 years at ambient temperature of storage.

Several factors impact the performance of MBP. These include:

- Polymer type
- Drug loading
- Microprecipitation parameters (MBP manufacturing)
- Galenical processing/additives
- Packaging and storage

Impact of these factors on the ability to manufacture MBP and subsequent impact on its performance is presented in the following section with several case studies.

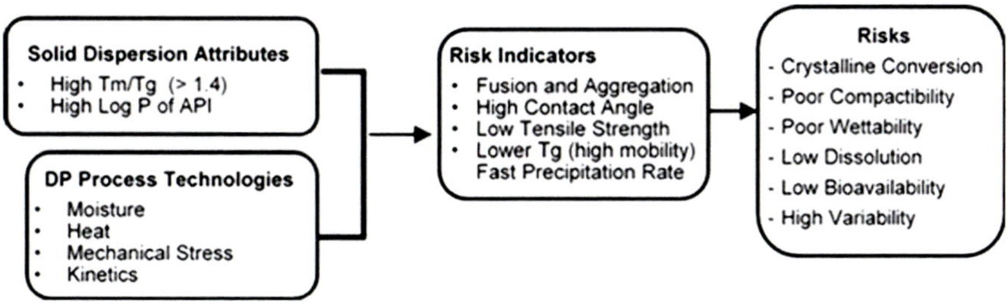

Fig. 12.2 Factors affecting processability of MBP and the associated risk factors

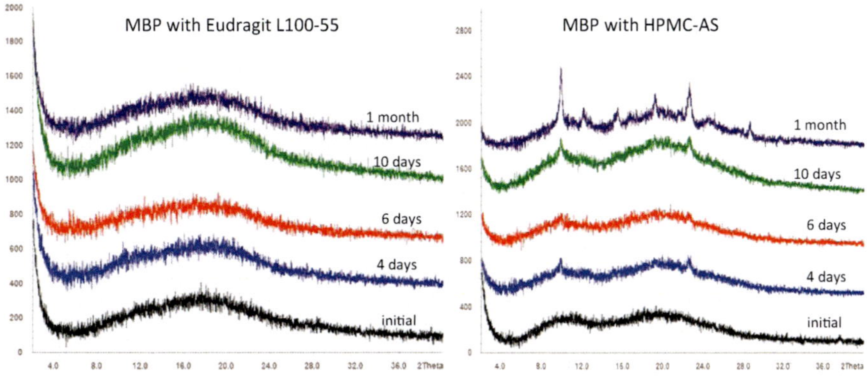

Fig. 12.3 Effect of polymer selection on stability of MBP of Compound "T"

12.4.1 *Polymer Selection and Drug Loading*

The MBP material contains amorphous drug dispersed in a polymeric carrier matrix either as molecular dispersion or as stabilized microdomains of drug. The matrix stabilizes the amorphous drug by various means such as vitrification of drug resulting in immobilized glass and drug–polymer-specific interactions. The dissolution characteristics of solid dispersions depend to a large extent on the physical state (amorphous), drug dispersivity (molecular dispersion), and particle size. The selection of polymer, drug loading, matrix composition, and preparation technique dictate the initial state of supersaturation (Urbanetz and Lippold 2005). The stability profile and ease of processing are dependent on the specific interactions between drug and polymer. Effect of API–polymer interaction on the MBP stability is shown in Fig. 12.3 for a poorly soluble Compound "T". The MBP was prepared using two different polymers, Eudrgait L100–55 and hypromellose acetate succinate (HPMC-AS). Both polymers provided amorphous material initially, but MBP with Eudragit L-100 55 exhibited greater stability than HPMC-AS when stored for 30 days at 40 °C/100 % RH, further proving the point that specific drug–polymer interactions between drug and polymer play a critical role in stabilization of amorphous state.

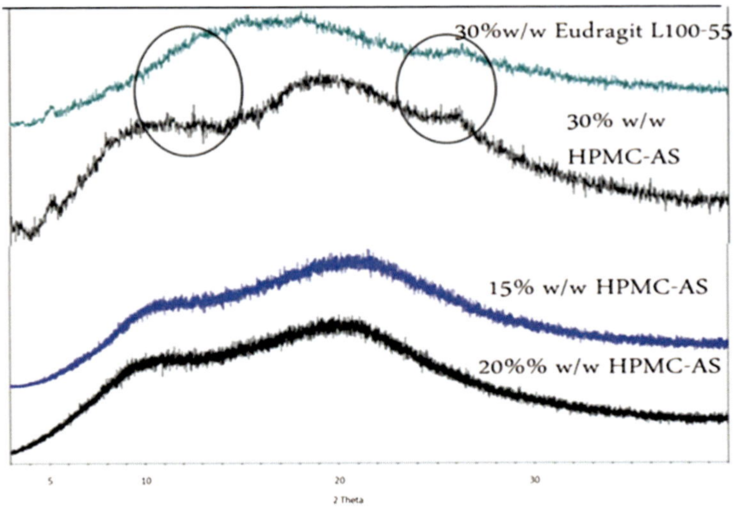

Fig. 12.4 MBP of Compound "X" showing effect of drug loading and polymer type

In the case of Compound "T," Eudragit L-100 55 provided stronger interaction, therefore better physical stability over HPMC-AS. However, a similar but opposite effect was observed with Compound "W." MBP with Eudragit L-100 55 did not yield completely amorphous material, whereas MBP prepared with HPMC-AS produced amorphous material that was stable for more than 2 years, further attesting that generation and stabilization of amorphous form are likely related to specific drug and polymer interaction.

In a study reported by Sertsou et al. (2002a, b) using the anti-solvent precipitation method, the impact of formulation and processing factors including drug loading, mixing speed, and anti-solvent pH was evaluated on the amorphous content of coprecipitated ASD. The effect of drug loading was found to be significant and the results were explained based on two competing phenomena influencing the amorphous content, i.e., crystallization inhibition by polymer and plasticization by solvent/anti-solvent. Similar results were obtained for MBP process as shown by application of MBP to Compound "X." As shown in Fig. 12.4, the drug loading up to 20 % provided a completely amorphous ASD, while drug loading above 30 % showed residual crystallinity for HPMC-AS.

12.4.2 Effect of Processing Technologies on MBP Stability

The handling of MBP requires consideration of heat, moisture, and shear stress that may destabilize the MBP. In a study of solid dispersion of Compound "Y," the MBP ASD prepared by MBP process provided a uniform and homogeneous solid solution of amorphous drug embedded in polymer while the same composition prepared by

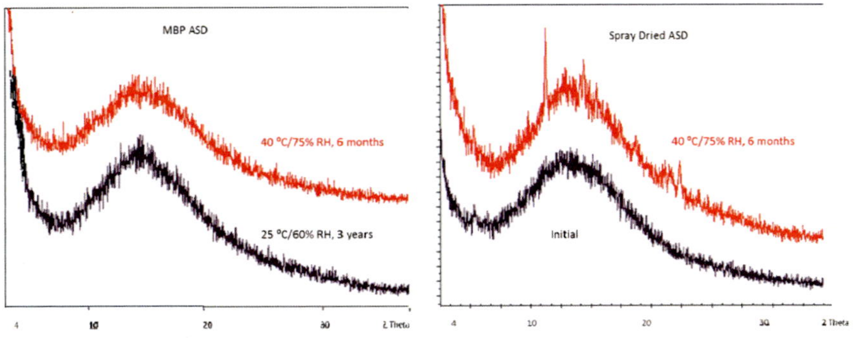

Fig. 12.5 Effect of processing technology on the crystallinity of ASD of Compound "Y" produced by MBP and spray drying with Eudragit L100 at 30 % drug loading

spray-drying process resulted in a two-phase dispersion that exhibited phase separation of drug and polymer. The particle size of the amorphous drug embedded in the ASD was determined by dissolving the polymer in an aqueous system, thereby separating the amorphous particles from the polymer matrix and leaving a suspension of amorphous API particles. Because of the destructive nature of the test, it is possible that some changes in the particle size could have occurred during the testing; regardless, the particle size of the recovered drug was found to be significantly different for the two processes. The $d_{90\%}$ of ASD from MBP process was found to be 0.9 μm, while that of spray-dried dispersion was 7.8 μm with a biphasic distribution, indicating that spray-dried dispersions may have a heterogeneous distribution of drug in the matrix resulting in higher variability during dissolution.

Further differentiation was observed in the stability of the ASD upon storage at accelerated stress conditions of temperature and humidity for 6 months. Crystalline peaks were observed for spray-dried dispersion, whereas ASD by MBP process remained amorphous as seen from Fig. 12.5 (Shah et al. 2012). Corresponding to this observation, the bioavailability of ASD by MBP was 100 %, while that of spray-dried ASD was 52 % when evaluated in a dog PK study.

As part of dosage form development, MBP ASD is often milled and densified for handling and filling operations. The densification is usually performed by either wet granulation or dry granulation processes. The aqueous wetting and massing of MBP granules can adversely affect its stability profile. As part of granulation process selection studies, the stability of wet granules of MBP in comparison with dry granules was studied. The adverse effect of wet granulation on physical stability was observed after long-term storage as shown in Fig. 12.6. Figure 12.6 shows the result of stress testing of tablets compressed from two types of granules, one by wet granulation and the other by dry granulation (roller compaction). Both tablets were shown to be amorphous at initially, but traces of crystalline peaks were observed from the wet granulated tablet after storage at accelerated conditions of 40 °C/75 % RH for 6 months.

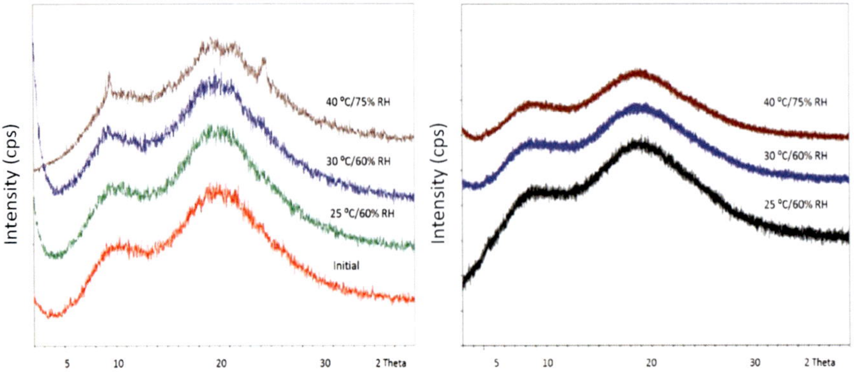

Fig. 12.6 pXRD profiles of wet and dry granulated MBP stored under various levels of stress (Compound Z)

Since the relaxation of amorphous state occurs over long periods of time, initial observations of amorphicity do not necessarily ensure long-term physical stability. Moreover, the relaxation may occur locally in microdomains instead of throughout the entire ASD during wet granulation. As such, multiple galenical technologies coupled with representative stability need to be evaluated to determine the robustness of the lead technology.

12.4.3 MBP Particulate Properties: Effect on Mechanical Properties, Downstream Processing, and Dissolution

The molecular state of API in the microstructural domains of MBP depends on the physicochemical properties of API, the polymer, and the specific interactions between API and polymer during precipitation. While the molecular state of API in ASD is important, the bulk particulate properties of MBP govern the critical galenical processes: material handling, flow properties, compaction behavior and performance, and dissolution. These particulate properties are closely related to the precipitation conditions such as shear, solvent–anti-solvent ratio, mode of addition, filtration efficiency, drying method, and milling. The final MBP powder is often milled and densified in order to minimize variability in the bulk particulate properties and to provide suitable flow and compactibility.

It is clear that the particle size of MBP can influence the downstream processing as well as performance of the final product. This is illustrated in Fig. 12.7 that shows the effect of MBP particle size on the particulate properties (bulk density and particle size) of the densified material, following the roller compaction process.

As discussed in Chap. 10 of this book, MBP particles are highly porous microstructures in general. This microporosity provides a number of benefits for enhanced

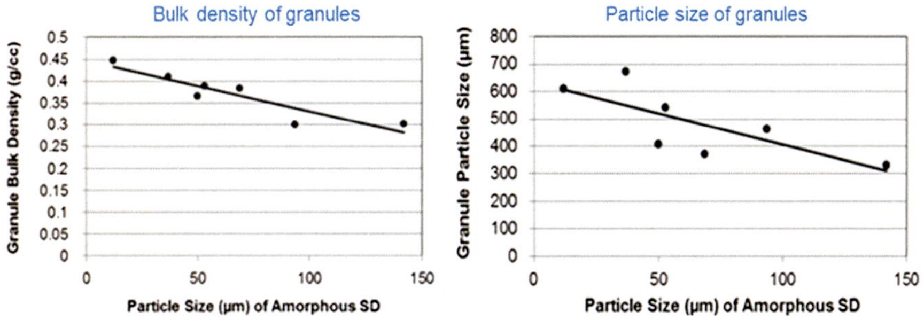

Fig. 12.7 Effect of primary particle size of MBP on bulk properties of roller compacted granules (Compound Z)

galenical processing, providing better compactibility, particle bonding, and densification. For example, the smaller particle sizes of the MBP powder, after roller compaction and milling, provided larger granules with high bulk density. Since MBP particles are porous, under compaction, they bond together efficiently, which results in densification. As expected, the extent of bonding between smaller particles is greater than larger ones. This occurs if, during roller compaction, the smaller MBP particles are compressed to the point of bonding, resulting in ribbons with high tensile strength. Such ribbons, upon milling, provide granules with a larger mean particle size and higher density or strength.

During product development, it is prudent to systematically evaluate the effect of MBP particulate properties on the properties of granules, tablet compaction, and dissolution preferably based on a statistically controlled experimental protocol to discern the interplay of the relevant interactions (Fig. 12.8). Due to the high polymeric content, the primary mode of compaction with amorphous materials is plastic deformation and as such the final compaction is sensitive to the dry granulation conditions (Herting and Kleinebudde 2008).

The impact of particle size on dissolution can be illustrated as follows: three different particle sizes of MBP powders were produced, in the range of 10–100 µm. The dissolution profile of these three MBP "as is" particles was monitored as shown in Fig. 12.9. As expected from the Noyes–Whitney equation, the dissolution of finer particles was faster than the dissolution of larger particles. Using these three particle sizes of MBP powders, tablets were prepared to the similar hardness value to 200 KN. The dissolution profile of these three tablets is shown in Fig. 12.10. Surprisingly, tablets made of smaller MBP particles dissolve slower than the tablet made of large particles.

The confounding effect of MBP particle size on the dissolution of tablet is attributed to the bonding of smaller particles during compression, particularly during roller compaction, where particle bonding is more pronounced with smaller particle sizes of MBP than larger particle sizes of MBP. These observations further support the hypothesis that particle size of MBP influences aggregation and bonding of the amorphous particles when subject to mechanical and/or thermal stress. The smaller

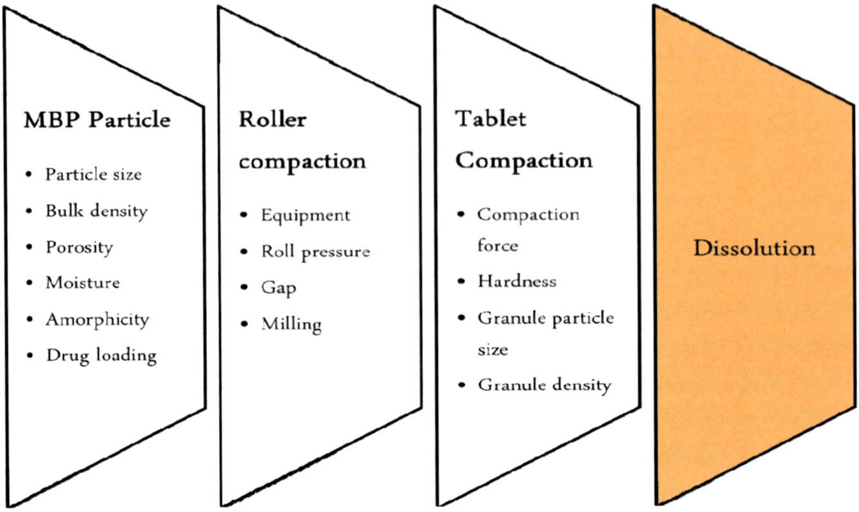

Fig. 12.8 Effect of MBP particulate properties on primary and secondary compression as well as dissolution

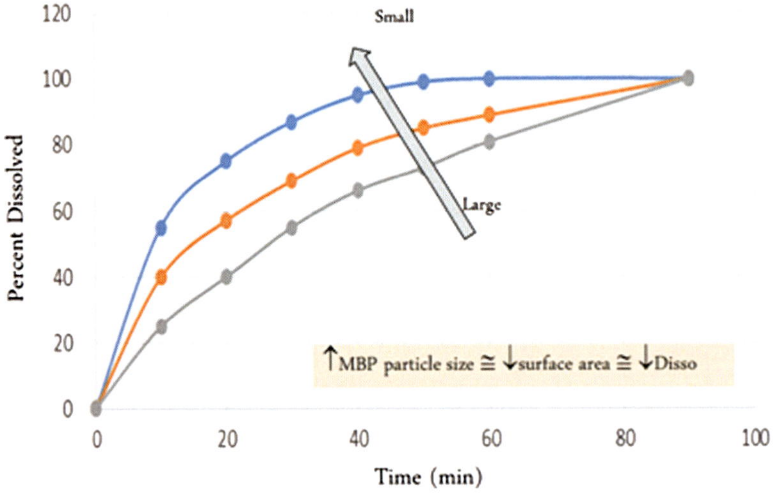

Fig. 12.9 An illustration of the effect of MBP particle size "as is" on dissolution

particles tend to exhibit a greater degree of sensitivity to external stress, resulting in comparably slower dissolution.

Both densification and compaction, being energy intensive processes, are sensitive to the physico-mechanical properties of the powders, more so in case of amorphous form. Bruno and colleagues showed that the dynamic indentation hardness of compacts of amorphous drug particles was approximately 30 % higher than that of

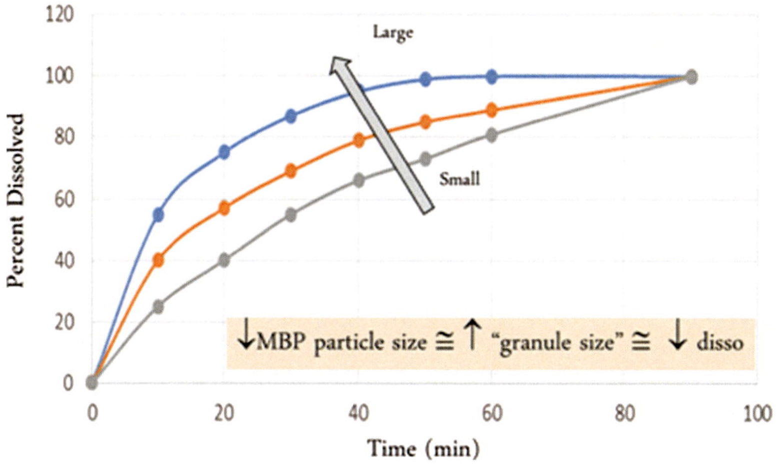

Fig. 12.10 An illustration of the effect of MBP particle size on dissolution of "Final Tablet"

crystalline particles of the same drug (Hancock et al. 2002). This suggests that the amorphous particles are prone to aggregation and fusion under mechanical stress. It may be hypothesized that when subjected to a high degree of compressive stress, the fused amorphous particles could form a hard surface that resists indentation. By the same token, it is possible that the ASD manufactured by different technologies can exhibit similar amorphous stability but behave differently under mechanical stress. The differences in the compactibility of ASD manufactured by spray drying versus melt extrusion have been recognized and studied extensively with a goal to improve the compaction properties of the melt-extruded products. Interestingly, similar results were observed when MBP was compared to spray-dried solid dispersions. In a compaction comparison study, two ASDs of an investigational compound were manufactured by MBP and by spray drying processes using the same polymer and drug loading. The tablets manufactured using MBP ASD exhibited several fold higher hardness than the tablets manufactured using spray-dried ASD. The difference in the mechanical properties can be attributed to the porosity of the MBP ASD.

12.4.4 *Effect of Moisture Content and Crystallinity on Dissolution of MBP*

The effect of moisture on the physical stability of ASD and its impact on dissolution is one of the most widely researched topics in the ASD literature (Simonelli et al. 1969; Hancock and Zografi 1994; Rumondor and Taylor 2010; Raina et al. 2013; Sarode et al. 2013). Reversion of amorphous systems to crystalline state occurs primarily as a function of temperature, water content, and storage time. Moisture can adversely impact stability of amorphous materials by lowering glass transition

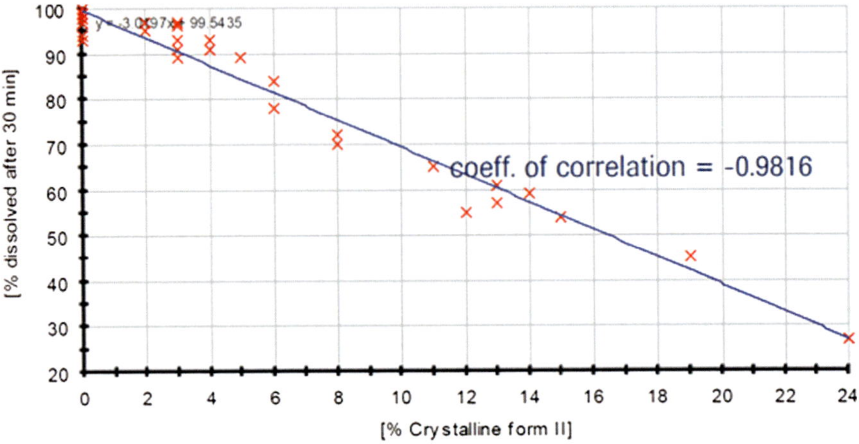

Fig. 12.11 Effect of crystalline content of Compound "Z" MBP on drug release (Shah et al. 2012)

temperature, thereby inducing mobility of the drug leading to phase separation, nucleation, and eventually crystallization. The moisture content in the MBP ASD is primarily controlled by its initial moisture content, the storage, and packaging conditions. Similar to the effect of aqueous wet granulation on product stability discussed in previous section, the MBP ASD also shows sensitivity to moisture during storage. A good correlation was observed between water content and the crystallinity in MBP ASD. In the authors' experience (unpublished work), the percent crystallinity of ASD was seen to increase with the moisture content up to a certain threshold value in a nonlinear fashion. Depending on the hygroscopicity and the crystallization tendency of the compound, the threshold moisture content above which MBP is significantly destabilized is generally between 3 and 10 %. The percent crystallinity in the ASD in turn is related to the dissolution performance of the product. For example, to investigate the effect of crystallinity, the percent drug dissolved at 30 min was plotted against percent crystallinity determined by pXRD (Fig. 12.11). As shown for Compound "Z," the percent dissolved was 90% or higher at crystallinity up to 4 %, but decreased linearly with percent increase in crystallinity beyond that level (Shah et al. 2013).

12.5 Case Studies of MBP of Poorly Soluble Drugs

The MBP process has been applied to numerous research compounds, enabling progression from preclinical to clinical stage. Examples of a few cases are presented in this section. The compounds were unsuitable for processing into ASD by spray drying or hot-melt extrusion due to either high melting point, thermal instability, or inadequate solubility in volatile solvents. The excerpts from these case studies

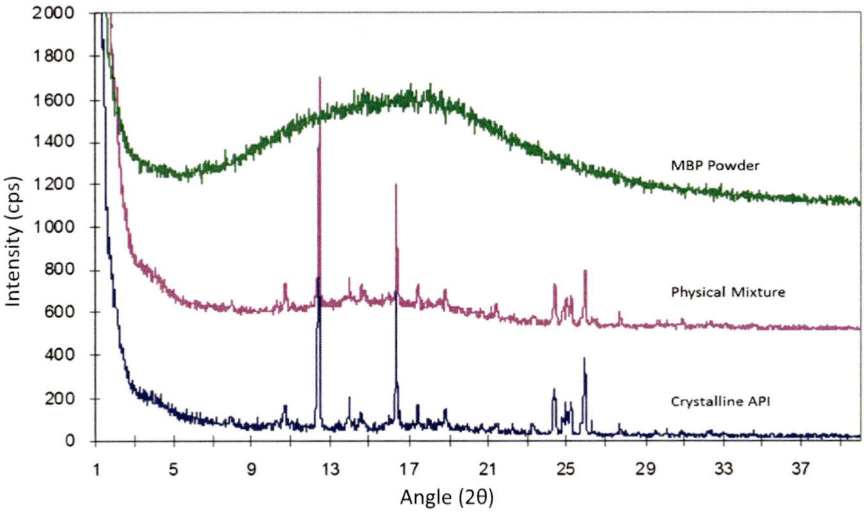

Fig. 12.12 pXRD of crystalline drug, physical mixture, and MBP of Drug A

are presented to demonstrate the application of MBP technology and highlight the relevance of various factors that ensure successful implementation of processes.

12.5.1 MBP Case A

Drug A has a high permeability but has very poor water solubility of < 1 μg/mL. Bioavailability in preclinical animal models was very low, 4 % in dogs and 9 % in rats. Nanomilling and lipid formulation did not provide acceptable exposures to move forward. Amorphous formulation approach using spray drying and hot-melt extrusion turned out not to be readily amenable to these processes, owing to poor solubility and thermal instability. Microprecipitation technology was employed to make amorphous solid dispersion of Drug A. MBP prescreening with polymers identified Eudragit L-100 as the best match for the physicochemical properties of the drug. Amorphicity by pXRD of MBP ASD is shown in Fig. 12.12 together with crystalline API and physical mixture of the same composition for comparison (Fig. 12.12). A drug load of up to 50 % was achieved, which was quite remarkable (Shah et al. 2012). Ability to achieve high drug loading was attributed to good miscibility of drug with Eudragit L-100 polymer, enhanced drug–polymer interaction, and the inherent versatility of the MBP process.

As Drug A was relatively non-hygroscopic compared to the polymer, the moisture sorption behavior of the MBP was in between that of pure drug and polymer as seen in Fig. 12.13.

The drug release profile of MBP ASD was compared to that of crystalline form using the United States Pharmacopoeia (USP) dissolution apparatus at a pH of 6.8.

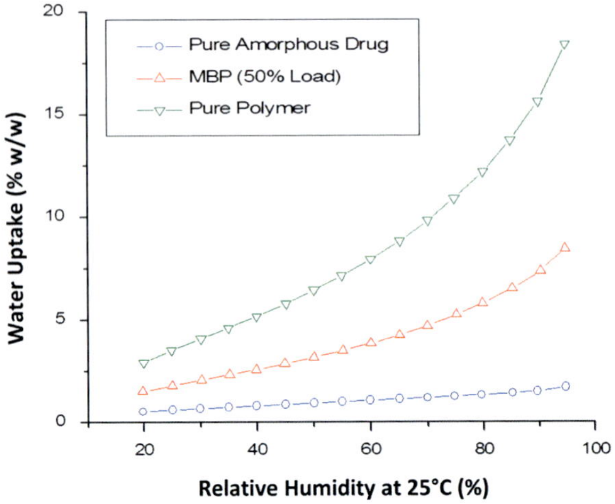

Fig. 12.13 Sorption Isotherm of MBP, polymer, and free form of Drug A. (Shah et al. 2012)

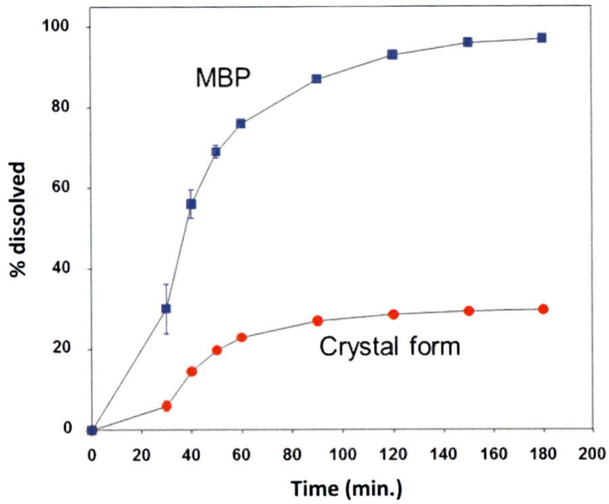

Fig. 12.14 Dissolution profile of crystalline form and MBP of Drug A. (Shah et al. 2012)

More than 80 % of drug was released from MBP ASD in 30 min, whereas about 30 % of drug was released from crystalline drug in the same time period. Moreover, MBP ASD maintained supersaturation for more than 3 h as demonstrated in Fig. 12.14.

In a dog PK study, the MBP provided 85 % bioavailability compared to 10 % for a crystalline nanosuspension formulation at an oral dose of 10 mg/kg (Shah et al. 2012). This product was evaluated in several clinical studies and was shown to provide a prolonged plasma release profile with MBP resulting in improved tolerability (Salazar et al. 2004), suggesting that the slow release of drug from the enteric polymeric matrix provides sustained release.

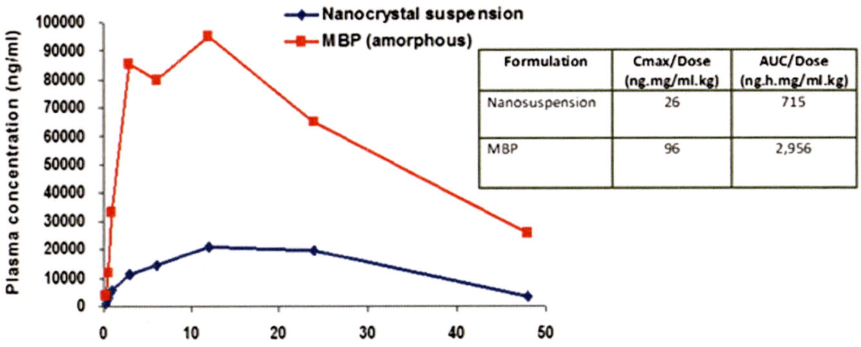

Fig. 12.15 Rat PK profile of nanocrystal formulation and MBP of Drug B

12.5.2 MBP Case B

Drug B has an aqueous solubility in the range of 3–10 μg/mL and provided inadequate exposures during preclinical studies. MBP was developed with an enteric polymer and compared against a nanocrystal suspension formulation in a rat PK study at 1000 mg/kg. A higher than fourfold increase in absorption was observed with MBP as compared to the crystalline form (Fig. 12.15).

12.5.3 MBP Case C

The dosage form development of drug C was very challenging because not only was the solubility poor with resultant poor bioavailability but also the plasma exposure levels of drug were very sensitive to the dosage regimen and frequency of dosing. An MBP formulation was developed and compared against a nanocrystalline suspension at a dose of 30 mg/kg in rat. The area under the curve (AUC) of MBP was about tenfold higher than crystalline form and was comparable to that of a cyclodextrin-based solution formulation.

Interestingly, MBP ASD prepared with Eudragit L-100 55 provided threefold higher AUC than that of MBP ASD prepared with HPMC-AS under similar dosing levels as shown in Fig. 12.16. This can be attributed to the specific drug and polymer interaction amongst other factors.

12.5.4 MBP Case D

Drug D had a high log P resulting in good lipophilicity for absorption; however, its solubility was extremely low at $<<1$ μg/mL resulting in the need for formulation

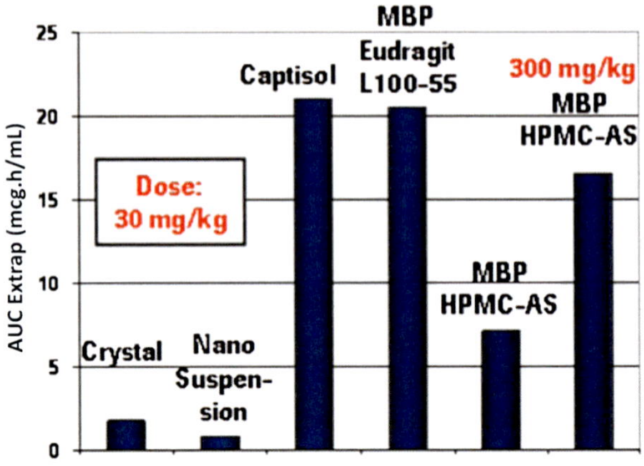

Fig. 12.16 Rat PK data of crystalline formulations and MBP of Drug C

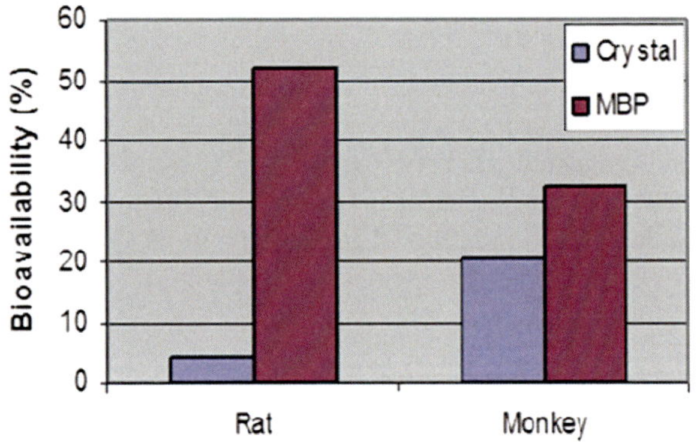

Fig. 12.17 Bioavailability of crystalline form and MBP of Drug D in rats and monkey

intervention. An MBP ASD was developed using HPMC-AS as the stabilizing polymer and it was evaluated in rat and monkey. The bioavailability was increased in the rat by more than tenfold and in the monkeys by more than 1.5 times compared to the crystalline form as shown in Fig. 12.17.

12.5.5 MBP Case E

Drug E is practically water insoluble (< < 1 μg/mL) with a melting point above 270 °C. The solubility in common organic solvents such as acetone, alcohol, and acetonitrile was poor < 5 mg/mL at 25 °C, but in DMA the solubility was exceptionally high > 500 mg/mL. The development of MBP and its impact on bioavailability has been published elsewhere (Shah et al. 2013).

Amorphous solid dispersion using spray drying and hot-melt extrusion was not readily applicable due to the poor organic solubility and high melting point. MBP technology was applied to make the ASD. MBP prescreening identified HPMC-AS

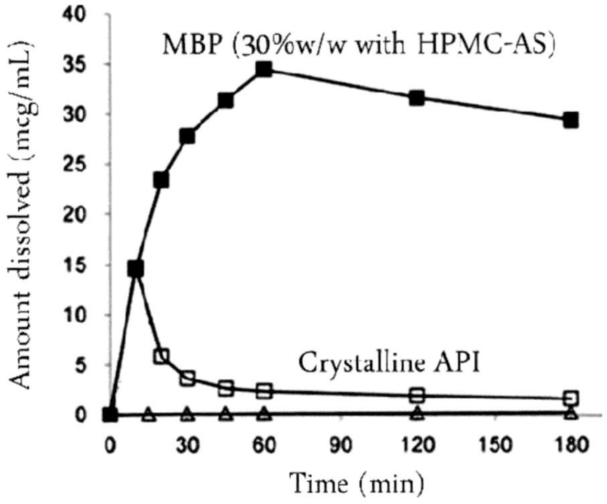

Fig. 12.18 Dissolution profile of crystalline form and MBP of Drug E (redrawn from Shah et al. 2013)

as a suitable polymer for Drug E. Further, miscibility study identified operable drug loading in the range of 30–40 %. ASDs with HPMC-AS were prepared using MBP technology and the resulting products were found to be pXRD amorphous upon preparation and storage under accelerated stress stability conditions of 40 °C/75 % RH for up to 6 months. The T_g values of amorphous Drug E and HPMCAS were 107 and 119 °C, respectively, while MBP ASD of 30 % Drug E exhibited a single T_g in the range of 100–110 °C, depending on residual moisture content in ASD (Shah et al. 2013).

The drug release profile from MBP ASD was compared against crystalline form (unstable crystalline form 1) using the USP dissolution apparatus and 900 mL of a pH 6.8 phosphate buffer medium (Fig. 12.18). A concentration of 35 μg/mL was achieved within 60 min and a supersaturation concentration of 30 μg/mL was maintained up to 3 h. The crystalline form (unstable crystalline form 1), on the other hand, exhibited an initial spike in concentration, which was immediately followed by a drop in concentration to the stable value of 1 μg/mL. Thus, about 20- to 30-fold increase in solubility (compared to unstable form 1) and maintenance of saturation levels was achieved with MBP (Shah et al. 2013).

In a relative bioavailability study comparing MBP formulations of Drug E against crystalline form (unstable crystalline form 1), the MBP formulations exhibited much higher exposures after a single dose of MBP compared to crystalline formulation, as seen in Fig. 12.19 (Shah et al. 2013). The relative bioavailability of the MBP formulations was four to fivefold higher than the crystalline formulation (capsule formulation). Furthermore, unlike the crystalline capsule formulation that reached a plateau at 600 mg dose, the exposure for MBP was dose-linear from up to 1200 mg.

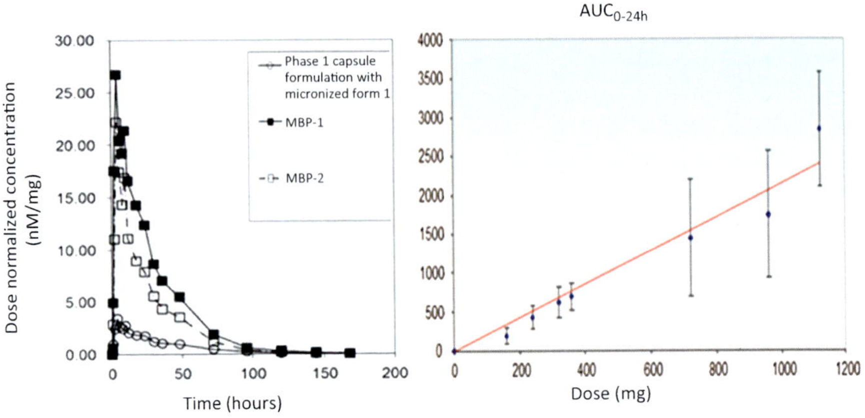

Fig. 12.19 Pharmacokinetic profile and dose-dependent increase in exposure observed with crystalline and MBP formulations of Drug E (Shah et al. 2013)

12.6 Summary and Conclusions

The MBP technology is well suited for compounds with poor solubility and high melting point, particularly when alternate ASD technologies such as spray drying and hot-melt extrusion are not readily applicable. This MBP technology has been employed to manufacture ASD of a number of poorly soluble drugs using stabilizing ionic polymers, mainly Eudragit L-100, L-100 55, and HPMC-AS. Drugs with high molecular weight, high melting point, low solubility (< 10 mcg/mL), and log P of greater than 3 seem to be highly suitable for MBP process. In all cases of MBP ASDs, the higher dissolution rate of the drug in MBP was translated into higher bioavailability and exposure in preclinical and clinical studies.

Application of MBP technique to diverse compounds has demonstrated the utility and the versatility of this technique. The MBP technique is highly adaptable to various manufacturing scales, from milligram quantities during preclinical research to hundreds of kilogram quantities in production phase with > 90 % recovery. As discussed in previous chapters, it is a material-sparing tool that can provide reproducible ASDs with superior performance, in some sense, compared to other ASD technologies for certain type of compounds. The material-sparing aspect is very important in the early stages of drug development to support animal studies when the drug supply is limited. Several research compounds have been scaled up from few milligrams to 100–1000 kg demonstrating that it is a scalable and robust process. The MBP technology can present a remarkable opportunity to advance certain poorly soluble compounds that otherwise would be considered undevelopable.

References

Albano A, Phuapradit W, Sandhu H, Shah N (2002) Amorphous form of cell cycle inhibitor having improved solubility and bioavailability. U.S. Patent. 6,350,786

Amidon GL, Lennernas H, Shah VP, Crison JR (1995) A theoretical basis for a biopharmaceutic drug classification: the correlation of in vitro drug product dissolution and in vivo bioavailability. Pharm Res 12(3):413–420

Bahl D, Hudak J, Bogner RH (2008) Comparison of the ability of various pharmaceutical silicates to amorphize and enhance dissolution of indomethacin upon co-grinding. Pharm Dev Technol 13(3):255–269

Bley H, Fussnegger B, Bodmeier R (2010) Characterization and stability of solid dispersions based on PEG/polymer blends. Int J Pharm 390(2):165–173

El-Badry M, Fathy M (2006) Enhancement of the dissolution and permeation rates of meloxicam by formation of its freeze-dried solid dispersions in polyvinylpyrrolidone K-30. Drug Dev Ind Pharm 32(2):141–150

Evonik (2014) Information from Evonik Industries website Retrieved from http://eudragit.evonik.com/product/eudragit/en/products-services/eudragit-products/enteric-formulations/l-100–55/pages/default.aspx. Accessed 14 Jan 2014

Faure B, Salazar-Alvarez G., Ahniyaz A, Villaluenga I, Berriozabal G, Miguel YRD, Bergström L (2013) Dispersion and surface functionalization of oxide nanoparticles for transparent photocatalytic and UV-protecting coatings and sunscreens. Sci Technol Adv Mater 14(2):023001

FDA, GfI (2000). Waiver of in vivo bioavailability and bioequivalence studies for immediate-release solid oral dosage forms based on a biopharmaceutics classification system. Guidance for Industry, U.S. Department of Health and Human Services, Food and Drug Administration, Center for Drug Evaluation and Research (CDER) August 2000

Hancock BC, Zografi G (1994) The relationship between the glass transition temperature and the water content of amorphous pharmaceutical solids. Pharm Res 11(4):471–477

Hancock BC, Carlson GT, Ladipo DD, Langdon BA, Mullarney MP (2002) Comparison of the mechanical properties of the crystalline and amorphous forms of a drug substance. Int J Pharm 241:73–85

Hancock BC, Shamblin SL, Zografi G (1995) Molecular mobility of amorphous pharmaceutical solids below their glass transition temperatures. Pharm Res 12(6):799–806

Herting MG, Kleinebudde P (2008) Studies on the reduction of tensile strength of tablets after roll compaction/dry granulation. Eur J Pharm Biopharm 70(1):372–379

Hu Q, Choi DS, Chokshi H, Shah N, Sandhu H (2013) Highly efficient miniaturized coprecipitation screening (MiCoS) for amorphous solid dispersion formulation development. Int J Pharm 450 (1–2):53–62

Kai T, Akiyama Y, Nomura S, Sato M (1996) Oral absorption improvement of poorly soluble drug using solid dispersion technique. Chem Pharm Bull (Tokyo) 44(3):568–571

Keck CM, Muller RH (2006) Drug nanocrystals of poorly soluble drugs produced by high pressure homogenisation. Eur J Pharm Biopharm 62(1):3–16

Kim EJ, Chun MK, Jang JS, Lee IH, Lee KR, Choi HK (2006) Preparation of a solid dispersion of felodipine using a solvent wetting method. Eur J Pharm Biopharm 64(2):200–205

Kohri N, Yamayoshi Y, Xin H, Iseki K, Sato N, Todo S, Miyazaki K (1999) Improving the oral bioavailability of albendazole in rabbits by the solid dispersion technique. J Pharm Pharmacol 51(2):159–164

Leuner C, Dressman J (2000) Improving drug solubility for oral delivery using solid dispersions. Eur J Pharm Biopharm 50(1):47–60

Liu R (2008) Water-insoluble drug formulation. CRC, Boca Raton

Loftsson T, Brewster ME (1996) Pharmaceutical applications of cyclodextrins. 1. Drug solubilization and stabilization. J Pharm Sci 85(10):1017–1025

Miller MA, DiNunzio J, Matteucci ME, Ludher BS, Williams RO, Johnston KP (2012) Flocculated amorphous itraconazole nanoparticles for enhanced in vitro supersaturation and in vivo bioavailability. Drug Dev Ind Pharm 38(5):557–570

Moser JD, Broyles J, Liu L, Miller E, Wang M (2008) Enhancing bioavailability of poorly soluble drugs using spray dried solid dispersions. Am Pharm Re, Drug Deliv 1 (9–10):68–71
Noyes A, Whitney W (1897) The rate of solution of solid substances in their own solutions. J Am Chem Soc 19:930–934
Okimoto K, Miyake M, Ibuki R, Yasumura M, Ohnishi N, Nakai T (1997) Dissolution mechanism and rate of solid dispersion particles of nilvadipine with hydroxypropylmethylcellulose. Int J Pharm 159(1):85–93
Rabinow BE (2004) Nanosuspensions in drug delivery. Nat Rev Drug Discov 3(9):785–796
Raina SA, Zhang GG, Alonzo DE, Wu J, Zhu D, Catron ND, Gao Y, Taylor LS (2013) Enhancements and Limits in Drug Membrane Transport Using Supersaturated Solutions of Poorly Water-Soluble Drugs. J Pharm Sci 103(9):2736–48
Rolfes H, Van Der Merwe TL, Truter PA (2001) Method of making controlled release particles of complexed polymers, Google Patents.
Rumondor AC, Taylor LS (2010) Effect of polymer hygroscopicity on the phase behavior of amorphous solid dispersions in the presence of moisture. Mol Pharm 7(2):477–490
Salazar R, Bissett D, Twelves C, Breimer L, DeMario M, Campbell S, Zhi J, Ritland S, Cassidy J (2004) A phase I clinical and pharmacokinetic study of Ro 31–7453 given as a 7- or 14-day oral twice daily schedule every 4 weeks in patients with solid tumors. Clin Cancer Res 10(13):4374–4382
Sarode AL, Wang P, Obara S, Worthen DR (2013) Supersaturation, nucleation, and crystal growth during single- and biphasic dissolution of amorphous solid dispersions: Polymer effects and implications for oral bioavailability enhancement of poorly water soluble drugs. Eur J Pharm Biopharm 86(3):351-360
Serajuddin AT (1999) Solid dispersion of poorly water-soluble drugs: early promises, subsequent problems, and recent breakthroughs. J Pharm Sci 88(10):1058–1066
Sertsou G, Butler J, Hempenstall J, Rades T (2002a) Solvent change co-precipitation with hydroxypropyl methylcellulose phthalate to improve dissolution characteristics of a poorly water-soluble drug. J Pharm Pharmacol 54(8):1041–1047
Sertsou G, Butler J, Scott A, Hempenstall J, Rades T (2002b) Factors affecting incorporation of drug into solid solution with HPMCP during solvent change co-precipitation. Int J Pharm 245(1–2):99–108
Shah N, Sandhu H, Phuapradit W, Pinal R, Iyer R, Albano A, Chatterji A, Anand S, Choi DS, Tang K, Tian H, Chokshi H, Singhal D, Malick W (2012) Development of novel microprecipitated bulk powder (MBP) technology for manufacturing stable amorphous formulations of poorly soluble drugs. Int J Pharm 438(1–2):53–60
Shah N, Iyer RM, Mair HJ, Choi DS, Tian H, Diodone R, Fahnrich K, Pabst-Ravot A, Tang K, Scheubel E, Grippo JF, Moreira SA, Go Z, Mouskountakis J, Louie T, Ibrahim PN, Sandhu H, Rubia L, Chokshi H, Singhal D, Malick W (2013) Improved human bioavailability of vemurafenib, a practically insoluble drug, using an amorphous polymer-stabilized solid dispersion prepared by a solvent-controlled coprecipitation process. J Pharm Sci 102(3):967–981
Shin SC, Cho CW (1997) Physicochemical characterizations of piroxicam-poloxamer solid dispersion. Pharm Dev Technol 2(4):403–407
Simonelli AP, Mehta SC, Higuchi WI (1969) Dissolution rates of high energy polyvinylpyrrolidone (PVP)-sulfathiazole coprecipitates. J Pharm Sci 58(5):538–549
Urbanetz NA, Lippold BC (2005) Solid dispersions of nimodipine and polyethylene glycol 2000. Eur J Pharm Biopharm 59(10):107–118
Williams IRO, Watts ABA, Miller DA (2010) Formulating Poorly Soluble Drugs. Aapspress and Springer, New York
Williams HD, Trevaskis NL, Charman SA, Shanker RM, Charman WN, Pouton CW, Porter CJ (2013) Strategies to address low drug solubility in discovery and development. Pharmacol Rev 65(1):315–499
Wu K, Li J, Wang W, Winstead DA (2009) Formation and characterization of solid dispersions of piroxicam and polyvinylpyrrolidone using spray drying and precipitation with compressed antisolvent. J Pharm Sci 98(7):2422–2431

Chapter 13
Downstream Processing Considerations

Susanne Page and Reto Maurer

13.1 Introduction

Independent of the manufacturing technology used to create the amorphous solid dispersion and the anticipated route of administration, downstream processing is required to convert it into the final dosage form. For an oral application of a drug, tablets are the preferred solid dosage form followed by capsules. The development objective is captured in the target product profile (TPP) and the quality target product profile (QTPP) as these define the intended release mode, the type and acceptable size of dosage form, the anticipated shelf life, and storage conditions. They therefore set the framework for the formulation scientists, to help define the most important critical quality attributes such as dissolution kinetics, chemical and physical stability, appearance, and mechanical properties.

Dissolution Kinetics: In most cases, the drug substance is desired to be released immediately from the tablets or capsules, and only on rare occasions, an extended or pH-dependent drug release is envisaged. In both cases, the mechanism for achieving the supersaturation potential of the solid amorphous dispersion needs to be maintained in the final dosage form. Therefore, the formulator needs to have a clear understanding of all the factors that have an influence on the dissolution kinetics and the supersaturation potential. The amount of drug released determined in a standard dissolution test is the sum of different processes occurring all at the same time, namely release of the drug from the amorphous solid dispersion, nucleation, and crystal growth. The amount of drug released per time ($\mathrm{d}W/\mathrm{d}t$) is directly proportional to the diffusion coefficient (D), the surface area of the solid (A), the difference between the saturation concentration (c_s) and the concentration of drug in solution at time t (c_t), and indirect proportional to the diffusion layer thickness (L) as by Noyes and Whitney and Nernst and Brunner (Dokoumetzidis and Macheras 2006). The

S. Page (✉) · R. Maurer
Formulation Research and Development, F. Hoffmann-La Roche Ltd., Basel, Switzerland
e-mail: susanne.page@roche.com

N. Shah et al. (eds.), *Amorphous Solid Dispersions*,
Advances in Delivery Science and Technology, DOI 10.1007/978-1-4939-1598-9_13

impact of the surface area on dissolution kinetic was illustrated by comparing the dissolution rate of cefdinir tablets obtained from spray drying (SD) or a supercritical anti-solvent (SAS) process (Park et al. 2010), and by another example of solid dispersions containing indomethacin (IMC) and polyethylene glycol 6000 (PEG 6000; Ford and Elliott 1985). Interestingly, the fastest release is not always obtained from the smallest particles (Ford and Elliott 1985). This observation could be explained by the fact that not only the amount of drug substance released per time is quite high for very small particles but also nucleation and subsequently crystal growth are fast. In particular, this aspect is important when the solid dispersion needs to be milled/micronized prior to the downstream processing. The surface area or particle size distribution of such systems might be a critical material attribute (CMA), and needs to be carefully investigated and subsequently controlled. Besides the particle size of the amorphous solid dispersion, the selection of type and amount of excipients present in the final dosage form will also have a major effect on both the downstream processing itself and the drug release profile. As an example, the investigations from Lepek et al. illustrate this point. They compared the effect of lactose and microcrystalline cellulose (MCC) during direct compression of formulations containing amorphous telmisartan, and showed improved flowability as well as compressibility for the final blends containing microcrystalline cellulose. Further investigations showed that the disintegration time ranged from 30 s for the formulation containing sodium carboxymethyl starch, 1 min for croscarmellose sodium and starch glycolate, up to 30 min for formulations with potato starch (Lepek et al. 2013). The dissolution kinetic of the final drug product also depends on the ratio between the solid dispersion and the excipients in the drug product, and on the processing conditions during downstream manufacturing, like roller compaction and compression force. Furthermore, the dissolution of the final product might change upon storage as described below.

When an amorphous component is exposed to high temperature and/or relative humidity above glass transition temperature (T_g), the amorphous system transforms to a rubbery state. Molecules in rubbery state are mobile and can form interparticle bridges or form particle aggregates, often resulting in particle fusion (Descamps and Palzer 2007). This effect of particle fusion can be magnified in amorphous solid dispersion particles with large surface area (Matteucci et al. 2007; Alonzo et al. 2011; Shah et al. 2013).

Amorphous solid dispersions, particularly amorphous microprecipitated bulk powder (MBP) or spray-dried powder, have large surface area. For example, a surface area of 23–24 m^2/g was obtained for amorphous MBP solid dispersion of vemurafenib (Shah et al. 2013). Similarly, surface areas in the range of 23–51 m^2/g were obtained for itraconazole (ITZ) MBP with drug loading in the range of 33–50 % (Matteucci et al. 2007). In general, the large surface area powder dissolves faster than small surface area powder following the equation from Noyes and Whitney and Nernst and Brunner. However, in the case of amorphous dispersion, dissolution phenomenon is not straightforward. Because of potential particle fusion under stress condition, amorphous solid dispersion with large surface area can actually dissolve slower than small surface area if amorphous solid dispersion fused together to hard aggregates.

The fusion of amorphous polymer particles above their T_g is a slow coalescence process most likely driven by surface energy, which reduces the free volume and the total surface area (Rosenzweig and Narkis 1983; Palzer 2011). Since amorphous dispersions often have a polymeric component, the sintering of amorphous polymer colloids occurs at/or above the T_g which is dependent on the particle size and packing fraction within the polymer as determined by the melt viscosity (Mazur et al. 1997).

Chemical and Physical Stability: Due to the fact that the drug substance is present in a high-energy amorphous form in the formulation, the chemical stability of the system might be altered compared to formulations containing the crystalline drug substance. Furthermore, the polymer itself can interact with and trigger instability with the active pharmaceutical ingredient (API) as shown by Dong and Choi for hydroxypropyl methylcellulose acetate succinate (HPMCAS). This polymer may undergo hydrolysis under harsh processing conditions (e.g., heating at 140 °C up to 5 h) with the generation of succinic acid and acetic acid, which can form ester bonds with the hydroxyl groups present in the API. This was shown by the authors for model compound A, where the succinate esters of the model compound and its epimer were found in the product, as well as for dyphylline (Dong and Choi 2008).

Recrystallization of the drug substance can be triggered by absorption of water/presence of moisture, and energy input in the form of heat or mechanical stress. The absorption of water by the amorphous solid dispersion leads to a decrease in the T_g of the system as the T_g of water is very low (− 137 °C), and an increase in molecular mobility due to disruption of intermolecular hydrogen bonds (Ahlneck and Zografi 1990). Water can be present in the amorphous solid dispersion itself in the form of residual moisture, or it can be introduced into the system by the moisture bound to excipients and water used as a processing liquid, for instance, during film coating. Energy is transferred to the amorphous solid dispersion upon milling, compaction, and compression of the material. A direct energy transfer occurred when jet or high peripheral-speed pin mills (e.g., shock action mills) were used for milling (Colombo et al. 2009). Further down, the material is either directly filled into capsules, compressed into tablets (if the bulk density is reasonably high), or dry granulated. In case of dry granulation and tablet compression, the pressure applied to the system can induce amorphous–amorphous phase separation in these systems (Ayenew et al. 2012a), and increase the extent of crystallization (Ayenew et al. 2012b). Recrystallization of the drug substance can also be triggered when a film coat is applied; besides the aforementioned effect of water, the temperature used in the process should be selected carefully. Finally, the packaging configuration for the final drug product needs to be selected to maintain stability over the shelf life of the product.

Physical and Mechanical Properties: Depending on the manufacturing technology used for the manufacture of the amorphous solid dispersion, the material will have different physical properties clearly impacting flow and compression behavior of the material. Comparing spray-dried powders, for instance, with milled extrudates will reveal the difference of both materials. The smaller spray-dried particles have a higher tendency towards cohesion and thus impacting powder flow. On the other hand, the material has a higher porosity compared to the dense extrudates, making tablet compression easier. In addition to the physical and mechanical properties of the materials, those properties of the material subject to an applied stress are clearly of importance for the whole downstream processing process. Iyer et al. (2013) compared polymers, such as copovidone and HPMCAS, with other excipients such as lactose, MCC, or dibasic calcium phosphate anhydrous (DCP-A) and showed that the polymers itself exhibit compression pressures on the low, but acceptable end of the spectrum (solid fraction; SF = 0.85). MCC and lactose are in the typical range, and DCP-A requires extremely high compression pressures in order to obtain the same solid fraction (for DCP-A the value was extrapolated to SF = 0.85). Modification of the polymers by either SD or hot-melt extrusion (HME) led to an increase in the compression pressure by 24 % in case of spray-dried HPMCAS, 61 % in case of melt-extruded HPMCAS and a decrease of approximately 10 % upon extrusion of copovidone. The same authors also investigated the tensile strengths of the materials at a solid fraction of 0.85, and observed a decrease of tensile strength for the melt-extruded polymers compared to the polymer as is, whereas an increase in tensile strength was observed for the spray-dried HPMCAS, indicating an enhanced ability to form strong compacts. Further tests showed that in addition to compression pressure and tensile strength, other parameters such as dynamic hardness, brittle fracture index and dynamic bonding index were also altered by either SD or HME (Iyer et al. 2013).

The following sections will give a detailed overview on the downstream processing of the amorphous solid dispersions manufactured by different technologies. Overall, it can be concluded that the downstream process should avoid the use of water, higher temperatures and pressures as much as possible.

13.2 Downstream Processing of Hot-Melt Extrudates

HME is a fusion-based technology widely used for manufacturing of amorphous intermediates where the API is molecularly dispersed in a stabilizing polymer matrix. The number of polymers approved for pharmaceutical applications is limited and a preferred polymer may have a number of inherent challenges in terms of processability, dissolution performance, and downstream processing. For example, commonly used polymers like Eudragit® L100-55 and povidones of higher molecular weights show a high melt viscosity and can only be extruded with the incorporation of adequate plasticizers. In some cases, the API itself has a significant plasticizing effect on the polymer.

The high-energy amorphous state of the HME intermediate is facilitated by applying shear stress and thermal energy to a physical powder blend in order to overcome the crystal lattice of the drug and to soften the polymer allowing fusion with excipients. The extrusion process is followed by an immediate solidification or cooling step in order to freeze the glassy state and immobilize the incorporated API molecules.

The obtained extruded intermediates is normally not considered to be the final drug product as it usually appears as solid strands or films of undefined length which either have to be shaped, cut, or milled in order to process it further into the desired final solid dosage form. Contrary to other manufacturing technologies for amorphous solid dispersions, extrudates are dense particles, having a high bulk density, enabling capsule filling with a powder blend or direct compression.

13.2.1 Powder Blends and Direct Compression

Direct Shaping of Extrudates: In the literature, the most prominent dosage forms described for solid dispersions made by HME are tablets and capsules. The simplest approach to produce tablets is to cut the solidified strands manually into cylindrical mini-matrices (Bruce et al. 2007; Schilling et al. 2008; Read et al. 2010; Dierickx et al. 2012). This direct shaping process is well suited for small batch sizes where the weight uniformity of the particles can be assured by individual weight check. For large batch sizes, direct shaping and automated calendaring is much more challenging as it must be preceded by a steady and nonpulsatile HME process facilitating consistent strand or film dimensions.

Milling of Extrudates: Feng et al. (2012) and Jijun et al. (2010) prepared tablets simply by blending a defined sieve fraction of the pulverized extrudates with functional excipients and directly compressing it with a single-punch press. Jijun et al. (2010) showed that the particle size of the milled extrudates had an impact on the dissolution kinetics as well as the flowability of the final powder blend for direct compression. They showed that dissolution of finer particles was inferior compared to the performance of the coarser sieve fraction and the flowability declined with particle size reduction (Jijun et al. 2010). The same group also investigated the downstreaming process by comparing the quality and performance of direct compressed tablets with tablets generated via wet granulation of the milled extrudates (Jijun et al. 2011). The wet granulation process initiated recrystallization upon storage resulting in a different dissolution release profiles.

Deng et al. produced HME strands of 2 mm diameter which were resized using a Fitz® Mill comminutor. The resulting powder was mixed with filler, disintegrants, and lubricants and directly compressed to tablets. Different superdisintegrants were used in order to investigate the impact on the dissolution kinetics. Alternatively, they prepared pellets of 1 mm thickness by utilizing a pelletizer. The pellets were simply filled into hard gelatin capsules and used for dissolution testing as well (Deng et al. 2013). In another example, Kindermann et al. milled the HME strands with an ultracentrifugal mill and used the 355–500-μm sieve fraction to prepare double-layer tablets with tailor-made release profiles (Kindermann et al. 2012).

Read et al. used a cryogenic mill for the amorphous ketoprofen extrudates to produce particle with a mean size of 15–250 μm. The milled fraction reached 100 % dissolution after approximately 600 min while the manually cut rods eroded slowly resulting in only 50 % dissolution in the same time span. Cryogenic milling can be of advantage in cases where low-melting polymers have to be milled or when obtained strands show lack of brittleness (Read et al. 2010).

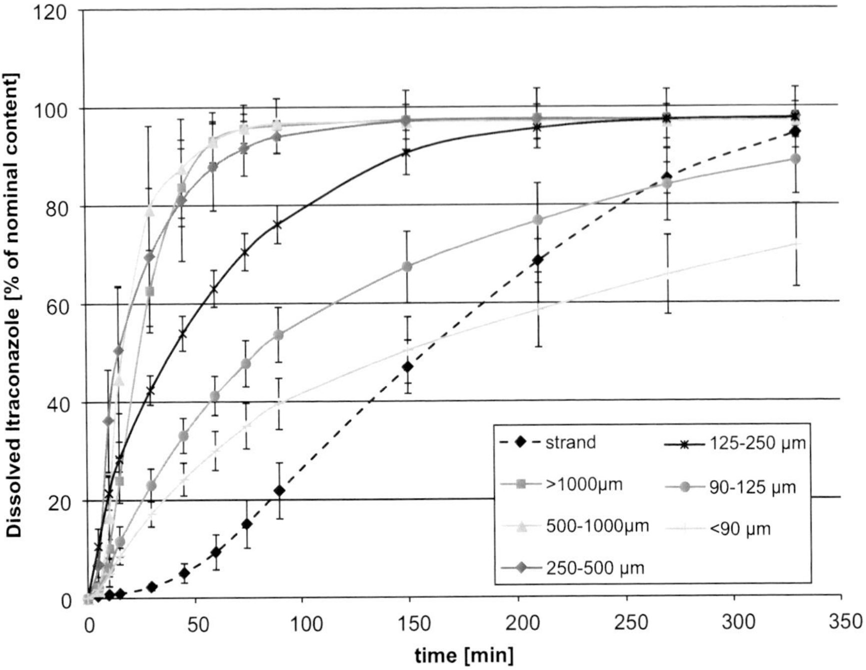

Fig. 13.1 Itraconazol (ITZ) release from hot-melt extrudates in dependence of particle size. (USP2 basket, simulated gastric fluid)

The impact of particle size on the dissolution performance was also observed for amorphous ITZ–Soluplus® extrudates. Samples of defined sieve fractions as well as samples of the unmilled strands were compared (simulated gastric fluid, $n = 6$). The results (Fig. 13.1; data not published) clearly showed that after 100 min, the API from milled particles with a mean particle size above 250 µm was fully dissolved, whereas the dissolution from finer particles was significantly slower and incomplete. The unmilled strands were steadily eroding resulting in almost 100 % dissolution after 330 min. For this formulation, coarser particles were more beneficial than finer grades, potentially as a result of polymer swelling or concurrent recrystallization effects.

Effect of Excipients: The composition and powder properties of the initial blend of API, polymer, and additional excipients not only impact the extrusion process but also the downstream process. The formulation has to exhibit sufficient flowability in order to facilitate a steady extrudate flow and screw fill degree ensuring uniform HME strands and an acceptable throughput. Many of the available pharmaceutical polymers show poor flowability and therefore the addition of flow aids like silicone dioxide or application of free-flowing polymers and excipients have to be considered. Godderis et al. investigated the flow properties and bulk density of ternary systems containing drug substance, Eudragit® E100, and surfactant tocopheryl polyethylene glycol succinate (TPGS). By estimating the Hausner ratio and the Carr index, the best excipient ratio was evaluated with respect to the flowability (Goddeeris et al. 2008). The need for free-flowing excipients for pharmaceutical HME is also recognized by the polymer manufacturers. For example, Soluplus® was developed as dissolution and solubility enhancer especially for HME applications and shows favorable flow properties.

Another important aspect is the bulk density of the initial blend. If the bulk density is too low, the feeding unit, which is usually based on a twin-screw system, becomes the limiting part of the equipment with regard to throughput and screw fill degree. Therefore, the formulator is requested to balance between sufficient flow properties, bulk density, and particle size distribution which should be similar for all ingredients in order to avoid demixing.

The advantages of a drug substance formulated as a solid dispersion is sometimes not reflected in the final dosage form as many hydrophilic polymers tend to swell and form gel matrices when exposed to aqueous environment. This effect can change the dissolution kinetics from the targeted immediate-release profile to a slower erosion process. One option to improve the dissolution rate of such formulations is to reduce the solid dispersion content in the final solid dosage form by adding large amounts of a diluent like fine MCC (DiNunzio et al. 2012) or lactose (Jijun et al. 2011) as external phase to the milled extrudates before direct compression. These excipients act as spacers increasing the porosity and prevent formation of slowly disintegrating gel structures. Here the formulator has to be aware that the drug load in solid dispersion is already reduced since normally it contains substantial amounts of stabilizing polymer.

A different approach to overcome polymer gelling is to add inorganic salts like potassium bicarbonate. They can facilitate dehydration and precipitation of the polymers in aqueous environment which is reflected by an improved dissolution profile (Hughey et al. 2013).

Another possibility to adjust the dissolution kinetics is to incorporate specific excipients directly into the extrudate. Water-soluble pore former like mannitol (Deng et al. 2013), citric acid, and sucrose (Schilling et al. 2008) as well as water-soluble polymers like hypromellose, polyethylene oxide (Read et al. 2010), or poloxamers (Zhu et al. 2006) can be extruded together with the API and stabilizing polymer and these have also been shown to improve the dissolution rate.

13.2.2 Film Coating

As a last step in the downstream process, tablet coating should be given due consideration specifically in regard to preventing hygroscopic solid dispersions from moisture uptake or facilitating a targeted release profile. Jijun et al. described coating of HME-based tablets with an Opadry® amb coat (Jijun et al. 2010, 2011). In this case, the physicochemical properties of the amorphous API should be carefully monitored in order to exclude recrystallization during the coating procedure.

13.3 Downstream Processing of Spray-Dried Powders

Spray-dried powders often exhibit relatively small particles representing a large surface area, low bulk density, and often show poor flowability. Therefore, often a pre-compaction step is needed during downstream processing in order to increase the bulk density to a level which later allows tablet compression or capsule filling. At the same time, this reduces the surface area of the material which could have an impact on the drug release rate, especially when appropriate attention is not given to the formulation composition.

The powder characteristic, e.g., flowability, particle size, and bulk density, of the spray-dried material can be improved using an internal fluid bed in the spray dryer. Alternatively, other manufacturing technologies such as fluid bed coating/layering or spray granulation, where the organic solution containing the drug substance and the polymer is sprayed onto a carrier, can be used.

13.3.1 Powder Blends, Dry Granulation and Compression

Effect of Excipients: One issue which is often described in the literature for unformulated or poorly formulated capsule or tablet formulations containing spray-dried powders is the formation of hard plugs that inhibit dissolution. Langham et al. investigated the dissolution behavior of spray-dried amorphous solid dispersions of felodipine and copovidone, and showed that compaction leads to a significant decrease in the rate and extent of dissolution, which is drug-load dependent (Langham et al. 2012). Fakes et al. showed that the dissolution rate for amorphous material relative to the crystalline drug slowed down upon exposure to the aqueous dissolution medium, and that this may be attributed to the initial rapid conversion of the amorphous to crystalline material. The deposition of the crystalline drug on the insoluble excipient MCC, which was used as filler in the formulation, formed a hard plug at the surface of the capsules thus inhibiting dissolution. The addition of fast dissolving or readily dispersible fillers should therefore improve the disintegration/initial dissolution rate, which was demonstrated by Fakes et al. by comparing the effect of lactose and MCC as fillers, clearly showing that the dissolution rate significantly improved for the lactose-containing formulation (Fakes et al. 2009).

The underlying physical processes that resulted in poor dissolution performance of an encapsulated amorphous solid dispersion consisting of amorphous celecoxib, polyvinyl pyrrolidone (PVP), and meglumine was investigated by Puri et al. (2011). They concluded that rapid hydration of the capsule in aqueous media leads to leaching out of meglumine; resulting in decreased ratio of amorphous celecoxib to PVP and the interaction in the solid dispersion causing hydrophobization of PVP. The water-mediated H-bond interlinked in the amorphous solid dispersion promoted interparticle cohesivity and formation of a nondispersible plug (Puri et al. 2011). In order to circumvent undesired interfacial interactions, they proposed surface modification by particle coating, reduction in exposed surface area, and use of high-specific surface area and/or surface-adsorptive excipients as spacers in the formulation blend as effective measures to improve the dissolution behavior. In order to improve the wettability of the spray-dried material, surfactants can be integrated into the spray-dried solid dispersions. This approach needs careful consideration as the presence of surfactants in spray-dried amorphous solid dispersions can significantly affect the compressibility of the material, resulting in decreased tablet strengths, increased elastic deformation, and capping (Roberts et al. 2011).

Effect of Compaction/Compression: The effect of compression on the phase behavior of amorphous solid dispersion was first investigated in detail by Ayenew et al. They showed that compression can result in amorphous–amorphous phase separation in solid dispersions, and that this effect is more pronounced in metastable compositions of solid dispersions (Ayenew et al. 2012a). Further evidence that compression can lead to increased crystallinity upon storage can be found by Leane et al. (2013). They compared the crystallinity of tablets prepared from roller-compacted granules, tablets prepared by direct compression and blends filled into capsules without compression under accelerated stability testing. The degree of crystallinity increased with increasing number of compression steps (Leane et al. 2013).

In this connection, the effect of fillers on the physical stability of the compressed tablets was investigated by several research groups. Leane et al. compared the effect of MCC, mannitol, and lactose, and showed that greater physical stability was observed for formulations containing MCC, which is attributed to the fact that MCC deforms primarily by plastic deformation, whereas lactose and mannitol deform by brittle fracture (Leane et al. 2013). Based on the investigations done by Schmidt et al. (2003), carrageenan has the potential of protecting drugs from polymorphic transformation during tablet compression. Dhumal et al. (2007) also investigated the effect of carrageenan (Gelcarin® GP-379) on the physical stability of amorphous spray-dried dispersions upon compression and storage. Physical stability of the tablets improved when carrageenan was co-precipitated with the solid dispersion in the SD process compared to a physical mixture (PM) with carrageenan or to the solid dispersion alone. This effect may be attributed to the cushioning action provided by carrageenan, which releases mechanical stress by expansion and stores less stress in the tablet. It is assumed that the better stability of the co-precipitate is related to the close proximity of carrageenan, celecoxib, and PVP in the co-precipitate compared to the PM (Dhumal et al. 2007).

13.3.2 Fluid Bed Coating/Layering and Spray Granulation

The low bulk density and the bad flowability associated with spray-dried powders can be circumvented by manufacturing the amorphous solid dispersion either by fluid bed coating/layering or granulation, in particular spray granulation. In both cases an organic solution, consisting of the drug substance, the polymer and potential other excipients, is applied on a filler which could either be a spherical pellet (fluid bed layering) or a conventional excipient. The development of barrier coated drug layered particles is described by Puri et al. After applying a methanolic solution containing celecoxib, PVP and meglumine (solid content: 10 % w/v) onto MCC in a Wurster process, a film coat is applied to the particles in order to avoid formation of an agglomerated capsule-shaped mass upon dissolution, and to improve the physical stability of the system. The authors investigated three different materials for the film coat, namely inulin, polyvinyl alcohol (PVA) and polyvinyl acetate phthalate (PVAP), and finally selected PVA for the accelerated stability study (Puri et al. 2012). Oshima et al. applied an organic solution of ITZ, polysorbate 80 and either hypromellose or hypromellose phthalate onto a powder blend consisting mainly of Ceolus RC® (a colloidal grade of MCC, the surface of which is covered with carmellose sodium) in a fluid bed coating process. The flowability of the obtained granules was optimized by adding 0.1 % of light anhydrous silicic acid as surface modifier. The amount of disintegrant was optimized and 2 % of croscarmellose sodium was finally selected (Oshima et al. 2007). A similar approach was tested by Chowdary and Rao who applied an organic ITZ solution onto three different superdisintegrants, or lactose or MCC and investigated the dissolution behavior (Chowdary and Rao 2000). The solid dispersions in superdisintegrants gave much higher rates of dissolution than the dispersions in other excipients (Ac-Di-Sol > Kollidon CL > Primojel > MCC > lactose).

13.3.3 *Film Coating*

The moisture uptake during an aqueous film coating process can lead to increased level of crystallinity as shown during stability testing. There was no difference observed between a moisture-barrier-containing formulation (Opadry® amb) and an Opadry® II system (Leane et al. 2013). However, reduced moisture uptake and improved physical stability was observed when a hygroscopic, amorphous solid dispersion-containing tablet was coated with hydroxypropyl methylcellulose phthalate (HPMCP) using organic solvents (Reven et al. 2013).

13.4 Downstream Process of MBP

MBP are usually free-flowing powders. MBP exhibit a certain porosity and surface roughness as compared to particles prepared by HME (Dong et al. 2008). A comparison of the specific surface area (BET) showed that the MBP particles had 47 times larger specific surface area, even so the true density was comparable (1.33 and 1.30 g/cm^3 for the MBP and the HME product, respectively) (Dong et al. 2008). In case of vemurafenib, the MBP is described as spongy network having pores in the range of 50–200 nm and some bigger bubbles in the range of 3–10 μm (Shah et al. 2013). The properties of MBP particles are dependent on and can be tracked back to the process parameters used for precipitation. Especially, the amount of API and polymer dissolved in the organic phase as well as the solvent to anti-solvent ratio are critical process parameters in the MBP processes (Shah et al. 2012) and have a direct impact on the porosity, surface roughness, size, and bulk density of the particles.

A comparison of two different capsule formulations containing 40 mg of vemurafenib as MBP, one obtained by dry blending and the other one by a wet-mixing process, showed that the mean values of AUC_{0-inf} (86.2 ± 52.1 μMh and 79.8 ± 42.8 μMh) and the C_{max} were comparable with each other in a single 160 mg dose human bioavailability study. The AUC and C_{max} in each case were greater than that for the crystalline reference formulation (Shah et al. 2013).

13.4.1 Powder Blend, Dry Granulation and Compression

Milling of MBP: Depending on the particle size of the MBP, a milling step might be recommended prior to any downstream processing of the material. Milling could be done using different kinds of mills, like jet mills, pin mills, hammer mills, and so on. The selection of the type of mill as well as the process parameters used will have an impact on the final particle size obtained. The particle size itself will not only have an effect on the dissolution behavior (Shah et al. 2013), but might also effect the entire downstream processing as the flowability of jet-milled material might not be adequate for robust downstream manufacturing later on.

The selection of the downstream processing itself strongly depends on the bulk density of the obtained material. In case of sufficiently high bulk density, a direct compression approach could be used, otherwise it is recommended to increase the bulk density using a roller compaction process.

Effect of Excipients: Depending on the wetting behavior of the MBP particles, which strongly depends on the drug to polymer ratio, the lipophilicity of the compound as well as the hydrophilicity of the polymer used (Shah et al. 2012), additional excipients need to be added during downstream processing. Functional excipients like wetting agents, glidants, fillers, and disintegrants can be added intra- or extragranular to improve the wettability of the MBP, to improve the flowability of the powder blend, as well as to avoid sticking of the material onto the rolls/punches and enable a fast disintegration/dissolution behavior of the granules/tablets. The type and amount of the different excipients need to be carefully adjusted.

Effect of Compaction/Compression: Process parameters like compaction force, gap width, and screen size used for breaking the ribbons can have an impact on the properties of the granules and tablets. Compaction force and gap width should be selected in a way that the granules have a sufficiently high bulk density, but could still be compressed to tablets. In other words, overcompaction should be avoided as this leads to tablets with insufficient hardness.

The screen size used in the granulator unit needs to be selected based on the width of the anticipated tablet size of the lowest dose strengths. The granules should be characterized thoroughly in order to gain a good process understanding. Dissolution tests of the granules provide additional information on the process and help to link the measured attributes of the pure MBP with those of the final tablet. It is strongly recommended to investigate the effect of the roller compaction force, gap width, compression force, and other potential critical process parameters (pCPP) on the tablet properties like hardness, disintegration time, abrasion, dissolution, and so on. The selected process parameters should enable a robust manufacturing of MBP tablets at the end.

13.4.2 *Film Coating*

Depending on the desired product profile, a film coating may be applied to the tablets. The film coating parameters need to be selected carefully in order to avoid an uptake of water by the tablet kernels and exposure of the tablets to high temperatures. Both, water and temperature, might otherwise lead to increased molecular mobility in the amorphous system and the drug substance could recrystallize from the amorphous solid dispersion.

13.5 Downstream Processing of Mesoporous Silica-Based Systems

Mesoporous silica drug delivery systems are characterized by a unique pore structure in which the drug substance is entrapped at a molecular level. Upon contact with liquids, the drug substance is released from the pores at a certain rate, which depends among other factors on the pore size of the mesoporous silica and the degree of loading (Mellaerts et al. 2007). The rate limiting step for drug release seems to be the time needed for diffusion out of the internal pores, which is a function of the silica particle size and pore diameter, as shown by investigations of the drug release of ten physicochemically different drug molecules (Speybroeck et al. 2009). This indicates that the pore structure needs to be maintained during the downstream processing. Critical material attributes of mesoporous silica particles are the low bulk density (below 0.1 g/cm^3), poor compressibility, and flowability (Vialpando et al. 2011), which is a challenge for the development of tablets. Dry granulation or direct compression has primarily been investigated as downstream processing methods, but one reference mentions wet granulation as well.

13.5.1 Powder Blends, Dry Granulation and Direct Compression

Mesoporous Silica Loading Process: Mesoporous silica particles can be loaded with a broad range of different drug substances using different loading methods. Investigations showed that the loading method had an impact on the degree of loading, the degree of residual crystallinity, as well as bulk density (Limnell et al. 2011) therefore affecting the final drug product performance.

Effect of Excipients: The dissolution kinetic can be altered by adding additional excipients as shown by Limnell et al., who observed an increase in the amount of IMC released when mesoporous silica particles were blended with excipients. The increase in the release was attributed to PVP K30 in the excipient blend, which functioned as a precipitation inhibitor (Limnell et al. 2011). A systematic investigation on the addition of precipitation inhibitors was done by Speybroeck et al. (2010) by evaluating the in vitro and in vivo behavior of formulations consisting of ordered mesoporous silica SBA-15 loaded with ITZ and HMPC or HPMCAS, respectively. Due to the pH dependent solubility, HPMCAS was not able to prevent precipitation of ITZ in vitro at low pH and even upon transfer to FaSSIF, where rapid precipitation of ITZ occurred despite minimal or no HPMCAS being dissolved. Contrary to HPMCAS, HPMC was able to maintain the supersaturation in vitro and led to more than 60 % increase in absorption compared to ITZ-loaded SBA-15 particles in a rat pharmacokinetic (PK) study. The PM of the ITZ-loaded SBA-15 particles and HPMC (1:4:6) achieved 88 % of the AUC relative to Sporanox®, which was used as reference in this study (Speybroeck et al. 2010). Similar bioavailability of ITZ-loaded SBA-15 and Sporanox® was obtained in a PK study in rabbits and dogs. In this study, the loaded mesoporous silica system (49 %) was blended with croscarmellose (25 %), lactose (25 %), and sodium lauryl sulphate (SLS; 1 %) ensuring fast disintegration and good dispersion of the loaded ordered mesoporous silica system (Mellaerts et al. 2008).

The effect of adding disintegrants, low-hydroxypropylcellulose (L-HPC) or pregelatinized starch (PCS), to the tablet formulation was investigated by Takeuchi et al. The dissolution rate of IMC from the tablets was significantly improved and similar to the solid dispersion particles itself when an L-HPC was present in the formulation. Formulations containing PCS were also able to improve the dissolution and tableting properties, but the dissolution rate of IMC slightly decreased and the compaction property was slightly lower than that in the case of L-HPC (Takeuchi et al. 2005).

Ratio Between Drug-loaded Silica Particles and Excipients: Besides the influence of certain excipients on the drug release, several authors investigated the effect of the ratio between drug-loaded silica particles and excipients in the formulation. Limnell et al. (2011) used 25 % of IMC-loaded MCM-41 in their tablet formulations and obtained tablets with a fast release. Tahvanainen et al. further increased the amount of drug-loaded silica particles to 25, 30, and 35 % using IMC-loaded thermally oxidized mesoporous silicon microparticles (TOPSi-IMC), and observed a decrease in dissolution rate and permeability as a result of loss of unique pore structure due to deformation of the particles under compression (Tahvanainen et al. 2012).

Effect of Compression: Increasing the compression force applied to the PM of TOPSi-IMC and excipients leads to a decrease in the release of IMC (Tahvanainen et al. 2012). Further investigations linking the effect of compression force on the drug release were conducted (Limnell et al. 2011; Vialpando et al. 2011; Kiekens et al. 2012). Limnell et al. observed a slight decrease in the amount of drug released from tablets compared to capsules containing IMC-loaded MCM-41 particles. Nevertheless, the tablets retained their ability for fast release of IMC as no major alteration in the porous structures of the particles after tablet compression was observed (Limnell et al. 2011). Vialpando et al. investigated the effect of compression force on ITZ-loaded ordered mesoporous silica (SBA-15 and COK-12), and observed a decrease in the amount of drug released with increasing pressure. This was related to a reduction in the pore size and volume. A comparison of both silica materials showed that SBA-15 is more sensitive towards compression than COK-12. This was related to the slightly thicker walls and higher condensation degree of the silica framework of COK-12 (Vialpando et al. 2011). The addition of plastic deforming materials, such as MCC, was helpful in protecting the silica and improving the release rate following compression as shown by Vialpando et al. who added 30, 50, or 70 % of MCC to the drug-loaded SBA-15 and COK-12, compressed tablets at 120 MPa, and investigated their release profile. The dissolution profile of the tablets clearly improved, but was still slower in comparison to the PM (approx. 80 % of ITZ after 60 min). Addition of 4.8–5.1 % croscarmellose sodium further enhanced the drug release following compression (Vialpando et al. 2011). The effect of compression was also investigated by Kiekens et al. comparing a 5 and 10 mg tablets and a 5 mg capsule containing ezetimibe-loaded ordered mesoporous silica (OMS) with a 10 mg Ezetrol tablet (reference formulation) in vitro and in vivo. In vitro, both OMS tablets showed comparable, but improved dissolution behavior compared to the reference. This was not reflected in vivo (PK study in Beagle dogs), where the area under the curve (AUC) for the OMS tablets and Ezetrol tablet was comparable, and more than two times increase in AUC was observed for the 5 mg OMS capsule compared to the reference. As the 5 mg capsule and tablet had an identical composition, this result showed reduced bioavailability due to compression (Kiekens et al. 2012).

13.5.2 Wet Granulation as Alternative Granulation Technique

Wet granulation was investigated as an alternative downstream process for ordered mesoporous silica by Vialpando et al. (2012). COK-12 was used as model ordered mesoporous silica due to its thicker walls and higher degree of silica condensation, which results in higher resistance towards compression. The authors successfully demonstrated that wet granulation can improve powder flow and compactibility by increasing the particle size, bulk density, and smoothing of the surface of the

ITZ-loaded COK-12 particles. In order to achieve this, process parameters such as binder concentration, binder addition rate, and granulation temperature need to be carefully selected avoiding overwetting of the material and therefore premature drug release on one hand, and ensuring agglomeration on the other hand. The decrease in the release profile upon compression was compensated by extragranular addition of croscarmellose sodium (2.4 %). Overall, the amount of "drug-loaded silica particles" in the tablet could be increased by wet granulation compared to the dry processes. In addition to the investigations of process parameters, the application of the wet granulation process to COK-12-loaded silica particles loaded with ITZ, fenofibrate, naproxen, or ibuprofen revealed that the risk of premature drug release during wet granulation is primarily compound dependent (Vialpando et al. 2012).

13.5.3 Modification of the Release Profile

The drug release from tablets containing drug-loaded mesoporous silica systems can be modified by applying a functional coat on top of the tablets. A pH-dependent drug release systems for intestinal drug release were achieved by using Eudragit S100 (Xu et al. 2011) or HPMCP (Xu et al. 2009). The concentration in the coating solution, the coating thickness, and the drying temperature had an effect on the amount of drug released at pH 1.2, whereas the drug release kinetic at pH 7.4 was unchanged (Xu et al. 2009, 2011).

Modification of the release rate of the loaded mesoporous silica can for instance be achieved by surface functionalization of the mesoporous silica (Song et al. 2005). This can be achieved either by applying polyelectrolyte multilayer coatings (Zhu et al. 2005), or by ionic interaction of oppositely charged polycations and anionic SBA-15 (Yang et al. 2005) or by anchoring suitable polyamines on the external surface to obtain a pH and anion-controlled nano-supramolecular gate-like ensemble (Bernardos et al. 2008).

13.6 Summary and Conclusion

The usual downstream processing of amorphous solid dispersion involves generation of granules either directly through milling or via granulation. This is followed by blending, capsule filling or tablet compression, and film coating. Due to the fact that solid dispersions contain substantial amounts of a stabilizing polymer, the properties of the polymer will have an impact on disintegration behavior as well as compactibility. As a consequence, one important aspect of formulation development is the selection of suitable excipients especially fillers and disintegrants. Especially, the selection of fillers is quite controversial in the literature. Fast dissolving fillers like lactose or mannitol are preferred from a disintegration and dissolution perspective, but these can induce brittle fracture, whereas plastic deformable fillers like MCC or

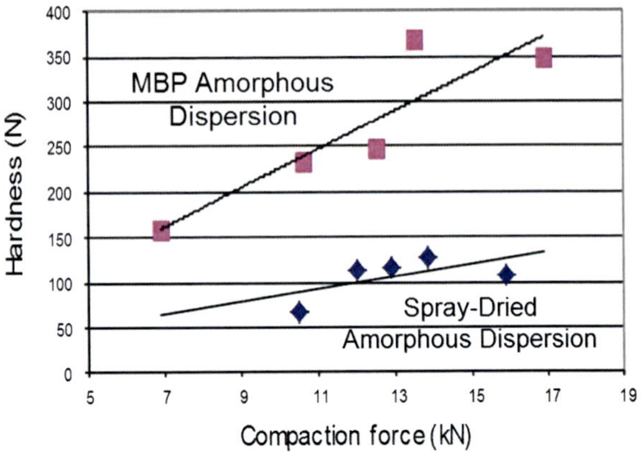

Fig. 13.2 Hardness–compression force profiles of MBP and spray-dried amorphous dispersions of vemurafenib

carrageen improve the physical stability of the system by protecting the amorphous solid dispersion during roller compaction/ tablet compression.

The value of proper compaction characterization, i.e., assessment of the tensile strength, compression pressure, solid fraction relationships leading to the compactability, tabletability, and compressibility (CTC) profiles, provides basic mechanical property information (Tye et al. 2005). Along with tensile strength, compression pressure and solid fraction, the elastic modulus, permanent deformation pressure, and brittleness of compacts are additional important properties (Hiestand and Smith 1984a, b) pharmaceutical scientists used to quantify the mechanical nature of materials. For example, mechanical properties of a compact are very much influenced by solid fraction, and even a change of 0.01 can influence mechanical property as much as 10–20 % (Amidon et al. 2009). These properties, therefore, are of significant interest in supporting tablet development in a scientific manner.

Hancock et al. reported (Hancock et al. 2002) a dynamic indentation hardness, 30 % higher for the amorphous form of a drug (178.4 MPa) compared to a crystalline form of a drug (230.3 MPa). Aggregation and fusion of compacted amorphous particles could form a harder surface that resists indentation, compared to crystalline particles. However, even among amorphous dispersions of similar composition, the techniques used to prepare these dispersions can impact the deformation attributes of the resulting product. At similar compression force levels, tablets of a given composition prepared by co-precipitation (containing amorphous MBP) had a significantly higher hardness than that of similar composition prepared by spray drying, as seen from Fig. 13.2.

In addition, tensile strength of compacts of amorphous solid dispersions is impacted by the level of polymers, the level of inclusion in the dispersion, and particularly dependent on the processing technique. Amorphous solid dispersion generated by HME process is quite different than amorphous solid dispersions prepared by other techniques. Powders processed by HME are subjected to elevated processing temperatures as well as high pressure. The reduced free volume retards molecular mobility and prevents further densification during tableting (Zhu et al. 2002; Young et al. 2005), which could adversely impact the product performance

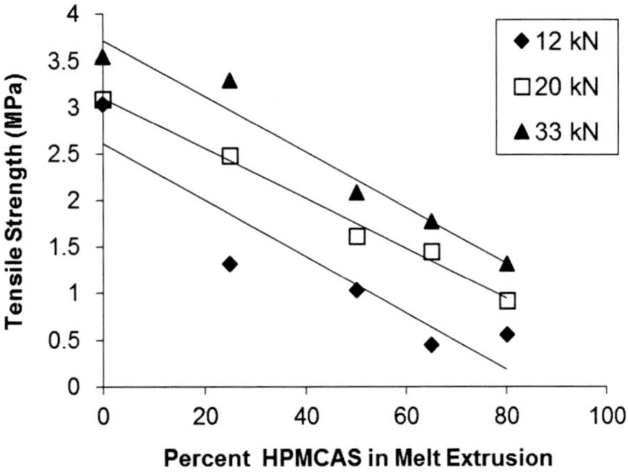

Fig. 13.3 Effect of HPMCAS level in melt-extruded solid dispersion on tablet tensile strength

such as drug dissolution. For example, solid dispersions containing HPMCAS prepared by melt extrusion exhibit decreasing tablet tensile strength with increasing level of HPMCAS, as shown in Fig. 13.3. The tensile strength was observed to decrease three- to fourfold when HPMCAS level was present at 80 % in the dispersion.

In addition, material properties such as dynamic hardness and tensile strength of amorphous solid dispersions are also governed by polymers and technologies used to prepare such dispersions (Iyer et al. 2013). The dynamic hardness of melt-extruded HPMCAS was greater than that of "as is" materials. However, both spray-dried HPMCAS and melt-extruded copovidone did not exhibit a significant change in dynamic hardness from that of "as is" materials, respectively, as seen from Fig. 13.4. The tensile strength of melt-extruded HPMCAS and copovidone decreased significantly compared to their respective native materials, as seen from Fig. 13.5 indicating that tablet development of melt-extruded solid dispersions could be challenging.

During each step of the downstreaming process, energy is applied to the system which can lead to amorphous–amorphous phase separation and can trigger recrystallization. In addition, a reduction in surface area occurs which might affect the dissolution behavior when the formulation is not adequately designed. Therefore, it is recommended to characterize each intermediate product as well as the final drug product. Figure 13.6 presents a schematic depiction of downstream process together with proposed analytical tests after each step as well as the critical quality attributes and critical process parameters. Especially the dissolution behavior and physical stability should be tested not only for the final drug product but also at the immediate level allowing identification of potential adverse effects earlier in the development chain. Several analytical methods, or combinations thereof, can help to gain a mechanistic insight into the dissolution behavior of tablets containing amorphous solid dispersions. Langham et al. combined the ultraviolet (UV) absorbance measurement of the re-circulating dissolution media from a flow cell with simultaneous acquisition of magnetic resonance images. The MR images showed the fundamental difference in the dissolution behavior of the investigated solid dispersions, and could thus explain the difference observed in the drug release (Langham et al. 2012).

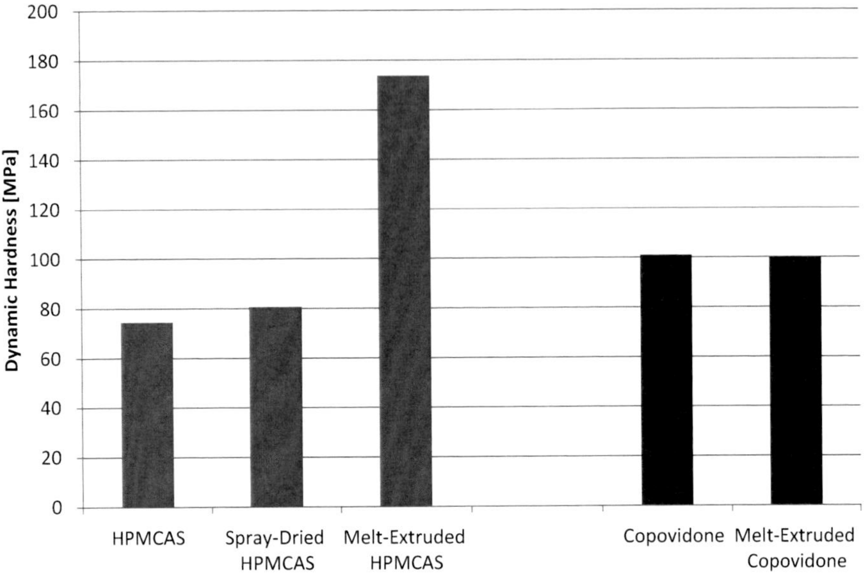

Fig. 13.4 Dynamic hardness of "as is" and processed HPMCAS and copovidone

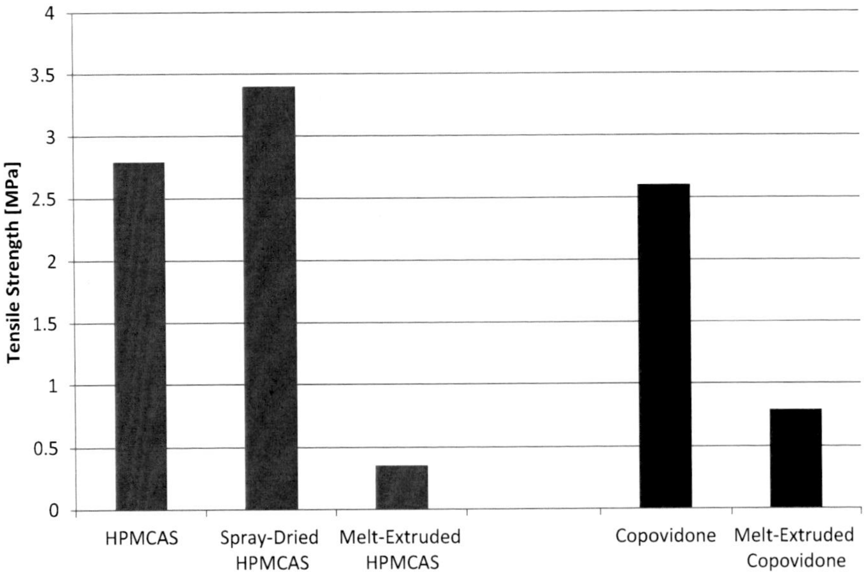

Fig. 13.5 Tensile strength of "as is" and processed HMPCAS and copovidone

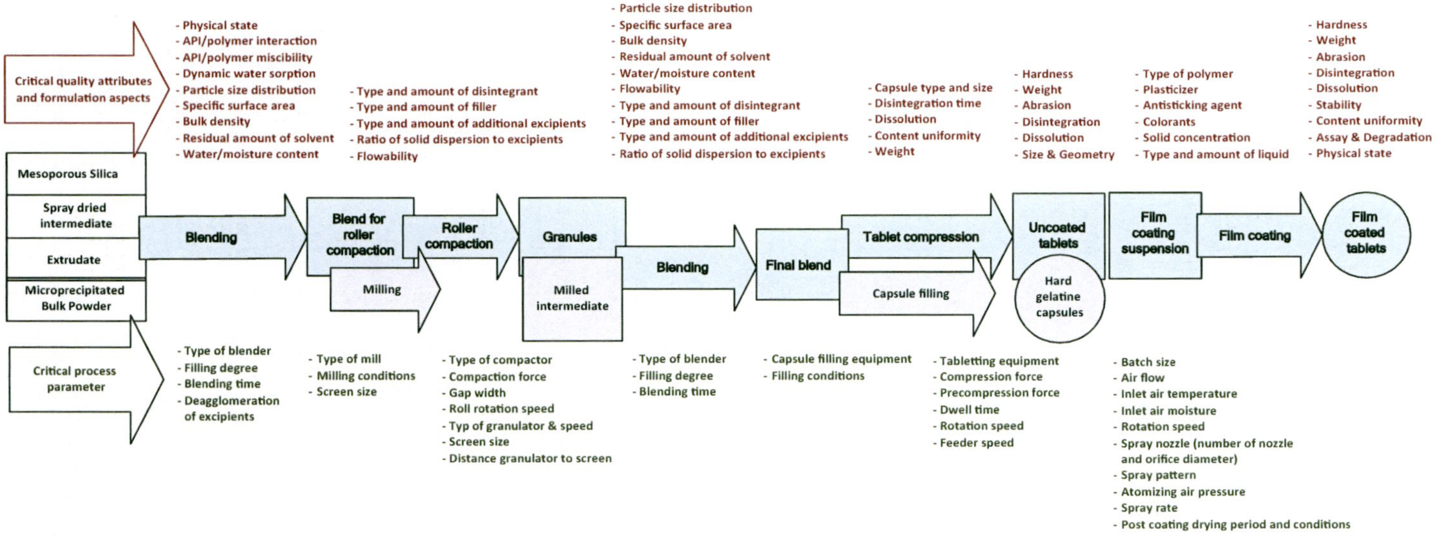

Fig. 13.6 Flow chart: amorphous solid dispersion downstream processing

In addition to the dissolution testing, analytical methods need to be established and validated in order to detect low amounts of crystalline material in the amorphous solid dispersion and/or the final drug product. Xie et al. described the development of such a method, and established a reliable multivariate curve resolution (MCR) method based on the second derivative Raman measurements for quantitative determination of the solid state forms of the drug substance and tablets (Xie et al. 2008).

Amorphous solid dispersions can be successfully transformed into an appropriate solid form like a tablet or a hard gelatin capsule by judiciously selecting the processing technologies. Critical process parameters must be identified and excipients carefully selected in order to obtain a final drug product with the desired quality attributes. Each of these factors plays a critical role in developing a successful commercial product from a solid amorphous dispersion.

References

Ahlneck C, Zografi G (1990) The molecular basis of moisture effects on the physical and chemical stability of drugs in the solid state. Int J Pharm 62:87–95

Alonzo DE, Gao Y et al (2011) Dissolution and precipitation behavior of amorphous solid dispersions. J Pharm Sci 100(8):3316–3331

Amidon GE, Secreast PJ et al (2009) Particle, powder, and compact characterization. In: Qiu Y, Chen Y, Zhang GGZ, Liu L, Porter WR (eds) Developing solid oral dosage forms: pharmaceutical theory and practice. Elsevier, New York, pp 176–173

Ayenew Z, Paudel A et al (2012a) Can compression induce demixing in amorphous solid dispersions? A case study of naproxen-PVP K25. Eur J Pharm Biopharm 81:207–213

Ayenew Z, Paudel A et al (2012b) Effect of compression on non-isothermal crystallization behaviour of amorphous indomethacin. Pharm Res 29:2489–2498

Bernardos A, Aznar E et al (2008) Controlled release of vitamin B2 using mesoporous materials functionalized with amine-bearing gate like scaffoldings. J Control Release 131:181–189

Bruce C, Fegely KA et al (2007) Crystal growth formation in melt extrudates. Int J Pharm 341:162–172

Chowdary KPR, Rao SS (2000) Investigation of dissolution enhancement of itraconazole by solid dispersion in superdisintegrants. Drug Dev Ind Pharm 26(11):1207–1211

Colombo I, Grassi G et al (2009) Drug mechanochemical activation. J Pharm Sci 98(11):3961–3986

Deng W, Majumdar S et al (2013) Stabilization of fenofibrate in low molecular weight hydroxypropylcellulose matrices produced by hot-melt extrusion. Drug Dev Ind Pharm 39(2):290–298

Descamps N, Palzer S (2007) Modeling of sintering of water soluble amorphous particles. PARTEC, Nürnberg

Dhumal R et al (2007) Development of spray-dried co-precipitate of amorphous celecoxib containing storage and compression stabilizers. Acta Pharm 57(3):287–300

Dierickx L, Saerens L et al (2012) Co-extrusion as manufacturing technique for fixed-dose combination mini-matrices. Eur J Pharm Biopharm 81:683–689

DiNunzio JC, Schilling SU et al (2012) Use of highly compressible Ceolus™ microcrystalline cellulose for improved dosage form properties containing a hydrophilic solid dispersion. Drug Dev Ind Pharm 38(2):180–189

Dokoumetzidis A, Macheras P (2006) A century of dissolution research: from Noyes and Whitney to the biopharmaceutics classification system. Int J Pharm 321:1–11

Dong Z, Choi DS (2008) Hydroxypropyl methylcellulose acetate succinate: potential drug–excipient incompatibility. AAPS Pharm Sci Tech 9(3):991–997

Dong Z, Chatterji A et al (2008) Evaluation of solid state properties of solid dispersions prepared by hot-melt extrusion and solvent co-precipitation. Int J Pharm 355(1–2):141–149

Fakes MG, Vakkalagadda BJ et al (2009) Enhancement of oral bioavailability of an HIV-attachment inhibitor by nanosizing and amorphous formulation approaches. Int J Pharm 370:167–174

Feng J, Xu L et al (2012) Evaluation of polymer carriers with regard to the bioavailability enhancement of bifendate solid dispersions prepared by hot-melt extrusion. Drug Dev Ind Pharm 38(6):735–743

Ford JL, Elliott PNC (1985) The effect of particle size on some in-vitro and in-vivo properties of indomethacin-polyethylene glycol 6000 solid dispersions. Drug Dev Ind Pharm 11(263):537–549

Goddeeris C, Willems T et al (2008) Dissolution enhancement of the anti-HIV drug UC 781 by formulation in a ternary solid dispersion with TPGS 1000 and Eudragit E100. Eur J Pharm Biopharm 70:861–868

Hancock BC, Carlson GT et al (2002) Comparison of the mechanical properties of the crystalline and amorphous forms of a drug substance. Int J Pharm 241:73–85

Hiestand E, Smith D (1984a) Indices of tableting performance. Powder Technol 38:145–159

Hiestand E, Smith D (1984b) Three indices for characterizing the tableting performance of materials. Adv Ceramic 9:47–57

Hughey JR, Keen JM et al (2013) The use of inorganic salts to improve the dissolution characteristics of tablets containing Soluplus-based solid dispersions. Eur J Pharm Sci 48:758–766

Iyer R, Hegde S et al (2013) The impact of hot melt extrusion and spray drying on mechanical properties and tableting indices of materials used in pharmaceutical development. J Pharm Sci 102:3604–3613

Jijun F, Lili Z et al (2010) Stable nimodipine tablets with high bioavailability containing NM-solid dispersions prepared by hot-melt extrusion. Powder Technol 204:214–221

Jijun F, Lishuang X et al (2011) Nimodipine (NM) tablets with high dissolution containing NM solid dispersions prepared by hot-melt extrusion. Drug Dev Ind Pharm 37(8):934–944

Kiekens F, Eelen S et al (2012) Use of ordered mesoporous silica to enhance the oral bioavailability of ezetimibe in dogs. J Pharm Sci 101(3):1136–1144

Kindermann C, Matthée K et al (2012) Electrolyte-stimulated biphasic dissolution profile and stability enhancement for tablets containing drug-polyelectrolyte complexes. Pharm Res 29:2710–2721

Langham ZA, Booth J et al (2012) Mechanistic insights into the dissolution of spray-dried amorphous solid dispersions. J Pharm Sci 101(8):2798–2810

Leane MM, Sinclair W et al (2013) Formulation and process design for a solid dosage form containing a spray-dried amorphous dispersion of ibipinabant. Pharm Dev Technol 18(2):359–366

Lepek P, Sawicki W et al (2013) Effect of amorphization method on telmisartan solubility and the tabletting process. Eur J Pharm Biopharm 83:114–121

Limnell T, Santos HA et al (2011) Drug delivery formulations of ordered and nonordered mesoporous silica: comparison of three drug loading methods. J Pharm Sci 100(8):3294–3306

Matteucci ME, Brettmann BK et al (2007) Design of potent amorphous drug nanoparticles for rapid generation of highly supersaturated media. Mol Pharm 4(5):782–793

Mazur S, Beckerbauer R et al (1997) Particle size limits for sintering polymer colloids without viscous flow. Langmuir 13(16):4287–4294

Mellaerts R, Aerts CA et al (2007) Enhanced release of itraconazole from ordered mesoporous SBA-15 silica materials. Chem Comm 13:1375–1377

Mellaerts R, Mols R et al (2008) Increasing the oral bioavailability of the poorly water soluble drug itraconazole with ordred mesoporous silica. Eur J Pharm Biopharm 69:223–230

Oshima T, Sonoda R et al (2007) Preparation of rapidly disintegrating tablets containing itraconazole solid dispersions. Chem Pharm Bulletin 55(11):1557–1562

Palzer S (2011) Agglomeration of pharmaceutical, detergent, chemical and food powders—similarities and differences of materials and processes. Powder Technol 206(1–2):2–17

Park J, Park HJ et al (2010) Preparation and pharmaceutical characterization of amorphous cefdinir using spray-drying and SAS-process. Int J Pharm 396:239–245

Puri V, Dantuluri AK et al (2011) Investigation of atypical dissolution behavior of an encapsulated amorphous solid dispersion. J Pharm Sci 100(6):2460–2468
Puri V, Dantuluri AK et al (2012) Barrier coated drug layered particles for enhanced performance of amorphous solid dispersion dosage form. J Pharm Sci 101(1):342–353
Read MD, Coppens KA et al (2010) Hot melt extrusion technology for the manufacture of poorly soluble drugs with controlled release dissolution profiles. ANTEC
Reven S, Homar M et al (2013) Preparation and characterization of tablet formulation based on solid dispersion of glimepiride and poly(ester amide) hyperbranched polymer. Pharm Dev Technol 18(2):323–332
Roberts M, Ehtezazi T et al (2011) The effect of spray drying on the compaction properties of hypromellose acetate succinate. Drug Dev Ind Pharm 37(3):268–273
Rosenzweig N, Narkis M (1983) Newtonian sintering simulator of two spherical particles. Polym Eng Sci 23(1):32–35
Schilling SU, Bruce CD et al (2008) Citric acid monohydrate as a release-modifying agent in melt extruded matrix tablets. Int J Pharm 361(1–2):158–168
Schmidt AG, Wartewig S et al (2003) Potential of carrageenans to protect drugs from polymorphic transformation. Eur J Pharm Biopharm 56(1):101–110
Shah N, Sandhu H et al (2012) Development of novel microprecipitated bulk powder (MBP) technology for manufacturing stable amorphous formulations of poorly soluble drugs. Int J Pharm 438:53–60
Shah N, Iyer RM et al (2013) Improved human bioavailability of vemurafenib, a practically insoluble drug, using an amorphous polymer-stabilized solid dispersion prepared by a solvent-controlled coprecipitation process. J Pharm Sci 102(3):967–981
Song S-W, Hidajat K et al (2005) Functionalized SBA-15 materials as carriers for controlled drug delivery: influence of surface properties on matrix-drug interactions. Langmuir 21:9568–9575
Speybroeck MV, Barillaro V et al (2009) Ordered mesoporous silica material SBA-15: a broad-spectrum formulation platform for poorly soluble drugs. J Pharm Sci 98(8):2648–2658
Speybroeck MV, Mols R et al (2010) Combined use of ordered mesoporous silica and precipitation inhibitors for improved oral absorption of the poorly soluble weak base itraconazole. Eur J Pharm Biopharm 75:354–365
Tahvanainen M, Rotko T et al (2012) Tablet preformulations of indomethacin-loaded mesoporous silicon microparticles. Int J Pharm 422:125–131
Takeuchi H, Nagira S et al (2005) Tabletting of solid dispersion particles consisting of indomethacin and porous silica particles. Chem Pharm Bull 53(5):487–491
Tye CK, Sun C et al (2005) Evaluation of the effects of tableting speed on the relationships between compaction pressure, tablet tensile strength, and tablet solid fraction. J Pharm Sci 94(3):465–472
Vialpando M, Aerts A et al (2011) Evaluation of ordered mesoporous silica as a carrier for poorly soluble drugs: influence of pressure on the structure and drug release. J Pharm Sci 100(8):3411–3420
Vialpando M, Backhuijs F et al (2012) Risk assessment of premature drug release during wet granulation of ordered mesoporous silica loaded with poorly soluble compounds itraconazole, fenofibrate, naproxen, and ibuprofen. Eur J Pharm Biopharm 81:190–198
Xie Y, Tao W et al (2008) Quantitative determination of solid-state forms of a pharmaceutical development compound in drug substance and tablets. Int J Pharm 362:29–36
Xu W, Gao Q et al (2009) pH-Controlled drug release from mesoporous silica tablets coated with hydroxypropyl methylcellulose phthalate. Mater Res Bull 44:606–612
Xu Y, Qu F et al (2011) Construction of a novel pH-sensitive drug release system from mesoporous silica tablets coated with Eudragit. Solid State Sci 13:641–646
Yang Q, Wang S et al (2005) pH-Responsive carrier system based on carboxylic acid modified mesoporous silica and polyelectrolyte for drug delivery. Chem Mater 17:5999–6003
Young CR, Dietzsch C et al (2005) Compression of controlled-release pellets produced by a hot-melt extrusion and spheronization process. Pharm Dev Technol 10(1):133–139

Zhu Y, Shah NH et al (2002) Influence of thermal processing on the properties of chlorpheniramine maleate tablets containing an acrylic polymer. Pharm Dev Technol 7(4):481–489

Zhu Y, Shi J et al (2005) Stimuli-responsive controlled drug release from a hollow mesoporous silica sphere/polyelectrolyte multilayer core-shell structure. Angew Chem Int Ed 44:5083–5087

Zhu Y, Shah NH et al (2006) Controlled release of a poorly water-soluble drug from hot-melt extrudates containing acrylic polymers. Drug Dev Ind Pharm 32:569–583

Part III
Characterization

Chapter 14
Structural Characterization of Amorphous Solid Dispersions

Amrit Paudel, Joke Meeus and Guy Van den Mooter

14.1 Introduction

Amorphous solid dispersions (ASD) consist of active pharmaceutical ingredient (API) molecules dispersed in stabilizing carrier(s) which are mostly amorphous polymers along with some functional excipients and are powder, extrudates, thin films, porous foams, surface-coated beads, etc. (Paudel et al. 2013). Hygroscopicity of amorphous materials as well as the carriers increase the analytical complexity (Palermo et al. 2012a, b). Additional excipients further complicate the characterization of amorphous systems in the finished dosage form. The comprehensive qualitative/quantitative characterization of molecular mobility, miscibility, phase separation, domain size, crystallinity, surface chemistry, moisture/solvent, molecular interactions in ASD requires a gamut of analytical techniques. Calorimetric techniques (differential scanning calorimetry (DSC), isothermal microcalorimetry (IMC), and localized thermal analysis) are common for the analysis of ASD (Baird and Taylor 2012). Dielectric spectroscopy and thermomechanical techniques are also increasingly used for the ASD analysis. Infrared, Raman, and solid-state nuclear magnetic resonance spectroscopy (SS-NMR) analyze the molecular interactions among the components of ASD and structural changes during phase separation/crystallization and quantify crystallinity (Vogt and Williams 2010). Polarized light microscopy, scanning/transmission electron microscopy, and atomic force microscopy (AFM) probe the morphological characteristics, spatial phase distribution, and crystallinity. Powder X-ray diffraction is selective for detecting/quantifying the crystallinity (Vogt and Williams 2010). X-ray photoelectron scattering (Dong and Boyd 2011), inverse gas chromatography (IGC) and time-of-flight secondary ion mass spectrometry (Ho and Heng 2013; Barnes et al. 2011) are highly sensitive

G. Van den Mooter (✉) · A. Paudel · J. Meeus
Drug Delivery and Disposition, University of Leuven, Leuven, Belgium
e-mail: Guy.Vandenmooter@pharm.kuleuven.be

A. Paudel
Research Center Pharmaceutical Engineering GmbH (RCPE), Graz, Austria

N. Shah et al. (eds.), *Amorphous Solid Dispersions*,
Advances in Delivery Science and Technology, DOI 10.1007/978-1-4939-1598-9_14

and/or selective to ASD surface analysis. Gravimetric vapor sorption (GVS) probes the hygroscopicity, crystallinity/crystallization and drug–polymer interactions (Burnett et al. 2009). Thermogravimetric analysis measures the moisture and/or volatile content in ASD. The use of multiple simultaneous measurement tools for the integrated information with a spatiotemporal resolution is increasing. This chapter focuses on the thermal, diffractometric, and moisture sorption analysis of amorphous pharmaceuticals.

14.2 Differential Scanning Calorimetry (DSC) Studies of ASD

Non-isothermal DSC involves the controlled heating and/or cooling of a material in a DSC sample holder (pan) along with a reference (usually an empty pan) inside a furnace supplied with a constant flow of inert gas and a cooling system. In a heat flux setup, the temperature difference between the sample and the empty reference pan placed inside the same furnace is measured by separate thermocouples as a function of temperature. In contrast, the instrumental output is the electrical power difference between the sample and the reference pan housed inside the isolated furnace in case of a power-compensation DSC. The heat flow evolved from or transferred in the sample is derived from the measured temperature or power difference as the end response. DSC also enables measurement of isothermal crystallization kinetics. Various endothermic events/transitions such as glass transition, melting, desolvation, enthalpy recovery and some degradation reactions absorb heat while the exothermic processes viz., crystallization, crystal perfection, and some thermal decomposition liberate heat from the sample. A DSC thermogram obtained by heating or cooling at linear rate includes the heat-capacity (C_P)-related transitions and kinetic events. Thus, the total heat flow (THF) signal recorded by DSC (*dQ*/*dt*) can be presented as:

$$\frac{dQ}{dt} = C_p.\beta + f(t, T), \tag{14.1}$$

where *t*, *T*, and β are the time, temperature, and heating rate, respectively. The first term in the right hand side is the C_p-related heat flow and the second the kinetic heat flow. The multiple transitions occurring concomitantly in DSC traces pose challenges in qualitative and quantitative interpretation. One way of improving this is superimposing a nonlinear heating/cooling program on the linear temperature program, prevalently called modulated temperature DSC (mDSC). A single frequency sinusoidal oscillation is the most used nonlinear heating in mDSC setup for pharmaceutical solid dispersions. The expression for *dQ*/dt, in case of mDSC, can be rewritten as:

$$\frac{dQ}{dt} = C_P.[\beta + A_T.\omega.\cos(\omega.t)] + f'(t, T) + A_K.\sin(\omega.t) \tag{14.2}$$

where A_T and ω($= 2\pi/\text{modulation period } (p)$) are respectively the amplitude and the angular frequency of temperature modulation and β is the linear heating rate.

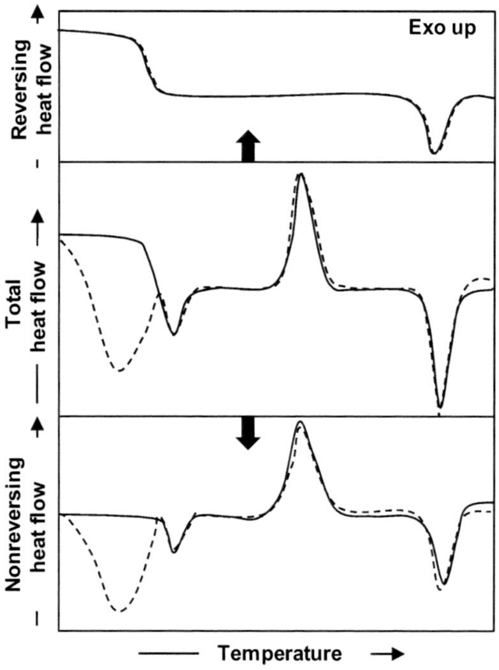

Fig. 14.1 Characteristic mDSC thermograms of amorphous solids undergoing non-isothermal crystallization (*solid line*) and the same of the amorphous solid containing moisture (*dashed line*)

Here, $f'(t, T)$ represents the kinetic component without temperature modulation while A_K is the amplitude of kinetic response to the temperature modulation. In this way, the measured frequency-dependent heat capacity in addition to the total heat capacity makes it possible to deconvolute the THF signal into the reversing (C_p related) and nonreversing (kinetic) heat flow components. Reversing heat flow (RHF) usually comprises all transitions that are thermodynamically reversible at the temperature and the time they are measured, e.g., glass transition and melting. In contrast, nonreversing heat flow (NRHF) which is obtained by subtraction of RHF from THF, consists of transitions nonreversible at the temperature and time of the measurement such as enthalpic recovery, cold crystallization, evaporation, desolvation, thermal decomposition, curing, etc. A typical mDSC thermogram for an amorphous material undergoing non-isothermal crystallization is depicted in Fig. 14.1. The dashed line represents an amorphous material containing some moisture and/or volatile solvent. The first event exhibiting the step jump in RHF is the glass transition of the material. The accompanying endothermic overshoot from the baseline in THF is originated from the overlapping enthalpy recovery signal and additional broad endotherm is from desolvation, both apparent in NRHF. The exothermic peak is of the cold crystallization and the final endotherm is of melting.

Various experimental conditions crucially affect the resultant heat flow measured by DSC, e.g., sample size, pan and purge gas type, sample–pan contact, the heating/cooling rate. The increase in the scan rate proportionally improves the sensitivity while inversely affect the resolution. Many times, fast heating or cooling rates are preferred to kinetically enhance the desirable event in DSC. Heating or cooling at a rate that is considerably faster than the time scale of the process of interest, using fast scan DSC, hyper DSC or flash DSC, can be helpful to obtain reliable calorimetric data about the initial material structure (Ford and Mann 2012). Fast cooling can be advantageous for in situ amorphization for rapidly crystallizing materials. Ideally, the fast heating enhances the weak glass transition signals or for detecting the glass transition of thermally unstable materials which start degrading prior to its T_g. For high scan rate, the sample size and thickness need to be extremely small and thermal contact should be excellent to avoid thermal gradient and lag (Zhuravlev and Schick 2010). The mDSC signals rely on the combinations of β, A_T, and ω (Santovea et al. 2010). For example, heat-iso amplitude ($\beta = A_T.\omega$) is suitable for studying melting/crystallization. For amorphous blends, where various transitions are expected heating–cooling mode ($\beta < A_T.\omega$) is more preferred. Great caution is needed while choosing the combinations of β, A_T, and p so that the applied program can be followed by the sample. At least four to six cycles across each transition ensure efficient signal deconvolution. The "p" is selected according to the C_P of a material and the width of transitions such that the lower the C_P, the lower is the period required while the wider the transition, the higher is the period required. Higher A_T generally increases sensitivity and decreases resolution. The plot of modulated heat flow as a function of temperature, propagating as a smooth regular sine wave indicates the optimal combination of mDSC parameters, while the distorted profile suggests the improper modulation parameters. Lissajous plot, the modulated heat flow versus the modulated heating rate, can diagnose the stability of modulation condition, the distortion indicating the uncontrolled condition. The width and the slope of a Lissajous coil represent the phase lag and the C_P, respectively. The amplitude of the periodic function changes across the C_P -related transition retracing the eccentricity of ellipse.

14.2.1 Glass Fragility, Molecular Mobility, and Enthalpy Recovery

The heat capacity difference measured between glass and supercooled liquid (ΔC_P), the position as well as the shape of T_g describe the dynamics of amorphous systems. In general, amorphous polymers behave as strong glasses that exhibit quasi-Arrhenius behavior of viscosity and structural relaxation time often yieldinga broad glass transition (larger T_g width; ΔT_g) with relatively small ΔC_P. The ratio of T_g to melting temperature (T_m) exceeds significantly to 2/3. On the other hand, most of small molecular weight APIs are fragile glasses that significantly deviate from the Arrhenius path in the T_g region, thus show a sharp T_g and T_g/T_m often below 2/3. The ΔT_g measured during heating or cooling in DSC can thus be correlated to the apparent activation energy for molecular motion at T_g (Moynihan 1993). Hancock et al. 1998 tested this relationship on various sugars, indomethacin, and polyvinylpyrrolidone (PVP) with different molecular weights. Kawakami (2011) estimated the size of the cooperatively rearranging region of glassy ribavirin as a function of sub-T_g annealing time using the ΔT_g.

The fragility (m), the extent of deviation from Arrhenius behavior, of a material can be estimated from a series of T_g's obtained under different heating/cooling rates. An empirical relation of m with DSC data ($=56 \times T_g \Delta C_P / \Delta H_m$) was proposed by Wang et al. (2002) for many glasses. The derived parameters from DSC data correlate with the crystallization tendency of some amorphous APIs (Kawakami et al. 2012). For indomethacin/PVP ASD, it was reported that the fragility parameters deducted from DSC data considerably differ for the dispersions prepared from different routes viz., the values of "m" were in the order of melt quenching > spray drying > ball milling (Ke et al. 2012). The configurational entropy was unable to predict nonequilibrium glass stability (Graeser et al. 2009).

DSC is the most used technique to study the molecular mobility of amorphous pharmaceuticals at and below T_g. The higher molecular mobility of the glassy state of amorphous systems approaches towards the virtual structural equilibrium through molecular densification and the loss of the free volume and enthalpy with time. The process is called structural or enthalpy relaxation. The primary process, referred to as α-relaxation, originates from the cooperative global mobility of entire structure or segmental motion. It has been shown that the α-process is the main precursor for crystallization (Caron et al. 2010). The temperature dependence of this global mobility below T_g can be described using the Adam–Gibbs–Vogel (AGV) expression, and the non-Arrhenius behavior of the same across glass transition region can be described using Vogel–Fulcher–Tammann (VFT) law (Greco et al. 2012). The Arrhenius-type secondary relaxations evolve from local motions mostly intramolecular, rotational motions of a nonrigid molecular fragment, the common being a β-process. This faster motion has been found responsible for the crystallization of some drugs below T_g and also as the precursor for global mobility (Bhattacharya and Suryanarayanan 2009). It is also possible for molecules to possess multiple secondary processes.

The molecular domains in amorphous structure behave like an ensemble of autonomous substates, each following unique relaxation kinetics during annealing (Kawakami and Pikal 2005). This relaxation distribution is often expressed using an empirical Kohlrausch–Williams–Watts (KWW) equation (Eq. 14.3):

$$\phi_{\mathrm{KWW}}(t,T) = \exp\left[-\left(\frac{t}{\tau(T)}\right)^{\beta_K}\right] = 1 - \frac{\Delta H(t,T)}{\Delta H(\infty,T)}, \qquad (14.3)$$

where $\phi_{\mathrm{KWW}}(t,T)$, τ, and β_K ($0 < \beta_K \leq 1$) are the extent of relaxation, average relaxation time, and distribution parameter, respectively, at time t and temperature T. The narrower relaxation distribution of substates tends towards unity of β_K while β_K approaches to zero for higher dynamic heterogeneity. Here, $\Delta H\ (t,T)$ and $\Delta H\ (\infty,T)(=\Delta C_P\ (T_g\text{–}T))$ are the values of enthalpy recovery at annealing time "t" and that virtually at the completion of process, respectively. The KWW approach is limited to the systems with the relaxation times significantly larger than the annealing. A modified stretched exponent (MSE) function exists for the systems with very short ($t \ll T_1$) relaxation times with respect to the annealing times. DSC is extensively used to study the effect of annealing to mimic the aging during storage of amorphous pharmaceuticals (Hancock and Shamblin 2001). DSC measures enthalpy recovery equivalent to the relaxation, provided the recovery is not accompanied by other physical or instrument-related factors. The recovered enthalpy at T_g is largely the consequence of α-relaxation. Precise integration of both the T_g and the superimposing enthalpy recovery measured by conventional DSC can be problematic. A direct correlation has been shown between the ratio of the height of recovery overshoot to ΔC_P and the β_K for some excipients (Pikal et al. 2004).

The fictive temperature (*temperature of the equilibrium supercooled liquid that is isoenthalpic to the glass*) progressively decreases towards the annealing temperature with time. This limits DSC annealing approach by the fact that relaxation time progressively increases while aging. Also, it is important to introduce the temperature correction for ΔC_P in view of the temperature dependence of the latter. The τ and β_K obtained by fitting KWW or MSE expressions are reported to be significantly higher and lower compared to their initial values, respectively. The annealing period greatly affect these values as well. The time constant, τ^{β_K}, has been found comparatively invariant to the annealing period and thus more reliable. Meaningful comparison of τ among systems requires similar β_K (Kawakami and Pikal 2005). Another concern is the possible overestimation due to extra annealing while heating at slow rate in mDSC. A frequency-related T_g shift in the RHF relative to the THF signal gives an endothermic signal in NRHF overlapping with enthalpy recovery. This effect can be subtracted from the actual signal by measuring the same in the cooling cycle in identical experimental condition (Kawakami and Pikal 2005). However, there would also be enthalpy loss while reaching to and mainly τ^{β_K} while residing at the annealing temperature. Therefore, it is advisable to correct the enthalpy recovery data by subtracting the signal obtained from frequency- and temperature-effect together from a time-zero sample.

There are fewer recent literature on the relationship of the ΔT_g and molecular mobility (Chieng et al. 2013a, b). As the increase in β_K tends the system to approach equilibrium faster, the larger β_K is associated with narrower ΔT_g. A study suggests that τ^{β_K} measured by mDSC for amorphous API, such as indomethacin and nifedipine, below T_g showed Arrhenius-type temperature dependence while that for ketoconazole showed typical VFT behavior (Bhugra et al. 2006). Caron et al. 2010 determined the value of τ^{β_K} for amorphous nifedipine and phenobarbital and their ASD in PVP by mDSC and found a correlation with sub-T_g crystallization. The relaxation time constant of indomethacin/PVP ASD obtained by mDSC is proposed to represent the bulk relaxation (Hasegawa et al. 2009). The different τ^{β_K} values of the differently prepared ASD of the same system indicate their diverse structural dynamics (Ke et al. 2012). The general utility of KWW to describe the molecular mobility of ASD is highly questionable unless the real chemical identities of the relaxing domains are known. The complex non-KWW behavior of enthalpy decay obtained by DSC has been reported for many ASD of wherein individual components show KWW profiles. For example, such observation for celecoxib/PVP has been attributed to their different composition-dependent H-bonding interactions (Bansal et al. 2010).

The enthalpy recovery in DSC of sub-T_g relaxations is often feeble and the relaxation time decreases upon annealing, and thus DSC is not sensitive enough to probe the isolated local mobility (Bhattacharya and Suryanarayanan 2009). In general, the β-process contributes more at the lower annealing temperature/longer time in DSC. Therefore, the sub-T_g recovery temperature increases with longer time and/or higher temperature of annealing due to the increasing involvement of the α-process. Vyazovkin and Dranca (2006, 2007) have published their works on the use of DSC for studying β-relaxation of various organic glasses including that of some APIs and polymers. Amorphous PVP, indomethacin, and ursodeoxycholic acid annealed at various temperatures below $0.8T_g$ in DSC and the enthalpy recovery peaks during subsequent analysis were only observed for samples annealed at or above a certain temperature (Vyazovkin and Dranca 2006). Interestingly, activation energy for β-relaxation obtained from heating rate dependence of the recovery temperature showed satisfactory correlation with T_g for PVP and APIs. The sub-T_g peak of the β-relaxation of maltodextrin was observed to be influenced by temperature and humidity (Descamps et al. 2009). DSC sometimes provides indirect information on surface versus bulk relaxation behavior of ASD. Puri et al. (2012) annealed celecoxib/PVP/meglumine ASD with two different thicknesses and found that the enthalpy recovery for ASD with relatively higher surface-to-volume ratio was approximately three times higher compared to those for the ASD with lower values.

14.2.2 Molecular Miscibility and Compositional Homogeneity

DSC historically stands at the forefront for studying the molecular miscibility and phase homogeneity in ASD. Despite many limitations inherent to this technique, it is the first-line technique to screen the feasibility of molecular dispersion formation and to study solid-state miscibility in ASD before and during stability studies (Baird and Taylor 2012). Deconvolution of overlapping signals by mDSC facilitates phase analysis of ASD over conventional DSC. Fast scanning DSC can analyze the materials that are difficult to amorphicize (Guns et al. 2010). Depending upon the composition, a single mixed T_g (T_{gm}) of a miscible binary ASD can normally be distinguished from that of the API or the polymer. However, strong drug–polymer intermolecular interactions such as ionic/polyelectrolytic interactions, salt formation, or others can increase the T_{gm} to be higher than that of individual components (Weuts et al. 2005) as recently shown for a hydrogen bonding (H-bonding) system (Calahan et al. 2013). A single T_{gm} is considered as an indicator of complete molecular mixing between drug and polymer. For completely miscible compositions, volume additivity expressions of the Gordon–Taylor or Couchman–Karasz approach (Couchman and Karasz 1978; Gordon and Taylor 1952) aid the calculation of T_{gm} assuming the equivalent strength of homo- and heteromolecular interactions in the system. Therefore, positive deviation of the experimental T_g from the predicted one implies enthalpic contributions owing either to the strong intermolecular H-bonding (e.g., MK-0591/PVP) or to the negative excess volume of mixing (e.g., itraconazole/polyvinylpyrrolidone vinyl acetate, PVPVA; Kalogeras 2011), while negative deviations point to the predominant effect of positive excess free volume of mixing over

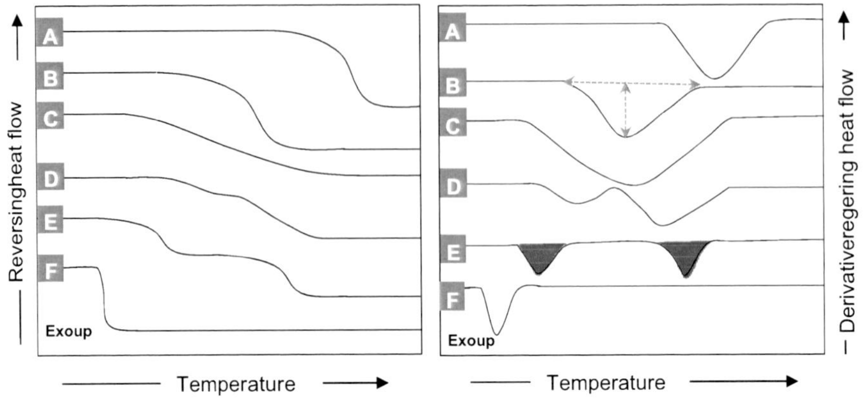

Fig. 14.2 An overlay of hypothetical mDSC thermograms depicting reversing heat flow (RHF; *left*) and derivate RHF (*right*) versus temperature. Trace *A*: polymer, *B–E*: various ASD, and *F*: API

either moderate-to-strong (e.g., felodipine ASD with PVP or hydroxypropyl methyl cellulose, HPMC) or weak-to-moderate (e.g., naproxen–PVP) heteromolecular interactions. Kalogeras (2011) modeled the experimental T_{gm} (composition) profiles of diverse ASD using an empirical expression. The H-bonding interaction has also been confirmed in a drug candidate/PVP ASD wherein the T_g measured by DSC overlapped with the predicted Gordon–Taylor profile. This suggests the need of cautious interpretation of such data from thermal analysis alone (Tobyn et al. 2009).

Hypothetical RHF signals and the corresponding first-derivative RHF signals with respect to temperature (dRHF) observable of ASD bearing diverse physical structures are illustrated in Fig. 14.2 (B–E). At least two thermal transitions are clearly exhibited in phase-separated ASD, each originating from one of the partially or completely separated phases. The separated phases rich in API and polymer exhibit respectively low and high T_g that are positioned distinctly apart provided sufficiently different composition (*E*). In such case, the ratio of ΔC_P of two phases or moreover the height or area of peaks in dRHF signals can be comparable with the phase composition (Paudel et al. 2013). On the other hand, when the composition of separated phases is not too different, borderline merger of two T_g's can be observed (*D*). For this, the corresponding dRHF signals can provide clearer identification. The lesser the difference between two T_g's, the more homogeneous the system is (Paudel and Van den Mooter 2012). Apart from the number and position of T_g, the shape of the transition can also provide information on the molecular state of ASD. For instance, the width and the symmetry of glass transition are very important features. Although the mid-positions of the glass transition seem similar for lines B and C, their shapes strikingly differ pointing the criticality of the shape/width in interpreting the compositional homogeneity. More precisely, the T_{gm} width of B is almost half that of C. As calorimetric techniques are reported to detect the phase separation into the larger domain size (> 30 nm), the wider T_g can be associated with heterogeneity possibly beyond the detection limit of DSC. The Raman chemical mapping has proven the fact that a single T_g is not necessarily an indication of molecular miscibility (Qian et al. 2010). There is potential danger of introducing integration errors for feeble and wider signals to obtain the consistent T_{gm} midpoint and ΔC_P as of the case given in Fig. 14.2 (trace C). Rather, the use of the width and the area of the peak in the dRHF signal are seemingly reliable to describe the extent of microheterogeneties. Recently, increasing interests on the use of dRHF responses can be witnessed for polymer composites to model the blend homogeneity (Shi et al. 2013). The semiquantitative value of this simple data analysis approach has been applied for pharmaceutical ASD as well (Paudel et al. 2013).

The most problematic cases in distinguishing a single versus two amorphous phases are: ASD consisting of drug and carrier with essentially similar T_g's such as felodipine and Eudragit® EPO (T_g difference $< 3\,°C$; Qi et al. 2010a, b). Locating T_{gm}'s belonging to drug-rich and polymer-rich domains was found ambiguous for the phase-separated compositions of itraconazole/Eudragit E100 ASD (T_g difference $\leq 5\,°C$; Janssens et al. 2010). Instead, the respective enthalpy- recovery endotherms in NRHF signals facilitated the distinction of the T_{gm} positions. Enthalpy recovery can also be apparent even for a very small fraction of separated phases in NRHF traces that could otherwise not be contrasted from the RHF baseline. Sometimes, the presence of enthalpy recovery of the separated phases indirectly ensures that phase separation had occurred in the original sample rather than the in situ separation induced by heating in DSC. Analysis by fast scanning DSC can be superior as the decrease in the exposure time at higher temperature provides kinetically less-affected miscibility data. The inaccuracy often associated with C_P measurement of heterogeneous systems limits its routine use for studying the phase analysis of ASD. The change in phase angle, the angle between the modulated heating rate and modulated heat flow, can be correlated to phase separation through sophisticated deconvolution (Pieters et al. 2006). Such advanced mDSC experiments can worth the miscibility study of challenging systems. The DSC interpretation of ASD of API in multiple polymers needs further details. The mDSC analysis of felodipine ASD prepared in a blend of immiscible polymers viz., Eudragit EPO and PVPVA, suggested that the increase in drug loading leads to the higher fraction of drug in the polymer in which the drug has higher solid solubility (Yang et al. 2013). Likewise, thermal characterization of ASD containing an API dispersed in a copolymer is complicated if the amorphous drug has diverse miscibility in the constituting monomers as evidenced from the miconazole dispersion in Kollicoat IR (*poly (ethylene glycol-co-vinyl alcohol)*; Litvinov et al. 2012).

14.2.3 Crystallization, Melting, Crystallinity, and Mixing Interactions in ASD

The use of DSC for studying isothermal/non-isothermal crystallization is extensively reported (Baird and Taylor 2012; Svoboda and Málek 2011). Although not always mutually exclusive, an entire crystallization process proceeds through an initial nucleation step followed by the growth of nuclei. A very small fraction of a sample mass actuates the nucleation process, often indistinguishable by DSC. The crystallization exotherm usually evolves at a temperature (T_C) following the T_g while heating an amorphous glass (Fig. 14.1) and it can appear while slowly cooling the melt as well. There are cases where the crystallization exotherm is detected before T_g during both scanning and isothermal measurements for physically unstable API. Crystallization from the cryomilled amorphous etravirine was detected at T_g −32 °C (Qi et al. 2010a, b). DSC studies have revealed the atypical bimodal crystallization of milled amorphous griseofulvin wherein the first exotherm occur prior to the T_g (Trasi and Byrn 2012; Trasi et al. 2010; Willart et al. 2012). Such behavior was also noticed for other milled amorphous form of felodipine, sulfamerazine, piroxicam, hydrochlorothiazide (Chattoraj et al. 2012). Interestingly, the first crystallization event originates from surface that is followed by bulk process beyond T_g. Highest surface crystal growth rates around T_g were confirmed for APIs exhibiting surface crystallization (Otte et al. 2012).

Experimental conditions such as thermal contact with the DSC pan, sample particle size, heating rate, etc. can markedly affect the crystallization enthalpy. The integrated area under a single and symmetrical crystallization exotherm can directly correlate to the crystallization enthalpy and the activation energy (Svoboda and Málek 2011). The activation energy of the non-isothermal crystallization (E_C) can be determined using the modified Kissinger method (Eq. 14.4) using the heating rate (β) dependence data of T_C from a series of DSC measurements:

$$\ln\left(\frac{\beta^n}{T_C^2}\right) = -\frac{mE_C}{RT_C} + \text{constant}. \tag{14.4}$$

Here, n is the order parameter and m is the dimensionality of growth. In mDSC analysis, the positive overall periodic minimum heating rate ensures proper deconvolution of crystallization exotherm. Grisedale et al. (2010) studied crystallization kinetics of amorphous salbutamol sulfate using mDSC and found that the spray-dried product had the higher T_C and E_C compared to that of the milled ones and for the latter these values increased with milling time.

The polymer in ASD significantly retards API crystallization such that T_c increases and H_F increases. For specifically interacting ASD of a drug candidate with PVPs, T_C was directly correlated with the T_g of the polymer (Khougaz and Clas 2000). The polymer can also selectively decelerate or inhibit surface crystallization. The presence of 20 % PVPVA in griseofulvin/PVPVA cryomilled ASD shifted the first T_C while this surface crystallization exotherm completely vanished upon increasing PVPVA (Chattoraj et al. 2012). While T_{gm} being unaltered, nifedipine/PVP ASD exhibited the particle-size-dependent crystallization exotherm with lower T_C for smaller particle fraction (Miyanishi et al. 2013).

The common method for modeling both isothermal and non-isothermal crystallization kinetics from amorphous solids is the Johnson–Mehl–Avrami -Kolmogorov

(JMAK) nucleation-growth model (Eq. 14.5; Weinberg et al. 1997):

$$\alpha_C = 1 - e^{[(K(t-t_{ind}))^m]}. \tag{14.5}$$

Here K is the nucleation rate constant, t_{ind} is nucleation induction period, and α_C is the crystallized fraction at time t. The values of K and m are obtained by regressing the experimental data using the double logarithmic linear form of Eq. 14.5. Originally derived for a homogeneous single component system, zero or constant nucleation rate in course of crystallization assumed by JMAK model can be an oversimplification of nucleation process for the process starting from ASD. Furthermore, this model is advocated to be inefficient in explaining crystal impingement at higher α_C (> 0.8; Sousa et al. 2010). The JMAK model is occasionally modified to account for the decrease in nucleation rate accompanied by the increase in the growth rate while crystallization progresses. Yang et al. (2010) proposed a better kinetic model (than JMAK) that considers nucleation rate to be inversely proportional to the crystallinity in order to study isothermal crystallization of etravirine–PVP ASD. Yoshihashi et al. (2006, 2010) studied isothermal crystallization of flurbiprofen, tolbutamide, and naproxen from ASD above T_{gm} using DSC. Their method typically involves heating the sample above T_m of API, cooling below T_{gm}, heating to the intended temperature, and keeping isothermal until the API crystallizes.

Non-isothermal DSC crystallization data can often enable the quantification of initial amorphicity. Assuming complete crystallization of the amorphous fraction present during DSC analysis, the degree of amorphicity is simply the ratio of area under the crystallization exotherm to that under the melting endotherm (Baird and Taylor 2012). These two events occurring at different temperature need correction for the temperature-dependent enthalpy. Equation 14.6 determines the initial crystallinity including the correction for the possible noncrystallizing fraction (α_{NC}; Grisedale et al. 2010):

$$Crystallinity = 1 - \left[\frac{\Delta H_C}{\Delta H_m - (T_m - T_C)\Delta C_P} \times \frac{1}{(1 - \alpha_{NC})}\right]. \tag{14.6}$$

The α_{NC} can be estimated as the ratio of melting enthalpy obtained on (partial) crystallization of fully amorphous material (during DSC analysis) to that of the pure crystalline material of same form. For materials undergoing degradation during melting, ΔH_m and T_m can be replaced by ΔH_C and T_C obtained for fully amorphous material, respectively. In case of API in pure amorphous form or in ASD undergoing complete crystallization during DSC, the ratio of ΔC_P at T_g of partially crystalline material to that of completely amorphous material gives the degree of amorphicity while that of ΔH_C of partially crystalline material to that of completely amorphous material gives the degree of crystallinity in the sample (Aso et al. 2009). Apart from simple crystallization and melting, the possible occurrence of multiple polymorphic or liquid transitions during heating of amorphous API or ASD complicates the direct determination of crystallinity from DSC data (Janssens et al. 2010).

Occasionally, unintentional residues of submicron or bulk crystals dispersed within fresh ASD is encountered or partial crystallinity may develop during storage (Janssens and Van den Mooter 2009). The melting transition(s) of API appears in the DSC thermogram of partially crystalline dispersion. However, multiple kinetic effects should be considered before thermodynamic interpretation. Below melting temperature of the drug, a fraction of crystals can dissolve in the matrix near and above the T_g or melting temperature of the polymer (Qi et al. 2010a, b). Also, the exothermic process of polymer dissolution in excessively molten API can affect the melting endotherm of API. The crystallinity estimated as the ratio of ΔH_m of the API in ASD to that in the corresponding physical blend can possibly correct these kinetic effects (Yang et al. 2010). The extent of crystallization in efavirenz/PVP ASD was computed as the ratio of melting enthalpy of the API in the ASD after time t to that after maximum crystallization of the same ASD. The complete crystallization may not necessarily occur as 40 % efavirenz crystallization was reported at plateau time (steady state) from PVP-based ASD.

There exist cases where a slow scanning mDSC thermogram discerns no melting for partially crystalline ASD (Bikiaris et al. 2005), as case of naproxen–PVP solid dispersions (Paudel and Van den Mooter 2012). Crystallites inhomogeneously dispersed in a polymeric matrix could experience different local compositions and hence variable local melting temperatures. This spans the overall melting event over a wide range. Fast scanning DSC can detect melting probably due to hindered drug–polymer dissolution kinetics and increased sensitivity. This also sacrifices the resolution among the multiple events and poses interruption from other kinetic signals retained such as desolvation. Disappearance/diminution of a drug melting endotherm can possibly attribute to the very fine crystallites distributed within ASD matrix. For the trace amount of nifedipine or griseofulvin embedded in PVP, crystallite size reduction to several nanometers resulted in more than 10 % drop in T_m (Liu et al. 2007). Also, the reduction in melting enthalpy of submicron or nano level crystallites in ASD results from the alteration of the bond energy of surface atoms of small crystals on the internal energy (Liang et al. 2002). In absence of melting, the crystallinity can be indirectly estimated by the C_P change at a temperature $> T_{gm}$ during a heat-cool-heat DSC method. The C_P at a temperature would be lesser in the first heating due to an initially present crystalline fraction as compared to that obtained during the following cooling or subsequent heating cycle. The heat-cool-heat mDSC program enabled the quantification of the API crystallinity of felodipine/Eudragit® EPO ASD (Qi et al. 2010a, b) and surface crystallinity of nifedipine–PVP ASD developed during aging, but with undetectable melting (Miyanishi et al. 2013).

Melting temperature of API in a solid dispersion obtained under quasi-equilibrium conditions also inherit a wealth of information about the drug–carrier mixing interaction (Sun et al. 2010). The overall chemical potential decreases when an API is dispersed in a polymer matrix leading to the depression in its equilibrium melting temperature. The depression is higher for stronger favorable (exothermic) interaction existing between mixing components. The composition–melting point profile for a drug/polymer system can yield the activity coefficient of molten drug in polymer, the Flory–Huggins interaction parameter and consequently the thermal phase diagram (composition-Gibbs free energy of mixing). These thermodynamic state functions are applicable at their best to the equilibrium data requiring the investigational system prepared with the least input of kinetic energy. DSC data of dispersions prepared by slow solvent evaporation (Paudel et al. 2012) are closer to the assumption (Caron et al. 2011) than by spray drying or milling. Extrapolation of heating-rate-dependent melting temperatures to 0 °C/min heating rate yields quasi-equilibrium melting temperatures, which minimize the kinetic effect and are more reliable for theoretical use. The crossing point of the melting line with the T_{gm} line as the function of composition provides the predicted solid solubility of API in polymer, which still remains experimentally unverified for many drug–polymer systems. The extrapolated endset melting temperature in DSC ideally indicates the completion of melting and thus is meaningful for miscibility analysis (Marsac et al. 2006).

14.3 Isothermal Microcalorimetry (IMC) Studies of ASD

IMC, a complementary technique to DSC, is a versatile and sensitive tool for non-specific thermal activity monitoring (TAM) of heat change occurring during any physical, chemical, or biological process (Ball and Maechling 2009). TAM possesses extreme sensitivity towards the heat flow (0.1 μW) and temperature change (10^{-4} °C), therefore suitable for quantitative analysis of several processes. The applicability of IMC has been extensively proven in monitoring subtle thermal events originating from processes such as mixing, chemical degradation, crystallization, and other phase transformations (Gaisford 2005). The TAM enables calorimetric experiments on various solid-state processes under controlled temperature and relative humidity (RH). The isothermal temperature is accurately maintained by a water bath surrounding a sample vial while a hygrostat containing saturated aqueous salt solution optionally placed along with a sample creates the designated RH at a set temperature (O'Neill and Gaisford 2011). The RH of the sample headspace can be continuously varied by passing the programmed composition of dry (0 %RH) and wet nitrogen (100 %RH) through a mass flowmeter. This RH perfusion microcalorimetry can probe various interactions through moisture-induced thermal activity traces (MITAT; Lechuga-Ballesteros et al. 2003). When operated as isothermal solution microcalorimetry (SC) or isothermal titration microcalorimetry (ITC), IMC measures solution or solution-mediated processes (Ehtezazi et al. 2000). The solid or liquid analyte is sealed in a glass vial (and equilibrated in a specific liquid) which can be broken to release the sample in the solution which results in the heat flow due to mixing and/or dissolution in SC while during an ITC experiment the heat flow generated from titration of continuously dispensed analyte in the liquid equilibrated inside the calorimeter is measured (Blandamer et al. 1998). A heat-conduction IMC measures heat flow (dq/dt) from ongoing processes in a sample. The detected electrical power (P) is the product of the voltage (U) generated by the temperature difference due to a thermal process in the sample and a calibration constant (ε_C). Generally, the exothermic process is represented by a positive signal and the endothermic event by negative signal in a heat flow curve. The integration of the heat flow curve (power data) over a particular time interval provides heat (q). Provided the total enthalpy of a process (Q) is known, the temporal calorimetric data generates a solid-state kinetics in the form of a fraction converted ($\alpha = q/Q$; Sousa et al. 2010).

In spite of superior sensitivity, various challenges associated with IMC, e.g., larger sample size (50–200 mg) and limited temperature range (10–80 °C) restrict the types of experiments possible on amorphous samples. The thickness and sample amount play vital role on the eventual onset of surface versus bulk induction kinetics (Gaisford et al. 2009). The analysis time including the preceding equilibration is so long that the initial data points for fast processes such as the induction of enthalpy relaxation, nucleation/crystallization can escape detection. In contrary, ceaseless heat flow originating from some of very slow processes might not return to the baseline within the experimental period. Another main concern about IMC is the nonspecificity of the recorded TAM signal as the outcome of the entire processes taking place during the experimental period. This can hinder the data interpretation of amorphous systems possessing temporally overlapping processes such as relaxation, phase separation, and nucleation/crystal growth. For elevated static humidity or RH perfusion microcalorimetry, the potential degradation of labile components or moisture sorption/desorption/condensation signals can present further aberration in the measured data (Buckton and Darcy 1999). The optimal experimental conditions such as small sample size, temperature, gas flow rate, and RH ramp rate should be maintained to obtain (quasi)equilibrium measuring environment and an acceptable signal to noise ratio. Since thermal history of a sample cannot be erased prior to IMC analysis, it is extremely crucial to have consistency of process history, formulation conditions, and more importantly residual solvent of samples. At positive end, the feasibility of "*as is*" sample analysis by IMC can have discriminative advantage to detect the subtle structural change in ASD inherited from formulation and/or manufacturing processes (Kawakami and Pikal, 2005). Some existing applications of IMC relevant to amorphous pharmaceuticals are presented below.

14.3.1 Enthalpy Relaxation Studies on Amorphous Pharmaceuticals by IMC

The quantitative application of IMC for the rate of enthalpy relaxation of a pure amorphous APIs and excipients are explicitly documented but unfortunately few examples are available hitherto for ASD (Caron et al. 2010). Exothermic heat flow is detected in TAM below T_g as a consequence of relaxation. During IMC measurement, the early data points are often excluded during data analysis to avoid noise resulting from sample positioning (Kawakami and Pikal 2005). However, IMC records more temporal data points during enthalpy relaxation and hence yields relaxation parameters from a single run when compared to DSC. Thus, the power–time profiles obtained in TAM can be directly treated with the power equations of relaxation models viz., KWW (Eq. 14.3) or MSE equation with respect to time. The derivative form of the MSE equation can describe the experimental relaxation data measured by IMC more consistent, especially those recorded at lower annealing temperature (Kawakami and

Pikal 2005). The study of Bhugra et al. (2006) on diverse amorphous APIs revealed that the relaxation time constants measured below T_g by IMC for indomethacin and ketoconazole were notably shorter than those measured by mDSC. However, the extrapolation of the data trends measured by both techniques conversed at T_g.

The IMC relaxation data for ASD are complex to understand as well. Alem et al. (2010) investigated enthalpy relaxation of amorphous mixtures of sucrose–lactose and sucrose–indomethacin. For the mixture components with comparable relaxation times, physically meaningful relaxation parameters were obtained by modeling the experimental data of binary mixtures with the derivative KWW and MSE expressions. As the individual relaxation times differ markedly, the KWW relaxation parameters lost the ability to confer a physical meaning while MSE parameters were still meaningful. The relaxation time constants of nifedipine/PVP and phenobarbital/PVP ASD below T_{gm} using IMC data were significantly lower compared to those from mDSC (Caron et al. 2010). Moreover, the lower crystallization onsets predicted by relaxation data from IMC than the experimental value for ASD was anticipated to stem from the possibility of IMC measuring the α-process as well as some fast relaxation irrelevant to crystallization and not measured by mDSC. Overall, β-value obtained using IMC is found to be smaller than obtained using DSC data of amorphous API as well of ASD (Bhugra et al. 2006; Caron et al. 2010). The reason for the same is still controversial whether it is due to the higher portion of β-process measured by IMC or owing to the different impact of time-dependent molecular mobility in the two techniques. Chieng et al. (2013a, b) found that the relaxation times measured by IMC for ASD of starch derivatives and disaccharides and/or polyols to be directly proportional to one-third power of that derived from ΔT_{gm} methodology. Moisture-induced enthalpy relaxations are reported for amorphous sodium indomethacin sucrose, lactose, raffinose, and PVP using RH perfusion IMC (Lechuga–Ballesteros et al. 2003). The MITAT of amorphous spray-dried raffinose revealed an exotherm after a threshold RH originated from α-relaxation triggered by moisture (Miller and Lechuga-Ballesteros 2006). This was tested by recording MITAT at different RH scanning rate and by using samples with different thermal history. The relaxation exotherm expectedly vanished after *ex situ* annealing prior to analysis. Data on T_g–RH (moisture content) relation can assist the interpretation of MITAT.

14.3.2 Crystallization Kinetics of and Crystallinity in Amorphous Systems by IMC

IMC has been ubiquitously utilized for studying the isothermal crystallization kinetics of amorphous pharmaceuticals owing to the ultimate sensitivity of the technique towards subtle heat flow (Gaisford 2012). Low quantification and detection limit of the technique lead to enhanced crystallization enthalpy and therefore superior S/N ratio as compared to DSC. Several studies report the inert environment in studying crystallization kinetics or at static RH or scanning RH using RH perfusion calorimetry (Yonemochi et al. 1999). Different modified kinetic models are fitted to temporal data on crystalline conversion to obtain crystallization parameters. For purely amorphous material undergoing complete crystalline conversion, the enthalpy of the crystallization exotherm from IMC equivalences the melting enthalpy obtained by subsequent DSC analysis (Buckton and Darcy 1999). The ratio of crystallization enthalpy of a partially crystalline sample to that of a completely amorphous reference provides the crystallinity. There are always chances to miss earlier induction points for rapidly crystallizing APIs especially in the absence of a crystallization inhibitor or while measuring at elevated RH. Sousa et al. (2010) estimated the total heat of crystallization by applying mathematical models on intermediate experimental IMC data points and found that the methods correctly predicted the heat of crystallization, in case of amorphous indomethacin. Thus, such models can enable the calculation of the total heat for missing data points of a very fast or slow process. The IMC study of Hédoux et al. (2009) on isothermal crystallization of amorphous cryomilled indomethacin exhibited an unusual broad exotherm with double peaks suggesting a possible overlap of a surface and bulk process. Interestingly, Bhugra et al. (2008) observed the early power curve essentially resembling the sub-T_g decay profile for amorphous indomethacin measured at $T_g + 20\,^{\circ}\mathrm{C}$ that was followed by a significantly longer nonzero baseline and a plateau before the crystallization exotherm. Since the expected relaxation time at this temperature ranges in ms–µs, the authors attribute the early decay to the decreasing nucleation rate triggered after the formation of stable nuclei population. The samples withdrawn from TAM before the exotherm and inspected by microscopy revealed that it already contained indomethacin crystals implying the lesser sensitivity of IMC for the crystallization induction. Crystallization studied using IMC of amorphous nifedipine after intensive mixing with glass beads showed markedly decrease of induction time in comparison to the untreated sample, the reason being the trace crystallinity developed during the longer mixing time (Song et al. 2005).

TAM signals recorded in humid condition must be interpreted cautiously. The crystallization enthalpy of lactose measured under elevated RH was markedly lower than the melting enthalpy by DSC (Hogan and Buckton 2001). Moisture triggers nucleation/crystallization beyond a critical RH by enhancing molecular mobility, often accompanied by desorption of water. The crystallization enthalpy obtained by IMC here (the exothermic heat of crystallization minus endothermic heat of desorption) would be lower than the melting enthalpy by DSC. The supply of the water of crystallization in RH perfusion calorimetry facilitates the hydrate crystallization of some materials that can restrain desorption. Crystallization enthalpy for the low mass of spray-dried amorphous raffinose recorded in TAM was equivalent to that of melting in DSC (Hogan and Buckton 2001). The sorption, evaporation, crystallization of salt, and sample wetting can contribute data recorded using aqueous salt solution hygrostat (Gaisford 2012). The additional plasticizer mixed with water is sometimes used to control headspace relative vapor pressure during TAM measurements. Yonemochi et al. (1999) studied isothermal crystallization of different amorphous ursodeoxycholic acid with the sample environment maintained using a varying ratio of ethanol–water. None of samples crystallized upon exposure to only humid air

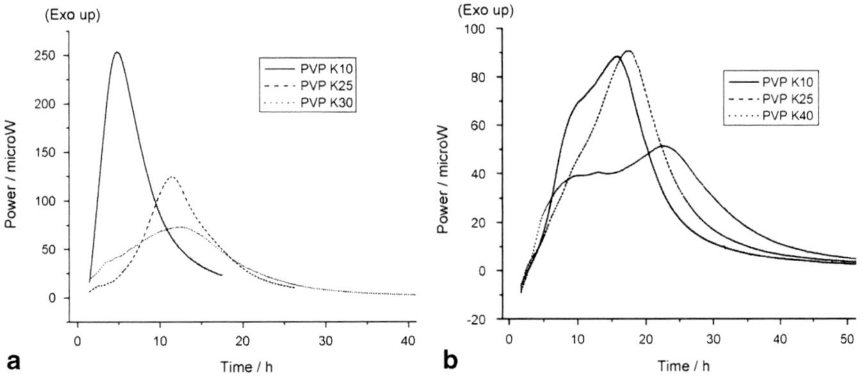

Fig. 14.3 Power–time data observed for indomethacin– PVP ASD films at 25 °C (**a**) and 37 °C (**b**). (Source: Gaisford et al. 2009, with permission from Elsevier)

while ground samples crystallized and quench-cooled samples did not crystallize in presence of ethanol vapor.

Although crystallization studies from ASD using IMC is challenging, Latsch et al. (2003, 2004) showed great potential of IMC in monitoring crystallization of amorphous steroidal drugs from dilute multicomponent transdermal patches. Transformation of amorphous estradiol dispersed in a matrix made up of acrylic polymer and polyethylene glycol (PEG) to hemihydrate was monitored using IMC (Latsch et al. 2004). Heat flow associated with drug crystallization expectedly increased with increase in drug loading in the formulation and was detected for drug loading even below 2 %w/w. In contrast to the case of amorphous indomethacin (Bhugra et al. 2008), the induction of estradiol crystallization from polymeric patches was detected quite earlier by IMC than by microscopy. Likewise, crystallization kinetics from a low amount of amorphous norethindrone acetate dispersed in a transdermal polymeric patch was studied using IMC (Latsch et al. 2003). The crystallization from the patch containing 4 %w/w drug was detected more by IMC than by microscopy. Crystallization from the patches with drug loading 4–10 %w/w was isokinetic while the rate was markedly accelerated from 12 % drug content. Crystallization kinetics from patches containing 2–14 %w/w mixtures of estradiol and norethindrone acetate dispersed in acrylic polymer and a plasticizer were studied using IMC (Latsch et al. 2004). The crystallization rate was the highest from patches containing PEG as a plasticizer. Gaisford et al. (2009) monitored the crystallization kinetics of indomethacin from ASD prepared with different PVP at 25 and 37 °C using IMC. As shown in Fig. 14.3, a crystallization exotherm was discernible at 25 °C while two adjoining peaks appeared at 37 °C. Post-TAM microscopy revealed that indomethacin entirely transformed into the stable γ-form at 25 °C while at 37 °C the early fraction (the first peak) transformed into the α-form and the second exotherm is attributed to the crystallization of the γ-form. No significant alteration in the overall crystallization profiles was apparent among the ASD containing different PVP. Urbanovici–Segal

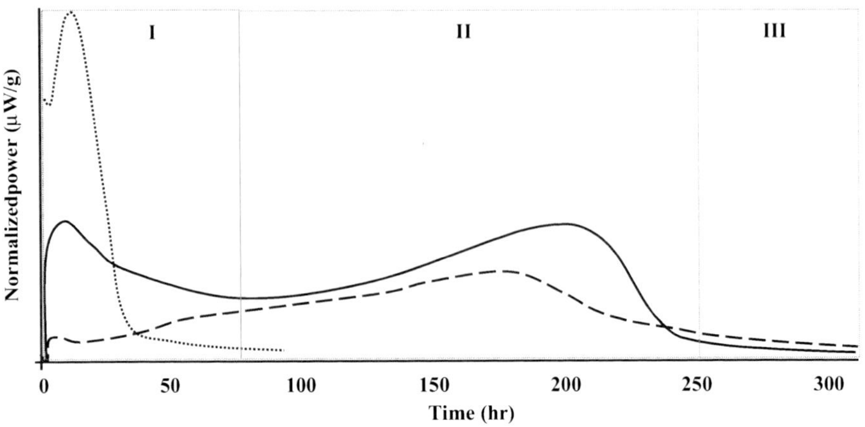

Fig. 14.4 TAM thermograms recording crystallization AMG 517 from pure amorphous state at 80 °C/25 %RH (. . .), from amorphous solid dispersions (ASD) with 50 %w/w hydroxypropyl methyl cellulose-acetate succinate (HPMC-AS) (- – -) and from ASD with 18 %w/w HPMC-AS (—). The region-I represents crystallization from drug-rich phase, region-II represents that from miscible ASD portion and region-III represents the absence of crystallization signal due to the amorphous fraction retained below solid solubility. (Adapted from Calahan et al. 2013)

models described experimental data better compared to the Avrami, Tobin model. Calahan et al. (2013) investigated the crystallization behavior of AMG 517, an investigational drug, from its ASD prepared in hydroxypropyl methyl cellulose-acetate succinate (HPMC-AS) using IMC at elevated temperature and RH. TAM thermograms (Fig. 14.4) exhibited the bimodal crystallization for ASD with higher drug load which turn into unimodal with decreasing drug content. The first exotherm of ASD overlapping the crystallization exotherm of the pure amorphous drug stems from the crystallization of amorphous drug clusters present in supersaturated ASD. The second exotherm originates from crystallization of drug from separated domains having miscible drug composition above the solid solubility while the remaining amorphous drug fraction is proposed to represent solid solubility of drug in the polymer.

IMC measurements under humid condition are also frequently used to quantify crystallinity in samples (Gaisford 2012). IMC is suitable for quantifying trace amorphicity ($<$ 1 %w/w) in bulk crystals while the quantification of the trace crystallinity in bulk ASD is relatively inferior (Giron et al. 1997). The heat flow data are converted to initial amorphous content using the crystallization enthalpy or calibration curve. The contribution of unwanted processes while using saturated salt solution hygrostat, especially endothermic evaporation of water released during crystallization hinders quantification in small samples (Gaisford 2012). The RH perfusion cell overcomes many of these adverse effects. Addition of a drying (at 0 %RH) and a rewetting step after the crystallization step during the measurement is shown to provide the correction data for the wetting signal (Gaisford 2012). These methodologies are only valid in absence of all moisture-induced solid-state processes except crystallization, which is rarely possible for ASD. SC is an alternative technique for quantification of amorphicity. The instantaneous heat flow signal from a solid in contact with a solvent constitute the disruption of solute–solute interactions (endothermic), solvent–solvent interactions (endothermic), and the solvent–solute interactions (exothermic; Van den Mooter 2012). The solute–solute interaction such as lattice energy significantly prevails in the crystalline substance over the amorphous from. Therefore, the dissolution enthalpy can be endothermic for a crystalline and exothermic for an amorphous form (Syll et al. 2012). Calibration curves in a suitable solvent can be used to quantify the amorphicity. Chadha et al. (2013) utilized SC to discriminate the structural features of amorphous hypoglycaemic agents prepared by different methods and to quantify the amorphous content in the sample. However, this methodology can again be limiting for ASD. The THF subsequent to the introduction of solid in a solvent will originate from different strength of drug–solvent, drug–carrier, and carrier–solvent interactions. Rather, this makes SC a useful technique to study the nature of interaction in binary systems and to estimate the heat of mixing between the dispersion components (Casarino et al. 1996).

14.4 Powder X-ray Diffractometry (pXRD) and X-ray Scattering

X-ray diffractometry (XRD) is a gold standard technique for studying crystalline materials. Several properties associated with long-range three-dimensional (3D) periodicity of crystals such as unit cell dimensions, lattice parameters, etc. are accessible from XRD data. The single crystal XRD involves the measurements on a precisely grown single crystal, with the incidence of X-rays on entire crystallographic planes through continuous rotation thus detecting a series of spots resulting from constructive interferences of the diffracted X-rays. The crystallographic interpretations of pharmaceuticals are detailed elsewhere (Ochsenbein and Schenk 2006). Also, the basic theories of X-ray diffraction and scattering are decipherable in several standard text books and are beyond the scope of this chapter. Most samples, often present in powder form, contain crystals of multiple orientations and the diffraction is detected as continuous Debye rings eventually resulting in powder patterns. XRD analysis of polycrystalline powders is called powder X-ray diffractometry (pXRD). Bragg's law of diffraction must be satisfied for the diffraction of incident X-ray onto a given crystallographic lattice plane through a constructive interference of coherently scattered waves that are "in-phase" in a certain direction (Eq. 14.7):

$$n\lambda = 2d \sin\theta \tag{14.7}$$

where n, λ, d, and θ are an integer (diffraction order), the incident X-ray wavelength, the spacing between a set of lattice planes, and the angle between the incident ray and the diffraction planes, respectively. Since d is a structural constant, either λ or θ needs to be varied to meet Bragg's criterion by all lattice planes present in a crystal. Most laboratory XRD measurements are carried out using a monochromatic X-ray source by continuously increasing θ until the entire coverage of d and thus Bragg condition is met once at a time for each plane. The common sources with Cu and Mo produce X-rays with the wavelength of 1.5418 and 0.71073 Å, respectively. Another alternative is Laue diffraction wherein incidence of X-rays containing a range of wavelengths at fixed θ allows Bragg condition simultaneously for many planes as recorded by synchrotron X-ray source. High flux of X-rays from synchrotron source generates the supreme data quality, although the accessibility is limited and expensive. Common pXRD configurations are reflection and transmission mode. The Bragg–Brentano θ–2θ setup is the classical reflection geometry and the Debye–Scherrer geometry is a transmission setup. Transmission mode is suitable for low absorption samples and liquids, suspensions, etc. (in capillary sample holder).

The signals at a particular scattering vector, $Q\,(=4\pi\sin\theta/\lambda)$, represent the spatial density fluctuations in real space. Diffractometer configurations are common covering different ranges of scattering angle viz., $> 10°$ for wide angle X-ray scattering (WAXS) capturing structural information of < 1 nm, 0.1–10° for small-angle X-ray scattering (SAXS) and 0.001–0.3° for ultrasmall-angle X-ray scattering (USAXS; Dong and Boyd 2011). SAXS/USAXS are suitable for the analysis of liquid crystals, mesophases, macromolecules, dispersions, molecular self-assemblies, pores, colloidal structures, etc. (< 1–200 nm). The scattering measured in the Guinier's region by USAXS yields the radius of gyration of macromolecules. Simultaneous SAXS/WAXS instrument facilities provide in situ monitoring of *ms* scale phase transitions. Instruments measuring simultaneous SAXS/WAXS and DSC also exists (Pili et al. 2010). Some instrumental configuration selectively measures the surface scattering at the defined surface depth of the sample, e.g., grazing incidence measurements (Koch and Bras 2008).

A powder diffractogram is thus a plot of the diffraction intensity against 2θ or Q. The peak related to a particular space group is designated as a Bragg peak and is the signature of the crystalline material. The peak positions are related to the metrics of unit cell (d-spacing and other lattice parameters) and thus useful for qualitative phase analysis. The intensity and area under the Bragg peak are quantitatively associated with crystal structure (atomic positions, occupancy, etc.) while peak width to the crystal defects (strain, disorders) and size of discrete domain of crystallites. The Scherrer equation expresses the volume averaged thickness normal to the reflecting plane of crystallites with uniform size and shape as the reciprocal of full width at half height of the Bragg peak in radians (Patterson 1939). Background signals from the sample, holder, air, etc. and instrument-related broadness from polarization and optical shadow factors should be identified and removed (Bates 2011).

An amorphous material exhibits translational, orientational, and/or conformational periodicity only at short range (Bates et al. 2006). The pXRD pattern of an isotropic amorphous material appears as a continuous halo without distinct Bragg peaks. A partially crystalline material exhibits the Bragg peaks superimposed over the amorphous halo. The X-ray amorphous pattern portrays the mean response of the average local order of an ensemble of short-range orders each constituting local energy minima and contributing the coherent X-ray diffraction (Bates 2011). Alteration of local structure during T_g hardly respond in pXRD recorded with laboratory X-ray source. Sometimes, the demarcation of short-range order and disorder is ambiguous, the end-ordered dimension (e.g., nanocrystallites) being the persistence of the unit cell characteristics. The domain size depends upon molecular size and the distance between the neighbors. The periodicity limited up to five basic units of atoms or molecules or unit cells in a sample leads to X-ray amorphous pattern which can be liquids, glasses, mesophases, nanocrystals (Bates et al. 2006). Each amorphous system exhibits unique peak maxima profiles of a halo and the area ratio of different halos, even unique to the preparation methods. Remarkable different pXRD halos of amorphous indomethacin prepared by melt quenching and by cryo-grinding has been reported (Karmwar et al. 2012). Although mechanistically unclear, the multiple glassy states of the same system with altered configurational entropies and microstructures correspond to different kinetic modifications. There are existing debates on such altered microstructures on amorphous states relating to polymorphism (Hancock et al. 2002). Water exists in at least three different X-ray amorphous forms differing in density and short-range H-bonding pattern (Winkel et al. 2009). Below some cases of the pXRD analysis of local structures, miscibility and crystallinity of amorphous pharmaceuticals are presented.

14.4.1 Local Structure of Amorphous Pharmaceuticals Using Total X-ray Scattering

The total scattering data refer to the collection of all coherent scattering measured over the entire Q-space treating Bragg and diffuse scattering with equivalent gravity (Billinge et al. 2010). It is the plot of total scattering structural function $S(Q)$, intensity normalized by incident flux per atom, versus Q. The X-ray counts divided by the scattering cross section of the sample yields the square of the atomic form factor, $f(Q)^2$. The weaker signals at higher Q get amplified with the small value of $f(Q)$ at higher Q eventually providing information even of the weaker signal. $S(Q)$ depends upon the magnitude of Q rather than on the direction for isotropic materials and is the sum of intra- and intermolecular scattering (Benmore et al. 2013). The pXRD data recorded in a shallow sample holder with low background or a capillary holder with larger step size and longer counting time are preferred.

Fourier transformation of $S(Q)$ profiles generates a real-space function $G(r)$ called the atomic pairwise distribution function (PDF) and is popular for the analysis of total scattering data (Billinge 2008, Farrow and Billinge 2009). The PDF profile ($G(r)$ vs. r) presents the probability of finding an atom at a distance r from any reference atom and the product of peak area and distance relates to the number of the particular atomic pairs distributed in real space. While low-r peaks (< 2.5 Å) correspond to intramolecular distances, the high-r signals to the overlap of intermolecular packing and thus sensitive to the short-range structures (Benmore et al. 2013). Radially averaged 3D structural information obtained from the PDF traces of pXRD data can serve as interatomic fingerprint. Radial distribution function and differential PDF provide further details of local order such as atomic coordination number, i.e., the number of neighboring atoms for a reference atom at distance r (Benmore et al. 2013, Farrow and Billinge 2009). Limiting r-region to 20 Å of amorphous pharmaceuticals avoids fitting artifacts (Newman et al. 2008). The PDF amplitude of amorphous materials falls rapidly along r compared to crystals and the r leasing to the virtual loss of $G(r)$ oscillation yields the size of coherent domains (Atassi et al. 2010). Benmore et al. (2013) characterized amorphous APIs using synchrotron X-ray diffraction and spallation neutron diffraction. Structural information was lost after 5 Å of both glassy and crystalline carbamazepine. Miconazole nitrate and clotrimazole consist of the altered structural motifs in amorphous states from the respective crystals while amorphous cinnarizine retains the crystalline conformational residue. The PDF traces suggested the weaker memory of γ-form present in the amorphous indomethacin prepared from both $\alpha$and γ forms (Bates et al. 2006). Multivariate analysis (MVA)-PDF can efficiently assess the degree and the nature of disorder during milling, storage time/temperature-mediated crystallization of milled amorphous material (Bøtker et al. 2011; Naelapää et al. 2012; Boetker et al. 2012). The rapid acquisition PDF using synchrotron radiation of Ca-ketoprofen uniquely discriminated amorphous and mesomorphous structures of the crystalline phase (Atassi et al. 2010). Total diffraction data of a drug candidate, GNE068, were used for the in-depth investigation of the disordered mesophase (Chakravarty et al. 2013).

One hurdle associated with meaningful PDF analysis is the need of the broad Q range. Evidently, liming the range of Q causes poor resolution and thus remarkable loss of structural information. Indeed, a synchrotron source is preferable to generate total scattering data with the highest resolution consisting of minimal statistical uncertainties. The PDF data with $Q_{max} < 18$ Å^{-1} constitute all structural features for pharmaceuticals (Atassi et al. 2010). Dykhne et al. (2011) found that the PDF fingerprint of amorphous carbamazepine is unambiguously correlated with that of crystalline form when recorded using synchrotron radiation ($Q_{max} = 20$ Å^{-1}), silver-anode ($Q_{max} = 15.9$ Å^{-1}) or molybdenum-anode ($Q_{max} = 12.5$ Å^{-1}) while statistically poor and suboptimal resolution was obtained with Cu-anode ($Q_{max} \leq 8$ Å^{-1}). The PDF of melt-quenched carbamazepine collected using a synchrotron source by Billinge et al. (2010) revealed the same to be nanocrystallites instead of being amorphous. PDF analysis of synchrotron data of amorphous indomethacin showed poor correlation with both crystal (Billinge 2008) that was apparent using a Cu source (Bates et al. 2006). Overall, discriminatory PDF analysis requires at least the data collected using a molybdenum source.

14.4.2 Molecular Miscibility in ASD Using pXRD and Computational Analysis

Traditional pXRD analysis ascertains amorphicity through the absence of Bragg peaks. As such, X-ray halo patterns of amorphous composites lack direct information on the miscibility compared to thermal methods. Various computational, chemometric, and statistical analysis treatments of the measured pXRD infer the qualitative and semiquantitative information on miscibility in ASD (Ivanisevic 2010). Figure 14.5 depicts the general approaches for the miscibility assessment.

Since an amorphous physical mixture can be considered as the representation of the immiscible state, the pXRD of ASD overlaying a physical mixture implies amorphous immiscibility and any mismatch indicates miscibility. However, the miscibility inferred from this comparison is susceptible to the physical mixture preparation and many other factors. Provided the availability of reproducible X-ray amorphous halo of pure drug and polymer, it is convenient to generate pXRD pattern by linearly combining the weighted powder patterns of individual components (Fig. 14.5I). This digital pXRD pattern serves as the data for virtual physical mixture (Bates 2011). The change in local drug–drug and polymer–polymer coordination occurs in miscible ASD. Measurement under inert environment and low background improve the data quality for modeling (Newman et al. 2008; Ivanisevic et al. 2009). Since proper normalization of pXRD data is necessary before generating the digital pXRD, the type of normalization scheme can also affect ultimate results (Newman et al. 2008).

Likewise, the dissimilarity between calculated PDF traces from pure components and PDF trace of ASD qualitatively infer the miscibility. PDF transformation autonormalize data in an absolute scale of electron units amplifying subtle differences that enables robust fitting. If the ratio of the scale factors conversing the simulated PDF pattern to the measured PDF of ASD results the theoretical drug/polymer ratio, the ASD is phase separated, otherwise miscible. Newman et al. (2008) contrasted the total percentage from composition calculated using PDF against the real composition for ASD of dextran–PVP, indomethacin–PVP, and trehalose–dextran dispersions. In agreement with T_g's of dextran and PVP exhibited by the DSC thermograms of ASD, the theoretical blend compositions could be correctly modeled by PDF. Higher deviation of PDF-derived total percentage from the true value is noticeable for the single T_{gm} indomethacin–PVP systems. However, the PDF data evidenced phase separation in dextran–trehalose ASD despite a single T_{gm}. Thus, PDF analysis detects phase separation in ASD with multiple T_{gm} as well as in the single T_{gm} system. Since PDF analysis experiences higher random errors compared to direct pXRD pattern analysis, statistical significance of the differences should be confirmed. Moore et al. (2010) utilized the error propagation in PDF analysis of miscibility. The sum of squares of the errors of the individual curves estimates the error in the difference between the measured and the digital PDF of ASD ($\Delta G(r)$ residuals). The statistical significance of the immiscibility/miscibility can be established through hypothesis testing. The inclusion of zero by the plot of $\pm 3\sigma$ interval around the residual plot throughout the r-range ensures the acceptable condition for inferring the immiscibility, e.g., terfenadine–PVP (Fig. 14.5III). While the exclusion of zero by the $\pm 3\sigma$ difference plot intervals indicates statistically significant dissimilarities between experimental PDF of ASD and the physical mixture indicating miscibility, e.g., felodipine–PVPVA 64 ASD.

Ivanisevic et al. (2009) applied pure curve resolution method (PCRM) and alternative least square (ALS) method to pXRD data of ASD for miscibility studies. The PCRM-based approach follows the analysis of the variance among the measured intensity points of pXRD patterns of ASD with varying drug loadings thus

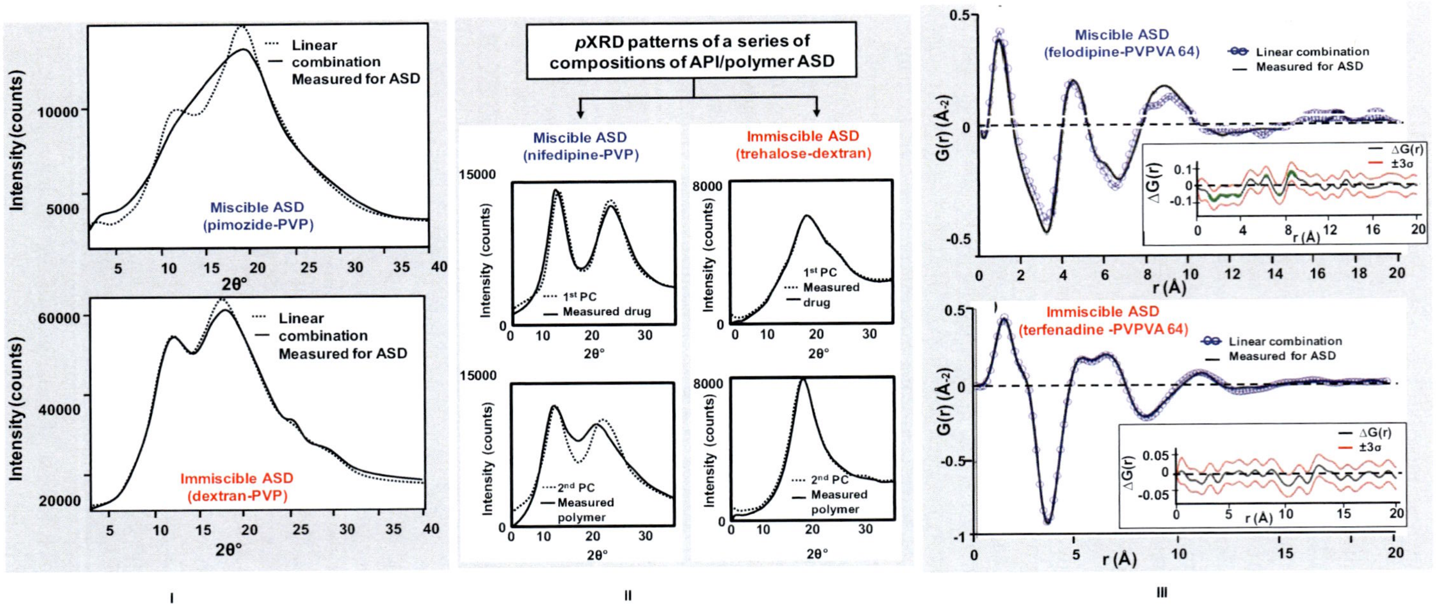

Fig. 14.5 Illustrations of miscibility assessment using powder X-ray diffractometry (*pXRD*) data: I. comparison of linear combination of drug and polymer pXRD patterns with measured pattern of *ASD*, II. comparison of pure component pXRD pattern extracted from multivariate curve resolution of ASD patterns with the measured patterns of pure components, and III. comparison of linear combination of drug and polymer pairwise distribution function (*PDF*) patterns with the PDF pattern derived from the measured pXRD data of ASD (residual plot with standard deviation is shown in inset). (Schemes are constructed based on Bates 2011; Ivanisevic et al. 2009; Moore and Wildfong 2011)

eventually allowing the extraction of pure curves (PCs), i.e., pXRD patterns of the constituting components (Fig. 14.5II). The variance of the measured powder patterns is describable by two extracted PCs representative of API and excipient for phase-separated systems representing the weighted drug–drug and polymer–polymer intermolecular interactions as per blend composition. The miscibility is diagnosed by the poor fit of extracted PCs with the measured powder patterns of pure components as drug–polymer coordination additionally contributes the orthogonal PCs of API and polymer (Fig. 14.5II, left panel). Additional component or the shift of the amorphous halo position(s) on the order $\geq 1°2\theta$ in extracted PCs can even indirectly indicate miscibility. The ratio of scale factors of PCs corresponds to the blend phase-separated composition. ALS approach can estimate the drug–polymer coordination number representing the degree of miscibility as the drug–polymer coordination substitutes the drug–drug and polymer–polymer coordination as a function of composition.

The homogeneity in ASD can be investigated within the range of 1–100 nm using SAXS (Laitinen 2009). The intensity and angular distribution of the scattered intensity in SAXS patterns can yield the size or surface area per unit volume of a constituting domain. Apart from the atomic structure, the electronic density variations of amorphous composites stem from the heterogeneity of the microstructure. Laitinen et al. (2009) studied the distribution of perphenazine in its ASD with PVP and PEG. The pair density distribution profiles as a function of cluster radius were extracted from SAXS data. The ASD exhibited a peak with maxima < 30 nm along with a minor peak at 70–90 nm, as for pure polymers. This was considered as the sign of molecular level dispersion formation. A subsequent SAXS study of the effect of storage at elevated humidity showed unaltered size of clusters pointing to the retention of molecular miscibility (Laitinen et al. 2010).

14.4.3 Crystallization Studies and Crystallinity of Amorphous Systems Using pXRD

The inherent specificity and quantitative power towards the crystalline state make pXRD as the technique of choice for studying crystallization from amorphous systems (Ochsenbein and Schenk 2006). Also, the instrumental flexibility allows in situ monitoring of crystallization as function of time, temperature, pressure, RH, or combinations. An API can crystallize to different polymorphs from ASD (Guns et al. 2011). The reference powder pattern enables the identification/quantification of the polymorphs developing in ASD (Ivanisevic et al. 2010). Preferred orientation of crystalline faces is a major source of error. Generally, analysis in transmission mode reduces the preferred orientation and avoid other instrumentally induced distortions and anisotropic shifts (Moore et al. 2009).

Quantification of the degree of crystallinity present initially, induced during storage or processing in amorphous pharmaceuticals by pXRD is well established. The powder pattern of a partially crystalline dispersion includes Bragg peaks associated with the prominent crystal faces superimposed to the amorphous halo. The intensity and area of Bragg peaks systematically increase with increases in crystallinity and in expense of the amorphous halo. An indirect quantitative method for crystallinity utilizes the ratio of intensity of Bragg peaks of the sample to that of the (internal/external) reference standard. Since the predominant amorphous fraction largely contribute to the diffraction pattern of ASD, an indirect method presents a major analytical challenge. The ratio of net areas of all Bragg peaks to the sum of areas of Bragg peaks and of the corresponding amorphous halos beneath at the respective 2θ obtained from the normalized pattern directly yield the degree of crystallinity in a sample (Paudel et al. 2013; Rumondor and Taylor 2010; and Zidan et al. 2012). This is a common method for solid dispersions and works well unless the powder pattern comprises many overlapping peaks. Constructing calibration curves of partially crystalline dispersion is tedious for this approach since it is hard to generate the calibrants with identical microstructures as that of the analyte. Preferably, calibrants must contain ASD mixed with the known quantity of crystalline API. Modern pXRD software allows integrating the area under a particular Bragg peak and amorphous halo below it separately. Rumondor and Taylor (2010) found for felodipine–PVP system that the predicted crystallinity by partial least-squares (PLS) method exhibited less error compared to the area ratio method. The use of intensity ratio of a peak or a subset of characteristic peaks is also ubiquitous in quantifying crystallinity in ASD (Paudel and Van den Mooter 2012; Paudel et al. 2013). It requires all measurements under identical conditions and no peak broadening due to lattice strain or particle size and the use of an internal standard. If an ample region in pXRD pattern lacks Bragg peaks, the diffraction intensities of the noncrystalline regions representing the short-range order can be used to estimate the amorphicity (Ivanisevic et al. 2010). The difference in X-ray absorption coefficients of API and polymer should be duly considered before the quantitative analysis by such approach (Madsen et al. 2011).

More challenging is to quantify trace (nano) crystallinity in ASD. The Scherrer broadening due to the small and imperfect crystallites present in a low volume fraction can possibly bring the diffracted Bragg intensity down to that of the diffuse halo of an amorphous fraction (Koch and Bras 2008). These scenarios can lead to the false impression of complete amorphicity. Total scattering pattern enables quantification of trace crystallinity, since the scattering vector (Q) is proportional to the crystalline fraction (Liu et al. 2012). Benmore et al. 2013 quantified the trace crystallinity in various amorphous APIs by comparing the Bragg scattering intensity to the diffuse scattering intensity measured using synchrotron X-ray source. The intensities of the halo of various ASDs are shown to alter with aging as for PVP-based dispersions of nifedipine, felodipine, indomethacin, etc. (Ivanisevic 2010).

Temporal pXRD measurements enable monitoring in situ isothermal crystallization kinetics from ASD (Ivanisevic et al. 2010). Relative crystallinity is often estimated using the area ratio method at different time periods. The crystallinity degree at each time point is normalized with that at the point where crystallization reaches a maximum and is stationary. The extents of crystallization can be plotted as a function of time and modeled using diverse models of crystallization. Since the fractals and particulate properties of growing virgin crystals are difficult to predict, only relative crystallinity can be estimated with such experiments and is sufficient for comparison of kinetics. Also, long pXRD-recording programs may enhance crystallization rate by X-ray energy leading to the overestimation of the rate as reported for APIs like nifedipine (Ivanisevic 2010). Paudel et al. (2013) monitored the kinetics of crystallization of naproxen at 94 %RH from naproxen–PVP ASD prepared by spray drying at various process parameters.

14.5 Gravimetric Vapor Sorption (GVS)

The moisture–solid interaction is an inevitable aspect of pharmaceutical development. Elucidation of moisture-induced physical alterations in amorphous pharmaceuticals is crucial, especially for ASD. Gravimetric measurement on the rate and extent of moisture gain (sorption) by or loss (desorption) from amorphous samples as a function of RH or as a function of time at a constant RH (isohumic condition) can provide a wealth of information of ASD. The key structural properties of ASD measureable by moisture sorption/desorption are drug–polymer interactions, moisture-induced glass transition, crystallization, hydrate formation/dehydration, etc. while that associated with particulate or bulk properties are hygroscopicity, diffusivity, pore size, surface area, etc. (Burnett et al. 2009).

The traditional way of installing the desired RH is by using a saturated aqueous solution of a particular salt or salt mixture (Greenspan 1977). Commercial instruments are now available, which can measure gravimetric moisture sorption and desorption while applying an RH ramp or at the selected humidity, known as dynamic or gravimetric vapor sorption analyzers (DVS or GVS; Penner 2013). Thus, moisture sorption/desorption isotherms can be expressed as a function of RH or time. The sorption isotherms may appear in various shapes for different types of samples and several theoretical models exist for describing different types of sorption isotherms, which is beyond the scope of our current discussion. The interested readers are referred to the standard books for the details (Reutzel-Edens and Newman 2006). In addition, moisture diffusivity in an amorphous solid matrix is an extremely complex process, which depends upon the hygroscopicity, texture, microstructure, porosity, surface chemistry and particle size, morphology, moisture-induced structural relaxation and phase transformations as well as the extent of water content. Different models have been proposed to account these phenomena, but none can provide complete account of all underlying processes (Perdana et al. 2014).

An interesting use of GVS, for amorphous dispersions, is detection and quantification of crystallinity. Due to higher hygroscopicity of hydrophilic polymers and amorphous API of ASD, the exposure of a sample to elevated humidity environment triggers the sorption of water molecules by polar/hydrophilic functional groups at the air–solid interface. The α-relaxation time of amorphous material often correlates with its water sorption potential (Bhardwaj and Suryanarayanan 2013). The gravimetric vapor sorption (GVS)–desorption led to the dissimilar structurally reversal of annealed amorphous trehalose when compared to that obtained by heating beyond T_g (Saxena et al. 2013). The sorbed water molecules ($T_g = -137\,°C$) increases the molecular mobility and plasticize ASD (Rumondor et al. 2009). The extent of depression of T_g due to moisture sorption depends upon the water—sample intermolecular interactions (Yuan et al. 2011; Crowley and Zografi 2002; Roos 1995). The moisture-induced glass-to-rubber transition drastically increases sample fluidity after a certain RH. With sufficient moisture sorption and time, the increase in molecular mobility facilitates the energy barrier and hence induces nucleation/crystallization. Crystallization of an amorphous phase sharply reduces the tendency of moisture sorption and this might be apparent from the change in the slope of the sorption isotherm from the particular RH (Hancock and Dalton 1999; Roe and Labuza 2005). The moisture-induced nucleation and crystal growth manifest drastic reduction in surface area (voids/spaces) and surface free energy of the sample. This appears as a sharp drop in the moisture sorption profile originating from the loss of excess water during crystallization. The plots of extent of mass loss as a function of time yield the crystallization kinetics profile at particular RH. Temperature also imparts the synergistic effect on the induction point (RH or time) of crystallization. The ratio of moisture induced reduced T_g to the onset of crystallization for amorphous lamotrigine mesylate was fairly constant over a wide range of RH and temperature combination (Schmitt et al. 1996). Interestingly, the logarithm of onset of crystallization time below T_{gm} (dry) was linearly related to the ratio of reduced T_{gm} to the experimental temperature (corresponding to the combination of RH/T) for ASD of the drug candidate SAR and HPMC phthalate (Greco et al. 2012). Yang et al. (2010) reported the linear relationship between the crystallization constant and RH for efavirenz–PVP ASD. Various methods have been reported for the quantification of amorphous content below 1 % using GVS that are based on the inherently higher hygroscopicity of amorphous phases compared to crystalline counterparts (Young et al. 2007). The extent of equilibrium moisture sorption at a particular RH for calibrants with known amorphicity enables the quantification. However, the possible contribution of various accompanying phenomena, such as moisture-induced phase separation, limits the quantitative application of GVS at higher amorphous content. Another hurdle in using GVS analysis for the precise quantification of crystallinity of an API in pure form or in ASD would be for the system wherein the incomplete crystallization occurs even by the end of the DVS cycle, which indeed implies the reliance on the complementary confirmation data post DVS analysis (Qi et al. 2010a, b).

Gravimetric moisture sorption data at a particular RH provide an estimate of the strength of binary intermolecular interactions among drug, polymer, and water molecules and hence the extent of moisture-induced destabilization possible in ASD (Crowley and Zografi 2002; Paudel et al. 2010). By applying Flory–Huggins–Vrentas expressions for drug–polymer–water systems, drug–polymer interaction parameters have been derived from the moisture sorption data of physical mixtures or ASD of many drug–polymer systems (Paudel et al. 2010; Rumondor et al. 2009). The non-covalent interactions such as H-bonding established by hydrophobic drug with the hydrophilic moiety of the hygroscopic polymer in ASD lead to the reduction of moisture sorption by the polymer in ASD compared to the pure state, the process sometimes known as hydrophobization. Therefore, stronger drug–polymer interactions makes the polymer more hydrophobized and hence leads to the larger negative deviation of moisture uptake by ASD as compared to the linear addition of weighted values for drug and polymer. The PVP/VA 64-Eudragit ASD markedly retarded extent and rate of water sorption compared to the corresponding physical mixture (Yang et al. 2013). The same polymer blend loaded with 10 % felodipine prepared by melt extrusion showed drastic improvement on the moisture protection for the amorphous system. With some geometric assumption, van Drooge et al. 2006 utilized a model to estimate the size of amorphous diazepam clusters dispersed in phase-separated PVP-based ASD using moisture sorption data. The thickness of the hydrophobized layer and volume of hydrophobized PVP were estimated assuming that the drug dispersed at molecular level reduces the hygroscopicity more than clusters having less drug–PVP contacts. The estimated size of amorphous drug clusters increased from 9.6 (265 molecules) to 20.8 nm (2674 molecules) while increasing the API content from 35 to 65 % in phase-separated ASD.

14.6 Other Techniques

Besides thermal, diffractometric, and moisture sorption techniques, there are ample other techniques, as aforementioned, that are commonplace for the analysis of the various structural features of ASD. Table 14.1 lists various techniques and brief descriptions for their measuring principles relevant to ASD. The details of working principle and various other applications of these techniques are available elsewhere. Some representative and relevant references are also included in Table 14.1. Tables 14.2–14.8 portray the structural characteristics measurable by various these techniques and also possible hurdles and interferences.

14.7 Regulatory Perspective of ASD Analysis

Chemistry, manufacturing and control of API, and dosage forms are strictly regulated by United States Food and Drug Administration (FDA; USFDA 2012). The current regulatory initiation towards the risk-based quality by design (QbD) approach encourages the implementation of a range of analytical tools varying in the depth of information, operational cost and time during different developmental stages and product life cycle (ICHQ8(R2) 2009). Many regulatory challenges thus apply on the analytical techniques characterizing and controlling the key quality attributes of these products. The inevitable qualities in any finished dosage form are the intended level of pharmaceutical performance and physical/chemical stability. Various molecular properties of intermediate and final product such as phase homogeneity, molecular mobility, miscibility, hygroscopicity, crystallinity, surface chemistry, etc. can principally govern the quality attributes of ASD-based products (in vitro dissolution, in vivo behavior, and physical stability). Thus, characterization and monitoring of these physical attributes during preformulation, manufacturing, product release, and stability studies enable the priori control on the ASD-based product quality ensuring the patients need and the expectations of regulatory agencies. This requires various analytical techniques that interface among physical chemistry, materials science, surface chemistry, etc.

Preformulation studies furnish the physicochemical data including relevant solid-state properties (amorphous stability, crystallization kinetics, hygroscopicity, etc.) of a drug candidate intended for ASD development (biopharmaceutics classification system (BCS) II/IV) during preclinical stage. Early screening for polymer(s), (optional) surfactant selection and drug loading includes high-throughput solvent casting or melt quenching with a series of compositions (Paudel et al. 2012). At this stage, rapid and in situ analysis by polarized light microscopy and vibrational spectroscopic probes (IR, Raman) enable identification of the crystallizing systems and compositions. Furthermore, DSC analysis will assist for miscibility studies and composition-phase analysis. *In sample,* micro-dissolution follows the solid-state analysis that evaluates the stability of aqueous supersaturation generated by ASD (Wyttenbach et al. 2013). Again, spectroscopic probe or microscopy can monitor the physical manifestation, especially nucleation/crystallization or phase separation while in contact with the aqueous medium. Accelerated stability studies under the elevated RH/temperature can be studied in situ using pXRD or spectroscopy. These preclinical formulation prototyping supported by promising animal exposure data assist the nomination of first-in-human ASD and together constitute the data for investigational new drug submission.

More detailed product characterization increasingly stipulates sophisticated analytical tools while the clinical studies on drug candidate progresses. The physicochemistry of drug candidate and excipient(s) of the elected formulation and characterization data obtained during preclinical stage will guide the manufacturing feasibility studies at the laboratory scale and process selection for intermediate ASD during phase I studies. Implementation of various non-destructive spectroscopic process analytical technology (PAT) probes such as Fourier transform infrared spectroscopy (FTIR), Raman, near-infrared spectroscopy (NIR), or terahertz (THz) fiber optics at multiple locations of processing equipment will provide real-time monitoring of amorphization and evolution of desired physical properties such as compositional homogeneity or stabilizing interactions in the resulting ASD. This *in-process* supervisory control of product quality attributes will ensure running of the process within the design space. Such PAT tools have been successfully applied to some ASD manufacturing by hot melt extrusion (Saerens et al. 2013a, b). Analytical tools highly sensitive to the minute alteration in product physical structure can only discriminate the process parameters effect on physical attributes of ASD relevant to the performance and physical stability. Comprehensive studies on solid-state solubility and kinetic miscibility of the selected drug–polymer combination require mDSC experiments and spectroscopic studies. Thermoanalytical data can be modeled using suitable theoretical models to extract physical stability descriptors such as heteromolecular interaction parameters. Furthermore, vibrational spectroscopy such as FTIR, NIR, Raman, and THz will help in identifying different stabilizing intermolecular interactions such as H-bonding. Often, these techniques can be insensitive to the subtle process-related microstructural variations. Therefore, use of total scattering (PDF) pXRD and SS-NMR spectroscopy/relaxometry is encouraged at this stage for the estimation of miscibility at molecular level. AFM-based techniques implemented to characterize molecular miscibility and stability of melt-quenched films has shown promising correlation to that obtained in the dispersion prepared by hot melt extrusion (Lauer et al. 2013). SS-NMR or sensitive pXRD measurements can verify various features such as the extent of molecular interaction, crystallinity, domain size, etc. achievable by faster techniques such as NIR, Raman. Detailed elucidation of various kinetic processes concomitantly occurring during bio-relevant in vitro drug dissolution such as nucleation/crystallization, plasticization, polymer swelling, aggregation, etc. can facilitate more rational correlation with in vivo human pharmacokinetic data. Various spectroscopy probes (IR, Raman, THz), microspectroscopic imaging (NIR, Raman, MRI, etc.), microscopy to scattering techniques (SAXS and SANS) can provide different level of information in such investigations.

Rigorous dosage from development and manufacturing activities commencing parallel to the phase II clinical studies further commend stronger analytical support. DVS moisture sorption/desorption analysis, surface analysis using time-of-flight secondary ion mass spectroscopy (TOF-SIMS), X-ray photoelectron spectroscopy (XPS), IGC, or localized thermal analysis will help in the process-related effect on the surface versus bulk microstructure of ASD that can be relevant to physical and chemical stability. Comprehensive elucidation of primary and various secondary molecular motions using DES, thermally stimulated discharge current (TSDC), or SS-NMR can help in predicting the physical stability against crystallization. The projected stability under stressed condition can be experimentally verified. Advanced multidimensional SS-NMR spectroscopy/relaxometry will provide the superior dynamic and structural information of ASD product that can be correlated with the process as well as performance. The intended products of most ASD are solid oral dosage forms such as tablets and capsules. This necessitates various particulate level characterizations (powder flow, micromeritics, etc.) as well as compatibility studies of API in ASD in presence of the intended adjuvants such as fillers, binders, lubricants, etc. With the further dilution of drug in the finished product, the analytical tool needs to more sensitive. The use of localized tests with spatiochemical specificity such as microspectroscopic imaging (NIR, Raman) and micro/nano-thermal analysis can be more helpful in elucidating content as well as phase distribution. Further, sound understanding on the effect of downstream processing such as granulation, compaction, coating, etc. on the product microstructure require surface sensitive and discriminatory techniques. Compression can also lead to molecular demixing in some systems (Ayenew et al. 2012). Dynamic TOF-SIMS revealed the orientation of pyrrolidone group of PVP towards the surface during compression to develop tablets containing PVP-based ASD (Leane et al. 2013). Coating-induced trace crystallization was detected by pXRD.

The commercial drug product manufacturing process with real-time monitoring is established by phase III clinical stage, since the intended commercial formulation is preferably used for these studies. Therefore, robust PAT tools and well-validated multivariate chemometric models need the finalization to set the mechanistic QbD framework of process monitoring in relation to the product specification (real-time release and stability). Solid-state chemical and physical stability are integral parts of the pre-approval accelerated and long-term stability reports for new drug application filing. Sensitivity of solid-state techniques should be adequate to precisely identify any alteration of physical structure (mobility, miscibility, crystallinity, etc.) in finished product in comparison to the previously characterized intermediate ASD. Based on the nature of the product and behavior in the stability environment, mDSC and/or pXRD and vibrational-spectroscopy-based methods can be established as stability monitoring methods. This becomes often challenging as API in ASD-based final product is further diluted. The analytical capabilities of these standard techniques can be inadequate in case of consolidated finished dosage tablets containing ASD equivalent to API content $< 1\,\%$. The limit of detection and quantitation of chemometric NIR method for crystallinity spiked in indomethacin–PVP/VA ASD was recently found the lowest compared to pXRD and Raman methods in tablets with total drug content below 0.5 % (Palermo et al. 2012a, b). This forecasts a great utility of NIR method for online stability monitoring in ASD-based tablets of potent medications. The microstructures of amorphous films in coated tablets can significantly alter during storage that can impact the drug release profile which can be monitored using PALS. In general, regulators seek more characterization data on finished products than on the intermediate ASD and they also make more sense in correlating quality attributes of the finished products with its stability and in vivo data. Unfortunately, there seems to be very few scientific contributions on characterization of ASD-based finished dosage form with a great detail in comparison to the intermediate ASD.

The post-approval life cycle management of ASD-based products can become more challenging compared to ordinary crystalline formulation. Any alteration in physical structure of the product related to technology transfer and scale-up should fall under the scientifically characterized design space and should clearly reflect in the analytical data generated. The FDA recalls of nearly 100 solid oral dosages of small molecular API in past 2 years due to the physical defect or instability implied to criticality of product stability control (Recalls 2011–2013). The major specifications of current regulations are towards the chemical stability and degradations in the drug substances and product while limited guidance exists on the physical attributes. Moreover, predictability of physical stability of ASD-based products in ambient condition (shelf life) by routine accelerated stability testing is quantitatively poor and challenging. Therefore, different line extension products of ASD systems may need analytical data, specific to the particular products, to file the abbreviated applications.

A schematic overview of various structural characterization tools implementable at different stages of ASD-based product development is provided in Fig. 14.6.

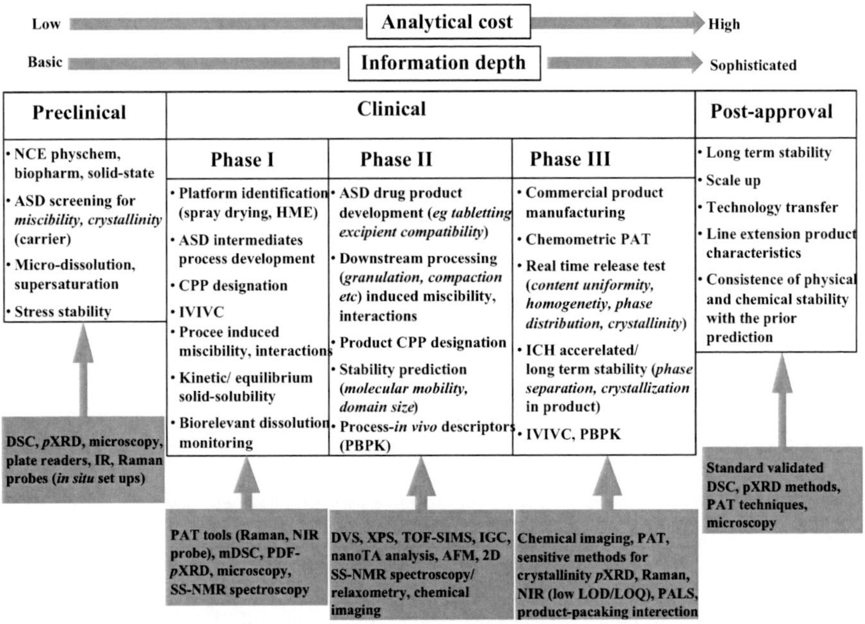
Low
Analytical cost
High
Basic
Information depth
Sophisticated
Preclinical
Clinical
Post-approval
• NCE physchem, biopharm, solid-state
• ASD screening for miscibility, crystallinity (carrier)
• Micro-dissolution, supersaturation
• Stress stability
Phase I
Phase II
Phase III
• Platform identification (spray drying, HME)
• ASD intermediates process development
• CPP designation
• IVIVC
• Procee induced miscibility, interactions
• Kinetic/ equilibrium solid-solubility
• Biorelevant dissolution monitoring
• ASD drug product development (eg tabletting excipient compatibility)
• Downstream processing (granulation, compaction etc) induced miscibility, interactions
• Product CPP designation
• Stability prediction (molecular mobility, domain size)
• Process-in vivo descriptors (PBPK)
• Commercial product manufacturing
• Chemometric PAT
• Real time release test (content uniformity, homogenetiy, phase distribution, crystallinity)
• ICH accerelated/ long term stability (phase separation, crystallization in product)
• IVIVC, PBPK
• Long term stability
• Scale up
• Technology transfer
• Line extension product characteristics
• Consistence of physical and chemical stability with the prior prediction
DSC, pXRD, microscopy, plate readers, IR, Raman probes (in situ set ups)
PAT tools (Raman, NIR probe), mDSC, PDF-pXRD, microscopy, SS-NMR spectroscopy
DVS, XPS, TOF-SIMS, IGC, nanoTA analysis, AFM, 2D SS-NMR spectroscopy/ relaxometry, chemical imaging
Chemical imaging, PAT, sensitive methods for crystallinity pXRD, Raman, NIR (low LOD/LOQ), PALS, product-pacaking interection
Standard validated DSC, pXRD methods, PAT techniques, microscopy

Fig. 14.6 Schematic presentation of utilization of various structural characterization tools implementable at different stages of ASD-based product development

The remaining techniques are summarized in Tables 14.1–14.9 at the end of the chapter.

Table 14.1 Description of various analytical tools utilized for the analysis of ASD

Techniques	Measuring principle/subtype
Dielectric relaxation spectroscopy (DRS) (Asami 2002; Kaatze 2013; Vassilikou-Dova and Kalogeras 2008)	The responses from orientational polarization within the analyte subjected to a varying external electric field Isochronal DRS: Constant frequency with temperature scanning, or varying time Isothermal frequency sweep broad-band DRS Varying frequency at constant temperature, measurement at different temperatures generates 3D profile
Thermally stimulated current (TSC) spectroscopy (Boutonnet-Fagegaltier et al. 2002; Gun'ko et al. 2007; Vassilikou-Dova and Kalogeras 2008)	Current released from analyte measured on linear reheating after polarization during quenching of the heated sample from the designated temperature under a constant DC electric field to a sub-ambient temperature ($\ll T_g$) Thermally stimulated depolarization current (TSDC) Thermal-windowing TSDC (TW-TSDC) Short temperature windows of polarization steps Thermally stimulated polarization current (TSPC) Current measured during first heating
Dynamic mechanical analysis (DMA) (Jones et al. 2012; Kalichevsky et al. 1992; Vassilikou-Dova and Kalogeras 2008)	Rheological behavior or mechanical stiffness (modulus) of the material subjected to an oscillating mechanical stress at selected frequencies Isochronal temperature sweep DMA (DMTA) More than 80 % application, constant frequency with temperature scanning, or varying RH Isothermal frequency sweep DMA Analogous to frequency sweep DRS
Solid-state vibrational spectroscopy/microscopy (Amigo 2010; Breitkreitz and Poppi 2012; Chen et al. 2011; Gendrin et al. 2008; Jørgensen et al. 2009; Kazarian and Ewing 2013; McIntosh et al. 2012; Pavia et al. 2001; Prats-Montalbán et al. 2012; Reich 2005; Van Eerdenbrugh and Taylor 2011; Zeitler et al. 2007)	Change in frequency of incident electromagnetic radiation due to the absorption or inelastic scattering of a portion resonant to the frequency of molecular, lattice or phonon vibration (stretching, bending, bending, rocking, wagging, scissoring, etc.) Middle infrared (MIR or IR) spectroscopy Asymmetric vibration of the polar bonds mediated by the alteration of the dipole moment Near infrared (NIR) spectroscopy Overtones and combinations of the fundamental vibrations Far infrared (terahertz, THz) spectroscopy Supramolecular vibrational, e.g., H-bonding, halogen-bonding or lattice/phonon vibrations Raman spectroscopy Scattering by non-isotropic change in molecular polarizability and leading to changes the rotational and vibrational motions Hyperspectral imaging/mapping Sequential collection of spectral data by locally scanning the selected areas within a static sample or simultaneous acquisitions of thousands of spectra resolved spatially PAT application Portable and temperature/pressure resistant fiber optics of NIR and Raman probes

Table 14.1 (continued)

Techniques	Measuring principle/subtype
Solid-state nuclear magnetic resonance (SS-NMR) techniques (Belton and Hills 1987; Courtier-Murias et al. 2012; Duer 2004; Fu and Sun 2011; Mantle 2011)	Interaction between a static magnetizing field and the spin angular momentum of spin active nuclei processing with angular frequency of a sample placed inside a radiofrequency (*rf*) coil ^{1}H SS-NMR spectroscopy Diverse local chemical environments makes a same proton nucleus experiences varying shielding effect against external field ^{13}C CP-MAS SS-NMR spectroscopy Magnetization transfer from abundant ^{1}H nuclei to rare ^{13}C(CP) and magic angle spinning (MAS) to nullify chemical shift anisotropy Heteronuclear (X = ^{15}N,^{19}F,^{17}O,^{31}P) SS-NMR spectroscopy Same as ^{13}C CP-MAS, but more selective 2D SS-NMR correlation spectroscopy Homo-nuclear (e.g.^{1}H-^{1}H) correlation (COSY), hetero-nuclear (^{1}H-X, X = ^{15}N,^{19}F,^{17}O,^{31}P) correlation (HETCOR) through- space and/or -bond dipolar coupling and nuclear interactions recorded by designed rf pulse sequences SS-NMR relaxometry Measurement, in time domain, of the rate of nuclear magnetization decay towards equilibrium following the excitation by an rf-pulse Longitudinal (spin-spin)-T_1(static)/$T_{1.}$ (rotating frame) relaxation time Transverse (spin lattice) - T_2 (from NMR line width) relaxation time NMR imaging Magnetic resonance (MRI), constant time (CTI) Spatial molecular mobility recorded via T_2 relaxation times of proton nuclei
Microscopic techniques (Carlton 2011; Dieing et al. 2011; Eddleston et al. 2010; Vitez et al. 1998)	Interaction of molecular solids with optical/electronic source resulting information on morphology, solid state Polarized light microscopy Identification of solid form based on birefringence and morphology Scanning electron microscopy (SEM) Focused electron beam scanned over the specimen surface under a high vacuum Penetrate the sample interior and the emitted electrons produces the specimen topography Transmission electron microscopy (TEM) Spatial variation of the inelastic interaction of electron beam while transmitting through a sample produces highly resolved image
Thermo-gravimetry (TGA; Giron et al. 2004)	Weight change of the sample as a function of temperature and/or time measured using a sensitive microbalance

Table 14.1 (continued)

Techniques	Measuring principle/subtype
Surface analytical techniques (Barnes et al. 2011; Brown and Vickerman 1984; Chehimi et al. 2011; Clark 1977; Harding et al. 2007; Ho and Heng 2013; Pollock and Hammiche 2001; Sitterberg et al. 2010; Thielmann and Levoguer 2002; Voelkel et al. 2009)	Atomic force microscopy (AFM) A sharp probe on a flexible cantilever tip (10–20 nm) scanned (contact mode, intermittent contact or tapping mode or non-contact mode) on a sample surface experiences attractive and repulsive forces thus yield high-resolution image and nano-mechanical information Localized thermal analysis (LTA; micro- or nano-thermal analysis) An AFM probe equipped with a thermal probe heats up sample resulting topographical imaging and thermal analysis of specific areas of the sample surface, sudden change in deflection of the heated cantilever indicates a thermal transition Time-of-flight secondary ion mass spectrometry (ToF-SIMS) Mass spectrometry of sample surface maps spatial compositional information: the distribution of compounds within 1–2 nm depth X-ray photoelectron spectroscopy (XPS) Low-energy X-ray irradiated on a sample surface under high vacuum allows quantitation of atomic concentration over a depth up to 10 nm Inverse gas chromatography (IGC) The retention behavior (volume, time) of a probe gas flowing at constant rate through a column packed with the solid sample yield information on surface interaction/energetics
Emerging techniques (Dlubek et al., 2007; Kissick et al., 2011; Magazu et al., 2011; Pansare et al., 2012; Zelkóa et al., 2006)	Positron annihilation lifetime spectroscopy (PALS) Molecular free volume (distribution) of a sample probed by the ortho positronium (o-Ps) formed by the parallel spin interaction of injected positron atom with the electron of the sample Second order nonlinear optical imaging of chiral crystals (SONICC) Selective images are produced by the detection of half the incident wavelength due to the occurrence of second harmonic generation by well-ordered chiral crystals Fluorescence resonance energy transfer (FRET) The extent of nonradiative energy transfer via long-range dipole–dipole coupling from excited fluorescent donor molecule to acceptor molecule is inversely related to the pair distance (3–10 nm) Neutron scattering Incoherent neutron scattering by the sample atoms provides information the self-dynamics such as mean square displacements and the onset of anharmonic motions of these atoms
Simultaneous tools (Dazzi et al. 2012; Feth et al. 2011; Ghita et al. 2008; Huang and Dali 2013; Pandita et al. 2012; Pili et al. 2010; Rahman et al. 2013)	Simultaneous multi-methodological measurement on a same sample, increases confidence on the data interpretation of complex systems DSC-TGA, DSC-XRD, DSC-Raman, DSC-FTIR, and DSC-NIR GVS-NIR AFM-IR

PAT process analytical technology

Table 14.2 Structural properties of ASD measurable by dielectric relaxation spectroscopy (DRS) and possible hurdles

Molecular mobility (Adrjanowicz et al. 2009, 2012; Bhugra et al. 2008; Bra's et al. 2008; Carpentier et al. 2006; Diogo and Moura-Ramos 2009; Havriliak and Negami 1967; Kaminski et al. 2011; Schönhals; Williams 2009)	Miscibility (Bhattacharya and Suryanarayanan 2011; Caron et al. 2010; Grzybowska et al. 2012; Kaminska et al. 2013; Korhonen et al. 2008; Roig et al. 2011; Tarek El and Roland 2007; Vassilikou-Dova and Kalogeras 2008; Zhang et al. 2004)	Crystallinty/crystallization (Alie et al. 2004; Bhattacharya and Suryanarayanan 2011; Bra's et al. 2008; Dantuluri et al. 2011; Gnutzmann et al. 2013; Grzybowska et al. 2012; Kaminski et al. 2011; Maheswaram et al. 2013; Wojnarowska et al. 2010)	Interference/limitations (Bhardwaj and Suryanarayanan 2011; Caron et al. 2010; Micko 2012; Vassilikou-Dova and Kalogeras 2008)
α-Process at T_g and other processes above T_g, and other sub-T_g relaxation processes (JG-, β-, γ-...)	Temperature scanning isochronal DRS exhibits composition-dependent T_g of blend as in DSC	Amorphous materials possess up to 10^8 times higher conductivity than crystalline	DC conductivity interferes the low frequency signals
Relaxation profiles obtained as loss peak in imaginary dielectric curve and as step jump of real permittivity line	Partial miscibility might exhibit multiple relaxation peaks overlaying or towards individual components, or display diverse relaxation rate even $> T_g$	Non-isothermal crystallization by isochronal temperature sweep DSC (cold crystallization and melting discernible as sudden drop and steep rise of static dielectric permittivity, respectively)	Crystallization onset predicted using DRS relaxation time in coupling model faster than the experimental time
Relaxation times are generally shorter than measured by thermal analysis	Molecularly mixed system show broader loss peak compared to individual peak	Isothermal crystallization above and below T_g can be studied by broadband DRS	Effect of water difficult to discern
Various model available to fit the non-Debye relaxation profiles	Width of loss peak also provide the measure of dynamic heterogeneity	Dielectric intensity/strength or α-peak decreases as crystallization progresses with time)	Exact identity of molecular relaxor needs complementary analysis
	Loss peak of some secondary processes can obscure with the stronger inter-component H-bonding in blend		Physical meaning of relaxation times parameters obtained by fitting DRS data by existing molecular mobility models still unclear
	DRS probes plasticization of selective motional process		

DSC differential scanning calorimetry

Table 14.3 Structural properties of ASD measurable by thermally stimulated current (TSC) spectroscopy and possible hurdles

Molecular mobility (Alie et al. 2004; Bhugra et al. 2008; Hirakura et al. 2007; Moura Ramos et al. 2002; Vassilikou-Dova and Kalogeras 2008; Viciosa et al. 2009)	Miscibility (Dong et al. 2008; Ghosh et al. 2011; Roig et al. 2011; Shmeis et al. 2004a, b)	Interference/limitations (Antonijevi'c et al. 2008; Dong et al. 2008; Jain et al. 2012; Shi et al. 2011)
Best resolution of the subtle sub-T_g local relaxation originating from rotational mobility of molecular dipoles	Increase in plasticization of loss peak with increasing a lower T_g drug content increases in ASD, T_g often resembles calorimetric T_g	Interference from space-charge peak (interfacial polarization signal), preconditioning can eliminate such effect
Peak of α-process at T_g, post-T_g signal from rigid amorphous fraction	Higher scatter of TW-TSDC peaks under a global peak indicates higher compositional heterogeneity	Non-Debye decay functions (e.g., KWW) is inappropriate to describe the TSDC relaxation process measured in temperature scanning mode
TW-TSDC deconvolutes a global peak of a particular motional process into ensemble of Debye peaks	TSPC would be able to measure initial structure of ASD with no need of thermal cleaning	Cleaning of thermal history disadvantageous to measure initial structure of sample by TSDC
Cooperative process identified by the convergence of individual relaxation times to a single point (compensation temperature)	Discrimination of free versus plasticizing water in ASD	

ASD amorphous solid dispersions, *TW-TSDC* thermal-windowing thermally stimulated discharge current

Table 14.4 Structural properties of ASD measurable by dynamic mechanical analysis (DMA) and possible hurdles

Molecular mobility (Andronis and Zografi 1997; Jones et al. 2012; Kalichevsky et al. 1992; Vyazovkin and Dranca 2006)	Miscibility (Abiad et al. 2010; Fadda et al. 2010; Hoppu et al. 2009; Jones et al. 2012; Labuschagne et al. 2010; Lamm et al. 2012; Liu et al. 2012; Poirier-Brulez et al. 2006; Suknunthaa et al. 2012; Yang et al. 2011)	Crystallinty/crystallization (Gupta and Bansal 2005; Soutari et al. 2012)	Interference/limitations (Abiad et al. 2010; Ayenew et al. 2012; Suknunthaa et al. 2012)
Detects from a relaxation to subtle sub-T_g processes that show otherwise negligible change in heat capacity (β,γ etc.)	Complementary to DSC, symmetrical and narrower loss peak for miscible system, multiple loss peaks for highly heterogeneous system	Employs the quantifiable difference in mechanical response of crystalline and amorphous state of a material	Need of a special sample geometry and amount, difficult to mount powder
$\hat{I}$'-peak not interfered (e.g., by DC in DRS)	Solid solubility estimation from melt rheology	Monitoring amorphous-to-crystalline transition as a function of temperature and/or time	Brittle ASD with higher drug content inappropriate for DMA analysis
Activation energy for α,β process between that measured by DRS and TSC	T_g measured higher than calorimetric T_g		ASD compact preparation may risk the alteration of the physical structure
	Composition-dependent hardness, reduced, and storage moduli obtained by nano-DMA provide miscibility data		

Table 14.5 Vibrational spectroscopic and microscopic analysis of ASD structure and possible hurdles

Miscibility and molecular interactions (Andrews et al. 2010; Breitkreitz and Poppi 2012; Hartshorn et al. 2013; Hédoux et al. 2011; Marsac et al. 2010; Padilla et al. 2010; Paudel et al. 2012; Qian et al. 2010; Rumondor et al. 2011; Wegiel et al. 2012; Zeitler et al. 2007)	Crystallinty/crystallization (Chan et al. 2004; Hédoux et al. 2011; Hédoux et al. 2009; Kao et al. 2012; Palermo et al. 2012a, b; Priemel et al. 2012; Sinclair et al. 2011; Vehring et al. 2012)	Interference/limitations (Gendrin et al. 2008; Jørgensen et al. 2009; Matero et al. 2013; Taylor et al. 2001)
IR, Raman scrutinize homo/heteromolecular interaction (H-bonding, dipolar interaction), higher sensitivity of NIR for H-bonding (OH overtones/combinations)	Amorphization generally weakens many intermolecular interactions while strengthens some	Overlapping peaks need deconvolution and multivariate data analysis for quantification, multi-component image analysis requires chemometric methods
Various structural information from band shift, relative peak intensities and/or width, e.g., drug–polymer H-bonding (red shift), monomer–multimer ratio of API	Relative crystallinity can be quantified from the ratio of peak intensities of vibrational bands unique to amorphous and crystalline forms or the ratio of peak width of partially to completely crystalline samples of IR or Raman spectra	Contribution from the alteration of the specific molar absorbance on newer bands (e.g., drug–polymer interactions) difficult to account for absolute quantification
Homo- (drug–drug, polymer–polymer) versus heteromolecular (drug–polymer) interactions, conformational states, dipolar self- and interassociation, phase-separation	Low-frequency (LF) Raman region sensitive to amorphous form preparation method (polyamorphism), associated to the excess quasi-harmonic vibrational density of states of amorphous materials representing the feeble phonon signature of underlying the crystal	Particle size, distribution and shape of sample require crucial control for reproducibility
Variable temperature/RH Raman or FTIR for heat/moisture-induced molecular demixing	Spectral monitoring of crystallization during storage, etc., LF Raman susceptibility presents superior sensitivity to the phonon peaks from lattice vibration thus detect nanocrystallites (30 Å) even in low API content (2–5 %)	MIR region highly susceptible to the interference from moisture peaks thus poses hurdle for hygroscopic samples
Time-domain THz measurement for molecular mobility studies		Raman signals are less affected by water peaks but confronted by fluorescence related artifacts from elastic scattering

Table 14.5 (continued)

Miscibility and molecular interactions (Andrews et al. 2010; Breitkreitz and Poppi 2012; Hartshorn et al. 2013; Hédoux et al. 2011; Marsac et al. 2010; Padilla et al. 2010; Paudel et al. 2012; Qian et al. 2010; Rumondor et al. 2011; Wegiel et al. 2012; Zeitler et al. 2007)	Crystallinty/crystallization (Chan et al. 2004; Hédoux et al. 2011; Hédoux et al. 2009; Kao et al. 2012; Palermo et al. 2012a, b; Priemel et al. 2012; Sinclair et al. 2011; Vehring et al. 2012)	Interference/limitations (Gendrin et al. 2008; Jørgensen et al. 2009; Matero et al. 2013; Taylor et al. 2001)
Vibrational (MIR, NIR, Raman, THz) microscopy (imaging/mapping) serves for fast and unambiguous quantification of the different solid forms of API in matrix, high signal-to-noise ratio and spatiochemical resolution Physical states and distribution API in matrix, studies of spatiotemporal (in situ) phase transitions and real-time structural evolution (e.g., during dissolution) Selective identification of the solid-state of API in formulations in THz spectra due to THz transparency of most pharmaceutical excipients Stronger scattering aromatic APIs than aliphatic polymers provides better structural contrast in Raman imaging and also lesser interference by water	Loss of phonon vibration results in the diffuse THz spectrum for the amorphous state enables identification and quantitation of crystal	NIR bands are often too diffused for quantitative use for higher number of constituents in ASD
NIR/Raman as PAT tools (Almeida et al. 2012; Alonzo et al. 2010; de Veij et al. 2009; Saerens et al. 2011, 2012; Smith-Goettler et al. 2011; Tumuluri et al. 2008; Wahl et al. 2013)		
Expeditious and non-destructive PAT tools for remote sampling *On-line* monitoring of solid-state phase transformations in dry *or su*spended samples during manufacturing (e.g., by hot melt extrusion) Monitoring solid- and solution-mediated transformations (e.g., nucleation, crystallization) during dissolution of ASD		

API active pharmaceutical ingredient, *FTIR* Fourier transform infrared spectroscopy, *NIR* near-infrared spectroscopy, *PAT* process analytical technology, *THz* terahertz

Table 14.6 Structural characterization of ASD by solid-state nuclear magnetic resonance (SS-NMR) spectroscopy and relaxometry and possible hurdles

Molecular mobility (Apperley et al. 2005; Aso et al. 2000; Benmore et al. 2013; Carpentier et al. 2006; Geppi et al. 2008; Ito et al. 2010)	Miscibility and drug–polymer interaction (Aso and Yoshioka 2006; Aso et al. 2002, 2007, 2009; Brettmann et al. 2012a, b; Chen et al. 2005; Forster et al. 2003; Geppi et al. 2008; Klama 2010; Pham et al. 2010; Tatton et al. 2013; Vogt et al. 2011, 2013)	Crystallinty/crystallization (Aso et al. 2009; Dahlberg et al. 2011; Geppi et al. 2008; Offerdahl et al. 2005; Seliger and Žagar 2013; Urbanova et al. 2013; Vogt and Williams 2012; Vogt et al. 2011)	Interference/limitations (Geppi et al. 2008; Pham et al. 2010; Schantz et al. 2009; Vogt et al. 2011)
Record motional process in a molecule within wider time scale (10^{-11}–10^{-4} s)	H-bonding results deshielding, intra and intermolecular interaction of API in matrix, H-bonding configurations (*groups, H-bond distance*)	Crystallinity determination based on different dynamics/spectral features of amorphous (faster, disordered) and crystalline (slower, ordered) fraction, T_1 relaxation highly sensitive to the crystalline state	Risk of crystallization of during longer measurement time necessary for T_1
Identification of dynamics of a particular molecular fractions with diverse mobility	Wider frequency range ^{13}C CP SS-NMR eases interpretation	Analysis of NMR line width versus delay time (T_2 profiles) using Gaussian–Lorentzian function (*Gaussian part: crystalline and Lorentzian: amorphous fraction*) for partially crystallinity	Ambiguity about the source of biphasic spin relaxation decay profile (due to *dynamic heterogeneity and/or heteronuclear coupling*)
T_1-from local/rapid, T_1-from α-like process and T_2- sensitive to local motions	Information on the H-bonding geometry nuclear quadrupolar moment and EFG at quadrupolar nuclei (spin > 1/2) such as ^{14}N, ^{35}Cl, ^{17}O	Variable temperature NMR relaxometry to study non-isothermal crystallization	Detection of heterogeneity based on relaxometry limited to the API and polymers having different independent relaxation times as (*analogous to identical T_g problem for DSC*)

Table 14.6 (continued)

Molecular mobility (Apperley et al. 2005; Aso et al. 2000; Benmore et al. 2013; Carpentier et al. 2006; Geppi et al. 2008; Ito et al. 2010)	Miscibility and drug–polymer interaction (Aso and Yoshioka 2006; Aso et al. 2002, 2007, 2009; Brettmann et al. 2012a, b; Chen et al. 2005; Forster et al. 2003; Geppi et al. 2008; Klama 2010; Pham et al. 2010; Tatton et al. 2013; Vogt et al. 2011, 2013)	Crystallinty/crystallization (Aso et al. 2009; Dahlberg et al. 2011; Geppi et al. 2008; Offerdahl et al. 2005; Seliger and Žagar 2013; Urbanova et al. 2013; Vogt and Williams 2012; Vogt et al. 2011)	Interference/limitations (Geppi et al. 2008; Pham et al. 2010; Schantz et al. 2009; Vogt et al. 2011)
Richer information from variable temperature and hetero-nuclear relaxometry (e.g.,^{19}F T_1/T_1 nearest neighbor distance)	Molecular mechanism of miscibility from 2D-correlation contours (strength, i.e., distance of drug–polymer H-bonding)	Convoluted spectral profile of partially crystalline dispersions due to overlapping sharp signals of crystalline fraction and the broad peaks of the amorphous fraction	Expensive, time and resource intensive technique
T_1- and T_1-relaxation times of abundant nuclei-dynamic heterogeneity	Hetero-nuclear correlation (e.g.,^{1}H-^{19}F) more selective,^{17}O SS-NMR susceptible to split of drug–drug H-bonding by drug–polymer mixing	Minimal MVA or gravimetric calibration for quantitatively specific high-resolution CP-NMR	Best results by NMR imaging only of static sample, dynamic dissolution conditions generate complexity
	Qualitative drug–polymer miscibility from the observed spectral alterations (*peak shift, shielding/deshielding, broadening*)	Up to 1 % API crystallinity detectable in polymeric mixtures	
	Estimates of the upper boundary of domain size of inhomogeneity in ASD from magnetization spin-diffusion length (from ^{1}H $T_1/T_{1\rho}$ times)	Crystallization monitoring > T_g alternatively by nuclear quadrupole double resonance	

Table 14.6 (continued)

Molecular mobility (Apperley et al. 2005; Aso et al. 2000; Benmore et al. 2013; Carpentier et al. 2006; Geppi et al. 2008; Ito et al. 2010)	Miscibility and drug–polymer interaction (Aso and Yoshioka 2006; Aso et al. 2002, 2007, 2009; Brettmann et al. 2012a, b; Chen et al. 2005; Forster et al. 2003; Geppi et al. 2008; Klama 2010; Pham et al. 2010; Tatton et al. 2013; Vogt et al. 2011, 2013)	Crystallinty/crystallization (Aso et al. 2009; Dahlberg et al. 2011; Geppi et al. 2008; Offerdahl et al. 2005; Seliger and Žagar 2013; Urbanova et al. 2013; Vogt and Williams 2012; Vogt et al. 2011)	Interference/limitations (Geppi et al. 2008; Pham et al. 2010; Schantz et al. 2009; Vogt et al. 2011)
	For phase-separated ASD exhibiting two distinct sets of relaxation times, separated domain size estimated from respective $T_1/T_{1\rho}$ times		
	Domain size of API in intimately mixed ASD (≤ 10 nm) deduced using T_2 relaxation times (*amorphous cluster versus molecular dispersion*)		
	2D-NMR through-space hetero-nuclear dipolar coupling distance to directly probe molecular proximity		
NMR imaging (*Dahlberg et al. 2011; Gladden and Sederman 2013; Langham et al. 2012; Mantle 2013*)			
Cross-section image from inside of an opaque (undisturbed) specimen Series of in situ temporal imaging of kinetic events during dissolution process viz., drug release and/or (re)crystallization, polymer mobilization, medium ingress, etc. Local distribution of nuclear spin density Spatial mapping of the associated component (API, polymer or dissolution medium) using T_2-relaxation times or self-diffusion coeffic$_i$ent of a specific nucleus within a sample volume Mutual interactions with other components and translation motions			

ASD amorphous solid dispersions, *API* active pharmaceutical ingredient, *MVA* multivariate analysis, *EFG* electric field gradient, *DSC* differential scanning calorimetry

Table 14.7 Light and electron microscopic analysis of morphology and microstructure of ASD

Microscopy	Structural information	Interference/limitations
Polarized light microscopy (Cai et al. 2011; Kestur and Taylor 2010, 2013; Konno and Taylor 2006; Yang and Gogos 2013)	Rapid detection of crystals based upon birefringence as completely amorphous sample lacks the same Nucleation and crystal growth kinetics (number, size and fractal) at variable temperature and/or RH Surface and bulk crystallization represented by uncovered and covered sample, respectively Morphology and aspect ratio of crystals, particle/crystal size, and distribution Detection of melting, dehydration, crystallization, degradation, etc., by hot stage microscopy	Insufficient illumination of trace/small crystallites dispersed in polymer matrix Isotropic crystal (e.g., cubic) non-birefringent Some amorphous polymer also exhibit virtual birefringence due to shear induced deformation Semiquantitative
Scanning electron microscopy (SEM) (Karavas et al. 2007a, b; Qi et al. 2010a, b)	Morphological inspections of nanostructures and amorphous systems with high (nm) spatial resolution under several thousand times magnification Rapid detection of surface crystal (size, number, and shape), suitable for the analysis of particulate and bulk level characteristics (particle size, crystal morphology, agglomerates) Surface topography, rugosity, etc. Chemical distribution map of the location and heterogeneity of different components using energy-dispersive X-ray spectroscopy (EDS)	Possible electronic ablation of labile sample surface Inadequate for the detailed analysis of amorphous phase structure (e.g. *phase separation, domain/cluster size/shape*) Inherent need of smooth/flat (microtomic) sample surface for quantitative applications by EDS Dissimilar drug–polymer elemental composition required for discriminatory EDS mapping
Transmission electron microscopy (TEM) (Bikiaris et al. 2005; De Zordi et al. 2012; Karavas et al. 2007a, b; Ma et al. 2013; Nakayama et al. 2009)	Ultimate spatial resolution for analyzing compositional heterogeneities, density variations, grain boundaries, etc. Reasonable contrast provided by hetero-elemental API for the measurement of the dimension of amorphous drug nano-clusters dispersed in the polymer matrix TEM/EDS for drug–polymer miscibility, i.e., molecularly mixed systems appear as a continuous matrix without elemental localization (lack of local elemental spikes) Spatial distinction of nanoscopic physical structure among amorphous, crystalline, and other mesophases in heterogeneous polymeric dispersion using selective negative staining of unsaturated API by heavy metal oxides (e.g., OsO_4) over excipients Characterization of polymorphs, nanocrystals, salts, co-crystals, etc.	Electron beam damage for some samples Impedingly tedious sample preparation Exotic elemental marker (considerable elemental dissimilarity between API and polymer) required for heterogeneity detection Ultimate safety required to handle the toxic staining agents

RH reversing heat

Table 14.8 Micro-/nano-structure analysis of ASD surface

Surface analytical tools	Structural information	Interference/limitations
Atomic force microscopy (AFM) (Lauer et al. 2011, 2013; Meeus et al. 2013; Qi et al. 2013a, b)	Surface nanoscopic phase imaging (*spatial phase/chemical nano-heterogeneities*) and nanometric topography imaging of ASD, dimension of surface phase separation Phase contrast in heterogeneous system due to more adhesive drug-rich than the polymer-rich domains Colloidal AFM for drug–carrier adhesion force measurement Local mechanical properties (viscoelasticity, Young's modulus, surface free energy, stiffness, etc.)	Damage of soft polymeric samples Smoother sample preferred as access to a very small area (*relatively smaller cantilever size than the particle curvature*) produce erroneous surface rugosity Interference of phase imaging results by the variations in sample-probe contact exerted by nano-topography
Localized thermal analysis (LTA) (Meeus et al. 2012; Meeus et al. 2013; Meng et al. 2012; Zhang et al. 2009)	Thermal transition mapping: softening temperature, T_g, melting at surface (â˜¼10–100 nm), markedly different T_g, T_m from bulk (by DSC; e.g., *reduced melting of surface crystals*) Local mapping of miscibility (spatial T_g distribution) and the physical state of sample surface, capturing early onset of instability (nano phase separation), 3D topology/thermal conductivity mapping at the surface Analysis on inter-particulate composition in individual powder particle	Due to faster heating rates than DSC difficult to compare with latter Very thin shell-core structures difficult to analyze (more noises) Semi-quantitative analysis, tedious calibration
Time-of-flight secondary ion mass spectrometry (ToF-SIMS) (Meeus et al. 2013; Scoutaris et al. 2012; Shard et al. 2009)	Analysis of surface chemical composition of a sample (*top 1–2 nm layer*) Mapping of spatial surface distribution of mixture components, 3D chemical distribution and miscibility, relative drug distribution in multi-polymeric system Depth profiling by surface sputtering (*removing defined surface monolayers*)	Spatial resolution limits the estimation of phase-separated domain

Table 14.8 (continued)

Surface analytical tools	Structural information	Interference/limitations
X-ray photoelectron spectroscopy (XPS) (Baer et al. 2011; Dahlberg et al. 2008; Maniruzzaman et al. 2013; Meeus et al. 2013)	Quantitative surface chemical composition analysis based on atomic concentration Profiling surface segregation (enrichment or depletion) of API and polymer in ASD particles Angle-resolved analysis of top surface layer, quantitative depth profiling of formulations Indication of drug–polymer interactions by spectral shift or new peaks	Spatial resolution around 1 μm for imaging Limited to drug–polymer with dissimilar elemental composition
Inverse gas chromatography (IGC) (Hasegawa et al. 2009; Ke et al. 2012; Lim et al. 2013; Miyanishi et al. 2013; Otte et al. 2012)	Surface molecular mobility, surface energy (dispersive, acidic, basic), and heterogeneity (disorder) Surface acidity/basicity (H-bond donor or acceptor number) Surface cohesion energy density, Flory–Huggins interaction parameter Surface crystallinity, RH-dependent surface T_g	Adsorption of probe gas on solid sample can interfere equilibrium data, reproducibility can be affected by the subtle change in sample porosity, texture, etc. Possible phase separation/crystallization during analysis (due to high column pressure) not clear

ASD amorphous solid dispersions, *API* active pharmaceutical ingredient, *AFM* atomic force microscopy, *RH* reversing heat

Table 14.9 Some emerging and potential analytical applications for the advanced characterization of ASD

Emerging/potential tools	Structural information	Interference/limitations
Positron annihilation lifetime spectroscopy (PALS) (Antal et al. 2013; Bölcskei et al. 2011; Chieng et al. 2013a, b; Gottnek et al. 2013; Szabó et al. 2011; Szente et al. 2009; Zelkóa et al. 2006)	Very sensitive analysis of local and global molecular mobility based on free volume change Change in free volume owing to structural cross linking by H-bond formation in ASD Miscibility derived from change in number, size (distribution) of molecular holes as a function of API content in ASD Water–solid interaction	Relatively novel application in ASD area Requires further exploration
Second order nonlinear optical imaging of chiral crystals (SONICC) (Kestur et al. 2012; Kissick et al. 2011; Toth et al. 2012; Wanapun et al. 2011)	Qualification of crystallinity and size distribution of trace chiral crystal in an amorphous matrix Extremely sensitive (4 ppm) of crystallinity for some chiral crystal systems Imaging of nanocrystal (90 nm) in polymeric matrix Potential utility for monitoring early stage nucleation/crystal growth from amorphous system	Limited to the chiral crystals Requires further exploration
Fluorescence resonance energy transfer (FRET) (van Drooge et al. 2006)	Study of miscibility in product with low drug content (0.5–1 %w/w) Estimation of domain size and identification of molecular cluster to dispersion possible Selective	Not applicable for high drug-loading system as API molecular proximity below FRET distance Fluorescent properties required and FRET should occur
Neutron scattering (Bordallo et al. 2012; Lerbret et al. 2012; Magazù et al. 2008; Magazù et al. 2010; Qi et al. 2013a, b)	Molecular dynamics and structural relaxation of multicomponent amorphous systems Small-angle neutron scattering for polymer–surfactant interaction during dissolution of ternary ASD	Sample preparation tedious Requires isotope labeling or enrichment

ASD amorphous solid dispersions, *API* active pharmaceutical ingredient

References

Abiad MG, Campanella OH, Carvajal MT (2010) Assessment of thermal transitions by dynamic mechanical analysis (DMA) using a novel disposable powder holder. Pharmaceutics 2:78–90

Adrjanowicz K, Wojnarowska Z, Wlodarczyk P, Kaminski K, Paluch M, Mazgalski J (2009) Molecular mobility in liquid and glassy states of Telmisartan (TEL) studied by broadband dielectric spectroscopy. Eur J Pharm Sci 38:395–404

Adrjanowicz K, Zakowiecki D, Kaminski K, Hawelek L, Grzybowska K, Tarnacka M, Paluch M, Cal K (2012) Molecular dynamics in supercooled liquid and glassy states of antibiotics: azithromycin, clarithromycin and roxithromycin studied by dielectric spectroscopy. Advantages given by the amorphous state. Mol Pharm 9:1748–1763

Alem N, Beezer AE, Gaisford S (2010) Quantifying the rates of relaxation of binary mixtures of amorphous pharmaceuticals with isothermal calorimetry. Int J Pharm 399:12–18

Alie J, Menegotto J, Cardon P, Duplaa H, Caron A, Lacabanne C, Bauer M, (2004) Dielectric study of the molecular mobility and the isothermal crystallization kinetics of an amorphous pharmaceutical drug substance. J Pharm Sci 93:218–233

Almeida A, Saerens L, De Beer T, Remon JP, Vervaet C (2012) Upscaling and in-line process monitoring via spectroscopic techniques of ethylene vinyl acetate hot-melt extruded formulations. Int J Pharm 439:223–229

Alonzo D, Zhang GZ, Zhou D, Gao Y, Taylor L (2010) Understanding the behavior of amorphous pharmaceutical systems during dissolution. Pharm Res 27:608–618

Amigo JM (2010) Practical issues of hyperspectral imaging analysis of solid dosage forms. Anal Bioanal Chem 398:93–109

Andrews GP, Abu-Diak O, Kusmanto F, Hornsby P, Hui Z, Jones DS (2010) Physicochemical characterization and drug-release properties of celecoxib hot-melt extruded glass solutions. J Pharm Pharmacol 62:1580–1590

Andronis V, Zografi G (1997) Molecular mobility of supercooled amorphous indomethacin, determined by dynamic mechanical analysis. Pharm Res 14:410–414

Antal I, Kállai N, Luhn O, Bernard J, Nagy ZK, Szabó B, Zelkó IK (2013) Supramolecular elucidation of the quality attributes of microcrystalline cellulose and isomalt composite pellet cores. J Pharm Biomed Anal 84:124–128

Antonijevi'c MD, Craig DQM, Barker SA (2008) The role of space charge formation in the generation of thermally stimulated current (TSC) spectroscopy data for a model amorphous drug system. Int J Pharm 353:8–14

Apperley DC, Forster AH, Fournier R, Harris RK, Hodgkinson P, Lancaster RW, Rades T (2005) Characterisation of indomethacin and nifedipine using variable-temperature solid-state NMR. Magn Reson Chem 43:881–892

Asami K (2002) Characterization of heterogeneous systems by dielectric spectroscopy. Prog Pol Sci 27:1617–1659

Aso Y, Yoshioka S (2006) Molecular mobility of nifedipine–PVP and phenobarbital–PVP solid dispersions as measured by ^{13}C-NMR spin-lattice relaxation time. J Pharm Sci 95:318–325

Aso Y, Yoshioka S, Kojima S (2000) Relationship between the crystallization rates of amorphous nifedipine, phenobarbital, and flopropione, and their molecular mobility as measured by their enthalpy relaxation and ^{1}H NMR relaxation times. J Pharm Sci 89:408–416

Aso Y, Yoshioka S, Zhang J, Zografi G (2002) Effect of water on the molecular mobility of sucrose and poly(vinylpyrrolidone) in a colyophilized formulation as measured by ^{13}C-NMR relaxation time. Chem Pharm Bull 50:822–826

Aso Y, Yoshioka S, Miyazaki T, Kawanishi T, Tanaka K, Kitamura S, Takakura A, Hayashi T., Muranushi N. (2007) Miscibility of nifedipine and hydrophilic polymers as measured by ^{1}H-NMR spin-lattice relaxation. Chem Pharm Bull 55:1227–1231

Aso Y, Yoshioka S, Miyazaki T, Kawanishi T (2009) Feasibility of ^{19}F-NMR for assessing the molecular mobility of flufenamic acid in solid dispersions. Chem Pharm Bull 57:61–64

Atassi F, Mao C, Masadeh AS, Byrn SR (2010) Solid-state characterization of amorphous and mesomorphous calcium ketoprofen. J Pharm Sci 99:3684–3697

Ayenew Z, Paudel A, Van den Mooter G (2012) Can compression induce demixing in amorphous solid dispersions? A case study of naproxen-PVP K25. Eur J Pharm Biopharm 81:207–213

Baer DR, Gaspar DJ, Nachimuthu P, Techane SD, Castner DG (2011) Application of surface chemical analysis tools for characterization of nanoparticles. Anal Bioanal Chem 983–1002

Baird JA, Taylor LS (2012) Evaluation of amorphous solid dispersion properties using thermal analysis techniques. Adv Drug Deliv Rev 64:396–421

Ball V, Maechling C (2009) Isothermal microcalorimetry to investigate non specific interactions in biophysical chemistry. Int J Mol Sci 10:3283–3315

Bansal SS, Kaushal AM, Bansal AK (2010) Enthalpy relaxation studies of two structurally related amorphous drugs and their binary dispersions. Drug Dev Ind Pharm 36:1271–1280

Barnes TJ, Kempson IM, Prestidge CA (2011) Surface analysis for compositional, chemical and structural imaging in pharmaceutics with mass spectrometry: a ToF-SIMS perspective. Int J Pharm 417:61–69

Bates S (2011) Amorphous materials: a structural perspective. Amorphous Workshop PPXRD–11

Bates S, Zografi G, Engers D, Morris K, Crowley K, Newman A (2006) Analysis of amorphous and nanocrystalline solids from their X-ray diffraction patterns. Pharm Res 23:2333–2349

Belton PS, Hills BP (1987) The effects of diffusive exchange in heterogeneous systems on N.M.R. line shapes and relaxation processes. Mol Phys 61:999–1018

Benmore CJ, R Weber JK, Tailor AN, Cherry BR, Yarger JL, Mou Q, Weber W, Neuefeind J, Byrn SR (2013) Structural characterization and aging of glassy pharmaceuticals made using acoustic levitation. J Pharm Sci 102:1290–1300

Bhardwaj SP, Suryanarayanan R (2011) Subtraction of DC conductivity and annealing: approaches to identify Johari-Goldstein relaxation in amorphous trehalose. Mol Pharmaceutics 8:1416–1422

Bhardwaj S, Suryanarayanan R (2013) Molecular mobility as a predictor of the water sorption by annealed amorphous trehalose. Pharm Res 30:714–720

Bhattacharya S, Suryanarayanan R (2009) Local mobility in amorphous pharmaceuticals—characterization and implications on stability. J Pharm Sci 98:2935–2953

Bhattacharya S, Suryanarayanan R (2011) Molecular motions in sucrose-PVP and sucrose-sorbitol dispersions: I. implications of global and local mobility on stability. Pharm Res 28:2191–2203

Bhugra C, Shmeis R, Krill S, Pikal M (2006) Predictions of onset of crystallization from experimental relaxation times I-correlation of molecular mobility from temperatures above the glass transition to temperatures below the glass transition. Pharm Res 23:2277–2290

Bhugra C, Shmeis R, Krill SL, Pikal MJ (2008a) Different measures of molecular mobility: comparison between calorimetric and thermally stimulated current relaxation times below Tg and correlation with dielectric relaxation times above T_g. J Pharm Sci 97:4498–4515

Bhugra C, Shmeis R, Krill SL, Pikal MJ (2008b) Prediction of onset of crystallization from experimental relaxation times. II. comparison between predicted and experimental onset times. J Pharm Sci 97:455–472

Bikiaris D, Papageorgiou GZ, Stergiou A, Pavlidou E, Karavas E, Kanaze F, Georgarakis M (2005) Physicochemical studies on solid dispersions of poorly water-soluble drugs: evaluation of capabilities and limitations of thermal analysis techniques. Thermochim Acta 439:58–67

Billinge SJL (2008) Nanoscale structural order from the atomic pair distribution function (PDF): there's plenty of room in the middle. J Solid State Chem 181:1695–1700

Billinge SJL, Dykhne T., Juhas P., Bozin E, Taylor R, Florence AJ, Shankland K (2010) Characterisation of amorphous and nanocrystalline molecular materials by total scattering. CrystEngComm 12:1366–1368

Blandamer MJ, Cullis PM, Engberts JBFN (1998) Titration microcalorimetry. J Chem Soc Faraday Trans 94:2261–2267

Boetker JP, Koradia V, Rades T, Rantanen J, Savolainen M (2012) Atomic pairwise distribution function analysis of the amorphous phase prepared by different manufacturing routes. Pharmaceutics 4:93–103

Bölcskei É, Süvegh K, Marekc GRT Jr, Pintye-Hódi K (2011) Testing of the structure of macromolecular polymer films containing solid active pharmaceutical ingredient (API) particles. Radiat Phys Chem 80:799–802

Bordallo HN, Zakharov BA, Boldyreva EV, Johnson MR, Koza MM, Seydel T, Fischer J (2012) Application of incoherent inelastic neutron scattering in pharmaceutical analysis: relaxation dynamics in phenacetin. Mol Pharm 9:2434–2441

Bøtker JP, Karmwar P, Strachan CJ, Cornett C, Tian F, Zujovic Z, Rantanen J, Rades T (2011) Assessment of crystalline disorder in cryo-milled samples of indomethacin using atomic pairwise distribution functions. Int J Pharm 417:112–119

Boutonnet-Fagegaltier N, Menegotto J, Lamure A, Duplaa H, Caron A, Lacabanne C, Bauer M (2002) Molecular mobility study of amorphous and crystalline phases of a pharmaceutical product by thermally stimulated current spectrometry. J Pharm Sci 91:1548–1560

Brás AR, Noronha JP, Antunes AMM, Cardoso MM, Schönhals A, Affouard Fdr, Dionísio M, Correia NIT (2008) Molecular motions in amorphous ibuprofen as studied by broadband dielectric spectroscopy. J Phys Chem B 112:11087–11099

Breitkreitz MC, Poppi RJ (2012) Trends in Raman chemical imaging. Biomed Spectros Imag 1:159–183

Brettmann B, Bell E, Myerson A, Trout B (2012a) Solid-state NMR characterization of high-loading solid solutions of API and excipients formed by electrospinning. J Pharm Sci 101:1538–1545

Brettmann BK, Myerson AS, Trout BL (2012b) Solid-state nuclear magnetic resonance study of the physical stability of electrospun drug and polymer solid solutions. J Pharm Sci 101:2185–2193

Brown A, Vickerman JC (1984) Static SIMS for applied surface analysis. Surf Interface Anal 6:1–14

Buckton G, Darcy P (1999) Assessment of disorder in crystalline powders-a review of analytical techniques and their application. Int J Pharm 179:141–158

Burnett D, Malde N, Williams D (2009) Characterizing amorphous materials with gravimetric vapour sorption techniques. Pharma Tech Eur 21

Cai T, Zhu L, Yu L (2011) Crystallization of organic glasses: effects of polymer additives on bulk and surface crystal growth in amorphous nifedipine. Pharm Res 28:2458–2466

Calahan JL, Zanon RL, Alvarez-Nunez F, Munson EJ (2013) Isothermal microcalorimetry to investigate the phase separation for amorphous solid dispersions of AMG 517 with HPMC-AS. Mol Pharm 10:1949–1957

Carlton R (2011) Thermal microscopy, pharmaceutical microscopy. Springer, New York, pp 65–84

Caron V, Bhugra C, Pikal MJ (2010) Prediction of onset of crystallization in amorphous pharmaceutical systems: phenobarbital, nifedipine/PVP, and phenobarbital/PVP. J Pharm Sci 99:3887–3900

Caron V, Tajber L, Corrigan OI, Healy AM (2011) A comparison of spray drying and milling in the production of amorphous dispersions of sulfathiazole/polyvinylpyrrolidone and sulfadimidine/polyvinylpyrrolidone. Mol Pharm 8:532–542

Carpentier L, Decressain R, Gusseme A, Neves C, Descamps M (2006) Molecular mobility in glass forming fananserine: a dielectric, NMR, and TMDSC investigation. Pharm Res 23:798–805

Casarino P, Lavaggi P, Pedemonte E (1996) Heat of mixing in polymer blends based on poly(vinyl acetate). J Therm Anal Calorim 47:165–170

Chadha R, Bhandari S, Arora P, Chhikara R (2013) Characterization, quantification and stability of differently prepared amorphous forms of some oral hypoglycaemic agents. Pharml Dev Technol 18:504–514

Chakravarty P, Bates S, Thomas L (2013) Identification of a potential conformationally disordered mesophase in a small molecule: experimental and computational approaches. Mol Pharm 10:2809–2822

Chan KLA, Fleming OS, Kazarian SG, Vassou D, Chryssikos GD, Gionis V (2004) Polymorphism and devitrification of nifedipine under controlled humidity: a combined FT-Raman, IR and Raman microscopic investigation. J Raman Spectrosc 35:353–359

Chattoraj S, Bhugra C, Telang C, Zhong L, Wang Z, Sun C (2012) Origin of two modes of non-isothermal crystallization of glasses produced by milling. Pharm Res 29:1020–1032

Chehimi MM, Djouani F, Benzarti K (2011) XPS studies of multiphase polymer systems. In: Boudenne A, Ibos L, Candau Y, Thomas S (eds) Handbook of multiphase polymer systems, 1st edn. Wiley, Chichester, pp 585–637

Chen J-Z, Ranade SV, Xie X-Q (2005) NMR characterization of paclitaxel/poly (styrene-isobutylene-styrene) formulations. Int J Pharm 305:129–144

Chen Z, Lovett D, Morris J (2011) Process analytical technologies and real time process control a review of some spectroscopic issues and challenges. J Process Control 21:1467–1482

Chieng N, Cicerone MT, Zhong Q, Liu M, Pikal MJ (2013a) Characterization of dynamics in complex lyophilized formulations: II. Analysis of density variations in terms of glass dynamics and comparisons with global mobility, fast dynamics, and Positron Annihilation Lifetime Spectroscopy (PALS). Eur J Pharm Biopharm 85:197–206

Chieng N, Mizuno M, Pikal M (2013b) Characterization of dynamics in complex lyophilized formulations: I. Comparison of relaxation times measured by isothermal calorimetry with data estimated from the width of the glass transition temperature region. Eur J Pharm Biopharm 85:189–196

Clark DT (1977) ESCA applied to polymers, molecular properties. Springer, pp 125–188

Couchman PR, Karasz FE (1978) A classical thermodynamic discussion of the effect of composition on glass-transition temperatures. Macromolecules 11:117–119

Courtier-Murias D, Farooq H, Masoom H, Botana A, Soong R, Longstaffe JG, Simpson MJ, Maas WE, Fey M, Andrew B, Struppe J, Hutchins H, Krishnamurthy S, Kumar R, Monette M, Stronks HJ, Hume A, Simpson AJ (2012) Comprehensive multiphase NMR spectroscopy: basic experimental approaches to differentiate phases in heterogeneous samples. J Magn Reson 217:61–76

Crowley KJ, Zografi G (2002) Water vapor absorption into amorphous hydrophobic drug/poly(vinylpyrrolidone) dispersions. J Pharm Sci 91:2150–2165

Dahlberg C, Millqvist-Fureby A, Schuleit M (2008) Surface composition and contact angle relationships for differently prepared solid dispersions. Euro J Pharm Biopharm 70:478–485

Dahlberg C, Dvinskikh SV, Schuleit M, Furó I (2011) Polymer swelling, drug mobilization and drug recrystallization in hydrating solid dispersion tablets studied by multinuclear NMR microimaging and spectroscopy. Mol Pharm 8:1247–1256

Dantuluri AKR, Amin A, Puri V, Bansal AK (2011) Role of a-relaxation on crystallization of amorphous celecoxib above T_g probed by dielectric spectroscopy. Mol Pharm 8:814–822

Dazzi A, Prater CB, Hu Q, Chase DB, Rabolt JF, Marcott C (2012) AFM-IR: combining atomic force microscopy and infrared spectroscopy for nanoscale chemical characterization. Appl Spectrosc 66:1365–1491

de Veij M, Vandenabeele P, De Beer T, Remon JP, Moens L (2009) Reference database of Raman spectra of pharmaceutical excipients. J Raman Spectrosc 40:297–307

De Zordi N, Moneghini M, Kikic I, Grassi M, Del Rio Castillo AE, Solinas D, Bolger MB (2012) Applications of supercritical fluids to enhance the dissolution behaviors of Furosemide by generation of microparticles and solid dispersions. Eur J Pharm Biopharm 81:131–141

Descamps N, Palzer S, Zuercher U (2009) The amorphous state of spray-dried maltodextrin: sub-sub-T_g enthalpy relaxation and impact of temperature and water annealing. Carbohydr Res 344:85–90

Dieing T, Hollricher O, Toporski J, Haefele T, Paulus K (2011) Confocal Raman microscopy in pharmaceutical development, confocal raman microscopy. Springer, Berlin, pp 165–202

Diogo HP, Moura-Ramos JJ (2009) Secondary molecular mobility in amorphous ethyl cellulose: aging effects and degree of co-operativity. J Polymer Sci Part B: Polymer Phys 47:820–829

Dlubek G, Shaikh MQ, Krause-Rehberg R, Paluch M (2007) Effect of free volume and temperature on the structural relaxation in polymethylphenylsiloxane: a positron lifetime and pressure-volume-temperature study. J Chem Phys 126:24906

Dong Y-D, Boyd BJ (2011) Applications of X-ray scattering in pharmaceutical science. Int J Pharm 417:101–111

Dong Z, Chatterji A, Sandhu H, Choi DS, Chokshi H, Shah N (2008) Evaluation of solid state properties of solid dispersions prepared by hot-melt extrusion and solvent co-precipitation. Int J Pharm 355:141–149

Duer MJ (2004) Solid-state NMR spectroscopy. Wiley-Blackwell

Dykhne T, Taylor R, Florence A, Billinge SL (2011) Data requirements for the reliable use of atomic pair distribution functions in amorphous pharmaceutical fingerprinting. Pharm Res 28:1041–1048

Eddleston MD, Bithell EG, Jones W (2010) Transmission electron microscopy of pharmaceutical materials. J Pharm Sci 99:4072–4083

Ehtezazi T, Govender T, Stolnik S (2000) Hydrogen bonding and electrostatic interaction contributions to the interaction of a cationic drug with polyaspartic acid. Pharm Res 17:871–877

Fadda HM, Khanna M, Santos JC, Osman D, Gaisford S, Basit AW (2010) The use of dynamic mechanical analysis (DMA) to evaluate plasticization of acrylic polymer films under simulated gastrointestinal conditions. Eur J Pharm Biopharm 76 493–497

Farrow CL, Billinge SJL (2009) Relationship between the atomic pair distribution function and small-angle scattering: implications for modeling of nanoparticles. Acta Crystallograph Sec A 65:232–239

Feth MP, Jurascheck J, Spitzenberg M, Dillenz J, Bertele G, Stark H (2011) New technology for the investigation of water vapor sorption-induced crystallographic form transformations of chemical compounds: a water vapor sorption gravimetry-dispersive raman spectroscopy coupling. J Pharm Sci 100:1080–1092

Ford JL, Mann TE (2012) Fast-scan DSC and its role in pharmaceutical physical form characterisation and selection. Adv Drug Deliv Rev 64:422–430

Forster A, Apperley D, Hempenstall J, Lancaster R, Rades T (2003) Investigation of the physical stability of amorphous drug and drug/polymer melts using variable temperature solid state NMR. Die Pharmazie 58:761

Fu W, Sun P (2011) Solid state NMR study of hydrogen bonding, miscibility, and dynamics in multiphase polymer systems. Front Chem China 6173–189

Gaisford S (2005) Stability assessment of pharmaceuticals and biopharmaceuticals by isothermal calorimetry. Curr Pharm Biotechnol 6:181–191

Gaisford S (2012) Isothermal microcalorimetry for quantifying amorphous content in processed pharmaceuticals. Adv Drug Deliv Rev 64 431–439

Gaisford S, Verma A, Saunders M, Royall PG (2009) Monitoring crystallisation of drugs from fast-dissolving oral films with isothermal calorimetry. Int J Pharm 380:105–111

Gendrin C, Roggo Y, Collet C (2008) Pharmaceutical applications of vibrational chemical imaging and chemometrics: a review. J Pharm Biomed Anal 48:533–553

Geppi M, Mollica G, Borsacchi S, Veracini CA (2008) Solid-state NMR studies of pharmaceutical systems. Appl Spectrosc Rev 43:202–302

Ghita OR, Beard MA, McCabe J, Bottom R, Richmond J, Evans KE (2008) A study into first and second order thermal transitions of materials using spectral-DSC. J Mater Sci 43:4988–4995

Ghosh I, Snyder J, Vippagunta R, Alvine M, Vakil R, Tong W-Q, Vippagunta S (2011) Comparison of HPMC based polymers performance as carriers for manufacture of solid dispersions using the melt extruder. Int J Pharm 419:12–19

Giron D, Remy P, Thomas S, Vilette E (1997) Quantitation of amorphicity by microcalorimetry. J Therm Anal 48:465–472

Giron D, Mutz M, Garnier S (2004) Solid-state of pharmaceutical compounds. J Therm Anal Calorim 77:709–747

Gladden LF, Sederman AJ (2013) Recent advances in flow MRI. J Magn Reson 229:2–11

Gnutzmann T, Kahlau R, Scheifler S, Friedrichs F, Rossler EA, Rademann K, Emmerling F (2013) Crystal growth rates and molecular dynamics of nifedipine. CrystEngComm 15:4062–4069

Gordon M, Taylor JS (1952) Ideal copolymers and the second-order transitions of synthetic rubbers. I. non-crystalline copolymers. J Appl Chem 2:493–500

Gottnek M, Süvegh K, Pintye-Hódi K, Regdon G Jr (2013) Effects of excipients on the tensile strength, surface properties and free volume of Klucel® free films of pharmaceutical importance. Radiat Phys Chem 89:57–63

Graeser KA, Patterson JE, Zeitler JA, Gordon KC, Rades T (2009) Correlating thermodynamic and kinetic parameters with amorphous stability. Eur J Pharm Sci 37:492–498

Greco SP, Authelin J-R, Leveder C, Segalini A (2012) A practical method to predict physical stability of amorphous solid dispersions. Pharm Res 29:2792–2805

Greenspan L (1977) Humidity fixed points of binary saturated aqueous solutions. J Res Nat Bur Stand Sect A 81:89–96

Grisedale LC, Jamieson MJ, Belton PS, Barker SA, M Craig DQ (2010) Characterization and quantification of amorphous material in milled and spray-dried salbutamol sulfate: a comparison of thermal, spectroscopic, and water vapor sorption approaches. J Pharm Sci 100:3114–3129

Grzybowska K, Paluch M, Wlodarczyk P, Grzybowski A, Kaminski K, Hawelek L, Zakowiecki D, Kasprzycka A, Jankowska-Sumara I (2012) Enhancement of amorphous celecoxib stability by mixing it with octaacetylmaltose: the molecular dynamics study. Mol Pharm 9:894–904

Gun'ko VM, Zarko VI, Goncharuk EV, Andriyko LS, Turov VV, Nychiporuk YM, Leboda R, Skubiszewska-Zięba J, Gabchak AL, Osovskii VD, Ptushinskii YG, Yurchenko GR, Mishchuk OA, Gorbik PP, Pissis P, Blitz JP (2007) TSDC spectroscopy of relaxational and interfacial phenomena. Adv Colloid Interface Sci 131:1–89

Guns S, Kayaert P, Martens JA, Van Humbeeck J, Mathot V, Pijpers T, Zhuravlev E, Schick C, Van den Mooter G (2010) Characterization of the copolymer poly(ethyleneglycol-g-vinylalcohol) as a potential carrier in the formulation of solid dispersions. Eur J Pharm Biopharm 74:239–247

Guns S, Dereymaker A, Kayaert P, Mathot V, Martens J, Mooter G (2011) Comparison between hot-melt extrusion and spray-drying for manufacturing solid dispersions of the graft copolymer of ethylene glycol and vinylalcohol. Pharm Res 28:673–682

Gupta P, Bansal AK (2005) Devitrification of amorphous celecoxib. AAPS PharmSciTech 6:E223–E230

Hancock BC, Dalton CR (1999) The effect of temperature on water vapor sorption by some amorphous pharmaceutical sugars. Pharma Dev Technol 4:125–131

Hancock BC, Shamblin SL (2001) Molecular mobility of amorphous pharmaceuticals determined using differential scanning calorimetry. Thermochim Acta 380:95–107

Hancock B, Dalton C, Pikal M, Shamblin S (1998) A pragmatic test of a simple calorimetric method for determining the fragility of some amorphous pharmaceutical materials. Pharm Res 15 762–767

Hancock BC, Shalaev EY, Shamblin SL (2002) Polyamorphism: a pharmaceutical science perspective. J Pharm Pharm 54:1151–1152

Harding L, King WP, Dai X, Craig DQ, Reading M (2007) Nanoscale characterisation and imaging of partially amorphous materials using local thermomechanical analysis and heated tip AFM. Pharm Res 24:2048–2054

Hartshorn CM, Lee YJ, Camp CH, Liu Z, Heddleston J, Canfield N, Rhodes TA, Hight Walker AR, Marsac PJ, Cicerone MT (2013) Multicomponent chemical imaging of pharmaceutical solid dosage forms with broadband CARS microscopy. Anal Chem 85:8102–8111

Hasegawa S, Ke P, Buckton G (2009) Determination of the structural relaxation at the surface of amorphous solid dispersion using inverse gas chromatography. J Pharm Sci 98:2133–2139

Havriliak S, Negami S (1967) A complex plane representation of dielectric and mechanical relaxation processes in some polymers. Polymer 8:161–210

Hédoux A, Paccou L, Guinet Y, Willart J-Fo, Descamps M (2009) Using the low-frequency Raman spectroscopy to analyze the crystallization of amorphous indomethacin. Eur J Pharm Sci 38:156–164

Hédoux A, Guinet Y, Descamps M (2011) The contribution of Raman spectroscopy to the analysis of phase transformations in pharmaceutical compounds. Int J Pharm 417:17–31

Hirakura Y, Yamaguchi H, Mizuno M, Miyanishi H, Ueda S, Kitamura S (2007) Detection of lot-to-lot variations in the amorphous microstructure of lyophilized protein formulations. Int J Pharm 340:34–41

Ho R, Heng JYY (2013) A review of inverse gas chromatography and its development as a tool to characterize anisotropic surface properties of pharmaceutical solids. Kona Powder Part J 30:164–180

Hogan SE, Buckton G (2001) Water sorption/desorption-near IR and calorimetric study of crystalline and amorphous raffinose. Int J Pharm 227:57–69

Hoppu P, Hietala S, Schantz S, Juppo AM (2009) Rheology and molecular mobility of amorphous blends of citric acid and paracetamol. Euro J Pharm Biopharm 71:55–63

Huang J, Dali M (2013) Evaluation of integrated Raman-DSC technology in early pharmaceutical development: characterization of polymorphic systems. J Pharm Biomed Anal 86:92–99

ICHQ8(R2) (2009) ICH harmonised tripartite guideline (2009). Pharmaceutical Development

Ito A, Watanabe T, Yada S, Hamaura T, Nakagami H, Higashi K, Moribe K, Yamamoto K (2010) Prediction of recrystallization behavior of troglitazone/polyvinylpyrrolidone solid dispersion by solid-state NMR. Int J Pharm 383:18–23

Ivanisevic I (2010) Physical stability studies of miscible amorphous solid dispersions. J Pharm Sci 99:4005–4012

Ivanisevic I, Bates S, Chen P (2009) Novel methods for the assessment of miscibility of amorphous drug-polymer dispersions. J Pharm Sci 98:3373–3386

Ivanisevic I, McClurg RB, Schields PJ, Gad SC (2010) Uses of X-ray powder diffraction in the pharmaceutical industry: drug discovery, development, and manufacturing, pharmaceutical sciences encyclopedia. Wiley, pp 1–42

Jain D, Chandra LSS, Nath R, Ganesan V (2012) Low temperature thermal windowing (TW) thermally stimulated depolarization current (TSDC) setup. Meas Sci Technol 23:025603

Janssens S, Van den Mooter G (2009) Review: physical chemistry of solid dispersions. J Pharm Pharmacol 61:1571–1586

Janssens S, Zeure A, Paudel A, Humbeeck J, Rombaut P, Mooter G (2010) Influence of preparation methods on solid state supersaturation of amorphous solid dispersions: a case study with itraconazole and eudragit E100. Pharm Res 27:775–785

Jones DS, Tian Y, Abu-Diak O, Andrews GP (2012) Pharmaceutical applications of dynamic mechanical thermal analysis. Adv Drug Deliv Rev 64:440–448

Jørgensen AC, Strachan CJ, Pöllänen KH, Koradia V, Tian F, Rantanen J (2009) An insight into water of crystallization during processing using vibrational spectroscopy. J Pharm Sci 98:3903–3932

Kaatze U (2013) Measuring the dielectric properties of materials. Ninety-year development from low-frequency techniques to broadband spectroscopy and high-frequency imaging. Meas Sci Technol 24:12005

Kalichevsky MT, Jaroszkiewicz EM, Ablett S, Blanshard JMV, Lillford PJ (1992) The glass transition of amylopectin measured by DSC, DMTA and NMR. Carbohydr Polym 18:77–88

Kalogeras IM (2011) A novel approach for analyzing glass-transition temperature vs. composition patterns: application to pharmaceutical compound + polymer systems. Eur J Pharm Sci 42:470–483

Kaminska E, Adrjanowicz K, Kaminski K, Wlodarczyk P, Hawelek L, Kolodziejczyk K, Tarnacka M, Zakowiecki D, Kaczmarczyk-Sedlak I, Pilch J, Paluch M (2013) A new way of stabilization of furosemide upon cryogenic grinding by using acylated saccharides matrices. The role of hydrogen bonds in decomposition mechanism. Mol Pharm 10:1824–1835

Kaminski K, Adrjanowicz K, Wojnarowska Z, Grzybowska K, Hawelek L, Paluch M, Zakowiecki D, Mazgalski J (2011) Molecular dynamics of the cryomilled base and hydrochloride ziprasidones by means of dielectric spectroscopy. J Pharm Sci 100:2642–2657

Kao JY, McGoverin CM, Graeser KA, Rades T, Gordon KC (2012) Measurement of amorphous indomethacin stability with NIR and Raman spectroscopy. Vib Spectrosc 58:19–26

Karavas E, Georgarakis E, Sigalas MP, Avgoustakis K, Bikiaris D (2007a) Investigation of the release mechanism of a sparingly water-soluble drug from solid dispersions in hydrophilic carriers based on physical state of drug, particle size distribution and drug-polymer interactions. Eur J Pharm Biopharm 66:334–347

Karavas E, Georgarakis M, Docoslis A, Bikiaris D (2007b) Combining SEM, TEM, and micro-Raman techniques to differentiate between the amorphous molecular level dispersions and nanodispersions of a poorly water-soluble drug within a polymer matrix. Int J Pharm 340:76–83

Karmwar P, Graeser K, Gordon KC, Strachan CJ, Rades T (2012) Effect of different preparation methods on the dissolution behaviour of amorphous indomethacin. Eur J Pharm Biopharm 80:459–464

Kawakami K (2011) Dynamics of ribavirin glass in the sub-T_g temperature region. J Phys Chem B 115:11375–11381

Kawakami K, Pikal MJ (2005) Calorimetric investigation of the structural relaxation of amorphous materials: evaluating validity of the methodologies. J Pharm Sci 94:948–965

Kawakami K, Usui T, Hattori M (2012) Understanding the glass-forming ability of active pharmaceutical ingredients for designing supersaturating dosage forms. J Pharm Sci 101:3239–3248

Kazarian SG, Ewing AV (2013) Applications of Fourier transform infrared spectroscopic imaging to tablet dissolution and drug release. Expert Opin Drug Deliv 10:1–15

Ke P, Hasegawa S, Al-Obaidi H, Buckton G (2012) Investigation of preparation methods on surface/bulk structural relaxation and glass fragility of amorphous solid dispersions. Int J Pharm 422:170–178

Kestur US, Taylor LS (2010) Role of polymer chemistry in influencing crystal growth rates from amorphous felodipine. CrystEngComm 12:2390–2397

Kestur US, Taylor LS (2013) Evaluation of the crystal growth rate of felodipine polymorphs in the presence and absence of additives as a function of temperature. Cryst Growth Des 13:4349–4354

Kestur US, Wanapun D, Toth SJ, Wegiel LA, Simpson GJ, Taylor LS (2012) Nonlinear optical imaging for sensitive detection of crystals in bulk amorphous powders. J Pharm Sci 101:4201–4213

Khougaz K, Clas S-D (2000) Crystallization inhibition in solid dispersions of MK-0591 and poly(vinylpyrrolidone) polymers. J Pharm Sci 89:1325–1334

Kissick DJ, Wanapun D, Simpson GJ (2011) Second-order nonlinear optical imaging of chiral crystals. Annu Rev Anal Chem (Palo Alto, Calif.) 4:419–437

Klama F (2010) NMR-studies of multi component solids. M. Sc. Dissertation, University of East Anglia, p 186

Koch MHJ, Bras W (2008) Synchrotron radiation studies of non-crystalline systems. Ann Rep Prog Chem Sect C 104:35–80

Konno H, Taylor LS (2006) Influence of different polymers on the crystallization tendency of molecularly dispersed amorphous felodipine. J Pharm Sci 95:2692–2705

Korhonen O, Bhugra C, Pikal MJ (2008) Correlation between molecular mobility and crystal growth of amorphous phenobarbital and phenobarbital with polyvinylpyrrolidone and L-proline. J Pharm Sci 97:3830–3841

Labuschagne PW, John MJ, Sadiku RE (2010) Investigation of the degree of homogeneity and hydrogen bonding in PEG/PVP blends prepared in supercritical CO_2: comparison with ethanol-cast blends and physical mixtures. J Supercrit Fluids 54:81–88

Laitinen R (2009) Title. University of Kuopio

Laitinen R, Suihko E, Toukola K, Björkqvist M, Riikonen J, Lehto VP, Järvinen K, Ketolainen J (2009) Intraorally fast-dissolving particles of a poorly soluble drug: preparation and in vitro characterization. Eur J Pharm Biopharm 71:271–281

Laitinen R, Suihko E, Bjorkqvist M, Riikonen J, Lehto V-P, Jarvinen K, Ketolainen J (2010) Perphenazine solid dispersions for orally fast-disintegrating tablets: physical stability and formulation. Drug Dev Ind Pharm 36:601–613

Lamm MS, Simpson A, McNevin M, Frankenfeld C, Nay R, Variankaval N (2012) Probing the effect of drug loading and humidity on the mechanical properties of solid dispersions with nanoindentation: antiplasticization of a polymer by a drug molecule. Mol Pharm 9:3396–3402

Langham ZA, Booth J, Hughes LP, Reynolds GK, Wren SAC (2012) Mechanistic insights into the dissolution of spray-dried amorphous solid dispersions. J Pharm Sci 101:2798–2810

Latsch S, Selzer T, Fink L, Kreuter Jr (2003) Crystallisation of estradiol containing TDDS determined by isothermal microcalorimetry, X-ray diffraction, and optical microscopy. Eur J Pharm Biopharm 56:43–52

Latsch S, Selzer T, Fink L, Horstmann M, Kreuter Jr (2004a) Use of isothermal heat conduction microcalorimetry, X-ray diffraction, and optical microscopy for characterisation of crystals grown in steroid combination-containing transdermal drug delivery systems. Eur J Pharm Biopharm 57:397–410

Latsch S, Selzer T, Fink L, Kreuter Jr (2004b) Determination of the physical state of norethindrone acetate containing transdermal drug delivery systems by isothermal microcalorimetry, X-ray diffraction, and optical microscopy. Eur J Pharm Biopharm 57:383–395

Lauer ME, Grassmann O, Siam M, Tardio J, Jacob L, Page S, Kindt JH, Engel A, Alsenz J (2011) Atomic force microscopy-based screening of drug-excipient miscibility and stability of solid dispersions. Pharm Res 28:572–584

Lauer ME, Siam M, Tardio J, Page S, Kindt JH, Grassmann O (2013) Rapid assessment of homogeneity and stability of amorphous solid dispersions by atomic force microscopy—from bench to batch. Pharm Res 30:1–13

Leane MM, Sinclair W, Qian F, Haddadin R, Brown A, Tobyn M, Dennis AB, (2013) Formulation and process design for a solid dosage form containing a spray-dried amorphous dispersion of ibipinabant. Pharm Dev Technol 18:359–366

Lechuga-Ballesteros D, Bakri A, Miller D (2003) Microcalorimetric measurement of the interactions between water vapor and amorphous pharmaceutical solids. Pharm Res 20:308–318

Lerbret A, Affouard Fdr, Hédoux A, Krenzlin S, Siepmann Jr, Bellissent-Funel M-C, Descamps M (2012) How strongly does trehalose interact with lysozyme in the solid state? Insights from molecular dynamics simulation and inelastic neutron scattering. J Phys Chem B 116:11103–11116

Liang LH, Zhao M, Jiang Q (2002) Melting enthalpy depression of nanocrystals based on surface effect. J Mater Sci Lett 21:1843–1845

Lim RTY, Ng WK, Widjaja E, Tan RBH (2013) Comparison of the physical stability and physicochemical properties of amorphous indomethacin prepared by co-milling and supercritical anti-solvent co-precipitation. J Supercr Fluids

Litvinov VM, Guns S, Adriaensens P, Scholtens BJR, Quaedflieg MP, Carleer R, Van den Mooter G (2012) Solid state solubility of miconazole in poly[(ethylene glycol)-g-vinyl alcohol] using hot-melt extrusion. Mol Pharm 9:2924–2932

Liu X, Yang P, Jiang Q (2007) Size effect on melting temperature of nanostructured drugs. Mater Chem Phys 103:1–4

Liu H, Zhang X, Suwardie H, Wang P, Gogos CG (2012a) Miscibility studies of indomethacin and Eudragit® E PO by thermal, rheological, and spectroscopic analysis. J Pharm Sci 101:2204–2212

Liu Y-M, Xu J-T, Fu Z-S, Fan Z-Q (2012b) Effect of phase separation on overall isothermal crystallization kinetics of PP/EPR in-reactor alloys. J Appl Polym Sci 127:1346–1358

Ma H, Choi D, Zhang Y-E, Tian H, Shah N, Chokshi H (2013) Evaluation on the drug-polymer mixing status in amorphous solid dispersions at the early stage formulation and process development. J Pharm Innov 8:163–174

Madsen IC, Scarlett NVY, Kern A (2011) Description and survey of methodologies for the determination of amorphous content via X-ray powder diffraction. Z Kristallogr 226:944–955

Magazù S, Maisano G, Migliardo F, Galli G, Benedetto A, Morineau D, Affouard Fdr, Descamps M (2008) Characterization of molecular motions in biomolecular systems by elastic incoherent neutron scattering. J Chem Phys 129:155103

Magazù S, Migliardo F, Affouard F, Descamps M, Telling MTF (2010) Study of the relaxational and vibrational dynamics of bioprotectant glass-forming mixtures by neutron scattering and molecular dynamics simulation. J Chem Phys 132:184512

Magazu S, Migliardo F, Benedetto A (2011) Elastic incoherent neutron scattering operating by varying instrumental energy resolution: principle, simulations, and experiments of the resolution elastic neutron scattering (RENS). Rev Sci Instrum 82:105111–105115

Maheswaram MP, Mantheni D, Perera I, Venumuddala H, Riga A, Alexander K (2013) Characterization of crystalline and amorphous content in pharmaceutical solids by dielectric thermal analysis. J Therm Anal Calorim 1–11

Maniruzzaman M, Morgan DJ, Mendham AP, Pang J, Snowden MJ, Douroumis D (2013) Drug-polymer intermolecular interactions in hot-melt extruded solid dispersions. Int J Pharm 443:199–208

Mantle MD (2011) Quantitative magnetic resonance micro-imaging methods for pharmaceutical research. Int J Pharm 417:173–195

Mantle MD (2013) NMR and MRI studies of drug delivery systems. Curr Opin Colloid Interface Sci 18:214–227

Marsac P, Shamblin S, Taylor L (2006) Theoretical and practical approaches for prediction of drug-polymer miscibility and solubility. Pharm Res 23:2417–2426

Marsac PJ, Rumondor ACF, Nivens DE, Kestur US, Stanciu L, Taylor LS (2010) Effect of temperature and moisture on the miscibility of amorphous dispersions of felodipine and poly(vinyl pyrrolidone). J Pharm Sci 99:169–185

Matero S, Den Berg Fv, Poutiainen S, Rantanen J, Pajander J (2013) Towards better process understanding: chemometrics and multivariate measurements in manufacturing of solid dosage forms. J Pharm Sci 102:1385–1403

McIntosh AI, Yang B, Goldup SM, Watkinson M, Donnan RS (2012) Terahertz spectroscopy: a powerful new tool for the chemical sciences? Chem Soc Rev 41:2072–2082

Meeus J, Chen X, Scurr DJ, Ciarnelli V, Amssoms K, Roberts CJ, Davies MC, Den Mooter Gv (2012) Nanoscale surface characterization and miscibility study of a spray-dried injectable polymeric matrix consisting of poly (lactic-co-glycolic-acid) and polyvinylpyrrolidone. J Pharm Sci 101:3473–3485

Meeus J, Scurr DJ, Amssoms K, Davies MC, Roberts CJ, Van den Mooter G (2013) Surface characteristics of spray-dried microspheres consisting of PLGA and PVP: relating the influence of heat and humidity to the thermal characteristics of these polymers. Mol Pharm 10:3213–3224

Meng J, Levina M, Rajabi-Siahboomi AR, Round AN, Reading M, Craig DQM 2012 The development of thermal nanoprobe methods as a means of characterizing and mapping plasticizer incorporation into ethylcellulose films. Pharm Res 29:2128–2138

Micko B (2012) title. University of Bayreuth

Miller D, Lechuga-Ballesteros D (2006) Rapid assessment of the structural relaxation behavior of amorphous pharmaceutical solids: effect of residual water on molecular mobility. Pharm Res 23:2291–2305

Miyanishi H, Nemoto T, Mizuno M, Mimura H, Kitamura S, Iwao Y, Noguchi S, Itai S (2013) Evaluation of crystallization behavior on the surface of nifedipine solid dispersion powder using inverse gas chromatography. Pharm Res 30:502–511

Moore MD, Wildfong PLD (2011) Informatics calibration of a molecular descriptors database to predict solid dispersion potential of small molecule organic solids. Int J Pharm 418:217–226

Moore M, Steinbach A, Buckner I, Wildfong PD (2009) A structural investigation into the compaction behavior of pharmaceutical composites using powder X-ray diffraction and total scattering analysis. Pharm Res 26:2429–2437

Moore M, Shi Z, Wildfong PD (2010) Structural interpretation in composite systems using powder X-ray diffraction: applications of error propagation to the pair distribution function. Pharm Res 27:2624–2632

Moura Ramos J, Correia Nl, Taveira-Marques R, Collins G (2002) The activation energy at T_g and the fragility index of indomethacin, predicted from the influence of the heating rate on the temperature position and on the intensity of thermally stimulated depolarization current peak. Pharm Res 19:1879–1884

Moynihan CT (1993) Correlation between the width of the glass transition region and the temperature dependence of the viscosity of high-T_g glasses. J Am Ceram Soc 76:1081–1087
Naelapää K, Boetker JP, Veski P, Rantanen J, Rades T, Kogermann K (2012) Polymorphic form of piroxicam influences the performance of amorphous material prepared by ball-milling. Int J Pharm 429:69–77
Nakayama S, Ihara K, Senna M (2009) Structure and properties of ibuprofen-hydroxypropyl methylcellulose nanocomposite gel. Powder Technol 190:221–224
Newman A, Engers D, Bates S, Ivanisevic I, Kelly RC, Zografi G (2008) Characterization of amorphous API:polymer mixtures using X-ray powder diffraction. J Pharm Sci 97:4840–4856
Ochsenbein P, Schenk KJ (2006) Crystallography for polymorphs, polymorphism. Wiley-VCH Verlag GmbH & Co. KGaA, pp 139–166
Offerdahl TJ, Salsbury JS, Dong Z, Grant DJW, Schroeder SA, Prakash I, Gorman EM, Barich DH, Munson EJ (2005) Quantitation of crystalline and amorphous forms of anhydrous neotame using ^{13}C CPMAS NMR spectroscopy. J Pharm Sci 94:2591–2605
O'Neill MAA, Gaisford S (2011) Application and use of isothermal calorimetry in pharmaceutical development. Int J Pharm 417:83–93
Otte A, Zhang Y, Carvajal MT, Pinal R (2012) Milling induces disorder in crystalline griseofulvin and order in its amorphous counterpart. CrystEngComm 14:2560–2570
Padilla AM, Ivanisevic I, Yang Y, Engers D, Bogner RH, Pikal MJ (2010) The study of phase separation in amorphous freeze-dried systems. Part I: Raman mapping and computational analysis of XRPD data in model polymer systems. J Pharm Sci 100:206–222
Palermo R, Short S, Anderson C, Tian H, Drennen J III (2012a) Determination of figures of merit for near-infrared, Raman and powder X-ray diffraction by net analyte signal analysis for a compacted amorphous dispersion with spiked crystallinity. J Pharm Innov 7:56–68
Palermo RN, Anderson CA, Drennen Iii JK (2012b) Review: use of thermal, diffraction, and vibrational analytical methods to determine mechanisms of solid dispersion stability. J Pharm Innov 7:2–12
Pandita SD, Wang L, Mahendran RS, Machavaram VR, Irfan MS, Harris D, Fernando GF (2012) Simultaneous DSC-FTIR spectroscopy: comparison of cross-linking kinetics of an epoxy/amine resin system. Thermochim Acta 543:9–17
Pansare VJ, Hejazi S, Faenza WJ, Prud'homme RK (2012) Review of long-wavelength optical and NIR imaging materials: contrast agents, fluorophores, and multifunctional nano carriers. Chem Mater 24:812–827
Patterson AL (1939) The Scherrer formula for X-ray particle size determination. Phys Rev 56:978–982
Paudel A, Mooter G (2012) Influence of solvent composition on the miscibility and physical stability of naproxen/PVP K 25 solid dispersions prepared by cosolvent spray-drying. Pharm Res 29:251–270
Paudel A, Van Humbeeck J, Van den Mooter G (2010) Theoretical and experimental investigation on the solid solubility and miscibility of naproxen in poly(vinylpyrrolidone). Mol Pharm 7:1133–1148
Paudel A, Nies E, Van den Mooter G (2012) Relating hydrogen-bonding interactions with the phase behavior of naproxen/PVP K 25 solid dispersions: evaluation of solution-cast and quench-cooled films. Mol Pharm 9:3301–3317
Paudel A, Worku ZA, Meeus J, Guns S, Van den Mooter G (2013a) Manufacturing of solid dispersions of poorly water soluble drugs by spray drying: formulation and process considerations. Int J Pharm 453:253–284
Paudel A, Loyson Y, Van den Mooter G (2013b) An investigation into the effect of spray drying temperature and atomizing conditions on miscibility, physical stability, and performance of naproxen–PVP K 25 solid dispersions. J Pharm Sci 102:1249–1267
Pavia DL, Lampman GM, Kriz GS (2001) Infrared spectroscopy in 'Introduction to spectroscopy': a guide for students of organic chemistry, 3rd edn. WB Saunders Co., Philadelphia

Penner EA (2013) Comparison of the new vapor sorption analyzer to the traditional saturated salt slurry method and the dynamic vapor sorption instrument. M. Sc. dissertation, University of Illinois

Perdana J, van der Sman RGM, Fox MB, Boom RM, Schutyser MAI (2014) Measuring and modelling of diffusivities in carbohydrate-rich matrices during thin film drying. J Food Eng 122:38–47

Pham TN, Watson SA, Edwards AJ, Chavda M, Clawson JS, Strohmeier M, Vogt FG (2010) Analysis of amorphous solid dispersions using 2D solid-state NMR and $^{1}H\ T_1$ relaxation measurements. Mol Pharm 7:1667–1691

Pieters R, Miltner HE, Van Assche G, Van Mele B (2006) Kinetics of temperature-induced and reaction-induced phase separation studied by modulated temperature DSC. Macromol Symp 233:36–41

Pikal MJ, Chang L, Tang X (2004) Evaluation of glassy-state dynamics from the width of the glass transition: results from theoretical simulation of differential scanning calorimetry and comparisons with experiment. J Pharm Sci 93:981–994

Pili B, Bourgaux C, Amenitsch H, Keller G, Lepê tre-Mouelhi S, Desmaële D, Couvreur P, Ollivon M (2010) Interaction of a new anticancer prodrug, gemcitabine-squalene, with a model membrane: coupled DSC and XRD study. Biochim Biophys Acta (BBA) Biomembr 1798:1522–1532

Poirier-Brulez F, Roudaut G, Champion D, Tanguy M, Simatos D (2006) Influence of sucrose and water content on molecular mobilityin starch-based glasses as assessed through structureand secondary relaxation. Biopolymers 81:63–73

Pollock HM, Hammiche A (2001) Micro-thermal analysis: techniques and applications. J Phys D Appl Phys 34:R23

Prats-Montalbán JM, Jerez-Rozo JI, Romaach RJ, Ferrer A (2012) MIA and NIR chemical imaging for pharmaceutical product characterization. Chemometr Intell Lab Syst 117:240–249

Priemel PA, Grohganz H, Gordon KC, Rades T, Strachan CJ (2012) The impact of surface- and nano-crystallisation on the detected amorphous content and the dissolution behaviour of amorphous indomethacin. Euro J Pharm Biopharm 82:187–193

Puri V, Dantuluri AK, Bansal AK (2012) Barrier coated drug layered particles for enhanced performance of amorphous solid dispersion dosage form. J Pharm Sci 101:342–353

Qi S, Belton P, Nollenberger K, Clayden N, Reading M, Craig DM (2010a) Characterisation and prediction of phase separation in hot-melt extruded solid dispersions: a thermal, microscopic and NMR relaxometry study. Pharm Res 27:1869–1883

Qi S, Weuts I, De Cort S, Stokbroekx S, Leemans R, Reading M, Belton P, Craig DQ (2010b) An investigation into the crystallisation behaviour of an amorphous cryomilled pharmaceutical material above and below the glass transition temperature. J Pharm Sci 99:196–208

Qi S, Moffat JG, Yang Z (2013a) Early stage phase separation in pharmaceutical solid dispersion thin films under high humidity: improved spatial understanding using probe-based thermal and spectroscopic nanocharacterization methods. Mol Pharm 10:918–930

Qi S, Roser S, Edler K, Pigliacelli C, Rogerson M, Weuts I, Dycke F, Stokbroekx S (2013b) Insights into the role of polymer-surfactant complexes in drug solubilisation/stabilisation during drug release from solid dispersions. Pharm Res 30:290–302

Qian F, Huang J, Zhu Q, Haddadin R, Gawel J, Garmise R, Hussain M (2010) Is a distinctive single T_g a reliable indicator for the homogeneity of amorphous solid dispersion? Int J Pharm 395:232–235

Rahman Z, Siddiqui A, Khan MA (2013) Assessing the impact of nimodipine devitrification in the ternary cosolvent system through quality by design approach. Int J Pharm 455:113–123

Recalls U (2011–2013) http://www.fda.gov/Safety/Recalls/ArchiveRecalls/default.htm

Reich G (2005) Near-infrared spectroscopy and imaging: basic principles and pharmaceutical applications. Adv Drug Deliv Rev 57:1109–1143

Reutzel-Edens SM, Newman AW (2006) Physical characterization of hygroscopicity in pharmaceutical solids, polymorphism. Wiley-VCH Verlag GmbH & Co. KGaA, pp 235–258

Roe KD, Labuza TP (2005) Glass transition and crystallization of amorphous trehalose-sucrose mixtures. Int J Food Prop 8:559–574
Roig Fdr, Dantras E, Dandurand J, Lacabanne C (2011) Influence of hydrogen bonds on glass transition and dielectric relaxations of cellulose. J Phys D Appl Phys 44:045403
Roos YH (1995) Chapter 4—water and phase transitions, phase transitions in foods. Academic Press, San Diego, pp 73–107
Rumondor ACF, Taylor LS (2010) Application of partial least-squares (PLS) modeling in quantifying drug crystallinity in amorphous solid dispersions. Int J Pharm 398:155–160
Rumondor ACF, Marsac PJ, Stanford LA, Taylor LS (2009) Phase behavior of poly(vinylpyrrolidone) containing amorphous solid dispersions in the presence of moisture. Mol Pharm 6:1492–1505
Rumondor AF, Wikström H, Van Eerdenbrugh B, Taylor LS (2011) Understanding the tendency of amorphous solid dispersions to undergo amorphous-amorphous phase separation in the presence of absorbed moisture. AAPS PharmSciTech 12:1209–1219
Saerens L, Dierickx L, Lenain B, Vervaet C, Remon JP, Beer TD (2011) Raman spectroscopy for the in-line polymer-drug quantification and solid state characterization during a pharmaceutical hot-melt extrusion process. Eur J Pharm Biopharm 77:158–163
Saerens L, Dierickx L, Quinten T, Adriaensens P, Carleer R, Vervaet C, Remon JP, De Beer T (2012) In-line NIR spectroscopy for the understanding of polymer-drug interaction during pharmaceutical hot-melt extrusion. Eur J Pharm Biopharm 81:230–237
Saerens L, Vervaet C, Remon JP, De Beer T (2013a) Process monitoring and visualization solutions for hot-melt extrusion: a review. J Pharm Pharmacol. 66:180–203
Saerens L, Vervaet C, Remon J-P, De Beer T (2013b) Visualization and process understanding of material behavior in the extrusion barrel during a hot-melt extrusion process using raman spectroscopy. Anal Chem 85:5420–5429
Santovea A, Piero MJ, Llabres M (2010) Comparison between DSC and TMDSC in the investigation into frozen aqueous cryoprotectants solutions. Drug Dev Ind Pharm 36:1413–1421
Saxena A, Jean YC, Suryanarayanan R (2013) Annealing effect reversal by water sorption–desorption and heating above the glass transition temperature—comparison of properties. Mol Pharm 10:3005–3012
Schantz S, Hoppu P, Juppo AM (2009) A solid-state NMR study of phase structure, molecular interactions, and mobility in blends of citric acid and paracetamol. J Pharm Sci 98:1862–1870
Schmitt E, Davis CW, Long ST (1996) Moisture-dependent crystallization of amorphous lamotrigine mesylate. J Pharm Sci 85:1215–1219
Schönhals A Dielectric spectroscopy on the dynamics of amorphous polymeric systems. Novocontrol, Application note. http://novocontrol.de/pdf_s/APND1.PDF. Accessed 27 June 2013
Scoutaris N, Hook AL, Gellert PR, Roberts CJ, Alexander MR, Scurr DJ (2012) ToF-SIMS analysis of chemical heterogenities in inkjet micro-array printed drug/polymer formulations. J Mater Sci Mater Med 23:385–391
Seliger J, Žagar V (2013) Crystallization of an amorphous solid studied by nuclear quadrupole double resonance. Chem Phys 421:44–48
Shard AG, Rafati A, Ogaki R, Lee JLS, Hutton S, Mishra G, Davies MC, Alexander MR (2009) Organic depth profiling of a binary system: the compositional effect on secondary ion yield and a model for charge transfer during secondary Ion emission. J Phys Chem B 113:11574–11582
Shi Y, Zhang X-Y, Gong LL (2011) Molecular motion and detrapping behavior of trapped space charges in polyvinyl pyrrolidone: a thermally stimulated depolarization current study. Polym Bull 67:1595–1604
Shi P, Schach Rg, Munch E, Montes Hln, Lequeux Fo (2013) Glass transition distribution in miscible polymer blends: from calorimetry to rheology. Macromolecules 46:3611–3620
Shmeis R, Wang Z, Krill S (2004a) A mechanistic investigation of an amorphous pharmaceutical and its solid dispersions, Part I: a comparative analysis by thermally stimulated depolarization current and differential scanning calorimetry. Pharm Res 21:2025–2030

Shmeis R, Wang Z, Krill S (2004b) A mechanistic investigation of an amorphous pharmaceutical and its solid dispersions, Part II: Molecular mobility and activation thermodynamic parameters. Pharm Res 21:2031–2039
Sinclair W, Leane M, Clarke G, Dennis A, Tobyn M, Timmins P (2011) Physical stability and recrystallization kinetics of amorphous ibipinabant drug product by Fourier transform raman spectroscopy. J Pharm Sci 100:4687–4699
Sitterberg J, Özcetin A, Ehrhardt C, Bakowsky U (2010) Utilising atomic force microscopy for the characterisation of nanoscale drug delivery systems. Eur J Pharm Biopharm 74:2–13
Smith-Goettler B, Gendron CM, MacPhail N, Meyer RF, Phillips JX (2011) NIR monitoring of a hot-melt extrusion process. Spectrosc Lett
Song M, de Villiers MM, Redelinghuys A-M, Liebenberg W (2005) Isothermal and dynamic microcalorimetry for quantifying the crystallization of an amorphous drug during interactive powder mixing. Particul Sci Technol 23:323–334
Sousa LAE, Alem N, Beezer AE, Oâ €™ Neill MAA, Gaisford S (2010) Quantitative analysis of solid-state processes studied with isothermal microcalorimetry. J Phys Chem B 114:13173–13178
Soutari N, Buanz ABM, Gul MO, Tuleu C, Gaisford S (2012) Quantifying crystallisation rates of amorphous pharmaceuticals with dynamic mechanical analysis (DMA). Int J Pharm 423:335–340
Suknunthaa K, Jonesb DS, Tantishaiyakula V (2012) Properties of felodipine-poly (vinylpyrrolidone) solid dispersion films and the impact of solvents. Sci Asia 38:188–195
Sun Y, Tao J, Zhang GGZ, Yu L (2010) Solubilities of crystalline drugs in polymers: an improved analytical method and comparison of solubilities of indomethacin and nifedipine in PVP, PVP/VA, and PVAc. J Pharm Sci 99:4023–4031
Svoboda R, Málek Jr (2011) Interpretation of crystallization kinetics results provided by DSC. Thermochim Acta 526:237–251
Syll O, Richard B, Willart JF, Descamps M, Schuck P, Delaplace G, Jeantet R (2012) Rehydration behaviour and ageing of dairy powders assessed by calorimetric measurements. Innov Food Sci Emerg Technol 14:139–145
Szabó B, Süvegh K, Zelkó R (2011) Effect of storage on microstructural changes of Carbopol polymers tracked by the combination of positron annihilation lifetime spectroscopy and FT-IR spectroscopy. Int J Pharm 416:160–163
Szente V, Süvegh K, Marek T, Zelkó R (2009) Prediction of the stability of polymeric matrix tablets containing famotidine from the positron annihilation lifetime distributions of their physical mixtures. J Pharm Biomed Anal 49:711–714
Tarek El G, Roland Bh (2007) Dielectric relaxation processes in solid and supercooled liquid solutions of acetaminophen and nifedipine. J Phys Condens Matter 19:205134
Tatton AS, Pham TN, Vogt FG, Iuga D, Edwards AJ, Brown SP (2013) Probing hydrogen bonding in cocrystals and amorphous dispersions using ^{14}N–^{1}H HMQC solid-State NMR. Mol Pharm 10:999–1007
Taylor LS, Langkilde FW, Zografi G (2001) Fourier transform Raman spectroscopic study of the interaction of water vapor with amorphous polymers. J Pharm Sci 90:888–901
Thielmann F, Levoguer C (2002) iGC-A new instrumental technique for characterising the phyisco-chemical properties of pharmaceutical materials. Application note 301
Tobyn M, Brown J, Dennis AB, Fakes M, Gao Q, Gamble J, Khimyak YZ, McGeorge G, Patel C, Sinclair W, Timmins P, Yin S (2009) Amorphous drug–PVP dispersions: application of theoretical, thermal and spectroscopic analytical techniques to the study of a molecule with intermolecular bonds in both the crystalline and pure amorphous state. J Pharm Sci 98:3456–3468
Toth SJ, Madden JT, Taylor LS, Marsac P, Simpson GJ (2012) Selective imaging of active pharmaceutical ingredients in powdered blends with common excipients utilizing two-photon excited ultraviolet-fluorescence and ultraviolet-second order nonlinear optical imaging of chiral crystals. Anal Chem 84:5869–5875

Trasi N, Byrn S (2012) Mechanically induced amorphization of drugs: a study of the thermal behavior of cryomilled compounds. AAPS PharmSciTech 13:772–784

Trasi N, Boerrigter SM, Byrn S (2010) Investigation of the milling-induced thermal behavior of crystalline and amorphous griseofulvin. Pharm Res 27:1377–1389

Tumuluri VS, Kemper MS, Lewis IR, Prodduturi S, Majumdar S, Avery BA, Repka MA (2008) Off-line and on-line measurements of drug-loaded hot-melt extruded films using Raman spectroscopy. Int J Pharm 357:77–84

Urbanova M, Brus J, Sedenkova I, Policianova O, Kobera L (2013) Characterization of solid polymer dispersions of active pharmaceutical ingredients by ^{19}F MAS NMR and factor analysis. Spectrochim Acta A 100:59–66

USFDA (2012) http://www.accessdata.fda.gov/scripts/cdrh/cfdocs/cfcfr/CFRsearch.cfm?CFRPart=314. Accessed 18 Jan 2014

Van den Mooter G (2012) The use of amorphous solid dispersions: a formulation strategy to overcome poor solubility and dissolution rate. Drug Discov Today Technol 9:e79–e85

van Drooge DJ, Braeckmans K, Hinrichs WLJ, Remaut K, De Smedt SC, Frijlink HW (2006a) Characterization of the mode of incorporation of lipophilic compounds in solid dispersions at the nanoscale using fluorescence resonance energy transfer (FRET). Macromol Rapid Commun 27:1149–1155

van Drooge DJ, Hinrichs WLJ, Visser MR, Frijlink HW (2006b) Characterization of the molecular distribution of drugs in glassy solid dispersions at the nano-meter scale, using differential scanning calorimetry and gravimetric water vapour sorption techniques. Int J Pharm 310:220–229

Van Eerdenbrugh B, Taylor LS (2011) Application of mid-IR spectroscopy for the characterization of pharmaceutical systems. Int J Pharm 417:3–16

Vassilikou-Dova A, Kalogeras IM (2008) Dielectric analysis (DEA), thermal analysis of polymers. Wiley, pp 497–613

Vehring R, Ivey J, Williams L, Joshi V, Dwivedi S, Lechuga-Ballesteros D, Dalby RN, Byron PB, Peart J, Suman JD (2012) High-sensitivity analysis of crystallinity in respirable powders using low frequency shift-Raman spectroscopy. Respir Drug Deliv 2:641–644

Viciosa T, Pires G, Ramos JJM (2009) Revisitation of the molecular mobility of the amorphous solid 4,4′-methylenebis (N, N-diglycidylaniline) (MBDA): new contributions from the TSDC technique. J Mol Liquids 148:114–119

Vitez IM, Newman AW, Davidovich M, Kiesnowski C (1998) The evolution of hot-stage microscopy to aid solid-state characterizations of pharmaceutical solids. Thermochim Acta 324:187–196

Voelkel A, Strzemiecka B, Adamska K, Milczewska K (2009) Inverse gas chromatography as a source of physiochemical data. J Chromatogr A 1216:1551–1566

Vogt FG, Williams GR (2010) Advanced approaches to effective solid-state analysis: X-ray diffraction, vibrational spectroscopy and solid-state NMR. Am Pharm Rev 13:1–17

Vogt F, Williams G (2012) Analysis of a nanocrystalline polymer dispersion of ebselen using solid-state NMR, Raman microscopy, and powder X-ray diffraction. Pharm Res 29:1866–1881

Vogt FG, Clawson JS, Strohmeier M, Pham TN, Watson SA, Edwards AJ, Gad SC (2011) New approaches to the characterization of drug candidates by solid-state NMR, Pharmaceutical Sciences Encyclopedia. Wiley

Vogt FG, Yin H, Forcino RG, Wu L (2013) ^{17}O Solid-state NMR as a sensitive probe of hydrogen bonding in crystalline and qmorphous solid forms of diflunisal. Mol Pharm 10:3433–3446

Vyazovkin S, Dranca I (2006) Probing beta relaxation in pharmaceutically relevant glasses by using DSC. Pharm Res 23:422–428

Vyazovkin S, Dranca I (2007) Effect of physical aging on nucleation of amorphous indomethacin. J Phys Chem B 111:7283–7287

Wahl PR, Treffer D, Mohr S, Roblegg E, Koscher G, Khinast JG (2013) Inline monitoring and a PAT strategy for pharmaceutical hot melt extrusion. Int J Pharm 455:159–168

Wanapun D, Kestur US, Taylor LS, Simpson GJ (2011) Single particle nonlinear optical imaging of trace crystallinity in an organic powder. Anal Chem 83:4745–4751

Wang L-M, Velikov V, Angell CA (2002) Direct determination of kinetic fragility indices of glass-forming liquids by differential scanning calorimetry: kinetic versus thermodynamic fragilities. J Chem Phys 117:10184

Wegiel LA, Mauer LJ, Edgar KJ, Taylor LS (2012) Mid-infrared spectroscopy as a polymer selection tool for formulating amorphous solid dispersions. J Pharm Pharmacol. doi:10.1111/jphp.12079

Weinberg MC, Birnie I DP, Shneidman VA (1997) Crystallization kinetics and the JMAK equation. J Non-Cryst Solids 219:89–99

Weuts I, Kempen D, Verreck G, Peeters J, Brewster M, Blaton N, Van den Mooter G (2005) Salt formation in solid dispersions consisting of polyacrylic acid as a carrier and three basic model compounds resulting in very high glass transition temperatures and constant dissolution properties upon storage. Eur J Pharm Sci 25:387–393

Willart J-F, Carpentier L, Danède F, Descamps M (2012) Solid-state vitrification of crystalline griseofulvin by mechanical milling. J Pharm Sci 101:1570–1577

Williams G (2009) Chain dynamics in solid polymers and polymerizing systems as revealed by broadband dielectric spectroscopy. Macromol Symp 286:1–19

Winkel K, Bowron DT, Loerting T, Mayer E, Finney JL (2009) Relaxation effects in low density amorphous ice: two distinct structural states observed by neutron diffraction. J Chem Phys 130:204502

Wojnarowska Z, Grzybowska K, Adrjanowicz K, Kaminski K, Paluch M, Hawelek L, Wrzalik R, Dulski M, Sawicki W, Mazgalski J, Tukalska A, Bieg T (2010) Study of the amorphous glibenclamide drug: analysis of the molecular dynamics of quenched and cryomilled material. Mol Pharm 7:1692–1707

Wyttenbach N, Janas C, Siam M, Lauer ME, Jacob L, Scheubel E, Page S (2013) Miniaturized screening of polymers for amorphous drug stabilization (SPADS): rapid assessment of solid dispersion systems. Eur J Pharm Biopharm 84:583–598

Yang M, Gogos C (2013) Crystallization of poly(ethylene oxide) with acetaminophen-A study on solubility, spherulitic growth, and morphology. Eur J Pharm Biopharm 85:889–897

Yang J, Grey K, Doney J (2010) An improved kinetics approach to describe the physical stability of amorphous solid dispersions. Int J Pharm 384:24–31

Yang M, Wang P, Suwardie H, Gogos C (2011) Determination of acetaminophen's solubility in poly(ethylene oxide) by rheological, thermal and microscopic methods. Int J Pharm 403:83–89

Yang Z, Nollenberger K, Albers J, Craig D, Qi S (2013) Microstructure of an immiscible polymer blend and its stabilization effect on amorphous solid dispersions. Mol Pharm 10:2767–2780

Yonemochi E, Inoue Y, Buckton G, Moffat A, Oguchi T, Yamamoto K (1999) Differences in crystallization behavior between quenched and ground amorphous ursodeoxycholic acid. Pharm Res 16:835–840

Yoshihashi Y, Iijima H, Yonemochi E, Terada K (2006) Estimation of physical stability of amorphous solid dispersion using differential scanning calorimetry. J Therm Anal Calorim 85:689–692

Yoshihashi Y, Yonemochi E, Maeda Y, Terada K (2010) Prediction of the induction period of crystallization of naproxen in solid dispersion using differential scanning calorimetry. J Therm Anal Calorim 99:15–19

Young PM, Chiou H, Tee T, Traini D, Chan H-K, Thielmann F, Burnett D (2007) The use of organic vapor sorption to determine low levels of amorphous content in processed pharmaceutical powders. Drug Dev Ind Pharm 33:91–97

Yuan X, Carter BP, Schmidt SJ (2011) Determining the critical relative humidity at which the glassy to rubbery transition occurs in polydextrose using an automatic water vapor sorption instrument. J Food Sci 76:E78–E89

Zeitler JA, Taday PF, Newnham DA, Pepper M, Gordon KC, Rades T (2007) Terahertz pulsed spectroscopy and imaging in the pharmaceutical—a review. J Pharm Pharmacol 59:209–223

Zelkóa Rn, Orbán Ãdm, Süvegh Kr (2006) Tracking of the physical ageing of amorphous pharmaceutical polymeric excipients by positron annihilation spectroscopy. J Pharm Biomed Anal 40:249–254

Zhang S, Painter PC, Runt J (2004) Suppression of the dielectric secondary relaxation of poly(2-vinylpyridine) by strong intermolecular hydrogen bonding. Macromolecules 37:2636–2642
Zhang J, Bunker M, Chen X, Parker AP, Patel N, Roberts CJ (2009) Nanoscale thermal analysis of pharmaceutical solid dispersions. Int J Pharm 380:170–173
Zhuravlev E, Schick C (2010) Fast scanning power compensated differential scanning nano-calorimeter: 2. Heat capacity analysis. Thermochim Acta 505:14–21
Zidan AS, Rahman Z, Sayeed V, Raw A, Yu L, Khan MA (2012) Crystallinity evaluation of tacrolimus solid dispersions by chemometric analysis. Int J Pharm 423:341–350

Chapter 15
Dissolution of Amorphous Solid Dispersions: Theory and Practice

Nikoletta Fotaki, Chiau Ming Long, Kin Tang and Hitesh Chokshi

15.1 Introduction

The dissolution test has long served as the scientific, industrial, and regulatory standard for measuring the rate of drug solubilization or drug release from solid oral drug products and other dosage forms. The monitoring of the rate of drug solubilization or release from the dosage form is particularly important for drug products in which drug release is the rate-determining step for absorption and pharmacodynamic effect. Therefore, variations in dissolution rate may result in potential variability in bioavailability and safety or efficacy concerns. The dissolution test is a holistic measure of critical material attributes of the composition of the drug product (drug substance and excipients), manufacturing process, and physicochemical changes in the drug product upon aging or exposure to stress (e.g., tablet hardening, loss of disintegrant functionality, physical form change, etc.). For drug products containing amorphous solid dispersions, the properties of the carrier and, consequently, its dissolution behavior will be important to the in vitro dissolution performance of the active drug/drug product. Indeed, the dissolution test, when properly designed, is a powerful tool to assist in the selection of the polymer carrier by virtue of its ability to discriminate between polymers with respect to its ability to facilitate dissolution, to maintain a supersaturated solution, and to influence solid-state stability of the amorphous state of the drug substance. It is also noted that the dissolution test is model dependent. That is, the results are influenced by test parameters such as test media pH, hydrodynamics, and test apparatus. As such, it is not surprising to note standardized test descriptions and apparatus described in the USP (USP 2009a), Ph. Eur.

N. Fotaki (✉) · C. M. Long
Department of Pharmacy and Pharmacology, University of Bath, Bath, UK
e-mail: n.fotaki@bath.ac.uk

K. Tang
Pharma Technical Regulatory, Genentech Inc., South San Francisco, CA, USA

H. Chokshi
Roche Pharma Research & Early Development, Roche Innovation Center, New York, NY, USA

N. Shah et al. (eds.), *Amorphous Solid Dispersions*,
Advances in Delivery Science and Technology, DOI 10.1007/978-1-4939-1598-9_15

(PhEur 2011), and JP (JP 2006), and various regulatory guidances published by the major health authorities, (FDA 1997) and (EMA 2010), covering dissolution testing. USP < 1092 > provides detailed requirements for the development and validation of dissolution methods (USP 2009b). As noted in the guidance, recommendations are made for sink condition to be established for the test whereby sink condition is defined as the volume of medium at least three times that required in order to form a saturated solution of drug substance. When sink conditions are present, it is more likely that dissolution results will reflect the properties of the dosage form. USP < 1092 > allows for non-sink conditions if the method is shown to be more discriminating or otherwise appropriately justified. The guidances indicate that a small amount of surfactant is justified as needed to establish sink condition for the test.

15.2 Regulatory Quality Control Dissolution Method

As amorphous characteristic is a critical quality attribute of the amorphous solid dispersion-based drug product, it must be monitored and controlled to ensure quality of the product for the desired in vivo product performance/bioavailability throughout its shelf life. The regulatory quality control (QC) dissolution method should be capable of discriminating substantial changes in the amorphous characteristic of a drug product, corroborating results obtained by spectroscopic methods such as X-ray powder diffraction (XRPD), near-infrared chemical imaging (NIR-CI), and infrared (IR). The QC dissolution method must be able to discriminate presence of the crystalline drug in the amorphous solid dispersion-based drug product. This discriminatory power of the dissolution method allows monitoring of batch-to-batch consistency as well as physical form stability of an amorphous solid dispersion-based drug product. The QC dissolution method should be validated as per the regulatory standards with the additional emphasis on the amorphous–crystalline discriminatory power.

Where dissolution or drug release is the rate-limiting step to drug absorption, it may be possible to establish an in vitro–in vivo correlation (IVIVC) whereby validated predictions of in vivo performance may be made from in vitro dissolution data.

For the dissolution testing of an amorphous drug product, any of the test apparatus typically used for testing of solid oral dosage forms may be selected based on its demonstrated capability to provide meaningful drug-release data and to discriminate for critical factors. Typically, the basket and paddle apparatus are the most commonly used. However, the reciprocating cylinder apparatus and the flow-through apparatus also may be used.

When the drug product contains an amorphous solid dispersion, it is important not only to characterize the dissolution rate but also to determine the ability of the formulation to maintain a supersaturated state upon dissolution of the drug. This prolongation of the supersaturated state is an important aspect of the dissolution test when applied towards the selection of polymers for the solid dispersion during the formulation development phase. Once the formulation has been established, a dissolution test which incorporates sink condition in the test for monitoring lot-to-lot differences in release rate is appropriate.

15.3 Amorphous State and Solid Dispersion

As described in previous chapters, solid forms are classified into crystalline and amorphous states based on the order of molecular packing (Raumer et al. 2006; Brittain et al. 1991; Ossi 2003). Molecules aggregate together with long-range order in the crystalline state, but this is not the case for the amorphous state (Shalaev and Zografi 2002; Raumer et al. 2006).

The dissolution and bioavailability of a poorly water-soluble drug can be enhanced when formulated as an amorphous form (Goldberg et al. 1966; Chiou and Riegelman 1970; Hancock and Parks 2000; Six et al. 2004; Leuner and Dressman 2000).

Solid dispersion is defined as one type of method to produce an amorphous compound by incorporating a hydrophobic drug into a hydrophilic carrier (Chiou and Riegelman 1971). It is one of the most studied methods to solubilize and to enhance dissolution rate of biopharmaceutical classification system (BCS) class 2 compounds. For instance, a solid dispersion of ritonavir (Law et al. 2001), ER-3421 (a dual 5-lipoxygenase/cyclooxygenase inhibitor; Kushida et al. 2002), was found to have a much higher dissolution rate than the crystalline counterpart and resulted in higher area under curve (AUC) and $C_{\max}$ in the in vivo study.

15.4 Theoretical Aspects of Dissolution Testing of Amorphous Drugs

15.4.1 Solubility and Dissolution of Amorphous Compounds and Solid Dispersions

The amorphous form has attracted increasing interest within the pharmaceutical field because of its higher kinetic solubility in aqueous solvents and faster dissolution rate, which may result in faster absorption rate and increased bioavailability of poorly water-soluble compounds (Leuner and Dressman 2000). The solubility increment of amorphous forms over crystalline states depends on the potential energy difference between these physical states (Hancock and Zografi 1997; Gupta et al. 2004a). It was estimated that 10–1600 folds of solubility increment can be achieved by applying the amorphous form (El-Zein et al. 1998; Fawaz et al. 1996; Ali and Gorashi 1984; Kohri et al. 1999; Hancock and Parks 2000). However, the measured solubility of amorphous compounds was much less, due to rapid crystallization of amorphous materials upon contact with water. Thus, it is important to store amorphous solids protected from moisture.

It is well established that amorphous solids will generally exhibit faster rates of dissolution and higher solubility than the stable crystalline form of the drug. Furthermore, the "enhanced" solubility observed with amorphous solids is transient. Over time, the concentration of dissolved drug will decrease to a concentration consistent with the thermodynamic solubility of the stable crystalline form in the test medium, which is attributed to the precipitation of the dissolved drug from the supersaturated solution as crystalline precipitate. Additionally, solid-state transformation from the amorphous state to the crystalline state may occur due to the effect of solvent on the glass transition temperature (T_g) of the amorphous solid. These characteristics are also typical of amorphous solid dispersions, although the dissolution rate, extent of dissolution, ability to maintain a supersaturated solution, and the T_g of the solid dispersion are largely dependent upon the choice of the polymer in the solid dispersion.

It is generally believed that a solid dispersion can enhance the dissolution rate of a poorly soluble drug through one or a combination of factors (Chiou and Riegelman 1971; Ford 1986; Bloch and Speiser 1987; Craig 1990; Torrado et al. 1996; Fawaz et al. 1996; Serajuddin 1999; Kohri et al. 1999; Leuner and Dressman 2000; Kushida et al. 2002; Gohel and Patel 2003; Miller et al. 2008b). In the case of amorphous solid dispersions, a homogeneous molecular dispersion of the drug in the hydrophilic carrier is formed. Since the drug is already present in the molecular state as a "solid solution," the step of drug dissolution is bypassed.

The dissolution rate of a solid is usually described by the Noyes and Whitney (1897) model (Noyes and Whitney 1897) and the Nernst (1904) model (Nernst 1904). However, the dissolution rate of a two-component system or binary mixture of a drug within a carrier is more complex which led Higuchi et al. (Higuchi et al. 1965) to investigate a uniform, intimate, non-disintegrating mixture of two dissolving compounds both in crystalline state. But the Higuchi model was claimed to not adequately explain the dissolution process in the amorphous solid dispersion model (Van Drooge et al. 2004).

A model proposed by Van Drooge et al. (2004) claimed to describe more accurately the dissolution of amorphous solid dispersions compared to Higuchi's model as it considers phase transition during dissolution from amorphous to crystalline phase (Van Drooge et al. 2004). Moreover, this model is able to deal with altered dissolution behavior due to crystallization before dissolution that was also reported by other researchers (Torrado et al. 1996; Moneghini et al. 1998; Forster et al. 2001). The model was based on slow release profile of diazepam from solid dispersion tablets with amorphous disaccharide as carrier. During the first phase, drug release is slow, but is gradually accelerated and both carrier and drug dissolve according to a nonlinear profile. In the second phase, a linear release at higher rate than during the first phase is observed that stops when the entire amount of carrier is dissolved. In this phase, dissolution of the solid dispersion results in transportation of drug to the bulk. During the third phase, the crystalline drug dissolves slowly. The geometry of the undissolved crystalline drug skeleton determines the release profile. Apparently, this is not enough to form a robust skeleton tablet: drug particles will be dispersed in the dissolution medium and dissolve according to a first-order release profile. On

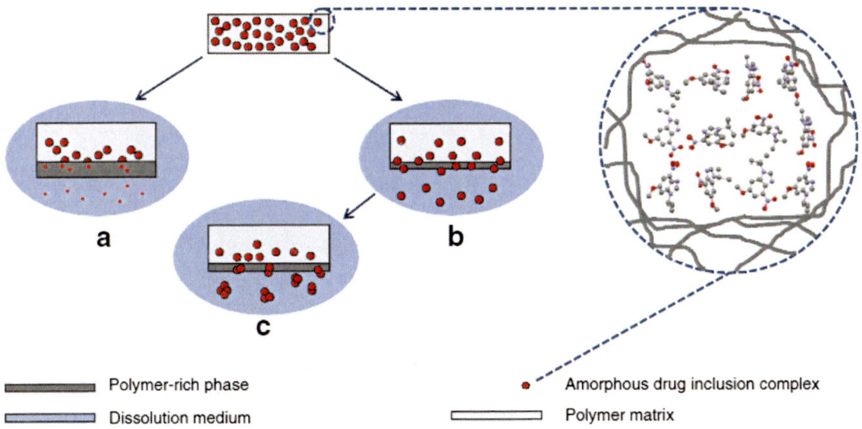

Fig. 15.1 Release modes of amorphous drug inclusion complexes during dissolution of a solid dispersion. (Reproduced with permission from Moore and Wildfong 2009)

the other hand, when the carrier is dissolved, over 80 or 90 % of the drug is still undissolved: a skeleton of crystalline drug is formed, yielding a zero-order release.

Similar behavior was reported previously by Allen and coworkers for tablets prepared from dispersions of small saccharides like sucrose, mannitol, and sorbitol (Allen et al. 1978). The faster release phase was attributed to a molecular incorporated fraction and the slow release to a crystalline fraction of the lipophilic drug, which was present due to incomplete dissolution of the drug in the molten sugar during preparation (Allen et al. 1977). The results imply that crystallization occurred during dissolution of disaccharide solid dispersions.

Another model was proposed to describe the dissolution behavior of solid dispersions and the process of crystallization (Craig 2002). It was suggested that the release behavior of drug molecules from a biphasic solid dispersion as being either carrier mediated or drug mediated (Fig. 15.1)

Initially, a polymer-rich diffusion layer is formed between the solid dispersion and the dissolution medium. In a carrier-mediated dissolution (Fig. 15.1a), after diffusion of drug into the polymer-rich phase, drug is further released into the dissolution medium either as solvated molecules or as amorphous particles at a rate dictated by the carrier. In a drug-mediated dissolution (Fig. 15.1b), the high solubility of the carrier in the dissolution medium does not allow the formation of a polymer-rich phase and drug is dissolved through diffusion of the amorphous complex from the dispersion to the dissolution medium at a rate proportional to the aqueous solubility of the amorphous drug. Agglomeration can occur, resulting in instability in the dissolution (Fig. 15.1c). When the amorphous molecules diffuse through the polymer-rich diffusion layer too rapidly and their dissolution rate in the aqueous medium is relatively slow, crystallization may occur, which creates a high-energy interfacial boundary that slows down the dissolution rate. The reduction in specific surface area due to agglomeration would slow the dissolution rate at a greater extent, permitting a longer

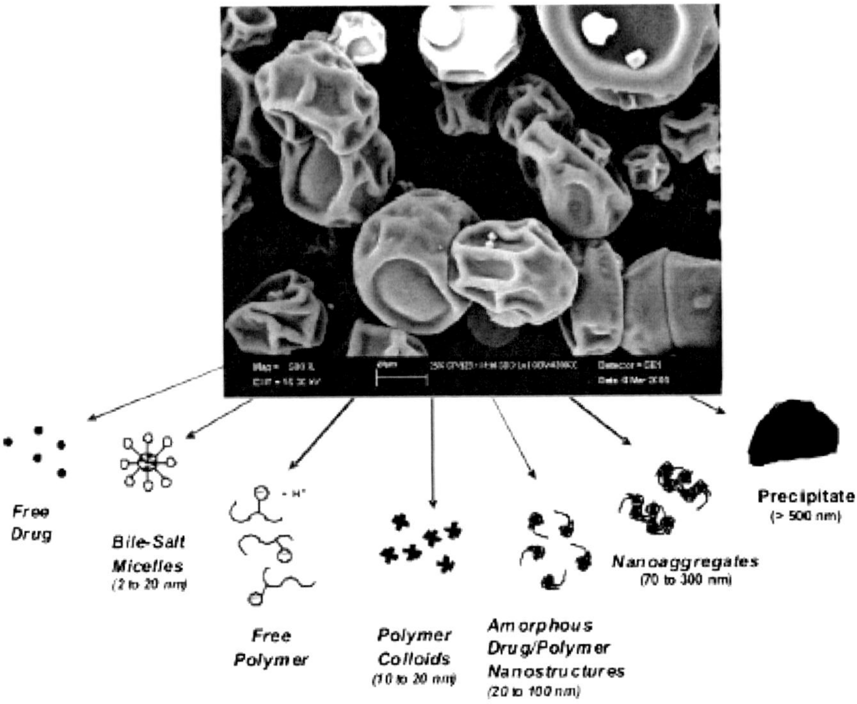

Fig. 15.2 Species formed when SDDs are added to aqueous solutions, simulating duodenal and intestinal contents. (Reproduced with permission from Friesen et al. 2008)

time frame for crystallization (Craig 2002). The model also implied that the relative aqueous solubility of the carrier component may induce physical instability and the effect of manufacturing methods will affect release behavior (critical manufacturing variables) (Moore and Wildfong 2009). Intermolecular interactions between drug and carrier on drug release were not taken into account in this model.

In biorelevant media, according to Friesen et al. (2008), solid drug dispersions (SDDs) rapidly dissolve and/or disperse and produce a wide variety of potential species—categorized based on their size and composition: (1) free or solvated drug, (2) drug in bile salt micelles, (3) free or solvated polymer, (4) polymer colloids, (5) amorphous drug/polymer nanostructures, (6) small aggregates of amorphous drug/polymer nanostructures (termed "nanoaggregates"), and (7) large amorphous particles/precipitate (Fig. 15.2; Friesen et al. 2008).

The rapid formation of drug/polymer nanostructures and nanoaggregates upon introduction of solid dispersion to an aqueous solution which are stable in aqueous suspension can justify the enhanced oral drug absorption. These nanostructures and nanoaggregates produce a free drug concentration higher than the solubility of crystalline drug which is sustained by replacing free drug as it is absorbed (Friesen et al. 2008).

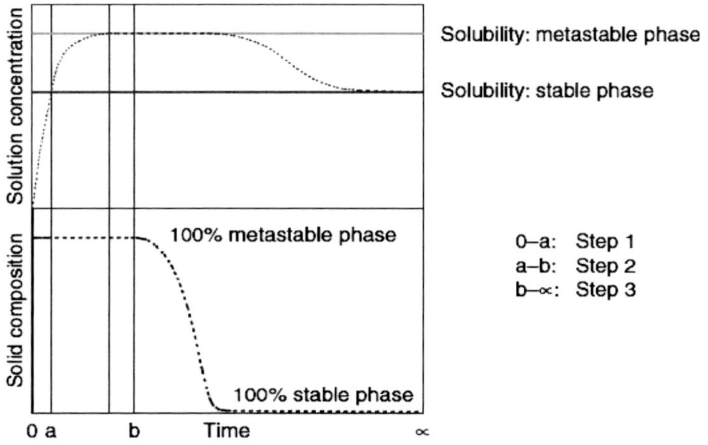

Fig. 15.3 Concentration in the solution and the solid compositions as a function of time during the solution-mediated phase transformation. (Reproduced with permission from Zhang et al. 2009)

15.4.2 Solution-Mediated Transformation During Dissolution

One of the issues relating to the stability of the amorphous state, particularly in vivo, is its solution-mediated transformation characteristic. Solution-mediated transformation of amorphous to crystalline state is the conversion of metastable solids such as amorphous solids to the crystalline state when the solids are exposed to a solvent, in this case water. The transformation to the thermodynamically stable crystalline state occurs at a higher rate in the presence of solvents than in the dry state because of higher molecular mobility in the presence of solvents.

Characterization of solution-mediated transformations in the amorphous state can give an insight into amorphous crystallization (Zhang et al. 2009). The importance of the phase transition kinetics, molecular interpretations, and process implications has been emphasized in numerous studies (Cardew and Davey 1985; Davey et al. 1986, 1997a, 1997b; Rodriguez-Hornedo et al. 1992; Blagden et al. 1998).

The presence of the solvent does not change the thermodynamics and stability relationship, unless a solvate/hydrate is formed with the solvent. However, owing to the much higher mobility in the solution state than in the solid, transformation to the stable phase is much faster. This process is analogous to the effect of catalysts for chemical reactions (Zhang et al. 2009). The schematic representation of concentrations in the solution, as well as the solid compositions as a function of time, is shown in Fig. 15.3 for a typical solution-mediated transformation process (Zhang et al. 2009).

As depicted in Fig. 15.3, three consecutive steps are involved in a solution-mediated transformation (Cardew and Davey 1985; Zhang et al. 2002; Rodriguez-Hornedo et al. 1992): (1) initial dissolution of the metastable phase into the solution to reach and exceed the solubility of the stable phase, (2) nucleation of the stable phase, and (3) crystal growth of the stable phase coupled with the continuous dissolution of the metastable phase. Step (2) or (3) is usually the slowest of the steps. For the overall solution-mediated transformation process, step (2) is usually the slowest. Therefore, this is the rate-determining step as nucleation is involved. When step (2) is the rate determinant, factors such as the solubility and the solubility difference between the phases, processing temperature, contact surfaces, agitation, and soluble excipients/impurities that affect nucleation will influence the overall transformation. When step (3) is the rate-controlling step, the kinetics of the conversion is determined by solubility difference, solid/solvent ratio, agitation, processing temperature, particle size of the original phase, and soluble excipients/impurities.

15.4.3 Crystallization During Dissolution

While it is acknowledged that orally administered amorphous formulations may supersaturate and crystallize in vivo (when exposed to the gastrointestinal (GI) fluids in the human body), the ability to study this phenomenon in the in vivo setting has been hampered by the dearth of analytical tools, the complexity of GI fluids, and the inaccessibility of the GI tract where the measurements need to be made. In contrast, there are many reported studies on proposed mechanisms and kinetics devoted to crystallization in vitro (Cardew and Davey 1985; Davey et al. 1986, 1997a, 1997b; Weissbuch et al. 1987; Rodriguez-Hornedo et al. 1992; Weissbuch et al. 1994; Zhu and Grant 1996; Zhu et al. 1996; Blagden et al. 1998; Davey et al. 2002).

The crystallization of amorphous drugs during dissolution could be shown by the nearly 20 % difference between the predicted and experimentally measured solubility ratios of amorphous drug. For instance, in the case of indomethacin (IMC) and griseofulvin particles, the predicted solubility ratio in water ranged from 25 to 104 and 38 to 441, respectively, whereas the experimentally measured value was only 4.5 (IMC) and 1.4 (griseofulvin; Hancock and Parks 2000; Matteucci et al. 2008; Elamin et al. 1994).

The concentration of the solution at equilibrium will decrease after a period of time to the level equivalent to the solubility of the most stable crystalline form. The duration of the period of increased (metastable) solubility is generally thought to be controlled by the rate of nucleation and, thus, the rate of growth of the more stable phase (Clarkson et al. 1992).

The suggested mechanism behind transformation from disordered structure to ordered thermodynamically stable structure is mainly through surface solid-state transition (Mosharraf et al. 1999). This would result in a very slow reduction in the apparent solubility plateau level down to the thermodynamically stable value. The investigation of the relationship between equilibrium solubility, the amount of solute added to the solvent, and the proportion of disordered or amorphous structures on the surface of the particles can provide valuable information which can be used to predict and control the solubility and dissolution behavior of sparingly soluble hydrophobic drugs (Mosharraf et al. 1999).

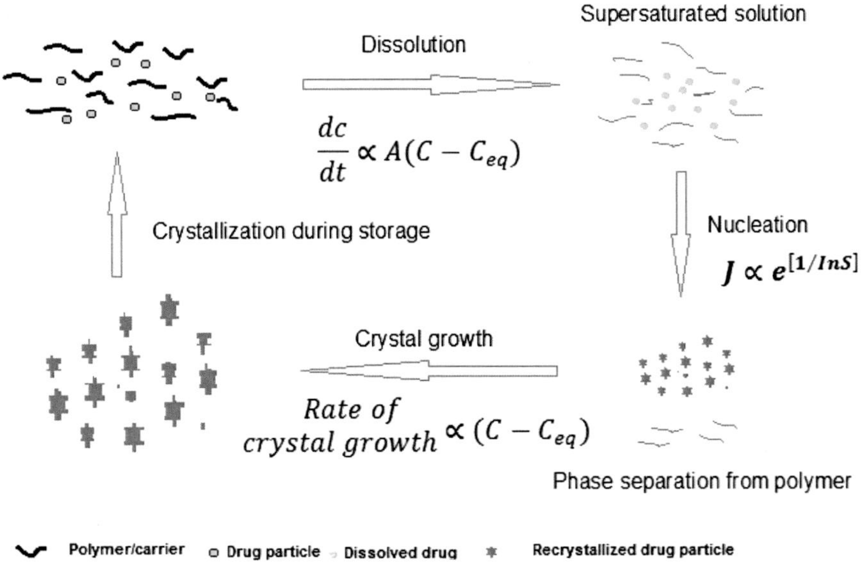

Fig. 15.4 Schematic illustration of the competition between dissolution and crystallization via the solid or solution state for amorphous systems. (Reproduced with permission from Alonzo et al. 2010)

Dissolution, precipitation, and crystallization that can occur during dissolution of an amorphous system are summarized in Fig. 15.4 (Alonzo et al. 2010). Modified Noyes and Whitney equation is used to describe the dissolution pathway, where dc/dt represents the dissolution rate which is directly proportional to the surface area (A) and the difference between the solution concentration (C) and the equilibrium concentration (C_{eq}; Alonzo et al. 2010). In the nucleation path, J represents the nucleation rate, which is proportional to the degree of supersaturation (S). For the growth path, the rate of crystal growth is also proportional to the difference between the actual solution concentration and the equilibrium concentration (Alonzo et al. 2010).

15.4.4 Supersaturation of Amorphous Systems During Dissolution

In recent pharmaceutical literature, the terms "equilibrium solubility" and "kinetic or apparent solubility" are often used for the systems with stable and metastable equilibria, respectively (Lipinski et al. 2001; Huang and Tong 2004).

A solid phase is crystallized from solution if the chemical potential of the solid phase is less than that of the dissolved component. A solution in which the chemical potential of the solute is the same as that of the corresponding solid phase and is in equilibrium with the solid phase under the given conditions (temperature, pH, and concentration) is referred to as a saturated solution (Gamsjager et al. 2008). In order for crystallization from solution to occur, this equilibrium concentration or solubility must be exceeded. This excess concentration or chemical potential, called the supersaturation, is the driving force for nucleation and crystal growth (Strickley et al. 2007).

For poorly water-soluble drugs, the maximum achievable intraluminal drug concentration may limit absorption. However, the intraluminal concentration of a drug is not necessarily limited by its solubility in GI fluids (Brouwers et al. 2009). Drugs may be in solution at a concentration above their saturation solubility, that is, in a state of supersaturation. The degree of supersaturation can be expressed by the supersaturation ratio, *S* (Eq. 15.1; Brouwers et al. 2009):

$$\mathbf{S} = \mathbf{C}/\mathbf{C}_{\mathbf{eq}} \tag{15.1}$$

with C and C_{eq} representing the solubility and equilibrium solubility (saturation), respectively.

A solution is defined as unsaturated, saturated, or supersaturated based on the following relationships: $S < 1$, $S = 1$, or $S > 1$, respectively (Brouwers et al. 2009).

Amorphous drugs are high-energy solid systems that are capable of reaching higher kinetic solubility values (supersaturation) than would be expected from the equilibrium solubility of a crystalline material (Wei-Qin 2009). A supersaturated drug solution is thermodynamically unstable compared with the equilibrium condition (saturation). Thus, it has the tendency to return to the equilibrium state by drug precipitation. The higher the supersaturation, the more precipitation will take place as the former is the driving force for the latter (Six et al. 2004). The accelerated dissolution and the higher apparent solubility provided by amorphous systems could induce the generation of supersaturated solutions in the GI lumen that can result in increased absorption (Brouwers et al. 2009).

The dissolution characteristics of solid dispersions depend to a large extent on the physical state (ideally: amorphous), drug dispersivity (ideally: molecular dispersion), and particle size (Brouwers et al. 2009). Supersaturation can be achieved through:

1. The in vivo GI conditions: effect of pH and content changes from stomach to intestine.
   Due to the pH gradient in the GI lumen (pH 1.5–2 in the stomach compared with pH 5–8 in the intestine), the solubility of weak bases in gastric fluid (ionized form) typically exceeds their solubility in the intestinal fluid (unionized form). Therefore, higher dissolution of poorly water-soluble weak bases in the stomach before transfer to the intestine may result in supersaturation (Brouwers et al. 2009; Bevernage et al. 2010).
2. High-energy and rapidly dissolving solid forms.
   As mentioned earlier, amorphous state requires less energy to dissolve, resulting in higher apparent solubility and increased dissolution rates (Hancock and Parks 2000). Methods such as particle size reduction through milling and co-grinding that can form an amorphous state may increase dissolution rates by enhancing the surface area available for dissolution (Sarode et al. 2013; Brouwers et al. 2009).

This higher initial solubility may be sufficient to ensure increased and more rapid absorption for a drug with good permeability such as BCS class 1 and 2 drugs. But a more thermodynamically stable form may crystallize at any time inside the GI tract, and the crystallization would have a major impact on the product performance in vivo (Strickley et al. 2007; Sarode et al. 2013). The higher dissolution rate and apparent solubility of an amorphous drug usually causes supersaturation during in vivo dissolution. It should be noted though that rapid dissolution and supersaturation could prove counterproductive in some cases, for example, for drugs that precipitate during transit from the stomach to upper small intestine where an increase in the pH is observed. This precipitation in the GI tract may compromise oral bioavailability (Brouwers et al. 2009; Overhoff et al. 2008).

If a crystallization-inhibitory polymer is incorporated into the amorphous solid dispersion, the in vivo precipitation may be delayed or completely eliminated, resulting in much improved oral absorption. It is ideal if the polymeric carrier can function as a precipitation (crystallization) inhibitor during in vivo dissolution (Zhang et al. 2009).

The commercially available Sporanox® capsule formulation is a solid dispersion relying on the principle of supersaturation to enhance the intestinal absorption of the antifungal itraconazole (ITR), a weak base (pKa = 4) with an extremely low and pH-dependent aqueous solubility (ca. 1 ng/mL in water, 6 mg/mL in 0.1 M HCl). This formulation comprises a molecular dispersion of ITR in a hydroxypropylmethylcellulose (HPMC) matrix, which is coated onto inert sugar spheres. Dissolution of HPMC in media simulating the gastric environment results in supersaturated concentrations which are maintained for at least 4 h. HPMC is believed to prevent ITR from precipitation in the stomach and in the intestine, resulting in significant absorption (maximum fraction absorbed ca. 85 %) and oral bioavailability (ca. 55 %; Brewster et al. 2008).

Yamashita et al. (2003) investigated the dissolution in acidic medium of solid dispersions containing the macrolide lactone tacrolimus in an amorphous state comparing three different polymers (HPMC, PVP, and PEG 6000) as the carrier. Rapid dissolution and supersaturated concentrations of tacrolimus up to 25-fold higher than the equilibrium solubility (2 mg/mL) were observed. Even though the polymer choice did not affect the maximum degree of supersaturation, it was only HPMC that could fully inhibit precipitation for up to 24 h.

The changes of pH in stomach and intestine and in fasted and fed state will affect the solubility and dissolution of weak bases, for example, ITR. The GI pH may also change the performance of precipitation inhibitors, especially their solubility and hydrogen bonding between the H donor and acceptor. Surface tension, viscosity of GI fluids, and presence of bile salts and phospholipids may influence the solubility and dissolution of the drug (Dressman et al. 2007) and subsequently the intraluminal supersaturation and drug-precipitation kinetics (Brouwers et al. 2009). Supersaturation may be affected by the hydrodynamics and the composition of the GI fluids (Brouwers et al. 2009). Thus, for the in vitro dissolution, testing the extent of supersaturation following acidic-to-neutral pH transition must be considered in order to correlate the in vitro dissolution with the in vivo absorption (Miller et al. 2008a).

Furthermore, ITR absorption mostly occurs in the proximal small intestine (Miller et al. 2007; Six et al. 2005); hence, immediate release (IR) formulations are provided with a small window for absorption because supersaturated levels of ITR in the gastric environment rapidly precipitate upon exit from the stomach (Miller et al. 2007; Six et al. 2005). A study was performed using various types of ITR amorphous formulations (including Sporanox®) and reported finding revealed inconsistency between the in vitro dissolution performance in simulated gastric fluid (SGF) and the in vivo absorption. It was suggested that faster release and increased supersaturation in the acidic medium along with differences in crystallization rate upon transfer to the small intestine correlated with the lower predicted bioavailability (Six et al. 2005). Precipitation can decrease the driving force for transport across the biological membrane and limit the time available for absorption which complicates development of IVIVC with amorphous compound (Overhoff et al. 2008). The importance of simulating the GI pH shift during supersaturation dissolution testing of amorphous compound to evaluate whether supersaturation is maintained in the small intestine was described (Six et al. 2005).

15.5 Factors That Influence Dissolution of Amorphous Drug Products

15.5.1 Formulation Factors

As with conventional formulations of IR solid oral drug products containing crystalline drug, dissolution behavior of amorphous drug products are greatly influenced by the composition of the formulation as well as the manufacturing process. Of note is the polymer used as the carrier for the solid dispersion. The dissolution rate of the dosage form is determined by the characteristic of the carrier (Leuner and Dressman 2000). Such carrier systems include cellulose-based polymers such as HPMC and its acidic derivative hydroxypropylmethylcellulose acetate succinate (HPMC-AS); polyethylene glycol (PEG), polyvinylpyrrolidone (PVP) and its copolymers; acrylate polymers (Eudragit), sugars and their derivatives; emulsifiers, organic acids and its derivatives. There are only a handful of oral pharmaceutical products containing amorphous active pharmaceutical ingredient (API) that have been successfully marketed despite several decades of effort in research and development (Chap. 3). However, recently launched products such as Incivek®, Kalydeco®, and Zelboraf® demonstrate a great versatility of the amorphous solid dispersions in increasing the rate of solubilization and bioavailability.

Whereas, for conventional formulations, where the dissolution of the drug may occur from the disintegrated granule (i.e., drug substance particle), dissolution of amorphous drug products occur from the solid dispersion consisting of the amorphous drug and the polymer. Therefore, the dissolution behavior of the polymer plays a key role in the dissolution of the amorphous drug product. Ionizable polymers, such as those of the acrylates or hypromellose acetate succinate will dissolve through salt formation at pH above the pKa. Additionally, the dissolution rates of different grades of ionizable polymers will differ depending on the degree of substitution. Unlike ionizable polymers, the dissolution rate of non-ionizable polymers (e.g., copovidone) is pH independent. The viscosity of the polymer will also influence the dissolution rate of the amorphous solid dispersion and the final drug product.

Drug load, expressed as the ratio of the drug to polymer in the amorphous solid dispersion, will also play a significant role in the dissolution and physical stability of the supersaturated solution. Lower drug load (i.e., higher polymer content) will generally result in enhanced dissolution and stability of the supersaturated solution.

15.5.2 Manufacturing Factors Which Influence Solid-State Properties

The process for the manufacture of the drug product, particularly the processing step for the amorphous solid dispersion and those that impact the solid-state properties of the solid dispersion, will have a major influence on the dissolution of the drug product. Amorphous processing technologies are described in a different chapter of this book and, therefore, will not be covered in detail here. It should be mentioned, however, that these technologies generally will result in material with different solid-state properties, most notably particle size, shape, porosity and density. In addition to the particulate properties, there may be potential differences in the type and extent of interactions of the drug and polymer as a function of the processing technology used to prepare the solid dispersion. Systematic research in this area is lacking; however, it is anticipated that solvent-based processes may be conducive to certain interactions that may not be feasible with non-solvent-based technologies such as hot-melt extrusion (HME).

Material produced from spray-drying processes are generally spherical and hollow due to process in which the solution of dissolved drug and polymer are sprayed as fine droplets and then rapidly dried in an inert stream of warm air. Spray-dried dispersions are oftentimes porous and fragile due to the escape of solvent through the solid matrix. Processing factors that influence the particle properties of the spray-dried dispersion (drying temperature, spray rate, droplet size, air flow, etc.) will influence the dissolution rate.

In contrast, there are no solvents used in HME processes for manufacturing amorphous solid dispersions. Instead, the mixture of crystalline drug and polymer are heated to a temperature at which the components melt or form a eutectic, and then flash cooled, resulting in a dense amorphous glassy solid. The solid is then milled to achieve uniform particle size distribution, which is then processed into the final formulation. Therefore, the milling step will determine particle size and surface area, which in turn is related to the dissolution rate of the solid.

A newer, innovative processing technology for the manufacture of amorphous solid dispersions is the microprecipitation method in which an organic solution of the drug and polymer is introduced into a miscible anti-solvent in which the drug and polymer are insoluble or less soluble (e.g., water), causing the precipitation of the drug and polymer as a microprecipitated bulk powder (MBP). As there is water and solvent involved in the process, a subsequent drying step and milling step are required. Materials made by the MBP process tend to be porous. Manufacturing processes that influence the porosity and particle size of the MBP (rate of precipitation, drying temperature, MBP milling) will significantly impact the dissolution of the MBP and the final drug product.

It is noted that discussion has centered on the processing technologies and parameters that will influence the MBP, and no mention is made of the drug substance. This should be inherently clear that the solid-state properties of the starting drug substance is not that important to the dissolution of the drug product, as the crystalline drug substance is converted to an amorphous solid dispersion, and it is the properties of the amorphous solid dispersion that is responsible for the dissolution behavior of the drug product. It should also be mentioned that the above factors that influence dissolution are discussed independently, whereas it is the combination of these factors, that is, composition and process factors which influence the chemical and physical properties of the amorphous solid dispersion, which will determine the dissolution rate of the solid dispersion.

15.6 Dissolution Case Studies to Guide Formulation Development

In addition to the physical form stability of the amorphous drug, a sound understanding of the chemical form of the drug in the amorphous solid dispersion and its behavior during and after dissolution are important elements of the quality-by-design approach to development of an amorphous solid dispersion drug product. The chemical form of a drug (weak acid, weak base, or neutral) and its pH-dependent solubility is known to impact in vivo performance of the amorphous solid dispersion product due to its interplay with the pH gradient of the GI tract. Weak acids may rapidly precipitate, crystallize, or gel in the stomach at low pH, while weak bases may rapidly dissolve at the lower pH of the stomach but precipitate at the higher pH in the lower GI upon transiting into the intestine. It is therefore very important to consider the impact of the drug's chemical form and pKa while designing the amorphous solid dispersion with respect to selection of stabilizing polymer (nonionic vs. ionic, and pH-dependent release), drug loading, manufacturing technology, as well as downstream processing (IR vs. eroding vs. enteric coating) to maximize the duration of in vivo supersaturation. In order to guide the design of amorphous solid dispersions of new chemical entities (NCEs) with respect to these elements, researchers have recently successfully utilized in vitro dissolution testing in non-sink and biorelevant media as well as two-stage media as a predictive tool for in vivo precipitation, kinetic solubility, and supersaturation. Dissolution medium representing 100 % saturation is preferred for formulation screening; however, several successful examples of use of less than 100 % saturation media during early design of amorphous solid dispersion have been published. Formulations that are able to sustain supersaturation for at least 2 h (physiologically relevant) would represent viable formulations that can be investigated in vivo. Formulation that can maintain supersaturation for less than 60 min would need careful evaluation as these may be more prone to higher pharmacokinetic (PK) variability. The following examples illustrate various in vitro dissolution methodologies utilized to guide amorphous formulation development with high predictive power for the desired in vivo performance.

15.6.1 Between Amorphous and Crystalline Phase

Solid-state changes that may occur during dissolution of amorphous carbamazepine (CBZ) were studied in phosphate buffer pH 7.2 at room temperature using in situ Raman spectroscopy (off-line; Savolainen et al. 2009). The findings of this study confirmed that the surface of the CBZ samples crystallize immediately upon contact with the dissolution medium. The transition from the amorphous form to crystalline anhydrate (form I) of CBZ and then a solution-mediated transformation from form I to dihydrate, as previously demonstrated by Murphy et al. (Murphy et al. 2002), was proposed. The transition from the amorphous form to the crystalline anhydrate is likely a solid-state transition as amorphous CBZ has been shown to crystallize to an anhydrate form in dry conditions (25 °C and < 10 % RH; Patterson et al. 2005). To validate the instant crystallization of CBZ, the dissolution test in phosphate buffer was performed and no significant improvement on the dissolution rates was noted (Savolainen et al. 2009). Dissolution from most of the amorphous samples was even slightly slower than from the dihydrate compacts. Analysis of the remaining sample after the dissolution experiment confirmed that the sample surface had converted to the dihydrate (crystalline form) during dissolution (Savolainen et al. 2009).

In another case, dissolution of crystalline and amorphous ciclesonide was studied (Feth et al. 2008). Crystalline and amorphous ciclesonide exhibit the same saturation solubility (Feth et al. 2008). For the crystalline ciclesonide, within the first 60 min of dissolution, the concentration measured (50 mg/L) was significantly lower than its saturation solubility (90.1 $\pm$ 2.2 mg/L). On the contrary, for the amorphous form, concentration increased almost instantaneously to values up to four times higher than the saturation solubility determined for amorphous ciclesonide (91.6 $\pm$ 5.2 mg/L). After 24 h, for both forms, concentrations were approaching the saturation concentration. It was suggested that the amorphous ciclesonide was able to form a stable supersaturated solution in water for at least 60 min due to the absence of inoculation crystals in the amorphous phase. On the other hand, in crystalline ciclesonide, the crystal itself acts as crystallization seed in the slurry which speeds up the process.

15.6.2 Between Different Temperatures and Polymers

Both temperature and polymer (which act as a crystallization inhibitor) can greatly affect the solubility of amorphous drugs (Alonzo et al. 2009). The concentration–time (dissolution) profiles attained with solid dispersions may be higher than those achieved with the pure amorphous API (Gupta et al. 2004b), suggesting that certain polymers are able to further enhance solution concentrations relative to pure amorphous drug. The increased solution concentrations observed following dissolution of amorphous solid dispersions have been attributed to the inhibition of API crystallization from the supersaturated solution by the polymer (Gupta et al. 2004b; Tanno et al. 2004) and increased equilibrium solubility of the API due to solution complexation with the polymer (Usui et al. 1997; Acartürk et al. 1992; Loftsson et al. 1996).

In order to guide design of solid dispersions, a simple supersaturation test/dissolution study for rank ordering of polymers and drug loading was reported without actually making solid dispersions (Konno et al. 2008). Dissolution of amorphous IMC at different temperatures (25 °C, 37 °C) in simulated intestinal fluid without pancreatin showed that IMC underwent solution-mediated transformation with higher concentration and lower temperature. The inclusion of polymer inhibited crystallization (Konno and Taylor 2008; Alonzo et al. 2009). For the amorphous felodipine, extensive crystallization was observed in 10–15 min at 25 °C, whereas instant crystallization was observed at 37 °C. A supersaturated solution was not formed when polymer was not included (Konno et al. 2008; Alonzo et al. 2009).

15.6.3 Dissolution of Salts

In the stomach and intestine, drug solubility can be enhanced by food and bile components such as bile salts, lecithin, and fatty acids. Supersaturation in the intestinal fluid is an important property that can play a significant role in drug absorption. For compounds with poor intrinsic solubility in the intestinal fluid, solubility is often a limiting factor for absorption. For many of these compounds, it may not be possible to enhance the saturation solubility to the extent required such that the whole dose is dissolved in the GI fluid. In this case, creating or maintaining supersaturation in the intestinal fluid can be an effective way to enhance absorption of these compounds (Wei-Qin 2009).

15.6.3.1 Weak Bases

In most cases, supersaturation is induced from solubilized formulations or formulations that contain a high-energy state of the drug. However, for weakly basic drugs, even intake of the crystalline powder may result in supersaturation in the small intestine. Due to the pH gradient in the GI lumen in fasted state conditions (pH 1.5–2 in the stomach vs. pH 5–8 in the intestine), the gastric solubility of weak bases (ionized form) typically exceeds their intestinal solubility (unionized form). Hence, after dissolution of poorly water-soluble weak bases in the stomach, transfer to the intestine may result in supersaturated concentrations and an increased flux across the intestinal mucosa (Wei-Qin 2009). By simulating the GI pH shift during dissolution experiments, this behavior can be monitored. For instance, Kostewicz et al. (Kostewicz et al. 2004) evaluated the behavior of three weakly basic drugs (dipyridamole, BIBU 104 XX, and BIMT 17 BS) in an in vitro system simulating both the pH gradient between stomach and intestine and the presence of bile salts and phospholipids in the intestine. Upon transfer of a solution of the drug in an acidic medium (pH 2, simulating fasted state gastric conditions), supersaturated concentrations of the weak bases were observed in both fasted- and fed-state simulated intestinal fluids (FaSSIF and FeSSIF). Presumably, this mechanism plays an important role in the intestinal absorption of various poorly water-soluble weak bases.

Depending on the properties of the salt and its corresponding base or acid, the fate of the salt in the GI tract may vary significantly. When the salt of the basic drug gets in the GI tract, it may dissolve in the stomach and remain either in solution or precipitate out as the free base when it is emptied into the intestine. It may also convert to the hydrochloride salt if the hydrochloride salt is less soluble, especially with the influence of the common-ion effect. In this case, the dissolution in the intestine is in reality the dissolution of the precipitated hydrochloride salt. When salt conversion happens in vivo, it can precipitate out as either the crystalline or the amorphous form with different particle size that will affect solubility and dissolution (Wei-Qin 2009; Li et al. 2005).

Since transport across the biological membrane of weak bases will be more pronounced in the small intestine (uptake of the unionized form), sufficient precipitation inhibition (polymer) is required upon transfer of the supersaturated solution to the intestine. Therefore, one cannot rely on dissolution studies at constant acidic pH to predict the performance of formulations of weak bases in vivo (Miller et al. 2007). For instance, Six et al. (2005) observed a discrepancy between the results of in vitro dissolution tests in acidic medium and in vivo absorption for four solid dispersions of ITR: faster release and increased supersaturation in acidic medium correlated with lower bioavailability. Presumably, this effect can be explained by differences in crystallization rate upon transfer to the small intestine (increased driving force for precipitation in case of higher supersaturation). Thus, it is crucial to simulate the GI pH shift during supersaturation dissolution testing of weak bases to evaluate whether supersaturation is maintained in the small intestine.

Use of a GI pH shift model was also employed in the development of a propriety weak base (compound A) which exhibited high solubility at gastric pH, but very low solubility at intestinal pH, and high PK variability in humans. A two-stage dissolution test in which the drug product is tested in pH 2 media (HCl) or pH 4.5 media (phosphate buffer) for 30 min followed by testing in a dissolution medium containing sodium taurocholate and lecitihin at pH 6.5 (fasted state simulated intestinal fluid, FaSSIF) was used to evaluate the performance of an IR tablet containing the thermodynamically stable crystalline-free base. The results showed that although very high solubility is observed at low pH (2), at elevated gastric pH (4.5) the drug exhibits very low solubility. Moreover, upon transitioning from pH 2 to pH 6.5 (FaSSIF), the solubility drops quickly to the measured solubility for the amorphous drug. These results suggested that significant PK variability observed in patients may be attributable to differences in gastric pH associated with the disease state and with concomitant gastric pH-elevating medications, and that an amorphous formulation may improve the PK variability while also potentially enhancing bioavailability. Two tablet formulations containing amorphous solid dispersion with 35 % drug load and 60 % drug load in HPMC-AS polymer made by spray drying were developed and evaluated using this two-stage dissolution test. The results shown (Figs. 15.5 and 15.6) support the hypothesis of improved variability and potential enhancement of bioavailability.

Two tablet formulations containing crystalline drug substance and 35 % amorphous solid dispersion were evaluated in humans. The results show substantial enhancement (1.7 fold) in the bioavailability of the drug and improved PK variability (Coefficient of Variation (CV) 80 % → 18 %) with the tablet made with amorphous solid dispersion (Roche in-house data). The results from this study illustrate the utility of dissolution test conditions that are carefully chosen to simulate physiological conditions (i.e., biorelevant) when applied to testing of a poorly water-soluble weak base and the potential for improved absorption through amorphous formulation technologies.

15.6.3.2 Weak Acids

Dissolution of the salt of an acidic compound has its own complications. The salt is likely to convert to the free acid. When this happens, the liberated free acid may coat the surface of the remaining drug particles or nucleate on other particle surfaces, leading to a slowdown of dissolution (Wei-Qin 2009). As described in the earlier sections, weak acids may rapidly precipitate/crystallize or gel in stomach before transit to the lower GI. It is therefore important to select amorphous solid dispersion and downstream technology, yielding a drug product with optimal supersaturation at physiologically relevant pH for absorption.

15.6.4 Biorelevant Dissolution Testing of Amorphous Solid Dispersions

To better predict the in vivo behavior after oral administration and estimate the impact of solubility, degree of supersaturation, and dissolution on absorption, the in vitro dissolution method should be physiologically relevant (biorelevant), taking into account the contents and the transit through the GI tract.

For example, in the case of in vitro dissolution of amorphous ITR, the initial burst in FaSSGF, a biorelevant medium containing sodium taurocholate and lecithin Vertzoni et al. 2005 was far greater than in SGF and led to higher dissolution

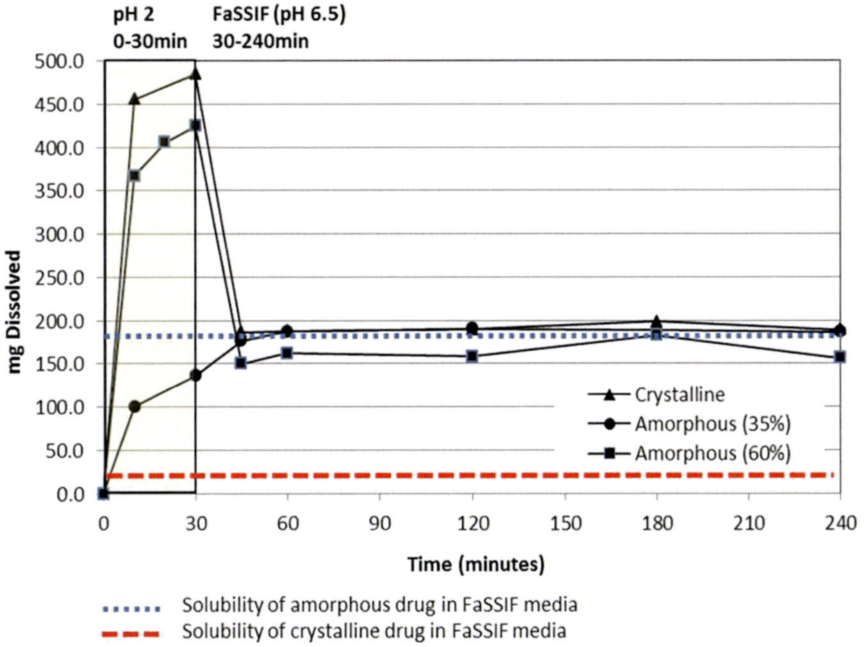

Fig. 15.5 Two-stage dissolution of tablet formulations made with crystalline drug substance and amorphous solid dispersion in media simulating normal gastric pH

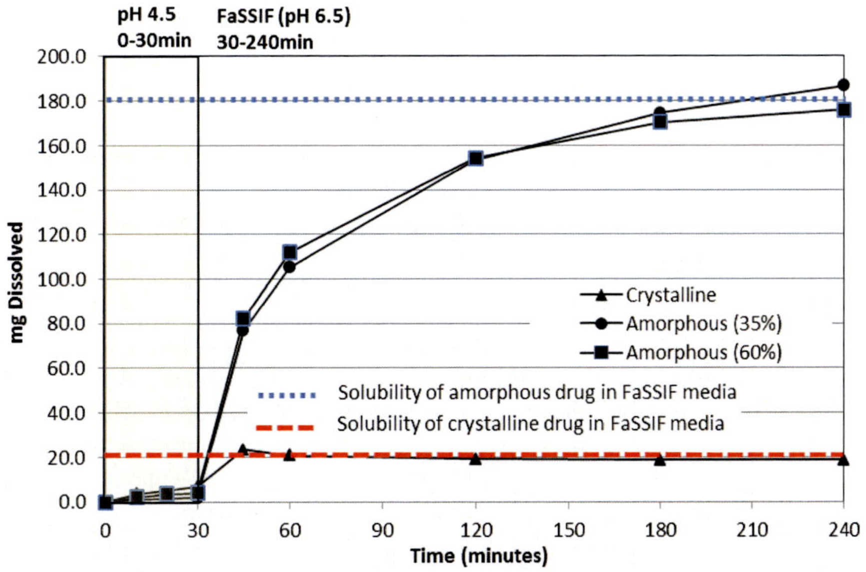

Fig. 15.6 Two-stage dissolution of tablet formulations made with crystalline drug substance and amorphous solid dispersion in media simulating elevated gastric pH

(Ghazal et al. 2009). Furthermore, the presence of lecithin and sodium taurocholate in FaSSIF and FeSSIF enhanced ITR's dissolution rate, with a greater increase in the lower pH and more bile salts and lecithin of FeSSIF, in accordance to the in vivo food effect of the drug.

The precipitation behavior of the amorphous JNJ-25894934 from three different liquid formulations in phosphate buffer, FaSSIF, and FeSSIF was investigated (Dai et al. 2007). The solubility of the drug was similar in FaSSIF and FeSSIF, but was eightfold lower in the phosphate buffer. Differences in precipitation were observed between the different media, especially for the fast-precipitating formulation. While precipitation from this formulation was immediate and complete in phosphate buffer, a metastable zone containing about 20 and 80 % dissolved drug, was maintained during 8 h in FaSSIF and FeSSIF, respectively, and precipitation continued after 8 h. As there was no difference in drug solubility in FaSSIF versus FeSSIF, enhanced concentrations in the metastable zone in FeSSIF clearly suggest specific precipitation-inhibiting interactions that are more effective in FeSSIF, presumably due to the increased bile salt/phospholipid concentrations. The in vitro precipitation profiles in FeSSIF correlated better with in vivo absorption than those in phosphate buffer and FaSSIF (Dai et al. 2007).

Another example of the usefulness of biorelevant media, in this case FaSSIF, could be seen in the development of Zelboraf® (vemurafinib), an IR tablet drug product made with amorphous solid dispersion (Shah et al. 2013). The initial clinical formulation was an IR capsule containing a fast dissolving metastable crystalline form of the vemurafinib. In order to achieve higher exposures in the clinical studies, IR capsule prototypes made with amorphous solid dispersion were prepared. Whereas dissolution results for the three formulations suggested comparable exposures; the actual exposures obtained in humans revealed the amorphous formulations to be far superior to the formulation made with the metastable crystalline drug substance (Fig. 15.7). However, when the formulations were tested using biorelevant media (FaSSIF), a clear distinction could be made between the formulations, which corresponded to the observed differences in vivo (Fig. 15.8). Further investigation into the underlying reason for the substantial differences found evidence of solid-state transformation of the metastable crystalline form of the drug substance (form 1) to its thermodynamically stable, insoluble crystalline form (form 2).

The same test conditions were used to demonstrate the effect of stressing on the amorphous formulation (Fig. 15.9), which showed lower rate and extent of dissolution of the amorphous formulation when stressed by heat and humidity. The diminished dissolution observed in the stressed sample was determined to be due to stress-induced crystallization of the amorphous product (Shah et al. 2013).

Although it is not possible to measure intraluminal supersaturation in the GI tract as mentioned previously, more reliable prediction of intraluminal supersaturation of amorphous solid dispersions can be gained by performing supersaturation/dissolution assays in biorelevant media. It is important to evaluate supersaturation not only in gastric but also in intestinal conditions, as discussed previously, and simulation of the GI pH gradient would be essential (Brouwers et al. 2009). A pH shift can be simulated by multi-compartmental dissolution techniques (Kostewicz et al. 2004;

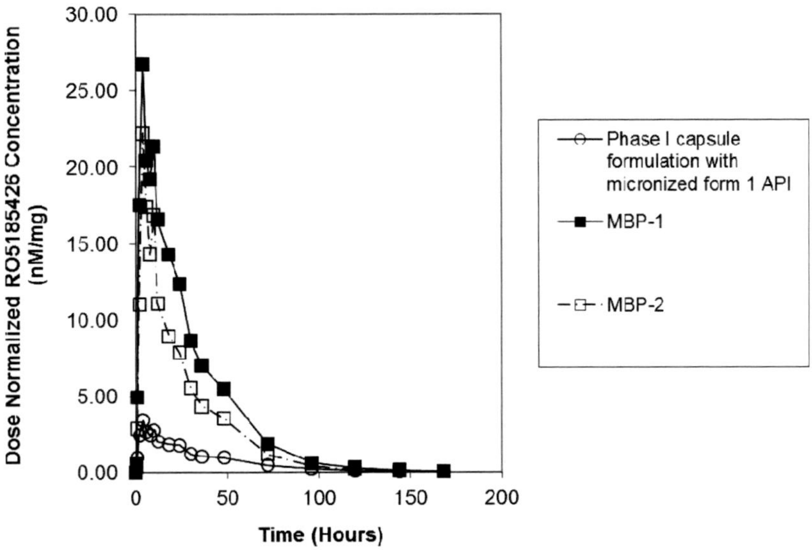

Fig. 15.7 Plasma concentration time profiles of vemurafinib, given as a single dose of capsule formulations made with metastable crystalline drug substance (form 1) or amorphous MBP (MBP-1 and MBP-2) to healthy volunteers. (Reproduced with permission from Shah et al. 2013)

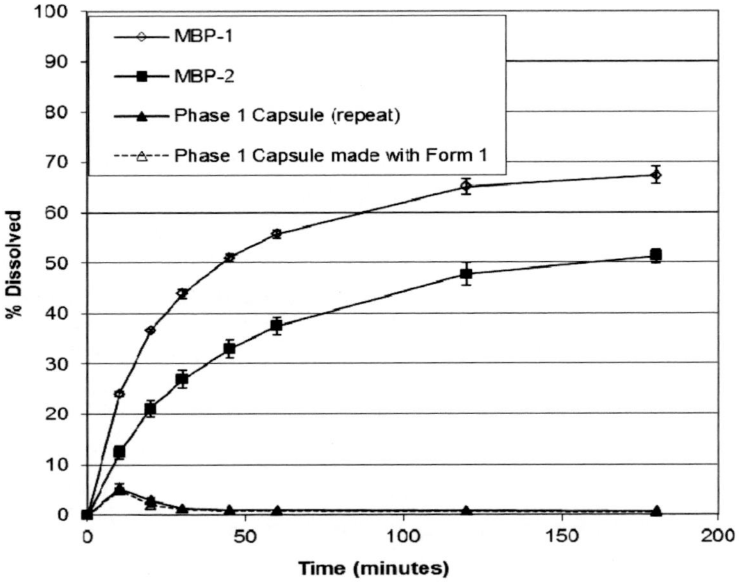

Fig. 15.8 Dissolution profiles of capsule formulations made with metastable crystalline drug substance (form 1) and amorphous solid dispersions (MBP-1 and MBP-2) obtained using paddle apparatus and fasted state simulated intestinal fluid (FaSSIF). (Reproduced with permission from Shah et al. 2013)

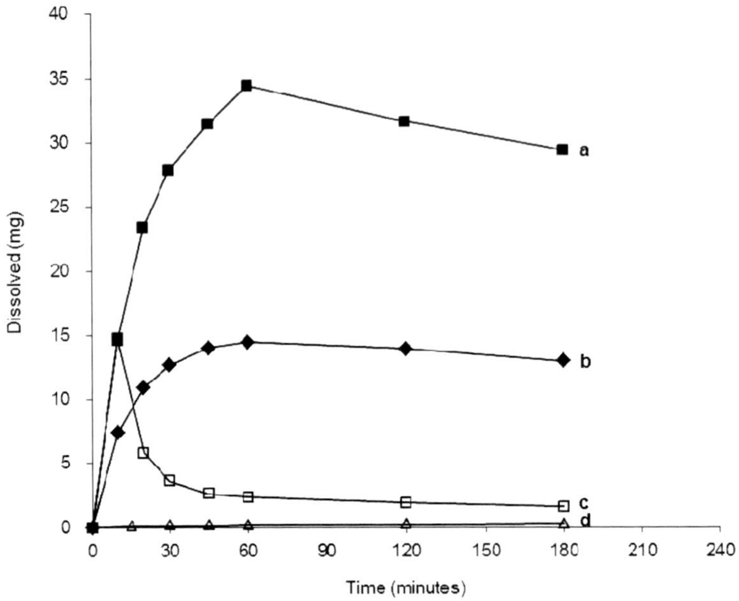

Fig. 15.9 Dissolution profiles of MBP and crystalline vemurafenib in USP apparatus 2: **a** unstressed vemurafenib MBP, **b** stressed vemurafenib MBP, **c** metastable crystalline vemurafenib, and **d** stable crystalline vemurafenib. (Reproduced with permission from Shah et al. 2013)

Gu et al. 2005) or simple manual transfer of gastric medium into intestinal medium (Mellaerts et al. 2008). Understanding of the intraluminal factors that affect supersaturation in both fasted and fed state would allow the development/modification of more appropriate biorelevant media for dissolution testing of amorphous solid dispersions. As endogenous and exogenous components such as bile salts, phospholipids, and food digestion products alter the dissolution and solubility of drugs (Dressman et al. 2007), more studies are needed in order to understand their influence on the rate and extent of intraluminal supersaturation as currently little is known regarding the impact of these factors (Brouwers et al. 2009). The discrepancy in supersaturation of amorphous compounds between in vitro studies and in vivo results (DiNunzio et al. 2008) presents another biopharmaceutical consideration in the development of IVIVCs.

15.7 IVIVCs of Amorphous Formulations

A few successful examples of IVIVC for amorphous formulation have been reported in the literature. A level A correlation was developed for disintegration-controlled amorphous nilvadipine matrix tablet (DCMT; Tanaka et al. 2006) (Fig. 15.10). Two DCMT amorphous nilvadipine formulations, DCMT-1 and DCMT-2, with different compositions of disintegrant were prepared and tested in vitro (USP apparatus 2 (100

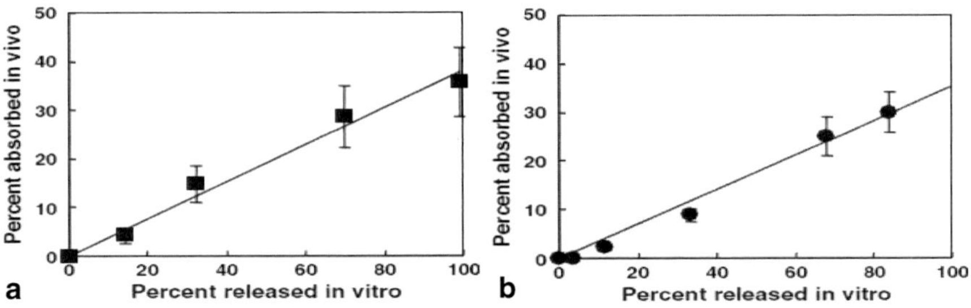

Fig. 15.10 Level A IVIVCs of DCMT-1 (**a**) and DCMT-2 (**b**) amorphous nilvadipine formulation. The reported linear regressions were $y = 0.995x$ (**a**) and $y = 0.9512x$ (**b**). (Reproduced with permission from Tanaka et al. 2006)

rpm) in Japanese Pharmacopoeia (JP) first medium (pH 1.2, 37 ± 0.5 °C)) and in vivo (fasted beagle dogs; in vivo absorption calculated with numerical deconvolution).

A successful multiple level C correlation between C_{max} and AUC with % released at 5 min (Q_{5min}) and at 60 min (Q_{60min}) was developed for poly(ethylene glycol) (PEG)–ritonavir amorphous solid dispersion formulations (Law et al. 2004). In this study, in vitro dissolution was conducted with a USP apparatus 1 (50 rpm, 37 ± 0.5 °C) in 900 mL of 0.1 N hydrochloric acid. The in vivo data was obtained from beagle dogs.

15.8 Conclusion

The dissolution test is still the only in vitro test that can potentially serve as a surrogate for in vivo performance. A well-designed test can serve as a valuable tool for the development of amorphous solid dispersions and the drug products containing amorphous solid dispersions. Appropriate test conditions may be developed to discriminate for composition, drug load, manufacturing process, and solid-state properties of the solid dispersion. Use of biorelevant and two-stage dissolution methods as well as supersaturation and sink versus non-sink conditions should be carefully considered during design of amorphous solid dispersion. The test may also offer insight into supersaturation and crystallization behavior as the drug is exposed to water during dissolution. The QC dissolution method must be able to discriminate between the presence and absence of the crystalline drug in the amorphous solid dispersion-based drug product. As a final note, consideration should be given to new developments in the area of integrated modeling and simulation, which offer the opportunity to integrate solubility, dissolution, precipitation, and absorption with simulations of dissolution behavior and impact on bioavailability of amorphous formulations.

References

Acartürk F, Kislal Ö, Çelebi N (1992) The effect of some natural polymers on the solubility and dissolution characteristics of nifedipine. Int J Pharm 85:1–6

Ali AA, Gorashi AS, (1984) Absorption and dissolution of nitrofurantoin from different experimental formulations. Int J Pharm 19:297–306

Allen LV Jr, Yanchick VA, Maness DD (1977) Dissolution rates of corticosteroids utilizing sugar glass dispersions. J Pharm Sci 66:494–497

Allen LV, Levinson RS, Martono DD (1978) Dissolution rates of hydrocortisone and prednisone utilizing sugar solid dispersion systems in tablet form. J Pharm Sci 67:979–981

Alonzo DE, Gao Y, Taylor LS (2009) Crystallization behavior of amorphous pharmaceuticals during dissolution. In: AAPS annual meeting and exposition. AAPS Journal, Los Angeles

Alonzo DE, Zhang GGZ, Zhou DL, Gao Y, Taylor LS (2010) Understanding the behavior of amorphous pharmaceutical systems during dissolution. Pharm Res 27:608–618

Bevernage J, Brouwers J, Clarysse S, Vertzoni M, Tack J, Annaert P, Augustijns P (2010) Drug supersaturation in simulated and human intestinal fluids representing different nutritional states. J Pharm Sci 99(11):4525–4534

Blagden N, Davey RJ, Rowe R, Roberts R (1998) Disappearing polymorphs and the role of reaction by-products: the case of sulphathiazole. Int J Pharm 172:169–177

Bloch DW, Speiser PP (1987) Solid dispersions—fundamentals and examples. Pharm Acta Helv 62:23–27

Brewster ME, Vandecruys R, Peeters J, Neeskens P, Verreck G, Loftsson T (2008) Comparative interaction of 2-hydroxypropyl-beta-cyclodextrin and sulfobutylether-beta-cyclodextrin with itraconazole: phase-solubility behavior and stabilization of supersaturated drug solutions. Eur J Pharm Sci 34:94–103

Brittain HG, Bogdanowich SJ, Bugay DE, Devincentis J, Lewen G, Newman AW (1991) Physical characterization of pharmaceutical solids. Pharm Res 8:963–973

Brouwers J, Brewster ME, Augustijns P (2009) Supersaturating drug delivery systems: the answer to solubility-limited oral bioavailability? J Pharm Sci 98:2549–2572

Cardew PT, Davey RJ (1985) The kinetics of solvent-mediated phase transformations. Proc Royal Soc Lond A 398:415–428

Chiou WL, Riegelman S (1970) Oral absorption of griseofulvin in dogs: increased absorption via solid dispersion in polyethylene glycol 6000. J Pharm Sci 59:937–942

Chiou WL, Riegelman S (1971) Pharmaceutical applications of solid dispersion systems. J Pharm Sci 60:1281–1302

Clarkson JR, Price TJ, Adams CJ (1992) Role of metastable phases in the spontaneous precipitation of calcium-carbonate. J Chem Soc Faraday Trans 88:243–249

Craig DQM (1990) Polyethyelene glycols and drug release. Drug Dev Ind Pharm 16:2501–2526

Craig DQM (2002) The mechanisms of drug release from solid dispersions in water-soluble polymers. Int J Pharm 231:131–144

Dai WG, Dong LC, Shi X, Nguyen J, Evans J, Xu Y, Creasey AA (2007) Evaluation of drug precipitation of solubility-enhancing liquid formulations using milligram quantities of a new molecular entity (NME). J Pharm Sci 96:2957–2969

Davey RJ, Cardew PT, Mcewan D, Sadler DE (1986) Rate controlling processes in solvent-mediated phase-transformations. J Cryst Growth 79:648–653

Davey RJ, Black SN, Goodwin AD, Mackerron D, Maginn SJ, Miller EJ (1997a) Crystallisation in polymer films: control of morphology and kinetics of an organic dye in a polysilicone matrix. J Mater Chem 7:237–241

Davey RJ, Blagden N, Potts GD, Docherty R (1997b) Polymorphism in molecular crystals: stabilization of a metastable form by conformational mimicry. J Am Chem Soc 119:1767–1772

Davey RJ, Allen K, Blagden N, Cross WI, Lieberman HF, Quayle MJ, Righini S, Seton L, Tiddy GJT (2002) Crystal engineering-nucleation, the key step. Cryst Eng Comm 4:257–264

Dinunzio JC, Miller DA, Yang W, Mcginity JW, Williams RO (2008) Amorphous compositions using concentration enhancing polymers for improved bioavailability of itraconazole. Mol Pharm 5:968–980

Dressman JB, Vertzoni M, Goumas K, Reppas C (2007) Estimating drug solubility in the gastrointestinal tract. Adv Drug Deliv Rev 59:591–602

El-Zein H, Riad L, El-Bary AA (1998) Enhancement of carbamazepine dissolution: in vitro and in vivo evaluation. Int J Pharm 168:209–220

Elamin AA, Ahlneck C, Alderborn G, Nyström C (1994) Increased metastable solubility of milled griseofulvin, depending on the formation of a disordered surface structure. Int J Pharm 111:159–170

EMA (2010) Guideline on the investigation of bioequivalence, appendix I: dissolution testing and similarity of dissolution EMA 2010 profiles CPMP/EWP/QWP/1401/98 Rev. 1/ Corr. http://www.ema.europa.eu/docs/en_GB/document_library/Scientific_guideline/2010/01/ WC500070039.pdf

Fawaz F, Bonini F, Guyot M, Bildet J, Maury M, Lagueny AM (1996) Bioavailability of norfloxacin from peg 6000 solid dispersion and cyclodextrin inclusion complexes in rabbits. Int J Pharm 132:271–275

FDA (1997) Guidance for industry: dissolution testing of immediate release solid oral dosage forms. Rockville, md: US department of health and human services, food and drug administration, center for drug evaluation and research (CDER)

Feth MP, Volz J, Hess U, Sturm E, Hummel RP (2008) Physicochemical, crystallographic, thermal, and spectroscopic behavior of crystalline and X-ray amorphous ciclesonide. J Pharm Sci 97:3765–3780

Ford JL (1986) The current status of solid dispersions. Pharm Acta Helv 61:69–88

Forster A, Hempenstall J, Tucker I, Rades T (2001) The potential of small-scale fusion experiments and the gordon-taylor equation to predict the suitability of drug/polymer blends for melt extrusion. Drug Dev Ind Pharm 27:549–560

Friesen DT, Shanker R, Crew M, Smithey DT, Curatolo WJ, Nightingale JAS (2008) Hydroxypropyl methylcellulose acetate succinate-based spray-dried dispersions: an overview. Mol Pharm 5:1003–1019

Gamsjager H, Lorimer JW, Scharlin P, Shaw DG (2008) Glossary of terms related to solubility. Pure Appl Chem 80:233–276

Ghazal HS, Dyas AM, Ford JL, Hutcheon GA (2009) In vitro evaluation of the dissolution behaviour of itraconazole in bio-relevant media. Int J Pharm 366:117–123

Gohel MC, Patel LD (2003) Processing of nimesulide-peg 400-pg-pvp solid dispersions: preparation, characterization, and in vitro dissolution. Drug Dev Ind Pharm 29:299–310

Goldberg AH, Gibaldi M, Kanig JL, Mayersohn M (1966) Increasing dissolution rates and gastrointestinal absorption of drugs via solid solutions and eutectic mixtures IV. Chloramphenicol–urea system. J Pharm Sci 55:581–583

Gu CH, Rao D, Gandhi RB, Hilden J, Raghavan K (2005) Using a novel multicompartment dissolution system to predict the effect of gastric ph on the oral absorption of weak bases with poor intrinsic solubility. J Pharm Sci 94:199–208

Gupta P, Chawla G, Bansal AK (2004a) Physical stability and solubility advantage from amorphous celecoxib: the role of thermodynamic quantities and molecular mobility. Mol Pharm 1:406–413

Gupta P, Kakumanu VK, Bansal AK (2004b) Stability and solubility of celecoxib-pvp amorphous dispersions: a molecular perspective. Pharm Res 21:1762–1769

Hancock BC, Parks M (2000) What is the true solubility advantage for amorphous pharmaceuticals? Pharm Res 17:397–404

Hancock BC, Zografi G (1997) Characteristics and significance of the amorphous state in pharmaceutical systems. J Pharm Sci 86:1–12

Higuchi WI, Mir NA, Desai SJ (1965) Dissolution rates of polyphase mixtures. J Pharm Sci 54:1405–1410

Huang LF, Tong WQ (2004) Impact of solid state properties on developability assessment of drug candidates. Adv Drug Deliv Rev 56:321–334

JP XV (2006) Japanese pharmacopoeia, Chapter 6.10 dissolution test.

Kohri N, Yamayoshi Y, Xin H, Iseki K, Sato N, Todo S, Miyazaki K (1999) Improving the oral bioavailability of albendazole in rabbits by the solid dispersion technique. J Pharm Pharmacol 51:159–164

Konno H, Taylor LS (2008) Ability of different polymers to inhibit the crystallization of amorphous felodipine in the presence of moisture. Pharm Res 25:969–978

Konno H, Handa T, Alonzo DE, Taylor LS (2008) Effect of polymer type on the dissolution profile of amorphous solid dispersions containing felodipine. Eur J Pharm Biopharm 70:493–499

Kostewicz ES, Wunderlich M, Brauns U, Becker R, Bock T, Dressman JB (2004) Predicting the precipitation of poorly soluble weak bases upon entry in the small intestine. J Pharm Pharmacol 56:43–51

Kushida I, Ichikawa M, Asakawa N (2002) Improvement of dissolution and oral absorption of er-34122, a poorly water-soluble dual 5-lipoxygenase/cyclooxygenase inhibitor with anti-inflammatory activity by preparing solid dispersion. J Pharm Sci 91:258–266

Law D, Krill SL, Schmitt EA, Fort JJ, Qiu YH, Wang WL, Porter WR (2001) Physicochemical considerations in the preparation of amorphous ritonavir-poly(ethylene glycol) 8000 solid dispersions. J Pharm Sci 90:1015–1025

Law D, Schmitt EA, Marsh KC, Everitt EA, Wang WL, Fort JJ, Krill SL, Qiu YH (2004) Ritonavir-peg 8000 amorphous solid dispersions: in vitro and in vivo evaluations. J Pharm Sci 93:563–570

Leuner C, Dressman J (2000) Improving drug solubility for oral delivery using solid dispersions. Eur J Pharm Biopharm 50:47–60

Li S, Wong S, Sethia S, Almoazen H, Joshi YM, Serajuddin ATM (2005) Investigation of solubility and dissolution of a free base and two different salt forms as a function of ph. Pharm Res 22:628–635

Lipinski CA, Lombardo F, Dominy BW, Feeney PJ (2001) Experimental and computational approaches to estimate solubility and permeability in drug discovery and development settings. Adv Drug Deliv Rev 46:3–26

Loftsson T, Fririksdóttir H, Gumundsdóttir TK (1996) The effect of water-soluble polymers on aqueous solubility of drugs. Int J Pharm 127:293–296

Matteucci ME, Miller MA, Williams RO, Johnston KP (2008) Highly supersaturated solutions of amorphous drugs approaching predictions from configurational thermodynamic properties. J Phys Chem B 112:16675–16681.

Mellaerts R, Mols R, Kayaert P, Annaert P, Van Humbeeck J, Van Den Mooter G, Martens JA, Augustijns P (2008) Ordered mesoporous silica induces ph-independent supersaturation of the basic low solubility compound itraconazole resulting in enhanced transepithelial transport. Int J Pharm 357:169–179

Miller DA, Mcconville JT, Yang W, Williams RO, Mcginity JW (2007) Hot-melt extrusion for enhanced delivery of drug particles. J Pharm Sci 96:361–376

Miller DA, Dinunzio JC, Yang W, Mcginity JW, Williams RO (2008a) Enhanced in vivo absorption of itraconazole via stabilization of supersaturation following acidic-to-neutral ph transition. Drug Dev Ind Pharm 34:890–902

Miller DA, Mcginity JW, Williams III RO (2008b) Solid dispersion technologies. In: Williams III RO, Taft DR (eds) Advanced drug formulation design to optimize therapeutic outcomes. Informa Healthcare, New York

Moneghini M, Carcano A, Zingone G, Perissutti B (1998) Studies in dissolution enhancement of atenolol. Part I. Int J Pharm 175:177–183

Moore MD, Wildfong PLD (2009) Aqueous solubility enhancement through engineering of binary solid composites: pharmaceutical applications. J Pharm Innov 4:36–49

Mosharraf M, Sebhatu T, Nystrom C (1999) The effects of disordered structure on the solubility and dissolution rates of some hydrophilic, sparingly soluble drugs. Int J Pharm 177:29–51

Murphy D, Rodriguez-Cintron F, Langevin B, Kelly RC, Rodriguez-Hornedo N (2002) Solution-mediated phase transformation of anhydrous to dihydrate carbamazepine and the effect of lattice disorder. Int J Pharm 246:121–134

Nernst W (1904) Theorie der reaktionsgeschwindigkeit in heterogenen systemen. Z Phys Chem 47:52–102

Noyes AA, Whitney WR (1897) The rate of solution of solid substances in their own solutions. J Am Chem Soc 19:930–934

Ossi PM (2003) Structural changes induced by swift heavy ions in non-metallic compounds. Beam interactions with materials and atoms. Nucl Instrum Methods Phys Res 209:55–61

Overhoff KA, Mcconville JT, Yang W, Johnston KP, Peters JI, Williams RO (2008) Effect of stabilizer on the maximum degree and extent of supersaturation and oral absorption of tacrolimus made by ultra-rapid freezing. Pharm Res 25:167–175

Patterson JE, James MB, Forster AH, Lancaster RW, Butler JM, Rades T (2005) The influence of thermal and mechanical preparative techniques on the amorphous state of four poorly soluble compounds. J Pharm Sci 94:1998–2012

PhEUR (2011) European pharmacopoeia, Chapter 2.9.3 Dissolution test for solid dosage forms, strasbourg, france, council of europe, european directorate for the quality of medicines and healthcare

Raumer MV, Hilfiker R, Blatter F (2006) Relevance of solid-state properties for pharmaceutical products. In: Hilfiker R (ed) Polymorphism in the pharmaceutical industry. Wiley VCH, Weinheim

Rodriguez-Hornedo N, Lechuga-Ballesteros D, Wu HJ (1992) Phase transition and heterogeneous/epitaxial nucleation of hydrated and anhydrous theophylline crystals. Int J Pharm 85:149–162

Sarode AL, Sandhu H, Shah N, Malick W, Zia H (2013) Hot melt extrusion for amorphous solid dispersions: temperature and moisture activated drug–polymer interactions for enhanced stability. Mol Pharm 10:3665–3675

Savolainen M, Kogermann K, Heinz A, Aaltonen J, Peltonen L, Strachan C, Yliruusi J (2009) Better understanding of dissolution behaviour of amorphous drugs by in situ solid-state analysis using raman spectroscopy. Eur J Pharm Biopharm 71:71–79

Serajuddin ATM (1999) Solid dispersion of poorly water-soluble drugs: early promises, subsequent problems, and recent breakthroughs. J Pharm Sci 88:1058–1066

Shah N, Iyer RM, Mair H-J, Choi DS, Tian H, Diodone R, Fähnrich K, Pabst-Ravot A, Tang K, Scheubel E, Grippo JF, Moreira SA, Go Z, Mouskountakis J, Louie T, Ibrahim PN, Sandhu H, Rubia L, Chokshi H, Singhal D, Malick W (2013) Improved human bioavailability of vemurafenib, a practically insoluble drug, using an amorphous polymer-stabilized solid dispersion prepared by a solvent-controlled coprecipitation process. J Pharm Sci 102:967–981

Shalaev E, Zografi G (2002) The concept of structure in amorphous solids from the perspective of the pharmaceutical sciences. Amorphous food and pharmaceutical systems. The Royal Society Of Chemistry, Cambridge

Six K, Verreck G, Peeters J, Brewster M, Van Den Mooter G (2004) Increased physical stability and improved dissolution properties of itraconazole, a class II drug, by solid dispersions that combine fast- and slow-dissolving polymers. J Pharm Sci 93:124–131

Six K, Daems T, De Hoon J, Van Hecken A, Depre M, Bouche MP, Prinsen P, Verreck G, Peeters J, Brewster ME, Van Den Mooter G (2005) Clinical study of solid dispersions of itraconazole prepared by hot-stage extrusion. Eur J Pharm Sci 24:179–186

Strickley RG, Oliyai R (2007) Solubilizing vehicles for oral formulation development. In: Augustijins P, Brewster ME (eds) Solvent systems and their selection in pharmaceutics and biopharmaceutics. Biotechnology: pharmaceutical aspects, vol VI. Springer, New York, pp 257–308

Tanaka N, Imai K, Okimoto K, Ueda S, Tokunaga Y, Ibuki R, Higaki K, Kimura T (2006) Development of novel sustained-release system, disintegration-controlled matrix tablet (DCMT) with solid dispersion granules of nilvadipine (II): in vivo evaluation. J Control Release 112:51–56

Tanno F, Nishiyama Y, Kokubo H, Obara S (2004) Evaluation of hypromellose acetate succinate (HPMCAS) as carrier in solid dispersions. Drug Dev Ind Pharm 30:9–17
Torrado S, Torrado S, Torrado JJ, Cadórniga R (1996) Preparation, dissolution and characterization of albendazole solid dispersions. Int J Pharm 140:247–250
USP (2009a) The United States pharmacopeia and the national formulary, <711> Dissolution. The Official Compendia of Standards USP 32–NF 27
USP (2009b) The United States pharmacopeia and the national formulary, <1092> The dissolution procedure: development and validation. The Official Compendia of Standards USP 32–NF 27
Usui F, Maeda K, Kusai A, Nishimura K, Keiji Y (1997) Inhibitory effects of water-soluble polymers on precipitation of RS-8359. Int J Pharm 154:59–66
Van Drooge DJ, Hinrichs WL, Frijlink HW (2004) Anomalous dissolution behaviour of tablets prepared from sugar glass-based solid dispersions. J Control Release 97:441–452
Vertzoni M, Dressman J, Butler J, Hempenstall J, Reppas C (2005) Simulation of fasting gastric conditions and its importance for the in vivo dissolution of lipophilic compounds. Eur J Pharm Biopharm 60:413–417
Wei-Qin T (2009) Salt screening and selection: new challenges and considerations in the modern pharmaceutical research and development paradigm. Developing solid oral dosage forms. Academic Press, San Diego
Weissbuch I Leisorowitz L, Lahav M (1994) Tailor-made and charge-transfer auxiliaries for the control of the crystal polymorphism of glycine. Adv Mater 6:952–956
Weissbuch I, Zbaida D, Addadi L, Leiserowitz L, Lahav M (1987) Design of polymeric inhibitors for the control of crystal polymorphism—induced enantiomeric resolution of racemic histidine by crystallization at 25-Degrees-C. J Am Chem Soc 109:1869–1871
Yamashita K, Nakate T, Okimoto K, Ohike A, Tokunaga Y, Ibuki R, Higaki K, Kimura T (2003) Establishment of new preparation method for solid dispersion formulation of tacrolimus. Int J Pharm 267:79–91
Zhang GGZ, Gu C, Zell MT, Burkhardt RT, Munson EJ, Grant DJW (2002) Crystallization and transitions of sulfamerazine polymorphs. J Pharm Sci 91:1089–1100
Zhang GGZ, Zhou D, Yihong Q, Yisheng C, Geoff GZZ, Lirong L, William RP (2009) Crystalline and amorphous solids. Developing solid oral dosage forms. Academic Press, San Diego
Zhu HJ, Grant DJW (1996) Influence of water activity in organic solvent plus water mixtures on the nature of the crystallizing drug phase 2. Ampicillin. Int J Pharm 139:33–43
Zhu HJ, Yuen CM, Grant DJW (1996) Influence of water activity in organic solvent plus water mixtures on the nature of the crystallizing drug phase 1. Theophylline. Int J Pharm 135:151–160

Chapter 16
Stability of Amorphous Solid Dispersion

Xiang Kou and Liping Zhou

16.1 Introduction

Presently, the majority of new chemical entities in a typical pharmaceutical company's pipeline are reported to be poorly water soluble (Siew and Arnum 2013). Amorphous solid dispersion (ASD) is one of the most attractive approaches to provide a supersaturating drug delivery system (SDDS; Newman et al. 2012). The solid dispersion approach has been used to increase the rate of dissolution and solubility of active pharmaceutical ingredients (API) without sacrificing their intestinal membrane permeability (Miller et al. 2012). It has also been used for controlled release of an API (Tran et al. 2011). Compared to crystalline solids, amorphous materials lack three-dimensional long-range order. They often have the desired pharmaceutical properties of dissolving faster and showing higher kinetic solubility when compared to corresponding crystals. A well-designed amorphous system can exist in a supersaturated state in vivo, thus improving the exposure of the drug.

Despite these advantages, ASDs also demonstrate poor chemical and physical stability, creating development challenges related to their commercialization as marketed products. Major contributors to instability and other unpleasant surprises, particularly in the late-phase development include (1) lack of understanding of the physicochemical properties of the API, the stabilizing polymer, other additives, and the interactions among all the ingredients; (2) shortage of reliable technologies for early prediction of the amorphous formulation stability with limited amount of material; and (3) disconnection between early formulation development and downstream processing technologies/methodologies due to the difference in manufacturing setups. To mitigate the risks associated with physical instability, a

X. Kou (✉)
Chemical and Pharmaceutical Profiling,
Novartis Pharmaceuticals, Shanghai, China

L. Zhou
Ipsen Biosciences, Cambridge, MA, USA
e-mail: liping.zhou@ipsen.com

N. Shah et al. (eds.), *Amorphous Solid Dispersions*,
Advances in Delivery Science and Technology, DOI 10.1007/978-1-4939-1598-9_16

comprehensive understanding of the molecular structure and the interactions among different ingredients is essential to the successful development of an ASD dosage form.

The fundamentals regarding ASD have already been introduced in the previous chapters and elsewhere (Taylor and Shamblin); readers are encouraged to use those for reference. This chapter introduces the mechanisms of instability and influencing factors, options to prevent instability, and methods to characterize API and other ingredients as well as to predict the instability. In addition, we introduce the typical stability programs applicable for early ASD development. It is important to note that this chapter mainly focuses on the physical stability aspects. Though not the focus of this chapter, chemical instability of ASD is of equal importance to physical instability (Pikal and Dellerman 1989). Chromatographic approach is the most widely adopted method for the assessment of chemical stability of both API and the key excipients.

16.2 Factors Affecting the Stability of Amorphous Solid Dispersion (ASD)

ASDs ideally can be imagined as amorphous solids being molecularly dispersed into an inert polymer matrix (Chiou and Riegelman 1971). Amorphous solids exist as a nonequilibrium phase lacking all long-range order symmetry. They exhibit short-range order over a few molecules, including nearest-neighbor or next-nearest-neighbor interactions. Sufficient nonbonded interactions enable amorphous material to present as a condensed phase and behave mechanically like crystalline solids. The lack of periodicity in nonbonded interactions results in high internal energy relative to the crystalline state, providing increased apparent aqueous solubility and enhanced dissolution. The enhancement in API absorption is further strengthened by the solubilization effects of excipients (polymer and surfactant). However, as a metastable state, there exists a thermodynamic potential for amorphous solids to spontaneously revert to a more stable crystalline state. The reduction in molecular mobility (Korhonen et al. 2008; Van den Mooter et al. 2001) and reduction in molecular coupling due to specific interactions between the drug and the polymer (Aso and Yoshioka 2006) are additional driving forces for form conversion from amorphous stage to crystalline stage. In addition, there is a balance between the thermodynamic driving force for nucleation and crystal growth and the kinetic factors, mainly the molecular mobility. Nucleation occurs at lower temperature more likely due to the thermodynamic driving force, whereas molecular mobility is mostly at higher temperature. The processing and storage conditions, and parameters such as temperature and relative humidity (RH) mainly, play significant roles on both the thermodynamic and kinetic aspects (Leuner and Dressman 2000; Serajuddin 1999). A thorough understanding of these ASD physical stability-governing aspects will elicit conscious evaluation of ASD formulation and facilitate rational drug product design and development.

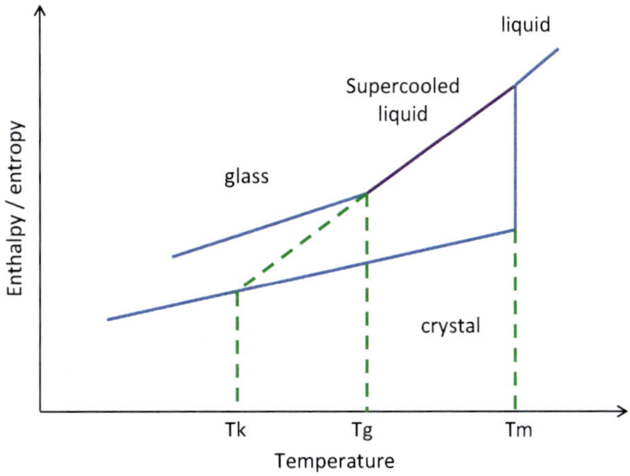

Fig. 16.1 Schematic diagram of the thermodynamic relationship of amorphous and crystalline states: T_m the melting point, T_g the glass transition point, and T_k the Kauzmann temperature

16.2.1 Thermodynamic Aspect

ASD is a glass solution of a poorly soluble API in a hydrophilic polymer carrier with a high glass transition temperature. The solid-state change is on a molecular level. The thermodynamic driving force for crystallization from an amorphous solid arises from the higher Gibbs free energy of the amorphous system relative to that of the crystal for all temperatures lower than the equilibrium melting point. The glass will relax toward lower configurational enthalpy and entropy. This driving force increases with the degree of supercooling; thus, the further from the crystal melting temperature, the greater the driving force for crystallization (Fig. 16.1). The difference in Gibbs free energy between the supercooled liquid and crystalline phase, ΔG, is given by Gibbs equation,

$$\Delta G(T) = \Delta H(T) - T\Delta S(T) \tag{16.1}$$

The enthalpy change can be estimated by,

$$\Delta H(T) = \Delta H_{\text{fus}} + \int_{T_m}^{T} \Delta C_p dT, \Delta S(T) = \Delta S_{\text{fus}} + \int_{T_m}^{T} \frac{\Delta C_p}{t} dT \tag{16.2}$$

Although the thermodynamic driving force increases with decreasing temperature, molecular mobility decreases on cooling, thereby increasing the kinetic barrier to crystallization. The kinetic barrier to crystallization is a consequence of the high viscosity and decreased molecular mobility of supercooled liquids and glassy systems, and is obviously strongly temperature dependent.

The reduction of free energy is modeled according to the classical nucleation theory, which describes factors influencing nucleation kinetics from supercooled liquids (Turnbull and Fisher 1949; James 1985). In this model, the rate of homogeneous nucleation (I) is governed by the free energy change occurring on formation of a nucleus with a critical size, ΔG^*, and the activation energy for transporting a molecule from the amorphous phase to the nucleus, ΔG_{a}:

$$I = A \exp \frac{-(\Delta G^* + \Delta G_{\mathrm{a}})}{kT} \tag{16.3}$$

In this equation, A is a constant, k is the Boltzmann constant, and T is the temperature. ΔG^* represents the balance between the energy penalties associated with creating a new surface. The thermodynamic driving force for nucleation rises with the degree of supercooling. Once a stable nucleus has been formed, an increase in the mass of crystalline material is achieved by crystal growth. Crystal growth starts with the diffusion of molecules from the bulk solution toward the liquid–solid interface, followed by the integration of the molecules into the crystal lattice. Crystal growth from viscous liquids is often described by the normal crystal growth model,

$$u = \tfrac{k}{\delta}\left[1 - \exp\left(\tfrac{\Delta G}{RT}\right)\right] \tag{16.4}$$

where ΔG is the free energy difference between the liquid and crystalline forms, k is the Boltzmann constant, independent of temperature, and δ is the viscosity. More complex modeling equations were reported elsewhere (Nascimento and Zanotto 2010).

The thermodynamic driving force of crystal growth lies in the metastable equilibrium at the interface of solid and liquid and is dictated by the degree of supercooling.

16.2.2 Kinetic Driving Force: Molecular Mobility

Amorphous material has higher apparent solubility and can remain in supersaturated state upon transit from gastric compartment to the intestinal compartment with or without the assistance of precipitation inhibitors. According to the model of API in polymer solubility (Marsac et al. 2006b), APIs have the tendency to crystallize to the more thermodynamically stable form. Inhibition of API crystallization in solid dispersion is attributed predominantly to kinetic stabilization (Marsac et al. 2006b).

Molecular mobility is the rotational and transitional movement of molecules. It is well known that the reduced stability of amorphous solid is due to its greater molecular mobility relative to that of the corresponding crystalline form. The role of molecular mobility in crystallization lies in the fact that it is necessary to allow diffusion and surface integration. Crystallization of the drug is generally preceded by phase separation and thus the formation of a drug-rich amorphous phase and polymer-rich phase. The molecular mobility of an amorphous compound is due to two main relaxations, global and local. The molecular mobility responsible for the glass transition is cooperative in nature and is also called the global mobility. Such molecular motions are also known as α-relaxations. Local mobility, otherwise known as β-relaxation or secondary relaxation, is noncooperative and much faster than global mobility (Bhattacharya and Suryanarayanan 2009). Both types of molecular mobility will impact the possibility of recrystallization.

16.2.2.1 Temperature Effect on Mobility

Temperature has a huge impact on the stability of an ASD. Thermodynamically, as clearly illustrated by Eqs. 16.1 and 16.2, the crystallization tendency is a function of temperature (Marsac et al. 2006a). Kinetically, the temperature affects the molecular mobility of the ASD system in a way that the temperature dependence of the molecular motions below the glass transition event is less than that the above T_g (Hancock et al. 1995). At elevated temperatures above T_g, with the transition from glass to supercooled liquid phase, structural relaxations happen rapidly leading to an enhanced phase separation and/or crystallization potential (Vasanthavada et al. 2004). In addition, thermal expansion of the API–polymer matrix may reduce the degree of interaction between its two components, affecting the hydrogen bonding, thereby decreasing the miscibility limit and contributing to the recrystallization as suggested elsewhere (Tang et al. 2002; Shibata et al. 2014).

Generally, it is recommended to store the amorphous material at least 50 °C below its glass transition temperature (Hancock et al. 1995). At lower temperatures, molecular mobility can practically be neglected due to the high viscosity of the system and the glass annealing (Qian et al. 2010). Albeit the low likelihood, recrystallization can still occur even with the storage temperature lower than $T_g - 50\,°C$ and has been observed by multiple research groups (Miyazaki et al. 2007; Yoshioka 1994). In a study on indomethacin, a β-relaxation was observed at $-20\,°C$ (Vyazovkin and Dranca 2005). The $T_g - 50\,°C$ rule is only suitable when the crystallization is driven by α-relaxation; however, local molecular mobility is noncooperative and requires less activation energy. At temperatures well below T_g, β-relaxation is the predominant driving force. An alternative recommendation is to keep the storage temperature at or below the Kauzmann temperature (T_k). As shown in Fig. 16.1, the entropy difference between the liquid and solid phase decreases as a liquid is supercooled. Kauzmann temperature (T_k) is a theoretical temperature value at which the potential for molecular rearrangement approaches a minimum value. The status at T_k is equal to that of the crystal. At this temperature and below, rotational and diffusive motions are improbable, even over extremely long timescale (Yu 2001; Craig et al 1999), T_k nominally represents a temperature below which many processes requiring molecular rearrangement will cease (Kauzmann 1948). It can be roughly estimated by $(T_g)^2/T_m$ (T_g, T_m expressed in kelvin scale). Materials having both T_g and T_k values below ambient temperature are often hard to produce and maintain in an amorphous state.

16.2.2.2 Moisture/Water Effect on Mobility

The influence of moisture or water on ASD stability is governed by the interaction of water with either API or polymer. The amorphous solids can interact with water via two mechanisms: the adsorption of the water molecules at the surface and the absorption of water into the bulk structure. Absorption is possible due to the lower density structure of amorphous solids whereby the free volume facilitates the sorption of water molecules, and hence, amorphous solids typically absorb considerably more water than their crystalline counterparts.

The water absorption by amorphous material is via hydrogen bonding. Water has a low glass transition temperature (T_g) at − 138 °C (Giovambattista et al. 2004) and has a plasticizing effect on the ASD, which lowers T_g of most pharmaceutical systems and increases the crystallization rate (Miller and Lechuga-Ballesteros 2006). Plasticizers change a number of material properties such as decrease in mechanical strength, viscosity, T_g, and increase in molecular mobility, all resulting in greater tendency for crystallization and chemical reactivity. A research showed that with the moisture content increased from nearly zero to 2.7 %, the molecular mobility of the quenched trehalose increased approximately by a factor of 6 (Liu et al 2002). It has been found that even with as low as 1–2 % of water content, molecular mobility can be significantly influenced (Andronis and Zografi 1998; Andronis et al. 1997). In addition, it has been found that mixtures of deliquescent API and deliquescent excipient are more hygroscopic than either the API or the excipient alone (Salameh and Taylor 2005). In another study, it was reported that a system with stronger drug–polymer interactions and a less hygroscopic polymer is less susceptible to moisture-induced phase separation, while more hydrophobic drugs are more susceptible to this phenomenon even at low levels of absorbed moisture (Rumondor and Taylor 2009; Bhugra and Pikal 2008)

Polar polymers including proteins, starches, cellulose, and other water-soluble formulation excipients may absorb significant amount of moisture when exposed to water vapor under various humidity and temperature conditions (Weuts et al. 2005). As moisture uptake increases, the polymer becomes plasticized and undergoes structural changes which may then affect the mobility of the dispersed API. The nature of hydrogen bonding between a water molecule and a polymer molecule has been investigated by Fourier transform (FT) Raman spectroscopy in combination with gravimetric water vapor absorption and differential scanning calorimetry (DSC; Taylor et al. 2001). Hydrogen bonding between water and the polymer has been confirmed by a shift in the carbonyl peaks toward lower wave numbers in FT Raman spectrum. The extent of the red shift has been noted to be polymer specific (Taylor et al. 2001). This is evidenced by a larger shift in the carbonyl peaks at equivalent water content. The carbonyl group is able to interact to a greater extent with the absorbed water molecules in its rubbery state.

The mechanism of moisture-induced relaxation remains an open question. One hypothesis is that an amorphous solid contains a finite amount of hydrophilic sites that become saturated at some critical moisture content. Above this moisture content value, adsorbed/absorbed water molecules have a greater molecular mobility. Thus, water exhibits different type of impact on the solid dispersion below and above this critical moisture content. This water-binding-site saturation hypothesis is supported by calorimetric measurements which clearly show that the heat of water sorption approaches the enthalpy of condensation of water above RH threshold (Miller and Lechuga-Ballesteros 2006).

16.2.2.3 Molecular Mobility and Phase Separation

One mechanism of solid dispersion instability has been proposed by Vasanthavada and coworkers (Vasanthavada et al. 2004). In the absence of moisture, freshly prepared dispersion mixtures have high T_g values and hence the molecular mobility is very low at the storage temperatures. The API remains in a kinetically frozen state of miscibility. Upon exposure to moisture, the solid dispersion gets plasticized, and the molecular mobility increases. The exact role of water is not yet clear. Water either may weaken the H-bond interaction by bridging with polymer and API structural units or may merely increase the molecular mobility by plasticizing the mixture. In either case, diffusion of API through the polymeric matrix can result in separation of the API into an amorphous phase, which subsequently crystallizes. As more and more API phase separation occurs, increasing amount of free polymer units are left to interact with the remaining API. Such units possibly orient around and arrest the nondiffused API molecules by satisfying their H-bond requirement. At equilibrium, localized pockets of API molecules are almost entirely bonded to the polymer, reaching solid solubility. Upon phase separation, amorphous API crystallizes.

There is another explanation of phase separation. The existence of local density gradients stimulates translational diffusion of molecules from high- to low-density areas. This would ultimately lead to equalizing the density throughout the whole system so that it would assume the more homogeneous structure as of a liquid. A glass consisting of both high- and low-density regions demonstrates low and high molecular mobility. High-density regions are primarily involved in slower rotational diffusion, whereas faster translational diffusion is associated with low-density regions. The latter, therefore, are the most likely regions for nucleation to occur. In the process of density equalization, the molecules from the structurally arrested regions diffuse into neighboring high-mobility areas, therefore increasing the fraction of molecules capable of nucleating (Vyazovkin and Dranca 2007). The increased molecular mobility allows the system to equilibrate to reach thermodynamic solid solubility in the amorphous phases and to form a crystalline fraction of the excess amount of the API (Vasanthavada et al. 2004).

As the phase-separated API crystallizes, the fraction of polymer in close contact with the crystalline surface increases till the heat capacity reaches a plateau. An example of gradual phase separation is shown in Fig. 16.2.

16.2.3 Processing Methods

Solid dispersion can be prepared by many methods, with solvent removal and hot-melt extrusion being the most widely used approaches (Bikiaris 2011; Alam et al. 2012). Readers can find details on all types of processing technologies in other related chapters in this book. In the authors' experience, the binary phase behavior in solid dispersions exhibits substantial processing method dependency. For solid

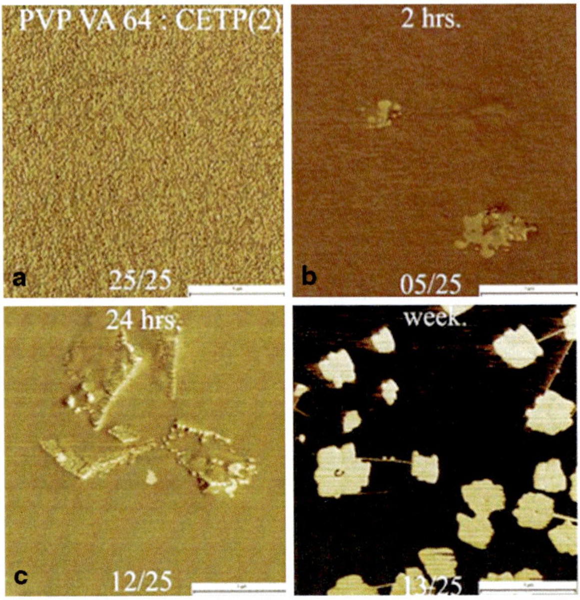

Fig. 16.2 Examples of phase separation. Phase maps recorded on fracture surface of cholesteryl ester transfer protein (CETP)/PVP VA64 film at different time points with exposure to stress conditions (RH = 75 %, $T = 40\,^{\circ}\mathrm{C}$): (**a**) $t = 0$, (**b**) $t = 2$ h, (**c**) $t = 24$ h, and (**d**) $t = 1$ week. (Reproduced with permission from Lauer et al. 2011))

dispersions with the same contents at the same drug load, often significant difference in physical stability and biopharmaceutical properties can result from different processing methods.

Structural relaxation is a multiexponential decay from the initial glassy state, where the configurations are "frozen in" during processing, toward the equilibrium supercooled liquid state. Therefore, the relaxation behavior would be expected to depend on processing methodology. It is found that quenched samples have longer relaxation times than the corresponding freeze-dried samples (Liu et al. 2002). Fourier transform infrared (FTIR) spectroscopy data indicate that the difference is attributed to two different types of H-bond populations (Paudel et al. 2012). The heterogeneity has also been observed by solid-state nuclear magnetic resonance (NMR) in the study of nifedipine–polyvinyl pyrrolidone (PVP) solid dispersions prepared by spray drying and melt quenching (Yuan et al. 2013). This is believed to be due to the different kinetic processes involved during the dispersion formation. Different processing methods result in different thermal histories, and this is reflected by the variations in molecular mobility. Amorphous simvastatin prepared via cryo-milled approach has been shown to have less stability than quench-cooled simvastatin (Graeser et al. 2009a). In another study, the physical stability of amorphous indomethacin has shown to be correlated to the cooling rate during the melt-quenching process, most likely due to the differences in nucleation under different conditions (Bhugra et al. 2008).

Table 16.1 Commonly used excipients in a solid dispersion formulation

PEG 4000	Cyclodextrin	Polyox®
PVP K30	SLS	Soluplus®
Eudragit® EPO	HPMCAS	Gelucire® 50/13
VA 64	Kollicoat® IR	Eudragit® RS 100
HPMC	Gelucire® 44/14	HPC

HPMC hydroxypropyl methyl cellulose, *HPMCAS* hydroxypropyl methyl cellulose acetyl succinate, *PEG* polyethylene glycol, *PVP* polyvinyl pyrrolidone, *SLS* sodium lauryl sulfate, *VA* vinyl acetate

16.2.4 Physicochemical Properties of the Additives: Polymer/Surfactant

Pharmaceutically acceptable polymers or copolymers are typically used as crystallization inhibitors in ASD. The selection of polymer and the drug load are vital not only to the creation of a good ASD but also to its physical stability. The glass transition temperature could be potentially modified, thus impacting the rate of crystallization (Yu 2001). In addition to the modification of T_g, polymers with good hydrogen-bond acceptors show more influence on the stabilization of the amorphous system (Wegiel et al. 2013). While working on ASD of acetaminophen, Miyazaki and coworkers observed that both acetaminophen–polyacrylic acid and acetaminophen–polyvinylpyrrolidone dispersions showed similar T_g values; however, the former had slower crystallization rate in the temperature range of 45–60 °C due to stronger hydrogen bonding (Miyazaki et al. 2004). Ghosh et al. reported phase separation in solid dispersion (SD) of NVS981 and hydroxypropyl methyl cellulose phthalate (HPMCP), but the dispersion in HPMC 3cps and hydroxypropyl methyl cellulose acetyl succinate (HPMC-AS) was stable (Ghosh et al. 2011). Many times, binary, ternary, or even quaternary ASD mixtures are ideal. Table 16.1 summarizes the polymers frequently used for ASD formulations.

In addition to inhibiting translational diffusion of APIs, steric hindrance either by polymers of different molecular weights or by a polymer of different concentrations may also slow crystallization mechanisms that proceed by rotational diffusion via a nearest substitution of a drug molecule with a polymer molecule. Either translational diffusion inhibition or steric hindrance may act to prevent drug molecule aggregation and/or interaction that are the precursors for crystallization (Matsumoto and Zografi 1999).

The appropriate use of polymer is very important to the physical stability of an ASD. Even at a low polymer concentration that results in a minor increase in T_g, enhanced drug stability has been observed (Matsumoto and Zografi 1999; Konno and Taylor 2006). This could be attributed to intermolecular interactions between the drug and the polymer in the dispersion. In the study with indomethacin, the drug–polymer interactions are indicated by peak shifts or peak intensity changes corresponding to specific vibrational modes of the functional groups involved in intermolecular interactions in FTIR (Taylor and Zografi 1997). PVP is commonly selected to stabilize indomethacin in its solid dispersions. Having no acidic protons, PVP does not self-associate through hydrogen bonding, although it can act as a proton acceptor through either the oxygen or nitrogen atom in the pyrrole ring. The carbonyl group is the least sterically hindered group and it is expected to be a more favorable site for interactions. It has been shown that solid dispersions of indomethacin-PVP produced FTIR spectra that showed an increase in the intensity of the band assigned to the nonhydrogen-bonded carbonyl and the PVP carbonyl stretch was shifted to a low wave number (Matsumoto and Zografi 1999). These results, not mirrored in the physical mixture data, suggested that hydrogen bonds between the indomethacin and PVP hydroxyl and carbonyl groups, respectively, were formed at the expense of indomethacin dimers. It was found that different types of hydrogen bonds were responsible for different behaviors in solid dispersions (Paudel et al. 2012).

In a study by Vasanthavada and coworkers, the data show that solid dispersion of griseofulvin–PVP had higher crystallization rate than that of the solid dispersion of indoprofen–PVP at the same drug load. This was attributed to the missing of hydrogen bonding between griseofulvin and PVP (Vasanthavada et al. 2005). In another study by Aso's group, it was revealed that the enthalpy relaxation time of amorphous nifedipine increased from 1 to 2 days in the absence of PVP to 18 days in the presence of 10 % PVP, whereas that of amorphous phenobarbital increased from 1.0 to 3.7 days in the absence and presence of 5 % PVP (Aso et al. 2004). This phenomenon has been attributed to the hydrogen bonding formed between PVP and the APIs based on data from carbon NMR (Aso and Yoshioka 2006).

Another type of excipient typically added to ASD is surfactant to improve the processibility of amorphous drug product (Ghebremeskel et al. 2006) and/or the in vivo performance (Qi et al. 2013). Comparing the impact of polymer on stabilizing ASD, the influence of surfactant is somewhat more complicated and can go in both directions. In a study on the impact of surfactants on crystal growth of amorphous celecoxib carried out by Dr. Taylor's group, the crystal growth rate increased with the presence of a surfactant alone (sodium lauryl sulfate (SLS), tocopherol polyethylene glycol succinate (TPGS), or sucrose palmitate); however, when polymer PVP was dispersed with one of the three surfactants mentioned above, the crystal growth rate was higher in the case of SLS and sucrose palmitate, but lower in the case of TPGS. It has also been observed that the rate was related to the amount of polymer used (Mosquera-Giraldo et al. 2013). In light of this observation, more attention is needed when addressing the physical stability of a more complex ternary solid dispersion system.

16.3 Principle and Techniques for Prediction of ASD Stability

Prediction methods of ASD stability can be categorized based on the stage of the formulation development: prediction during the formulation design and after formulating the ASD. The methods used in the ASD design phase are to measure the crystallization tendency of API alone or to test if a stable formulation is feasible by examining the API-additive miscibility. Once an ASD formulation is available, physical stability estimation can be carried out with methods to determine the kinetic factors influencing crystallization, for example, by determining molecular mobility or by measuring the phase separation. The appropriate application of the prediction approaches can mitigate the risks associated with the ASD as early as possible, thus improving the success rate.

16.3.1 API Crystallization Tendency

At the early stage of formulation development, formulation scientists can utilize some empirical rules to assess the suitability of an ASD for the API of interest. Simple estimations can be made using the ratio between glass transition temperature and melting point in kelvin as well as the Kauzmann temperature (T_k; Fig. 16.1). These estimations often provide quick educated guesses on the crystallization tendency of the API. Such information will be useful for estimating the difficulty of forming a stable ASD.

The glass-forming capability is an intrinsic property of organic materials. Upon solidification via either cooling or precipitation from solution, some organic materials prefer to attain crystallinity while others become amorphous. Thus, they are classified as good and poor glass formers, respectively. Over the past decades, it has been observed that the glass transition of a substance occurs in a temperature region that is approximately two thirds of the melting point in kelvin. If both T_g and T_m are known, then it is possible to estimate the deviation from the two-thirds rule, which is considered to reflect the temperature dependency of molecular motions in the region just above T_g. If T_g/T_m is significantly greater than 2/3, the material of interest is likely to have a greater than average temperature dependence of its molecular mobility in the region of T_g, and it is said to be "fragile," with a greater propensity to crystallize. Conversely, a material with a T_g/T_m ratio much less than 2/3 has a less than average temperature dependence of its molecular motions between these two temperatures (Hancock and Zografi 1997).

Though the two-thirds rule provides a quick estimation on the glass-forming ability of the material of interest, its predictivity is yet to be challenged. Compounds with similar T_g/T_m values may have very different crystallization behaviors. More sophisticated models using principal component analysis have been developed with better prediction power on material's glass-forming ability. Such models normally combine multiple parameters, including both experimental values and molecular intrinsic properties with the aim to develop a more predictive estimation than the simple two-thirds empirical rule. Baird et al. utilized thermodynamic parameters such as melting temperature, enthalpy, entropy, heat of fusion as well as molecular intrinsic properties like molecular weight and number of rotational bonds to build a model using 51 model compounds (Baird et al. 2010). Mahlin et al. built another model using multiple molecular descriptors, reflecting aromaticity, symmetry, flexibility, size, and distribution of electrons (Mahlin et al. 2011). These models provide high-quality alternatives for the prediction of ASD applicability and stability.

16.3.2 API and Additive Miscibility

"Miscibility" of an API and its additive(s) refers to a homogeneous system in which API and additives are mutually influencing the solid structures of each other. A miscible system indicates the API and the polymer are imbedded into each other's bonding network at a molecular level; hence, the system is stable and there is only one uniform phase. Miscibility is the key for the selection of a polymer carrier.

There are several parameters to consider for quick assessments. The detailed theoretical explanation can be found elsewhere. The most commonly used criteria are solubility parameter and lipophilicity (LogP). When the solubility parameter is used, a difference below 2.8 is empirically recognized as well miscible (Yoo et al. 2009). LogP value differences between the component pairs below 1.7 tend to stay amorphous; however, this rule is not strictly followed (Yoo et al. 2009). A more complicated miscibility estimation is to employ molecular dynamic (MD) simulation to predict the miscibility of a drug candidate and its carrier (Gupta et al. 2011). The MD approach can provide molecular level insight.

Besides the above-described ways for quick estimations, a prediction of miscibility using physical mixture of API and polymer is likely the most commonly used method. Different methodologies have been proposed for polymer selection (Van Eerdenbrugh and Taylor 2011), DSC being the most popular approach. The method is to test the T_g of the physical mixture, assuming one T_g indicates miscibility and hence a possible stable formulation. The physical mixture is first heated to pass the T_m of the API, followed by a quick cooling. During the modulated reheating process, the thermo activities are monitored. Albeit being a powerful and practical tool, a DSC also has some drawbacks. Formation of small domains (less than 15–30 nm) in a binary amorphous mixture containing more than one phase may result in failure to detect two distinct T_g events. Also during DSC measurements, the temperature of the sample is constantly changing, which in turn could result in a shifting miscibility of the system's components due to an increased or decreased miscibility with increasing temperature. Thus, the detection of a single T_g at temperatures higher than the T_g of the lowest individual component may not provide enough information to determine the number of amorphous phases present at room temperature (Rumondor et al. 2009).

Another method is to use rheological analysis (Liu et al. 2012). This technique is based on the fact that the curve of drug loading versus viscosity has a "V" shape, and drug load corresponding to the minimal viscosity indicates the API solubility in the polymer (dynamic frequency sweep). At lower drug load, with the increase in drug load, the viscosity decreases because of the plasticizer effect of the dissolved drug. Beyond the solubility point, the viscosity increases with the increasing drug load because the undissolved drug particles act as solid fillers (Liu et al. 2012).

Miscibility can be predicted by locating the minimal viscosity point. This method can also provide the minimum hot-melt extrusion processing temperature.

Additional approaches in place are the Flory–Huggins theory (Marsac et al. 2006b) and melting point depression approach (Marsac et al. 2006b; Marsac et al. 2009). Flory–Huggins theory investigates the solubility of the API in the polymer, while the melting point depression approach focuses on the change in melting point shift.

The Flory–Huggins lattice theory is originally developed to describe the solubility of a crystalline material in a solvent and can be derived to describe the relationship between drug and polymer, in which the polymer is the substitute of the solvent and the relationship is illustrated by Eqs. 16.5 and 16.6:

$$lna_1 = \frac{\Delta H_m}{R}\left(\frac{1}{T_m} - \frac{1}{T}\right) \tag{16.5}$$

where T_m is the melting point of the pure drug, ΔH_m is its molecular heat of melting, and T is the temperature at which the drug's solubility is measured (Zhu et al. 2010).

When the Flory-Huggins theory is applied to the drug–polymer, we get the following equation:

$$lna_1 = \ln \upsilon_1 + \left(1 - \frac{1}{x}\right)\upsilon_2 + \omega\upsilon_2^2 \tag{16.6}$$

where υ_1 and υ_2 are the volume fractions of the drug and the polymer respectively, x is the molar volume ratio of the polymer and the drug, and ω is the drug–polymer interaction parameter.

Volume fraction is the same as the weight fraction and the parameter x is the molecular weight ratio between the polymer and the drug (Sun et al. 2010; Tao et al. 2009). The drug–polymer interaction parameter ω is obtained by fitting the activity solubility relationship (Qian et al. 2012). If there is a net attraction between the two, the value of ω is negative. If there is a net repulsion, the ω value is positive. By comparing the ω value against the critical interaction parameter value for the binary system (0.5), the interaction between drug and polymer can be predicted (Rumondor et al. 2010).

A more complicated ternary system can also be predicted in a similar manner. For a ternary system, Flory–Huggins equation is,

$$ln\frac{p}{p_0} = ln\varphi_1 + (\varphi_2 + \varphi_3) - \frac{\varphi_2}{x_{12}} - \frac{\varphi_3}{x_{13}} + (\omega_{12}\varphi_2 + \omega_{13}\varphi_3)(\varphi_2 + \varphi_3) - \omega_{23}\frac{\varphi_2\varphi_3}{x_{12}} \tag{16.7}$$

Here the subscripts 1, 2, 3 refer to water, drug, and polymer, respectively; φ is volume fractions; ω is interaction parameters; ρ is the relative water vapor pressure and x is molecular size ratio parameter for the components (Rumondor et al 2010).

Finally, the melting point depression approach has been adapted to describe the API–polymer interaction, in which the thermal activities of the ASD are compared with those of the API alone. With the help of mathematic simulation using Eq. 16.8, a plot of $\left(\frac{1}{T_M^{mix}} - \frac{1}{T_M^{pure}}\right) \times \left(\frac{\Delta H_{fus}}{-R}\right) - \ln(\phi_{drug}) - \left(1 - \frac{1}{m}\right)\phi_{polymer}$ versus $\phi_{polymer}^2$ reveals the value of φ (Marsac et al. 2006b), which is the key interaction parameter for the understanding of drug–polymer interaction:

$$\left(\frac{1}{T_{\mathrm{M}}^{\mathrm{mix}}} - \frac{1}{T_{\mathrm{M}}^{\mathrm{pure}}}\right) = \frac{-R}{\Delta H_{\mathrm{fus}}}\left[\ln\phi_{\mathrm{drug}} + \left(1 - \frac{1}{m}\right)\phi_{\mathrm{polymer}} + \varphi\phi_{\mathrm{polymer}}^{2}\right] \quad (16.8)$$

where T_M^{mix} is the melting temperature of the drug in the presence of the polymer, T_M^{pure} ΔH_{fus} is the heat of fusion of the pure drug, m is the ratio of the volume of the polymer to that of the lattice site, ϕ, is the volume fraction, and φ, is the interaction parameter.

When using the melting point depression approach, melting event must precede chemical decomposition. Melting point of the drug should be sufficiently high so that the polymer is in a supercooled liquid-like state to interact and mix with the molten drug. This approach is more appropriate for polymers whose glass transition temperatures are significantly lower than the melting point of the drug. This method provides an estimation of the interaction parameter close to the melting point of the drug. The polymer drug ratio linearity is limited to relatively low polymer concentrations. The interaction parameter determined represents a composite value over this limited concentration range (Marsac et al. 2009).

16.3.3 Molecular Mobility Estimation of Formulated ASD

The molecular mobility of amorphous material, usually expressed as the relaxation time, τ, is typically evaluated through measurements of certain relaxation processes. From the stability perspective, storage conditions under which the materials exhibit relaxation times comparable to or greater than the timescale of the required shelf life are desirable. The general method in molecular mobility approach is to measure an indicating parameter under a certain condition, then to correlate this indicating parameter with relaxation time by different equations. This relaxation time could be an indication of the starting time of instability.

In an amorphous solid, relaxation may arise from different origins: enthalpy relaxation, volume relaxation, dielectric relaxation, and spin relaxation of protons and ^{13}C nuclei. Molecular mobility can be indirectly estimated using DSC and isothermal microcalorimetry (IMC) under the prerequisite that the estimated mobility is highly coupled to physical instability (Graeser et al. 2009b). Techniques such as dielectric spectroscopy and solid-state NMR directly measure the relaxation of a sample and provide information about the global α-relaxation as well as the local β-relaxation. However, solid-state NMR is not readily available. Hence, it is not widely used. Molecular mobility estimation using solid-state NMR will not be introduced in this chapter; readers can refer to literature elsewhere (Aso and Yoshioka 2006; Kojima et al. 2012; Ueda et al. 2012; Masuda et al. 2005). The molecular mobility can be estimated from the peak width obtained from NMR.

The molecular relaxation time value is methodology and instrumentation dependent. This is especially true when studying an amorphous system below its glass transition temperature since the sample may not be at true equilibrium with the experimental surroundings. It is preferable to utilize several complementary analytical tools to study the system of interest rather than relying on a single measuring technique. In addition, DSC method provides an estimate of the average molecular mobility that is associated with enthalpy changes in the sample, and other techniques (e.g., dielectric relaxation experiments) may provide slightly different absolute estimates of the average molecular relaxation time under similar time and temperature conditions. In all cases, the distribution parameter (β) indicates the extent to which the data deviate from a true exponential function, with a value of unity corresponding to an exponential function. Meaningful comparisons of average relaxation times from different experiments can only be made when the values of β are comparable (± 0.1) or when the impact of the non-exponential behavior on the average value of τ is fully understood. It should be noted that comparison of material properties should only be made using data generated under similar experimental conditions (Hancock and Zografi 1997).

In this section, we only cover the techniques that have been applied to solid dispersions; however, readers are advised that other techniques such as mechanical analysis have also been used for amorphous solids molecular mobility though not on solid dispersions (Andronis and Zografi 1997). In addition, it has been found that below T_g no clear relationship between the various factors and physical stability exists (Graeser et al. 2009b). It has been confirmed by a study with amorphous cephalosporin (Shamblin et al. 2006). Phase separation and crystallization involve diffusion and nucleation, both linked to molecular mobility. There are numerous studies attempting to correlate molecular mobility with physical stability. Though scientifically interesting and useful, readers should not relate ASD physical stability with molecular mobility solely, as molecular mobility is just one of the factors as shown in the earlier section on factors affecting physical stability. It is not surprising that the correlation between physical stability and molecular mobility is low. In addition, the correlation between the two is not complete. Most of the correlations only consider global mobility, not local mobility, though studies have shown that β-relaxation plays a key role in estimating the physical stability of solid dispersions (Vyazovkin and Dranca 2007). Hence, cautions should be taken when applying molecular mobility to predict stability.

A summary and comparison of different molecular mobility methods is shown in Table 16.2.

16.3.3.1 Molecular Mobility Measured by DSC

The most frequently employed approach for estimating molecular mobility in amorphous solids is that of enthalpy recovery experiments using DSC. The relaxation time measured by this approach is from samples stored at a temperature below T_g. Assessment of drug stability below T_g focuses on relating the measured recrystallization enthalpy of the samples to the recrystallization enthalpy of the fresh sample (Graeser et al. 2009b).

Table 16.2 Techniques used to determine molecular mobility

Method	Measured parameter	Description	Equation
DSC	Enthalpy (ΔH)	Determine the thermal events with temperature change	$1-(\frac{\Delta H}{\Delta H\infty})=\exp[-(\frac{t}{\tau})^{\beta}]$
Dielectric spectroscopy	Permittivity (ε)	Measure the dielectric properties of a medium as a function of frequency	$\varepsilon^{*}=\varepsilon_{\infty}+\frac{\varepsilon_{S}-\varepsilon_{\infty}}{(1+\mathrm{i}\omega\tau)^{\beta}}+\frac{\sigma_{dc}}{\mathrm{i}\omega\varepsilon_{S}}$
IMC	Heat powder (P)	Determine the thermal events with time change	$P=277.8\times\frac{\Delta H_{r}(\infty)}{\tau_{0}}$ $\times(1+\frac{\beta t}{\tau_{1}})\times(1+\frac{t}{\tau_{1}})^{\beta-2}$ $\times\exp\left[-(\frac{t}{\tau_{0}})\times(1+\frac{t}{\tau_{1}})^{\beta-1}\right]$ $\times\tau_{D}=\tau_{0}^{1/\beta}\times\tau_{1}^{(\beta-1)/\beta}$

DSC differential scanning calorimetry, *IMC* isothermal microcalorimetry

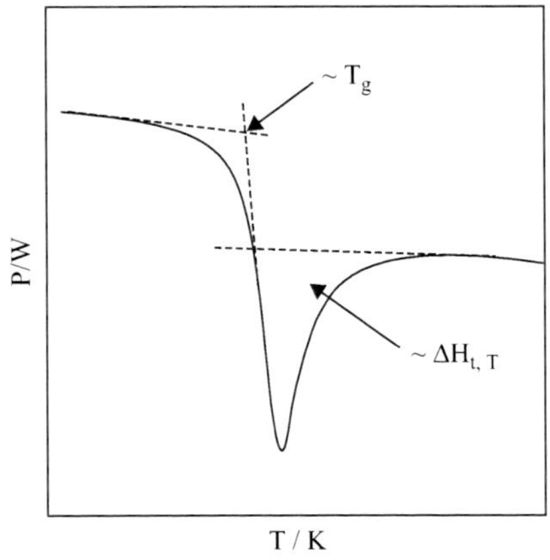

Fig. 16.3 A typical DSC thermogram from an enthalpy recovery ("relaxation") experiment. (Reproduced with permission from Hancock and Shamblin 2001)

In these experiments, the ASD is stored at different temperatures and different lengths of time, at a temperature below T_g. The sample is then reheated through T_g and the DSC trace at T_g is analyzed as shown in Fig. 16.3. The endothermic event is the recovery of the enthalpy that is lost by structural relaxation during the storage below T_g (also known as "aging"). The enthalpy associated with this endotherm ($\Delta H_{t,T}$) can be quantified in most cases by the subtraction of the response from a nonaged sample with an identical thermal history. Alternatively, it can be estimated by using the extrapolated supercooled liquid response to define an approximate baseline response as shown in Fig. 16.3. It is also possible to use modulated DSC (mDSC) as shown in Fig. 16.4.

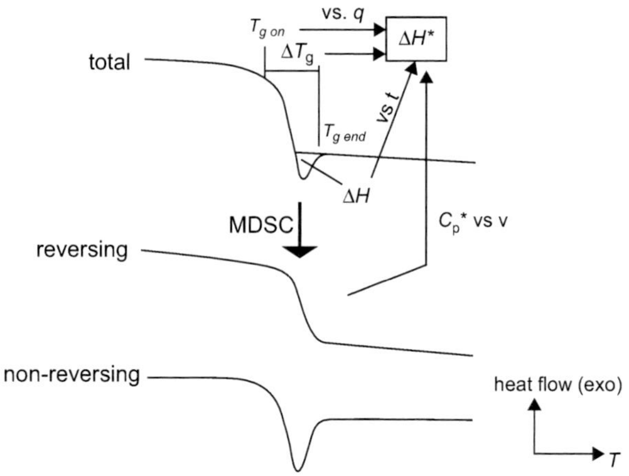

Fig. 16.4 Illustration of the use of DSC data for measuring T_g and ΔH^* (the activation energy for enthalpy relaxation). T_{gon}, T_{gend}, and ΔT_g indicate the onset, end, and width of the glass transition. Modulated DSC (mDSC) allows the separation of the total heat flow into reversing and nonreversing components. ΔH^* can be evaluated from (**a**) the dependence of T_{gon} on scanning rate *q*, (**b**) ΔT_g, (**c**) the dependence of the "relaxation enthalpy" ΔH (area of the "overshoot" on annealing time, and (**d**) the dependence of the complex heat capacity Cp^* (obtainable by mDSC) on modulation frequency *v*. (Reproduced with permission from Yu 2001)

The enthalpy change indicated by the measured endotherm peak corresponds to the extent that the sample is able to relax under the chosen storage conditions (temperature, humidity, and storage time). This enthalpy relaxation is directly related to the "average" molecular mobility of the material under those conditions. By repeating the experiment using a range of storage times, it is possible to determine the enthalpy change versus storage time, and from these data, the average molecular relaxation time can be estimated at any given temperature. The enthalpy change at each time point can then be expressed as a fraction of the total potential enthalpy relaxation, or the fraction relaxed expressed as $\frac{\Delta H_{t,T}}{\Delta H_{\max,T}}$; next the fraction relaxed is plotted as a function of storage time and, in most cases, a fit to the empirical Kohlrausch–Williams–Watts (KWW) equation. This is used to estimate the average relaxation time (τ) and the corresponding distribution parameter/stretched time function parameter (β), $\Delta H_{\max,T} = \Delta C_p^{T_g}(T_g - T)$ is the heat capacity change at T_g and T is the storage temperature:

$$\frac{\Delta H_{t,T}}{\Delta H_{\max,T}} = 1 - \exp\left(-\frac{t}{\tau}\right)^{\beta} \tag{16.9}$$

Once fitting the enthalpy data into the above equation, τ can be obtained. The fitted curve is shown in Fig. 16.5. In order for this curve-fitting procedure to provide meaningful data, at least five or six data points are needed. On the other hand, this method presents some limitations to its ability to describe the true relaxation behavior due to the assumption that the structural relaxation time τ is a constant throughout the

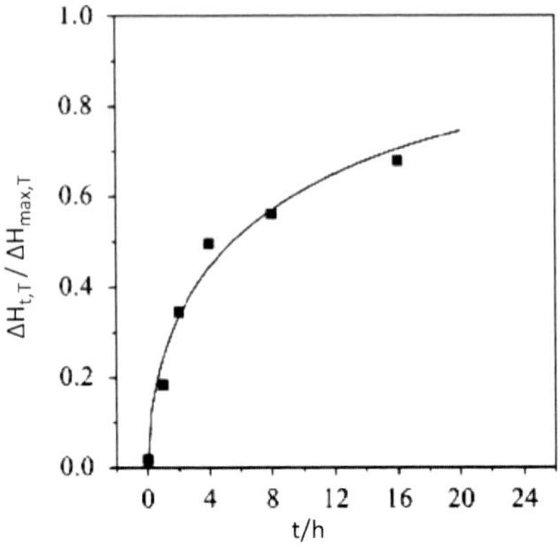

Fig. 16.5 Fraction-recovered enthalpy versus time for quench-cooled amorphous sucrose at 333 K. Line represents fit to the Kohlrausch–Williams–Watts equation. (Reproduced with permission from Hancock and Shamblin 2001)

relaxation process. In reality, τ tends to increase as relaxation of the glass progresses. As a consequence, if the change in τ during relaxation is considerably pronounced, it will lead to a situation in which no single τ value is sufficient to properly describe the relaxation process.

In case ΔH_{∞} is not easy to determine, it can be estimated by measuring the heat capacity (ΔCp) using mDSC (Six et al. 2001), where ΔH_{∞} can be obtained by $\Delta H_{\infty} = \Delta C_{\rm p}(T_{\rm g} - T_{\rm a})$, and $T_{\rm a}$ is the annealing temperature or the storage temperature (Hancock et al. 1995; Kakumanu and Bansal 2002).

In order to overcome the possibility of not being able to measure the value of ΔH_{∞}, a couple of alternate methods have been developed by Dr. Pinal's group using parameters determined by DSC (Mao et al. 2006a; Mao et al. 2006b) and the step-by-step procedures are listed in Table 16.3.

16.3.3.2 Molecular Mobility Measured by Dielectric Spectroscopy

Dielectric spectroscopy is a very promising tool to probe relaxation processes, especially since new broadband equipment can cover frequencies in the range of 10^{-4} to 10^{10} Hz. The theoretical introduction of this technique can be found in literature (Korhonen et al. 2008). Basically, electrodes of two different sizes are used, and the overall response, given by capacitance (C), depends on both the intrinsic characteristics (permittivity ε) and the geometry of the sample. The relationship between capacitance and sample geometry is given by:

$$C = \frac{A\varepsilon}{d} \tag{16.10}$$

where C is the capacitance, ε is the permittivity, and d is the geometry. When the experiment is carried out, stored sample is placed into a holder with electrodes connected to it. The permittivity (ε) is recorded. The permittivity data is fitted into

Table 16.3 Step-by-step instructions on the determination of relaxation time using DSC

Method 1 (82)	
1	Evaluating B and T_0 from scanning rate dependence of T_g using DSC $T_0 = T_g(1 - \frac{m_{min}}{m})$, $B = \frac{(\ln 10) T_g m^2{}_{min}}{m}$, $m_{min} = 16, m = \frac{\Delta H^*(T_g)}{(\ln 10) R T_g}$
2	Measuring of $\Delta C_p T_g$ and γ at T_g by DSC, $\gamma = \frac{\Delta C_p}{\Delta C_p^l}$
3	Calculating the initial enthalpy fictive temperature T_f^0 from T_g and γ, $T_f^0 = T_g^\gamma \times T_1^{(1-\gamma)}$
4	Evaluate time-dependent T_f by measuring recovered enthalpy using DSC after allowing samples to relax for given lengths of time, $T_f = T_f^0 \exp\left(-\frac{\gamma \Delta H_{relax}}{\Delta C_{p,T_g} T_g}\right)$
5	Calculate the time-dependent relaxation time, $\tau = \tau_0 \exp\left(\frac{B}{T(1-\frac{T_0}{T_f})}\right)$
Method 2 (83)	
1	Measure T_g as a function of heating rate (q)
2	Plot ln q versus $1/T_g$ (K), obtain the slope of the fitted line
3	Calculate the activation enthalpy, ΔH^* $(T_g) = 8.314$*slope
4	Calculate the fragility index, $m = \Delta H^* (T_g)/(2.303*8.314*T_g)$
5	Calculate D and T_0, $D = (2.303*m^2{}_{min})/(m - m_{min})$, $T_0 = T_g*(1 - m_{min}/m)$
6	Measure C_p of the liquid (l), glass (g), and crystalline (x) forms
7	Calculate the γ parameter, $\gamma = \frac{(C_p^l - C_p^g)}{(C_p^l - C_p^x)}$
8	Calculate the initial relaxation time, $\tau^0 = \tau_0 \exp \frac{DT_0}{T - T(\frac{T}{T_g})^\gamma}$, $\tau_0 = 10^{-14}$, $m_{min} = 16$

Eq. 16.11,

$$\varepsilon^* = \varepsilon_\infty + \frac{\varepsilon_S - \varepsilon_\infty}{(1 + i\omega\tau)^\beta} + \frac{\sigma_{dc}}{i\omega\varepsilon_S} \tag{16.11}$$

where ε^* is the complex permittivity, ε_S is the low-frequency limit of permittivity, ε_∞ is the high-frequency limit, ω is the frequency, τ is the relaxation time, β represents the distribution of relaxation times with a value close to unity indicating homogenously distributing species and value close to zero indicating broad distribution of relaxing substates, and σ_{dc} is the dc conductivity contribution (Bhugra et al. 2006). Different from the methods using DSC in which it is required to store the test sample at temperature below T_g, samples using dielectric spectrometry can be stored at temperature above T_g. This feature makes the test more useful as such conditions mimic the real ASD storage situations.

16.3.3.3 IMC

Isothermal microcalorimetry (IMC) monitors real-time heat change with elapsed time and has also been applied in ASD stability prediction. The typical sample size for each IMC measurement is about 200–1000 mg. A thermally inert material, glycerin in the crystalline state, is used as reference in the IMC experiment. The heat power (P), typically in the unit of μw/g, is monitored with time and the data are fitted into Eq. 16.12,

$$P=277.8\times\frac{\Delta H_r(\infty)}{\tau_0}\times\left(1+\frac{\beta t}{\tau_1}\right)\times\left(1+\frac{t}{\tau_1}\right)^{\beta-2}\times\exp\left[-\left(\frac{t}{\tau_0}\right)\times\left(1+\frac{t}{\tau_1}\right)^{\beta-1}\right] \tag{16.12}$$

where P is the heat power. The obtained value is then fitted into another equation,

$$\tau_D=\tau_0^{1/\beta}\times\tau_1^{(\beta-1)/\beta} \tag{16.13}$$

and τ_D is the relaxation time. This approach allows the stability to be predicted under different temperature conditions (Liu et al. 2002).

16.3.4 Prediction of Physical Stability by Detection of Phase Separation and Crystallization

In API–polymer systems, many times the physical instability is initiated by amorphous–amorphous phase separation ending in heterogeneous arrangements. Drug crystallization from a solid solution is the segregation of drug molecules from the polymers prior to nucleation and subsequent crystal growth. Hence, determination of the drug phase separation from the solid solution can provide an estimate for the kinetics of physical stability. The general practice is to place the sample at the designed temperature and humidity and to detect phase separation and then crystallization. The parameters that indicate the occurrence of phase separation or crystallization are then characterized at different time intervals. The results from different time intervals can then be plotted over the time period of test to predict when the crystallization start or the stability failed the requirement.

A summary of techniques used to study phase separation is shown in Table 16.4.

16.3.4.1 Phase Separation Detection by Atomic Force Microscopy

Atomic force microscopy (AFM) has been used for stability studies of solid dispersions. The presumption of using AFM to characterize phase separation is that freshly made solid dispersion is homogeneous and the surface is smooth, and changes in surface roughness are induced by phase separation. Once the phase separation occurs, surface roughness will be increased. An example of AFM image was shown in Fig. 16.2. Hence, solid dispersion is placed under AFM over certain period of time under stress conditions. The surface roughness is then monitored and plotted over time. The stability can thus be monitored and predicted (Lauer et al. 2011). A more advanced version is AFM coupled with IR (Van Eerdenbrugh et al. 2012), offering additional information on chemical stability. The disadvantage of AFM is the surface

Table 16.4 Techniques used to characterize phase separation

Method	Measured parameter	Description	Equations
AFM	Surface roughness	Fractured films are prepared by annealing and quench cooling	Roughness versus time
Raman/IR	Peak intensity	Sensitive approach to get the homogeneity of the sample	
DSC	T_g	Determine the number of glass transitions events with 1 T_g indicating no phase separation	$(1-\alpha)_t = 1-\frac{T_{g(\text{polymer})}-T_{g,T}}{T_{g(\text{polymer})}-T_{g(\text{initial})}}$, $[-ln(1-\alpha)] = kt$ it is the rate constant for solid-state transformation
XRD	Crystallinity and crystallization onset time (τ_{oc})	Measure the intensity of diffraction peaks. The crystallinity is indicated by the diffraction peaks' sharpness	$@\ T > T_g, \log(\tau_{oc}) = \log(\tau_{oc}^0) + \frac{1}{ln(10)} \times \frac{D'T_0}{T-T_0}$ $@\ T < T_g, \log(\tau_{oc}) = \log(\tau_{oc}^0) + \frac{1}{\ln(10)} \times \frac{\varepsilon DT_0}{T-T_0/T_f}$ When humidity is considered, ln $[\tau_{oc}(T, \text{RH})] = \log(\tau_{oc}^0) + A\frac{T_g(\text{RH})}{T}$
IGC	Retention volume (V)	Detect surface crystallization under different humidity	$V = [1 - \exp(-Bt^m)](V_{\max}-V_0)+V_0$ $\alpha = (V - V_0)/(V_{\max} - V_0)$ $[-\ln(1-\alpha)]^{1/2} = kt$ $\alpha = kt$ $\ln k = \ln A - E_a/RT$

AFM atomic force microscopy, *DSC* differential scanning calorimetry, *IR* infrared, *XRD* X-ray diffraction, *IGC* inverse gas chromatography

of the solid dispersion needs to be relatively smooth; otherwise, it cannot be detected by AFM unless surface polish is applied. Another drawback of AFM is the surface area for detection is very limited.

16.3.4.2 Phase Separation Detection by Raman Mapping and IR

Raman mapping has been shown to be able to detect phase separation in systems which do not exhibit glass transition detectable by calorimetry (Padilla et al. 2011). In other words, the sensitivity of this approach is superior to that of DSC. The phase separation indication is the deviation in the peak intensity of the compound over an area. It is measured by scanning 100 μm line map; an average of the peak intensity for the one component system is detected. By subtracting this average intensity from the intensity of the appropriate peak at each point across a map, a deviation from homogeneity is obtained. The deviation in the peak intensity at each point across the map is then weighted relative to the corresponding peak height. This deviation is mainly due to phase separation (Padilla et al. 2011) and plotted over time. The stability plot is then obtained. Same principle applies to the measurement by IR (Rumondor and Taylor 2009).

16.3.4.3 Phase Separation Detection by T_g Change

The kinetics of drug phase separation can be determined by monitoring the change in T_g of the solid dispersion. The time-dependent changes in T_g values are used in calculating the fraction of the drug phase separated $(1-\alpha)$ from the solid dispersions, where α is the fraction that remains miscible in the solid dispersion. The fraction of the drug that is phase separated at a specific storage time (t) is calculated by using the ratio shown below:

$$(1-\alpha)_t = 1 - \frac{T_{g(polymer)} - T_{g,T}}{T_{g(polymer)} - T_{g(initial)}} \qquad (16.14)$$

Here $T_{g(polymer)}$ is the T_g of the polymer, $T_{g,T}$ is the T_g of the solid dispersion under a certain condition, $T_{g(initial)}$ is the T_g of the freshly made solid dispersion. To estimate the rate constant of phase separation, the phase-separated fraction of the drug is plotted against the storage time and a linear fit obtained by using the equation,

$$[-\ln(1-\alpha)] = kt \qquad (16.15)$$

where k is the rate constant for the solid-state transformation. Hence, the instability time can be estimated (Vasanthavada et al. 2005). An illustration is shown in Fig. 16.6.

16.3.4.4 Crystallization Tendency Measured by X-Ray Diffraction

The principle of using X-ray (powder) diffraction (XRPD) to detect stability is to monitor the change of crystallinity over time under different temperature and humidity conditions (Greco et al. 2012). The advantage of this technique is that the temperature can be both above and below T_g. Small-angle X-ray diffraction (XRD) can also be used (Zhu et al. 2010). The onset of crystallization, τ_{oc}, is hence obtained.

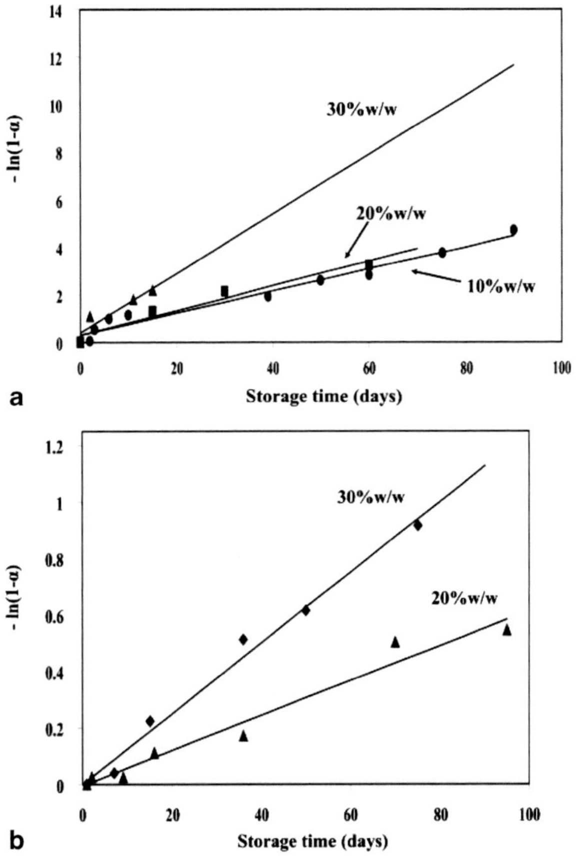

Fig. 16.6 An illustration of the estimation of the extent of phase separation of **a** griseofulvin and **b** indoprofen from the PVP solid dispersions. (Reproduced with permission from Vasanthavada et al. 2005)

By fitting τ_{oc} versus temperature data into different equations (Eqs. 16.16 and 16.17), according to the scenario, unknown parameter, τ_{oc}^0, $D' = \varepsilon D$ and T_0 can be derived. The long-term stability is then predicted by inputting the target temperature back to the equation with already known parameters:

$$\mathrm{Log}(\tau_{oc}) = \mathrm{Log}(\tau_{oc}^0) + \frac{1}{\ln(10)} \times \frac{D'T_0}{T - T_0} \tag{16.16}$$

$$\mathrm{Log}(\tau_{oc}) = \mathrm{Log}(\tau_{oc}^0) + \frac{1}{\ln(10)} \times \frac{\varepsilon D T_0}{T - T_0/T_f} \tag{16.17}$$

Once humidity effect is considered, equations will be changed into,

$$\ln(\tau_{oc}(T, \mathrm{RH})) = \mathrm{Log}(\tau_{oc}^0) + A\frac{T_g(\mathrm{RH})}{T} \tag{16.18}$$

Here A is a constant $\left(A = \frac{\varepsilon D T_0/T_g}{(1-\frac{T_0}{T_f})}\right)$, and T_g (RH) is the T_g at that RH.

16.3.4.5 Crystallization Tendency Measured by Inverse Gas Chromatography

Inverse gas chromatography (IGC) has been used to detect surface crystallization and to predict solid dispersion stability (Miyanishi et al. 2012). The principle of IGC is out of scope of this chapter; readers are encouraged to read elsewhere. In IGC, sample of the solid dispersion is placed in the capillary in the machine. Solvent vapor is continuously purged on the surface of the freshly made solid dispersion over a period of time; the volume of the solvent retained on the surface can be measured over time. A retention volume will change if the surface crystallization occurs. Therefore, the retention volume will be an indication of solid dispersion instability. A retention volume versus time plot can be drawn from the IGC results. A retention volume change versus temperature plot can also be obtained if the experiment was carried out at different temperature. The steps to obtain the crystallization rate and finally the prediction of instability are shown below:

By fitting the retention volume data into the equation, V_{max} could be obtained, which is the retention volume when crystallization at the surface is completed:

$$V = [1 - \exp(-Bt^m)](V_{max} - V_0) + V_0 \tag{16.19}$$

where V is the retention volume at time t, V_0 is the retention volume at $t = 0$, and V_{max} is the retention time when the crystallization at the surface is completed ($t = \infty$). m and B represent the reaction mechanism and reaction rate, respectively.

Once V_{max} is obtained, it is fit into the Eq. 16.20 to obtain α, which represents the percentage of crystallized drug at the surface of the solid dispersion at time t:

$$\alpha = \frac{(V - V_0)}{(V_{max} - V_0)} \tag{16.20}$$

Once α is obtained, retention volume data from the experiments are fitted into two equations, $[-\ln(1 - \alpha)]^{1/2} = kt$ and $\alpha = kt$, to determine which equation to use. The criterion is regression r. Once decided, k would be derivded from regression using one of the above equations.

Following that is to fit k into $\ln k = \ln A - E_a/RT$, in which A is a constant, E_a is the activation energy, R is the gas constant, and T is the temperature of the experiment at which solid dispersion crystallization is carried out. Hence, experiments at several temperatures will be needed for solid dispersion crystallization in order to get the equation. From the equation fitting, A and E_a can be derived.

Finally, the derived temperature of storage can be applied to the equation $\ln k = \ln A - E_a/RT$, to calculate the corresponding k. Once k is available, the crystallization rate can be obtained, and the stability prediction can be made based on that.

16.4 Stability Programs

Due to the complexity of the interplaying factors on ASD stability, studies are often carried out focusing on one or a few parameters at a time. As a consequence, there are different options on the prioritization of the different impacting factors to identify the underlying mechanism for instability and to evaluate approaches to inhibit crystallization. Here we want to share with readers our rationale for a sophisticated stability program which provides a rather general understanding.

As discussed in Sect. 18.3, several models are available and can be applied with limited amounts of experimental data in combination with calculated molecular descriptors. These models are extremely valuable at the early stage of development of the API. A quick decision can be made regarding if ASD is an applicable approach for the candidate. Once this question is answered, the selection of an appropriate polymer comes in as the next step. If other additives are needed, for example, a surfactant to improve the manufacturability or to enhance the pharmacokinetic (PK) performance, the implication on system stability has to be assessed. The different formulations developed can be rank ordered based on their stability under severe temperature and humidity conditions. When the composition of the ASD is finalized, a more systematic stress test is used to understand the recrystallization risk. Though the changes in physicochemical properties by and large appear in a nonlinear fashion, the stress tests normally are sufficient to assess the risks, thus providing the formulation scientists confidence to estimate the stability of the drug product.

There are many approaches that are in place for mechanistic insights of the instability mechanism from both thermodynamic and kinetic viewpoints as discussed previously. These approaches normally require solid understanding of the instrumentation involved and the data interpretation may not always be straightforward. In addition, not all of these approaches are readily available in the formulation laboratories in most of the pharmaceutical companies. Instead, they are more prone to academic research. In a more conservative approach, instead of predicting the stability, a stability program is carried out normally under stress conditions. In a typical stability program, suitable amorphous formulation should exhibit long-term stability with respect to solid-state properties as well as chemical integrity (Verreck et al. 2003). Finalized ASD can be produced via an appropriate processing method and the amorphous samples lightly ground prior to storage for evaluation. Although varying from laboratory to laboratory, company to company, a common set of stress conditions would generally include 4 °C, 25 °C, 50 °C, 80 °C, 25 °C/60 % RH, and 37 °C/75 % RH with time points ranging from one day up to six months or two years. At each time point under each stress condition, the sample is analyzed using XRPD, DSC, and/or FTIR as illustrated above. In the late-phase drug development, long-term stability is carried out for years to monitor the stability by either XRD or DSC. In this case, the objective is not to predict but to evaluate real-time stability.

In addition to the stability assessment of ASD with the goal for drug product development, the ASD formulation also needs to maintain supersaturation during the in vivo dissolution window so as to achieve the solubility enhancement and to optimize drug absorption.

In case of preclinical studies, suspension stability may also be important. The ASD suspension stability is conducted to support PK and/or toxicology studies. The preferred suspension vehicle is the one in which the API would remain amorphous for up to 4–6 h at room temperature.

Conclusions

ASD approach is of high interest due to the potential for oral bioavailability enhancement, particularly for an API with poor aqueous solubility. At the same time, these are challenging systems with regard to chemical and physical stability due to their high energy state as compared to their crystalline counterparts. Chemical and physical stability of ASD is the key for its successful development and commercialization. Fundamental understandings of both thermodynamic and kinetic aspects are essential to ensure sound rational formulation design. In this chapter, we have focused on the physical stability of ASD and discussed the factors that impact the physical stability, available prediction tools, and approaches applicable to improve the stability, as well as practical stability programs suitable for different stages of ASD development.

Although an ASD drug formulation has inherent stability risks, they present significant opportunities. The task for formulation scientists lies in the design of the ASD matrix with API, hydrophilic carrier, and/or surfactant that offers supersaturated API solution in the gastrointestinal tract with sufficient stability in addition to a stable formulation matrix under standard storage conditions. With an increased mechanistic understanding of the instability, advancement in accuracy and accessibility of modern analytical technologies, and downstream processing development, the intrinsic risks can be successfully mitigated to enable full exploitation of the solid dispersion formulation strategies.

Acknowledgements The authors would like to express their gratitude to Dr. Quanying Bao at Novartis for useful discussions and her help with the proof reading.

References

Alam MA, Ali R, Al-Jenoobi FI, Al-Mohizea AM (2012) Solid dispersions: a strategy for poorly aqueous soluble drugs and technology updates. Expert Opin Drug Deliv 9(11):1419–1440

Andronis V, Zografi G (1997) Molecular mobility of supercooled amorphous indomethacin, determined by dynamic mechanical analysis. Pharm Res 14(4):410–414

Andronis V, Zografi G (1998) The molecular mobility of supercooled amorphous indomethacin as a function of temperature and relative humidity. Pharm Res 15(6):835–842

Andronis V, Yoshioka M, Zografi G (1997) Effects of sorbed water on the crystallization of indomethacin from the amorphous state. J Pharm Sci 86(3):346–351

Aso Y, Yoshioka S (2006) Molecular mobility of nifedipine–PVP and phenobarbital–PVP solid dispersions as measured by 13 C-NMR spin-lattice relaxation time. J Pharm Sci 95(2):318–325

Aso Y, Yoshioka S, Kojima S (2004) Molecular mobility-based estimation of the crystallization rates of amorphous nifedipine and phenobarbital in poly(vinylpyrrolidone) solid dispersions. J Pharm Sci 93(2):384–391

Baird JA, Van Eerdenbrugh B, Taylor LS (2010) A classification system to assess the crystallization tendency of organic molecules from undercooled melts. J Pharm Sci 99(9):3787–3806

Bhattacharya S, Suryanarayanan R (2009) Local mobility in amorphous pharmaceuticals—characterization and implications on stability. J Pharm Sci 98(9):2935–2953

Bhugra C, Pikal MJ (2008) Role of thermodynamic, molecular, and kinetic factors in crystallization from the amorphous state. J Pharm Sci 97(4):1329–1349

Bhugra C, Shmeis R, Krill S, Pikal M (2006) Predictions of onset of crystallization from experimental relaxation times I-Correlation of molecular mobility from temperatures above the glass transition to temperatures below the glass transition. Pharm Res 23(10):2277–2290

Bhugra C, Shmeis R, Pikal MJ (2008) Role of mechanical stress in crystallization and relaxation behavior of amorphous indomethacin. J Pharm Sci 97(10):4446–4458

Bikiaris DN (2011) Solid dispersions, part I: recent evolutions and future opportunities in manufacturing methods for dissolution rate enhancement of poorly water-soluble drugs. Expert Opin Drug Deliv 8(11):1501–1519

Chiou WL, Riegelman S (1971) Pharm applications of solid dispersion systems. J Pharm Sci 60(9):1281–1302

Craig DQM, Royall PG, Kett VL, Hopton ML (1999) The relevance of the amorphous state to pharmaceutical dosage forms: glassy drugs and freeze dried systems. Int J Pharm 179(2): 179–207

Ghebremeskel AN, Vemavarapu C, Lodaya M (2006) Use of surfactants as plasticizers in preparing solid dispersions of poorly soluble API: stability testing of selected solid dispersions. Pharm Res 23(8):1928–1936

Ghosh I, Snyder J, Vippagunta R, Alvine M, Vakil R, Tong W-Q et al (2011) Comparison of HPMC based polymers performance as carriers for manufacture of solid dispersions using the melt extruder. Int J Pharm 419(1–2):12–9

Giovambattista N, Angell CA, Sciortino F, Stanley HE (2004) Glass-transition temperature of water: a simulation study. Phys Rev Lett 93(4):047801

Graeser KA, Patterson JE, Rades T (2009a) Applying thermodynamic and kinetic parameters to predict the physical stability of two differently prepared amorphous forms of simvastatin. Curr Drug Deliv 6(4):374–382

Graeser KA, Patterson JE, Zeitler JA, Gordon KC, Rades T (2009b) Correlating thermodynamic and kinetic parameters with amorphous stability. Eur J Pharm Sci 37(3–4):492–498

Greco S, Authelin J-R, Leveder C, Segalini A (2012) A practical method to predict physical stability of amorphous solid dispersions. Pharm Res 29(10):2792–2805

Gupta J, Nunes C, Vyas S, Jonnalagadda S (2011) Prediction of solubility parameters and miscibility of pharm compounds by molecular dynamics simulations. J Phys Chem B 115(9):2014–2023

Hancock BC, Shamblin SL (2001) Molecular mobility of amorphous pharmaceuticals determined using differential scanning calorimetry. Thermochimica Acta 380(2):95–107

Hancock BC, Zografi G (1997) Characteristics and significance of the amorphous state in pharmaceutical systems. J Pharm Sci 86(1):1–12

Hancock B, Shamblin S, Zografi G (1995) Molecular mobility of amorphous pharmaceutical solids below their glass transition temperatures. Pharm Res 12(6):799–806

James PF (1985) Kinetics of crystal nucleation in silicate glasses. J Non-Crystalline Solids 73(1–3):517–540

Kakumanu VK, Bansal AK (2002). Enthalpy relaxation studies of celecoxib amorphous mixtures. Pharm Res 19(12):1873–1878

Kauzmann W (1948) The nature of the glassy state and the behavior of liquids at low temperatures. Chem Rev 43(2):219–256

Kojima T, Higashi K, Suzuki T, Tomono K, Moribe K, Yamamoto K (2012) Stabilization of a supersaturated solution of mefenamic acid from a solid dispersion with EUDRAGIT® EPO. Pharm Res 29(10):2777–2791

Konno H, Taylor LS (2006) Influence of different polymers on the crystallization tendency of molecularly dispersed amorphous felodipine. J Pharm Sci 95(12):2692–2705

Korhonen O, Bhugra C, Pikal MJ (2008) Correlation between molecular mobility and crystal growth of amorphous phenobarbital and phenobarbital with polyvinylpyrrolidone and L-proline. J Pharm Sci 97(9):3830–3841

Lauer M, Grassmann O, Siam M, Tardio J, Jacob L, Page S et al (2011) Atomic force microscopy-based screening of drug-excipient miscibility and stability of solid dispersions. Pharm Res 28(3):572–584

Leuner C, Dressman J (2000) Improving drug solubility for oral delivery using solid dispersions. Eur J Pharm Biopharm 50(1):47–60

Liu J, Rigsbee DR, Stotz C, Pikal MJ (2002) Dynamics of pharmaceutical amorphous solids: the study of enthalpy relaxation by isothermal microcalorimetry. J Pharm Sci 91(8):1853–1862
Liu H, Zhang X, Suwardie H, Wang P, Gogos CG (2012) Miscibility studies of indomethacin and Eudragit® E PO by thermal, rheological, and spectroscopic analysis. J Pharm Sci 101(6):2204–2212
Mahlin D, Ponnambalam S, Hockerfelt MH, Bergstrom CA (2011) Toward in silico prediction of glass-forming ability from molecular structure alone: a screening tool in early drug development. Mol Pharm 8(2):498–506
Mao C, Chamarthy SP, Pinal R (2006a) Time-dependence of molecular mobility during structural relaxation and its impact on organic amorphous solids: an investigation based on a calorimetric approach. Pharm Res 23(8):1906–1917
Mao C, Prasanth Chamarthy S, Byrn SR, Pinal R (2006b) A calorimetric method to estimate molecular mobility of amorphous solids at relatively low temperatures. Pharm Res 23(10):2269–2276
Marsac P, Konno H, Taylor L (2006a) A comparison of the physical stability of amorphous felodipine and nifedipine systems. Pharm Res 23(10):2306–2316
Marsac P, Shamblin S, Taylor L (2006b) Theoretical and practical approaches for prediction of drug–polymer miscibility and solubility. Pharm Res 23(10):2417–2426
Marsac P, Li T, Taylor L (2009) Estimation of drug–polymer miscibility and solubility in amorphous solid dispersions using experimentally determined interaction parameters. Pharm Res 26(1):139–151
Masuda K, Tabata S, Sakata Y, Hayase T, Yonemochi E, Terada K (2005) Comparison of Molecular Mobility in the Glassy State Between Amorphous Indomethacin and Salicin Based on Spin-Lattice Relaxation Times. Pharm Res 22(5):797–805
Matsumoto T, Zografi G (1999) Physical properties of solid molecular dispersions of indomethacin with poly(vinylpyrrolidone) and poly(vinylpyrrolidone-co-vinyl-acetate) in relation to indomethacin crystallization. Pharm Res 16(11):1722–1728
Miyazaki T, Yoshioka S, Aso Y, Kojima S (2004) Ability of polyvinylpyrrolidone and polyacrylic acid to inhibit the crystallization of amorphous acetaminophen. J Pharm Sci 93(11):2710–2717
Miller D, Lechuga-Ballesteros D (2006) Rapid assessment of the structural relaxation behavior of amorphous pharmaceutical solids: effect of residual water on molecular mobility. Pharm Res 23(10):2291–2305
Miller JM, Beig A, Carr RA, Spence JK, Dahan A (2012) A win–win solution in oral delivery of lipophilic drugs: supersaturation via amorphous solid dispersions increases apparent solubility without sacrifice of intestinal membrane permeability. Mol Pharm 9(7):2009–2016
Miyanishi H, Nemoto T, Mizuno M, Mimura H, Kitamura S, Iwao Y, Noguchi S, Itai S (2012) Evaluation of Crystallization Behavior on the Surface of Nifedipine Solid Dispersion Powder Using Inverse Gas Chromatography. Pharm Res 30(2):502–511
Miyazaki T, Yoshioka S, Aso Y, Kawanishi T (2007) Crystallization rate of amorphous nifedipine analogues unrelated to the glass transition temperature. Int J Pharm 336(1):191–195
Mosquera-Giraldo LI, Trasi NS, Taylor LS (2013) Impact of surfactants on the crystal growth of amorphous celecoxib. Int J Pharm 461(1–2):251–257
Nascimento ML, Zanotto ED (2010) Does viscosity describe the kinetic barrier for crystal growth from the liquidus to the glass transition? J Chem Phys 133(17):174701
Newman A, Knipp G, Zografi G (2012) Assessing the performance of amorphous solid dispersions. J Pharm Sci 101(4):1355–77
Padilla AM, Ivanisevic I, Yang Y, Engers D, Bogner RH, Pikal MJ (2011) The study of phase separation in amorphous freeze-dried systems. part I: Raman mapping and computational analysis of XRPD data in model polymer systems. J Pharm Sci 100(1):206–222
Paudel A, Nies E, Van den Mooter G (2012) Relating hydrogen-bonding interactions with the phase behavior of naproxen/pvp k 25 solid dispersions: evaluation of solution-cast and quench-cooled films. Mol Pharm 9(11):3301–3317
Pikal MJ, Dellerman KM (1989) Stability testing of pharmaceuticals by high-sensitivity isothermal calorimetry at 25 °C: cephalosporins in the solid and aqueous solution states. Int J Pharm 50(3):233–252

Qi S, Roser S, Edler KJ, Pigliacelli C, Rogerson M, Weuts I et al (2013) Insights into the role of polymer-surfactant complexes in drug solubilisation/stabilisation during drug release from solid dispersions. Pharm Res 30(1):290–302

Qian F, Huang J, Hussain MA (2010) Drug-polymer solubility and miscibility: Stability consideration and practical challenges in amorphous solid dispersion development. J Pharm Sci 99(7):2941–2947

Qian F, Wang J, Hartley R, Tao J, Haddadin R, Mathias N et al (2012) Solution behavior of PVP-VA and HPMC-AS-Based amorphous solid dispersions and their bioavailability implications. Pharm Res 29(10):2766–2776

Rumondor ACF, Taylor LS (2009) Effect of polymer hygroscopicity on the phase behavior of amorphous solid dispersions in the presence of moisture. Mol Pharm 7(2):477–490

Rumondor AF, Ivanisevic I, Bates S, Alonzo D, Taylor L (2009) Evaluation of drug-polymer miscibility in amorphous solid dispersion systems. Pharm Res 26(11):2523–2534

Rumondor ACF, Konno H, Marsac PJ, Taylor LS (2010) Analysis of the moisture sorption behavior of amorphous drug–polymer blends. J Appl Polym Sci 117(2):1055–1063

Salameh AK, Taylor LS (2005) Deliquescence in binary mixtures. Pharm Res 22(2):318–324

Serajuddin ATM (1999) Solid dispersion of poorly water-soluble drugs: early promises, subsequent problems, and recent breakthroughs. J Pharm Sci 88(10):1058–1066

Shamblin SL, Hancock BC, Pikal MJ (2006) Coupling between chemical reactivity and structural relaxation in pharmaceutical glasses. Pharm Res 23(10):2254–2268

Shibata Y, Fujii M, Suzuki A, Koizumi N, Kanada K, Yamada M et al (2014) Effect of storage conditions on the recrystallization of drugs in solid dispersions with crospovidone. Pharm Dev Technol 19(4):468–474

Siew A, Arnum PV (2013) Industry perspectives: achieving solutions for the challenge of poorly water-soluble drugs. Pharma Technol 37(6):60–62, 66

Six K, Verreck G, Peeters J, Augustijns P, Kinget R, Van den Mooter G (2001) Characterization of glassy itraconazole: a comparative study of its molecular mobility below T(g) with that of structural analogues using MTDSC. Int J Pharm 213(1–2):163–173

Sun Y, Tao J, Zhang GGZ, Yu L (2010) Solubilities of crystalline drugs in polymers: An improved analytical method and comparison of solubilities of indomethacin and nifedipine in PVP, PVP/VA, and PVAc. J Pharm Sci. 99(9):4023–4031

Tang XC, Pikal MJ, Taylor LS (2002) The effect of temperature on hydrogen bonding in crystalline and amorphous phases in dihydropyrine calcium channel blockers. Pharm Res 19(4):484–490

Tao J, Sun Y, Zhang GGZ, Yu L (2009) Solubility of small-molecule crystals in polymers: d-Mannitol in PVP, Indomethacin in PVP/VA, and Nifedipine in PVP/VA. Pharm Res 26(4):855–864

Taylor LS, Shamblin SL (2009) Amorphous solids. In: Polymorphism in pharmaceutical solids, 2nd edn. Informa healthcare, New York, pp 587–630

Taylor L, Zografi G (1997) Spectroscopic characterization of interactions between PVP and Indomethacin in amorphous molecular dispersions. Pharm Res 14(12):1691–1698

Taylor LS, Langkilde FW, Zografi G (2001) Fourier transform Raman spectroscopic study of the interaction of water vapor with amorphous polymers. J Pharm Sci 90(7):888–901

Tran P-L, Tran T-D, Park J, Lee B-J (2011) Controlled release systems containing solid dispersions: strategies and mechanisms. Pharm Res 28(10):2353–2378

Turnbull D, Fisher JC (1949) Rate of nucleation in condensed systems. J of Chem Phys 17(1):71–73

Ueda K, Higashi K, Limwikrant W, Sekine S, Horie T, Yamamoto K et al (2012) Mechanistic differences in permeation behavior of supersaturated and solubilized solutions of carbamazepine revealed by nuclear magnetic resonance measurements. Mol Pharm 9(11):3023–3033

Van den Mooter G, Wuyts M, Blaton N, Busson R, Grobet P, Augustijns P et al (2001) Physical stabilisation of amorphous ketoconazole in solid dispersions with polyvinylpyrrolidone K25. Eur J Pharm Sci 12(3):261–269

Van Eerdenbrugh B, Taylor LS (2011) An ab initiopolymer selection methodology to prevent crystallization in amorphous solid dispersions by application of crystal engineering principles. CrystEngComm 13(20):6171–6178

Van Eerdenbrugh B, Lo M, Kjoller K, Marcott C, Taylor LS (2012) Nanoscale mid-infrared imaging of phase separation in a drug–polymer blend. J Pharm Sci 101(6):2066–2073

Vasanthavada M, Tong W-Q, Joshi Y, Kislalioglu MS (2004) Phase behavior of amorphous molecular dispersions i: determination of the degree and mechanism of solid solubility. Pharm Res 21(9):1598–1606

Vasanthavada M, Tong W-Q, Joshi Y, Kislalioglu MS (2005) Phase behavior of amorphous molecular dispersions ii: role of hydrogen bonding in solid solubility and phase separation kinetics. Pharm Res 22(3):440–448

Verreck G, Six K, Van den Mooter G, Baert L, Peeters J, Brewster ME (2003) Characterization of solid dispersions of itraconazole and hydroxypropylmethylcellulose prepared by melt extrusion—part I. Int J Pharm 251(1–2):165–174

Vyazovkin S, Dranca I (2005) Physical stability and relaxation of amorphous indomethacin. J Phys Chem B 109(39):18637–18644

Vyazovkin S, Dranca I (2007) Effect of physical aging on nucleation of amorphous indomethacin. J Phys Chem B 111(25):7283–7287

Wegiel LA, Mauer LJ, Edgar KJ, Taylor LS (2013) Crystallization of amorphous solid dispersions of resveratrol during preparation and storage-Impact of different polymers. J Pharm Sci 102(1):171–184

Weuts I, Kempen D, Decorte A, Verreck G, Peeters J, Brewster M et al (2005) Physical stability of the amorphous state of loperamide and two fragment molecules in solid dispersions with the polymers PVP-K30 and PVP-VA64. Eur J Pharm Sci 25(2–3):313–320

Yoo SU, Krill SL, Wang Z, Telang C (2009) Miscibility/stability considerations in binary solid dispersion systems composed of functional excipients towards the design of multi-component amorphous systems. J Pharm Sci 98(12):4711–4723

Yoshioka M, Hancock BC, Zografi G (1994) Crystallization of indomethacin from the amorphous state below and above its glass transition temperature. J Pharm Sci 83(12):1700–1705

Yu L (2001) Amorphous pharmaceutical solids: preparation, characterization and stabilization. Adv Drug Deliv Rev 48(1):27–42

Yuan X, Sperger D, Munson EJ (2013) Investigating miscibility and molecular mobility of nifedipine-pvp amorphous solid dispersions using solid-state NMR spectroscopy. Mol Pharm 11(1):329–337

Zhu Q, Taylor LS, Harris MT (2010) Evaluation of the microstructure of semicrystalline solid dispersions. Mol Pharm 7(4):1291–1300

Chapter 17
Regulatory Considerations in Development of Amorphous Solid Dispersions

Ziyaur Rahman, Akhtar Siddiqui, Abhay Gupta and Mansoor Khan

17.1 Introduction

Amorphous solid dispersion (ASD) provides an opportunity to enhance the bioavailability and therapeutic performance by improving the physicochemical properties of poorly aqueous soluble drugs. Generally, an amorphous form of the drug has higher solubility as well as dissolution rate and extent than its crystalline counterpart due to its ability to produce high supersaturation in the aqueous conditions (Ivanisevic et al. 2009; Tobyn et al. 2009). On the other hand, noncrystalline drug delivery strategies such as cosolvent or self-emulsifying drug delivery system are liquid or semi-solid dosage forms which have high manufacturing cost and a undesirably high level of surfactants and/or solvents, which sometimes may not be acceptable to regulatory authorities (Fatouros et al. 2007; Gao and Morozowich 2005). Although amorphous materials possess crystal-like short-range order arrangements, they lack long-range translational oriental symmetry characteristics of the crystalline molecules. This makes amorphous systems inherently metastable or thermodynamically unstable. Due to this reason, these systems have a tendency to phase transform to their stable crystalline form during storage, usage, dissolution testing, and/or upon oral ingestion (Ivanisevic et al. 2009; Tobyn et al. 2009). Thus, crystalline reversion negates the solubility and dissolution advantages of the amorphous system. This is a major challenge for the pharmaceutical manufacturers to keep the drug in the amorphous form in the solid dispersion in order to meet the quality specifications and to ensure consistent in vivo response. This challenge is compounded for low-dose ASD such as tacrolimus (Prograf; Janssens and Van den Mooter 2009; Van den Mooter 2009) because it is difficult to monitor devitrification in the final dosage forms. It is a regulatory challenge to monitor and control such a pharmaceutical product since often the product meets the quality specifications at the time of release but later fails quality

M. Khan (✉) · Z. Rahman · A. Siddiqui · A. Gupta
Division of Product Quality and Research, Center of Drug Evaluation and Research,
U.S. Food and Drug Administration, MD, USA
e-mail: Mansoor.Khan@fda.hhs.gov

N. Shah et al. (eds.), *Amorphous Solid Dispersions*,
Advances in Delivery Science and Technology, DOI 10.1007/978-1-4939-1598-9_17

control tests such as dissolution, as the product ages during the storage or use. The reason for discrepancy in the quality control test results is due to the fact that the product may be fully amorphous or contain higher amorphous content when tested initially but its amorphous percentage changes with time as ASD products are thermodynamically unstable (Ivanisevic et al. 2009; Tobyn et al. 2009). Additionally, low-dose drugs may also have blending and content uniformity problems. This may make such drug products unsafe and inefficacious if not understood and processed well. In the past, various ASD products were recalled by FDA due to safety and efficacy issues. In 2012, A pharmaceutical company recalled its commercial tacrolimus capsule (generic version of Prograf) due to its failure to meet quality specification. Similarly, there were instances of lyophilized product recalled due to particles in the reconstituted product. These particles were the crystallized drug which did not dissolve on reconstitution (Guo et al. 2013; FDA recalls 2013). A structured approach to product development using quality by design (QbD) should result in a good quality product. QbD is a new product development paradigm where quality is built into the product rather than confirmed by quality control tests. It is all about understanding, monitoring, and controlling the factors that could affect the critical quality attributes (CQAs) of the product by utilizing novel technologies and mathematical tools (multivariate analysis; Rosencrance 2011). This chapter reviews regulatory aspects of the ASD.

17.2 NDA Versus ANDA

The application for a new drug product is called a new drug application (NDA) and submitted to the Office of New Drug Quality Assessment (ONDQA). Similarly, the application for generic drug product is called an abbreviated new drug application (ANDA) and submitted to the Office of Generic Drugs (OGD). A generic product is the equivalent or copy-cat version of the original drug product, reference listed drug (RLD) approved by the FDA. Generic drug manufacturers have to show that their product is pharmaceutically equivalent and bioequivalent, and hence therapeutically equivalent to the RLD. The generic products should also be appropriately labeled and manufactured in compliance with the current good manufacturing practices (cGMPs; FDA-ANDA 2013). Therapeutic equivalency information of generic drug products is available on electronic Orange Book (FDA-Orange Book 2013). The major differences between NDA (full) and ANDA (abbreviated) relate to preclinical and clinical data which are not required for the generic drug application. Since NDA has already established the safety and efficacy of the drug, the ANDA sponsor does not have to repeat these studies (FDA-NDA 2013; FDA-ANDA 2013). This is based on the assumption that when a therapeutically equivalent generic product is administered, it will be safe, effective, and bioavailable (rate and extent) as to the RLD. The data submission requirements of ANDAs and NDAs are the same beside differing requirement on preclinical and clinical data. These include chemistry, manufacturing, controls, testing, and labeling. It is the responsibility of the ANDA sponsors to demonstrate that their product meets the same quality standard as that of RLD. Drug application for RLD and generic products are submitted in common technical document (CTD) format as devised by the International Conference of Harmonization (ICH). CTD contains five modules and each module deals with the specific information of the application. Module 1 contains administrative and prescribing information such as label information, module 2 contains overviews and summaries, module 3 contains information on the manufacturing and quality control aspects of the drug products, module 4 contains nonclinical study data and module 5 contains clinical study reports (FDA-eCTD 2003).

17.3 Quality by Design

A pharmaceutical quality product is "a product free from contamination and reproducibly delivers the therapeutic benefits promised in the label to the consumer" (Woodcock 2004). The traditional approach of pharmaceutical development may be called quality by testing (QbT) in which qualities of pharmaceuticals are ensured by a battery of quality control tests. FDA ensures product quality by tightening the specifications which are based on the properties of the exhibit or clinical batches. This approach is based on the assumption that tighter specification will be able to detect differences among batches if there are changes in the formulation and/or processes parameters. Tighter specifications entail increased number of tests in order to meet the quality standard. However, FDA understood that increased testing does not necessarily increase or improve the quality of the product. In a newer approach for pharmaceutical development, FDA is emphasizing that quality should be built into the product starting at the product development stage rather than quality confirmed by testing at the end of manufacturing stage. This newer approach is called quality by design (QbD) as defined by ICH Q8 (R2) document "a systematic approach to development that begins with predefined objectives and emphasizes product and process understanding and process control, based on sound science and quality risk management"(ICH-Q8(R2) 2009). The overall objective of QbD concept is to develop quality product from the conception stage of the product development. Predefined quality of the product can be achieved and maintained throughout the life cycle of the product using QbD principle. This involves thorough understanding, monitoring, and controlling of drug substance, excipients, processes, and packaging system using QbD tools such as risk assessment, design of experiment, process analytical technology (PAT) and multivariate analysis. Additionally, enhanced knowledge about product quality and performance can be obtained using these tools that allow studying the range of material attributes, manufacturing methods, and process parameters. Knowledge gained through QbD helps in establishing design space, specifications, and manufacturing controls. Those aspects of drug substance, excipients, container closure systems, and manufacturing processes that are critical to product quality should be determined and controlled. Similarly, critical formulation and process parameters that may contribute to variation in the product quality and hence in vivo performance should be understood and controlled. As illustrated in the ICH-Q8(R2) document, the element of QbD include:

- Quality target product profile
- Risk assessment
- Critical quality attributes
  - Drug substance
  - Polymers and/or excipients
  - Amorphous/crystalline ratio
  - Dissolution
- Manufacturing process
- Quality risk management
- Design space
- Control strategy

Some of these elements and their subparts will be described in detail in the following sections.

17.3.1 Quality Target Product Profile

According to ICH-Q8 (R2), QTPP is "a prospective summary of the quality characteristics of a drug product that ideally will be achieved to ensure the desired quality, taking into account safety and efficacy of the drug product" (ICH-Q8(R2) 2009). It is an essential element of QbD that elaborates the design criteria of an intended drug product. Furthermore, it forms the basis for development of CQA, critical process parameter (CPP) and control strategy. It also forms the basis of strategies to be adopted to ensure quality, safety, and efficacy of the drug product. Following are the QTPP as per ICH-Q8(R2): clinical use, route of administration, dosage form, delivery systems, dosage strength, container-closure system, therapeutic moiety release or delivery, attributes affecting pharmacokinetic characteristics (e.g., dissolution, aerodynamic performance) and drug product quality criteria (e.g., sterility, purity, stability, and drug release) appropriate for the intended marketed product. Basically, QTPP lays the foundation of product and process design and optimization. It also includes patient-relevant product performance characteristics such as assay, dissolution, amorphous/crystalline (A/C) ratio, content uniformity, impurities and stability profile, etc. Interestingly, QbD implementation difference on NDA and ANDA products are visible at this stage. QTPP of NDA is usually determined before or during the drug product development progress. On the other hand, QTPP has already been established from the characterization, labeling, and clinical data of the RLD for the ANDA. Moreover, it is expected that generic product would have same QTPP as that of RLD but may implement and achieve it using different formulation or design approaches.

17.3.2 Critical Quality Attributes

Critical quality attribute (CQA) according to ICH-Q8(R2) document is a "physical, chemical, biological, or microbiological characteristic that should be within the appropriate limit range or distribution to ensure the desired product quality" (ICH-Q8(R2) 2009). Identification of CQAs is done through quality risk assessment which is the part of quality risk management as outlined in ICH-Q9 document (ICH-Q9 2009). Prior product knowledge is also considered in the risk assessment and forms the basis of linking CQAs to its safety and efficacy. Based on quality risk assessment, CQAs may be arranged in order of their importance. CQAs identify and link critical materials attribute (CMAs) and process parameters to the QTPP of the drug product. It covers both aspects of product performance and determinants of product performance. In general, CQAs are considered to be the attributes of the drug substance, excipients (polymers), and final drug products but may also include CQA of the drug product intermediates. The CQAs of a drug product may include those characteristics that affect purity, strength, drug release, and stability, e.g., assay, impurity profile, accelerated stability, dissolution, and A/C ratio, etc. More specifically, for ASD, important CQAs are dissolution and A/C ratio. These CQAs may also be important for intermediate ASD in order to control and maintain quality of final product over its intended shelf life. Intermediates ASD could be primary granules containing a solid dispersion of drugs and polymers and/or excipients before mixing with other ingredients of the formulation such as diluents, lubricants, and glidants, etc. to formulate into tablet/capsule.

17.3.2.1 Drug Substance

Drug substance properties should be carefully evaluated. These property will dictate the excipients, method, and process selection for ASD product. The properties to be considered are solubility in organic and aqueous solvents, miscibility with polymers, melting point, particle size, and thermal stability. These properties determine manufacturability, product performance, and long-term stability. For example, hot melt extrusion cannot be used for thermally labile drug molecules (Forster et al. 2001; Vasconcelos et al. 2007; Leuner and Dressman 2000). On the other hand, for spray drying process, drug solubility determines the selection of the solvent as well as inlet process temperature (Vasconcelos et al. 2007; Leuner and Dressman 2000).

17.3.2.2 Polymers and/or Excipients

Polymers or matrix formers are the major ingredients of ASD formulations and should meet regulatory requirements. They should be food or pharmaceutical-grade materials and include materials in the "Generally Regarded As Safe" (GRAS) category. List of safe excipients/polymers and their percentage level that can be safely used in the formulation are found in FDA inactive ingredient database (FDA-IIG

Table 17.1 Commercial amorphous solid dispersion products

Commercial name	Polymer	Drug	Dosage form	Manufacturers
Kaletra®	Copovidone	lopinavir/ritonavir	Tablet	Abbott Laboratories
Gris-PEG®	Polyethylene glycol	Griseofulvin	Tablet	Pedinol Pharmacal Inc.
Cesamet™	Polyvinyl pyrollidone	Nabilone	Capsule	Valeant Pharmaceuticals
Sporanox®	Hydroxypropylmethyl cellulose	Itraconazole	Capsule	Janssen Pharmaceutica
Intelence®	Hydroxypropylmethyl cellulose	Etravirine	Tablet	Tibotec Pharmaceuticals
Certican®	Hydroxypropylmethyl cellulose	Everolimus	Tablet	Novartis
Isoptin SR-E	Hydroxypropylmethyl cellulose-Hydroxypropyl cellulose	Verapamil	Tablet	Abbott
Prograf®	Hydroxypropylmethyl cellulose	Tacrolimus	Capsule	Astellas

list 2013). The polymers used in the commercial ASD products (Table 17.1) includes copovidone (Chandwani and Shuter 2008; Janssens and Van den Mooter 2009; FDA-Kaletra® 2013), polyethylene glycol (Janssens and Van den Mooter 2009; FDA-Gris-PEG® 2013), polyvinyl pyrollidone (PVP; Janssens and Van den Mooter 2009; FDA-Cesamet™ 2013), hydroxypropylmethyl cellulose (HPMC; Janssens and Van den Mooter 2009; Vajna et al. 2011; FDA-Intelence® 2013; FDA-Sporonox® 2013; EMA Assessment Report 2011; FDA-Prograf® 2013) and hydroxypropyl cellulose (HPC; Vajna et al. 2011; Janssens and Van den Mooter 2009). The selection of the polymer depends also on physicochemical properties of the drug, manufacturing process, and its manufacturability as it is one of the determinant of CQAs of the ASD. The properties of a polymer to be evaluated as a component of the ASD formulation are polymer type, molecular weight, polydispersity, concentration or amount, number of the polymer in the formulation, melting point and/or glass transition temperature (T_g), extent of miscibility with the drug, solvent solubility, particle size, hygroscopicity, compatibility with the drug and other excipients of the formulation, presence or absence of the intermolecular interactions (chemistry of the polymer), mechanical properties and chemical stability (Konno and Taylor 2006; Ingkatawornwong et al. 2001; Van Eerdenbrugh and Taylor 2010; Padden et al. 2011).

Polymer conformations have shown to influence the ASD performance (Al-Obaidi et al. 2009; Six et al. 2004). Similarly, molecular weight of the polymer influences the solubility, dissolution (rate and extent), physical and chemical stability (Wu et al.

2011). Correlation is reported between decreasing molecular weight of the polymer and increasing dissolution rate of the drug. This is related to polymer swelling, viscosity, and diffusion layer thickness (Tantishaiyakul et al. 1999; Hilton and Summers 1986). Higher dissolution rate would be observed in lower molecular weight polymer due to thin diffusion layer whereas higher molecular weight give lower dissolution rate which provide thick diffusion layer around the particles which act as a physical barrier. On the contrary, higher molecular weight polymer prevents the physical transformation to stable crystalline form during dissolution testing (Kogermann et al. 2013). Similarly, drug–polymer miscibility characteristics are very important as it will impact devitrification propensity of the system. The probability of crystallization or phase separation of the drug is lower in a miscible system as compared to a partially miscible or immiscible system (Ivanisevic 2010; Qian et al. 2010). Miscibility also depends upon intermolecular interactions between the drug and polymers. Additionally, these intermolecular interactions further increase the physical stability of amorphous systems (Tobyn et al. 2009). The propensity of the material to form a miscible solid dispersion could be predicted from the chemical structure of the drug and polymers/excipients. Other factors affecting miscibility are temperature and presence of other ingredients in the formulation including water. Miscible systems are indicated by a single transition in the DSC studies but there are reports in the literature of the phase-separated system showing single thermal transition (Lodge et al. 2006; Krause and Iskandar 1978). It is important to confirm miscibility by other analytical tools such as powder X-ray diffraction and computational models (Ivanisevic et al. 2009; Newman et al. 2008). Most commonly used polymers for the ASD are hydrophilic and hygroscopic. Hygroscopicity of the polymer affects the long-term stability of the ASD, e.g., ASD of felodipine with hydroxypropylmethyl cellulose acetate succinate (HPMCAS) is more stable than PVP. This is probably due to higher hygroscopicity of PVP, which increases the molecular mobility and hence crystallization (Rumondor et al. 2009). Water acts as a plasticizing agent (Hancock and Zografi 1994; Zhang and Zografi 2001) and changes the driving force of crystallization. It also competes for H-bonding and intermolecular interaction between the drug and polymers/excipients (Marsac et al. 2008).

Drug to polymer ratio is selected based on polymer properties and manufacturing method. The selected ratio should be convenient to process and allow for the intermediate to be processed into tablet or capsule dosage form. The biggest challenge of a solid dispersion is to maintain the drug in the amorphous form. This can be achieved to some extent by using low drug and high polymer levels. On the other hand, if selected polymers/excipients promote chemical interactions, its high level will produce extensive chemical interactions, poor stability, and degradations. Thus, balance must be maintained to produce a physically and chemically stable system, as thermodynamics of crystallization/destabilization driving forces depend on the drug loading, drug–polymer solubility and miscibility, and its T_g (Marsac et al. 2006a, b).

Amorphous drugs are thermodynamically unstable and revert to stable crystalline form. They are characterized by unique thermal event called glass transition temperature, T_g. These systems exhibit higher molecular mobility above their T_g. One method of increasing physical stability of an amorphous system is to increase the T_g of the drug by incorporating high T_g polymers/excipients. At temperatures below T_g, amorphous systems have low molecular mobility and thus a lower chance of molecular reorientation, generation of crystal nuclei, and growth. In general, crystallization involves grouping of critical mass of the drug, alignment, and reorientation to form crystal lattice (Marsac et al. 2006a). Mixing of drugs with polymers prevents and/or decreases the crystallization by one or more of the following mechanism: increase T_g of the drug, decrease the chemical potential of the drug (Hancock et al. 1995), increase activation energy for crystallization (Marsac et al. 2006b), provide kinetic barrier against crystallization (Crowley and Zografi 2003), and/or impede the transport of the drug to the crystallization phase (Marsac et al. 2006a). These mechanisms influence the driving force for drug crystallization. It is proposed that if the storage temperature is at least 50 K below the T_g, amorphous phase will be stable for many years (Newman et al. 2012; Hancock et al. 1995). However, it is difficult to predict stability of ASD based on T_g alone, e.g., amorphous indomethacin is more stable than amorphous phenobarbital, although both have same T_g (Fukuoka et al. 1989) and similar behavior is shown by amorphous felodipine which is more stable than amorphous nifedipine (Marsac et al. 2006b). Another very important factor to consider in predicting the devitrification potential of the ASD is the chemistry of the polymer such as the type and number of the chemical group capable of forming ionic and/or hydrogen bonds with drugs. Polymers/excipients and drugs could interact by ionic bond (between acid and base, Yoo et al. 2009), hydrogen bond, van der Waals forces, and k–k stacking (Ghebremeskel et al. 2006). The intermolecular interactions between the components are strong enough to overcome crystal lattice energy to ensure stable and homogeneous system. For example, PVP is a commonly used polymer due to its high glass transition temperature and ability to form hydrogen bond, which prevents drug crystallization in the ASD (Taylor and Zografi 1997). Based on the same principle, co-amorphous approach has been proposed to increase the physical stability in which two drugs or a drug and small molecule are co-processed to produce an amorphous system and stabilized by the intermolecular interactions (Chieng et al. 2009; Lobmann et al. 2011).

Bicomponent versus multicomponent ASD also influence physical stability. Multicomponent offers advantage in terms of enthalpy by maximizing the chance of the intermolecular interactions (hydrogen bonds, ion–ion, and ion–dipole forces) and their intensity. It also lowers entropy, thus improving the physical stability. The mechanism of the improvement in the physical stability involves anti-plasticizing effect, which is indicated by an increase in T_g, and intermolecular interactions. Another advantage of multicomponent ASD is improvement of solubility/dissolution as one polymer stabilizes the solid state and the other prevents solution-mediated crystallization during dissolution or in gastrointestinal fluid (Sakurai et al. 2012a) or processability. The selection of additional component is based on the assessment of individual and binary component. The other component could be another polymer or excipient (surfactants, wetting agents, organic acid, organic base or diluents; Ghebremeskel et al. 2006; Sakurai et al. 2012a; Huang et al. 2008; Al-Obaidi et al. 2011). One of the requirements of multicomponent ASD ingredients is mutual compatibility. They should neither promote drug degradation nor increase impurity generation that will impact the product safety and/or efficacy. The commonly used surfactants in the multicomponent ASD are Tween® 80, Span® 80, vitamin E polyethylene glycol 1000 succinate (D-α-tocopheryl polyethylene glycol 1000 succinate), or Cremophor® (Shamblin et al. 1998; Newman et al. 2008).

17.3.2.3 Amorphous/Crystalline Ratio

Superior in vivo performance of the ASD drug product is due to the presence of a completely or partially amorphous drug in these products when compared to its crystalline drug product. Furthermore, in vivo performance of the ASD may be related to A/C ratio, and maintaining that ratio throughout its shelf life would ensure consistent pharmacological response. Therefore, it is very important to understand the formulation and process parameters that could possibly change A/C ratio in the final product. At the same time, monitoring of this ratio is also critical during product development as it can point out what formulation and/or process factors need to change and control. Similarly, post approval monitoring of the ratio is also important as it can predict when product becomes unsafe/inefficacious to use. The A/C ratio monitoring and measurement should be done using appropriate analytical tools. This is one of the specification of ASD-based products in NDA/ANDA submission to the FDA as a measure of safety and efficacy of the product. Powder X-ray diffraction is the most commonly used technique because it is most definitive in identifying and providing a quantitative method to measure the crystalline drug in the amorphous system. The technique is very simple and nondestructive, with the ability to detect crystallinity at the level as low as 5 %. Crystalline material will show strong diffractions corresponding to the molecular arrangement in the crystalline lattice whereas amorphous material shows an amorphous halo and diffuse diffraction pattern due to lack of crystalline order at the molecular level (Shah et al. 2006). In case of the ASD intermediate product, it is mixed with other ingredients of the formulation, and there is a dilution in the final drug product leading to challenges in the determination of the A/C ratio, especially for the low-dose drug. Another problem arises due to overlapping of the crystalline peaks from the other formulation components that may interfere with the drug peaks. In such cases, it may be easier to determine the A/C ratio in the intermediate than the final product or to use other techniques of spectroscopy such as Fourier infrared, near infrared, or Raman spectroscopy in conjunction with chemometric methods such as principle component analysis and partial least square analysis, which may amplify and separate the peaks of the drug from the excipients (Zidan et al. 2012).

17.3.2.4 Dissolution

Dissolution is a very important CQA of the ASD and can be used in setting QTPP of the product. Although dissolution does not simulate the actual in vivo conditions the dosage form would encounter after oral ingestion, it can be used as an in vitro quality control test with an indirect link to the clinical performance. Change in the dissolution rate and extent indicates a change in the product during the shelf life or change in the formulation and/or process parameters. It is a good indicator of in vivo performance of ASD product. For example, dissolution behavior could predict the potential of amorphous drug transformation under in vivo conditions. There is always a risk of drug precipitation/crystallization due to the supersaturation phenomenon exhibited by an amorphous drug leading to a portion of the drug not being bioavailable. This is related to intraluminal supersaturation of the drug. Suitably formulated ASD should maintain the intraluminal supersaturated concentration for sufficient time so that sufficient epithelial absorption can take place (Hancock 2002; Hancock and Parks 2000; Bikiaris 2011).

Dissolution method of the ASD product should have discriminating ability. Dissolution methods commonly used for solid dispersions are USP apparatus 1 (paddle method) and apparatus 2 (basket method). Apparatus 3 (reciprocating cylinder) and apparatus 4 (flow through cell) can also be used as they provide sink conditions for low-solubility drugs and have a provision to change the pH of the media during the dissolution to simulate gastrointestinal conditions (Newman et al. 2012). In general, the agitation speed used is 50 rpm for paddle method and 100 rpm for basket method. Lower or higher agitation speed can be used as long as it can discriminate the effect of formulation, process, manufacturing variables, and change in amorphous/crystalline proportion of the drug in the ASD product. Furthermore, dissolution is usually conducted under sink condition, which is defined as three times the volume of dissolution media required to achieve saturation of the drug (USP34-NF29 2006). Another way of selecting the dissolution media type and volume is by dose–solubility ratio. It is dose of the drug soluble in 250 ml of media (Xia et al. 2010). It is also possible to run dissolution in non-sink condition when the drug is poorly soluble since solid dispersion provides supersaturated condition but the dissolution media should be able to maintain the supersaturation for sufficiently long time to allow for the analysis of the dissolution samples. The dissolution media used should be aqueous-based and biorelevant, simulating the physiological pH range of 1.2–7.5. Additionally, performing dissolution in biorelevant media helps in justifying dissolution conditions to the regulatory agency. The most commonly used dissolution media are 0.1 N HCl, water, simulated gastric fluid without enzyme, simulated intestinal fluid without enzyme, and pH 6.8 buffer (Jantratid et al. 2009; Newman et al. 2012). It is also recommended to run the dissolution in two media that can predict in vivo performance of ASD, most preferably in simulated gastric and simulated intestinal fluid (Gao et al. 2010; Polster et al. 2010). However, when possible, surfactants and organic solvents-based dissolution media should be avoided as they would diminish the discriminating ability of the dissolution method. For example, FDA recommends aqueous media adjusted to pH 4.5 containing 0.005 % w/v HPC for the tacrolimus solid dispersion capsule (FDA-Dissolution method 2013). This dissolution media is able to detect increase in the crystalline portion in the product with age/storage conditions as reflected by decrease in the percent drug dissolved. When this dissolution media is used with surfactant, it loses its discriminating ability (Zidan et al. 2012). If dissolution cannot be conducted without a surfactant, the surfactant level should be kept to bare minimum to keep the drug in solution and to allow for its analysis. At the same time, the dissolution method should still be able to detect the changes in the product due to process, formulation, or storage conditions. Attention should also be paid to solubility characteristics of the polymer in the dissolution media. The selected media should not only support the dissolution of the drug but also the dissolution and/or swelling of the polymer of the ASD product, otherwise complete dissolution of the drug will not take place, leading to misinterpretation of the results. For example, ASD-containing pH-dependent polymer such as Eudragit® E100 or HPMCAS or hydroxypropylmethyl cellulose phthalate (HPMCP) are insoluble in gastric pH. For such products, dissolution should be conducted in alkaline pH (Newman et al. 2012). Volume of the media could be 500–900 ml or lower depending upon the properties of ASD. It is also important to study the effect of media composition, volume, and stirring speed to get product profiles as dissolution parameters change. The physical characteristics of the ASD product such as gel formation, floating of the dispersion or particles on the top of the dissolution vessels, and particles clinging to shaft or vessels should also be investigated early in the development of the dissolution method (Puri et al. 2011).

Dissolution of ASD product involves various simultaneous processes such as dissolution of the amorphous drug, nucleation, and crystal growth of stable form of the drug and polymer dissolution. Crystallization of amorphous drugs could take place during in vitro and/or in vivo dissolution (Alonzo et al. 2010; Greco and Bogner 2012). Crystallization of metastable amorphous drugs could be from solid state, solution-mediated or simultaneously by both methods (Greco and Bogner 2012). In the solid-state devitrification, dissolution media act as a plasticizing agent that increases the mobility of molecules by decreasing T_g of the product which is the prerequisite for crystallization (Zhang et al. 2004). In such cases where crystallization is faster than dissolution, super-saturation would not be achieved and dissolution profile looks similar to its crystalline counterpart. In the solution-mediated crystallization, supersaturated solution produced during dissolution acts as a driving force for crystallization (Greco and Bogner 2012). The dissolved drug may crystallize on the surface of the matrix from supersaturated solution, e.g., crystallization of amorphous carbamazepine into carbamazepine dihydrate crystals (Savolainen et al. 2009). Crystal growth from solution-meditated crystallization consists of two steps: (1) diffusion of the drug from the supersaturated solution to crystal interface and (2) integration of the molecule into the crystal lattice which is accompanied by desolvation of the dissolution media. Several factors influence extent and kinetics of this transformation such as degree of solubility and dissolution enhancement of amorphous form compared to crystalline form and properties of the molecules under consideration (Greco and Bogner 2012). These transformations may have significant impact on the product performance as the crystallized drug present as thin layer on the surface or inside the matrix, would dramatically alter the dissolution profile of the product (Zhang et al. 2004). Ideally, stable amorphous system should be able to maintain supersaturated condition for long enough time for effective absorption to take place. Otherwise, the amorphous phenomenon advantage would be lost and the product would behave like a crystalline drug product.

Crystallization in the dissolution vessels could be prevented by various methods, e.g., increasing agitation speed, dissolution in large media volume, and adding polymers and/or surfactants in the dissolution media or a combination of these methods. Surfactants and polymers should be used at their lowest possible level and should not affect the drug solubility and/or dissolution behavior of the product, e.g., hydroxypropylmethyl cellulose (HPMC) prevented the crystallization of amorphous 9,3-diacetylmidecamycin during dissolution testing (Sato 1981). Since crystallization of drugs occurs at the surface through plasticizing effect of the dissolution media, polymers inhibit or delay surface crystallization by interacting with the surface (Wu et al. 2007). Furthermore, the polymer type also determines its ability to prevent crystallization from the supersaturated solution. For example, HPMC is more effective than polyvinyl pyrrolidone (PVP) in preventing the crystallization of amorphous indomethacin and felodipine during dissolution testing because HPMC acts both as the solid state and super saturated state stabilizer (Alonzo et al. 2010).

17.3.3 Manufacturing Process

The selected method and process critically influence the ASD CQAs. Commonly used methods and processes for the ASD are solvent evaporation, precipitation, milling, freeze drying, fluid bed drying, spray drying, hot melt extrusion, and recently introduced solvent co-precipitation variant called microprecipitated bulk powder "MBP" technology. Selected manufacturing method and process parameters will impact the characteristics of the final product, ultimately influencing the in vitro and in vivo performance. For example, physical stability of the troglitazone solid dispersion produced by solvent evaporation and milling method is due to differences in short range order (Alonzo et al. 2010). Similarly, rapid freezing in the freeze-drying method favors the formation of an amorphous drug while introducing annealing step promotes the formation of a crystalline drug (Pikal 1994). Process and method affect the morphology, performance, and stability. For example, spray drying process produces bulky product while hot melt extrusion produces dense product (Van den Mooter 2009). On the other hand, hot melt extrudate ASD have low moisture content and thus have better physical and chemical stability (Sakurai et al. 2012b). It is critically important to fully understand and control process and method variables by utilizing sophisticated and online tools such as near infrared and Raman spectroscopy.

Spray drying and fluid bed processes are based on solvent evaporation method. The main requirement for solvent evaporation method is the property of the solvent to dissolve the drug and polymers to produce single phase. The solvent should preferably be ICH class III solvent because of its low toxic potential (ICH-Q3(R5) 2011). Solvent type and volume (percentage solid) also influence the performance of the ASD (Al-Hamidi et al. 2010). The most critical processing parameters (CPP) are drying temperature, nozzle configuration, nitrogen/air flow rate, pump speed, temperature, and air volume (Marsac et al. 2010; Patterson et al. 2005; Wu et al. 2011). The most critical factors affecting the amorphous to crystalline ratio in the final product is the solvent evaporation rate. Solvent evaporation rate should be faster in order to prevent drug crystallization. In other word, drying time should be less than crystallization time (Wu et al. 2011; Leuner and Dressman 2000; Van den Mooter 2012). Shorter drying time prevents molecules to arrange into crystal lattice. The major advantage of solvent evaporation-based processes over hot melt extrusion processes is their ability to process a heat labile drug, and they can utilize high T_g or melting point polymers without compromising their heat lability. Since solvent evaporation-based processes use solvent and even a trace amount of the residual solvent in the final product can itself act as a plasticizer, increasing the drug mobility by decreasing its T_g and promote physical instability (Van den Mooter 2009), so it is very critical to reduce the solvent level in the ASD product to lowest possible level to maximize the physical stability.

Some advantages of the hot melt extrusion process include fast, simple, solvent free, continuous, and scalable operation. The first step is the selection of a heat-stable drug and polymer for the process. In an ideal situation, the drug should dissolve or be miscible in the polymer, not degrade in combination with it and/or other excipients and on exposure to high temperature. At the same time, the polymer should minimize the molecular mobility of the drug in the extruded product. Additionally, the polymer selected should be thermoplastic and have high fragility (Crowley et al. 2007; Van den Mooter 2009; Rauwendaal 1986). It is desirable that a polymer must have higher T_g than API to facilitate miscibility of a drug into a polymer. Plasticizers are commonly used in the hot melt extrusion process to improve thermal processability, modify drug release, and improve mechanical properties and surface appearance (Repka and McGinity 2000). When used in combination with a polymer, plasticizer improves the workability and flexibility of the polymer by increasing intermolecular separation of the molecules (Wang et al. 1997), and allows lower thermal processing (Zhu et al. 2002). Hot melt processing is usually carried out above T_g or melting temperature of the polymer to soften and lower melt viscosity so that it can easily flow through the extruder. Addition of plasticizer lowers the T_g of the polymer due to intermolecular interaction with the other molecules and allows processing at lower temperature (Repka and McGinity 2000). Surfactants are commonly used as plasticizers in the hot melt extrusion. Some examples include Tween® 80, docusate sodium, vitamin E polyethylene glycol 1000 succinate (D-α-tocopheryl polyethylene glycol 1000 succinate), or Cremophor® (Repka and McGinity 2000; Ghebremeskel et al. 2006; Shamblin et al. 1998; Newman et al. 2008). Drug to polymer homogeneity in the extrudate is a critical attribute that affects not only performance but also stability of the system. Nonhomogeneous system with a drug or polymer-rich region could lead to instability and variability in the product performance. T_g provides an indication of homogeneity or non-homogeneity of the solid dispersion (Guo et al. 2013).

Recently, solvent co-precipitation MBP technology was successfully utilized to prepare a stable ASD of a brick-dust drug to provide significantly higher exposure in humans. MBP technology is reported to be useful when conventional technologies such as hot melt extrusion and spray drying technologies are not suitable to form solid dispersions due to the very high melting point and very low solubility in low-boiling solvents. In this MBP process, drugs and ionic polymers (e.g., HPMCAS,

HPMCP, Eudragit) are dissolved in a super solvent such as dimethyl acetamide (DMA). The DMA solution is then introduced into acidic solution such as 0.01 N HCl maintained at 2–5 °C. The resulting precipitate is filtered, washed with cold dilute acid followed by cold water to remove DMA. The wet precipitate is dried to provide the ASD. Overall, the drug is precipitated in an ionic polymer in an amorphous form (microprecipitated bulk powder, MBP) at the nano-size or molecular level, thus significantly enhancing its bioavailability compared with micronized form of the drug. The stabilization of the amorphous form is imparted by the ionic nature of the polymers due to their high molecular weight and high glass transition temperature (Shah et al. 2012).

17.3.4 Design Space

According to ICH Q8(R2) document, design space is "a multidimensional combination and interaction of input variables (e.g., material attributes) and process parameters that have been demonstrated to provide assurance of quality" (ICH-Q8(R2) 2009). It is proposed by a sponsor and subject to regulatory review and approval by FDA. Working within design space is not considered a change because product will meet the defined quality. However, moving out of the design space is considered a change and initiates regulatory post approval process. The wider the designs space is, the more robust the product would be, as it would accommodate wider variation in the process and/or formulation parameters. Risk assessment, multivariate experimental design, literature, and prior experience/knowledge contribute in defining the design space. Design space of the product could be defined by range of material attributes and process parameters or through complex mathematical relationships. The process parameters studied should be critical process parameters (CPPs) that have significant effect on the CQAs. On the other hand, material attributes studied should be critical attributes of the drug substance (particle size, polymorphs, impurity, etc.) and excipients (moisture level, particle size, molecular weight, etc.) that would directly/indirectly affect the CQAs of the ASD. CPPs help in defining and controlling the design space. Sponsor could propose multiple design spaces of individual unit operation or a single design space encompassing multiple unit operations. Practically, it may be easier to develop and control design space of individual unit operations of a multistep process, and this approach would also provide greater operational flexibility.

17.3.5 Control Strategy

The objective of control strategy is to make sure that the product of required quality is produced consistently. The control strategy includes controls of in-process variables, input materials (drug substance and excipients), intermediates (in-process materials), packaging system, and drug product attributes. These controls should be based on the sound understanding of product formulation and process and should include the control of CPPs and CQA.

17.4 Conclusion

Introduction of QbD principles as laid out in the ICH documents Q8, Q8 (R2) and Q9 allow for a rational product development with well-controlled product intermediate and final product quality in the QbD paradigm. Solid dispersion products are highly amenable to the utilization of modern technologies with respect to the drug crystalline reversion and content uniformity throughout the shelf life. Post-marketing changes and product failures with recalls indicate that a thorough understanding of the product and process with validated and well-controlled analytical procedures are needed. The analytical methods should include methods with a discriminating ability to monitor crystalline reversion and the resultant lowering of dissolution upon storage, particularly in a high-temperature and humidity environment.

Disclaimer The views and opinions expressed in this chapter are only those of the authors and do not necessarily reflect the views or policies of the FDA.

References

Al-Obaidi H, Brocchini S, Buckton G (2009) Anomalous properties of spray dried solid dispersions. J Pharm Sci 98(12):4724–4737

Al-Hamidi H, Edwards AA, Mohammad MA, Nokhodchi A (2010) To enhance dissolution rate of poorly water-soluble drugs: glucosamine hydrochloride as a potential carrier in solid dispersion formulations. Colloids Surf B Biointerfaces 76(1):170–178

Al-Obaidi H, Ke P, Brocchini S, Buckton G (2011) Characterization and stability of ternary solid dispersions with PVP and PHPMA. Int J Pharm 419(1–2):20–27

Alonzo D, Zhang G, Zhou D, Gao Y, Taylor L (2010) Understanding the behavior of amorphous pharmaceutical systems during dissolution. Pharm Res 27(4):608–618

Bikiaris DN (2011) Solid dispersions, part I: recent evolutions and future opportunities in manufacturing methods for dissolution rate enhancement of poorly water-soluble drugs. Expert Opin Drug Deliv 8(11):1501–1519

Chandwani A, Shuter J (2008) Lopinavir/ritonavir in the treatment of HIV-1 infection: a review. Ther Clin Risk Manage 4(5):1023–1033

Chieng N, Aaltonen J, Saville D, Rades T (2009) Physical characterization and stability of amorphous indomethacin and ranitidine hydrochloride binary systems prepared by mechanical activation. Eur J Pharm Biopharm 71(1):47–54

Crowley K, Zografi G (2003) The effect of low concentrations of molecularly dispersed poly(vinylpyrrolidone) on indomethacin crystallization from the amorphous state. Pharm Res 20(9):1417–1422

Crowley MM, Zhang F, Repka MA, Thumma S, Upadhye SB, Kumar Battu S, Mcginity JW, Martin C (2007) Pharmaceutical applications of hot-melt extrusion: part I. Drug Dev Ind Pharm 33(9):909–926

EMA Assessment Report (2011) EMA/646111/2011 committee for medicinal products for human use (CHMP). http://www.ema.europa.eu/docs/en_GB/document_library/EPAR_-_Public_assessment_report/human/002311/WC500112240.pdf. Accessed 16 Nov 2013

Fatouros DG, Karpf DM, Nielsen FS, Mullertz A (2007) Clinical studies with oral lipid based formulations of poorly soluble compounds. Ther Clin Risk Manage 3(4):591–604

FDA-ANDA (2013) Food and Drug Administration, Abbreviated New Drug Application (ANDA): generics. http://www.fda.gov/Drugs/DevelopmentApprovalProcess/HowDrugsareDevelopedandApproved/ApprovalApplications/AbbreviatedNewDrugApplicationANDAGenerics/. Accessed 16 Nov 2013

FDA-Cesamet™ (2013) Food and Drug Administration, Cesamet™ Capsule label. http://www.accessdata.fda.gov/drugsatfda_docs/label/2006/018677s011lbl.pdf. Accessed 16 Nov 2013

FDA-Dissolution method (2013) Food and Drug Administration, Tacrolimus capsule. http://www.accessdata.fda.gov/scripts/cder/dissolution/dsp_SearchResults_Dissolutions.cfm. Accessed 16 Nov 2013

FDA-eCTD (2003) Food and Drug Administration, Guidance for industry M2 eCTD: electronic common technical document specification, Center for Drug Evaluation and Research, Center for Biologics Evaluation and Research. http://www.fda.gov/downloads/RegulatoryInformation/Guidances/UCM129624.pdf. Accessed 16 Nov 2013

FDA-Gris-PEG® (2013) Food and Drug Administration, Gris-PEG® Tablet label. http://www.accessdata.fda.gov/drugsatfda_docs/label/2010/050475s054lbl.pdf. Accessed 16 Nov 2013

FDA-IIG list (2013) Food and Drug Administration, Inactive ingredient search for approved drug products. http://www.accessdata.fda.gov/scripts/cder/iig/. Accessed 16 Nov 2013

FDA-Intelence® (2013) Food and Drug Administration, Intelence® Tablet label. http://www.accessdata.fda.gov/drugsatfda_docs/label/2013/022187s011s012s014lbl.pdf. Accessed 16 Nov 2013

FDA-Kaletra® (2013) Food and Drug Administration, Kaletra® Tablet Label. http://www.accessdata.fda.gov/drugsatfda_docs/label/2013/021251s045,021906s038lbl.pdf Accessed 16 Nov 2013

FDA-NDA (2013) Food and Drug Administration, New Drug Application (NDA). http://www.fda.gov/Drugs/DevelopmentApprovalProcess/HowDrugsareDevelopedandApproved/ApprovalApplications/NewDrugApplicationNDA/. Accessed 16 Nov 2013

FDA-Orange Book (2013) Food and Drug Administration, Orange Book: Approved Drug Products with therapeutic equivalence evaluations. http://www.accessdata.fda.gov/scripts/cder/ob/default.cfm. Accessed 16 Nov 2013

FDA-Prograf® (2013) Food and Drug Administration, Prograf® Capsule label. http://www.accessdata.fda.gov/drugsatfda_docs/label/2013/050708s043,050709s036lbl.pdf. Accessed 16 Nov 2013

FDA-Recalls (2013) Food and Drug Administration, Archive for recalls, market withdrawals & safety alerts. http://www.fda.gov/Safety/Recalls/ArchiveRecalls/default.htm. Accessed 16 Nov 2013

FDA-Sporonox® (2013) Food and Drug Administration, Sporonox® Capsule label. http://www.accessdata.fda.gov/drugsatfda_docs/label/2012/020083s048s049s050lbl.pdf. Accessed 16 Nov 2013

Forster A, Hempenstall J, Tucker I, Rades T (2001) The potential of small-scale fusion experiments and the gordon-taylor equation to predict the suitability of drug/polymer blends for melt extrusion. Drug Dev Ind Pharm 27(6):549–560

Fukuoka E, Makita M, Nakamura Y (1989) Glassy state of pharmaceuticals. IV: studies on glassy pharmaceuticals by thermomechanical analysis. Chem Pharma Bull 37(10):2782–2785

Gao P, Morozowich W (2005) Development of supersaturatable self-emulsifying drug delivery system formulations for improving the oral absorption of poorly soluble drugs. Expert Opin Drug Deliv 3(1):97–110

Gao Y, Carr RA, Spence JK, Wang WW, Turner TM, Lipari JM, Miller JM (2010) A pH-Dilution method for estimation of biorelevant drug solubility along the gastrointestinal tract: application to physiologically based pharmacokinetic modeling. Mol Pharm 7(5):1516–1526

Ghebremeskel A, Vemavarapu C, Lodaya M (2006) Use of surfactants as plasticizers in preparing solid dispersions of poorly soluble api: stability testing of selected solid dispersions. Pharm Res 23(8):1928–1936

Greco K, Bogner R (2012) Solution-mediated phase transformation: significance during dissolution and implications for bioavailability. J Pharm Sci 101(9):2996–3018

Guo Y, Shalaev E, Smith S (2013) Physical stability of pharmaceutical formulations: solid-state characterization of amorphous dispersions. Trends Anal Chem 49:137–144

Hancock BC (2002) Disordered drug delivery: destiny, dynamics and the Deborah number. J Pharm Pharmacol 54(6):737–746

Hancock B, Parks M (2000) What is the true solubility advantage for amorphous pharmaceuticals? Pharm Res 17(4):397–404

Hancock B, Zografi G (1994) The relationship between the glass transition temperature and the water content of amorphous pharmaceutical solids. Pharm Res 11(4):471–477

Hancock B, Shamblin S, Zografi G (1995) Molecular mobility of amorphous pharmaceutical solids below their glass transition temperatures. Pharm Res 12(6):799–806

Hilton JE, Summers MP (1986) The effect of wetting agents on the dissolution of indomethacin solid dispersion systems. Int J Pharm 31:157–164

Huang J, Wigent RJ, Schwartz JB (2008) Drug-polymer interaction and its significance on the physical stability of nifedipine amorphous dispersion in microparticles of an ammonio methacrylate copolymer and ethylcellulose binary blend. J Pharm Sci 97(1):251–262

ICH-Q8(R2) Guideline (2009) International conference of harmonization, tripartite guideline Pharmaceutical development. Q8 (R2)

ICH Q9 Guideline (2009) International conference of harmonization, tripartite guideline, Quality risk management Q9

ICH-Q3(R5) Guideline (2011) International conference of harmonization, tripartite guideline, impurities: guideline for residual solvents

Ingkatawornwong S, Kaewnopparat N, Tantishaiyakul V (2001) Studies on aging piroxicam-polyvinylpyrrolidone solid dispersions. Die Pharmazie 56(3):227–230

Ivanisevic I (2010) Physical stability studies of miscible amorphous solid dispersions. J Pharm Sci 99(9):4005–4012

Ivanisevic I, Bates S, Chen P (2009) Novel methods for the assessment of miscibility of amorphous drug-polymer dispersions. J Pharm Sci 98(9):3373–3386

Janssens S, Van Den Mooter G (2009) Review: physical chemistry of solid dispersions. J Pharm Pharmacol 61(12):1571–1586

Jantratid E, De Maio V, Ronda E, Mattavelli V, Vertzoni M, Dressman JB (2009) Application of biorelevant dissolution tests to the prediction of in vivo performance of diclofenac sodium from an oral modified-release pellet dosage form. Eur J Pharm Sci 37(3–4):434–441

Kogermann K, Penkina A, Predbannikova K, Jeeger K, Veski P, Rantanen J, Naelapaa K (2013) Dissolution testing of amorphous solid dispersions. Int J Pharm 444(1–2):40–46

Konno H, Taylor LS (2006) Influence of different polymers on the crystallization tendency of molecularly dispersed amorphous felodipine. J Pharm Sci 95(12):2692–2705

Krause S, Iskandar M (1978) Phase separation in styrene-methyl styrene block copolymers. In: Klempner D, Frisch K (eds), Polymer alloys. Springer, US, pp 231–243

Leuner C, Dressman J (2000) Improving drug solubility for oral delivery using solid dispersions. Eur J Pharm Biopharm 50(1):47–60

Lobmann K, Laitinen R, Grohganz H, Gordon KC, Strachan C, Rades T, (2011) Coamorphous drug systems: enhanced physical stability and dissolution rate of indomethacin and naproxen. Mol Pharm 8(5):1919–1928

Lodge TP, Wood ER, Haley JC (2006) Two calorimetric glass transitions do not necessarily indicate immiscibility: the case of PEO/PMMA. J Pol Sci Part B: Pol Phys 44(4):756–763

Marsac P, Konno H, Taylor L (2006a) A Comparison of the Physical stability of amorphous felodipine and nifedipine systems. Pharm Res 23(10):2306–2316

Marsac P, Shamblin S, Taylor L (2006b) Theoretical and practical approaches for prediction of druggçôpolymer miscibility and solubility. Pharm Res 23(10):2417–2426

Marsac P, Konno H, Rumondor A, Taylor L (2008) Recrystallization of nifedipine and felodipine from amorphous molecular level solid dispersions containing poly(vinylpyrrolidone) and sorbed water. Pharm Res 25(3):647–656

Marsac PJ, Rumondor ACF, Nivens DE, Kestur US, Stanciu L, Taylor LS (2010) Effect of temperature and moisture on the miscibility of amorphous dispersions of felodipine and poly(vinyl pyrrolidone). J Pharm Sci 99(1):169–185

Newman A, Engers D, Bates S, Ivanisevic I, Kelly RC, Zografi G (2008) Characterization of amorphous API: polymer mixtures using X-ray powder diffraction. J Pharm Sci 97(11):4840–4856

Newman A, Knipp G, Zografi G (2012) Assessing the performance of amorphous solid dispersions. J Pharm Sci 101(4):1355–1377

Padden BE, Miller JM, Robins T, Zocharski PD, Prasad L, Spence JK, LaFountaine J (2011) Amorphous solid dispersions as enabling formulations for discovery and early development. Am Pharm Rev 1(66):73

Patterson JE, James MB, Forster AH, Lancaster RW, Butler JM, Rades T (2005) The influence of thermal and mechanical preparative techniques on the amorphous state of four poorly soluble compounds. J Pharm Sci 94(9):1998–2012

Pikal MJ (1994). Freeze-drying of proteins. formulation and delivery of proteins and peptides. Am Chem Soc: pp 120–133

Polster CS, Atassi F, Wu SJ, Sperry DC (2010) Use of artificial stomach-duodenum model for investigation of dosing fluid effect on clinical trial variability. Mol Pharm 7(5):1533–1538

Puri V, Dantuluri AK, Bansal AK (2011) Investigation of atypical dissolution behavior of an encapsulated amorphous solid dispersion. J Pharm Sci 100(6):2460–2468

Qian F, Huang J, Hussain MA (2010). Drug-polymer solubility and miscibility: stability consideration and practical challenges in amorphous solid dispersion development. J Pharm Sci 99(7):2941–2947

Rauwendaal C (1986) Polymer extrusion. Munchen, Hanser Publishers pp 20–25

Repka MA, Mcginity JW (2000) Influence of Vitamin E TPGS on the properties of hydrophilic films produced by hot-melt extrusion. Int J Pharm 202(1–2):63–70

Rosencrance S (2011) QbD Status update generic drugs. http://www.fda.gov/downloads/Drugs/DevelopmentApprovalProcess/HowDrugsareDevelopedandApproved/ApprovalApplications/AbbreviatedNewDrugApplicationANDAGenerics/UCM292666.pdf. Accessed 16 Nov 2013

Rumondor A, Stanford L, Taylor L (2009) Effects of polymer type and storage relative humidity on the kinetics of felodipine crystallization from amorphous solid dispersions. Pharm Res 26(12):2599–2606

Sakurai A, Sakai T, Sako K, Maitani Y (2012a) Polymer combination increased both physical stability and oral absorption of solid dispersions containing a low glass transition temperature drug: physicochemical characterization and in vivo study. Chem Pharm Bull 60(4):459–464

Sakurai A, Sako K, Maitani Y (2012b) Influence of manufacturing factors on physical stability and solubility of solid dispersions containing a low glass transition temperature drug. Chem Pharm Bull 60(11):1366–1371

Sato T, Okada T, Sekiguchi K, Tsuda Y (1981) Difference in physicopharmaceutical properties between crystalline and noncrystalline 9, 3-diacetylmidecamycin. Chem Pharm Bull (29):2675–2682

Savolainen M, Kogermann K, Heinz A, Aaltonen J, Peltonen L, Strachan C, Yliruusi J (2009) Better understanding of dissolution behaviour of amorphous drugs by in situ solid-state analysis using Raman spectroscopy. Eur J Pharm Biopharm 71(1):71–79

Shah B, Kakumanu VK, Bansal AK (2006) Analytical techniques for quantification of amorphous/crystalline phases in pharmaceutical solids. J Pharm Sci 95(8):1641–1665

Shah N, Sandhu H, Phuapradit W, Pinal R, Iyer R, Albano A, Chatterji A, Anand S, Choi DS, Tang K, Tian H, Chokshi H, Singhal D, Malick W (2012) Development of novel microprecipitated bulk powder (MBP) technology for manufacturing stable amorphous formulations of poorly soluble drugs. Int J Pharm 438(1–2):53–60

Shamblin SL, Taylor LS, Zografi G (1998) Mixing behavior of colyophilized binary systems. J Pharm Sci 87(6):694–701

Six K, Verreck G, Peeters J, Brewster M, Mooter GVD (2004) Increased physical stability and improved dissolution properties of itraconazole, a class II drug, by solid dispersions that combine fast- and slow-dissolving polymers. J Pharm Sci 93(1):124–131

Tantishaiyakul V, Kaewnopparat N, Ingkataworwong S (1999) Properties of solid dispersions of piroxicam in polyvinylpyrrolidone. Int J Pharm 181(2):143–151

Taylor L, Zografi G (1997) Spectroscopic characterization of interactions between PVP and indomethacin in amorphous molecular dispersions. Pharm Res 14(12):1691–1698

Tobyn M, Brown J, Dennis AB, Fakes M, Gao Q, Gamble J, Khimyak YZ, Mcgeorge G, Patel C, Sinclair W, Timmins P, Yin S (2009) Amorphous drug-PVP dispersions: application of theoretical, thermal and spectroscopic analytical techniques to the study of a molecule with intermolecular bonds in both the crystalline and pure amorphous state. J Pharm Sci 98(9):3456–3468

USP34-NF29 (2006) United States Pharmacopeia 34–National Formulary 29, General Chapter <1092> the dissolution procedure: development and validation. United States Pharmacopeial Convention, Inc., Rockville, Maryland

Vajna B, Pataki H, Nagy Z, Farkas I, Marosi GR (2011) Characterization of melt extruded and conventional Isoptin formulations using Raman chemical imaging and chemometrics. Int J Pharm 419(1–2):107–113

Van Den Mooter G (2009) Solid dispersions as a formulation strategy for poorly soluble compounds. 20th Annual Symposium of the Finish Society of Physical Pharmacy, Vithi, Finland, Jan 28–29

Van Den Mooter G (2012) The use of amorphous solid dispersions: a formulation strategy to overcome poor solubility and dissolution rate. Drug Discov Today 9(2):e79–e85

Van Eerdenbrugh B, Taylor LS (2010) Small scale screening to determine the ability of different polymers to inhibit drug crystallization upon rapid solvent evaporation. Mol Pharm 7(4):1328–1337

Vasconcelos TF, Sarmento B, Costa P (2007). Solid dispersions as strategy to improve oral bioavailability of poor water soluble drugs. Drug Discov Today 12(23–24):1068–1075

Wang CC, Zhang G, Shah H, Infeld MH, Waseem Malick A, Mcginity JW (1997) Influence of plasticizers on the mechanical properties of pellets containing Eudragit RS 30 D. Int J Pharm 152(2):153–163

Woodcock J (2004) The concept of pharmaceutical quality. Am Pharm Rev 7(6):10–15

Wu T, Sun Y, Li N, De Villiers MM, Yu L (2007) Inhibiting surface crystallization of amorphous indomethacin by nanocoating. Langmuir 23(9):5148–5153

Wu JX, Yang M, Berg FVD, Pajander J, Rades T, Rantanen J (2011) Influence of solvent evaporation rate and formulation factors on solid dispersion physical stability. Eur J Pharm Sci 44(5):610–620

Xia D, Cui F, Piao H, Cun D, Piao H, Jiang Y, Ouyang M, Quan P (2010) Effect of crystal size on the in vitro dissolution and oral absorption of nitrendipine in rats. Pharm Res 27(9):1965–1976

Yoo SU, Krill SL, Wang Z, Telang C (2009). Miscibility/stability considerations in binary solid dispersion systems composed of functional excipients towards the design of multi-component amorphous systems. J Pharm Sci 98(12):4711–4723

Zhang J, Zografi G (2001). Water vapor absorption into amorphous sucrose-poly(vinyl pyrrolidone) and trehalose-poly(vinyl pyrrolidone) mixtures. J Pharm Sci 90(9):1375–1385

Zhang GGZ, Law D, Schmitt EA, Qiu Y (2004) Phase transformation considerations during process development and manufacture of solid oral dosage forms. Adv Drug Deliv Rev 56(3):371–390

Zhu Y, Shah NH, Malick AW, Infeld MH, Mcginity JW (2002) Solid-state plasticization of an acrylic polymer with chlorpheniramine maleate and triethyl citrate. Int J Pharm 241(2):301–310

Zidan AS, Rahman Z, Sayeed V, Raw A, Yu L, Khan MA (2012) Crystallinity evaluation of tacrolimus solid dispersions by chemometric analysis. Int J Pharm 423(2):341–350

Part IV
Emerging Technologies

Chapter 18
KinetiSol®-Based Amorphous Solid Dispersions

Dave A. Miller and Justin M. Keen

18.1 Background

KinetiSol has recently emerged as a novel technology in the field of amorphous solid dispersion (ASD) processing, and has been demonstrated in the pharmaceutical literature to produce ASDs with the most challenging compounds and compositions. Similar to hot-melt extrusion (HME) and spray drying, the fundamentals of the KinetiSol technology were established in another industry, specifically, commercial plastics processing, prior to its adaptation for pharmaceutical manufacturing. The viability of the KinetiSol process for production of pharmaceutical ASD systems was first established on commercial-scale plastics processing equipment (Miller 2007) and then subsequently scaled down to accommodate pharmaceutical development. A photograph of a model TC-254B batch-mode GMP KinetiSol compounder is shown in Fig. 18.1.

18.2 KinetiSol Fundamentals

KinetiSol is a fusion-based process that utilizes frictional and shear energies to rapidly transition drug and polymer blends into a molten state. Concurrent to the molten transition, KinetiSol rapidly and thoroughly mixes the active ingredient with its excipient carrier(s) on a molecular level to achieve a single-phase ASD system. The real-time temperature of the composition within the KinetiSol chamber is monitored by a computer-controlled module, and upon reaching the user-defined end point, molten material is immediately ejected from the process. Total processing times are generally less than 20 s, and elevated temperatures are observed for typically less than 5 s before discharge and cooling. On a laboratory scale, the process is

D. A. Miller (✉) · J. M. Keen
DisperSol Technologies LLC, Georgetown, TX, USA
e-mail: dave.miller@dispersoltech.com

N. Shah et al. (eds.), *Amorphous Solid Dispersions*,
Advances in Delivery Science and Technology, DOI 10.1007/978-1-4939-1598-9_18

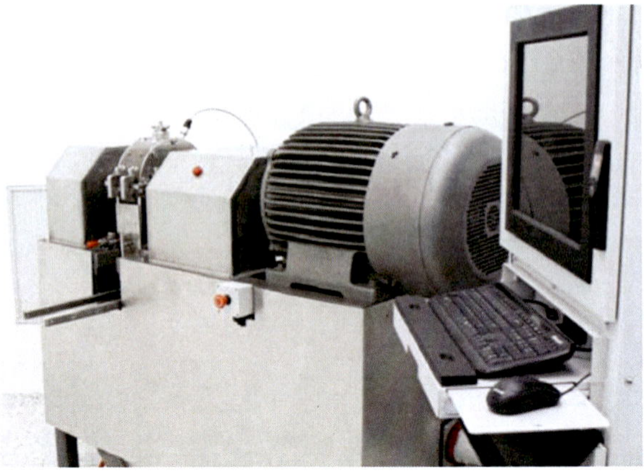

Fig. 18.1 Batch-mode GMP KinetiSol compounder model TC-254B

designed to operate in batch mode, whereas in commercial processing, it is operated semicontinuously, achieving product throughput as high as 1000 kg/h.

With its unique attributes, the KinetiSol process is providing novel solutions to emerging problems associated with ASD processing. KinetiSol's very brief processing times enable production of ASD systems with thermally sensitive active pharmaceutical ingredients (API) and excipients. The high rates of shear inherent to KinetiSol accelerate solubilization kinetics of drug compounds in molten polymers, which typically results in processing temperatures that are well below the melting point of the API. Consequently, the production of ASD systems with high melting point compounds (> 225°C) in a broad spectrum of concentration-enhancing polymers is routinely achieved with KinetiSol. The KinetiSol process is not torque limited, and hence processing of highly viscous/non-thermoplastic/high molecular weight polymers can be easily accomplished without the use of plasticizers. The capabilities of KinetiSol enable the use of unique drug/excipient combinations to create solubility-enhanced compositions that cannot be reproduced or manufactured at large scales by other technologies.

18.3 Process Development and Manufacturing with KinetiSol

KinetiSol process development and early-stage manufacturing are conducted in the batch-mode equipment configuration with batch sizes ranging from 50 to 300 g. In this configuration, the formulation is fully developed, the optimum batch size is determined, the influence of processing parameters on critical product quality attributes is characterized, and the processing parameters are optimized. Early development can be completed in a matter of days with API consumption on the order of just a few hundred grams. Once formulation and process development are complete, good manufacturing practice (GMP) production of amorphous intermediate can commence in batch mode to meet clinical trial material (CTM) demands of 10 kg or less for use in proof-of-principle clinical trials and stability analysis.

When larger product volume is required, the batch process is transferred to the semicontinuous KinetiSol equipment configuration. While it is possible to geometrically increase the size of the process (much larger machines are in use for plastics), a TC-254C continuous compounder produces commercial quantities using the same process geometry as the batch-mode compounder. The TC-254C is identical to the batch-mode equipment; however, the feeding and quenching operations are automated, allowing for the production of a sub-batch approximately every 30 s–2 min. Typical product throughputs on a TC-254C KinetiSol unit range from 20 to 30 kg/h. Due to the small footprint of the machinery, multiple compounders can be readily operated in parallel to provide greater throughputs when required. Because the processing geometry of the semicontinuous unit is identical to that of the batch-mode equipment, the starting parameters and design space developed in batch mode can be directly transferred to the production machine with little additional process development work. This attribute not only decreases the workload associated with process scale-up but also reduces the time and API required for the development of a high-volume production-ready process.

18.4 Novel Attributes of KinetiSol for the Processing of ASDs

By virtue of its unique characteristics, the KinetiSol process is providing novel solutions to emerging problems associated with ASD processing. KinetiSol's very brief processing times enable production of ASD systems with thermally sensitive APIs and excipients that meet potency, purity, and functionality requirements. The high rates of shear inherent to KinetiSol accelerate the solubilization kinetics of a compound in a molten polymer, which typically results in peak processing temperatures that are well below the melting point of the API. Consequently, the production of ASD systems with high melting point compounds (> 225°C) in a broad spectrum of concentration-enhancing polymers at high drug loadings is routinely achieved with KinetiSol. The KinetiSol process offers high torque output, and hence processing of highly viscous, non-thermoplastic, high molecular weight, and cross-linked polymers can be easily accomplished without the use of plasticizers. These capabilities of KinetiSol enable unique drug/excipient combinations which give rise to solubility-enhanced compositions that cannot be reproduced or manufactured at large scales by other technologies.

18.4.1 KinetiSol Processing of High-Melting-Point Drugs

Poorly water-soluble drugs with very high melting points (> 225°C) are emerging from drug discovery with greater frequency in recent years. High melting point compounds present significant challenges to thermal processing for the production of

ASD systems, namely polymer degradation and drug loading limitations. The advantage of KinetiSol for thermal processing of high melting point compounds is twofold: (1) the mechanical energy input typically renders crystalline compounds amorphous well below their melting points and (2) the process's high shear rates significantly accelerate solubilization kinetics of APIs in molten polymers. Consequently, KinetiSol can render high melting point APIs amorphous at temperatures well within the thermal-processing limits of most pharmaceutical polymers. Additionally, the rapid solubilization kinetics provided by KinetiSol allow for the achievement of high amorphous drug loading requirements.

An example of this unique benefit of KinetiSol was demonstrated by Bennett et al. (2013) in the generation of ASDs of 3-acetyl-11-keto-beta-*boswellic acid* (AKBA). AKBA is a botanical extract and experimental oncology compound that has demonstrated compelling efficacy in aggressive cancer models (Park et al. 2011; Yadav et al. 2012). The melting point of AKBA is 295°C and efforts to solubilize this molecule at target amorphous drug loadings using HME have been unsuccessful (Bennet et al. 2013). KinetiSol processing was successfully applied to produce ASD compositions of AKBA with four different polymeric carrier systems: HPMCAS-LF, HPMCAS-MF, Eudragit® L100-55, and Soluplus® in combination with Eudragit® L100-55, each also containing a surfactant. The KinetiSol process temperature profiles for each composition are shown in Fig. 18.2. It can be seen that all compositions were rendered amorphous at peak temperatures as much as 160 °C below the melting point of the drug. Also, processing times at elevated temperature were less than 5 s for all compositions, demonstrating the rapid solubilization kinetics achieved with KinetiSol. Potency analysis revealed that the AKBA content in all formulations was in excess of 99 %.

Pharmacokinetic studies in male Sprague-Dawley rats were conducted to compare the oral absorption of AKBA from the best KinetiSol compositions versus the micronized crystalline drug. As seen in Fig. 18.3, all KinetiSol compositions generated substantial improvements in systemic concentrations of AKBA over the pure drug. The results of KinetiSol processing with AKBA illustrate the capability of the process to achieve target amorphous drug loadings of a very high melting point compound in various solubility-enhancing polymeric carriers to dramatically improve bioavailability.

18.4.2 KinetiSol Processing of Thermally Labile Drugs

Owing to increasing molecular complexity (Keseru and Makara 2009), there is a growing percentage of poorly water-soluble compounds in contemporary development pipelines that are also highly thermally sensitive. Thermal degradation of a material is the result of total heat exposure, which is a function of both temperature and duration. KinetiSol- processing limits elevated processing temperatures to just a few seconds, and peak process temperatures are typically well below the melting point of the compound. KinetiSol therefore offers significant advantages for the production of ASD compositions with thermally labile compounds as it significantly reduces total heat exposure.

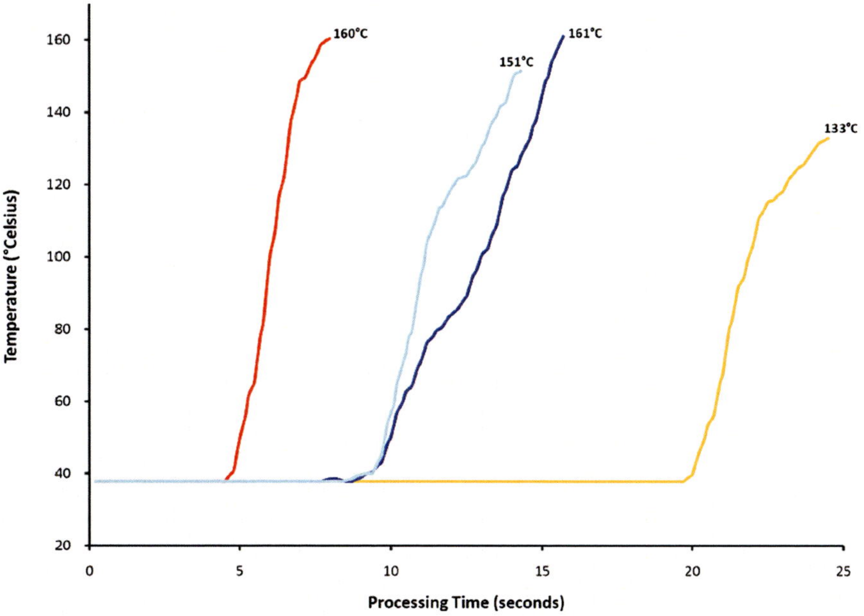

Fig. 18.2 KinetiSol-processing profiles for AKBA with four polymeric carriers

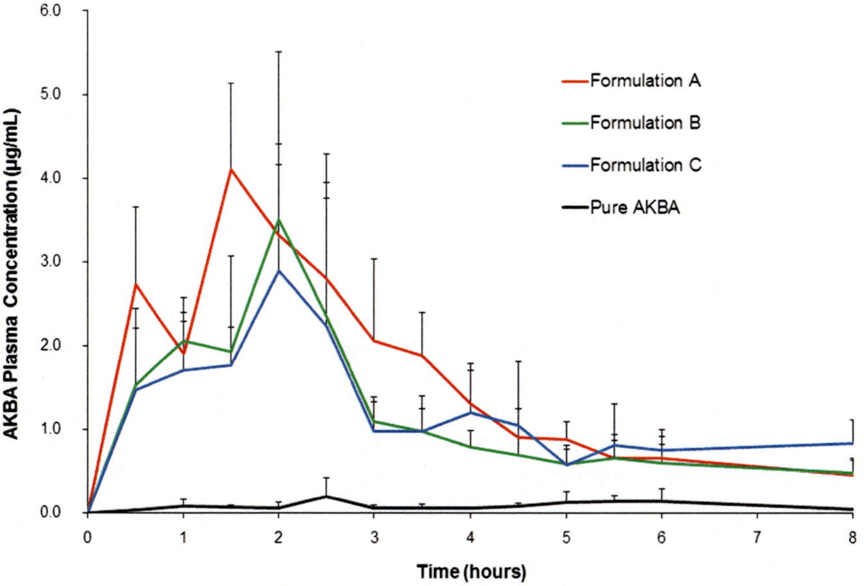

Fig. 18.3 AKBA plasma concentration profiles from oral dosing of three KinetiSol compositions and pure crystalline drug in male Sprague-Dawley rats (50 mg/kg)

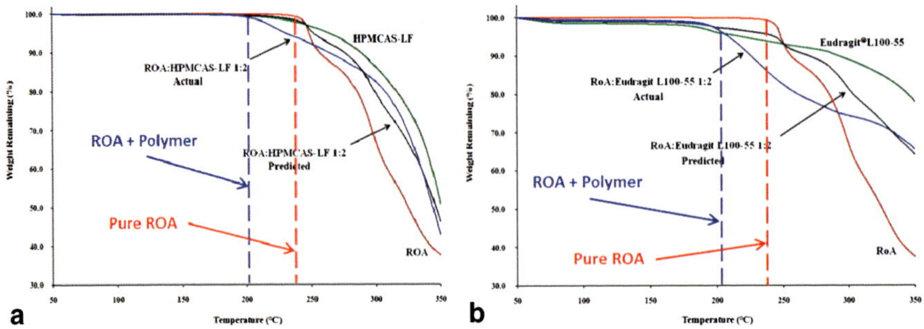

Fig. 18.4 Thermogravimetric analysis of: **a** ROA with HPMCAS-LF and **b** ROA with Eudragit L100-55. (Reprinted with permission from Hughey et al. 2010)

Hughey et al. (2010) demonstrated the advantages of KinetiSol over HME processing with a heat labile compound identified as ROA. Preformulation solubility enhancement studies identified Eudragit® L100-55 and HPMCAS-L as the optimum polymer carriers for an ASD of ROA. Thermogravimetric analysis (TGA) studies revealed that ROA degraded at an accelerated rate at elevated temperatures when in the presence of the anionic polymers (Fig. 18.4), which was attributed to an incompatibility with the acidic functional groups. The acceleration of thermal degradation in the presence of the anionic polymers thus presented even greater challenges to the thermal processing of ROA.

The KinetiSol-processing profiles for the ROA-Eudragit L100-55 and ROA-HPMCAS-LF compositions are shown in Fig. 18.5. As seen from these profiles, the processing time at elevated temperature for both compositions was limited to a few seconds and both ejection temperatures were less than 120°C; thus, demonstrating that KinetiSol was effective in rendering ROA substantially amorphous at processing temperatures far below its melting point of 230°C. When processed by HME, residence times were on the order of minutes and processing temperatures of 140 and 170°C were required to render ROA substantially amorphous in the Eudragit L100-55 and HPMCAS-LF compositions, respectively.

Potency analysis of the amorphous dispersion systems revealed that drug recovery was significantly higher with KinetiSol versus HME for comparable process feeds. In the case of Eudragit L100-55, the mean ROA potency value was 70.9 % for KinetiSol and 22.7 % for HME. The mean potency value for the HPMCAS-LF composition was 99.4 % with KinetiSol and 70.9 % for HME. A comparative summary of processing parameters and chemical analysis of the KinetiSol and HME products is provided in Table 18.1. The results demonstrated that KinetiSol was an effective method for producing ASDs where HME was not feasible, owing to the compound's thermal instability.

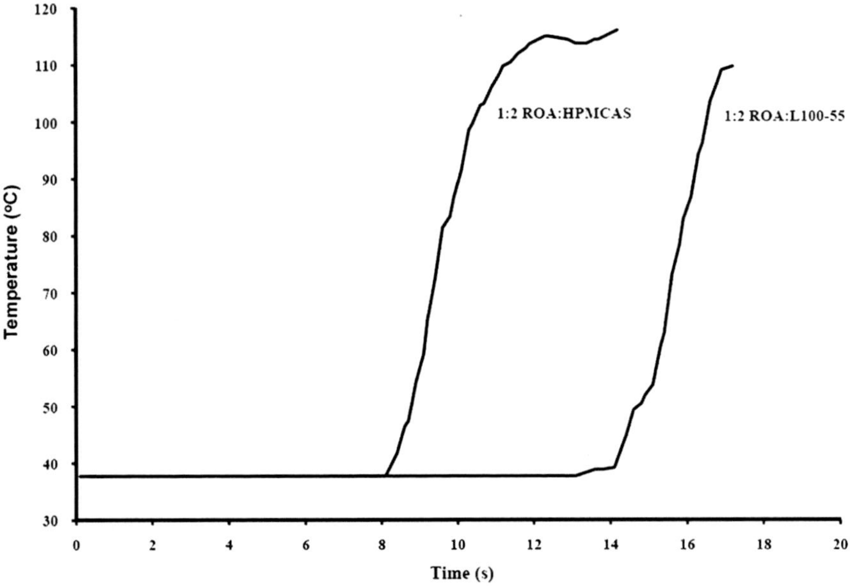

Fig. 18.5 KinetiSol-processing profiles of ROA:HPMCAS-LF and ROA:Eudragit L100-55 solid dispersion systems. (Reprinted with permission from Hughey et al. (2010))

Table 18.1 Summary of processing parameters and chemical analysis of ROA compositions with Eudragit L100-55 and HPMCAS-LF produced with KinetiSol and HME. (Source: Adapted from Hughey et al. 2010)

KinetiSol compositions						
Polymer	Particle size	Speed (RPM)	Temp.°C	Potency (%)	Impurities (%)	
Eudragit® L100-55	Unmicronized	1450	100	70.9 ± 0.8	12.9	
HPMCAS	Unmicronized	2400	112	99.4 ± 1.2	1.6	
Hot-melt extrusion compositions						
Polymer	Particle size	Processing temp.°C	Screw speed (rpm)	Recirculation time (min)	Potency (%)	Impurities (%)
Eudragit® L100-55	Unmicronized	140	300	2	22.7 ± 0.5	55.9
HPMCAS	Unmicronized	170	300	2	70.9 ± 0.3	10.2

18.4.3 KinetiSol Processing of Thermally Sensitive, Highly Viscous Polymer Systems Without Plasticizers

Many of the polymers commonly used in ASD systems present challenges for fusion-based processing with respect to thermal sensitivity, high viscosity, or both. Often, plasticizers and other processing aids are required to reduce polymer T_gs to enable thermal processing or facilitate thermal processing at lower temperatures. Because KinetiSol is not torque limited, highly viscous polymer systems are straightforward to process without plasticizers. Furthermore, KinetiSol reduces thermal stress on the polymer, and hence reduction of process temperatures by incorporating a plasticizer is not required. The elimination of plasticizers leads to increased composite T_gs and improved physical stability, in most cases.

DiNunzio et al. (2010a) evaluated this aspect of KinetiSol in comparison with HME in the production of ASD systems with itraconazole (ITZ) and Eudragit L100-55. Eudragit L100-55 is a heat labile polymer, which undergoes thermal degradation at approximately 155°C by decomposition of carboxylic acid side groups, followed by chain decomposition above 180°C (Lin and Yu 1999). Its thermal sensitivity coupled with high molten viscosity make Eudragit L100-55 a particularly challenging polymer to process thermally, and hence plasticizers are almost always required when processing by HME. This was the case in the study by DiNunzio et al., where Eudragit L100-55-based compositions prepared by HME required the aid of a plasticizer to achieve the necessary viscoelastic characteristics to enable processing. However, enablement of HME processing was achieved to the detriment of matrix rigidity as these materials exhibited a reduced compositional T_g of 54°C.

Conversely, KinetiSol processing allowed for the production of ASDs without the aid of a plasticizer, resulting in an amorphous ITZ-Eudragit L100-55 (1:2) composition with a measured T_g of 101°C. Examination of side group functionality of KinetiSol-processed L100-55 compositions showed similar functional levels to that of the unprocessed polymer, indicating that the polymer was not degraded during processing despite being processed above its degradation onset temperature.

This difference in composite T_g*s* between plasticized HME compositions and un-plasticized KinetiSol compositions was found to directly correlate with physical stability. ITZ crystallization was identified with the plasticized HME composition after just 1 month storage at 40°C/75 % RH; whereas, the un-plasticized KinetiSol composition remained physically stable for the entire 6-month study duration. With these results, this study established the unique ability of KinetiSol to process thermally sensitive, highly viscous polymers without use of a plasticizer and the corresponding improvement in physical stability of the resulting amorphous dispersion.

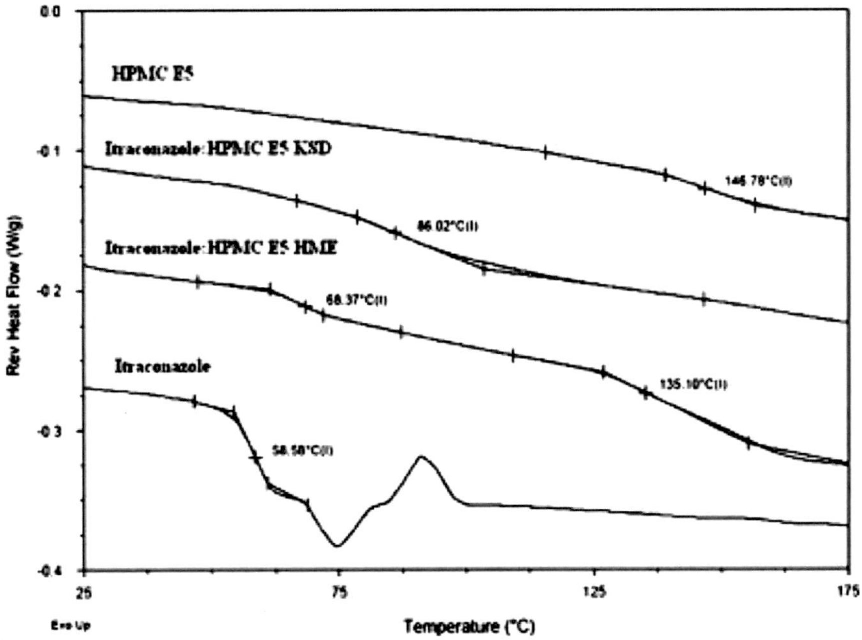

Fig. 18.6 Reversing heat-flow profiles of amorphous ITZ, HPMC E5, and ITZ:HPMC E5 solid dispersions produced by HME and KinetiSol. (Reprinted with permission from DiNunzio et al. 2010b)

18.4.4 KinetiSol Processing With a Non-Thermoplastic Polymer for Improved Homogeneity

Cellulosic polymers such as hypromellose (HPMC) and hypromellose acetate succinate (HPMCAS) have proven to be among the most effective concentration-enhancing polymers, yielding extensive and prolonged supersaturation from various ASD systems (Friesen et al. 2008; Miller et al. 2008; Curatolo et al. 2009). However, cellulosics are non-thermoplastic and as such are difficult to process thermally; often leading to heterogeneous, multiphase systems (Six et al. 2003; Verreck et al. 2003).

In a study conducted by DiNunzio et al. (2010), thermal processing of ITZ and HPMC by HME and KinetiSol was investigated. DiNunzio and coworkers determined that ITZ:HPMC E5 (1:2) ASD systems produced by HME were two phase, while the same composition processed by KinetiSol yielded a single-phase ASD with a single T_g of 86.02°C. A comparison of the thermograms for the HME and KinetiSol processed systems is shown in Fig. 18.6. The authors attributed the improved homogeneity of the KinetiSol processed system to the higher shear rates which provided improved mixing characteristics over HME.

KinetiSol processing of high-viscosity HPMC was established in a research published by Hughey et al. (2012). In this study, the authors successfully produced ASDs with a 50 cP grade of HPMC, demonstrating single-phase systems with enhanced performance with respect to ASDs comprising lower-viscosity HPMC. Further, it was shown that KinetiSol processing resulted in substantially less polymer degradation than HME processing conducted on a low-shear, small-scale extrusion system. The results of KinetiSol processing with HPMC presented by Hughey et al. and DiNunzio et al. illustrate the novelty and utility of KinetiSol with non-thermoplastic and highly viscous pharmaceutical polymers.

18.5 Summary

KinetiSol is a novel processing technology for the production of pharmaceutical solid dispersions that was adapted from the plastics processing industry. KinetiSol offers unique capabilities that provide ASD-processing solutions for challenging compounds and compositions. KinetiSol enables thermal processing of heat-labile APIs and polymers by substantially reducing total heat exposure. The rapid solubilization kinetics inherent to the KinetiSol process enables the achievement of ASD systems with very high melting point compounds in a broad spectrum of polymer carriers. KinetiSol enables processing of highly viscous, non-thermoplastic polymers without the use of plasticizers or other processing aids for the production of single-phase, high drug load ASDs. The streamlined development-to-production approach enabled by batch-mode development followed by semicontinuous mode production significantly simplifies the establishment of a commercial-ready process. With these unique attributes, KinetiSol is providing commercially relevant solutions to the problems presented by the challenging poorly water-soluble compounds clogging contemporary development pipelines.

References

Bennett RC, Brough C, Miller DA, O'Donnell KP, Keen JM, Hughey JR, Williams III RO, McGinity JW (2013) Preparation of amorphous solid dispersions by rotary evaporation and KinetiSol dispersing: approaches to enhance solubility of a poorly water-soluble gum extract. Drug Dev Ind Pharm 1–16. doi: 10.3109/03639045.2013.866142

Curatolo W, Nightingale J, Herbig S (2009) Utility of hydroxypropylmethylcellulose acetate succinate (HPMCAS) for Initiation and maintenance of drug supersaturation in the GI milieu. Pharm Res 26:1419–1431

DiNunzio JC, Brough C, Miller DA, Williams Iii RO, McGinity JW (2010a) Applications of KinetiSol® dispersing for the production of plasticizer free amorphous solid dispersions. Eur J Pharm Sci 40:179–187

DiNunzio JC, Brough C, Miller DA, Williams RO, McGinity JW (2010b) Fusion processing of itraconazole solid dispersions by kinetisol® dispersing: a comparative study to hot melt extrusion. J Pharm Sci 99:1239–1253

Friesen DT, Shanker R, Crew M, Smithey DT, Curatolo WJ, Nightingale JAS (2008) Hydroxypropyl methylcellulose acetate succinate-based spray-dried dispersions: an overview. Mol Pharm 5:1003–1019

Hughey J, DiNunzio J, Bennett R, Brough C, Miller D, Ma H, Williams R, McGinity J (2010) Dissolution enhancement of a drug exhibiting thermal and acidic decomposition characteristics by fusion processing: a comparative study of hot melt extrusion and KinetiSol® dispersing. AAPS PharmSciTech 11:760–774

Hughey JR, Keen JM, Miller DA, Brough C, McGinity JW (2012) Preparation and characterization of fusion processed solid dispersions containing a viscous thermally labile polymeric carrier. Int J Pharm 438:11–19

Keseru GM, Makara GM (2009). The influence of lead discovery strategies on the properties of drug candidates. Nat Rev Drug Discov 8:203–212

Lin S-Y, Yu H-L (1999) Thermal stability of methacrylic acid copolymers of Eudragits L, S, and L30D and the acrylic acid polymer of carbopol. J Polym Sci Part A: Polym Chem 37:2061–2067

Miller DA (2007) Improved oral absorption of poorly water-soluble drugs by advanced solid dispersion systems. PhD Dissertation, Division of Pharmaceutics, The University of Texas at Austin, Austin, TX, p 312

Miller DA, DiNunzio JC, Yang W, McGinity JW, Williams RO (2008) Enhanced in vivo absorption of itraconazole via stabilization of supersaturation following acidic-to-neutral pH transition. Drug Dev Ind Pharm 34:890–902

Park B, Prasad S, Yadav V, Sung B, Aggarwal BB (2011) Boswellic acid suppresses growth and metastasis of human pancreatic tumors in an orthotopic nude mouse model through modulation of multiple targets. PloS One 6:e26943

Six K, Berghmans H, Leuner C, Dressman J, Van Werde K, Mullens J, Benoist L, Thimon M, Meublat L, Verreck G, Peeters J, Brewster M, Van den Mooter G (2003) Characterization of solid dispersions of itraconazole and hydroxypropylmethylcellulose prepared by melt extrusion, part II. Pharm Res 20:1047–1054

Verreck G, Six K, Van den Mooter G, Baert L, Peeters J, Brewster ME (2003) Characterization of solid dispersions of itraconazole and hydroxypropylmethylcellulose prepared by melt extrusion—part I. Int J Pharm 251:165–174

Yadav VR, Prasad S, Sung B, Gelovani JG, Guha S, Krishnan S, Aggarwal BB (2012) Boswellic acid inhibits growth and metastasis of human colorectal cancer in orthotopic mouse model by downregulating inflammatory, proliferative, invasive and angiogenic biomarkers. Int J Cancer 130:2176–2184

Chapter 19
Amorphous Solid Dispersion Using Supercritical Fluid Technology

Pratik Sheth and Harpreet Sandhu

19.1 Introduction

The enhancement of solubility and oral bioavailability of poorly water-soluble drugs continue to be the most challenging aspect of drug development. Among the various formulation approaches, amorphous solid dispersion (ASD) is considered to be top ranked in terms of achieving the highest possible advantage (Chiou and Riegelman 1971; Stegemann et al. 2007). Various technologies are cited in the literature for preparation of ASD (Bikiaris 2011; Williams et al. 2013), which commonly involve mixing of drug with carrier polymer(s). Solid dispersion preparation methodology can be grouped into two main categories: fusion method and solvent evaporation method. Fusion method, also called melt method, requires high processing temperatures, at which many active pharmaceutical ingredients (API) and excipients may degrade. Solvent evaporation method, although requires milder processing conditions, has a drawback of using excessive volume of organic solvents and subsequent difficulties of removing the solvents below an acceptable level from the finished product. There are three subcategories of solvent evaporation method: (1) conventional solvent evaporation, (2) spray drying, and (3) supercritical fluid (SCF) technologies.

Solvent evaporation is commonly carried out in a rotary evaporator under reduced pressure and elevated temperature. A typical problem encountered in this process is the slow rate of solvent removal. This is due to the fact that as the "drying" proceeds, the viscosity of solution becomes higher which further slows down the evaporation rate. An improved version of the solvent evaporation is spray drying. In its basic form, spray drying is a very simple process where droplets or particles are dried while suspended in the drying gas, turning a liquid feed into a dry powder in a single continuous process step. Spray drying is recognized as one of the preferred

P. Sheth (✉)
Forum Pharmaceuticals, Inc., North Grafton, MA, USA
e-mail: pratiksheth@hotmail.com

H. Sandhu
Merck & Co., Inc., Summit, NJ, USA

N. Shah et al. (eds.), *Amorphous Solid Dispersions*,
Advances in Delivery Science and Technology, DOI 10.1007/978-1-4939-1598-9_19

Table 19.1 Physical properties of selected supercritical fluids

Fluid	Critical temp (°C)	Critical pressure (atm)	Critical density (g/cc)
Carbon dioxide	31.3	72.9	0.47
Water	374.1	217.8	0.32
Nitric oxide	36.5	72.5	0.45
Ammonia	132.5	112.5	0.24
Methane	– 79.6	45.4	0.16
Ethane	32.3	48.1	0.20
Propane	96.8	41.9	0.22
Ethylene	9.4	49.7	0.22
Methanol	239.6	79.8	0.27
Ethanol	240.9	60.6	0.28
Acetone	235.1	46.4	0.28

method for the production of ASDs. However, the spray drying process generally produces powders with low bulk density which requires downstream processing for the densification. In addition, product recovery and dust collection increase the cost of spray drying. These limitations as well as a search for an efficient and environmentally safe technique has led the researchers to explore SCF technology for the production of the ASDs (York et al. 2004).

SCF- based technologies have been increasingly used in the pharmaceutical industry for the preparation of micro- and nano-sized solid particles over the past several decades (York 1999). From a conceptual perspective, any material can exist in the supercritical phase depending on temperature and pressure conditions; however, for practical purposes only few of these conditions are feasible. Table 19.1 (Bartle 1988, Poling et al. 2001) lists the relevant properties of some fluids. Low critical temperature (31.2 °C) and pressure (72.9 atm) of carbon dioxide coupled with environmentally safe handling and noninflammable nature make it ideal for processing pharmaceuticals including heat-sensitive materials such as biologicals.

The solubilization properties of SCF are attributed to high diffusivity, low density, low viscosity, and low surface tension (see Table 19.2) compared to other states of matter. Based on the extensive application of SCF in chromatographic and extraction processes, the solvent polarity of $scCO_2$ is generally estimated to be similar to hexane. The solvent characteristics of $scCO_2$ can be further modulated by changes in the state variables such as temperature and pressure as well as using cosolvents. It is generally observed that an increase in density due to an increase in pressure induces more fluid-like properties and helps improve solubility of organic compounds.

Depending on the application, carbon dioxide can act either as a solvent, an antisolvent, or as a solute in the SCF processes. SCF technologies are further classified based on the particle growth mechanisms and their collection environment. Rapid expansion of supercritical solutions (RESS), gas antisolvent (GAS) precipitation,

Table 19.2 Relative differences in the solvent properties of different states of fluid

State of matter	Density (g/cc)	Viscosity (μPa s)	Diffusivity (mm^2/s)
Gas	0.001	10	1–10
Supercritical fluid	0.10–1.0	50–100	0.01–0.1
Liquids	1.00	500–1000	0.001

supercritical antisolvent (SAS) precipitation, precipitation with compressed fluid antisolvent (PCA), solution-enhanced dispersion by supercritical fluids (SEDS), and precipitation from gas-saturated solutions (PGSS) are the main variants of SCF-based technologies (Foster et al. 2003; Knez and Weidner 2003; Valle and Galan 2005). These techniques have successfully been used in the production of micro- and nanoparticles as well as ASDs to achieve particle size control and solubility. The ability to produce narrow particle size distribution with desired morphology has been extensively used in designing drug products for inhalation (York 1999; Lobo et al. 2005; Badens et al. 2009; Ayad et al. 2013).

19.2 Operations Where SCF Acts as a Solvent

19.2.1 Rapid Expansion of Supercritical Solvent (RESS)

This process is schematically illustrated in Fig. 19.1. The process is used when the solute (polymer, drug, or drug–polymer matrix) has high solubility in the SCF. The RESS process involves the saturation of SCF with drug or drug–polymer matrix, followed by depressurization of the solution by passing through a heated nozzle into a low-pressure chamber. The rapid decompression of the SCF containing the drug and/or polymers drives the nucleation and the particle formation. During decompression, the SCF solution experiences a Joule–Thomson cooling due to a large volumetric expansion.

RESS is an attractive process as it is a single-step process which requires minimum to no organic solvent. During rapid expansion, the solute experiences simultaneous pressure and temperature drop that considerably promotes the particle formation process.

Particle agglomeration is a common issue in RESS process. It becomes worse if residual cosolvent remains in the processed material. Several researchers have used various particle collection systems to overcome the agglomeration issues. Additionally, RESS process is only applicable to the compounds with good solubility in $scCO_2$. Unfortunately, many poorly soluble compounds with high molecular weights and polar bonds are also poorly soluble in $scCO_2$ at moderate temperatures (less than 60 °C) and pressures (less than 300 bar). Cosolvents such as methanol can be added to carbon dioxide to enhance solubility of the drug. However, inclusion of solvent to

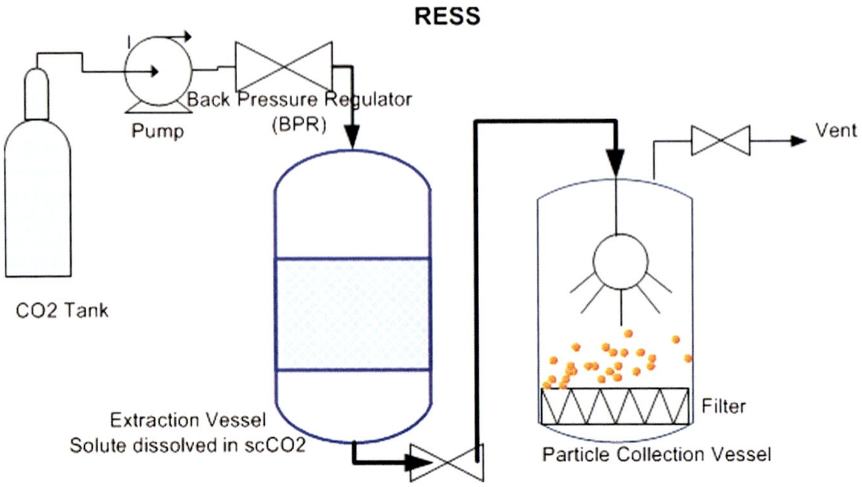

Fig. 19.1 Simplified schematic representation of RESS equipment setup showing components such as CO_2 tank, pump, back pressure regulator, extraction vessel, and particle collection vessel

aid in the solubility mandates the removal of residual solvent from the final product and adds significantly to the cost and complexity.

19.2.2 RESS Applications

Over the past several decades, RESS technology has been successfully used in the production of crystalline as well as amorphous particles of drugs and polymers (Perrut 2000; Yeo and Kiran 2005). A few examples from the literature are presented to illustrate the versatility of the technology and to explain the modification of basic RESS process to minimize agglomeration.

Using RESS process, nanoparticles of naproxen were produced with and without polylactic acid (PLLA). It was shown that the coating of naproxen particles with PLLA stabilized the particles against aggregation (Gadermann et al. 2009). Using this technique, another research group produced amorphous cefuroxime axetil nanoparticles (Varshosaz et al. 2009). In addition to polymer, they studied the effect of nozzle temperature and the extraction port temperatures on the particle morphology and size. Amorphous nanoparticles were obtained in all cases and the particle size decreased from 450 to 150 nm with increase in the drying efficiency.

The basic RESS process has been further modified to produce nanoparticles with tight particle size control and minimal agglomeration. Rapid expansion of a supercritical solution into a liquid solvent (RESOLV; Pathak et al. 2004) and rapid expansion from supercritical to aqueous solution (RESAS; Young et al. 2004; Tozuka et al. 2010) are the most notable ones. In RESOLV, the supercritical solution is allowed to expand into a liquid medium instead of an air medium. Pathak et al. showed that the RESOLV technique can successfully produce individual and spherical particles of naproxen and ibuprofen in nanoscales when the product is expanded into aqueous media containing polyvinylpyrrolidone (PVP). In the absence of PVP, the particles obtained were nanosized; however, they were agglomerated and nonspherical. It is attributed that the liquid at the receiving end of the rapid expansion in RESOLV suppresses the particle growth in the expansion jet, thereby allowing production of nano-sized and round particles.

The expansion of supercritical solution through an orifice or tapered nozzle into aqueous solution containing a stabilizer(s) is the operating principle of RESAS processes. This arrangement minimizes the particle agglomeration during free jet expansion. The stabilizers used in this process are mainly surfactants such as polysorbates, poloxamers, and lecithins or hydrophilic polymers. The presence of a stabilizer minimizes the particle aggregation by rapidly reducing the surface free energy of the primary particles together with steric stabilization. Indomethacin nanoparticles were produced using 1 % polyvinyl alcohol via the RESAS process (Tozuka et al. 2010).

In another variation of the RESS process, referred to as RESS-SC, a solid cosolvent such as menthol was used to modulate the solubility of the drug (phenytoin) in the polymer (Thakur and Gupta 2006). It was shown that when drug particles were mixed with solid cosolvent such as menthol, the presence of solid cosolvent inhibited particle–particle interactions between drug particles, thereby hindering the crystal growth. As illustrated in Fig. 19.2, phenytoin particles are surrounded by menthol, which reduced interparticulate interactions with other phenytoin particles. The co-solvent (menthol) is then removed by downstream processing such as sublimation or lyophilization.

Although RESS process is better suited to produce crystalline nanoparticles, it is amenable to the production of ASDs especially for compounds with low crystallization tendency. However, the successful application of RESS for ASD is also limited by the solubility of drug and polymer in the SCF. It is commonly observed that most organic compounds with poor solubility also exhibit low solubility in the SCF.

19.3 Operations Where SCF Acts as an Antisolvent

SCF-based processes such as GAS process, SAS process, aerosol solvent extraction system (ASES), SEDS address low solubility of the compounds in $scCO_2$. In these processes, the drug, polymer, or both are dissolved in an organic solvent to form a solution. Solvents used may include dimethyl sulfoxide, N-methyl pyrrolidone, methanol, ethanol, acetone, chloroform, or isopropanol. To successfully produce ASD, the drug and polymer should exhibit limited solubility in SCF and the organic solvent should be miscible with carbon dioxide. Collection of the precipitated particles in the antisolvent is carried out in the same vessel where solvent extraction takes place. The particles are collected on a filter unit located at the bottom of the vessel.

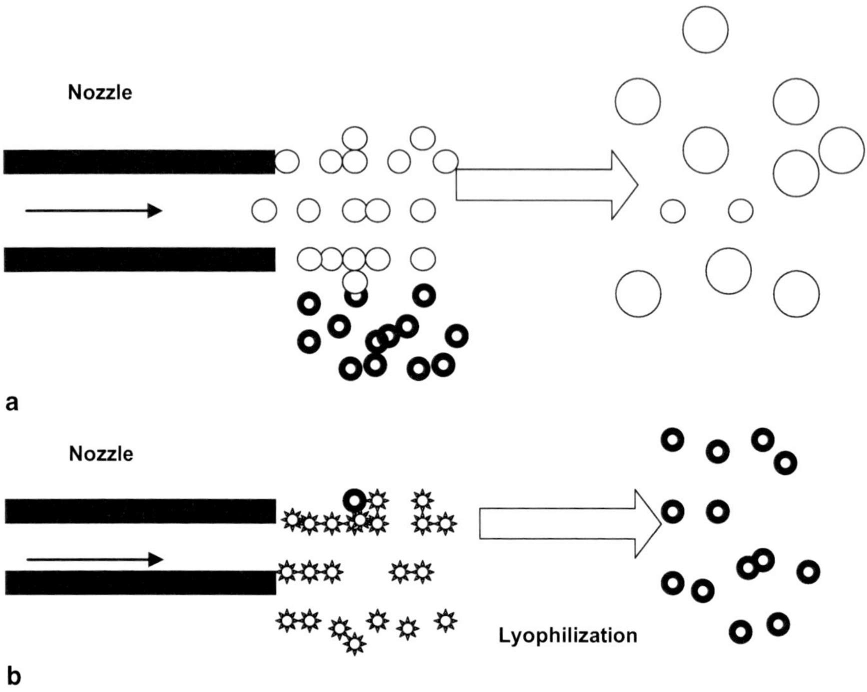

Fig. 19.2 Schematic representation of particle formation in expansion zone during conventional RESS process (**a**) with RESS-SC process (**b**)

19.3.1 Mechanism of Particle Formation in GAS, SAS/ASES

Figure 19.3a is the schematic representation of the GAS process. In this process, the solute is first dissolved in a liquid organic solvent or solvent mixture, and carbon dioxide gas is introduced to precipitate the solutes. The CO_2 gas used as the antisolvent does not have to be at supercritical condition. It is injected into the solution in a closed chamber, preferably from the bottom, in order to obtain uniform mixing. As a result of dissolution of CO_2 gas in organic solvent, the solubilization power of the solvent is reduced causing the precipitation of solutes. The particles produced are washed with additional antisolvent to remove the remainder of the solvent.

GAS processes are batch or semicontinuous operations. They do not work under constant pressure. The pressure varies continuously from 1 bar to the final pressure. GAS is favored by some researchers as it is a rather slow process that allows the growth of the particles in a controllable manner. Consequently, it is not considered suitable for ASD.

Figure 19.3b provides the schematic representation of the SAS process. Unlike GAS, this technique utilizes CO_2 gas in its supercritical stage as an antisolvent of the solute. The solute is first dissolved in a liquid solvent and then this solution is

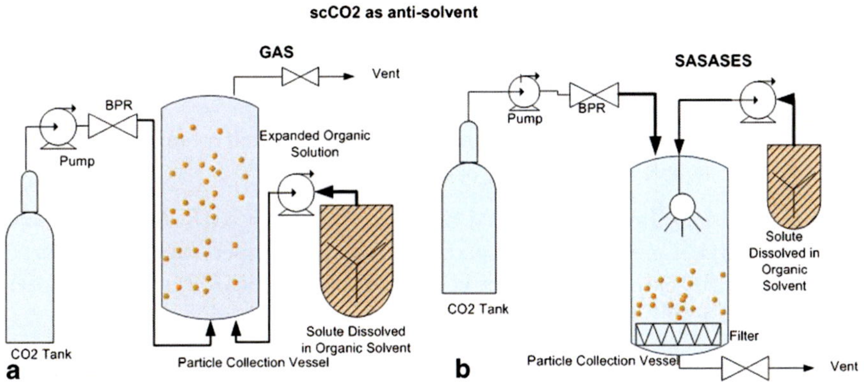

Fig. 19.3 Simplified schematic representation of GAS (**a**) and SAS/ASES (**b**) equipment set up showing components such as CO_2 tank, pump, back pressure regulator, extraction vessel, and solution of API and/or polymer in organic solvent

sprayed onto a chamber containing SCF (antisolvent). As organic solution droplets are diluted by SCF, the solubilization power of the organic solvent is reduced. As a consequence, the liquid mixture becomes supersaturated against the drug molecules, causing precipitation. Unlike GAS, this technology has produced favorable results during scale up to industrial scale (Jung et al. 2003). ASES process is the same process as SAS in principle.

The SAS- based processes are considered more amenable for the production of ASD. Besides the formulation composition (API, polymer, stabilizer, cosolvent, solvent, etc.), the key tunable variables that can affect the product attributes include pressure and temperature in the precipitation zone, atomization and feed rate of solution, flow rate of the SCF, and final drying/extraction time.

19.3.2 Applications of GAS, SAS and ASES

The earlier application of SCF in particle design has been focused on improving the morphology and particle size distribution of pure drug substance. Few examples are listed below:

Amorphous Nanoparticles of Amoxicillin Using SAS technology, Kalogiannis et al. produced amorphous nanoparticles of amoxicillin. The crystalline drug was dissolved in dimethysulfoxide (DMSO), or mixtures of DMSO with ethanol or methanol. Partially replacing the DMSO with ethanol or methanol further helped to reduce the particle size (Kalogiannis et al. 2005).

Rifampicin Similarly, Reverchon et al. produced amorphous particles of rifampicin using DMSO as the solvent and SAS as the processing technology (Reverchon et al. 2002).

Atorvastatin Calcium By modifying the processing conditions such as temperature and pressure during precipitation, concentration of drug in the feed solution, and ratio of feed solution and SCF, the amorphous particles of atorvastatin calcium were also obtained (Kim et al. 2008)

Among all the SCF-based processing, SAS has been most promising to produce ASD. A few examples are presented below:

Itraconazole ASD Itraconazole solid dispersion was prepared with hypromellose (HPMC 2910; Lee et al. 2014). The particle size of solid dispersion prepared ranged from 100 to 500 nm. It was further confirmed that itraconazole was molecularly dispersed in HPMC 2910 as an amorphous form.

Paclitaxel ASD ASD of paclitaxel was prepared with SAS using series of polymers and surfactant as stabilizers (Woo et al. 2006). The drug, polymer, and stabilizer were dissolved in an organic solvent such as ethanol and dichloromethane. This solution was then sprayed into a vessel containing SCF, to produce a highly uniform, nanoscale ASD. The solubility of the paclitaxel solid dispersion prepared by the SAS process was significantly higher than the untreated paclitaxel.

Phenytoin ASD prepared with PVP K 30 using compressed antisolvent process was compared to ASD produced by spray drying process (Muhrer et al. 2006). After ensuring that both were amorphous, dissolution profile of both were compared. Based on in vitro dissolution, the performance of the SAS- based material was at least 25 % better than the spray-dried material which can be attributed to better control of particulate properties such as morphology and particle size distribution, in SCF.

19.4 Solution-Enhanced Dispersion by Supercritical Fluids (SEDS)

This patented technique uses SCF as an antisolvent as well as a dispersing agent (Hanna and York 1998). The contact of the liquid solution containing the drug and the polymer with the SCF generates a finely dispersed mixture which leads to rapid particle precipitation. This technology is generally used to produce microspheres. The most important feature of the SEDS is the nozzle type. In this process, two types of coaxial nozzles are used: the first one is a nozzle with two channels which allows introduction of the SCF and the drug solution or drug polymer mixture at the same time. The second type of nozzle has three channels which allow introduction of three different fluids at once providing more choices in operating variables. For example, injection of the drug in an organic solvent, the polymer in an aqueous solution, and SCF at the same time is possible. The experimental arrangement of the SEDS process is shown in Fig. 19.4a. Figure 19.4b shows the schematic representation of the three-channeled coaxial nozzle.

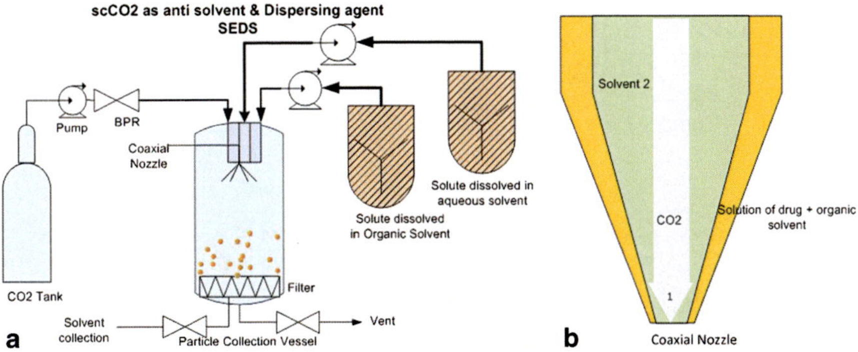

Fig. 19.4 Schematic drawing of SEDS (**a**) process and a simplified arrangement showing three-channeled coaxial nozzle (**b**) used in SEDS process

SEDS process was used to produce particles of salmeterol xinafoate with a polymer matrix (Beach et al. 1999). Two separate solutions of the active substance and the polymer (hydroxypropylcellulose) were prepared by dissolving in acetone, and co-introduced with supercritical CO_2 in a precipitator, using a three-passage nozzle. Analysis confirmed the inclusion of drug into the polymer matrix. Using a similar setup, spherical microparticles of hydrocortisone were entrapped within the biodegradable polymer poly (D, L-lactide-co-glycolide; DL-PLG) by using a combination of supercritical N_2 and CO_2 (Ghaderi et al. 2000). The use of N_2 together with CO_2 improved the homogeneity of mixing which led to more efficient integration of the polymer and the drug. This technique was also applied to produce PLLA-coated microparticles of puerarin (Chen et al. 2009).

19.5 Operations Where SCF Acts as Solute (Particles from Gas Saturated Solution)

Figure 19.5 provides a schematic illustration of PGSS equipment setup. As discussed earlier, it is difficult to dissolve high molecular weight or polar compounds in CO_2, which is a nonpolar solvent even in a supercritical state in the absence of the cosolvent. However, $scCO_2$ has the ability to diffuse into organic compounds, particularly into polymers. When $scCO_2$ diffuses into the polymer, it lowers the glass transition temperature and decreases its viscosity. These characteristics are bases of the PGSS process.

In the PGSS operations, the physical mixtures of the drug and the polymer are first exposed to SCF. In the presence of SCF and elevated pressure, the mixture starts to plasticize and melt. Following initial melt, introduction of the $scCO_2$ continues to dissolve the mixtures and the viscosity is decreased further. This nonviscous solution is then sprayed into a receiver using a nozzle and a pressure controlling valve. As

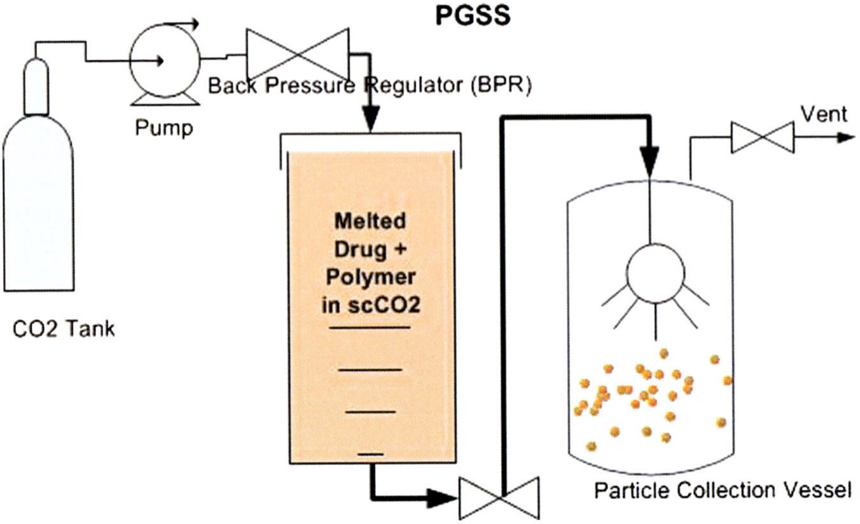

Fig. 19.5 Schematic representation of equipment setup of PGSS process

a result of rapid decompression, the dissolved SCF escapes from polymer matrix leading to the formation of composite microcapsules. This process suits well with materials that absorb SCF at high concentrations like PVP, polyethylene glycol, polyethylene, polyester, D, L-PLA, and poly(lactic-co-glycolic acid) PLGA.

Sencar-Bozic et al. made composite microparticles of nifedipine and polyethylene glycol (PEG 4000) using the PGSS process (Senčar-Božič et al. 1997). They showed that the solid dispersions had increased dissolution rates of nifedipine. Similar results were reported for the felodipine (Kerc et al. 1998). Rodrigues et al. prepared the microparticles of theophylline with hydrogenated palm oil (HPO) by the PGSS process demonstrating that the process enabled coprocessing of API with the excipient (Rodrigues et al. 2004). Particle size obtained was about 3 μm in diameter. Spherical morphology with a regular surface was obtained at higher expansion pressures. Although used to process crystalline material, this technology has successfully been scaled up to industrial scale (Weidner 2009).

The PGSS process can be performed without using any organic solvents. It usually requires lower operational pressures and gas consumption than the RESS process. One problem associated with the PGSS process is potential separation of the ingredients at pressure drop zone. However, PGSS can be modified to overcome the agglomeration, ingredient separation, and nonuniform particle size distribution problems. For example, using two separate mixing chambers, drug and polymer solution can be sequentially diluted with SCF to reduce viscosity, improve flow property, and atomization efficiency thereby produce more controlled particles (Shekunov et al. 2006).

Application of $scCO_2$ in melt extrusion process as a processing aid may function in either RESS or PGSS mode depending on the mutual solubility of drug and polymer

Table 19.3 Overview of various SCF process modalities evaluated for pharmaceutical product design

Criteria/process	RESS	GAS	ASES	SAS/modified SAS	PGSS
Mechanism	Rapid expansion of SCF	SCF introduced into solution	Solution introduced into SCF	Solution introduced into SCF	Solid incubated with SCF
Role of SCF	Solvent	Antisolvent	Antisolvent	Antisolvent	Plasticizer/solvent
Pressure changes	Variable	Variable	Variable	Fixed	Fixed
Compound/polymer solubility in SCF	Required	Not required	Not required	Not required	Preferred
Particle control	Limited	Controlled by SCF flow rate	Controlled by solution addition rate	Limited	Limited (similar to RESS)

ASES aerosol solvent extraction system, *GAS* gas antisolvent precipitation, *PGSS* precipitation from gas-saturated solutions, *RESS* Rapid expansion of supercritical solutions, *SAS* supercritical antisolvent, *SCF* supercritical fluid

in $scCO_2$. The pressure drop experienced by the extrudates at the die port can change the material characteristics depending on its interaction with $scCO_2$ (Verreck et al. 2005).

19.6 Conclusions

SCF-based technology has made tremendous progresses over the past several decades from the simple RESS process to a number of modified processes to accommodate various needs. The key features of some commonly used technologies are summarized in Table 19.3.

Analogous to melt extrusion and spray drying, SCF technology offers tremendous flexibility with respect to process design and operation. Despite a number of advantages, its application in the ASD has been limited, and that can be attributed to the lack of robust process understanding. However, continuous improvements over the past two decades have resulted in better equipment design and robust scale-up of such systems (Jung et al. 2003). Its use in the melt extrusion process as temporary plasticizers has contributed significantly to the development of ASDs. Further control of particulate properties with respect to physical stabilization, agglomeration, and densification will continue to grow in the future thus leading to more value-added processes and products.

References

Ayad MH, Bonnet B, Quinton J, Leigh M, Poli SM (2013) Amorphous solid dispersion successfully improved oral exposure of ADX71943 in support of toxicology studies. Drug Dev Ind Pharm 39(9):1300–1305

Badens E, Majerik V, Horváth G, Szokonya L, Bosc N, Teillaud E, Charbit G (2009) Comparison of solid dispersions produced by supercritical antisolvent and spray-freezing technologies. Int J Pharm 377(1–2):25–34

Bartle KD (1988) Theory and principles of supercritical fluid chromatography. Supercritical fluid chromatography. The Royal Society of Chemistry, S. R. M. London, pp 1–28

Beach S, Latham D, Sidgwick C, Hanna M, York P (1999) Control of the physical form of salmeterol xinafoate. Org Process Res Dev 3(5):370–376

Bikiaris DN (2011) Solid dispersions, part I: recent evolutions and future opportunities in manufacturing methods for dissolution rate enhancement of poorly water-soluble drugs. Expert Opin Drug Deliv 8(11):1501–1519

Chen AZ, Li Y, Chau FT, Lau TY, Hu JY, Zhao Z, Mok DK (2009) Microencapsulation of puerarin nanoparticles by poly(l-lactide) in a supercritical CO(2) process. Acta Biomater 5(8):2913–2919

Chiou WL, Riegelman S (1971) Increased dissolution rates of water-insoluble cardiac glycosides and steroids via solid dispersions in polyethylene glycol 6000. J Pharm Sci 60(10):1569–1571

Foster NR, Fariba D, Charoenchaitrakool Kiang M, Warwick B (2003) Application of dense gas techniques for the production of fine particles. AAPS PharmSciTech 5(2):32–38

Gadermann MKS, Al-Marzouqi A, Signorell R (2009) Formation of naproxen/polylactic acid nanoparticles by pulsed rapid expansion of supercritical solutions. Phys Chem Chem Phys 11:7861–7868

Ghaderi R, Artursson P, Carlfors J (2000) A new method for preparing biodegradable microparticles and entrapment of hydrocortisone in DL-PLG microparticles using supercritical fluids. Eur J Pharm Sci 10(1):1–9

Hanna M, York P (1998) Method and apparatus for the formation of particles, Google Patents

Jung J, Clavier JY, Perrut M (2003) Gram to kilogram scale up of supercritical anti-solvent process. Proceedings of the 6th international symposium on supercritical fluids, Versailles, France

Kalogiannis CG, Pavlidou E, Panayiotou CG (2005) Production of amoxicillin microparticles by supercritical antisolvent precipitation. Ind Eng Chem Res 44(24):9339–9346

Kerc J, Srcic S, Kofler B (1998) Alternative solvent-free preparation methods for felodipine surface solid dispersions. Drug Dev Ind Pharm 24(4):359–363

Kim MS, Jin SJ, Kim JS, Park HJ, Song HS, Neubert RH, Hwang SJ (2008) Preparation, characterization and in vivo evaluation of amorphous atorvastatin calcium nanoparticles using supercritical antisolvent (SAS) process. Eur J Pharm Biopharm 69(2):454–465

Knez Z, Weidner E (2003) Particles formation and particle design using supercritical fluids. Curr Opin Solid State Mater Sci 7(4–5):353–361

Lee TW, Boersen NA, Hui HW, Chow SF, Wan KY, Chow AH (2014) Delivery of poorly soluble compounds by amorphous solid dispersions. Curr Pharm Des 20(3):303–324

Lobo JM, Schiavone H, Palakodaty S, York P, Clark A, Tzannis ST (2005) SCF-engineered powders for delivery of budesonide from passive DPI devices. J Pharm Sci 94(10):2276–2288

Muhrer G, Meier U, Fusaro F, Albano S, Mazzotti M (2006) Use of compressed gas precipitation to enhance the dissolution behavior of a poorly water-soluble drug: generation of drug microparticles and drug-polymer solid dispersions. Int J Pharm 308(1–2):69–83

Pathak P, Meziani MJ, Desai T, Sun YP (2004) Nanosizing drug particles in supercritical fluid processing. J Am Chem Soc 126(35):10842–10843

Perrut M (2000) Supercritical fluid applications: industrial developments and economic issues. Ind Eng Chem Res 39(12):4531–4535

Poling BE, Prausnitz JM, John Paul OC, Reid RC (2001) The properties of gases and liquids. McGraw-Hill, New York

Reverchon E, De Marco I, Della Porta G (2002) Rifampicin microparticles production by supercritical antisolvent precipitation. Int J Pharm 243(1–2):83–91

Rodrigues M, Peiriço N, Matos H, Gomes de Azevedo E, Lobato MR, Almeida AJ (2004) Microcomposites theophylline/hydrogenated palm oil from a PGSS process for controlled drug delivery systems. J Supercrit Fluids 29(1–2):175–184

Senčar-Božič P, Srčič S, Knez Z, Kerč J (1997) Improvement of nifedipine dissolution characteristics using supercritical CO_2. Int J Pharm 148(2):123–130

Shekunov BY, Chattopadhyay P, Seitzinger JS (2006) Method and apparatus for enhanced size reduction of particles using supercritical fluid liquefaction of materials, Google Patents

Stegemann S, Leveiller F, Franchi D, de Jong H, Linden H (2007) When poor solubility becomes an issue: from early stage to proof of concept. Eur J Pharm Sci 31(5):249–261

Thakur R, Gupta RB (2006) Formation of phenytoin nanoparticles using rapid expansion of supercritical solution with solid cosolvent (RESS-SC) process. Int J Pharm 308(1–2):190–199

Tozuka Y, Miyazaki Y, Takeuchi H (2010) A combinational supercritical CO2 system for nanoparticle preparation of indomethacin. Int J Pharm 386(1–2):243–248

Valle EMMD, Galan MA (2005) Supercritical fluid technique for particle engineering: drug delivery applications. Rev Chem Eng 21:33

Varshosaz J, Hassanzadeh F, Mahmoudzadeh M, Sadeghi A (2009) Preparation of cefuroxime axetil nanoparticles by rapid expansion of supercritical fluid technology. Powder Technol 189(1): 97–102

Verreck G, Decorte A, Heymans K, Adriaensen J, Cleeren D, Jacobs A, Liu D, Tomasko D, Arien A, Peeters J, Rombaut P, Van den Mooter G, Brewster ME (2005) The effect of pressurized carbon dioxide as a temporary plasticizer and foaming agent on the hot stage extrusion process and extrudate properties of solid dispersions of itraconazole with PVP-VA 64. Eur J Pharm Sci 26(3–4):349–358

Weidner E (2009) High pressure micronization for food applications. J Supercrit Fluids 47(3): 556–565

Williams HD, Trevaskis NL, Charman SA, Shanker RM, Charman WN, Pouton CW, Porter CJ (2013) Strategies to address low drug solubility in discovery and development. Pharmacol Rev 65(1):315–499

Woo JS, Kim HJ, Kim Y (2006) Method for the preparatin of paclitaxel solid dispersion by using the supercritical fluid process and paclitaxel solid dispersion prepared thereby, Google Patents

Yeo S-D, Kiran E (2005) Formation of polymer particles with supercritical fluids: a review. J Supercrit Fluids 34(3):287–308

York P (1999) Strategies for particle design using supercritical fluid technologies. Pharm Sci Technol Today 2(11):430–440

York P, Kompella UB, Shekunov BY (2004) Supercritical fluid technology for drug product development. CRC

Young TJ, Johnson KP, Pace GW, Mishra AK (2004) Phospholipid-stabilized nanoparticles of cyclosporine A by rapid expansion from supercritical to aqueous solution. AAPS PharmSciTech 5(1):E11

Part V
Material Advances

Chapter 20
Supersolubilization by Using Nonsalt-Forming Acid-Base Interaction

Ankita Shah and Abu T. M. Serajuddin

20.1 Introduction

It is estimated that at least two thirds of new chemical entities (NCE) synthesized in the pharmaceutical industry fall in the category of poorly water-soluble drug and mostly belong to the class II of the biopharmaceutical classification system (BCS). Even when the compounds are considered to have poor or low aqueous solubility, the degree of how low the solubility could be differs greatly. Most of the drugs that were considered to be poorly water-soluble in the 1970s and 1980s, and exhibited bioavailability issues, had solubility in the range of 20–100 μg/mL. Two of the first drugs that had solubility of < 10 μg/mL were lovastatin and simvastatin, which were introduced in the market in the late1980s; their reported solubility was ~ 7 μg/mL (Serajuddin et al. (1991). The situation has changed so much during the past two decades that the solubility of < 1 μg/mL is now very common. The aqueous solubility of as low as 4–5 ng/mL has been reported for itraconazole (Six et al. 2005), and another drug, ziprasidone, has such a low aqueous solubility that it could not be determined for its unionized species (Kim et al. 1998). Therefore, depending on how low their solubility are, the same formulation strategies do not apply to all poorly water-soluble compounds, and the applicable technologies to enable their development into bioavailable oral dosage forms differ greatly. Williams et al. (2013) published an excellent article reviewing different strategies that may be applied to the formulation of drugs with low aqueous solubility. They include salt formation, particle size reduction, solid dispersion, and lipid-based drug delivery. Each of the technologies has its own advantages and limitations.

Among the technologies mentioned above, much research has been done on solid dispersion for over 50 years (Chiou and Riegelman 1969; Vasanthavada et al. 2008).

A. T. M. Serajuddin (✉) · A. Shah
Department of Pharmaceutical Sciences, College of Pharmacy and Health Sciences, St. John's University, Queens, NY, USA
e-mail: serajuda@stjohns.edu

N. Shah et al. (eds.), *Amorphous Solid Dispersions*,
Advances in Delivery Science and Technology, DOI 10.1007/978-1-4939-1598-9_20

However, despite its great promise, only a very limited number of products, formulated based on the solid dispersion principles, have been marketed. One of the major limitations of using these systems is that there could be physical and chemical instability of drugs in polymeric carriers (Serajuddin 1999). There is also the risk of drug precipitation in vivo due to the inability of the system to maintain a supersaturated state in the gastrointestinal (GI) fluid after oral administration. In case of weakly acidic and basic drugs, which exhibit pH-dependent solubility profiles, the drug release from conventional solid dispersions could also be pH dependent, and it may not be possible to maintain a steady drug release from dosage forms at different pH environments of the GI tract (Doherty and York 1989; Tran et al. 2010a).

The objective of this chapter is to present how weak acids and weak bases can be used as carriers in solid dispersions to modulate microenvironmental pH, and improve the dissolution rate of acidic or basic drugs. There are, however, limits on how much improvement in dissolution rate may be achieved by simply mixing acidic and basic excipients with, respectively, weakly basic and weakly acidic drugs. Recently, Singh et al. (2013) developed a novel approach of greatly increasing the solubility of basic drugs by interacting with weak acids that would not normally form salts with the drugs used. The aqueous solubility of haloperidol, which has an intrinsic solubility of 2.5 μg/mL, could be increased as high as > 300 mg/g of solution by using such weak acids as malic acid and tartaric acid. The authors called it the supersolubilization of drug. Since these acids did not form salts with haloperidol, the drug could be converted to amorphous forms by drying the highly concentrated drug solutions. In this chapter, the development of solid dispersions by such an acid–base interaction will be presented. This chapter will specifically contain the following topics:

- A review of how acidic and basic excipients have been used to modulate microenvironmental pH and improve the dissolution of drugs.
- Theory of supersolubilization by acid–base interaction.
- Superiority of supersolubilization over salt formation and conventional pH modulation
- Development of solid dispersion by using the supersolubilization principle.

20.2 Modulation of Microenvironmental pH

20.2.1 Microenvironmental pH

As acidic and basic drugs demonstrate pH-dependent solubility, their dissolution rates in aqueous media having different pH values may differ greatly. The Noyes–Whitney equation describing the relationship between dissolution rate and solubility (Noyes and Whitney 1897) is given below:

$$J = KA(C_s - C) \tag{20.1}$$

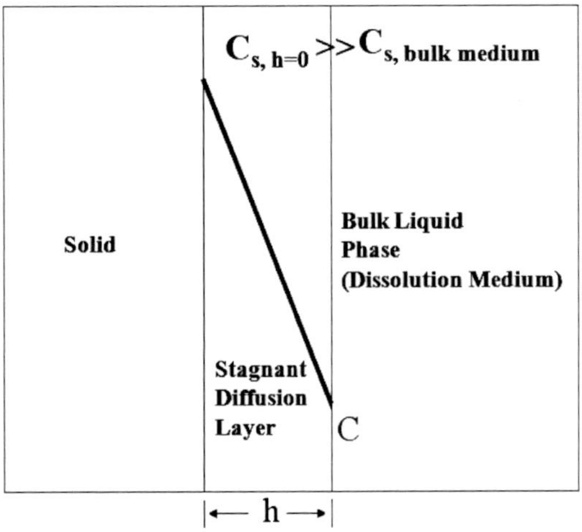

Fig. 20.1 Schematic diagram of diffusion layer on dissolving solid surface, where h represents diffusion layer thickness, C_s is the saturation solubility of solid, and C is the concentration in the bulk medium

where J is the dissolution rate, K is a constant, A is the surface area of the dissolving solid, C_s is the saturation solubility of the compound in the dissolution medium, and C is the concentration of the drug in the medium at different time points during dissolution. The difference between C_s and C in the equation represents the concentration gradient in the dissolution medium. The Noyes–Whitney equation was modified to

$$J = \frac{DA}{h}(C_s - C), \tag{20.2}$$

by considering that the solid drug dissolves instantly in a thin unstirred layer of dissolution medium surrounding the dissolving solid. This is shown schematically in Fig. 20.1. The transfer of the drug from the surface of the solid to the bulk dissolution medium outside this layer occurs through the diffusion of the solute from the solid surface through this layer, and this layer is also called the diffusion layer. The thickness of the diffusion layer and the diffusion coefficient of the drug through this layer are denoted in Eq. 20.2 by h and D, respectively, and at the nearest to the surface of solid drug where the thickness of the diffusion layer approaches zero, the concentration of drug in the diffusion layer approaches the saturation solubility C_S. Since the dissolution rate of a drug is dependent on its saturation solubility in the diffusion layer, it has been reported that the solubility of a drug under the pH condition of the diffusion layer rather than that in the bulk medium dictates the drug dissolution rate (Serajuddin and Jarowski 1985a,1985b). The pH in the diffusion layer, especially at the surface of the solid, represents the microenviromental pH (Pudipeddi et al. 2008). It is essentially the pH of the saturated solution in the immediate vicinity of drug particles.

20.2.2 Modulation of Microenvironmental pH by Salt Formation

According to the Noyes–Whitney equation described in the previous section, the dissolution rate of a drug is directly related to the C_s or saturation solubility. Therefore, if the pH of the diffusion layer is modified or modulated such that it becomes more favorable to an increase in drug solubility, the dissolution rate of the drug will increase accordingly.

The most common method of modulating microenvironmental pH is the salt formation. Since the solubility of salts are usually much higher than those of their corresponding base or acid forms, as shown in Fig. 20.1, the solubility of the drug in the diffusion layer at the surface of the salt is much higher than those in the bulk medium. As a result, the dissolution rate of the drug increases due to the salt formation (Serajuddin 2007). Using phenazopyridine hydrochloride and its free-base form, Serajuddin and Jarowski (1985a) demonstrated good correlation between the saturation solubility and the dissolution rate when the solubility under the microenvironmental pH, rather than the solubility under the bulk pH condition, was used. Since the acidic and basic drugs exhibit the pH-dependent solubility, there is a potential that the drug may precipitate out in the GI fluids after dissolution. The free acid may precipitate out from salts at the relatively lower pH of the stomach (pH 1–3), while the free base may precipitate out at the relatively higher pH of the intestine (pH 5.5–7.5). However, the precipitation usually occurs as fine particles with a large surface area, which redissolve rapidly as the dissolution continues in the GI tract (Serajuddin and Jarowski 1993).

Despite the potential advantage of salt formation, many compounds do not form salts. Even when salts are formed, they may have limited aqueous solubility. For example, aqueous solubilities of hydrochloride and mesylate salt forms of the weakly basic drug ziprasidone were 80 and 890 μg/mL, respectively, and, because of such low solubility, the dissolution of salts was incomplete, and the human bioavailability was variable (Thombre et al. 2011). There is also the potential that the salts may convert to the less soluble free base or acid form in the diffusion layer, thus coating the surface of the dissolving solid, and preventing further dissolution of the salt (Serajuddin 2007).

20.2.3 Modulation of Microenvironmental pH by Using pH Modifiers

Another approach to increase dissolution rates of poorly water-soluble acidic and basic drugs is to use microenvironmental pH modifiers (Badawy and Hussain 2007; Phuong et al. 2011; Tran et al. 2009). Generally, weak organic acids, such as ascorbic, adipic, citric, fumaric, glutaric, succinic, malic, tartaric acids, are used to decrease microenvironmental pH of weakly basic drugs. Various weak bases, such as sodium carbonate, sodium bicarbonate, magnesium oxide, calcium carbonate, potassium phosphate, sodium phosphate, and so forth, are used to increase the microenvironmental pH and thereby enhance the dissolution rate of poorly water-soluble weakly acidic drugs (Bi et al. 2011b; Tran et al. 2009, 2011). The pH modifiers may also prevent or retard conversion of salts into their unionized forms and, therefore, the dissolution rates remain high (Badawy et al. 2006). In addition to acids and bases, many researchers used polymeric carriers to modulate the local pH of dissolving solid and maintain supersaturation (Rao et al. 2003; Tatavarti et al. 2004; Tatavarti and Hoag 2006). The extent of pH modulation may be estimated by measuring microenvironmental pH using different methods described by Pudipeddi et al. (2008). Various physicochemical properties of pH modifiers, such as solubility, dissolution rate, and pK_a, would influence pH modulation and thereby dissolution rates of drugs. To obtain the complete dissolution of the drug from dosage form, the pH modifier may need to coexist with the drug in the formulation until the drug is completely dissolved. Accordingly, the excipients used and the manufacturing methods applied would influence the dissolution of drugs from the pH-modified solid dosage forms.

20.2.4 Use of pH Modifiers in Solid Dispersion

Many of the published reports in the literature on the use of pH modifiers in solid dosage forms are for controlling the drug release from modified-release dosage forms, such as matrix tablets and coated tablets and beads (Gohel et al. 2003; Kranz et al. 1969; Naonori et al. 1991; Siepe et al. 2006; Streubel et al. 2000; Tatavarti and Hoag 2006; Thoma and Zimmer 1990). The pH modifiers have also been used to prevent disproportion of salts in tablets (Zannou et al. 2007). The use of pH modifiers in solid dispersion to enhance dissolution rate of drugs is, however, rather limited.

As early as in 1969, Chiou and Riegelman (1969) reported the use citric acid in forming solid dispersion of griseofulvin, a poorly soluble drug, by mixing the components together in their molten state at high temperature. Later, Timko and Lordi (1979) also used citric acid to prepare solid dispersions of benzoic acid and phenobarbital by applying a similar method. However, in these studies, citric acid was used only as the water-soluble carrier, and not for their ability to modulate microenvironmental pH.

More recently, Tran et al. (2010b) evaluated the effect of four acidifiers, namely, citric acid, fumaric acid, glycolic acid, and malic acid, on the dissolution rate of a poorly soluble weakly basic drug, isradipine, from the solid dispersion system containing polyvinylpyrrolidone (PVP) as the polymeric carrier. The solubility of the drug at pH 1.2 and pH 7 (deionized water) were 114 and 7 μg/mL, respectively. By considering the higher drug solubility at a lower pH, it was expected that the incorporation of an acidifier in the solid dispersion would increase the dissolution rate of the drug. Each 100-mg tablet contained 5 mg of drug and10 mg of an acidifier. The pH of an aqueous suspension of the tablet components was 2.2–2.3, representing the microenvironmental pH of the tablets. Two sets of tablet formulations were prepared, where, in one case, the components were physically mixed before compressing them into tablets; and in the second case, solid dispersions of the drug, acidifiers, and PVP were prepared by the solvent evaporation method prior to compressing them into tablets. Overall, the use of acidifiers increased the dissolution rate of the drug as compared to the systems without acidifiers. The solid dispersion containing fumaric acid showed more than 90 % drug release in 60 min while the physical mixtures had only 50 % drug release at the same period of time.

Tran et al. (2009) also explored the importance of microenvironmental pH-modulating alkalizers in the preparation of self-emulsifying solid dispersions. Aceclofenac, a weakly acidic drug (pK_a 4–5), having extremely low solubility in acidic media and good solubility at higher pH, was used as the model drug. Ternary solid dispersions were prepared using the drug, Gelucire® 44/14 (the matrix) and an alkalizer at the weight ratio 70:70:28. The incorporation of an alkalizer increased the dissolution rate of the drug as compared to the dissolution when no alkalizer was present. Among various alkalizers used, Na_2CO_3 showed the highest drug release. The authors concluded that the three main parameters for enhancement of dissolution rate of aceclofenac from the solid dispersion were modulation of microenvironmental pH, change of drug substance from the crystalline to the amorphous form, and the self-emulsifying property of the matrix.

Schilling et al. (2008) reported that citric acid monohydrate can be used as the modifying agent for drug release from tablets prepared by melt extrusion of diltiazem HCl in Eudragit® RS PO. Being highly water-soluble, citric acid increased pore formation in the tablet matrix during dissolution, and thus enhanced the drug release. The drug converted to the amorphous form when a large amount of citric acid was used. The acid also served as the plasticizing agent during melt extrusion. In another study, the solid dispersion containing sibutramine, HPMC, gelatin, and citric acid showed the highest drug solubility of 5 mg/mL in water as compared to neat base which had the solubility of 0.01 mg/mL (Lim et al. 2010).

As mentioned earlier, the dissolution rate of basic drugs may also be increased by pH modulation. Telmisartan is a practically insoluble weakly acidic drug and shows pH-dependent solubility. Marasini et al. (2013) reported that its solubility, dissolution rate, and bioavailability in a rat model can be increased substantially by forming solid dispersion using PVP K30 as a carrier and Na_2CO_3 as an alkalizer.

For a weakly acidic drug, AMG009 with poor intrinsic solubility (0.6 μg/mL), Bi et al. (2011a, 2011b) prepared solid dispersions by using various nucleation inhibitors, such as HPMCE5 LV, HPMC K100 LV, plasdone K-17 (PVP K17); pH-modifiers, such as sodium carbonate, sodium bicarbonate, tromethamine, sodium acetate, sodium phosphate; and sodium citrate dihydrate. Both the nucleation inhibitor and the pH modifier were necessary to increase the dissolution rate and maintain supersaturation of the drug in the dissolution media.

As the above examples demonstrate, there have been several attempts to increase the dissolution rates of basic and acidic drugs by solid dispersion in the presence of pH modifiers. Any increase in solubility and dissolution rate of formulations due to pH modulation has, however, been rather modest. It was dependant on how much change in pH in the microenviroment could be achieved by pH modulation and what would be the increase in drug solubility due to such a change in pH. In

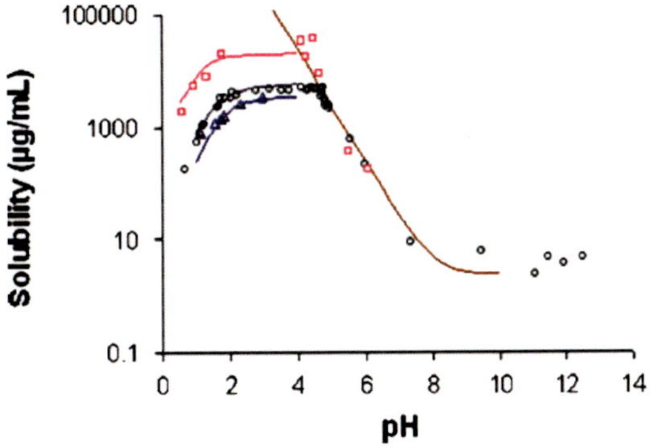

Fig. 20.2 pH-solubility profile of haloperidol at 37 °C determined by using methanesulfonic □, hydrochloric ○, and phosphoric Δ acids to adjust pH. (Adapted from Li et al. 2005b)

most of such solid dispersions, the increase in dissolution rates of drugs was due to the conversion of drugs to their amorphous forms. Unless the drug substances were converted to their amorphous forms by melting, melt extrusion, solvent evaporation, lyophilization, etc., the effect of the pH modifier would be limited.

20.3 Supersolubilization and Amorphization by Acid–Base Interaction

In 2013, Singh et al. (2013) reported a novel approach of increasing the solubility of basic drugs by acid–base interaction, which was different from the salt formation. They used a model drug, haloperidol, which had the intrinsic free-base solubility of 2.5 μg/mL and, as shown in Fig. 20.2, the maximum solubility of 1, 4, and 30 mg/mL, for, respectively, phosphate, hydrochloride, and mesylate salt forms (Li et al. 2005a, 2005b). No other salts for haloperidol could be prepared. By adding malic, tartaric, and citric acids to aqueous solutions, Singh et al. (2013) could increase the aqueous solubility of haloperidol to >300 mg/g of solution. In terms of amounts of water present in saturated solutions, the high solubility of 1.1, 1.3, and 0.8 g/mL of water could be obtained in the presence of malic, tartaric, and citric acids, respectively. The increase was 300,000–500,000 times larger than the solubility of the haloperidol free base, which the authors called supersolubilization. When such highly concentrated solutions were dried, the drug and most of the added acids converted to amorphous forms.

20.3.1 Theory of Supersolubilization

Singh et al. (2013) developed the theory of supersolubilization based on pH–solubility interrelationships for basic and acidic drugs, which were originally developed to explain their salt formation. The interrelationship of the solubility of a

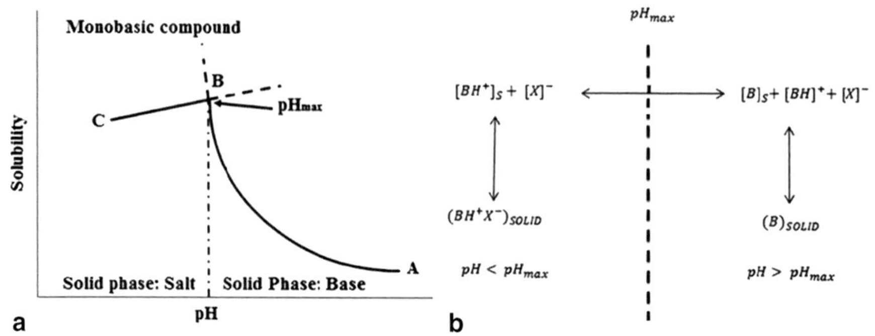

Fig. 20.3 Schematic representation of the interrelationship between the solubility of the base and its salt form as a function of pH. **a** *Line* AB represents free base as the equilibrium species *above* the pH_{max} and *line* BC represents salt as equilibrium species *below* the pH_{max}. **b** Different species formed at pH *above* and *below* the pH_{max}, when the pH of the suspension of a basic drug, B, is lowered by the addition of an acid, HX. The subscript "s" in $[BH^+]_s$ and $[B]_s$ indicates saturation. (Reprinted with the permission from Serajuddin and Pudipeddi 2002)

base, B, and its salt form, BH^+, as a function of pH was described by Kramer and Flynn (1972) by two independent curves, one where the free base is the equilibrium species and the other where the salt is the equilibrium species. This relationship is given in Fig. 20.3a, where the line A to B indicates that when an acid is added gradually to the aqueous suspension of a basic drug, the pH decreases and the solubility of the drug increases. However, when the pH decreases below a certain pH, which is called pH_{max} or the pH of maximum solubility, a phase transition occurs and the basic drug crystallizes into its salt form. This phenomenon is depicted by the line B→C in Fig. 20.3a. Thus, the free base and the salt are, respectively, the equilibrium species at pH above and below the pH_{max}, and only at the point of pH_{max}, both of them may coexist. Various acid-based equilibriums involved with the salt formation may be illustrated schematically in by Fig. 20.3b. Additionally, the solubility expressed by the line A→B above the pH_{max} and the line B→C below the pH_{max} may be depicted, respectively, by the following equations:

$$\begin{aligned} S_{T\text{base(pH>pHmax)}} &= [B]_S + [BH^+] = [B]\left(1 + \frac{H_3O^+}{K_a}\right), \\ &= [B]_S\left(1 + 10^{pK_a - pH}\right) \end{aligned} \tag{20.3}$$

$$\begin{aligned} S_{T\text{ salt (pH<pHmax)}} &= [BH^+]_S + [B] = [BH^+]\left(1 + \frac{K_a}{H_3O^+}\right) \\ &= [BH^+]_S\left(1 + 10^{pH - pK_a}\right) \end{aligned} \tag{20.4}$$

Singh et al. (2013) explored whether any deviation from or nonconformity with the classical pH–solubility interrelationships described by Fig. 20.3 and Eqs. 20.3 and 20.4 may be beneficially applied to the development of drug products. As shown in Fig. 20.3a, the acid added must remain fully ionized (X^-) below the pH_{max} to form a salt with the protonated base (BH^+). Singh et al. (2013) questioned what would happen if, say, the acid added to form the counterion is not fully ionized and remains only partially ionized at $pH < pH_{max}$. In such a situation, the acid–base equilibrium described in Fig. 20.3b will no longer exist as it will lead to three species in solution ($[BH^+]$, $[X^-]$, and [HX]) rather than two ($[BH^+]$ and $[X^-]$) necessary to form a salt (Serajuddin and Pudipeddi 2002). Consequently, no salt will be formed. Since Eq. 20.3 is essentially a modified version of the classical Henderson–Hasselbach equation, the solubility of a basic compound should increase indefinitely according to this equation if the pH is gradually lowered and no salt is crystallized out. In other words, the line A→B in Fig. 20.3a would extend indefinitely in the direction of B as the pH is decreased. Indeed, Singh et al. (2013) observed such an increase in solubility when weak nonsalt-forming acids were used to decrease pH and took advantage of this phenomenon in the supersolubilization of haloperidol.

20.4 Development of Solid Dispersion by Supersolubilization

Based on the solubility versus pH principles described, Singh et al. (2013) investigated the effect of adding weak organic acids on the solubility of a basic drug haloperidol (pK_a: 8). Malic acid (pK_a: 3.40, 5.13), tartaric acid (pK_a: 3.11, 4.80), and citric acid (pK_a: 3.12, 4.76, 6.39) were used as the acids, and, in all cases, the pH decreased by the addition of acids, and the aqueous solubility of haloperidol increased greatly. The results are shown in Fig. 20.4. Since, as mentioned earlier, the acids added did not form salts with haloperidol, the solubility of haloperidol continued to increase according to Eq. 20.3. It should be noted that the pH remained practically constant when it dropped to around 1.0–1.5; however, more drug dissolved as more acid was added to the solution and, ultimately, the aqueous solubility of the acid was the limiting factor in how much drug could be dissolved in a solution. In case of malic acid, > 0.30 g of haloperidol dissolved per gram of solution. Similar high solubility of haloperidol in presence of tartaric acid and citric acid were also observed (~ 0.30 and 0.25 g, respectively, per gram of solution). Each gram of such a solution contained drug, acid and water. Therefore, in terms of water, these were extremely high solubility as compared to the solubility of haloperidol salts shown in Fig. 20.4. For example, the maximum haloperidol solubility of 0.33 g per gram of solution was observed in presence of malic acid, where the amounts of malic acid and water present were, 0.37 and 0.30 g, respectively. Thus, when the solubility was calculated in terms of the water present, the solubility was 1.1 g/g of water. Considering that the highest solubility for any haloperidol salt observed was 30 mg or 0.03 g/mL of solution, this was a very high increase in drug solubility. The drug dissolved simply in the aqueous medium of water when the weak organic acids were added, without necessitating the addition of any organic solvents, complexing agents, etc. When the concentrated solutions were dried, they neither formed crystalline salts nor did they convert into crystalline free-base forms of the drug. Rather, the drug converted into

Fig. 20.4 pH-solubility profiles of haloperidol using weak organic acids. (Reprinted with the permission from Singh et al. 2013)

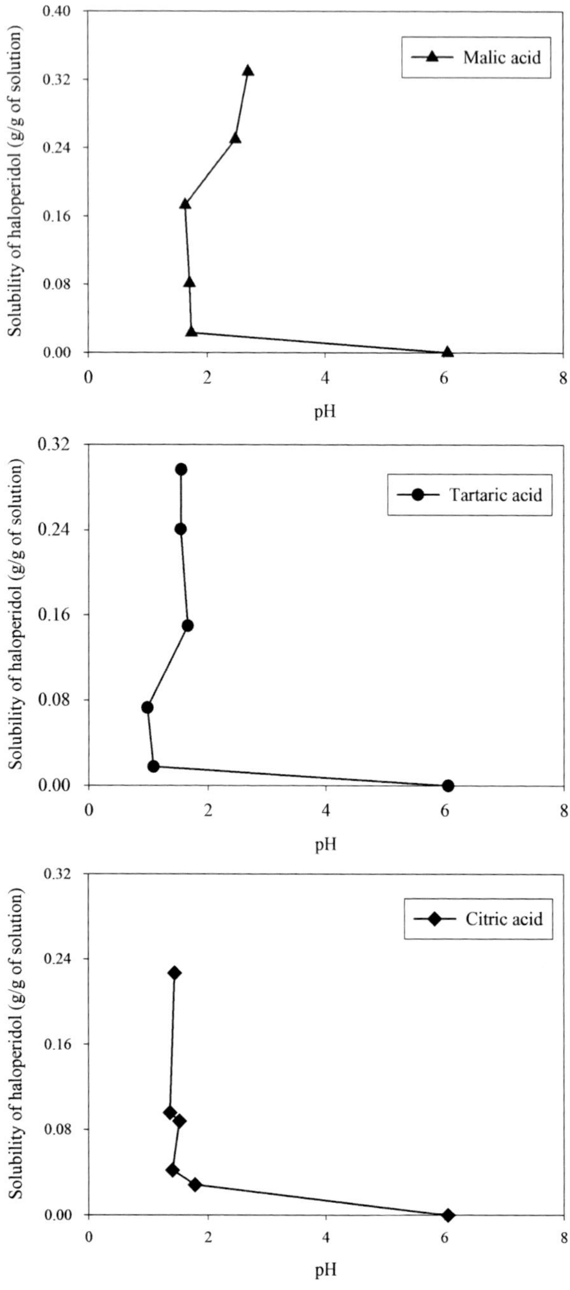

the amorphous form, which did not crystallize upon exposure to different accelerated stability conditions. These findings were exploited by Singh et al. (2013) in the development of novel solid dispersion systems for the model basic drug haloperidol.

Fig. 20.5 Powder XRD patterns of **a** haloperidol, **b** malic acid, and haloperidol–malic acid combination with molar ratios of **c** 0.05:1, **d** 0.1:1, **e** 0.14:1, **f** 0.16:1, **g** 0.18:1, and **h** 0.29:1 showing conversion of crystalline haloperidol and malic acid into amorphous form with the increase in molar ratio. (Reprinted with the permission from Singh et al. 2013)

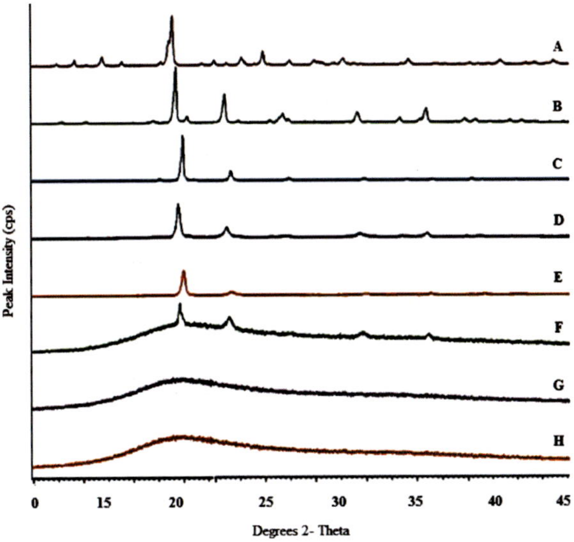

To prepare solid dispersions, Singh et al. (2013) dissolved haloperidol and weak organic acids (malic, tartaric, and citric acids) in water, where the molar ratios between the drug and the acid were increased gradually. The highest molar ratios of the drug to weak acids in solutions were 0.29:1, 0.24:1, 0.12:1, respectively, for malic acid, tartaric acid, and citric acid. All aqueous solutions were dried under vacuum for 7 days at 40 °C. Physical states of the dried solid dispersions of haloperidol in the acids were characterized by powder X-ray diffraction analysis (PXRD) and differential scanning calorimetry (DSC). PXRD and DSC analyses of haloperidol solid dispersions and malic acid are given in Figs. 20.5 and 20.6, respectively, as representative examples. At all drug to acid ratios, no crystallinity of the drug was observed by PXRD and DSC analyses. There was a gradual decrease in crystallinity of weak organic acids with the increase in drug concentration in dry solids with complete amorphization of both drug and acid at high drug to acid molar ratios. Thus, amorphous solid dispersions of the basic drug haloperidol in weak organic acids could be obtained by the supersolubilization of the drug in water followed by drying.

Figure 20.7 shows the results of the multi-step dissolution testing of the solid dispersions of haloperidol prepared using malic acid, tartaric acid, and citric acid. In the first step, the dissolution was carried out in 250 mL of a pH 2 dissolution medium (0.01N HCl) for 120 min, which was then followed for another 30 min by changing the pH to 4.5, and finally the pH was changed to 6.8 for another 90 min of dissolution testing. In all cases, dissolution rates of dry solids were much faster than that of the control haloperidol HCl powder. The solid dispersions showed > 85 % drug dissolution in 15 min at low molar ratios between haloperidol and an acid, indicating rapid drug dissolution. However, incomplete drug release was observed for all three acids when the ratios between haloperidol to weak acid were high. The authors reported that the physical nature of solid dispersions was responsible for incomplete

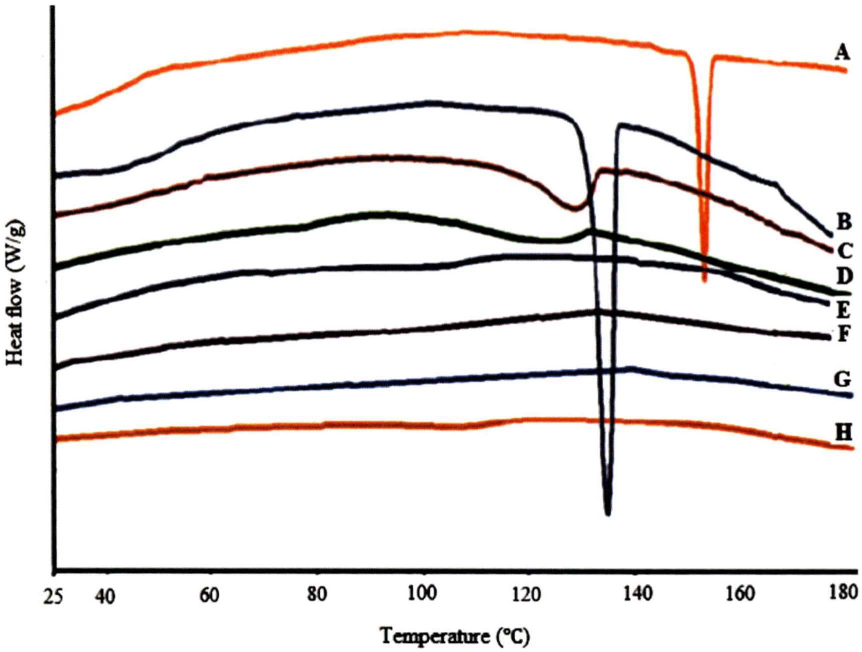

Fig. 20.6 DSC scans of **a** haloperidol, **b** malic acid, and haloperidol–malic acid combinations of **c** 0.05:1, **d** 0.1:1, **e** 0.14:1, **f** 0.16: 1, **g** 0.18:1, and **h** 0.29:1 molar ratios between haloperidol and malic acid. (Reprinted with the permission from Singh et al. 2013)

release. The dry solids with high amount of haloperidol were semisolid and viscous in nature, and thus difficult to disperse in the dissolution medium. Although the drug release was incomplete, the solid dispersions exhibited higher dissolution rates than that of its hydrochloride salt form and maintained high supersaturation even after the change in pH from 2 to 6.8. This can be observed by comparing Fig. 20.7 with 20.2, where the later figure shows that the solubility of haloperidol at pH 6.8 is extremely low.

Shah and Serajuddin (2014) conducted further studies to address the issue of incomplete drug release from solid dispersions of haloperidol in weak organic acids reported by Singh et al. (2013) and demonstrated in Fig. 20.7. They adsorbed the solid dispersions at 1.5:1 w/w ratios onto Neusilin® US2, which is chemically a magnesium aluminometasilicate, to convert the semisolid and viscous solid dispersions into free-flowing powders (Fig. 20.8). Earlier, Gumaste et al. (2013a, b) successfully loaded liquid microemulsion preconcentrates onto Neusilin® US2 to convert them into free-flowing and tabletable powders. Shah and Serajuddin (2014) determined that the solid dispersion could be adsorbed on to Neusilin® US2 at the ratio of as high as 1.5:1 solid to the silicate ratio and the powders were still free flowing and tabletable. The granules were vacuum dried, milled, and sieved before the compression into tablets. The multi-step dissolution of these tablets showed complete release

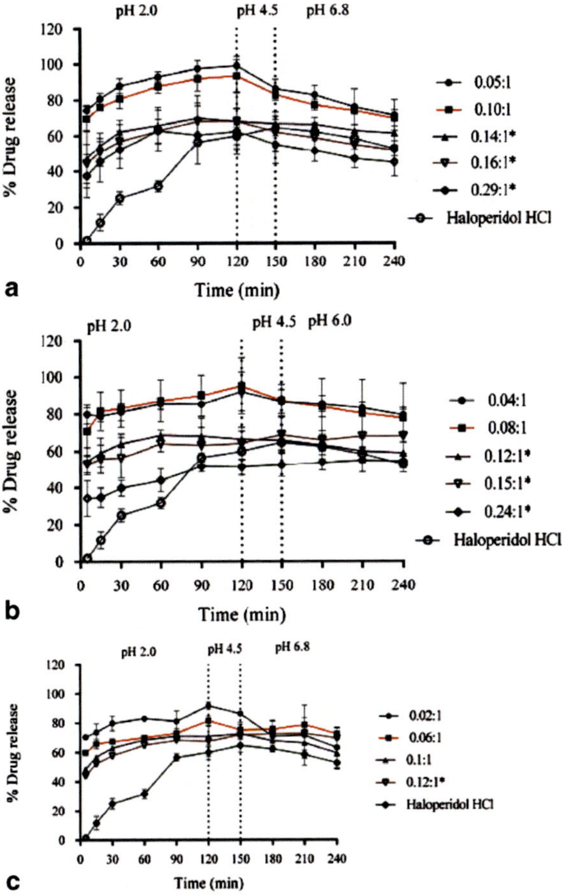

Fig. 20.7 Comparative step dissolution profiles of amorphous haloperidol from combinations with **a** malic acid, **b** tartaric acid, and **c** citric acid. Results for different haloperidol to acid ratios are given, and the dissolution profile of haloperidol HCl is shown for comparison. The *asterisk* represents the molar ratios that existed as sticky viscous mass during dissolution testing. (Reprinted with the permission from Singh et al. 2013)

of the drug as compared to only 40–50 % release with dry solids prepared by Singh et al. (2013) (Fig. 20.9). Thus, the supersolubilization of the basic drug did not only lead to the development of amorphous and stable solid dispersions, the materials could also be converted successfully into tablets by adsorbing them onto solid carriers. Shah and Serajuddin (2014) further demonstrated that a similar approach can also be applied to the formulation of solid dispersions for acidic drugs by using weak bases as the supersolubilizing agents (data not shown).

20.5 Concluding Remarks

As reported in various chapters of this book, there are many different strategies to prepare solid dispersions of poorly water-soluble drugs. In most cases, the drugs are converted into amorphous forms to increase solubility, and, thereby, their dissolution

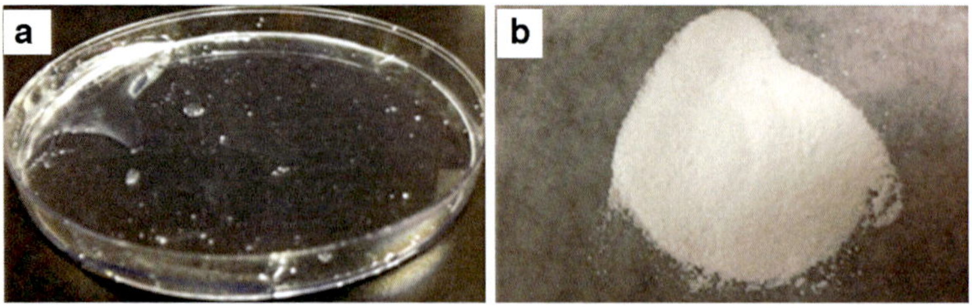

Fig. 20.8 Haloperidol–malic acid (molar ratio 0.29:1) dry solid: **a** without Neusilin® US2, **b** with Neusilin® US2

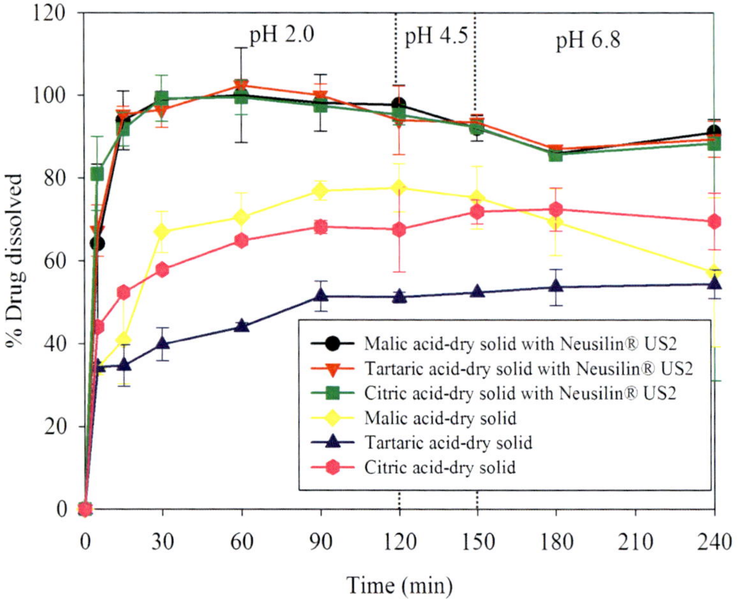

Fig. 20.9 Comparative multi-step dissolution profiles of haloperidol–weak acid dry solids with and without Neusilin® US2. Drug amount was kept 150 mg in each profile

rates from dosage forms. This chapter reviews how the microenvironmental pH modulation may be applied to the development of solid dispersions. In most cases reported in the literature, drugs were converted into amorphous forms by conventional means and pH modifiers were used to modulate microenvironmental pH such that the weakly acidic and basic drugs continued to dissolve even when the bulk pH conditions were unfavorable for dissolution. The overall impact of such microenvironmental pH modulation in increasing the dissolution rates of drugs has, however, been limited. For example, if the intrinsic solubility of a basic or acidic drug is increased tenfold from 5 to 50 μg/mL by microenvironmental pH modulation, the increase may not be able to ensure complete drug release. In some cases, the pH modifiers were used in combination with the conversion of drugs to amorphous forms. Physical stability issues of the amorphous forms, however, still remain.

In the present report, a novel supersolubilization technique to greatly increase solubility and dissolution rates of basic drugs by adding nonsalt-forming weak acids has been described. For a basic drug, haloperidol, the increase in its aqueous solubility in presence of malic, tartaric, and citric acids was as high as 300,000–500,000 times greater compared to the free-base solubility of 2.5 μg/mL. The important consideration was that the acids used to solubilize basic drugs had to have high aqueous solubility. When the concentrated solutions of haloperidol were dried, they formed physically stable solid dispersions, where the drug loads could be as high as 40–50 % w/w. Different types of chemical interactions could be involved in stabilizing the systems, such as electrostatic interactions, ion-pair interaction, H-bonding formation between the drug and the multifunctional acidic or basic excipients (Kadoya et al. 2008). The increased dissolution rates of such solid dispersion is due to a very high drug solubility when they come in contact with the aqueous media.

When the drug load was very high, the solid dispersion could, however, exist as a viscous semisolid mass and may not be processable into tablets. Such materials may not also disperse in aqueous media and, therefore, may exhibit incomplete drug release. The issues were resolved by adsorbing the solid dispersion on a metasilicate (Neusilin® US2). Further studies showed that the supersolubilization technology has general application in the development of solid dispersion systems for both basic and acidic drugs.

References

Badawy SIF, Hussain MA (2007) Microenvironmental pH modulation in solid dosage forms. J Pharm Sci 96:948–959

Badawy SIF, Gray DB, Zhao F, Sun D, Schuster AE, Hussain MA (2006) Formulation of solid dosage forms to overcome gastric pH interaction of the factor Xa inhibitor, BMS-561389. Pharm Res 23:989–996

Bi M, Kyad A, Alvarez-Nunez F, Alvarez F (2011a) Enhancing and sustaining AMG 009 dissolution from a bilayer oral solid dosage form via microenvironmental pH modulation and supersaturation. AAPS Pharm Sci Tech 12:1401–1406

Bi M, Kyad A, Kiang Y, Alvarez-Nunez F, Alvarez F (2011b) Enhancing and sustaining AMG 009 dissolution from a matrix via microenvironmental pH modulation and supersaturation. AAPS Pharm Sci Tech 12:1157–1162

Chiou WL, Riegelman S (1969) Preparation and dissolution characteristics of several fast-release solid dispersions of griseofulvin. J Pharm Sci 58:1505–1510

Doherty C, York P (1989) Microenvironmental pH control of drug dissolution. Int J Pharm 50:223–232

Gohel MC, Patel TP, Bariya SH (2003) Studies in preparation and evaluation of pH-independent sustained-release matrix tablets of verapamil HCl using directly compressible Eudragits. Pharm Dev Technol 8:323–333

Gumaste SG, Dalrymple DM, Serajuddin AT (2013a) Development of solid SEDDS, V: Compaction and drug release properties of tablets prepared by adsorbing lipid-based formulations onto Neusilin® US2. Pharm Res 30:3186–3199

Gumaste SG, Pawlak SA, Dalrymple DM, Nider CJ, Trombetta LD, Serajuddin AT (2013b) Development of solid SEDDS, IV: Effect of adsorbed lipid and surfactant on tableting properties and surface structures of different silicates. Pharm Res 30:3170–3185

Kadoya S, Izutsu K, Yonemochi E, Terada K, Yomota C, Kawanishi T (2008) Glass-state amorphous salt solids formed by freeze-drying of amines and hydroxy carboxylic acids: effect of hydrogen-bonding and electrostatic interactions. Chem Pharm Bull 56:821–826

Kim Y, Oksanen DA, Massefski Jr W, Blake JF, Duffy EM, Chrunyk B (1998) Inclusion complexation of ziprasidone mesylate with β-cyclodextrin sulfobutyl ether. J Pharm Sci 87:1560–1567

Kramer S, Flynn G (1972) Solubility of organic hydrochlorides. J Pharm Sci 61:1896–1904

Kranz H, Guthmann C, Wagner T, Lipp R, Reinhard J (2005) Development of a single unit extended release formulation for ZK 811 752, a weakly basic drug. Eur J Pharm Sci 26:47–53

Li S, Wong S, Sethia S, Almoazen H, Joshi YM, Serajuddin AT (2005a). Investigation of solubility and dissolution of a free base and two different salt forms as a function of pH. Pharm Res 22:628–635

Li S, Doyle P, Metz S, Royce AE, Serajuddin A (2005b) Effect of chloride ion on dissolution of different salt forms of haloperidol, a model basic drug. J Pharm Sci 94:2224–2231

Lim H, Balakrishnan P, Oh DH, Joe KH, Kim YR, Hwang DH et al (2010) Development of novel sibutramine base-loaded solid dispersion with gelatin and HPMC: physicochemical characterization and pharmacokinetics in beagle dogs. Int J Pharm 397:225–230

Marasini N, Tran TH, Poudel BK, Cho HJ, Choi YK, Chi S et al (2013) Fabrication and evaluation of pH-modulated solid dispersion for telmisartan by spray-drying technique. Int J Pharm 441: 424–432

Naonori K, Hiroshi Y, Ken I, Katsumi M (1991) A new type of a pH-independent controlled release tablet. Int J Pharm 68:255–264

Noyes AA, Whitney WR (1897) The rate of solution of solid substances in their own solutions. J Am Chem Soc 19:930–940

Phuong HT, Tran TT, Lee SA, Nho VH, Chi S, Lee B (2011) Roles of MgO release from polyethylene glycol 6000-based solid dispersions on microenvironmental pH, enhanced dissolution and reduced gastrointestinal damage of telmisartan. Arch Pharm Res 34:747–755

Pudipeddi M, Zannou EA, Vasanthavada M, Dontabhaktuni A, Royce AE, Joshi YM et al (2008) Measurement of surface pH of pharmaceutical solids: a critical evaluation of indicator dye-sorption method and its comparison with slurry pH method. J Pharm Sci 97:1831–1842

Rao VM, Engh K, Qiu Y (2003) Design of pH-independent controlled release matrix tablets for acidic drugs. Int J Pharm 252:81–86

Schilling SU, Bruce CD, Shah NH, Malick AW, McGinity JW (2008) Citric acid monohydrate as a release-modifying agent in melt extruded matrix tablets. Int J Pharm 361:158–168

Serajuddin A (1999) Solid dispersion of poorly water-soluble drugs: early promises, subsequent problems, and recent breakthroughs. J Pharm Sci 88:1058–1066

Serajuddin A (2007) Salt formation to improve drug solubility. Adv Drug Deliv Rev 59:603–616

Serajuddin A, Jarowski CI (1985a) Effect of diffusion layer pH and solubility on the dissolution rate of pharmaceutical bases and their hydrochloride salts I: phenazopyridine. J Pharm Sci 74:142–147

Serajuddin A, Jarowski CI (1985b) Effect of diffusion layer pH and solubility on the dissolution rate of pharmaceutical acids and their sodium salts II: salicylic acid, theophylline, and benzoic acid. J Pharm Sci 74:148–154

Serajuddin ATM, Jarowski CI (1993) Influence of pH on release of phenytoin sodium from slow-release dosage forms. J Pharm Sci 82:306–310

Serajuddin ATM, Pudipeddi M (2002) Solubility and dissolution of weak acids, bases, and salts. In: Stahl PH, Wermuth CG (eds) IUPAC handbook of pharmaceutical salts: properties, selection, and use. Verlag Helvetica Chimica Acta, Zurich, pp 19–39

Serajuddin A, Ranadive SA, Mahoney EM (1991) Relative lipophilicities, solubilities, and structure-pharmacological considerations of 3-hydroxy-3-methylglutaryl-coenzyme a (HMG-COA) reductase inhibitors pravastatin, lovastatin, mevastatin, and simvastatin. J Pharm Sci 80:830–834

Shah A, Serajuddin A (2014) Personal communication. (Submitted to Journal of Excipients and Food Chemicals, an Open Access journal)

Siepe S, Herrmann W, Borchert H, Lueckel B, Kramer A, Ries A et al (2006) Microenvironmental pH and microviscosity inside pH-controlled matrix tablets: an EPR imaging study. J Control Release 112:72–78

Singh S, Parikh T, Sandhu HK, Shah NH, Malick AW, Singhal D et al (2013) Supersolubilization and amorphization of a model basic drug, haloperidol, by interaction with weak acids. Pharm Res 30:1561–1573

Six K, Daems T, de Hoon J, Van Hecken A, Depre M, Bouche M et al (2005) Clinical study of solid dispersions of itraconazole prepared by hot-stage extrusion. Eur J Pharm Sci 24:179–186

Streubel A, Siepmann J, Dashevsky A, Bodmeier R (2000) pH-independent release of a weakly basic drug from water-insoluble and -soluble matrix tablets. J Control Rel 67:101–110

Tatavarti AS, Hoag SW (2006) Microenvironmental pH modulation based release enhancement of a weakly basic drug from hydrophilic matrices. J Pharm Sci 95:1459–1468

Tatavarti AS, Mehta KA, Augsburger LL, Hoag SW (2004) Influence of methacrylic and acrylic acid polymers on the release performance of weakly basic drugs from sustained release hydrophilic matrices. J Pharm Sci 93:2319–2331

Thoma K, Zimmer T (1990) Retardation of weakly basic drugs with diffusion tablets. Int J Pharm 58:197–202

Thombre AG, Herbig SM, Alderman JA (2011) Improved ziprasidone formulations with enhanced bioavailability in the fasted state and a reduced food effect. Pharm Res 28:3159–170

Timko RJ, Lordi NG (1979) Thermal characterization of citric acid solid dispersions with benzoic acid and phenobarbital. J Pharm Sci 68:601–605

Tran TT, Tran PH, Lee B (2009) Dissolution-modulating mechanism of alkalizers and polymers in a nanoemulsifying solid dispersion containing ionizable and poorly water-soluble drug. Eur J Pharm Biopharm 72:83–90

Tran PH, Tran TT, Lee K, Kim D, Lee B (2010a) Dissolution-modulating mechanism of pH modifiers in solid dispersion containing weakly acidic or basic drugs with poor water solubility. Expert Opin Drug Deliv 7:647–661

Tran TT, Tran PH, Choi H, Han H, Lee B (2010b) The roles of acidifiers in solid dispersions and physical mixtures. Int J Pharm 384:60–66

Tran PH, Tran TT, Park J, Min DH, Choi H, Han H et al (2011) Investigation of physicochemical factors affecting the stability of a pH-modulated solid dispersion and a tablet during storage. Int J Pharm 414:48–55

Vasanthavada M, Tong W, Serajuddin ATM (2008) Development of solid dispersion for poorly water-soluble drugs. In: Liu R (ed) Water-insoluble drug formulations, 2nd edn. Informa Healthcare, New York, pp 149–184

Williams HD, Trevaskis NL, Charman SA, Shanker RM, Charman WN, Pouton CW, Porter CJ (2013) Strategies to address low drug solubility in discovery and development. Pharmacol Rev 65:315–499

Zannou EA, Ji Q, Joshi YM, Serajuddin A (2007) Stabilization of the maleate salt of a basic drug by adjustment of microenvironmental pH in solid dosage form. Int J Pharm 337:210–218

Chapter 21
Stabilized Amorphous Solid Dispersions with Small Molecule Excipients

Korbinian Löbmann, Katrine Tarp Jensen, Riikka Laitinen, Thomas Rades, Clare J. Strachan and Holger Grohganz

21.1 Introduction to Small Molecules as Carriers in Solid Dispersions

Amorphous polymeric glass solutions are the most often investigated system for stabilizing amorphous drugs and improving their solubility and dissolution properties. However, this approach has drawbacks which have limited the number of pharmaceutical products on the market based on amorphous solid dispersions. Amorphous drug–polymer mixtures are often hygroscopic and hence absorb moisture, which reduces the glass transition temperature (T_g) and promotes phase separation and recrystallization (Lu and Zografi 1998; Rumondor and Taylor 2009; Vasconcelos et al. 2007). In addition, manufacturing solid dispersions into dosage forms can be challenging (Srinarong et al. 2011). Due to the oftentimes limited solubility of some drugs in the polymers, large quantities of polymer are often essential for stabilizing

T. Rades (✉) · K. T. Jensen · K. Löbmann · H. Grohganz
Department of Pharmacy, University of Copenhagen, Copenhagen, Denmark
e-mail: thomas.rades@sund.ku.dk

K. T. Jensen
e-mail: katrine.jensen@sund.ku.dk

K. Löbmann
e-mail: korbinian.loebmann@sund.ku.dk

H. Grohganz
e-mail: holger.grohganz@sund.ku.dk

R. Laitinen
School of Pharmacy, University of Eastern Finland, Kuopio, Finland
e-mail: riikka.laitinen@uef.fi

C. J. Strachan
Division of Pharmaceutical Chemistry and Technology, University of Helsinki, Helsinki, Finland
e-mail: clare.strachan@helsinki.fi

N. Shah et al. (eds.), *Amorphous Solid Dispersions*,
Advances in Delivery Science and Technology, DOI 10.1007/978-1-4939-1598-9_21

the drug load required, leading to large bulk volumes of the final dosage forms (Newman et al. 2012; Serajuddin 1999). However, the number of poorly water-soluble new chemical entities (NCEs) in the drug discovery process is growing constantly (Williams et al. 2013). Thus, the relatively slow entry of products based on solid dispersion technology to the market means that polymeric glass solutions alone are unlikely to cope with the growing need to produce stable amorphous formulations for commercial use. Thus, interest towards developing alternatives to polymeric amorphous dispersions has increased.

One such alternative is the use of binary amorphous systems with low molecular weight molecules as excipients instead of polymers. It has been reported that small molecules, such as citric acid, sugars, phospholipids, urea, and nicotinamide, can be used as carriers in solid dispersions (Ahuja et al. 2007; Hussain et al. 2012; Kumar and Gupta 2013; Lu and Zografi 1998; Masuda et al. 2012). The stabilizing effect of these binary amorphous systems has been attributed to an increase in the T_g of the mixture compared to those of the individual components as well as (amorphous) salt formation. However, not all of the excipients listed above are always able to create fully amorphous solid dispersions, e.g., solid dispersions with urea and nicotinamide often show at least partial crystallinity remaining for the drug (Aggarwal and Jain 2011; Arora et al. 2010a, 2010b; Chen et al. 2012; Samy et al. 2010).

Chieng et al. (2009) introduced the term "co-amorphous" in order to differentiate amorphous mixtures containing two low molecular weight components from the term "glass solutions" that nowadays is mainly linked to drug–polymer mixtures. Contrary to the above mentioned low molecular weight systems, the increase of the T_g is not seen as the main stabilizing mechanism. The co-amorphous systems are based on mixing at the molecular level and defined molecular interaction of the components, such as hydrogen bonding or π–π interactions. Ionic interaction, thereby salt formation, is a possibility but not a prerequisite, which would lead to the formation of a co-amorphous salt. Initially, this approach was applied for a combination of two pharmacologically relevant drugs. Promising candidates for combination therapy formulations could be produced (Allesø et al. 2009; Chieng et al. 2009; Löbmann et al. 2011, 2012b). These co-amorphous systems have the potential to avoid many of the disadvantages of polymeric glass solutions (such as the large bulk volumes and hygroscopicity). Co-amorphous systems have been found to provide high stability through stabilizing intermolecular interactions between the two drugs and enhanced dissolution rates due to synchronized release (Allesø et al. 2009; Löbmann et al. 2011). However, it is clear that the amount of applicable drug pairs is limited due to the limited amount of pharmacologically relevant drug pairs that could be used in combination therapy. Thus, novel co-amorphous drug–amino acid combinations have been recently introduced as more universally applicable systems. These combinations can potentially offer high amorphous stability and enhanced dissolution for poorly soluble drugs (Löbmann et al. 2013a, b).

In this chapter, the potential of small molecular weight carriers to stabilize amorphous drugs is discussed through several case studies. First, examples of citric acid and sugar molecules are described. These are followed by case studies of co-amorphous drug–drug combinations and the most recent findings on co-amorphous drug–amino acid combinations.

21.2 Case Studies

21.2.1 Combinations with Citric Acid

Citric acid can be used as an acidifying component in traditional polymeric dispersions for modulating the pH in dosage forms, and thus modifying the release rate of pH-dependent and ionizable drugs (Tran et al. 2010). However, citric acid alone has also been used as a carrier in solid dispersions. Lu and Zografi (1998) prepared amorphous binary mixtures of citric acid and indomethacin by cooling melts. They observed that citric acid and indomethacin were miscible up to a 0.25 weight fraction of citric acid as indicated by a single T_g. Miscibility was explained by the presence of hydrogen bonds between the carboxylic acid and hydroxyl groups of the individual components. Phase separation occurred in the systems above the miscibility limit, i.e., more than 25 % (w/w) of citric acid in the blend, as indicated by two T_g values. These phase-separated systems were described as a mixture of a saturated citric acid–indomethacin amorphous phase and another phase containing only citric acid. However, adding polyvinylpyrrolidone (PVP; 0.3 weight fraction in the system), which forms nonideally miscible systems with either citric acid or indomethacin at all compositions, produced a completely miscible system at all citric acid–indomethacin compositions. This was explained by the disruption of the intermolecular hydrogen bonding between indomethacin and citric acid when formulated together with PVP. The resulting self-association of citric acid and indomethacin enhanced their mutual miscibility.

In a further study, citric acid has been found to form an amorphous mixture with the antiviral drug acyclovir upon co-precipitation from N, N-dimethylformamide (Masuda et al. 2012). The molar ratio of acyclovir and citric acid was determined to be 1:2 in the co-precipitates. The miscibility of the components in the amorphous mixture was indicated by a single T_g value (68 °C) and the presence of hydrogen bonding interactions between acyclovir and citric acid. The formation of the amorphous mixture offered a significant improvement of skin permeation flux of the amorphous form from a polyethylene glycol (PEG) ointment *in vitro* compared to crystalline acyclovir.

Physically stable amorphous blends of paracetamol and citric acid anhydrate were prepared by melt quenching the binary mixtures. A long melting time and a high temperature in the melting process slightly lowered the T_g values due to higher amounts of degradation products being formed. Although the mixtures showed low T_g values, which decreased even further on addition of citric acid anhydrate, stable mixtures were formed. The 1:1 (w/w) mixture was the most stable mixture, remaining amorphous for 2 years when stored at dry ambient conditions (Hoppu et al. 2007, 2009a). The increased stability of the mixtures was attributed to hydrogen bond interactions between the phenol group of paracetamol and the carboxylic acid group of citric acid anhydrate and the alcohol group of citric acid anhydrate with paracetamol, as suggested by ^{13}C NMR measurements (Schantz et al.2009). Extrusion and ultrasound-mediated cutting of the sticky and amorphous mixtures had no effect on either the physical or chemical stability of the systems (Hoppu et al. 2009b).

The studies above revealed that a low molecular excipient such as citric acid can be used for the stabilization of an amorphous drug, even though its own T_g (11 °C) is rather low (Hoppu et al. 2007). Specific molecular interactions seemed to be the main stabilization mechanism. However, it was also shown that limited solid-state miscibility between drug and citric acid can lead to phase separation, which can be the starting point of crystallization.

21.2.2 Combinations with Sugar Molecules

Descamps et al. (2007) co-milled mixtures of lactose with mannitol and lactose with budesonide, respectively, with a ball mill. The differences in the T_g values of the raw materials in the two mixtures were large, i.e., mannitol 13 °C/lactose 111 °C and lactose 111 °C/budesonide 90 °C. Molecular-level mixing was achieved in both cases, as indicated by a single T_g observed for the mixtures. Lactose and budesonide were miscible at all budesonide molar fractions studied (the highest experimentally applied molar fraction of budesonide in the mixture was 0.65). This behavior was consistent with the fact that the two compounds could be made fully amorphous when separately milled. The experimental T_g of the mixture as a function of the molar fraction of budesonide was found to obey the Gordon–Taylor relationship, indicating the presence of ideal miscible systems with the absence of specific intermolecular interactions between the components in the mixture. Similarly, miscibility for lactose–mannitol mixtures was only obtained with mannitol fractions lower than 0.44. Interestingly, for larger mannitol fractions, the state obtained after milling was a mixture of lactose–mannitol glass solutions and crystalline α-mannitol. Extrapolation of T_g versus mannitol fraction revealed that for mannitol fractions > 0.50 the T_g of the lactose–mannitol glass solution became close to the milling temperature. It was suggested that in these conditions, the molecular mobility of the glass solution was high enough to induce the crystallization of mannitol, which acts as a plasticizer in the mixture during the milling process. When the T_g of the mixture reached the milling temperature, further amorphization of mannitol was not possible since it was counterbalanced by a rapid recrystallization to the α-form, leading to a steady-state concentration of mannitol in the glass solution. Thus, the process conditions and ratio of the components in the mixtures seemed to have a crucial impact on the successful formation of amorphous composites.

Recently, saccharin was used as a co-amorphous former for the model drug repaglinide (Gao et al. 2013). The initial aim of the authors was to prepare a co-crystal of these components since saccharin is a commonly used co-crystal former (Blagden et al. 2007), having one secondary amine (NH) group as the effective hydrogen donor, which may form a co-crystal through intermolecular hydrogen bonding to the drug. However, upon dissolving different drug–saccharin molar ratios (3:1, 2:1, 1:1, 1:2, and 1:3) in methanol and subsequent slow solvent evaporation at room temperature, a homogeneous co-amorphous precipitate was obtained at the molar ratio of 1:1 with a single T_g (51.9 °C). Fourier transform infrared spectroscopy (FTIR) measurements revealed intermolecular interactions between the NH group of repaglinide and the C = O group of saccharin in the co-amorphous mixture. This mixture was found to be stable for 3 months when stored at 40 °C/75 % RH. The co-amorphous mixture showed great improvement in solubility, e.g., a nearly 20-fold increase in distilled water for the drug. In addition, the dissolution rate of the drug was enhanced under sink conditions in various media. Furthermore, it was found that the co-amorphous blend supersaturated under nonsink conditions and saccharin was able to maintain this supersaturated state of the drug for more than 30 h (a maximum supersaturation level of 4.51 times the equilibrium solubility was reached). This finding may be explained by the presence of a repaglinide and saccharin complex with a higher apparent solubility in solution, i.e., interaction between both components in the supersaturated solution, rather than dissociation into individually dissolved molecules with lower solubility. Thus, supersaturation might have been achieved with respect to repaglinide but not with respect to the repaglinide–saccharin complex. The formation of such a complex might also explain the co-precipitation process in contrast to an individual recrystallization process. Although the authors did not investigate interactions between the components in solution, the investigated co-amorphous system again suggests the importance of molecular interactions between the components with respect to the formation of an amorphous composite, its stabilization, solubility, dissolution rate, and potential for supersaturation in comparison to the individual drug.

21.2.3 Co-Amorphous Drug–Drug Combinations

The trend of combining drugs from different classes for more efficient drug treatment and the increasing interest towards formulating poorly water soluble drugs with better dissolution properties lead to the development of co-amorphous drug–drug formulations. This formulation approach offers the opportunity not only to enhance bioavailability but also to reduce the size of the dosage form, since one drug is the dissolution-enhancing and amorphous-stabilizing excipient for the other drug molecule in the co-amorphous mixture and vice versa. This strategy can be seen as a potential way of preparing candidates for new formulations intended for combination therapy.

21.2.3.1 Co-Amorphous Drug Composites for Pulmonary Delivery

Amorphous particles in pulmonary delivery can be seen as advantageous as they are believed to have a higher adsorption to the lung tissue, and thus lower exhalation and better drug delivery. Amorphous composites of salbutamol sulfate together with ipratropium bromide, intended for asthma combination therapy, were produced in the weight ratios 10:1, 5:1, and 2:1 (w/w) by co-spray drying from aqueous and ethanol solution, in order to optimize powder morphology and shape for lung delivery to the respiratory region (Corrigan et al. 2006). It was possible to obtain amorphous ipratropium bromide when formulated together with salbutamol sulfate. This was not possible when spray drying ipratropium bromide alone.

Similar results were also obtained when co-spray drying budesonide together with formoterol at their therapeutic ratios of 100:6, 200:6, and 400:12 (w/w; Tajber et al. 2009a). The mixtures had a single T_g at approximately 90 °C and even though the amount of formoterol is limited, a small increase (approximately 1 °C) in the T_g could be obtained in the co-amorphous blends. This was in agreement with findings from FTIR studies, which indicated the presence of intermolecular interactions between both drugs. A follow-up study found a 2.6-fold increase for the respirable powder fraction of the spray-dried formulation at the ratio 100:6, when compared to a physical mixture of micronized powders (Tajber et al. 2009b). In addition, the amorphous composite between budesonide and formoterol had the advantage that drugs are delivered at a constant drug ratio, whereas a physical mixture has the risk of uneven dose uniformity due to demixing and particle separation.

Overall, these studies showed that formulating two drugs intended for pulmonary delivery into amorphous composites can result in better particle characteristics, a higher respirable fraction, reduced demixing, and thus better drug performance.

21.2.3.2 Co-Amorphous Indomethacin–Ranitidine HCl Systems

When Chieng et al. (2009) introduced the term "co-amorphous" in order to differentiate amorphous mixtures containing two low molecular weight components from drug–polymer mixtures, they deliberately combined the two drugs indomethacin and ranitidine hydrochloride, with the aim to obtain amorphous blends with improved amorphous properties. The choice of these two drugs was explained by their complimentary pharmacological profile, i.e., the histamine H_2-receptor antagonist ranitidine HCl is given together with nonsteroidal anti-inflammatory drugs (NSAIDs), such as indomethacin, in order to reduce gastrointestinal side effects induced by the NSAIDs. In addition, an amorphous formulation approach seemed reasonable in order to improve the performance of the biopharmaceutics classification system (BCS) class 2 drug indomethacin.

Using vibrational ball milling, it was possible to prepare co-amorphous blends of indomethacin and ranitidine HCl at weight ratios of 2:1, 1:1, and 1:2. A combination of the drugs resulted in a much faster crystalline to amorphous transformation as compared to individual milling of the drugs. Additionally, the co-amorphous blends were single-phase systems, indicated by a single T_g, and molecular interactions between indomethacin and ranitidine HCl could be identified. Both findings were used to explain the improved physical stability of the co-amorphous blends when compared to the individual amorphous drugs. Interestingly, increasing physical stability was not observed with increasing T_g, as is usually expected from amorphous systems. The co-amorphous blend at the 1:1 (w/w) ratio exhibited the highest physical stability but had an intermediate T_g between the 2:1 and 1:2 blends. In the latter two mixtures, the excess component recrystallized first (Fig. 21.1). This finding was

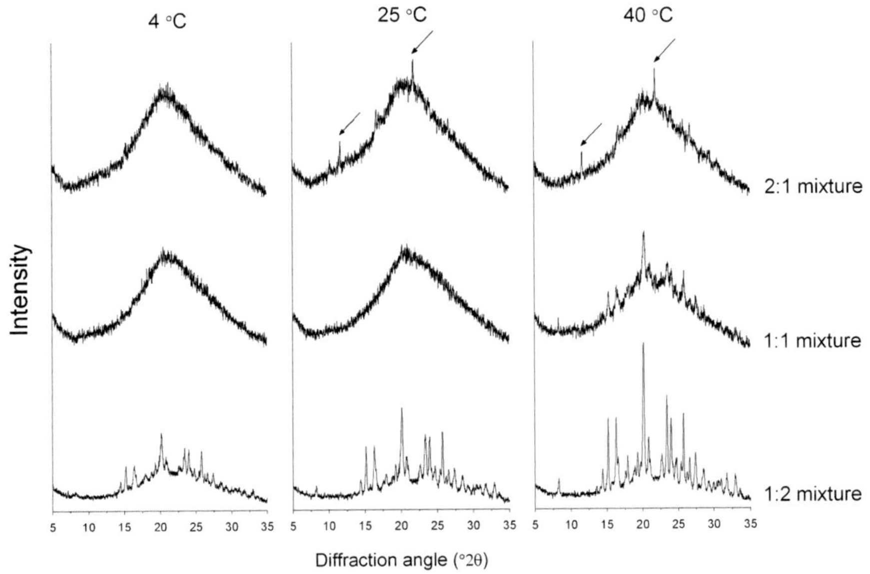

Fig. 21.1 X-ray diffractograms of the co-amorphous indomethacin–ranitidine HCl mixtures at the ratios 2:1, 1:1, and 1:2, after a storage period of 30 days, at 4, 25, and 40 °C under dry conditions. The *arrows* in the 2:1 mixtures indicate the initial recrystallization of indomethacin. The diffraction peaks in the 1:2 mixture belong to ranitidine HCl. (Reprinted from Chieng et al. 2009, with permission from Elsevier)

explained by the presence of molecular interactions on the bulk level that could be most pronounced at a similar weight ratio, i.e., the 1:1 blend. However, a connection to specific molecular interactions between both drugs at a 1:1 molar level was not made in this study.

21.2.3.3 Co-Amorphous Naproxen–Cimetidine Systems

A similar combination between a poorly soluble NSAID and an antihistaminic drug was investigated by Allesø et al. (2009). Their aim was to prepare co-amorphous blends of naproxen and cimetidine, with molar ratios (and not weight ratios) of 2:1, 1:1, and 1:2 to study potential intermolecular interactions and stability advantages at a molar level. The co-amorphous blends were then compared to the individual drugs with respect to intrinsic dissolution, physical stability, and ease of amorphization.

It was found that milling the two drugs together can significantly enhance the amorphization of naproxen, which could not be prepared as an individual amorphous component. This could be explained by the formation of homogeneous single-phase co-amorphous mixtures that had a significantly higher T_g, as compared to the theoretical ones for amorphous blends obtained by the Gordon–Taylor equation, and in particular to that of amorphous naproxen. A positive deviation to the Gordon–Taylor equation, in general, suggests the presence of intermolecular interactions between components in an amorphous mixture. This finding was confirmed using Raman spectroscopy, which indicated solid-state interactions between the carboxylic acid group of naproxen and the imidazole moiety of cimetidine.

The physical stability study revealed again that the excess component in the 1:2 and 2:1 co-amorphous mixtures primarily recrystallized, whereas the 1:1 co-amorphous mixture remained physically stable over a period of at least 6 months, regardless of its intermediate T_g between that of the 1:2 and 2:1 mixture. This finding was assigned to the molecular interactions found in the Raman spectroscopic study.

Based on the above findings, the authors investigated the dissolution behavior of the co-amorphous naproxen–cimetidine mixture at the 1:1 molar ratio. The intrinsic dissolution of both drugs was increased fourfold and twofold for naproxen and cimetidine, respectively, when compared to the individual crystalline and amorphous (only cimetidine) drugs (Fig. 21.2). Furthermore, a pair-wise (synchronized) release was observed for both drugs in the 1:1 co-amorphous formulation. It was suggested that the interactions between naproxen and cimetidine are strong enough to result in a pair-wise solvation, i.e., with every molecule of cimetidine, one molecule of naproxen is dissolved simultaneously into the dissolution medium. Overall, the co-amorphous formulation approach could be depicted as a promising approach to enhance the dissolution of poorly soluble drugs and at the same time increase the physical stability of the amorphous form of that drug.

21.2.3.4 Co-amorphous Indomethacin–Naproxen Systems

Löbmann et al. (2011) prepared co-amorphous binary blends of the two BCS class 2 drugs indomethacin and naproxen with the main aim to further investigate the type of molecular interactions in co-amorphous blends and their influence on recrystallization, dissolution and physicochemical properties. They were able to successfully produce co-amorphous blends at the molar ratios 2:1, 1:1, and 1:2 using quench cooling. Similar to the results obtained by Allesø et al. (2009), naproxen showed poor physical stability in its pure amorphous form. However, in combination with indomethacin, stable co-amorphous single-phase systems could be obtained.

Comparable results were also obtained in the physical stability study and for the intrinsic dissolution rates of co-amorphous indomethacin and naproxen at the 1:1 molar ratio. In spite of an intermediate T_g, this blend remained stable whereas the two other blends (1:2 and 2:1) showed recrystallization events of the excess component. The dissolution study revealed an increased dissolution rate, compared to the individual crystalline and amorphous (only indomethacin) drugs, and a synchronized release of both drugs in the co-amorphous 1:1 blend (Fig. 21.3).

The synchronized release of the co-amorphous 1:1 mixture suggested again a pair-wise release of both drugs at the same rate. In this context, the presence of molecular interactions was investigated and identified as peak shifts in the FTIR spectra. The presence of two carboxylic acid groups in both drugs and their possibility to form a carboxylic acid heterodimer was taken as starting point for the calculation

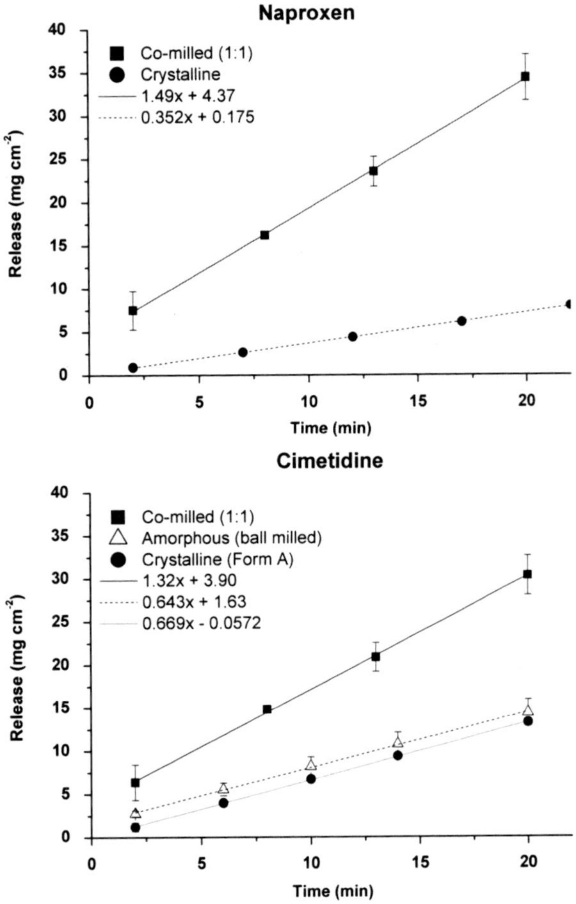

Fig. 21.2 Intrinsic dissolution profiles of naproxen (**a**) and cimetidine (**b**) showing the dissolution enhancement from the co-amorphous (co-milled) mixture compared to the crystalline and amorphous (only cimetidine) references. (Reprinted from Allesø et al. 2009, with permission from Elsevier)

of theoretical FTIR spectra using quantum mechanical tools, i.e., density functional theory (Löbmann et al. 2012a). The theoretical spectrum of the co-amorphous 1:1 indomethacin–naproxen heterodimer was compared to the theoretical spectra of the drug homodimers and the experimentally obtained FTIR spectra. It was found that the peak shifts from the experimental spectra were in line with those from the calculated spectra, which strongly suggested that the hypothesized heterodimer is formed in the co-amorphous 1:1 blend. In addition, the presence of indomethacin and naproxen homodimers in the individual amorphous drug form could be identified.

These results were used to explain the behavior of the co-amorphous blends in the physical stability study. The individual amorphous drugs comprise a near range molecular order in form of the homodimer which shows fast recrystallization kinetics. In the 1:1 co-amorphous blends, this microstructure will change towards the formation of a heterodimer between naproxen and indomethacin. In order for the heterodimer to recrystallize, the hydrogen bonds between the unlike molecules

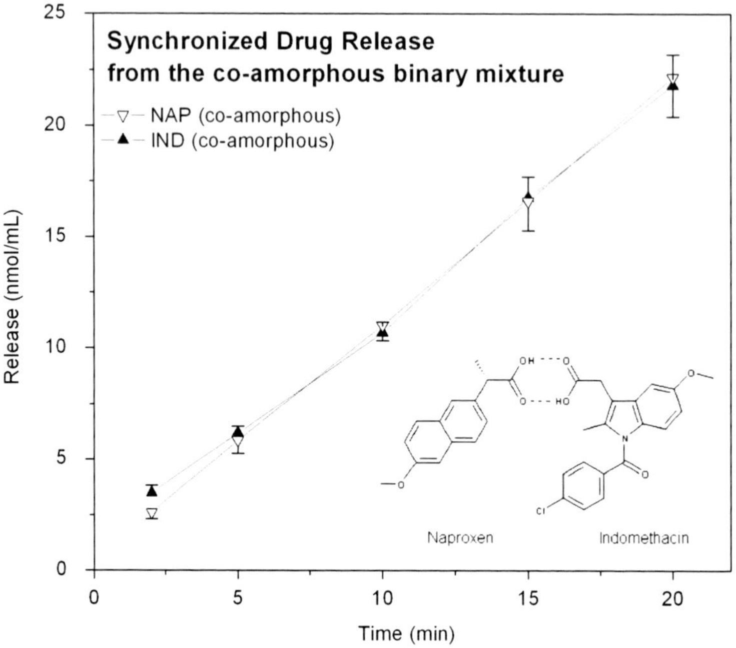

Fig. 21.3 Intrinsic dissolution profile (nmol/mL/min) of the co-amorphous 1:1 mixture showing the pair-wise (synchronized) release of the heterodimer of indomethacin and naproxen (depicted in the *lower right* corner), at the same rate. (Reprinted with permission from Löbmann et al. 2011, American Chemical Society)

have to break initially, followed by a reorientation towards like molecules, which in turn can then form a homodimer, leading to crystallization. In the 2:1 and 1:2 co-amorphous blends, both species, i.e., the heterodimer and excess drug homodimers, will be present. In this regard, the excess homodimers will be readily able to recrystallize, whereas the heterodimers will follow the recrystallization path of the 1:1 co-amorphous mixture.

On the basis of the results above, a new approach of using the Gordon–Taylor equation, to predict the T_g values of amorphous blends that show some degree of molecular interaction, was proposed. The experimentally observed T_g of the co-amorphous indomethacin and naproxen mixtures showed generally a negative deviation from those predicted by the Gordon–Taylor equation. It could be shown that this was related to an increased free volume found in the co-amorphous mixtures compared to the individual amorphous drugs, and was most pronounced in the co-amorphous 1:1 blend. Thus, the largest deviation from the theoretical glass transition values was also found for the co-amorphous 1:1 blend. It was thus suggested that the ratio between the heterodimer and homodimer microstructures, rather than the ratio between indomethacin and naproxen molecules, should be taken into account when calculating the T_g via the Gordon–Taylor equation. Therefore, the heterodimer was inserted as

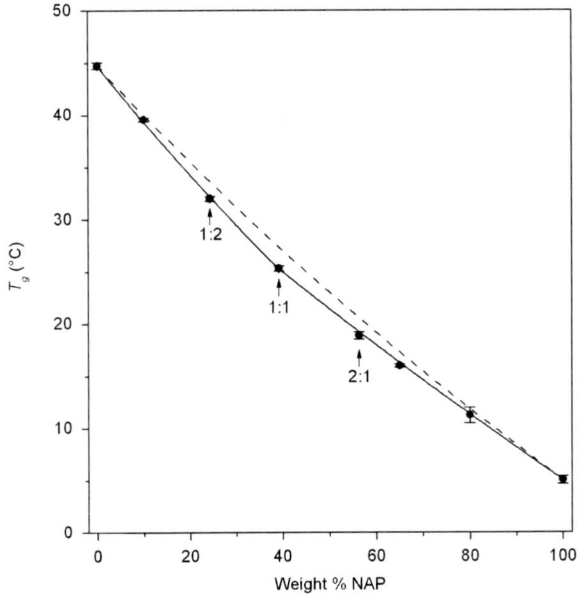

Fig. 21.4 Experimental (dots) and theoretical T_g values for various co-amorphous indomethacin–naproxen mixtures. The *dashed line* represents the T_g values using the Gordon–Taylor equation in its conventional way, i.e., each individual drug represents one part of the mixture. The *solid line* shows the T_g values using the Gordon–Taylor equation in its modified way as suggested by Löbmann et al. (2011), i.e., the indomethacin–naproxen heterodimer at the 1:1 molar ratio is one component in the equation and the excess drug the second. The modified Gordon–Taylor approach is in excellent agreement with the experimentally observed T_g values. (Reprinted with permission from Löbmann et al. 2011, American Chemical Society)

individual component into the Gordon–Taylor equation, with the homodimer being the additional excess component. Using this approach, the experimentally obtained T_g values were in excellent agreement with those theoretically calculated (Fig. 21.4). This behavior strongly supported the presence of a heterodimer as suggested above.

In an additional study, the individual drugs and the co-amorphous 1:1 mixture were characterized with respect to their glass-forming ability according to the amorphous classification system (Baird et al. 2010). Briefly, this system is based on the recrystallization of an organic component during a differential scanning calorimeter (DSC) measurement upon cooling from the melt and reheating above its melting point. In such a setup, components can be classified into three classes, where class 1 drugs show recrystallization already in the cooling phase, class 2 drugs show recrystallization upon reheating, and class 3 drugs do not show any recrystallization event during the DSC measurement. In this regard, class 1 drugs are the least favorable compounds to form an amorphous form as they possess the highest driving force towards crystallization, class 2 drugs possess intermediate properties, and class 3 drugs are the most likely components to result in a stable amorphous system. According to this system, indomethacin and naproxen can be classified as class 3 (Baird et al. 2010) and class 1 (Shimada et al. 2013) drugs, respectively. However, when formulating a co-amorphous blend between indomethacin and naproxen, the latter changes from an unfavorable class 1 drug to a favorable class 3 drug in the co-amorphous mixture (Shimada et al. 2013). Thus, the amorphization properties of organic components can be positively changed upon formulating an interacting co-amorphous blend such as the indomethacin–naproxen mixture.

Overall, it was concluded that the micro-structure found in co-amorphous drug–drug formulations, e.g., formation of a heterodimer, can significantly influence the behavior of these systems with respect to dissolution (improved and synchronized), amorphous stabilization, and physicochemical properties (e.g., free volume, glass-forming ability).

21.2.3.5 Co-amorphous Glipizide–Simvastatin Systems

A study of co-amorphous mixtures of the two BCS class 2 drugs glipizide and simvastatin gave the opportunity to study further factors that influence the physicochemical properties within these systems. Both drugs are commonly used for the treatment of metabolic diseases and because of their poor aqueous solubility, formulating a combination of both drugs into a co-amorphous drug delivery system seemed meaningful. Similar to the studies above, co-amorphous mixtures at the molar ratios 2:1, 1:1, and 1:2 were prepared using vibrational ball milling (conventional and cryo-milling). However, the individual drugs could only be transferred into their amorphous counterpart using cryo-milling. It is well known that mechanical activation at lower temperatures has a higher tendency to result in successful amorphization of a compound due to the more brittle nature of the material at these temperatures (Descamps et al. 2007). Thus, preparing a co-amorphous blend can significantly enhance the tendency of the two components to transform into their amorphous form as it was possible to prepare the co-amorphous glipizide–simvastatin mixtures at conventional milling conditions.

The resulting blends were homogeneous and single phased, with T_g values in agreement with those estimated by the Gordon–Taylor equation. This behavior suggested ideal miscibility without the presence of specific intermolecular interactions between the components in the mixture, which was further confirmed by FTIR. In agreement with these findings, the powder dissolution experiments showed no improvement in dissolution for the co-amorphous mixtures compared to the individual amorphous drugs. This in turn suggested that some form of specific intermolecular interactions is beneficial for dissolution improvement as seen from the studies described above. Interestingly, even though molecular interactions were lacking, the physical stability of the co-amorphous glipizide–simvastatin mixtures was higher than that of the individual amorphous drugs or a physical mixture of those. It was suggested that the intimate mixing of both drugs at a molecular level improved the physical stability of these co-amorphous systems even in the absence of directional molecular interactions.

Overall, this study revealed two interesting findings in co-amorphous systems. First, the amorphization tendency can be enhanced even though an interacting nature between the components could not be identified. And second, the homogeneous molecular-level mixing achieved through the milling process resulted in a positive effect on the physical stability of these systems. However, the presence of molecular interactions was still thought to be of higher importance to successfully prepare highly stable co-amorphous blends, mainly because of the recrystallization mechanism described for the indomethacin–naproxen heterodimer.

21.2.4 Co-Amorphous Drug-Amino Acid Formulations

The concept of a co-amorphous drug formulation approach comprising a combination of two pharmacologically relevant drugs intended for combination therapy was discussed in the previous sections. Many factors influencing the behavior of these systems have been identified and discussed. Generally, these combinations revealed that preparing co-amorphous drug–drug mixtures can facilitate drug amorphization, enhanced drug dissolution, and increased physical stability. However, it is not always necessary to administer more than one drug at the same time or simply not possible to find a suitable partner molecule for a co-amorphous drug–drug formulation. In these cases, it can be relevant to create co-amorphous mixtures of the poorly water-soluble drug with a low molecular weight excipient without biological activity. As outlined in the Sects. 21.2.1 and 21.2.2, a few of these carriers have already been investigated. However, excipients intended solely for a universal application in co-amorphous mixtures with poorly water soluble drugs have not been presented. In this context, the excipients should ideally be generally safe for the use in humans. For the performance of the co-amorphous formulations the excipient needs to form strong molecular interactions with the drug in the mixture.

Amino acids are small molecules that play an essential role in organisms as they are the building blocks for proteins, enzymes, and cell receptors, precursors for neurotransmitters and generally important as part of biosynthesis. Because of this, amino acids (in the form of proteins) are one of the main parts of human daily nutrition and therefore, generally well tolerated. Furthermore, they play a crucial role in the interaction with drugs at the active site of the receptor, and thus possess the possibility of molecular interactions with the drug *in vivo*. In addition, amino acids can improve the solubility of drugs due to their ability to act as salt-forming agents (Berge et al. 1977).

Amino acids were investigated for their potential to decrease the amount of cyclodextrin necessary to dissolve the drug naproxen (Mura et al. 2003). For this purpose the authors investigated the amino acids valine, leucine, arginine, or lysine in solution with hydroxypropyl-β-cyclodextrin and naproxen. They found that lysine and especially arginine further improved the solubility of the poorly water-soluble drug naproxen when dissolved together with hydroxypropyl-β-cyclodextrin. It was suggested that arginine established electrostatic interactions with the carboxyl group in naproxen as well as hydrogen bonds to the cyclodextrin. These interactions resulted in a ternary complex that is crucial for the solubility improvement. FTIR measurement indicated a salt formation between the acidic naproxen and the basic amino acid arginine upon co-milling and co-precipitation at an equimolar binary ratio, both with and without the presence of cyclodextrin (Mura et al. 2005). The co-precipitated ternary mixture naproxen–arginine–cyclodextrin furthermore showed a 15-fold increased dissolution rate for naproxen compared to the physical mixture and the pure drug. The improved performance of the ternary co-precipitate was attributed to a synergistic effect between the salt formation with arginine, cyclodextrin inclusion, and the specific role of arginine connecting both, drug and cyclodextrin, through interactions.

The interest in improving the solubility of drugs with amino acids without cyclodextrins was introduced along with the concept of the arginine-assisted solubilization system (AASS), which showed arginine hydrochloride to improve the solubility of alkyl gallates containing an aromatic ring when prepared together in a solution (Hirano et al. 2010). In comparison, lysine hydrochloride only had a marginal effect on the solubility of the alkyl gallates. The authors found that the guanidinium group in the side chain of arginine resulted in more favorable interactions with the alkyl gallates when compared to the amino group of lysine. These specific interactions were able to disrupt the alkyl gallate molecules leading to a better solubilization. However, for universal application of amino acids as co-amorphous excipient, further investigations were necessary, especially with respect to exploring further amino acids and non-salt-forming mixtures with respect to their applicability to form co-amorphous mixtures with neutral drug molecules.

21.2.4.1 Receptor Amino Acids as Stabilizers of Co-amorphous Mixtures

As outlined above, amino acids are an integral part of biological receptors. In the active site, they interact with the drug *in vivo* and thus are able to interact with a drug on a molecular level. Therefore, it was suggested that the amino acids from the active site of a given biological receptor should be a good starting point when selecting amino acids as excipients in co-amorphous formulations (Löbmann et al. 2013a). Similar to the studies above, the given potential of forming strong molecular interactions between the drug and the receptor amino acids that can prevent recrystallization in the co-amorphous blend were the main motivation for these investigations.

Indomethacin, for example, is a cyclooxygenase (COX) inhibitor and binds molecularly to the amino acids arginine and tyrosine in the active site of the receptor. In particular, the carboxylic acid group of indomethacin interacts with the phenol group of tyrosine and the guanidine group of arginine in the binding site of the COX-2 receptor (Rowlinson et al. 2003).

With this background, the feasibility of arginine and tyrosine to form co-amorphous binary and ternary mixtures with indomethacin by ball milling the molar ratios 1:1 and 1:1:1 was investigated (Löbmann et al. 2013a). X-ray powder diffraction (XRPD) and DSC measurements revealed that only the mixture of indomethacin in combination with arginine can be successfully prepared into a co-amorphous formulation under the applied ball-milling conditions. The obtained co-amorphous mixture between indomethacin–arginine had a markedly improved dissolution rate, resulting in an increase of approximately 200-fold when compared to the dissolution rate of both crystalline and amorphous indomethacin. Furthermore, this blend was physically stable for more than at least 6 months at 40 °C and dry storage conditions. In a follow-up study, it could be shown that the carboxylic acid group of indomethacin and the guanidine moiety of arginine form a salt during the milling process, as indicated by specific peak shifts of these groups in the FTIR spectrum of the co-amorphous mixture compared to spectra of amorphous indomethacin and crystalline arginine (Löbmann et al. 2013b). As mentioned previously, arginine is a well-known salt-forming agent for acidic drugs, and the presence of a salt was used to explain the much higher dissolution rate. In the presence of tyrosine, the XRPD diffractograms in the binary and ternary mixtures showed remaining crystalline reflections of tyrosine and it was argued that tyrosine in general had poor amorphization properties.

A similar approach was applied to carbamazepine that binds to the amino acids phenylalanine and tryptophan in neuronal Na^{+} channels (Yang et al. 2010). Co-amorphous mixtures were obtained for the binary mixture carbamazepine and tryptophan and in the ternary mixture including both receptor amino acids. The chemical structures of carbamazepine, phenylalanine, and tryptophan are given in Fig. 21.5. In the co-amorphous blends, a modest increase in the dissolution rate of carbamazepine was observed, with the ternary mixture showing the highest increase is dissolution rate. However, these effects were not significantly different from the dissolution of crystalline carbamazepine. The minor improvement in dissolution rate was explained by the ability of the amorphous carbamazepine to re-crystallize on the surface of the disc during the intrinsic dissolution experiment, which in turn reduces the dissolution rate (Löbmann et al. 2013a). However, the incorporation of amino acids into a co-amorphous system with carbamazepine resulted in highly stable amorphous formulations, remaining stable for more than 6 months at 40 °C under dry conditions whereas pure amorphous carbamazepine recrystallized within a week.

In the molecular interaction study of these co-amorphous blends, interactions between the tryptophan carboxylic acid and amide moieties of carbamazepine as well as additional H-bonding and π–π interactions were suggested, as the aromatic and amide vibrations were shifted in the respective FTIR spectra (Löbmann et al. 2013b). Comparing the co-amorphous spectrum of carbamazepine and tryptophan to amorphous carbamazepine prepared by quench cooling indicated that the interactions between carbamazepine and tryptophan are stronger than those found between the carbamazepine dimers in pure amorphous carbamazepine. Thus, the carbamazepine dimer present in the pure amorphous drug is disrupted by the presence of tryptophan, suggesting a similar stabilization mechanism as found in the co-amorphous example with indomethacin and naproxen (Sect. 21.2.3.4). When adding phenylalanine to the carbamazepine and tryptophan mixture, additional alterations in the spectra

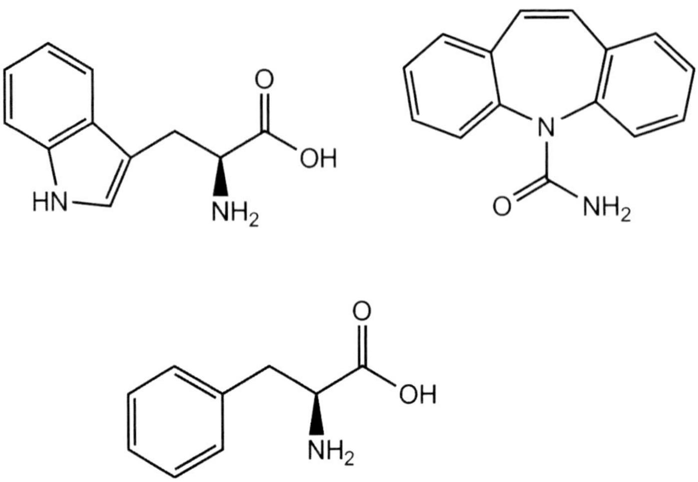

Fig. 21.5 The chemical structures of the drug carbamazepine and the receptor amino acids phenylalanine and tryptophan

were observed indicating further interactions between phenylalanine and either carbamazepine or tryptophan or both. It was not possible to prepare an amorphous mixture of carbamazepine and phenylalanine.

Overall, the receptor amino acids could be used to some degree to prepare co-amorphous blends with the drugs carbamazepine and indomethacin. The successfully prepared co-amorphous mixtures were physically stable and showed an increased dissolution rate of the drugs. In both cases, molecular interactions between the drugs and amino acids were observed as initially proposed for amino acids taken from the receptor binding site. However, differences in the type of interaction in the co-amorphous formulation and the receptor were observed. At the active site *in vivo*, only the amino acid side chains are able to interact with the drug whereas the amine and carboxylic acid head group are involved in the peptide backbone of the receptor. In comparison, it was found that the whole amino acid molecule, i.e., side chain and head group, interacted with the drug in the co-amorphous blends.

21.2.4.2 Combinations of Various Amino Acids to Improve the Dissolution Rate

To further explore amino acids as co-amorphous excipients, binary and ternary combinations of carbamazepine or indomethacin were investigated choosing amino acids beyond those present at the active site. For this purpose, both drugs were ball milled in varying combinations with the amino acids arginine, tryptophan, phenylalanine, and tyrosine (Löbmann et al. 2013a, b). From the 16 investigated blends, it was possible to produce 8 co-amorphous mixtures (Table 21.1). When analyzing the XRPD diffractograms of all these mixtures, a general trend for tyrosine being an

Table 21.1 Samples prepared by vibrational ball milling for 90 min at 30 Hz, their molar and weight ratio, success in amorphization, glass transition temperature (T_g), presence of molecular interactions in the co-amorphous mixtures, their physical stability at 40 °C under dry storage conditions, and intrinsic dissolution rate

Sample content	Molar ratio	Weight ratio	Success in amorphization	T_g (°C)	Molecular interactions	Physical stability at 40 °C/dry conditions	Intrinsic dissolution rate (mg cm^{-2} min^{-1})
CBZ	–	–	No[b]	$56.1 \pm 0.2^{(quenchcooled)}$	–	< 7 days$^{(quenchcooled)}$	0.034 (crystalline)
CBZ–ARG	1:1	1:0.74	No[b,c]	–	–	–	–
CBZ–PHE	1:1	1:0.70	No[b,d]	–	–	–	–
CBZ–TRP	1:1	1:0.86	*Yes*	81.0 ± 0.6	y	> 6 months	0.037
CBZ–TYR	1.1	1:0.77	No[b,a]	–	–	–	–
CBZ–ARG–TYR	1:1:1	1:1.50	No[b,a]	–	–	–	–
CBZ–PHE–TRP	1:1:1	1:1.56	*Yes*	75.1 ± 1.1	y	> 6 months	0.041
CBZ–ARG–TRP	1:1:1	1:1.60	*Yes*	65.4 ± 1.1	y	> 6 months	0.047
CBZ–TRP–TYR	1:1:1	1:1.63	No[a]	–	–	–	–

Table 21.1 (continued)

IND	–	–	*Yes*	36.7 ± 0.8	–	< 7 days	0.040 (crystalline) 0.064 (amorphous)
IND–ARG	1:1	1:0.49	*Yes*	64.1 ± 1.4	y	> 6 months	12.25
IND–PHE	1:1	1:0.46	*Yes*	47.8 ± 2.9	n	3–6 months	0.175
IND–TRP	1:1	1:0.57	*Yes*	68.7 ± 2.6	n	> 6 months	0.096
IND–TYR	1:1	1:0.51	No[a]	–	–	–	–
IND–ARG–TYR	1:1:1	1:0.99	No[a]	–	–	–	–
IND–PHE–TRP	1:1:1	1:1.03	*Yes*	63.1 ± 0.8	n	> 6 months	0.134
IND–ARG–PHE	1:1:1	1:0.95	*Yes*	77.9 ± 3.4	y	> 6 months	11.19
IND–TRP–TYR	1:1:1	1:1.08	No[a]	–	–	–	–

Amorphous carbamazepine was prepared by quench cooling to determine its T_g and physical stability
Remaining crystalline reflections in the XRPD diffractograms for:
[a] TYR
[b] CBZ
[c] ARG
[d] PHE

unsuitable excipient in co-amorphous formulations was found, as none of the mixtures including tyrosine resulted in an amorphous product. In contrast, tryptophan revealed particularly good amorphization properties as all the mixtures including tryptophan were amorphous, except those also including tyrosine. In total, three co-amorphous mixtures containing carbamazepine and five co-amorphous mixture containing indomethacin could be prepared (Table 21.1). In the carbamazepine mixtures, the presence of tryptophan seemed to be essential for a successful co-amorphization. Furthermore, only the ternary blend with arginine and tryptophan had a nonreceptor amino acid included. Interestingly, this mixture also showed the highest dissolution rate of carbamazepine. From FTIR measurements, H-bonds and π–π interactions were identified in all the investigated co-amorphous carbamazepine blends. In contrast, no specific interactions were observed between indomethacin and the nonreceptor amino acids tryptophan and phenylalanine in any of the co-amorphous mixtures thereof. Even though no specific interactions were identified between the amino acids and indomethacin, the nonsalt co-amorphous mixtures had a significantly improved dissolution rate compared to amorphous and crystalline indomethacin.

All the co-amorphous mixtures had a significantly higher T_g compared to the individual drugs when examined with DSC (Table 21.1). Furthermore, all the co-amorphous mixtures were physically stable for at least 6 months. The stability was correlated to the presence of a mixture on molecular level, elevated T_g, and molecular interactions. The best stability, however, was achieved for the mixtures in which specific molecular interactions occurred between the drug and the amino acids. Overall, it was concluded that amino acids are promising excipients for the stabilization of a poorly water-soluble drug in an amorphous form. A strong increase in dissolution rate can be achieved, when choosing the most advantageous amino acids with respect to the given drug. As molecular interactions were only observed, when at least one amino acid of the active site was included, those amino acids give a good starting point for the formulation of a co-amorphous blend. However, as seen from the example with tyrosine, this is not a general guarantee for successful co-amorphization, as the physicochemical properties of the amino acids towards amorphization also have to be taken into account. The low amount of excipient required in these formulations can be seen as an additional significant advantage when compared to solid dispersions prepared with polymers.

21.3 The Possibility of Co-crystal Formation with Low Molecular Weight Excipients

When preparing co-amorphous binary mixtures, the possibility of a co-crystal formation should be kept to mind. The saccharin example in Sect. 21.2.2 highlights the possibility of formation of an amorphous solid dispersion while attempting the preparation of a co-crystal using a small molecular weight excipient. One could easily think that the opposite, i.e., formation of a co-crystal upon attempting to prepare an amorphous solid dispersion or when storing the amorphous drug–excipient mixture under certain conditions, could also be likely to happen. For example, nicotinamide is a known co-crystal former (Blagden et al. 2007) and in addition to co-crystals with different drugs (Maheshwari et al. 2009; Seefeldt et al. 2007) it has been also observed to form at least partly crystalline solid dispersions depending on the preparation conditions (Aggarwal and Jain 2011).

Co-crystals are multicomponent crystals based on hydrogen bonding interactions without the transfer of hydrogen ions to form salts (Blagden et al. 2007; Lu and Rohani 2009). Furthermore, pharmaceutical co-crystals can be defined as crystalline materials composed of an active pharmaceutical ingredient (API) and one or more unique co-crystal formers which are solid at room temperature, in a stoichiometric ratio (Vishweshwar et al. 2006). Currently, little is known about the relation between co-amorphous systems and co-crystals (Laitinen et al. 2012). Both result from formation of interactions, such as hydrogen bonds, between the components. If these interactions are energetically more favorable than those between the molecules of one component, then either a co-amorphous system or a co-crystal can be formed. Prediction of the outcome is currently not possible and this question must be answered experimentally instead. The probability for co-crystal formation may be rationalized to some extent, for example by understanding the degree of conformational flexibility, as well as the supramolecular chemistry of the functional groups present in the given molecules and selection of appropriate co-crystal formers (Etter 1990; Vishweshwar et al. 2006). For example, acyclovir has been reported to form an amorphous mixture with anhydrous citric acid by co-precipitation through hydrogen bond interactions (as discussed in Chap. 2.1), but in the same study it formed a co-crystal with L-tartaric acid when processed with the same method (Masuda et al. 2012). The molar ratio of acyclovir and citric acid in the amorphous mixture was determined to be 1:2 while for the co-crystal the ratio was found to be 1:1.

In addition, the fate of the amorphous binary mixtures when stored in conditions that could induce phase separation is currently not known. After phase separation has occurred, the system might proceed to nucleation and crystal growth. These transformational stages are governed by both thermodynamic and kinetic factors that must be taken into account for the successful development of a stable amorphous binary system. However, it is possible that crystallization initiates the formation of co-crystals, which might be desirable if the physicochemical properties of the co-crystals are advantageous compared to those of the pure components (Pajula et al. 2010). Spontaneous co-crystal formation has been reported to occur when storing physical mixtures under extreme conditions, e.g., at high relative humidity (Arora et al. 2011; Jayasankar et al. 2007; Maheshwari et al. 2009). One example is the formation of a carbamazepine–nicotinamide co-crystal as a consequence of storage (Maheshwari et al. 2009). To our knowledge co-crystal formation has never been reported as a consequence of the storage of co-amorphous systems, but this possibility should be kept in mind and investigated, especially knowing that drug–drug co-crystals have been reported to exist (Aitipamula et al. 2009; Cheney et al. 2011; Lee et al. 2011; Nugrahani et al. 2007). Usually, co-amorphous systems have been observed to recrystallize to pure components, with the excess component crystallizing out first (Allesø et al. 2009; Chieng et al. 2009; Löbmann et al. 2011). However, the amorphous form can act as an intermediate state before formation of co-crystals (Seefeldt et al. 2007). Depending on the conditions (heating rate in DSC), amorphous mixtures of carbamazepine and nicotinamide were found to generate either co-crystals through a metastable co-crystalline phase (low heating rate) or co-crystals after initial recrystallization as individual components (high heating rate).

21.4 Conclusion and Future Perspectives

Interest towards the potential of binary amorphous systems, having small molecules as excipients instead of polymers, has increased. By combining a drug molecule with a low molecular weight excipient in a co-amorphous drug formulation, some drawbacks connected with polymeric dispersions, such as large bulk volumes, may be overcome. Amorphous composites were prepared using citric acid and sugar molecules, such as mannitol and saccharin, as carriers, and the combination of citric acid and paracetamol was found to be physically stable for 2 years when stored at dry ambient conditions. It is to be expected that the recently introduced co-amorphous drug–drug and drug–amino acid combinations could hold an even greater potential in stabilizing the amorphous form and in improving the dissolution rate of poorly soluble drugs.

The studies on co-amorphous drug–drug combinations have given an insight into the basic understanding of the physicochemical properties of these mixtures, their mechanism of amorphous stabilization and drug dissolution. Regardless of the preparation method and the initial glass-forming ability of a drug, the amorphization tendency of those drugs increased when they were formulated together with a second drug into a co-amorphous system rather than preparing an individual amorphous drug. It could be shown, that the physical stability of co-amorphous systems is dependent on several factors, presumably the degree of conformational flexibility of the components, molecular-level mixing and intermolecular interactions between the drug molecules, with the latter having a greater impact. Molecular interactions on a bulk and molecular level and the resulting microstructures, e.g., formation of a heterodimer, in these systems seemed to be crucial with respect to their (synchronized) dissolution behavior, recrystallization mechanism, and physicochemical properties, such as the free volume. Furthermore, these molecular interactions played a greater role in the physical stabilization of co-amorphous formulations than solely the presence of a high T_g, as shown in the studies on co-amorphous indomethacin-ranitidine HCl, naproxen–cimetidine, and indomethacin–naproxen.

Amino acids have proven to be useful excipients in co-amorphous drug delivery systems and may form the basis of a new potential platform technique to overcome challenges in the stabilization of the amorphous form of poorly soluble drugs. Though tryptophan has a potential as an overall good stabilizer of co-amorphous systems, it did not show specific molecular interactions with the model drug indomethacin in the co-amorphous blends. Receptor amino acids may be a useful starting point when exploring new co-amorphous stabilizers for a given drug in order to achieve strong interactions. Combining amino acids with other drug excipients such as polymers, cyclodextrins, or salt-forming agents can result in new formulations with improved capabilities. So far, the co-amorphous drug delivery approach concentrated only on the preparation and characterization of the pure co-amorphous mixture. Further studies, investigating the preparation of final dosage forms such as tablets, can give insight into the suitability of co-amorphous blends towards processability. Furthermore, the bioavailability of co-amorphous systems has not yet been explored and it will be interesting to see if further investigations including animal studies will confirm the use of amino acids as a useful approach in improving dissolution rate and bioavailability. Many new ideas may rise soon within the field of co-amorphous formulations as it is very likely that only a small fraction of its potential has been revealed thus far.

References

Aggarwal AK, Jain S (2011) Physicochemical characterization and dissolution study of solid dispersions of ketoconazole with nicotinamide. Chem Pharm Bull 59:629–638

Ahuja N, Katare OP, Singh B (2007) Studies on dissolution enhancement and mathematical modeling of drug release of a poorly water-soluble drug using water-soluble carriers. Eur J Pharm Biopharm 65:26–38

Aitipamula S, Chow PS, Tan RBH (2009) Trimorphs of a pharmaceutical cocrystal involving two active pharmaceutical ingredients: potential relevance to combination drugs. Cryst Eng Comm 11:1823–1827

Allesø M, Chieng N, Rehder S, Rantanen J, Rades T, Aaltonen J (2009) Enhanced dissolution rate and synchronized release of drugs in binary systems through formulation: amorphous naproxen-cimetidine mixtures prepared by mechanical activation. J Control Release 136:45–53

Arora S, Sharma P, Irchhaiya R, Khatkar A, Singh N, Gagoria J (2010a) Development, characterization and solubility study of solid dispersions of azithromycin dihydrate by solvent evaporation method. J Adv Pharm Technol Res 1:221–228

Arora S, Sharma P, Irchhaiya R, Khatkar A, Singh N, Gagoria J (2010b) Development, characterization and solubility study of solid dispersions of Cefuroxime Axetil by the solvent evaporation method. J Adv Pharm Technol Res 1:326–329

Arora KK, Tayade NG, Suryanarayanan R (2011) Unintended water mediated cocrystal formation in carbamazepine and aspirin tablets. Mol Pharm 8:982–989

Baird JA, Van Eerdenbrugh B, Taylor LS (2010) A classification system to assess the crystallization tendency of organic molecules from undercooled melts. J Pharm Sci 99:3787–3806

Berge SM, Bighley LD, Monkhouse DC (1977) Pharmaceutical salts. J Pharm Sci 66:1–19

Blagden N, de Matas M, Gavan PT, York P (2007) Crystal engineering of active pharmaceutical ingredients to improve solubility and dissolution rates. Adv Drug Deliv Rev 59:617–630

Chen L, Dang Q, Liu C, Chen J, Song L, Chen X (2012) Improved dissolution and anti-inflammatory effect of ibuprofen by solid dispersion. Front Med 6:195–203

Cheney ML, Weyna DR, Shan N, Hanna M, Wojtas L, Zaworotko MJ (2011) Coformer selection in pharmaceutical cocrystal development: a case study of a meloxicam aspirin cocrystal that exhibits enhanced solubility and pharmacokinetics. J Pharm Sci 100:2172–2181

Chieng N, Aaltonen J, Saville D, Rades T (2009) Physical characterization and stability of amorphous indomethacin and ranitidine hydrochloride binary systems prepared by mechanical activation. Eur J Pharm Biopharm 71:47–54

Corrigan DO, Corrigan OI, Healy AM (2006) Physicochemical and in vitro deposition properties of salbutamol sulphate/ipratropium bromide and salbutamol sulphate/excipient spray dried mixtures for use in dry powder inhalers. Int J Pharm 322:22–30

Descamps M, Willart JF, Dudognon E, Caron V (2007) Transformation of pharmaceutical compounds upon milling and comilling: the role of Tg. J Pharm Sci 96:1398–1407

Etter MC (1990) Encoding and decoding hydrogen-bond patterns of organic compounds. Account Chem Res 23:120–126

Gao Y, Liao J, Qi X, Zhang J (2013) Coamorphous repaglinide–saccharin with enhanced dissolution. Int J Pharm 450:290–295

Hirano A, Kameda T, Arakawa T, Shiraki K (2010) Arginine-assisted solubilization system for drug substances: solubility experiment and simulation. J Phys Chem B 114:13455–13462

Hoppu P, Jouppila K, Rantanen J, Schantz S, Juppo AM (2007) Characterisation of blends of paracetamol and citric acid. J Pharm Pharmacol 59:373–381

Hoppu P, Hietala S, Schantz S, Juppo AM (2009a) Rheology and molecular mobility of amorphous blends of citric acid and paracetamol. Eur J Pharm Biopharm 71:55–63

Hoppu P, Virpioja J, Schantz S, Juppo AM (2009b) Characterization of ultrasound extrudated and cut citric acid/paracetamol blends. J Pharm Sci 98:2140–2148

Hussain MD, Saxena V, Brausch JF, Talukder RM (2012) Ibuprofen–phospholipid solid dispersions: improved dissolution and gastric tolerance. Int J Pharm 422:290–294

Jayasankar A, Good DJ, Rodríguez-Hornedo N (2007) Mechanisms by which moisture generates cocrystals. Mol Pharm 4:360–372

Kumar S, Gupta S (2013) Effect of excipients on dissolution enhancement of aceclofenac solid dispersions studied using response surface methodology: a technical note. Arch of Pharm Res 37:1–12

Laitinen R, Löbmann K, Strachan CJ, Grohganz H, Rades T (2012) Emerging trends in the stabilization of amorphous drugs. Int J Pharm doi:10.1016/j.ijpharm.2012.04.066

Lee HG, Zhang GGZ, Flanagan DR (2011) Cocrystal intrinsic dissolution behavior using a rotating disk. J Pharm Sci 100:1736–1744

Lu J, Rohani S (2009) Polymorphism and crystallization of active pharmaceutical ingredients. Curr Med Chem 16:884–905 (APIs)

Lu Q, Zografi G (1998) Phase behavior of binary and ternary amorphous mixtures containing indomethacin, citric acid, and PVP. Pharm Res 15:1202–1206

Löbmann K, Laitinen R, Grohganz H, Gordon KC, Strachan C, Rades T (2011) Coamorphous drug systems: enhanced physical stability and dissolution rate of indomethacin and naproxen. Mol Pharm 8:1919–1928

Löbmann K, Laitinen R, Grohganz H, Strachan C, Rades T, Gordon KC (2012a) A theoretical and spectroscopic study of co-amorphous naproxen and indomethacin. Int J Pharm. doi:10.1016/j.ijpharm.2012.05.016

Löbmann K, Strachan C, Grohganz H, Rades T, Korhonen O, Laitinen R (2012b) Co-amorphous simvastatin and glipizide combinations show improved physical stability without evidence of intermolecular interactions. Eur J Pharm Biopharm 81:159–169

Löbmann K, Grohganz H, Laitinen R, Strachan C, Rades T (2013a) Amino acids as co-amorphous stabilizers for poorly water soluble drugs—Part 1: Preparation, stability and dissolution enhancement. Eur J Pharm Biopharm 85:873–881

Löbmann K, Laitinen R, Strachan C, Rades T, Grohganz H (2013b) Amino acids as co-amorphous stabilizers for poorly water-soluble drugs—Part 2: Molecular interactions. Eur J Pharm Biopharm 85: 882–888

Maheshwari C, Jayasankar A, Khan NA, Amidon GE, Rodriguez-Hornedo N (2009) Factors that influence the spontaneous formation of pharmaceutical cocrystals by simply mixing solid reactants. Cryst Eng Comm 11:493–500

Masuda T, Yoshihashi Y, Yonemochi E, Fujii K, Uekusa H, Terada K (2012) Cocrystallization and amorphization induced by drug–excipient interaction improves the physical properties of acyclovir. Int J Pharm 422:160–169

Mura P, Maestrelli F, Cirri M (2003) Ternary systems of naproxen with hydroxypropyl-β-cyclodextrin and aminoacids. Int J Pharm 260:293–302
Mura P, Bettinetti GP, Cirri M, Maestrelli F, Sorrenti M, Catenacci L (2005) Solid-state characterization and dissolution properties of Naproxen-Arginine-Hydroxypropyl-[beta]-cyclodextrin ternary system. Eur J Pharm Biopharm 59:99–106
Newman A, Knipp G, Zografi G (2012) Assessing the performance of amorphous solid dispersions. J Pharm Sci 101:1355–1377
Nugrahani I, Asyarie S, Soewandhi SN, Ibrahim S (2007) The antibiotic potency of amoxicillin-clavulanate co-crystal. Int J Pharm 3:475–481
Pajula K, Taskinen M, Lehto V-P, Ketolainen J, Korhonen O (2010) Predicting the formation and stability of amorphous small molecule binary mixtures from computationally determined Flory–Huggins interaction parameter and phase diagram. Mol Pharm 7:795–804
Rowlinson SW, Kiefer JR, Prusakiewicz JJ, Pawlitz JL, Kozak KR, Kalgutkar AS, Stallings WC, Kurumbail RG, Marnett LJ (2003) A novel mechanism of cyclooxygenase-2 inhibition involving interactions with Ser-530 and Tyr-385. J Biol Chem 278:45763–45769
Rumondor ACF, Taylor LS (2009) Effect of polymer hygroscopicity on the phase behavior of amorphous solid dispersions in the presence of moisture. Mol Pharm 7:477–490
Samy AM, Marzouk MA, Ammar AA, Ahmed MK (2010) Enhancement of the dissolution profile of allopurinol by a solid dispersion technique. Drug Discov Ther 4:77–84
Schantz S, Hoppu P, Juppo AM (2009) A solid-state NMR study of phase structure, molecular interactions, and mobility in blends of citric acid and paracetamol. J Pharm Sci 98:1862–1870
Seefeldt K, Miller J, Alvarez-Núez F, Rodríguez-Hornedo N (2007) Crystallization pathways and kinetics of carbamazepine–nicotinamide cocrystals from the amorphous state by in situ thermomicroscopy, spectroscopy, and calorimetry studies. J Pharm Sci 96:1147–1158
Serajuddin ATM (1999) Solid dispersion of poorly water-soluble drugs: early promises, subsequent problems, and recent breakthroughs. J Pharm Sci 88:1058–1066
Shimada Y, Goto S, Uchiro H, Hirota K, Terada H (2013) Characteristics of amorphous complex formed between indomethacin and lidocaine hydrochloride. Colloids Surf B Biointerfaces 105:98–105
Srinarong P, de Waard H, Frijlink HW, Hinrichs WL (2011) Improved dissolution behavior of lipophilic drugs by solid dispersions: the production process as starting point for formulation considerations. Expert Opin Drug Deliv 8:1121–1140
Tajber L, Corrigan DO, Corrigan OI, Healy AM (2009a) Spray drying of budesonide, formoterol fumarate and their composites—I. Physicochemical characterisation. Int J Pharm 367:79–85
Tajber L, Corrigan OI, Healy AM (2009b) Spray drying of budesonide, formoterol fumarate and their composites—II. Statistical factorial design and in vitro deposition properties. Int J Pharm 367:86–96
Tran TT-D, Tran PH-L, Choi H-G, Han H-K, Lee B-J (2010) The roles of acidifiers in solid dispersions and physical mixtures. Int J Pharm 384:60–66
Vasconcelos T, Sarmento B, Costa P (2007) Solid dispersions as strategy to improve oral bioavailability of poor water soluble drugs. Drug Discov Today 12:1068–1075
Vishweshwar P, McMahon JA, Bis JA, Zaworotko MJ (2006) Pharmaceutical co-crystals. J Pharm Sci 95:499–516
Williams HD, Trevaskis NL, Charman SA, Shanker RM, Charman WN, Pouton CW, Porter CJH (2013) Strategies to address low drug solubility in discovery and development. Pharm Rev 65:315–499
Yang Y-C, Huang C-S, Kuo C-C (2010) Lidocaine, carbamazepine, and imipramine have partially overlapping binding sites and additive inhibitory effect on neuronal Na+ channels. Anesthesiology 113:160–174

Chapter 22
Mesoporous ASD: Fundamentals

Alfonso Garcia-Bennett and Adam Feiler

22.1 Introduction

Silica is one of the most abundant chemicals at the earth's surface and mainly exists as one of its most stable polymorphs in the form of crystalline quartz, which is used extensively to manufacture optically pure window glazing and is an important part of today's lifestyle (Sosman 1965). Nature has perfected the synthesis of many other forms of crystalline architectures, for example, those of the zeolite family. Zeolites, natural and synthetic, are *micro*porous silicates of crystalline framework consisting of tetrahedral SiO_4 units, possessing pore openings and cavities in the region of ~ 10 Å with high specific surface areas. These are widely used in the catalysis industry and as water softeners in very large quantities, due to their ability to selectively direct many industrial reactions (such as petrochemical cracking) and to perform cation exchange (Rhodes 2010). Zeolites have found great utility in their ability to select between small molecules and different cations, hence their generic name of *molecular sieves*, but are limited by the small cavity size. To some, silica is only associated with its crystalline forms, but it exists both as crystalline and amorphous phases, and almost always as tetrahedral SiO_4 units.

Extending the use of zeolites into larger dimensions, say to catalyse enzymatic reactions and for the purification of colloidal precious metals, was the aim of researchers at Mobil Corporation (USA), who in 1992 discovered a viable and versatile synthetic procedure to prepare *meso*porous materials, i.e. materials with ordered porosity in the range between 20 and 500 Å (2–50 nm; Kresge et al. 1992). Their first material was termed MCM-41 (Mobil Composition of Matter) and the mechanisms involved templating of a silica sol-gel synthesis by an amphiphilic surfactant,

A. Garcia-Bennett (✉)
Department of Materials and Environmental Chemistry,
Arrhenius Laboratory, Stockholm University, Stockholm, Sweden
e-mail: alfonso@mmk.su.se

A. Feiler
Nanologica AB, Stockholm, Sweden

N. Shah et al. (eds.), *Amorphous Solid Dispersions*,
Advances in Delivery Science and Technology, DOI 10.1007/978-1-4939-1598-9_22

namely cetyltrimethylammonium bromide surfactant using tetraethyl orthosilicate as the silica source, in aqueous alkaline medium. This new family of materials was not crystalline but amorphous in composition, a fact that was later found to seriously impede their progress in petrochemical catalysis due to their poor hydrothermal stability. However, order was observed in the structure of the pores which were characterised as having a hexagonal arrangement, with very-well-defined pore-size distributions; even higher surface areas than the zeolites (above 1000 m^2/g) and very large pore volumes.

The discovery of mesoporous materials led to a research explosion in many academic and industrial areas (IMMS 2006). Surfactant science and inorganic chemistry merged in many PhD theses during the early 1990s, as researchers tried to control and develop these new types of materials. The increasing curve in publications mirrors similar trends in the patent literature. At the turn of the century, new and exciting applications in far-reaching areas such as sensors and biomedical devices were realised using mesoporous materials as the primary functional material (Raman et al. 1996). It is in the latter area and the pharmaceutical sector that the authors would like to concentrate in this chapter. We aim to review some of the seminal works that have led to the consideration of mesoporous materials as a functional excipient. We aim also to highlight some of the fundamental characterisation tools that have been used (and still are) in the development of mesoporous materials in the pharmaceutical sector, as well as to highlight the challenges ahead in order to see the implementation of these still exciting materials that keep surprising us and give us hope for more efficient drug delivery.

22.2 Ordered Mesoporous Materials: From MCM-41 to NFM-1

As mentioned, the discovery of ordered mesoporous materials is widely assigned to the work at Mobil laboratories (now Exxon Mobil) in the USA. However, it must be noted that researchers at Waseda University (Japan) led by Yanagisawa et al. 1990 also synthesised regularly ordered pore silicates composed of amorphous walls using surfactants as the pore-forming template. The Japanese work, whose publication predates the first Mobil patent filing date (Beck and Princeton 1991), is based on the use of the microporous mineral Kanemite and the combination of a hydrothermal treatment in the presence of surfactants to generate the mesoporosity through the intercalation of silicate sheets by the micellar surfactant. Mesoporous materials formatted through this mechanism are termed folded sheet materials (FSM-n). In the early years, controversy regarding to whom the first report of the materials was accredited ensued, and in time both mechanisms have been widely researched and developed, albeit with the Mobil-templating method becoming the most versatile and controllable approach to achieve different mesostructured products. Ironically, there is prior art in the patent literature for the formation of mesoporous materials. A method disclosing the synthesis of a material comparable to MCM-41 was reported, (Chiola et al. 1971) but those researchers did not foresee the full context and

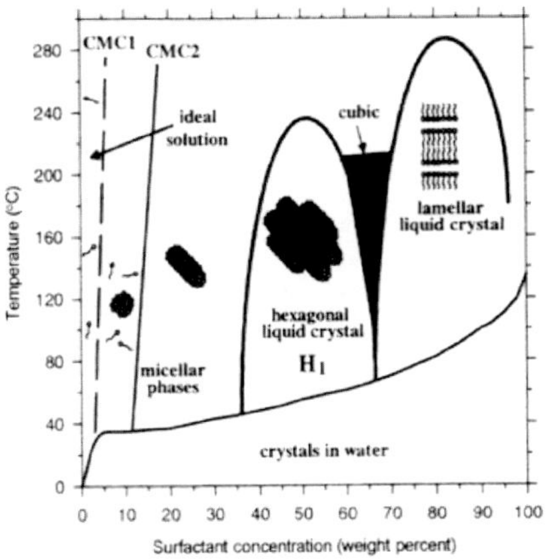

Packing Parameter, g	Micellar shape	Mesophase
< ⅓	Cone	Spherical micelles leading to close packing
⅓ - ½	Truncated cone	Cylindrical micelles, columnar structures
½ - 1	Truncated cone	Flexible bilayers
~1	Cylinder	Planar bilayers, and lamellar structures

Fig. 22.1 Phase diagram for the micellar packing of the surfactant cetyltrimethylammonium bromide changing with respect to temperature and surfactant concentration. (Reprinted with permission from ACS journals!)

potential of the material produced. The synthesis mechanism developed by Kresge et al. at Mobil Corporation is based on the self-assembly of surfactant molecules. The first molecules to be developed in this format were with cationic and polymeric surfactants. These molecules self-assemble in solution to form micellar structures which can undertake a larger variety of mesophases based on rod-like micelles to spherical micelles depending on factors such as the concentration of the surfactant in solution, the temperature of the system, its pH and the presence of additives (solvents, ions, etc.; Israelachvili 1991). Figure 22.1 shows the common shapes that micelles may take in solution above their critical micellar concentration (*cmc*), and a phase diagram for cationic surfactant cetyltrimethylammonium bromide, CTAB, $[CH_3(CH_2)_{15}N^+(CH_3)_3Br^-]$ which is employed for the synthesis of MCM-41.

Increases in the surfactant concentration lead to changes in the geometric conformation and packing of the micelles in solution. A direct relation between micellar shape and *meso*phase (surfactant phase in terms of structural properties) has been correlated, and it is known as the packing parameter, g, which describes the geometric parameters of the hydrophobic and hydrophilic sections of the amphiphilic surfactant molecule, and is described as follows:

$$g = \frac{v}{a_0 l}, \tag{22.1}$$

where v is the total volume of the hydrocarbon chain and a_0 is the head group area at the micelle surface and l the length of the hydrocarbon chain. Some predictive values of g are tabulated in Fig. 22.1. As g increases, the surface curvature of the micellar unit decreases from spherical to lamellar packing. While the effect of adding a silica source (such as tetraethyl orthosilicate, TEOS) to a surfactant solution adds a new complexity to the understanding of the packing parameter, this is a good initial design tool to design mesoporous structures of different parameters. For example, a high a_0 value will tend to form spherical micelles, leading to cubic and hexagonal (or indeed tetragonal; Garcia-Bennett et al. 2005a) close packing of the spherical units. Interaction of the surfactant head-group with the silicate species will lead to the eventual formation of *meso*caged porous materials (Huo et al. 1994), instead of straight channels which are typically formed with smaller a_0 surfactant values which are arranged into rod-like or cylindrical mesophases.

Thus, early synthetic development during the 1990s, focused on the versatility of the synthesis method and establishing the fundamental mechanism for the co-assembly of silica sources with surfactant-derived mesophases. Early on, it was noted that ordered mesoporous structures were formed when TEOS was added to surfactant solutions below their *cmc*, suggesting that mesophase order arises due to the interactions of the silica source with the surfactant prior to the interaction between the surfactant molecules (Chen et al. 1993). This leads to the proposition of a co-operative formation mechanism whereby multicharged inorganic species in an aqueous medium interact with the surfactant species and *co-operatively* form a liquid-crystalline arrangement through charge matching before silica condensation occurs (Huo et al. 1994). A number of other mechanisms have been proposed to justify the formation of mesoporous materials in acidic media and with a variety of surfactant types, and the reader is directed to the many excellent reviews describing this work (Wan and Zhao 2007; Berggren et al. 2005). With the knowledge of sol–gel and silica chemistry well established through decades of research into zeolite and ceramic materials (Brinker and Scherer 1990), the number of reports describing new mesoporous structures flourished to include lamellar; 2D–hexagonal, 2D–trigonal, 2D–orthorhombic (two dimensional) porous structures; and 3D–cubic, 3D–hexagonal and 3D–tretagonal, amongst others (Atluri et al. 2008; Hodgkins et al. 2005; Huo et al. 1994; Vartuli et al. 1994).

From the authors' perspective, there are several milestone synthetic papers that are worth noting. In 1998, Zhao et al. synthesised ordered mesoporous silica particles through the use of polymer surfactants enabling the extension of pore sizes by a significant amount (up to 35 nm; Zhao et al. 1998; Kim et al. 1995), in comparison to the smaller molecular surfactants. Using triblock copolymer such as poly(ethylene glycol)-block-poly(propylene glycol)-block-poly(ethylene glycol; known as P123) as template, some of the largest pore sizes were attained in a variety of ordered structures. These syntheses are performed in acidic conditions. The mechanism of silica formation is based on the interaction between the hydrophilic head group (S_0), acidic protons (H^+), anions (Cl^- in this case) and silica species (I^+) by electrostatic charge matching: denoted as (S_0H^+)(Cl^-I^+). The development of in situ and post-calcination procedures for functionalisation of mesoporous surfaces additionally allowed not only the large silanol surfaces produced after calcination but also to derive different functionalisations through the covalent bonding of other organosilane groups including hydrophobic silylating agents (see Fig. 22.2 below; Davidsson 2002).

In early synthetic development, anionic surfactants had been explored to direct the formation of mesoporous materials early in the synthetic development, but

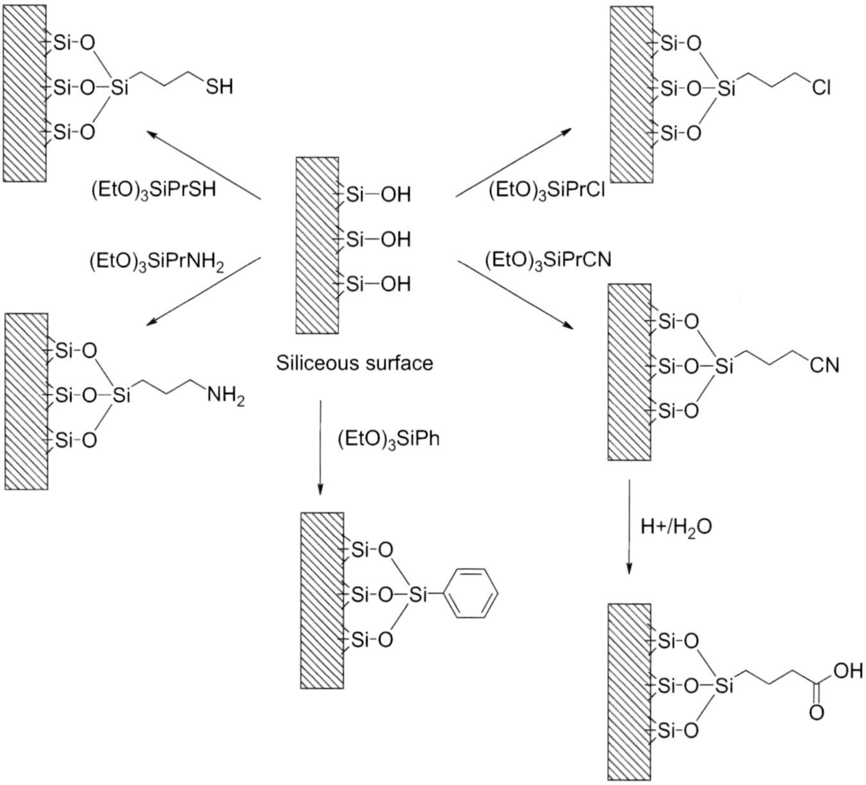

Fig. 22.2 Functionalisation strategies for mesoporous silanol surfaces based on post-calcination reaction with organosilanes

these preparations resulted only in lamellar and disordered structures (Wong and Ying 1998). Anionic surfactants are considered advantageous due to their non-toxic and non-irritant properties, which are widely exploited in the cosmetic industry (Seddon and Templer 1995). Additionally, families such as those of the amino acid-derived anionic surfactants (e.g. *N*-lauroyl-L-glysine; Sakamoto 2001), offer rich and biologically relevant liquid crystal behaviour, and could potentially increase the number and structure of unique mesoporous materials. The problem in replicating these mesophases into inorganic structures was to establish compatible charge-matching interactions between the surfactant head group (negatively charged) and the silica species. The latter are also negatively charged at alkaline pHs and hence an inadequate interaction was achieved in the synthesis gel. Che et al. were capable of solving this conundrum by introducing organoalkoxysilane *co-structure directing agents* (CSDAs) into the synthesis (Che et al. 2003). The negatively charged carboxylic headgroup of the anionic surfactant and the ammonium site of the organoalkoxysilane CSDA *N*-trimethoxysilylpropyl-*N, N,N*-trimethylammonium chloride (TMAPS) and 3-aminopropyl trimethoxysilane (APS).

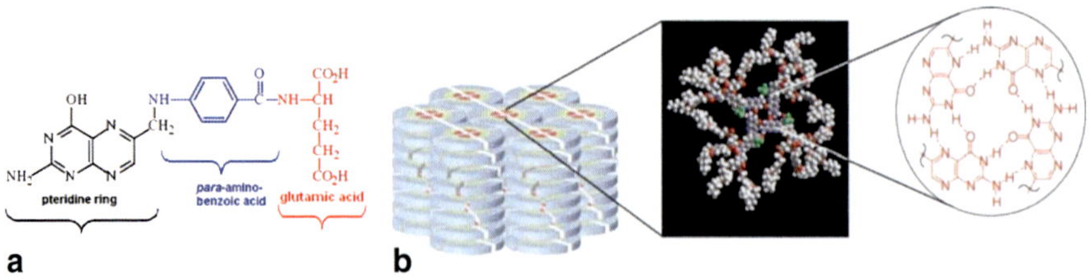

Fig. 22.3 **a** Structure formula of folic acid. **b** Schematic representation of self-assembled columnar stacking of pterin (or pteridine)-rings in folic acid

This family of materials has been named AMS-*n* (anionic surfactant-templated mesoporous silicas) and indeed allowed for the synthesis of various unique symmetries (Garcia-Bennett et al. 2004, 2005b; Che et al. 2004). The mechanism has been generalised as $S^-N^+\sim I^-$, where S^- is the surfactant, N^+ represents the positively charged amine moiety and I^- the inorganic silica species, both from the CSDA. An unexpected advantage of the AMS-*n* route is that removal of the surfactant via solvent extraction procedure results in an optimal coverage of the high surface area of the mesoporous material with amine groups (from the covalently grafted CSDA) reaching as high as several mmols/g (or a propylamine group for every two or three silicon atoms; Han et al. 2007).

A recent exciting development that extends the CSDA approach is the formation of ordered mesoporous materials through the use of non-surfactant-based mesophases, relying on biologically relevant supramolecular assemblies such as folates or nucleotide derivatives (Fig. 22.3). These form Hoogsteen-bonded tetrads and pentamers through H-donor and H-acceptor groups, capable of inducing self-organization to form columnar and hexagonal mesophases (Davis and Spada 2007). The biological importance of such macromolecular structures is exemplified by the assembly of guanosine-rich groups of telomere units with implications in chromosomal replication (Cong et al. 2002), or by folate derivatives: studied for the development of anticancer agents as increased folate receptors in tumour cells have led to several strategies for targeted drug delivery and treatment (Bacchi et al. 1980). Folic acid is composed of a pterin group, chemically and structurally similar to guanine, conjugated to the L-glutamate moiety via a p-amino benzoic acid.

Addition of CSDAs and silica sources to a solution of folic acid or guanosine monophosphate allows replicating their chiral mesophases to form ordered mesoporous materials with hexagonal phases. These materials have been denoted nanoporous folic acid material (NFM-1; Atluri et al. 2009). The chiral arrangement of the template within the pores was recently verified through circular dichroism and electron diffraction (Atluri et al. 2013).

Early applications of mesoporous materials, due to the origin of the initial discoveries, focused in and around their uses as catalysts. Both silica and silica–alumina (which offer acidic catalytic sites) compositions in particular were extensively explored in propene oligomerisation (gasoline and middle distillate production) due to their potential to have C9–C12 hydrocarbon selectivity (Bellussi et al. 1994). Other compositions or heterogeneously substituted compositions such as titania containing silica were explored for selective oxidation reactions and bimetallic hydrogenation (Thomas 1994; Thomas et al. 2003). For a full review of the uses of mesoporous materials in catalytic applications, see Schueth et al. (Taguchi and Scheuth 2004). Applications of these materials as adsorbents were realised soon, taking advantage of the precise control in pore size in combination with surface chemistry, exemplified by the remediation of heavy metal ions by propylthiol-functionalised mesoporous structures (Feng et al. 1997). Applications as adsorbents were supported by developments in the formation of ordered mesoporous carbon materials (Ryoo et al. 1999). Soon after these areas were covered from both intellectual property and proof-of-concept studies in the academic literature, more technologically advanced application was investigated in fibre optics and as electronic insulators taking advantage of their low-*k* dielectric properties of silica (Marlow et al. 1999; Doshi et al. 2003).

It is surprising that the link between pharmaceutical excipients and mesoporous silica was not made until 10 years after their discovery, an indication of the slow transfer of knowledge between research fields. Furthermore, drug delivery and formulation problems are very much the order of the day in the pharmaceutical industry, as it struggles to meet formulation challenges associated with drug discovery and repositioning of pharmaceutical actives (Cuatrecasas 2006; Wood 2006). Poor bioavailability from poorly soluble compounds, which make up a large proportion of lead compounds in pharmaceutical companies. Even their poor dissolution in suitable media for simple toxicology studies are major reasons for the rejection of lead compounds even before clinical trials (Takagi et al. 2006). The improvement of bioavailability and subsequent increases in potency or selectivity, the mitigation of toxicity, improving patient compliance through more patient-friendly administration routes or reduced number of doses and improving drug stability or reducing first pass effects are some of the important reasons that drive research within new drug delivery vehicles. Vallet-Regi et al. made this connection in 2001, and pioneered the field of pharmaceutical drug release using mesoporous materials (Yongde and Mokaya 2003; Vallet-Regi et al. 2001; Horcajada et al. 2004). As functional excipients for the pharmaceutical industry, ordered mesoporous silica particles have been shown in vitro and in vivo to: load large pay loads of single or multiple active molecules (Mellaerts et al. 2008); tailor the pharmacokinetic release profiles through diffusion or other mechanisms (Brohede et al. 2008); target the release of pharmaceutical products to specific cell types (Lu et al. 2010); increase the bioavailability of pH-sensitive drug candidates (Xia et al. 2012a); enhance the solubility of hydrophobic pharmaceutical actives with high partition constants (van Speybroeck et al. 2010); act as adjuvants in immunotherapies; act as a diagnostic and theranostic particles (Vallhov et al. 2007); and enhance the growth of apatite layers in tissue generation and bone implants (Wu and Chang 2012).

22.3 Characterisation of Mesoporous Materials

The characterisation of mesoporous solids play an important role in determining not only the type of materials produced (structurally and texturally) but also the purity. It is important to note that since the materials are amorphous, where the silica wall does not possess atomically ordered atoms, there is an added complication to their characterisation. Hence, the fundamental characterisation of mesoporous silicas must be conducted through a variety of techniques, and is mainly conducted through diffraction methods using electron microscopy and X-ray diffraction (XRD) which give details of the mesophase structure, type and quality, including the formation of any defects. Nitrogen adsorption measurements give indications of how well defined the porous system is including details of surface areas, pore volumes and pore size distributions. Functionalised silicas including tethered organic groups have been characterised by elemental analysis and spectroscopic methods (Ultraviolet-visible (UV), infrared (IR), nuclear magnetic resonance (NMR)) and the incorporation of precious metals and transition metal complexes are typically monitored by high resolution transmission electron microscope/energy-dispersive X-ray analyzer (HRTEM/EDX) and electron spin resonance (ESR) respectively.

X-ray diffractograms can reveal the *crystalline* (here referred to the long-range order of the mesopores) properties of the materials. Mesoporous materials have diffractograms with few peaks present at low 2θ angles—typically below $8°2\theta$—reflecting their large unit cells in comparison to microporous solids due to their amorphous silica composition (Thomas and Thomas 1997). Long-range order (with respect to the pores) of mesoporous silica is observed in the form of a broad peak (at around $20°2\theta$), hence the determination of atomic coordinates is not possible. In theory, there are not sufficient peaks in the diffractograms to resolve structures using XRD of mesoporous solids and it may also be difficult to determine the phase purity because of peak broadening and a lack of definition in higher angle peaks. Other additional practical problems arise due to the peak positions at very low angles (below $1°2\theta$) for very high unit cell materials such as SBA-15 (prepared with polymeric surfactants). In these cases, positioning of the detector is critical to reduce exposure of the material to the direct beam which may damage the detector and hide the desired peaks (see Fig. 22.4). In practice, some information can be derived if sufficient peaks can be recognised and indexed, such as the crystal class (hexagonal, cubic, etc.). However, for a full structural characterisation, modelling or further work using electron microscopy-based methods is required.

Transmission electron microscopy (TEM) is used to understand the local structure of a sample and in mesoporous solids, high-resolution TEM is used to elucidate the structure, together with XRD. There has, therefore, been extensive research on mesoporous solids using high-resolution TEM to help solve their structures, supported by techniques such as scanning electron microscopy (SEM), XRD and sorption studies (see later). Many structures can be modelled based on simulated Fourier transform diffractograms obtained from the high-resolution TEM images that represent specific

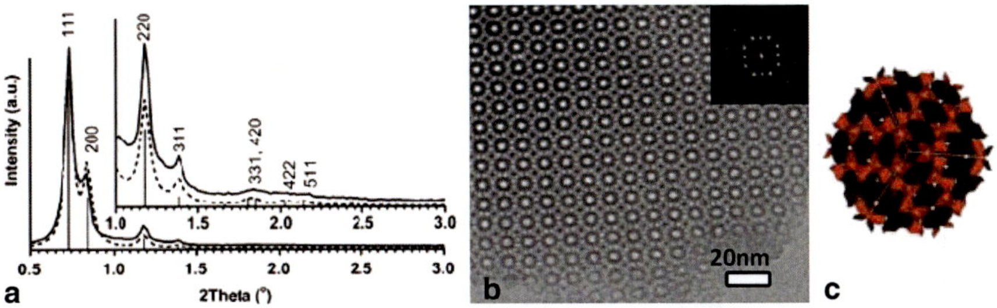

Fig. 22.4 **a** Simulated XRD pattern of bicontinuous cubic KIT-5 where *hkl* values may be indexed accordingly (Kleitz et al. 2003). **b** High-resolution TEM image of caged-type cubic mesoporous structure with *Pm3n* symmetry, viewed along the (Amidon et al. 1995) orientation and its corresponding Fourier transform diffractogram. **c** Unit cell model derived from TEM studies of a cage-type structure with *Fd3m* symmetry, viewed along the (Ritger and Peppas 1987) orientation. (Part **a** reproduced with permission from ACS journals!)

surface topologies and one can observe the presence or absence of local structural defects (Sakamoto et al. 2000).

Mesoporosity, in a strict sense, can only be identified by conducting sorption experiments since microscopic images only represent differences in contrast between the silica wall and *assumed* pore space (or pore template). Mesoporous materials adsorb according to type IV isotherm (in the standard Brunauer and IUPAC classification of isotherm curves; Brunauer et al. 1944), where the initial monolayer coverage is built upon by multilayer adsorption at higher adsorbate pressures, in an exactly similar way to that observed in type II. At the monolayer coverage, one can apply the Brunauer–Emmett–Teller (BET) equation to derive the specific surface area (Brunauer et al. 1938). At higher relative pressures, a steeper upward slope forms as a direct result of capillary condensation of nitrogen within the restrictive mesopores, characteristic of the type IV isotherm. The pressure, at which the capillary condensation increase occurs, is directly related to nitrogen filling of the pores and hence can be related to the diameter of the mesopores, derived traditionally using the Barrett–Joyner–Halenda (BJH) model which is based on the Kelvin equation (Barret et al. 1951), and expressions for the multilayer adsorbate film thickness as a function of adsorbate pressure. More recently, pore size distributions have more accurately been calculated using density functional theory (Ravikovitch et al. 1997), which does not introduce errors stemming from the application of Kelvin equation to smaller-size mesopores. The internal surface and volume decrease upon multilayer adsorption as critical pressure is reached at which capillary condensation occurs to fill the remaining volume preferentially over any larger-diameter pores. Upon pore filling, the isotherm curve reaches a plateau representing the upper limit of adsorption governed by the total pore volume (see Fig. 22.5). In large-pore mesoporous materials where pores or cavities are connected by smaller windows, de-gassing the nitrogen adsorbate from the material is governed by the size of the smaller connecting windows resulting in a lower relative pressure needed for the adsorbate to leave the pore. This effect can be followed in the desorption branch of the isotherm and gives

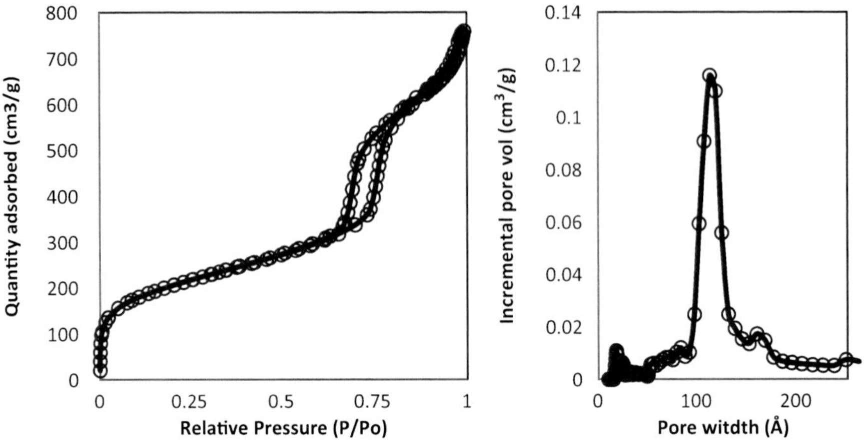

Fig. 22.5 Typical type IV nitrogen adsorption–desorption isotherm for mesoporous material SBA-15, showing a steep (denoting a sharp pore size distribution) capillary condensation rise at relative pressure of approximately 0.7. Its corresponding pore size distribution is shown. Note the presence of a hysteresis loop

rise to a hysteresis loop. Such effects are the subject of ongoing research in order to determine the shape of the pore geometry (Ravikovitch and Neimark 2002).

Thermogravimetric analysis (TGA) is commonly used as part of the characterisation toolbox for mesoporous materials. TGA measures mass change of a sample as a function of increasing temperature, which can be related to physical and chemical changes. In most cases, the measurements detect mass loss due to vaporisation or desorption. TGA is routinely used to quantify the mass loss of the templating agents from mesoporous particles and to measure the mass of material adsorbed in mesoporous particles after loading. The technique does not give specific chemical information directly and so, for unknown samples, complementary analysis needs to be done. TGA is often coupled to mass spectrometry or IR as a means to obtain chemical fingerprinting of the desorbing material. However, for rapid screening, the method is useful for measuring total masses and some general characterisations can be made, for example, mass loss of adsorbed water occurs close to 100 °C and volatile solvents below 100 °C.

When it comes to characterising the specific interactions between encapsulated molecules within the pores of mesoporous silica, common methods include differential scanning calorimetry (DSC) and various spectroscopic techniques. DSC measures changes in heat of a sample relative to a reference as a function of temperature. The technique is very sensitive to physical changes and this may be an endothermic processes such as melting or an exothermic processes such as crystallisation. DSC is also sensitive to more subtle phase changes like glass transitions. A DSC measurement comprises a scan of increasing and then decreasing temperature and can give clear indication of changes in physical properties. The enthalpy of transition is calculated from the DSC curve as a product of the area under the curve and the calorimetric constant. Some processes are reversible such as melting and glass transition whereas others are irreversible such as crystallisation, evaporation oxidation, degradation or denaturation. DSC is particularly useful for studying drug compounds to detect polymorphism and melting temperatures. When a drug molecule is confined in the narrow pores of mesoporous silica particles, the absence of a clear crystallisation peak is evidence that the drug is maintained in an amorphous form which is one of the major interests in using mesoporous material for drug formulation.

Various spectroscopy techniques are suitable for measuring the physico-chemical properties of mesoporous materials themselves and also investigating the interactions of molecules adsorbed onto and into the mesoporous materials. Spectroscopic techniques measure the interaction of matter with radiated energy. The energy may vary dramatically in magnitude and frequency and take the form of, for example, visible light, X-rays and microwaves. IR spectroscopy is very commonly used to identify chemical species of molecules through characteristic vibrations of molecules as they stretch, rotate and bend in the influence of adsorbing radiation. Every compound has a unique fingerprint spectra relating to the net interaction of atoms in the structure. Chemical or physical adsorption of a compound to a surface or trapped within a narrow pore results in changes to the IR spectra which can be used to identify specific interactions of the molecules with the surface. The stretching of the –OH bond gives a characteristic broad band in IR spectroscopy which is often used to detect the presence of water. For the case of mesoporous silica particles, the prominence of the OH stretching peak gives a measure of the degree of hydroxylation of the terminal silanol groups of the surface.

Needless to say, there are a battery of other analytical techniques suitable for the characterisation of mesoporous materials regarding aspects of composition and functionalisation.

22.4 Properties of Mesoporous Materials Relevant to Life Sciences

As mentioned earlier, mesoporous materials possess special physical properties such as high surface areas above $1000\,m^2/g$, high pore volumes above $1.0\,cm^3/g$ and controlled pore sizes between 2 and 50 nm. Apart from these physical properties, mesoporous materials offer other unique properties that make them suitable for applications in pharmaceutical sciences and, in particular, as functional excipients. It is also clear that the tensile strength of mesoporous materials under different tableting pressures as well as any thermal stresses will vary considerably depending on particle size and shape, as has been observed by Vialpando et al. (2011).

Control of morphological properties (particle size and shape) is critical in the behaviour of mesoporous materials in biological media. The interaction of SBA-15 and MCM-41 mesoporous silica particles with human red blood cells was investigated by Zhao et al. (2011). Larger SBA-15 particles (~ 531 nm, characterised by

SEM) were taken up by red blood cells to a greater extent than MCM-41 particles (~ 122 nm, by SEM). MCM-41 particles were adsorbed onto the cell surface without disturbing the cell membrane while the adsorption of SBA-15 particles to the cell surface led to greater local membrane deformation, internalization and subsequent rupture of the red blood cells (haemolysis). Hence, it is evident that small particle sizes require major membrane modifications in order to be taken up; this is due to the smaller contact area with the cell membrane and hence more energy is required to bind than for larger particles with larger contact area (Cedervall et al. 2007). Particle shape can also influence the interaction with cells by affecting the uptake mechanism. They additionally affect how particles circulate in the blood stream and how they filter through kidneys and spleen. In general, particles with different shapes experience different hydrodynamic forces in biological media (e.g. blood serum) where non-spherical particles have a higher tendency to move towards the vessel walls than spherical particles. (Donaldson et al. 2011). Non-spherical micron-sized particles can pass through the spleen when having at least one dimension smaller than 200 nm. Biopersistent nanofibres with high aspect ratio and fibre lengths $> 15\ \mu m$ (e.g. asbestos and carbon nanotubes) cannot be engulfed by macrophages leading to frustrated phagocytosis and to a non-effective clearance of the particles (Chithrani et al. 2006). However, the shape effect on cell uptake can be composition dependent. For example, in other type of nanoparticles (spherical gold nanoparticles), it has been shown that elongated particles are internalised faster by HeLa cells (Chithrani et al. 2006), while in a subsequent study, the opposite effect has been observed for hydrogel nanoparticles (cationic poly(ethylene glycol); Gratton et al. 2008). The immunological behaviour of mesoporous materials has also been identified to be particle shape and size dependent. (Vallhov et al. 2007).

The surface chemistry of mesoporous silica particles can influence the reactivity, solubility, interaction and agglomeration of particles in different environments, as well as their accumulation in organs and tissues. Surface chemistry will also strongly influence the nature of adsorbed protein corona and the interaction strength with it (Lynch et al. 2009).

Functionalisation of the particle surface with different chemical groups can be engineered to modulate the interaction with biological systems. For example, SBA-15 particles with or without propylcyanide functionalisation lead to stimulation of T cell development in different directions (Vallhov et al. 2012). The z-potential value for SBA-15-PrCN measured in serum supplemented cell culture media was -8.43 mV, and the corresponding value for the free SBA-15 was -10.0 mV. Whilst this may appear to be a small difference, SBA-15-PrCN had no effect on viability in comparison to controls and did not activate dendritic cells in contrast to SBA-15. Slowing et al. studied the uptake of MCM-41 mesoporous silica particles by HeLa cells expressing folate receptors. The more positive charged particles (with z-potential between -3.2 and $+12.8$ mV) were located inside endosomes in larger numbers than more negative charged ones (-4.7 and -34.7 mV; Slowing et al. 2006). These examples exemplify the effect of surface chemistry on both uptake, endosomal entrapment and the processing of particles within the cell (i.e. exposure or lack of it to lysozymes in the endolysosomes). Furthermore, functionalisation of mesoporous particles with polyethylene glycol (PEG) has been shown to decrease protein absorption leading to longer circulation times and altered biodistribution (Lin et al. 2011). On a practical note, surface functionalisation and changes to the surface chemistry of a mesoporous silica particle can improve its compatibility with different solvents and drug compounds being loaded within its pores, as well having effects on processing parameters such as tableting and its eventual administration route (Vallet-Regi et al. 2007).

22.5 Toxicological Implications of Mesoporous Materials

Regulatory approval of new pharmaceutical excipients are by definition complex and uncharted. It is often deemed necessary for novel excipients to be subjected to the same rigour as new drug compound or in combination with already existing drugs in which the excipient will form part of the final formulation (Kolter 2011). Pre-clinical toxicity studies for drug compounds must include: acute toxicity, sub-chronic and chronic assays, genotoxicity, reproductive toxicity, pharmacokinetics and pharmacology for absorption, distribution, metabolism and excretion. As mesoporous silica are potential excipients with some function (e.g. controlling drug dose as adjuvants or in diagnostics), their toxicity has been carefully assessed, despite ample data already existing on the toxicological properties of amorphous colloidal silica particles (already approved and widely used as food additives; British Pharmacopoeia 2009). This is a necessary barrier in the development of nanoparticles for clinical use, irrespective of the advantages that they may bring to the nanomedicine and pharmaceutical sector. There is a clear need for unification of toxicological assessment techniques and standards, which may facilitate the comparison of studies and endpoints in order to speed up this process (Fadeel and Garcia-Bennett 2010). Careful and accurate characterization of material properties, in terms of their in vivo particle size (i.e. in the relevant media to be used in assays or clinical context), surface chemistry, composition, textural properties (morphology, surface area, pore volume and pore size), degree of opsonisation and potential for the formation of reactive oxygen species (ROS), must be assessed. These are just as important as a battery of cellular studies. Secondly, analytical methods should be chosen to avoid their interactions with tested particles. The suitability of using the MTT assay to evaluate biocompatibility in vitro has been put into question for several nanoparticles including mesoporous silica (Laaksonen et al. 2007). Material–dye interactions and dye adsorption may interfere with the experimental outcome. When assessing the effect of mesoporous silica on cell viability, the MTT colorimetric assay overestimates cell death compared to other techniques such as flow cytometry due to an enhanced formazan exocytosis. Moreover, formazan can form crystals within the pores of carbon nanotubes limiting its solubilisation which result in a decreased colour and absorbance in the analysed solution and an underestimation of the cell viability. For this reason, other techniques than MTT, such as Trypan blue exclusion assay, propidium iodide (PI) and Annexin-V staining, should be used in conjunction (Worle-Knirsch et al. 2006). Toxicological and biocompatibility evaluations of mesoporous particles have been performed in several in vitro and in vivo systems via several administration routes.

We recently reported an in vivo oral toxicology study with two different types of mesoporous particles, AMS-6 and NFM-1 (with 300 nm and 1.5 μm particle size, 3 and 5 nm pore size respectively) which were administrated by oral gavage to rats (Kupferschmidt et al. 2013). The maximum tolerated dose (MTD) in the study could not be reached, i.e. no signs of toxicity were observed in response to the particles at any of the doses tested, between 30 and 2000 mg/kg for NFM-1 and 40 and 1200 mg/kg for AMS-6. A study of even higher doses was not possible due to ethical reasons. The results nonetheless confirm the high tolerance of the particles after oral administration, and a lack of systemic adsorption of mesoporous silica particles of sizes above 300 nm. All animals showed normal weight development and there were no differences in the haematology or the biochemistry parameters from animals treated with NFM-1 or AMS-6 compared to the control group. In a separate study, we investigated the effect of mesoporous silica materials in the diet using murine obesity model (Kupferschmidt et al. 2014a). Mesoporous silica particles with different pore sizes (20 and 110 Å for NFM-1 and SBA-15, respectively) were mixed in the high-fat (HF) diet of C57BL/6 J obese mice. No particles were added to the diet of control animals. The particle content in the HF diet did not result in differences between the animal groups when this was the only dietary source. However, obese mice receiving a standard diet with additional HF complement containing SBA-15 twice a week significantly decreased in weight and body fat composition. The reduction in body weight and body fat composition exerted by SBA-15 was attributed to the 110-Å pore size and its ability to encapsulate lipases, suggesting that pore size of mesoporous silica can be tailored to achieve reduction in body weight and fat composition. As a toxicological evaluation, this study supports the non-toxic classification of mesoporous silica for oral administration in time periods up to ~3 months. However, our results also suggest that the interaction or adsorption of drug and gastrointestinal substances must be investigated further. Accidental encapsulation of oral drugs into the particles may lead to difficulties in dose control. Obese patients usually suffer co-morbidities such as diabetes and cardiovascular diseases. Hence, pharmaceutical agents used to improve glycaemic control such as metformin (Riomet; Davidson 2012) or to treat high blood pressure such as beta-blockers may be affected by oral administration of mesoporous silica.

Intravenous administration (i.v.) results in systemic distribution and it offers the possibility of targeting via passive or active targeting. This is particularly important in the treatment of cancer with nanoparticles. Passive targeting occurs as a result of the tumour's leaky vasculature and reduced lymphatic drainage system. Particles of up to ~300 nm in diameter accumulate in the tumour via the enhanced permeability and retention effect (EPR; Adiseshaiah et al. 2010). Active targeting uses specific ligands attached on the particle surface to bind to specific or overexpressed surface molecules on the target tissues. Increased accumulation in the target can improve the effectiveness of toxic drugs and vaccines. This may result in the use of lower quantities of toxic drug compounds and decreased adverse effects. Mesoporous silica particles have been shown to be well tolerated when delivering efficient doses of anticancer drugs to tumours in human cancer xenograft mice after intraperitoneal and intravenous injection. Lu et al. have found that ten daily doses of > 100 mg/kg of ~ 100–130 nm particles administered by intraperitoneal injection (i.p.) or i.v. were well tolerated by mice (Lu et al. 2010). In a recent publication, the MTD after i.v. administration of mesoporous particles (spheres of 120 nm and rods of 36 × 1028 nm) and non-porous silica particles (150 nm) were respectively 30–65 mg/kg and 450 mg/kg (Adiseshaiah et al. 2010). For comparison, the MTD of i.v. administered amorphous non-porous silica particles (15 and 55 nm) and crystalline silica particles (400 nm, quartz) has been determined in rats to be respectively; ~ 50, ~ 125 and 100 mg/kg (Downs et al. 2012). Hudson et al. reported that i.p. and i.v. injections of ~ 1.2 g/kg doses of mesoporous silca materials (MCM-41, SBA-15 and MCF particles with particles sizes of 29–140, 800 nm and 4 μm and pore sizes of 3, 7, 16 nm respectively) were lethal or lead to euthanasia of mice. Dose reduction to 40 mg/kg was safe (Hudson et al. 2008). The silica content in different organs after injection has been shown to decrease over time indicating that the particles can be biodegraded. Low toxicity by other mesoporous silica particles after i.v. injection has been described in several publications (Huang et al. 2011). Complete particle clearance from the organism after ~ 4 weeks with no signs of toxicity after a single dose (500 mg/kg) and daily doses (14 days, 80 mg/kg) has been reported (Liu et al. 2011).

From a toxicological point of view, it is important to study the immune-modulatory properties of mesoporous particles, since they may be considered as a potential pathogen at worse, and may additionally interact with immune-competent cells causing an immune response. Immunological studies performed on mesoporous silica with different particle size and pore sizes showed that particle uptake by human dendritic and macrophage cells (immune competent cells) did not impair their functions or affect their cell viability. Smaller (AMS-6, 300 nm particle size, 4.5 nm pore size) particles were encapsulated into vesicular compartments while the larger (AMS-8, 3 μm particle size, 3.0 nm pore size) particles were found directly in the cytosol within human monocyte-derived dendritic cells (MDDC; Vallhov et al. 2007). Witasp et al. have shown that the uptake of AMS-6 and AMS-8 by MDDCs occurs through an energy-dependent (active) mechanism since none of the particles were taken up when cells were incubated at 4 °C, which inhibits active internalization (Witasp et al. 2009). Endosomal entrapment has also been shown for MCM-41 (150 nm particle size) in studies with HeLa cells (Slowing et al. 2006). In a recent study, ovalbumin (OVA)-sensitized mice were treated with SBA-15-OVA or alum-OVA, before challenge with OVA. The treatment with the particles was shown to skew the existing allergic inflammation (Th2-type response) towards a more Th1-like response. This was indicated by higher IgG2a serum levels and INF-γ production, measured in splenocytes after ex vivo stimulation with OVA, compared to alum adjuvant. Moreover, the OVA-specific IgE levels were lower than those in the alum-OVA treated group (Kupferschmidt et al. 2014b). These results make mesoporous silica particles suitable adjuvant candidates for vaccine/immunotherapy when a more Th1-like immune response is preferred over Th2. No symptoms of local toxicity or granuloma formation at the site of administration were observed in accordance with previous studies using SBA-15 as adjuvant (Mercuri et al. 2006).

22.6 Loading Drugs into Mesoporous Materials

While there have been many written reports on the role that mesoporous materials could play in sustained drug release, (Rosenholm et al. 2010) triggered (Garcia-Bennett 2011), or targeted release (Wang et al. 2010a), the first applications for these materials may be in improving the bioavailability of poorly soluble substances. The main reason for this is that while there are competing new and established technologies that tackle issues of targeting or sustained release, there are only a few industrial methods to improve the solubility of newly discovered drug compounds with poor solubility profiles, the so called class II (low solubility and high permeability) and class IV (low solubility and low permeability) APIs according to the Biopharmaceutics Classification System (BCS; Amidon et al. 1995). These established techniques include: (1) physical procedures including drug-particle size reduction through micronisation or the formation of nanosuspensions; modification of crystal habits or polymorph behaviour through re-crystallisation procedures; formation of the amorphous form or co-crystallisation; and (2) chemical procedures such as derivatization, complexation or formation of salts (Kerns and Di 2003). There is a common consensus amongst academics and industrial pharmaceutical experts that there is a need for new techniques for improving drug solubility. This is driven by the need to increase efficiency as well as optimise cost.

From the perspective of the patients, oral administration of drugs is undoubtedly the most patient-compliant and convenient method for ingestion of drug substances. It is cost-effective and offers flexibility in dosing. This flexibility is at the centre of the success of generic companies and the ease at which they can conduct bioequivalent studies (Chiou and Riegelman 1971). Ideally, a drug substance should emerge through lead optimisation stages in drug discovery with predictable and adequate solubility. Unfortunately, it is often the case that APIs that have good potency present formulation challenges due to poor solubility (Serajuddin 1999).

Loading or encapsulation of APIs into mesoporous silica has traditionally been conducted via impregnation of the solubilised drug followed by evaporation-induced capillary action. In essence, if you can dissolve the drug in a suitable solvent, you can load it into the desired mesoporous silica material by mixing the drug solution with the solid particles and evaporating the solvent in a rotary evaporator. The commonly used solvent in laboratory experiments include ethanol, methanol, acetic acid and acetone. The solid remaining will typically be composed of the mesoporous silica loaded with drugs, with a small fraction of drug remaining outside the pores and on the particle surface. Particularly for poorly soluble compounds, this method results in the formation of drug crystals outside of the pores (Xia et al. 2012a). Drug remaining on the surface will possess different dissolution behaviour than that loaded within the pores (van Speybroeck et al. 2010). While this may be of no consequence in some instances because of the small concentration of drug remaining on the exterior, it is still simpler to predict the behaviour of single-phase systems. There are several techniques that can be used in order to reduce the amount of drug remaining or crystallizing on the exterior of the particle. One may wash the loaded particles by choosing a suitable solvent in which the API has a lower solubility. One may additionally reduce the amount of loaded drug and repeat the loading procedure several times. Again, this is preferably avoided due to environmental and economic considerations. One may choose a large particle size material, in order to reduce the ratio of internal/external surface so as to minimise the drug remaining on the exterior of the pores. One may also avoid evaporation all together and filter the drug-loaded particles; in which case, a low amount of drug will be loaded but ensuring that all of it is contained within the pores. In practice, adequate loading procedures have to be established on a drug-by-drug basis taking into account solubility of the API in different solvents, the formation of solvates and different crystal polymorphs, the desired amount to be loaded, the pore size and particle size of the mesoporous particle being utilised. The loading of drug molecules into porous materials under confinement has been recently reviewed by Jiang and Ward (2014). Classical nucleation theory shows that crystal growth will occur when a nucleus of a certain size, r_{crit}, surpasses a maximum free energy change ΔG_{cryst}, or activation energy of crystallisation associated with the onset of crystal growth. Nuclei smaller than r_{crit} will remain in solution and not lead to crystallisation. This critical size is dependent on many factors such as concentration of the drug in solution, temperature, aromaticity, symmetry, electronegative atoms, formation of supramolecular interactions and molecular size. To complicate matters further, the formation of various polymorphs of API will yield different crystallisation behaviours (different r_{crit} and ΔG_{cryst} values; Roy et al. 2012). In essence, the mechanism of loading into mesoporous materials relies on preventing the drug molecules from reaching r_{crit}, and hence inducing a stable amorphous state of the drug compound within the pores. This is, of course, only possible if the mesopore size in question is below r_{crit}, i.e. below a certain critical pore diameter d^*. This model of suppression of crystallisation can be expressed thermodynamically according to the following equation:

$$d^* = 4\sigma cl T_{\text{m}}/(T_{\text{m}\infty} T)\Delta Hm\, \rho c, \qquad (22.2)$$

where σcl is the surface energy between crystal and melt, ΔHm is the heat of melting, T_{m} is the bulk melting temperature and ρc the crystal density. Further to thermodynamic concerns, nanoconfinement will decrease the crystallisation kinetics of the compound through a dispersion of nucleation sites within the porous structure and slow down the diffusion of growth nutrients to the growing crystal surface. Mesoporous materials offer exquisite control over pore size and distribution at ranges (of particular interest is the range between 2 and 8 nm) not attainable through other synthetic mechanisms, such as those Vycor-type glasses produced through etching techniques. In the latter, crystallisation often occurs within the pores of the silica matrix, while in our experience, no crystallisation has been observed within the pores of mesoporous silica materials < 5 nm, and it is extremely difficult with pores below < 10 nm. The reason is that organic crystal nuclei are considered to be in the order of nanometres to tens of nanometres, hence are restricted by the confinement of the meso-scaled pores. Some typical physical parameters from drugs loaded into mesoporous materials are shown in Table 22.1. Drug compounds loaded into mesoporous materials below approximately 100 Å are hence loaded in their amorphous state, which can be easily inferred through a variety of techniques, such as combination calorimetric techniques (TGA and DSC), diffraction and sorption-based methods. The precise nature of the amorphous state remains unknown and the formation of different degrees of non-crystallinity must be expected. One characteristic of the amorphous state is a lack of long-range translational and conformational order and a subsequent low-density non-crystalline packing of the drug compound within the pores. Short-range interactions, for example hydrogen bonding drug–drug interactions or electrostatic silica wall–drug interactions, must be playing a role within the pores. In short, more work is required in this area in order to fully elucidate the full effects of confinement of the pores on the amorphous nature of pharmaceutical compounds with a variety of physico-chemical parameters.

22.7 Fundamentals of Release

In molecular terms, the release of drug molecules from the internal space of mesoporous material can be considered as a random walk of molecules inside the pore system of the carrier. The first 60 % of release from dissolution curves has been shown to be adequately described by a semi-empirical power-law expression (Higuchi 1963):

$$Q = a + b \times t^{\kappa}. \tag{22.3}$$

This is applicable to various carrier symmetries such as planar, cylindrical and spherical. Here Q is the amount of molecules released per unit exposed area of the carrier, t denotes time and a, b and k are constants. This power-law function is related to the Weibull function that has been suggested as a universal tool for describing release from both Euclidian and fractal systems, and may be considered as a short-time approximation of the latter (Kosmidis et al. 2003). The constant a takes initial delay and burst effects into account, and b is a kinetic constant (Jamzad et al. 2005). The power law exponent, k, also called the transport coefficient, characterises the diffusion process and equals 0.5 for ordinary case I (or carrier controlled) diffusion in systems for which no swelling of the carrier material occurs, which can be expected for mesoporous material (Ritger and Peppas 1987). Diffusion-controlled release from a planar system, in which the carrier structure is inert, may be described by the Higuchi square-root-of-time law:

$$Q = \sqrt{D_{\text{eff}} C_S (2C_m - \varepsilon C_S) \times t}. \tag{22.4}$$

Here D_{eff} is the effective diffusion coefficient of the drug inside the carrier, and C_{m} is the total amount of drug present in the carrier per unit volume, ε is the total porosity of the carrier, defined as the volume fraction of pores when the drug material has been removed (here referred to as the porosity of the calcined samples) and C_{s} is solubility of the drug in the release medium. This theoretical framework has been shown to predict somewhat the release from mesoporous materials with different structures, as well as their diffusion coefficients (Brohede et al. 2008; Strømme et al. 2009a, b). A general theory for predicting the release properties is still difficult to transfer from drug to drug. This is primarily due to the myriad of drug properties, administration routes, formulation problems and translational issues that one encounters in the pharmaceutical sector. This should not deter the user, as the field is advancing rapidly and much progress has been made to understand where mesoporous materials are most advantageous. One such area is the observed improvements in bioavailability that can be attained by creation of an amorphous state of the drug when this is loaded into mesoporous materials (see examples in Table 22.1).

22.8 Matters of High Activity Antiretroviral Therapy (HAART): A Case Study

In an effort to discover how improvements in dissolution behaviour translate into actual oral bioavailability, studies must be performed in animals. This is illustrated by a study of the compound atazanavir (ATV), an antiretroviral protease inhibitor used for the treatment of infection by human immunodeficiency virus (HIV). Worldwide, over 40 million people are infected with HIV. The high activity antiretroviral therapy (HAART) combines at least three antiretroviral drugs and has been used to extend the lifespan of HIV-infected patients. Chronic use of HAART is needed to control HIV infections. The frequent administration of several drugs in relatively high doses is the main cause of patient noncompliance and a hurdle toward the efficacy of the pharmacotherapy (Sosnik et al. 2009). ATV is a lipophilic drug with partition constant (logP) value of 5.20. Despite four violations of the Lipinski's rule of five, its bioavailability is between 60 and 68 % and it has a half-life of 6.5 h when administered orally. However, the bioavailability of ATV is severely hampered, by as much as 78 % reduction in plasma concentration, when it is co-administered with proton-pump inhibitors, leading to a significant decrease in its effectiveness (Reyataz 2004). Proton-pump inhibitors are administered to HIV patients to treat the symptoms of heartburn and stomach pains that are common secondary effects of HIV medication. This decrease in bioavailability is caused by precipitation of the drug under the less acidic conditions caused by the proton-pump inhibitors. Hence, development of a formulation capable of maintaining high bioavailability when ATV is co-administered with proton pump inhibitors is likely to lead to a considerable improvement in patient comfort, as well as effectiveness of the treatment.

Table 22.1 Some examples of drug compounds loaded into mesoporous materials, with in vivo data available confirming enhancements in apparent solubility

Drug	Tm Free drug (°C)[a]	MW	logP[a]	Loading method	Meso-structure pore size	S_{BET}	Loading (wt%)	Amorphous state	*In vitro* enhancement of solubility and supersaturation state	DSC on loaded sample	AUC_0–8 h (μg ml−1 h−1)	$T_{1/2}$ (h)	C_{max} (μg/ml)	Ref
Aceclofenac *non-steroidal anti-inflammatory*	152	354.1	4.1	Solvent evaporation	MCM-41 2D-cylindrical 1.5 nm (BJH)	894	91.8	Crystalline peaks visible by XRD suggest drug on the exterior of the particle	Yes $Q_{85=}$ 30 min vs. 363 min for free drug	~5 °C melting depression when compared to crystalline drug	*Reference-formulation* 9.11 ± 7.54 *Meso-formulation* 17.30 ± 7.52	2.14 1.17	2.19 ± 0.73 4.05 ± 1.77	(Kumar et al. 2014)
Atazanavir *Anti-retroviral*	195	802.9	5.2	Solvent evaporation	NFM-1 2D-cylindrical 2.6 nm (DFT)	793	31.5	No peaks visible on XRD diffractograms	Yes Q_{87} = 7 h un-easurable for free drug.		*Reference-formulation* 0.065 *Meso-formulation* 0.271	– –	0.183 0.852	(Van Speybroeck et al 2010)
Celecoxib *nonsteroidal anti-inflammatory*	162	381.4	3.9	Diffusion and filtering	Carbon SBA-15 2D-cylindrical 4.4 nm (BJH)	722	39.2	No peaks visible on XRD diffractograms	Yes Q_{85} = ~15 min not reached for free drug	No melting depression is detected	*Reference-formulation* 330 ± 43 *Meso-formulation* 473 ± 45	9.60 ± 1.32 10.36 ± 1.13	2.43 ± 0.48 4.09 ± 0.18	(Wang et al. 2010b)

Table 22.1 (continued)

Drug	Tm Free drug (°C)[a]	MW	logP[a]	Loading method	Meso-structure pore size	S_{BET}	Loading (wt%)	Amorphous state	*In vitro* enhancement of solubility and super-saturation state	DSC on loaded sample	AUC_{0-8h} (μg ml−1 h−1)	$T_{1/2}$ (h)	C_{max} (μg/ml)	Ref
Glibenclamide *antidiabetic drug*	169	494.0	4.3	Solvent impregnation and evaporation	SBA-15 2D-cylindrical 7.2 nm (BJH)	606	22.5		Yes $Q_{85} = \sim 10$ min. not reached for commercial formulation	Did not exhibit a melting peak, nor did they exhibit a glass transition	*Reference-formulation* 1.413 *Meso-formulation* 6.236	$3.5 (T_{max})$ $2.5 (T_{max})$	0.292 2.255	(van Speybroeck et al. 2011)
Telmisartan *hypertension*	261	514.6	7.7	Solvent impregnation and drying	SBA-15 2D-cylindrical 11.3 nm (BJH)	917	27.5	No peaks visible on XRD diffractograms	Yes $Q_{85=}$ 30 min vs. 45 min for commercial tablet	Did not exhibit a melting peak	*Reference-formulation* 4.930 ± 0.381 *Meso-formulation* 6.487 ± 0.683	1 ± 0.2 (T_{max}) $1.42 \pm 0.2 (T_{max})$	2.102 ± 0.25 1.891 ± 0.27	(Jamzad et al. 2005)

[a]Data obtained from Chemspyder

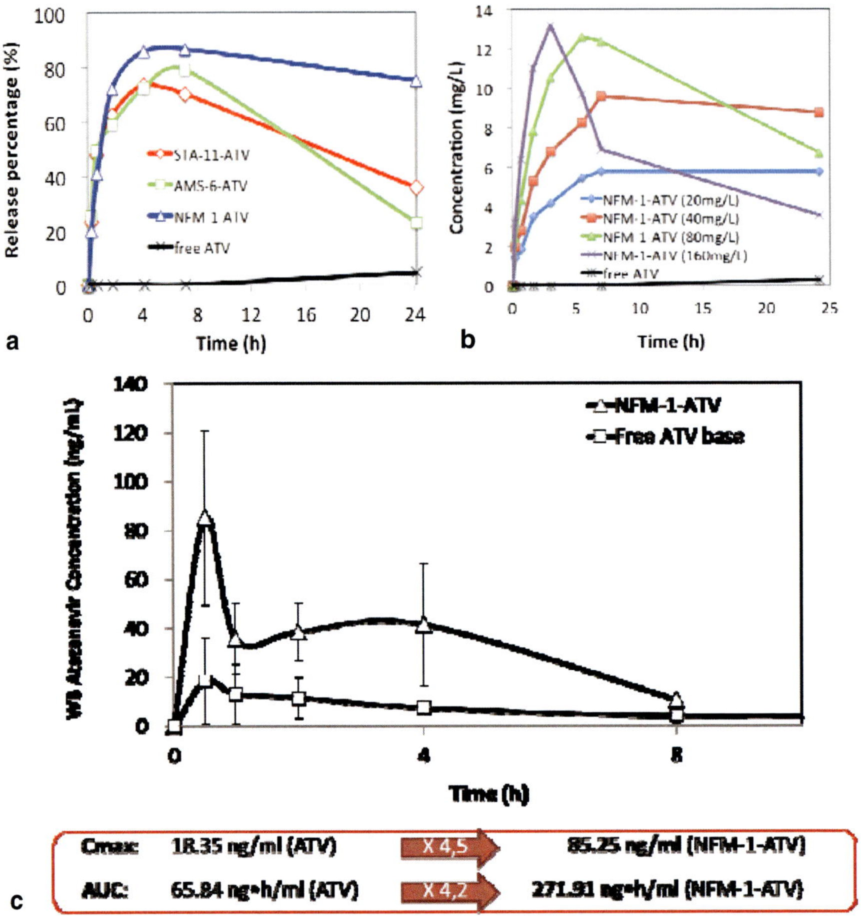

Fig. 22.6 **a** Dissolution curves of free ATV and loaded ATV into different mesoporous structures (simulated Intestine fluid, SIF, performed under sink conditions). **b** Kinetic release curve of different particles to SIF ratio. Varied amount of NFM-1-ATV was added into 500 mL SIF, respectively, corresponding to a series of particles-to-SIF ratio: 20, 40, 80 and 160 mg/L. **c** ATV plasma concentration, Sprague Dawley rats were administered with Omeprazole (100 mg/kg) 8 h prior to the administration of free atazanavir and NFM-1 Silica loaded with atazanavir (10 mg/Kg)

After loading ATV into mesoporous materials with varying structures ranging from 3D-cylindrical (4.0 nm pores, AMS-6) to 2D-cylindrical pores (7.0 nm pores, SBA-15), a significant enhancement in dissolution behaviour is observed for all mesoporous materials loaded with drugs (see Fig. 22.6a, b, c). In accordance with nano-confinement effects of drugs within cylindrical channels, the smaller pore size material (in the case NFM-1) results in the highest enhancement in apparent solubility (Xia et al. 2012b). To investigate further the extent of the solubility enhancement using mesoporous silica carriers, dissolution curves were obtained for NFM-1-ATV at different particles-to-SIF ratios: 20, 40, 80 and 160 mg L^{-1} (Fig. 22.6b). With increased particle concentration, the dissolution rate appears to decrease as a percentage of amount loaded. However, the actual solubility suggests that a higher concentration of NFM-1-ATV particles (160 mg L^{-1}) provides greater solubility, whereby the maximum solubility reached was 18 mg L^{-1} after 4 h, in comparison to 0.254 mg L^{-1} after a similar time for free ATV. In the case of the highest particle concentration, a supersaturation state is maintained for 4 h, after which time the amount of solubilised ATV decreases rapidly. Dissolution parameters can be fitted for both the power law and Higuchi model, supporting a diffusion-based mechanism of release of ATV from NFM-1 particles (2D-cylindrical with 3.0 nm pores). A single oral dose of 10 mg/kg ATV or 50 mg/kg NFM-1-ATV (20 wt% loading of ATV) was given to female Sprague Dawley rats ($n = 3$), fasted overnight prior to the experiment, approximately 5 h after administration of a proton-pump inhibitor (omeprazole, 100 mg/kg). The pharmacokinetic profile of whole-blood plasma ATV concentration for rats treated with NFM-1-ATV in comparison to those treated with free ATV is shown in Fig. 22.6c. A very pronounced improvement in ATV absorption is observed during the first hour of the study. Overall, a statistically significant improvement in bioavailability was observed for NFM-1-ATV in comparison with the free drug. The maximum concentration achieved (C_{max}) for free ATV in the present study was 18.35 ng mL^{-1}. In contrast, NFM-1-ATV results in a C_{max} value of 85.25 ng mL^{-1}. The total area under the curve (AUC) for an 8-h period shows a similar contrast in bioavailability, with values of 271.91 and 65.84 ng $h^{-1}mL^{-1}$ for NFM-1-ATV and ATV, respectively.

22.9 Conclusions and Future Perspectives

Some 20 years after the first synthetic mesoporous materials were produced there is growing acceptance and awareness of the potential use of these materials for pharmaceutical formulations. Encapsulation of molecules within the nanometre-sized pores leads to suppression of crystallisation, and stabilisation of an amorphous form which has been demonstrated to be effective at increasing the solubility of poorly water-soluble compounds. An added benefit of encapsulation often leads to protection from oxidation, hydrolysis and other degradation processes simply by nature of restriction of the molecules from the degrading environment. Hundreds of research articles on mesoporous silica particles have hailed these materials as ideal agents for controlled release delivery. Variation in the internal pore size and structure and interconnectivity of the mesoporous materials leads to significant and tunable differences in diffusion kinetics from different types of mesoporous materials. Thus, a platform of various materials can be envisaged to meet the needs of specific formulations. However, as more applied research is carried out for specific drug formulations, it is clear that simple generalisations of release kinetics from different materials are not adequate to predict a priori one material over another, and drug-specific interactions with the silica materials need to be investigated on a case-by-case basis.

Regarding the safety and regulatory issues, the plethora of studies addressing toxicity of nano- and microparticles are beginning to converge on a general acceptance that spherical silica particles larger than several hundred nanometres generally do not induce negative immune response in cells causing adverse reactions. Although the internal pore sizes of mesoporous materials are, by definition, in the nanometre-sized range and therefore fit into the classification scheme as nanomaterials, the particle size and morphology tends to dominate the biological response to these materials. Amorphous mesoporous silica particles have the same surface chemistry and physical and mechanical properties to fumed silica used currently in nearly all pharmaceutical tablets as well as many food product albeit at low concentrations and are therefore now accepted as having GRAS (generally regarded as safe) status. Accepting that due diligence is carried out for all new formulations and the necessary toxicity studies are conducted, the evidence is mounting that mesoporous silica particles will find a place in the formulators toolbox.

References

Adiseshaiah PP, Hall JB, McNeil SE (2010) Wiley Interdiscip Rev Nanomed Nanobiotechnol 2(1):99–112

Amidon GL, Lennernäs H, Shah VP, Crison JR (1995) A theoretical basis for a biopharmaceutic drug classification: the correlation of in vitro drug product dissolution and in vivo bioavailability. Pharm Res 12:413–420

Atluri R, Hedin N, Garcia-Bennett AE (2008) Chem Mater 20(12):3857–3866

Atluri R, Hedin N, Garcia-Bennett AE (2009) J Am Chem Soc 131:3189

Atluri R, Iqbal M, Bacsik Z, Hedin N, Villaescusa LA, Garcia-Bennett AE (2013) Langmuir 29:12003–12012

Bacchi CJ, Nathan HC, Hutner SH (1980) Science 210(4467):332–334

Barret EP, Oyner LG, Halendar PH (1951) J Am Chem Soc 73:373

Beck JS, Princeton NY (1991) US Patent No. 5 057 296

Bellussi G, Perego C, Carati A, Peratello S, Massara EP, Perego G (1994) Stud Surf Sci Catal 84:85

Berggren A, Palmqvist AEC, Holmberg K (2005) Soft Matter 1:219–226

Brinker CJ, Scherer GW (1990) Sol-Gel science: the physics and chemistry of sol-gel processing. Academic Press, London

British Pharmacopoeia (2009) Volume I and II monographs: medicinal and pharmaceutical substances Colloidal Anhydrous Silica

Brohede U, Atluri R, Garcia-Bennett AE, Stromme M (2008) Curr Drug Deliv 5(3):177–185

Brunauer S (1944) Physical adsorption of gases and vapours. Oxford University Press, Oxford

Brunauer S, Emmett PH, Teller E (1938) J Am Chem Soc 60:309

Cedervall T et al (2007) Understanding the nanoparticle-protein corona usingmethods to quantify exchange rates and affinities of proteins fornanoparticles. Proc Natl Acad Sci U S A 104(7): 2050–2055

Che S, Garcia-Bennett AE, Yokoi T, Sakamoto K, Kunieda H, Terasaki O, Tatsumi T (2003) Nature Mater 2:801

Che S, Liu T, Ohsuna K, Sakamoto O, Terasaki T (2004) Tatsumi Nature 429:281

Chen CY, Burkett SL, Li HL, Davis ME (1993) Microporous Mater 2:27

Chiola V, Ritsko JE, Vanderpool CD (1971) US Patent No. 3556725

Chiou WL, Riegelman S (1971) Pharmaceutical applications of solid dispersion systems. J Pharm Sci 60:1281–1302

Chithrani BD, Ghazani AA, Chan WCW (2006) Determining the size and shape dependence of gold nanoparticle uptake into mammalian cells. Nano Letters 6(4):662–668

Cong YS, Wright WE, Shay JW (2002) Microbiol Mol Biol Rev 66(3):407–425

Cuatrecasas P (2006) J Clin Invest 116(11)

Davidson MH (2012) Am J Cardiol 110(9 Suppl):43B–49B

Davidsson A (2002) Current opinion in colloid and interface. Science 7:92–106

Davis JT, Spada GP (2007) Chem Soc Rev 36(2):296–313

Donaldson K et al (2011) Identifying the pulmonary hazard of high aspect ratio nanoparticles to enable their safety-by-design. Nanomedicine (Lond) 6(1):143–156

Doshi DA, Gibaud A, Goletto V, Lu M, Gerung H, Ocko B, Han SM, Brinker CJ (2003) J Am Chem Soc 125:11646

Downs TR et al (2012) Mutat Res 745(1–2):38–50

Fadeel B, Garcia-Bennett AE (2010) Adv Drug Deliv Rev 62(3):362–374

Feng X, Fryxell GE, Wang LQ, Kim AY, Lui J, Kemmer KM (1997) Science 276:923

Garcia-Bennett AE (2011) Synthesis, toxicology and potential of ordered mesoporous materials in nanomedicine. Nanomedicine (Lond) 6:867–877

Garcia-Bennett AE, Che S, Terasaki O, Tatsumi T (2004) Chem Mater 16:813

Garcia-Bennett AE, Kupferschmidt N, Sakamoto Y, Che S, Terasaki O (2005a) Angewandte Chemie Int Ed 117(3):5451–5456

Garcia-Bennett AE, Kupferschmidt N, Sakamoto Y, Che S, Terasaki O (2005b) Angew Chem Int Ed 44:2

Gratton SEA et al (2008) The effect of particle design on cellular internalization pathways. Proc Natl Acad Sci U S A 105(33):11613–11618

Han L, Ruan L, Li Y, Terasaki O, Che S (2007) Chem Mater 19(11):2860–2867

Higuchi T (1963) J Pharm Sci 52:1145

Hodgkins RP, Garcia-Bennett AE, Wright PA (2005) Micropor Mesopor Mat 79:241–252

Horcajada P, Rámila A, Pérez-Pariente J, Vallet-Regı M (2004) Micropor Mesopor Mat 68(1–3):105–109

Huang X et al (2011) ACS Nano 5(7):5390–5399

Hudson SP et al (2008) 29(30):4045–4055

Huo Q, Margolese DI, Ciesla U, Demuth DG, Feng P, Gier TE, Sieger P, Firouzi A, Chmelka BF, Schuth F, Stucky GD (1994) Chem Mater 6:1176–1191

Israelachvili JN (1991) Intermolecular and surface forces, 2nd edn. Academic press, New York

Jamzad S, Tutunji L, Fassihi R. (2005) Int J Pharm 292:75

Jiang Q, Ward MD (2014) Chem Soc Rev 43:2066–2079

Kerns EH, Di L (2003) Drug Discov Today 8:316–323

Kim JM, Kwak JH, Jun S, Ryoo RJ (1995) Phys Chem 99:16742–16747

Kleitz F, Liu D, Anilkumar GM, Park IS, Solovyov LA, Shmakov AN, Ryoo R (2003) J Phys Chem B 107:14296

Kolter (2011) Pharm Tech Eur 23(10):

Kosmidis K, Argyrakis P, Macheras P (2003) J Chem Phys 119:6373

Kresge CT, Leonowicz ME, Roth J, Vartuli JC, Beck JS (1992) Nature 359:710–712

Kumar ID, Chirravuri SVS, Shastri NR (2014) Int J Pharm 461:459–468

Kupferschmidt N, Xia X, Hanoi R, Atluri R, Ballel L, Garcia-Bennett AE (2013) Nanomedicine 8(1):57–64

Kupferschmidt N, Csikasz R, Ballel L, Bengtsson T, Garcia-Bennett AE (2014a) Nanomedicine. doi:10.2217/nnm.13.138

Kupferschmidt N, Qazi K, Kemi C, Vallhov H, Garcia-Bennett AE, Gabrielsson S, Scheynius A (2014b) Nanomedicine. doi:10.2217/nnm.13.170

Laaksonen T et al (2007) Chem Res Toxicol 20(12):1913–1918

Lin YS, Abadeer N, Haynes CL (2011) Stability of small mesoporous silica nanoparticles in biological media. Chem Commun 47(1):532–534

Liu T et al (2011) Biomaterials 32(6):1657–1668

Lu J, Liong M, Li Z, Zink JI, Tamanoi F (2010) Small 6(16):1794–1805
Lynch I, Salvati A, Dawson KA (2009) Protein-nanoparticle interactions: what does the cell see? Nat Nanotechnol 4(9):546–547
Marlow F, McGehee MD, Zhao D, Chmelka BF, Stucky GD (1999) Adv Mater 11:632
Mellaerts R, Mols R, Jammaer JA, Aerts CA, Annaert P, Humbeeck JV, Van den Mooter G, Augustijns P, Martens JA (2008) Eur J Pharm Biopharm 69:223–230
Mercuri LP et al (2006) Small 2(2):254–256
Prescribing information for atazanavir sulfate (Reyataz) capsules (2004) Bristol–Myers Squibb, Princeton
Raman NK, Anderson MT, Brinker CJ (1996) Chem Mater 8:1682–1701
Ravikovitch P, Neimark A (2002) Langmuir 18:1550
Ravikovitch PI, Wei D, Chueh WT, Haller GL, Neimark AV (1997) J Phys Chem B 101:3671
Recent progress in mesostructured materials. Proceedings of the 5th international mesostructured materials symposium (IMMS 2006) Shanghai, China, 5–7 August 2006. Edited by D. Zhao
Rhodes CJ (2010) Properties and applications of zeolites. Sci Prog 93:223–284
Ritger PL, Peppas NA (1987) J Control Release 5:23
Rosenholm J, Sahlgren C, Linden M (2010) Nanoscale 2:1870–188
Roy S, Quiñones R, Matzger AJ (2012) Cryst Growth Des 12(4):2122–2126
Ryoo R, Joo SH, Jun S (1999) J Phys Chem B 103:7743
Sakamoto K (2001) In: Xia J, Nnanna IA (eds) Surfactant science series vol 101. Protein-based surfactants Ch. 10 261–280 Dekker, New York
Sakamoto Y, Kaneda M, Terasaki O, Zhao DY, Kim JM, Stucky G, Shin HJ, Ryoo R (2000) Nature 408:449
Seddon JM, Templer RH (1995) In: Lipowsky R, Sackmann E (eds) Polymorphism of lipid-water systems. Handbook of Biological Physics, vol 1. Elsevier, New York
Serajuddin AT (1999) Solid dispersion of poorly water-soluble drugs: early promises, subsequent problems, and recent breakthroughs. J Pharm Sci 88:1058–1066
Slowing I, Trewyn BG, Lin VSY (2006) J Am Chem Soc 128(46):14792–14793
Sosman RB (1965) The phases of silica. Rutgers University Press, New Brunswick
Sosnik A, Chiappetta DA, Carcaboso AM (2009) J Control Release 138:2–15
Strømme M, Brohede U, Atluri R, Garcia-Bennett AE (2009a) Nanomedicine Nanotechnology 1:140–148
Strømme M, Atluri R, Brohede U, Frenning G, Garcia-Bennett AE (2009b) Langmuir 25(8): 4306–4310
Taguchi A, Schueth F (2004) Micropor Mesopor Mat 77:1–45
Takagi T et al (2006) Mol Pharm 3(6):631–643
Thomas JM (1994) Nature 368:289
Thomas JM, Thomas WJ (1997) Principles and practice of heterogeneous catalysis. VCH, Weinheim
Thomas JM, Raja R, Johnson BFG, O'Connell TJ, Sankar G, Khimyak T (2003) J Chem Soc Chem Commun 10:1126
Vallet-Regi M, Balas F, Arcos D (2007) Mesoporous materials for drug delivery. Angew Chem Int Edit 46(40):7548–7558
Vallet-Regi M, Ramila A, Real RP, Perez-Pariente (2001) J Chem Mater 13:308–312
Vallhov H, Gabrielsson S, Strömme M, Scheynius A, Garcia-Bennett AE (2007) Mesoporous silica particles induce size dependent effects on human dendritic cells. Nano Lett 7:3576–3582
Vallhov H, Kupferschmidt N, Gabrielsson S, Paulie S, Strømme M, Garcia-Bennett AE, Scheynius A (2012) Small 8 No 13:2116–2124
Van Speybroeck M et al (2010) Eur J Pharm Biopharm 21:623–630
Van Speybroeck M et al (2011) J PharmSci 100(11)
Vartuli JC, Kresge CT, Leonowicz ME, Chu AS, McCullen SB, Johnson ID, Sheppard EW (1994) Chem Mater 6:2070–2077
Vialpando M, Aerts A, Persoons J, Martens J, Van Den Mooter G (2011) J Pharm Sci 100(8):3411
Wan Y, Zhao DY (2007) Chem. Rev 107:2821–2860

Wang et al (2010a) J Control Release 145:257–263

Wang F, Barnes TJ, Prestidge CA (2010b) Mesoporous Biomaterials 1(1–16):2300–2271

Witasp E et al (2009) Toxicol Appl Pharmacol 239:306–319

Wong MS, Ying JY (1998) Chem Mater 10:2067

Wood AJJ (2006) N Engl J Med 355:618–623

Worle-Knirsch JM, Pulskamp K, Krug HF (2006) Nano Letters 6(6):1261–1268

Wu C, Chang J, (2012) Interface Focus 2(3):292–306

Xia X, Zhou C, Ballel L, Garcia-Bennett AE (2012a) Chem Med Chem 1:43–48

Xia X, Zhou C, Ballell L, Garcia-Bennett AE (2012b) Chem Med Chem 7:43–48

Yanagisawa T, Shimizu T, Kuroda K, Kato C (1990) Bull Chem Soc Jpn 63:988

Yongde X, Mokaya R (2003) J Phys Chem B 107(29):6954–6960

Zhao D et al (1998) Triblock copolymer syntheses of mesoporous silica with periodic 50–300 angstrom pores. Science 279:548–552

Zhao Y et al (2011) Interaction of mesoporous silica nanoparticles with humanred blood cell membranes: size and surface effects. ACS Nano 5(2):1366–1375

Chapter 23
Mesoporous Silica Drug Delivery Systems

Yogesh Choudhari, Hans Hoefer, Cristian Libanati, Fred Monsuur and William McCarthy

23.1 Introduction

Solid dispersion (SD) is a well-accepted method to increase solubility of poorly soluble molecules and to improve amorphous state stability (Serajuddin 1999; Chiou 1971). The use of mesoporous silicas (MPS) for this technique has gained the attention of formulators due to its tunable porosity, high surface area, inertness, and good biocompatibility which makes it a suitable excipient in drug delivery (Mai Khanfar). The porous structure of silica itself can decrease the melting point and crystallinity of entrapped drug (Takeuchi et al. 2004). MPS has good flow properties and additional steps such as milling or sizing prior to tableting, and capsule filling can often be simplified when using these materials. This results in high recovery with minimum process loss, and it also minimizes the chances of the processed drug converting to its crystalline state (Takeuchi et al. 2004). MPS-based SD have become an important topic for further investigation as a drug delivery technology. Research is ongoing and many materials and techniques have been developed and studied. This chapter is meant as a basic overview of the background, theory, materials, and methods used by a growing number of researchers in the field. For the purposes of this chapter, we will maintain our focus on macroparticulate MPS -based drug delivery, particularly MPS-based SDs for oral dosage forms.

F. Monsuur (✉)
W. R. Grace and Company, Columbia, MD, USA
Tel.: +32 9 340 65 65
e-mail: fred.monsuur@grace.com

Y. Choudhari · H. Hoefer · C. Libanati · W. McCarthy
W. R. Grace and Company, Columbia, MD, USA

N. Shah et al. (eds.), *Amorphous Solid Dispersions*,
Advances in Delivery Science and Technology, DOI 10.1007/978-1-4939-1598-9_23

23.2 History

First use of MPS to increase the dissolution profile of a drug was reported in 1972 by Monkhouse and Lach (1972) using non-ordered mesoporous silica gel (MSG) and was further elaborated by Yang et al. (1979). The mechanism probably is based on the stabilization of the amorphous state in the pores. Amorphous drugs dissolve more easily compared to their crystalline form. However, the crystal form is more stable due to its lower energy and, generally, drugs tend to convert to the crystalline state. Stability of the amorphous form (physical and chemical stability) has remained a major concern in developing delivery systems based on drugs in amorphous state (Datta and Grant 2004; Andronis et al. 1997). A major challenge is to prevent the amorphous drug in a SD from reverting to its more stable crystalline state in the final dosage form over the shelf life of the product. Presence of a porous carrier may prevent this transformation and could ensure stability of the amorphous state. According to Takeuchi et al. (2005a), tablets of indomethacin-loaded porous silica were superior when compared to its physical mixture. This provides evidence that SD of a drug using porous silica has utility in formulating solid dosage forms like tablets and capsules.

In 1992, a new family of MPS was invented by The Mobil Corporation Laboratories named MCM-X (Mobil Crystalline Material; Kresge et al. 1992). These ordered mesoporous silicas (OMSs) were initially developed for catalyst applications and were only later studied as a drug delivery technology. These silicas were synthesized from surfactant micelles under basic conditions. They have unique properties such as their pore diameter, high surface area (up to 1500 m^2/g), large pore volumes (up to 1.5 cm^3/g), and silanol-rich surface which can be functionalized. Recently, the first pharmacompliant OMS materials have been developed.

Based on the pore structures, various grades are synthesized such as MCM-41, MCM-48, and MCM-50 having hexagonal, cubic, and lamellar pore shapes. In 1971, the same method of producing OMS was attempted by Chiola et al. (1971) using cationic surfactants. In 1995, Tanev and Pinnavaia (1995), Attard et al. (1995), and Bagshaw et al. (1995) reported the synthesis of MSU silica, and furthermore, in 1996, Ryoo et al. reported method for synthesis of KIT silica. Yu et al. (2000) and Chen et al. (2003) presented new OMS named as FDU and AMS, respectively. In 1998, Zhou et al. synthesized Santa Barbara Amorphous-X (SBA, where X correspond to specific pore structure and surfactant) using nonionic triblock polymers. SBA-15 possesses hexagonally ordered cylindrical pores synthesized from Poloxomer P123 as the surfactant, while pores of SBA-16 are spherical with a centered cubic structure. SBA-16 was synthesized from Lutrol F127 (Zhao et al. 1998b, 1998). Initially, application of ordered MPS in drug delivery was not clearly understood, but significant subsequent research has focused on these materials for drug delivery. MCM and SBA are the most studied OMS for drug delivery (Beck et al. 1992). Efforts are ongoing to advance the use of OMS materials in SDs and to characterize their regulatory compliance attributes for pharmaceutical applications.

Table 23.1 Classification of porous media

Types of pore	Mean pore diameter (nm)
Micropore	Less than 2
Mesopore	Between 2 and 50
Macropore	Greater than 50

Recently, applications of MPS in nanoparticulate drug delivery have been studied extensively by various researchers. Nanoparticles could provide an advantage in the improvement of controlled release, targeted delivery, and therapeutic efficiency. MPS nanoparticles (MSN) can be synthesized by self-assembly of surfactants (micelle) which acts as structure-directing agents, providing a template for silica species to congregate and grow to generate controlled-size nanoparticles. Pores in the MSN are developed after removal of structure-directing agents, and the nanosize of the pores can be achieved by varying process conditions like reactants, temperature, aging time, etc. These materials generally have high surface area and pore volume.

23.3 MPS Materials

According to International Union of Pure and Applied Chemistry of Porous Media, porous materials are classified according to their pore diameter as presented in Table 23.1 below (Sing et al. 1985).

Macroporous silica possesses wide pores, whereas the pores of microporous silica are so small that the opposite sidewall of the pore will overlap due to the proximity of walls. The pores can be of different shapes such as spherical or cylindrical with varying arrangements. Some structures may have large pores (more than 50 nm) in one dimension, but the width of the same pore may be in mesorange, and hence material can be considered as mesoporous. Non-ordered MPS are characterized by randomly oriented, interconnected pores with a representative pores size distribution, whereas OMSs are characterized by ordered pore orientation and size.

23.4 Synthesis of MPS

Pharmaceutical grades of MPS can be prepared by various methods. At present, on industrial scale, non-ordered MPS is made via the solgel process and the precipitation process. Ordered MPS employ templates helping to form oriented structures. Later, such templates have to be removed from the silica structure. Common to all MPS types is their three-dimensional structure built by SiO_4-tetrahedra linked via their tips. The inner surfaces of such structures contain silanol groups with varying concentrations and configurations depending on the synthesis history, temperature, and water vapor partial pressure. Approximately, five terminal silanols per square nanometer is a typical silanol group density on a silica surface. These terminal silanol groups play a major role in silica–drug interaction during amorphization (Qian and Bogner 2012).

23.4.1 Synthesis of Non-Ordered MPS (MSG)

The most common production processes for non-ordered MPS, MSG, employ the solgel reaction and a precipitation reaction. In both cases, aqueous solutions of alkaline silicates are used as silica sources. When mixed with a strong acid, a reaction takes place, forming a mixture of silica, alkaline salt, and water according to the exemplary formula:

$$A_2O \times 3.3\,SiO_2 + H_2B \rightarrow 3.3\,SiO_2 + A_2B - H_2O$$

with A = alkaline ion and B = acid residue. Typical starting materials are $Na_2O \times 3.3SiO_2$ and H_2SO_4.

The product mixture of silica, salt, and water is further processed by liquid–solid separation, followed by washing and drying using various techniques. In some cases, the silica can be structurally modified by washing with caustic. Metal silicates such as calcium silicates, magnesium silicates, and aluminum silicates can be produced by adding metal salts to the reaction mixture. Various sol–gel and precipitation techniques with modified key parameters such as reaction temperature, pH value, electrolyte concentration, and duration of reaction lead to silica with different pore structures and surface properties. In the case of precipitated silica, fine silica particles are generated which require sophisticated agglomeration and sizing steps. In contrast, the much coarser silica from the solgel process, as shown in Fig. 23.1, has to be carefully milled and classified prior to its further use.

Non-ordered MPS have randomly oriented pores with characteristic pore size distributions, which, in most cases, are accessible for molecules having sizes similar to the pore dimensions as shown in Fig. 23.2.

23.4.2 Synthesis of OMS

The OMSs are differentiated by their ordered porosity, which can be achieved by the self-assembly of amphiphilic molecules. The latter molecules act as organic templates during the hydrolysis of SiO_4 building blocks, which condense around such template and form amorphous silica walls. After completion of the formation of the three-dimensional silica framework, the organic template has to be removed by either calcination or solvent extraction, creating the ordered open mesoporous structure. The silica species may originate from different sources such as sodium silicate, or the alkoxydes, tetraethylorthosilicate (TEOS) and tetramethylorthosilicate (TMOS). Silica source, nature of template, ionic strength, pH value, composition of the reaction

Fig. 23.1 SEM picture of MPS (silica gel) particles. (Courtesy by W. R. Grace & Co)

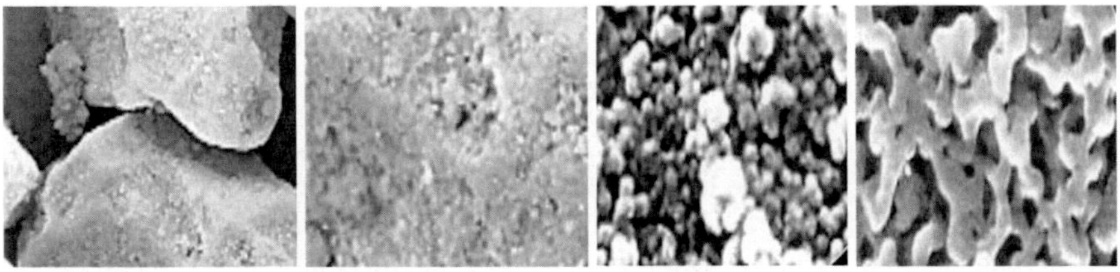

Fig. 23.2 Non-ordered MPS particles and surface under increasing magnification (left lower magnification, right higher magnification) showing the random orientation of intereconnected pores. (Courtesy by W. R. Grace & Co)

mixture, and temperature development during synthesis are essential to the control of pore size, pore diameter, pore volume, and wall thickness of the synthesized OMS (Giraldo et al. 2007).

MCM is synthesized by the liquid crystalline templating method (Kresge et al. 1992; Beck et al. 1992; Beck and Vartuli 1996). Initially, surfactants like alkyltrimethylammonium are dissolved in water to generate cylindrical micelles, in which hydrophobic carbon chains make up the central part, while polar groups direct themselves toward the surrounding water. The generated micelles interact with the SiO_4 building blocks, forming the OMS; after completion of the reaction, the mixture is filtered, and OMS is washed and calcined. The resulting MCM-41 has a hexagonally ordered channel-like pore structure. The corresponding pore size depends on the dimensions of the structure-providing micelle and can be controlled by the nature of the surfactant and its concentration in the reaction mixture. The pore diameter ranges from 3 to 10 nm when using surfactants of low molecular weight.

Zhao et al. (1998a, b) reported the use of amphiphilic triblock polymers such as polyethylene oxide–polypropylene oxide–polyethylene oxide in combination with tetraethoxysilane as the silica source. The OMS structure obtained by this method is known as SBA-15 with a pore size of around 30 nm and a wall thickness similar to MCM-41. SBA-15 is proven to be comparatively better for tablet dosage forms because it is better at withstanding compression forces encountered in tableting (Qian and Bogner 2012).

23.5 Applications of MPS in Drug Delivery

MPS have been widely used as excipients in pharmaceutical applications. Non-ordered MPS have also been utilized as functional excipients. As these application expanded, it was recognized that these materials could have interactions with the active drug substance. This property led to a new era of MPS uses in drug delivery applications. MPS drug delivery systems provide several functional applications when used in these systems. Figure 23.3 below describes this application space.

23.5.1 Delivery of Poorly Soluble Drugs

MPS proved to be promising substrates for enhancing dissolution and bioavailability of poorly soluble drugs. The formation of crystalline material is prevented by the confined space of the pores, which are slightly larger than the drug molecule and entraps the drug in its amorphous/noncrystalline, disordered form. High surface area and hydrophilicity of MPS enhances wettability which results in faster dissolution. It is also reported that MPS can help improve permeability of large hydrophilic molecules in presence of permeation enhancers (Foraker et al. 2003). Table 23.2 summarizes a wide body of research conducted by various investigators on several active drug substances using different types of MPS.

23.5.2 Amorphous Stability Improvement

In order to get the benefit from high solubility and dissolution, the challenge of poor stability of amorphous compounds has to be addressed seriously. MPS with high surface area and pore volume can improve the stability of amorphous compounds.

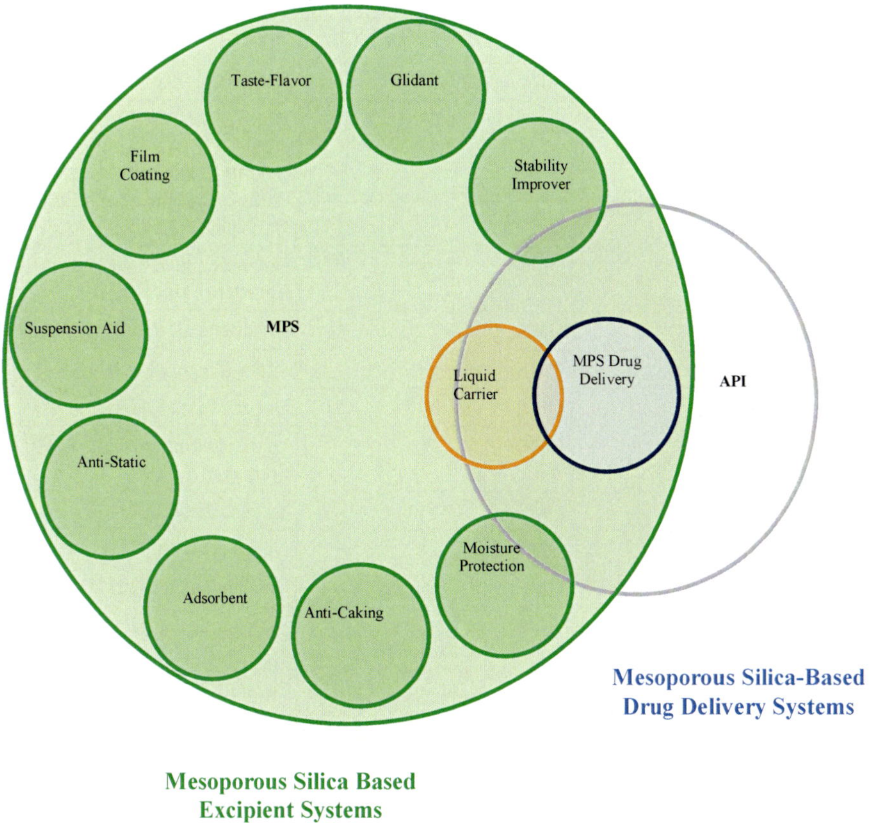

Fig. 23.3 Silica functionalities as they relate to excipient applications and interactions with active pharmaceutical ingredients

Upon adsorption, the high surface energy of MPS reduces the Gibbs free energy and improves the stability of amorphous drugs. In addition, molecules are incorporated in the small pores of MPS and chances of nucleation and crystal growth are reduced from the constrained space. Various researchers have claimed stability improvement of drugs by incorporation into MPS. Limnell et al. (2011a) found that stability of indomethacin improved significantly in MCM-41 and SBA-15 systems. The stability as well as dissolution of indomethacin was found to be satisfactory after prolonged storage at stressed conditions. Furthermore, a study conducted by Kinnari et al. (2011) observed that the amorphous state of itraconazole was maintained for 3 months at 40 °C, 75 % relative humidity (RH) in MPS (MSG) such as Syloid® 244FP silica. In contrast, complete degradation of itraconazole was found when stored on silica (Kinnari et al. 2011). Furthermore, the stability of poorly soluble drug K-832 was found to be satisfactory when K-832-loaded MPS (MSG; Sylysia® 740 and Sylysia® 350) was stored in open and closed condition at 60 °C, 80 % RH (Miura et al. 2011).

Table 23.2 Overview of APIs studied on ordered mesoporous silicas (OMS) and non-ordered mesoporous silica gel (MSG)

Drug studied	Silica used	Reference
Ibuprofen	OMS	Wang et al. (2009); Mortera et al. (2010); Aiello et al. (2002); Vallet-Regi et al. (2001); Ramila et al. (2003); Horcajada et al. (2004)
Naproxen	OMS	Halamova et al. (2010); Cavallero et al. (2004)
Vancomycin	OMS	Doadrio et al. (2010)
Amoxicillin	OMS	Vallet-Regi et al. (2004)
Gentamycin	OMS	Doadrio et al. (2004)
Ibuprofen	OMS	Izquierdo-Barba et al. (2009); Song et al. (2005)
Nimodipine	OMS	Yu and Zhai (2009)
Sertraline	OMS	Nunes et al. (2007)
Itraconazole	OMS	Vialpando et al. (2011)
Itraconazole, fenofibrate, naproxen, ibuprofen	OMS	Vialpando et al. (2012)
Aspirin, ibuprofen	OMS + MSG	Delle Piane et al. (2013)
Polythiazide	MSG	Sheth and Jarowski (1990)
Prednisolone, digoxin, griseofulvin	MSG	Yang et al. (1979)
Corticosteroids	MSG	Liao and Jarowski (1984)
Ezetmide	OMS	Kiekens et al.
Sulfthiazole	MSG	Patel
Indomethacine	MSG	Improving solubilty of poorly water soluble drug indomethacin by incorporating porous material in solid mDISPERSION
Ibuprofen, itraconazole	OMS	Kiekens et al. (2012)
Indomethacine	OMS + MSG	Limnell et al. (2011b)
Itraconazole	MSG + Silicon	Kinnaria et al. (2011)
Fenofibrate	OMS	Van Speybroeck et al. (2010)
Itraconazole, ibuprofen	OMS	Kiekens et al. (2012)
Tadalafil	MSG	Mehanna et al. (2011)
Itraconazole	OMS	Mellaerts et al. (2008)
Ibuprofen	MSG	Aerts et al. (2010); Verraedt et al. (2011); Aerts et al. (2007)

Table 23.2 (continued)

Drug studied	Silica used	Reference
Glebenchamide, indomethacin	OMS + MSG	Garcia-Bennet et al. (2013)
Atazanavir	OMS + MSG	Xia et al. (2012)
Carvedilol	OMS + MSG	Hu et al. (2012)
Telmisartan	MSG	Zhang et al. (2010)
Carbamazapine, cinnarazine, danazol, griseofulvin, ketoconazole, phenylbutazone, nifedipine	OMS	Van Speybroeck et al. (2009)

API active pharmaceutical ingredient, *MSG* mesoporous silica gel, *OMS* ordered mesoporous silicas

23.5.3 Controlled/Modified Release and Targeted Drug Delivery

Drug release from MPS is dependent on various factors like surface area, pore diameter, pore volume, surface silanol groups, etc. All of these parameters can easily be controlled during the synthesis of MPS, and the release of drug can be controlled by tailoring physicochemical properties of MPS. If the pores are cylindrical and pore opening is narrow, the time required for drug diffusion is long and release will be prolonged. Furthermore, the release profile can also be controlled via different surface treatments of the materials, leading to desired interactions between the porous substrate and the loaded substance. Here, the role of silanol groups is crucial. Calcined MPS with reduced number of silanol groups are ideal systems for controlled drug release (Andersson et al. 2004). pH-sensitive drug release from the MPS can be achieved by modification with functional groups. Doxorubicin hydrochloride loaded on poly(glutamic acid)-grafted MPS shows higher drug release at pH 5.5 than 7.4 (Zheng et al. 2013). Furthermore, surface-coated polyelectrolytes also aid pH-sensitive drug release for cancer therapy (Sun et al. 2014). Drug release from tablets of drug-loaded MPS can be controlled by use of excipients like stearic acid and cellulose polymers (hydroxyl propyl cellulose). Moritz and Laniecki (2012) suggest the principle involved in prolonged drug release is blockage of pores and hydrophobic characteristic of stearic acid. Controlled release of natural antibacterial, allyl isothiocyanate (AITC), was achieved by modulating the pore width and pore volume of SBA-15 (Park and Pendleton 2012). The release of ibuprofen from dimethylsilyl-modified MCM-41 is retarded due to presence of hydrophobic groups on the surface and further drug release can be controlled by the extent of surface modification (Tang et al. 2010). The use of MPS for targeted drug delivery for cancer therapy has been reported by researchers. The surface of silica can be modified with carboxyl groups and further functionalized with folate moiety. Since the surface of cancer tumor is enriched with folate receptor, the folate-modified silica can be captured by the receptors present on the tumor. As a result, anticancer drugs like doxorubicin can be targeted efficiently with reduced toxicity (Xie et al. 2013). Most recently, Porta et al. (2011) demonstrated the use of peptide valves on MPS to obtain controlled release functionalities.

23.5.4 Protein Drug Delivery

The stability of protein and peptides is a key consideration in developing drug delivery systems. The desired concentration of the protein/peptide molecules has to reach the site of action within a desired time frame. Various techniques have been reported, and some are even commercialized for the successful delivery of proteins and peptides. Recently, the use of MPS in protein and peptide delivery has attracted the attention of researchers due to their unique drug release properties. The strong interest is based on inertness of these materials, their biocompatibility, low toxicity, high surface area, adsorption capacity, controlled pore size, and possibilities of surface modification. Oral delivery of protein and peptides is challenging due to chances of protein degradation by pepsin and low gastric pH in the gastrointestinal (GI) tract. It is suggested that encapsulating proteins in a protective material can help maintain their activity in the harsh conditions of the GIT. MPS are good candidates for this purpose since the protein activity can be maintained inside its pores together with a high loading efficiency. However, there is a limitation that small pore diameters can't accommodate large molecules. As a result, proteins are adsorbed on the surface instead of deep in the pores. Pore diameter has therefore to be considered while selecting MPS for proteins and peptides. Loading of cytochrome C could be increased in the MPS as a function of pore size (Gu et al. 2013). Similarly, surface chemistry also influences loading of proteins on MPS. It was found that vancomycin has an overall negative charge with the exception of a positive charge head, and native SBA-15 also has negative charge; consequently, vancomycin is not adsorbed in the pore channel except the positively charged head (Doadrio et al. 2010). Several organizations are working on silica-based technologies to design systems for oral delivery of proteins. Recently, Slowing et al. used MPS for the intracellular delivery of cellular membrane-impermeable proteins. Furthermore, feasibility of MPS surface functionalization can also be used to generate stimuli-sensitive intracellular protein delivery. Such a system includes inorganic valves with CdS (Lai et al. 2003), Au (Vivero-Escoto et al. 2009; Aznar et al. 2009), or Fe_3O_4 (Giri et al. 2005) which can be dissolved under redox conditions and the nanovalves can be activated with light, redox reactions, temperature (Nguyen et al. 2005), or as a function of pH. However, since these valves are hydrophobic, they fail to achieve desired in vivo efficacy. Hence, hydrophilic valves like polymers, peptides, DNA/dendrimer complexes, and lipid bilayers are preferred. MPS surface modified with peptide functions as stimuli responsive and also ensures the cellular uptake of particles. Similarly, Porta et al. (2011) developed MPS modified with oligopeptide for the targeted and controlled delivery of drugs. The protein is released from pH-hydrolyzable citraconic acid-functionalized MPS when it reaches the endosomal pH environment (Park et al. 2010). MPS can also be used to form composites with other functionalized polymers. The composite of MPS with poly(D, L-lactide-co-glycolide) enables controlled delivery of a prime-boost vaccine via the encapsulation of plasmid DNA (pDNA) and protein in different compartments (Ho et al. 2010). Formulations using MPS are also designed for the delivery of small interfering (siRNA). However, the major hurdle for such systems is electrostatic repulsion between MPS and siRNA due to similar charges (negative). Hence, it is preferred to functionalize surface of MPS with amino groups to get maximum siRNA loading. It is important to note that loading of siRNA will be only on the surface of MPS and not in the pores. Xia and coworkers attached siRNA onto polyethyleneimine-capped MPS and achieved safe siRNA delivery (Xia et al. 2009). Recently, magnetic MPS have been successfully developed which accommodate siRNA in the pores, thus providing protection from degradation by enzymes (Li et al. 2011). Furthermore, controlled release of proteins can also be achieved by loading the proteins during the synthesis of MPS itself.

23.6 Theory of Drug Release from MPS

The improved bioavailability of poorly soluble drugs observed when an API is formulated in MPS is due to supersaturation conditions created when the amorphous form of the drug is released in the GI fluids. Typically, amorphous forms of materials, in particular organic materials such as drugs, have a higher solubility than their crystalline counterparts, often several folds higher in concentration. In vitro, when the drug is released from the carrier into a fixed volume of dissolution medium, the drug's concentration in the dissolution media increases and will exceed the saturation concentration of the more stable crystalline forms of the drug.

Applying the formalism of nucleation and growth that describes crystallization, or preparation of colloidal dispersion by condensation, to oral drug delivery systems (Brouwers et al. 2009; Sarode et al. 2014), the concentration of the drug in a fixed volume of dissolution medium over time will follow a profile characterized by LaMer's diagram (LaMer and Dinegar 1950; Fig. 23.4). As thermodynamics dictates, the ultimate concentration of the drug in the medium will converge toward the drug concentration, C_{eq}, which is in equilibrium with the more stable crystalline form of the drug. However, the rate of release of the drug from the MPS is controlled mostly by diffusion of the active molecules through the pore network and across the particle boundary layer at the surface of the MPS particle into the bulk dissolution fluid. The shape of the curve, and the generation of supersaturation of the drug, results from the fact that, initially, the kinetics of drug release from the MPS is faster than the nucleation and growth of crystalline forms of the drug from solution. Eventually, the nucleation and growth become dominant and the concentration will drop toward the saturation concentration for the crystalline forms that were condensed. This general principle applies to the in vivo drug profile *as* well, with the added sink for dissolved drug, that is, the absorption through the membranes of the GI tract. The extent of supersaturation is therefore determined by the interplay between the kinetics of drug release from the MPS and the kinetics and thermodynamics of formation of stable crystal isomorphs of the drug from solution. The latter is predicated mostly by the nature of the drug itself and its interaction with the dissolution medium. The physical and chemical properties of the delivery system will have little to no influence in the recrystallization of the drug. On the other hand, the properties of the delivery

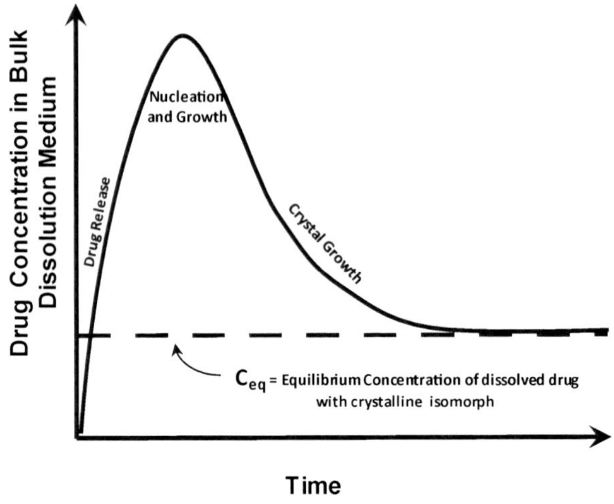

Fig. 23.4 LaMer diagram depicting the characteristic concentration profile versus time for the release of an amorphous form of a poorly soluble drug from an MPS system into a fixed volume of an aqueous dissolution medium. Three stages are evident: (1) at short times, release of the drug from the delivery system is the dominant factor and dissolved drug concentration increases past the saturation concentration (C_{eq}); (2) nucleation of the crystalline form of the drug and growth of the incipient crystals take place while dissolved drug concentration is sufficiently high and the drug falls out of solution; and (3) growth of crystals continues once nucleation stops and consumes the dissolved drug in excess of the saturation concentration

system, and in particular the physical and chemical structures of the MPS material, are fundamental determinants of the kinetics of release of the drug and therefore are key to the shape of the curve in Fig. 23.4 and the magnitude and duration of supersaturation of the drug in the GI tract. With regard to the MPS materials, it is important to understand the effect of particle size, pore structure, and surface chemistry on the kinetics of drug release in order to optimize drug delivery systems.

A mathematical model for the dissolution of solids was first proposed by Noyes and Whitney (1897) at the turn of the twentieth century. In the intervening century, and since the influential work of Higuchi (Siepmann and Siepmann 2013) on drug release from films, the models have been refined and adapted to the specifics of drug delivery. A recent review article provides a good survey of the models for diffusion controlled drug delivery systems (Siepmann and Siepmann 2013). The earliest and best-known descriptor of the rate of dissolution of a substance from a solid form, the Noyes–Whitney model, considers a simple process of diffusion from solid particles to bulk solution with empirical parameters to quantify the diffusion rate. The Nernst–Brunner formalism[6] introduces the concept of diffusion through a boundary layer around the solid particle, thus connecting the empirical parameters in the original equation of Noyes and Whitney (1897) to physical parameters of the solid. The equation that describes the rate of dissolution, *dM/dt,* is:

$$\frac{dM}{dt} = \frac{SD}{\delta}(C_s - C_t),$$

where C_s is the solubility of the substance, C_t is the concentration of the substance in solution at time *t*, *S* is the surface area available for diffusion, δ is the thickness of the boundary layer, and *D* is the diffusion coefficient of the substance through boundary layer. The surface area and diffusion parameters were all considered to be independent of time and dependent only on the properties of the diffusing molecule and the dissolution medium by the original investigators.

In most system used for drug delivery, *S* and *D* will depend on the properties of the system being considered and can vary as dissolution progresses. Hixson and Crowell (1931) extended the model to account for the change in surface area as drug dissolves from a solid particle with changing radius.

For delivery systems formulated using MPS, the pore size and pore structure as well as the particle diameter will affect the diffusion coefficient and the surface area (Van Speybroeck et al. 2010; Horcajada et al. 2004; Shen et al. 2011). The surface area can be calculated from geometric considerations, and the diffusion coefficient can be estimated by using traditional approaches for mass transfer in porous particles. Mortera et al. (2010) detail a good illustration of this approach where *D* was determined using the Stokes–Einstein equation to derive the drug diffusivity in the dissolution media (D_{dm}) and the Renkin equation to correct the value for the steric hindrance (δ) and the constrictivity (ω_r) in the pores:

$$D = \frac{D_{\mathrm{dm}}\delta\omega_r}{\tau},$$

$$D = \frac{D_{\mathrm{dm}}(1 - a/r)^2(1 - 2.1(a/r) + 2.09(a/r)^3 - 0.95(a/r)^5)}{\tau},$$

where δ and ω_r are both functions of the ratio between the drug and the pore radii (a/r) and τ is the tortuosity, which is assumed equal to one if pores are cylindrical, as is the case of ordered MPS. This modified equation shows much better agreement with experimental data than the original Noyes–Whitney equation.

The theoretical construct as well as the mathematical models provide a means to understand the effect of the properties of MPS on the performance of the delivery systems, in particular pore structure and particle size. More importantly, they can efficiently direct the design of new and improved materials and technologies to increase the bioavailability of poorly soluble drugs.

23.7 Methods of Drug Loading

MPS possesses high drug loading capacity which varies from 10 to 34 %, and it may go up to 60 % in some cases (Qu et al. 2006; Heikkilä et al. 2007). The type of method used for drug loading is important in understanding the efficacy and future problems

Table 23.3 Examples of drug and their loading efficiency using different solvents

Drug used	Solvent	Drug:silica ratio	% drug content, loading efficiency	Reference
Tadalafil	Methanol	Multistep	42	Mehanna et al. (2011)
Indomethacin	Ethanol	8:1	28.9	Limnell et al. (2011b)
Itraconazole	Methylene chloride	1:3	20	Mellaerts et al. (2008)

associated with the process. Various forces like hydrogen bonding, dispersive, and hydrophobic interactions are involved in the adsorption of drug molecule in the MPS pores. Hydrogen bonding is the vital force which ensures the drug is entrapped in the pores of MPS. The extent of hydrogen bonding is proportional to the amount of silanol groups present on the surface of the MPS and their interaction with functional groups of the drug molecules. In case there are fewer silanol groups on the MPS and the drug is hydrophobic, then the London type of interaction (dispersive) plays an important role (Delle Piane et al. 2013). Most studied methods of drug loading on MPS use organic solvents to solubilize drug and then load on the MPS. Solvents can often require processing steps to remove it prior to use in pharmaceutical products, and as a result, alternate methods involving solvent-free technology are under investigation. Therefore, we have broadly categorized the methods of drug loading on MPS as solvent-based and solvent-free technology.

23.7.1 Solvent-Based Methods

23.7.1.1 Solvent Immersion Method

Drug loading onto porous silica can be accomplished by adsorption from a drug solution in an appropriate solvent. A defined quantity of porous silica is suspended in a solution of the drug dissolved in a volatile solvent. The suspension is stirred for 1–2 h followed by filtration through micron filters to obtain drug-loaded silica. The wet mass is dried in an oven to get drug-loaded silica powder. A mixture of two or more solvents can also be used to ensure maximum drug loading. Once the solvent is evaporated, the same process can be repeated to increase percentage drug loading. Kovačič et al. (2011) prepared a porous silica-based SD of carvedilol using a solution of the drug in the solvent tetrahydrofuran (THF). The author observed that the crystallinity, amorphicity, and dissolution behavior of carvedilol can be controlled using various levels of drug content and methods of SD preparation. The dissolution behavior of the SD improved for the SD with lower level of drug content (Kovačič et al. 2011).Examples of drug and their loading efficiency using different solvents are shown in Table 23.3.

23.7.1.2 Solvent Drying Method

Drug solution is poured on the silica and the prepared suspension is stirred for a defined time. The solvent is evaporated using a rotary evaporator and all the dried material is recovered at the end of process. As the solvent starts evaporating, the drug concentration in the loading solution increases slowly which creates a concentration gradient and drug loading in the silica pores is initiated (HeikkiÏ et al. 2007). This method is preferred because of higher drug loading when compared to solvent immersion method. In the study by Limnell et al. (2011b), 27.0 % of indomethacin was loaded on non-ordered MPS using rotavapor method, while only 11.6 % was obtained by immersion method. Similarly, attempts have been made to evaporate the solvent using fluidized bed processors by spraying the suspension of the drug and silica dispersed in an organic solvent in the dryer. In the fluidized bed method, the author observed slightly increased indomethacin loading of 28.6 %; this may be due to the additional force of compressed air assisting drug penetration. Crystalline fraction of 5.6 % was observed during the calorimetry study of indomethacin drug-loaded OMS (Limnell et al. 2011b).

23.7.1.3 Incipient Impregnation Method

In this method, a concentrated drug solution near its saturation point is prepared. MPS is mixed with a volume of the concentrated drug solution which is equivalent to its pore volume. The solution is loaded on the MPS by capillary action. This method of drug loading is preferred when only a small quantity of drug is available. The amount of drug loaded can be determined easily; however, there may be issues regarding drug uniformity and crystallization of drug on the surface of silica (Xu et al. 2012). Liao and Jarowski (1984) loaded corticosteroids like prednisone, prednisolone, and hydrocortisone on Syloid® 244FP silica using a solvent mixture of *N*-dimethyl acetamide–polyethylene glycol 400 (7:3). The method resulted in significant improvement of corticosteroid dissolution (Liao and Jarowski 1984).

23.7.1.4 Spray Drying Method (Takeuchi et al. 2004, 2005b)

Spray drying has a close resemblance to the solvent immersion method. The main difference is the method of solvent removal; in this case, solvent is removed by a spray drying process. Additionally, extra spherical particle shape is achieved which is useful to aid flow and compression of the powder (Takeuchi et al. 2004). Among the different methods used for drug loading, spray drying with MPS yields high drug loading of the amorphous form along with enhanced dissolution properties (Vogt et al. 2008). According to Shen et al. (2011), dissolution of ibuprofen increased significantly after spray drying with MPS. Untreated crystalline ibuprofen showed 20 % drug release in the first 20 min, while it was 90 % for ibuprofen-loaded MPS.Superiority of spray drying method over solvent evaporation was proved by Takeuchi et al. (2004) while working on SD of tolbutamide with MPS. Crystals of tolbutamide were observed in the scanning electron microscopy (SEM) of SD prepared using the evaporation method. However, the particle size of the SD (4.29 ± 0.16 μm) remained same for both methods. The author also concluded that the heating rate in spray drying affects the properties and characteristics of a final SD. In case of tolbutamide, the rapid rate of drying facilitates the formation of the metastable form which may impact the solubility of the drug (Takeuchi et al. 2004).

23.7.1.5 Supercritical Fluid Method (Smirnova et al. 2003; Miura et al. 2010)

In this method, the drug is loaded on the MPS with the help of supercritical carbon dioxide as the solvent. A defined ratio of drug and MPS is placed in a closed pressurized chamber. The calculated quantity of supercritical CO_2 is purged through an orifice in a chamber at the desired pressure and temperature to initiate the drug loading in the pores of MPS. Although this method seems to be simple, it has not achieved the popularity of other methods. Reports suggest that only microparticles are generated through this method, and even these particles are dispersed all over the chamber due to the pressurized gas which results in low yield. In addition, the efficiency of solubility enhancement is also lower than the solvent immersion method (Miura et al. 2010). The mechanism involved in drug loading is by hydrogen bonding between drug and MPS. The experiments of Smirnova et al. (2003) found that solubility of both ketoprofen and miconazole increased significantly by adsorbing them on aerogel using supercritical carbon dioxide.

23.7.2 Solvent-Free Methods

23.7.2.1 Melt Mixing Method (Takeuchi et al. 2005b)

This method involves physical mixing of crystalline drug and MPS in a defined ratio, followed by heating to amorphize the drug. Capillary forces then trap the amorphous form in the pores of MPS. It was first reported by Chio and Riegelman in 1971 and further studied by various researchers. Watanabe et al. (2001) concluded that the extent of indomethacin amorphization is less in melted crystalline indomethacin without MPS as compared to its melted mixture with MPS. Also, it was observed that if the drug is melted without MPS, it starts converting to its original crystalline state on storage (Watanabe et al. 2001). The rate of recrystallization is dependent on the mixing time and varies inversely with it. This method is advantageous due to being a solvent-free technology; however, the heating step involved in this method makes it inappropriate for thermolabile drugs.

23.7.2.2 Co-grinding/Co-milling Method

Among various methods of amorphization, co-milling is easy and cheap in terms of large-scale manufacturing (Stein et al. 2010). The dry crystalline drug and MPS can be mixed with defined intensity and ratio to ensure amorphization of drug. The mechanism of amorphization is by interaction between drug and MPS to generate hydrogen bonds. Various technologies like planetary, oscillatory ball mill, turbulla mill, roller compaction, etc. were used to load drugs into pores of MPS. The extent of drug loading and degree of amorphization is influenced by various factors like time and intensity of mixing, reduced pressure, humidity, and silanol content. In most of the earlier studies, the extent of amorphization is found to be proportional to time and intensity of mixing (Konno and Kinuno 1989; Konno et al. 1986). However, contradictory observations were made with a study conducted by Pan et al. (2008) where the amorphization of indomethacin was found to be independent of both parameters, which is explained by the low glass transition temperature (T_g) of the drug. It was observed that dry blending increases the surface temperature of the drug due to particle–particle interaction, and if temperature increases above T_g, the drug starts to recrystallize. This could be the reason why the extent of amorphization is independent of the time and intensity of mixing. In addition, the drug loses its crystallinity via sublimation of the drug. Accordingly, drugs with higher vapor pressure have a more rapid rate of amorphization (Konno et al. 1986). The chance of drug entrapment in the pores is increased if the quantity of MPS is high. Also, if the surface area for particle–particle interaction is high, the extent of amorphization is increased. If the particle size of drug and MPS is decreased, surface area is increased, and this increases amorphization.

23.7.2.3 Microwave-Assisted Drug Loading (Waters et al. 2013)

Simple physical mixing of drug and silica often will not achieve the desired drug loading in the pores of silica and dissolution enhancement cannot be achieved. Microwave irradiation can be used to increase the drug loading in the silica pores. Waters et al. (2013) found that the extent of amorphization in the microwave-irradiated sample was significantly higher compared to simple physical mixture (from grinding). As a result, an improvement in the dissolution properties is observed.

23.7.2.4 Drug Loading During Synthesis (Solgel Process)

In this process, drug is dispersed in the colloidal silicate dispersion (sol) and gellation is induced by chemical means to convert into a porous network (gel; Ahola et al. 2000). The efficiency of encapsulation for hydrophilic drugs is > 85 % (Finnie et al.). Release can also be controlled by modulating internal pore structure which restricts diffusion of encapsulated drug. Further, release can also be controlled by using an alkoxide mixture consisting of varying ratios of methyltrimethoxysilane (MTMS) and tetramethoxysilane (TMOS) which modify the internal pore structure of the particles. Use of increased MTMS content adds methyl groups in the structure which provides flexibility to the Si–O–Si network, leading to gels with smaller pores than gels synthesized from pure TMOS (Scherer and Brinker 1990).

Similarly, Finnie et al. (2006) have claimed a method for the encapsulation of proteins and other biomolecules in the porous silica particles. Encapsulation efficiency ranged from 75 to 97 % while protein loading was between 5.7 and 6.6 %. The release of protein could be observed for 20 h which indicates a controlled release character of the MPS (Finnie et al. 2006).

23.8 Factors Affecting Drug Loading and Dissolution

23.8.1 Effect of Solvent

Solvent polarity (or dielectric constant) affects drug loading in silica-based carriers. In adsorption methods, drug loading is increased with hydrogen bonding between the drug and the surface of the adsorbent (MPS). Highly polar solvents will compete with drug for hydrogen bonding sites and will result in poor drug loading. Charnay et al. (2004) have studied the effect of solvent on ibuprofen loading by evaluating various solvents like dimethylsulfoxide (DMSO), dimethyl formamide (DMF), dimethylecetamide (DMA), ethanol, and hexane. In highly polar DMA, no drug loading was observed; conversely, 37 % of ibuprofen could be loaded using low-polarity hexane (Charnay et al. 2004). Furthermore, Fernandez-Nunez elaborated the concept using the polarizable continuum model (PCM) proposed by Tomasi (2004), and a similar relationship between drug loading on MPS and polarity of the solvent was observed. Free energy of solvation and radius of solvent (R_{sov}) also influences the drug loading. R_{sov} is directly proportional to the drug loading observed on SBA-15 and MCM-48 (Fernández-Núñez et al. 2009).

23.8.2 Effect of Pore Dimensions, Particle Size, and Surface Area

Pore size determines how large a molecule can be loaded into the silica and is generally dependent on the surface area and pore volume of the particle. As a rule of thumb, molecular size should be less than pore size to ensure easy and maximum drug loading. According to Horcajada et al. (2004), ibuprofen with less than 1 nm size can be loaded up to 19 % on 3.6-nm silica while loading decreases to 11 % if silica of 2.5 nm is used. Similarly, Zhang et al. (2010) observed loading of telmisartan was increased from 48.9 to 59.7 % when pore size increased from 3.6 to 12.9 nm. Pore size also influences the drug release significantly. Jin and Liang (2010) reported that MPS with pore size of 7.3 nm showed higher release of adsorbed ibuprofen than silica with 4.6 nm pore size. In addition, release of itraconazole from SBA-15 increased as pore diameter increased from 4.5, 6.4, 7.9 to 9.0 nm (Mellaerts et al. 2007).

Kinnari et al. (2011) compared two different pore sizes of MPS, Syloid® AL1FP silica (3.5 nm) and Syloid® 244FP silica (20 nm). Faster release was observed from Syloid® AL1FP silica despite its lower pore size. Kinnari et al. (1989) suggest the reason for this observation may be the higher surface area of the Syloid® AL1FP silica which could have improved dissolution of entrapped drug. It can be concluded that different surface areas of MPS also influence dissolution significantly.

While literature examples suggesting faster release for materials with wider pores are abundant, the benefits of modifying the pore size should be balanced against the risk of drug recrystallization in the mesopores. A number of studies have shown that drug recrystallize in wider pores, while confinement in comparatively smaller pores resulted in effective suppression of recrystallization.

Shorter pore channel lengths can speed dissolution. Chen et al. (2003) compared various silicas such as MCM-41, SBA-15, and SBA-15 LP and found that SBA-15 LP (with longer pores) showed comparatively lower release than MCM-41 and SBA-15. This suggests that the longer pore length reduces rate of drug release.[119] Smaller particles can reduce the pore length of MPS and subsequently increase drug release rates. Limnell et al. did comparative dissolution study of tablets prepared from indomethacin-loaded OMS, ordered (MCM-41) MSG, and non-ordered MPS (Syloid® 244FP silica). The faster drug release was observed from Syloid® 244FP silica tablets compared to MCM-41 which may be because of the larger pores and shorter distance for the diffusion in the small-sized Syloid® 244FP silica (Qu et al. 2006).

The different pore structure of OMS compared to MSG has also to be considered (previously shown in Fig. 23.2). One can conclude that a smaller pore size distribution of MSG is beneficial and should be controlled. Pore volume also affects release of entrapped drug, and it is observed that drug release is directly proportion to the pore volume (Izquierdo-Barba et al. 2005). Drug adsorbed on silica may be in a monolayer or multilayer format depending on the concentration of the drug solution and its interaction with silica surface. If concentration is high, the drug will be adsorbed as a multilayer and loading will be directly proportional to both pore volume and surface area. Alternatively, if adsorption is in a monolayer form, then the amount of drug loading is directly proportional to its surface area, and pore volume has a very negligible effect (Xu et al. 2012).

23.8.3 Effect of Humidity or Water Content

Water in trace amounts modifies the surface chemistry of MPS and can affect drug loading and release. The content of adsorbed water should be considered while loading drug in the pores of MPS. Moisture in the drug and MPS can reduce drug loading in the MPS pores. Water molecules interact with silanol groups and can reduce the groups available for hydrogen bonding with the drug. Pan et al. (2008)

found decreased amorphization when water content was less than 7 %; however, this reversed when water increased above 7 %. The reason may be that there is higher diffusion of drug into pores through absorbed water and, correspondingly, increased hydrogen bonding (Pan et al. 2008). Furthermore, the presence of free water in the dosage form enhances the chemical reactivity, and drugs are more prone to degradation. However, if the surface area of adsorbents like MPS is high, water tends to form a monolayer on the surface and water is not available for interaction with the drug (Waterman and Macdonald 2010).

It is important to consider the effect of humidity if the drug-loaded MPS is stored at higher humidity conditions. At high humidity, the surface of silica undergoes hydroxylation and surface hydrophilicity is increased. When drug-loaded MPS comes in contact with aqueous dissolution media, drug release is facilitated.

This concept is supported by the observations of the Mellaerts et al. (2010) who studied the effect of itraconazole release from SBA-15 OMS stored at 0, 52, and 97 % humidity. The highest release was observed in the sample stored at 97 % humidity (Mellaerts et al. 2010). Similar observations were reported by Gupta et al. (2002) who observed increased drug release, on storage at 75 % RH for weeks, from SD composed of Gelucire® 50/13 and Neusilin® (magnesium aluminum silicate). The reason cited was migration of the drug from Gelucire® 50/13 to the surface of Neusilin® at high humidity (Gupta et al. 2002). These results are contradictory to SD of hydrophilic polymer where drug release decreases with storage (Kalaiselvan et al. 2006).

23.8.4 Surface Chemistry

The surface of MPS is enriched with silanol (Si–OH) and siloxane (Si–O–Si) groups (Zhuravlev 2000). Inside the silica molecule, there are structurally bound (chemically adsorbed) water molecules denoted as internal silanol groups. The silanol and siloxane groups interact with the loaded drug to generate hydrogen bonding. Both physical and chemical interactions influence drug loading on MPS particles. Physical interactions are reversible and preferred if immediate release is desired. For hydrophobic drugs, physical interaction occurs with silica by hydrogen bonding and electrostatic and hydrophobic interaction. Hydrogen bonding is predominant if the number of surface silanol groups is high; otherwise, dispersive forces are predominant (Salonen et al. 2008). Ugliengo et al. did computational study with hydrophobic drugs aspirin and ibuprofen, using two different MCM-41 silica with 4.5 and 1.5 OH/nm^2. It was found that dispersive forces are strong in MCM-41 with less silanol groups (Delle Piane et al. 2013).

The number of OH groups per unit surface area of the silica materials varies between different silica materials and is affected by post-synthetic treatments like calcinations, time, and temperature. The value for MSG amorphous silica gel is 4.2–5.7, with decreasing values for OMS from SBA-15 (2.8–5.3) to MCM-41 (1.4–3) materials (Bahl and Bogner 2006; Kozlova and Kirik 2010). Silanol groups can exist in three different forms: isolated, vicinal, and geminal. Isolated and geminal silanol groups can be used as grafting templates, for example, for amino- or dendrimer-functionalized silica (Zhao and Lu 1998a; Muoz et al. 2003). Such functionalization also affects drug loading based on the properties of the drug. Pan et al. (2008) observed that drug loading in functionalized MPS was reduced significantly while it was increased when MPS was treated with dichlorosiloxane. The reason may be reduced silanol content due to functionalization and less hydrogen bonding with drug. Alternatively, treatment with dichlosiloxane increases the silanol content and increases drug loading (Pan et al. 2008). Similarly, drug release is affected by surface modifications. When the surface of silica is hydrophilic, there are more hydroxyl groups and a correspondingly stronger interaction between drug and MPS. Accordingly, drug release is slowed down. Alternatively, if surface groups are modified with hydrophobic groups like octadecyl silane (C18), drug interaction is poor and drug release is rapid. Therefore, it can be concluded that surface polarity of silica influences drug release proportionally. These assumptions are supported by observations by Izquierdo-Barba et al. (2009).

Mehanna et al. (2011) showed 93.45 % of tadalafil could be released from Syloid® 244FP silica-based SD in 10 min, while its physical mixture with silica could release only 21.78 %, and no release was observed from pure native drug. From the observations of dissolution studies, one can predict the efficiency of silica to improve dissolution of poorly soluble drugs. Takeuchi et al. (2004) used dissolution studies to compare SD of tolbutamide prepared with hydrophilic silica and hydrophobic silica. Hydrophobic silica showed much slower dissolution of tolbutamide, confirming the role of hydrophilic silica in increasing the rate of drug dissolution (Takeuchi et al. 2004).

23.9 Techniques of SD Evaluation (Kovačič et al. 2011)

23.9.1 Particle Size and Morphology

Size of the drug and SD particles can be determined by differential laser scattering methods. Measurement of particle size may provide insights for understanding amorphization and could be used as a tool for estimating the degree of amorphization in SD. Kovacic et al. (2011) showed that particle size of crystalline carvedilol was reduced from 41.5 to 10 μm after preparation of the SD. Also, SEM images of the carvedilol SD reveal no traces of free carvedilol, indicating complete encapsulation of the drug in the pores of the silica.[107] Similar observations were reported by Waters et al. (2013) while preparing fenofibrate SD using amorphous silica.

23.9.2 *Differential Scanning Calorimetry and Thermogravimetric Analysis*

A primary objective of using MPS is to convert crystalline drug into amorphous form and maintain its stability for a long period of time. It is necessary to ensure drug remains in amorphous form for as long as possible. Thermal studies are complementary techniques frequently used for determination of polymorphs. Various polymorphs can be differentiated by different melting temperatures and heats of fusion. In the case of amorphous compounds, there is no lattice energy and there will be no distinct melting point. However, glass transition temperature can be its identifying characteristic. The differential scanning calorimetry (DSC) curve of crystalline carvedilol showed an endothermic peak at 115 °C, dH 128 J/g, indicating characteristic melting point while SD of carvedilol with MPS showed no distinct melting point; however, a glass transition temperature can be observed at 38 °C. This confirms that crystalline carvedilol has been converted to the amorphous form in the silica-based SD (Brinker 2005).

23.9.3 *X-ray Diffractometry*

Mehanna et al. (2011) used X-ray diffractometry (XRD) technique to confirm the loading of tadalafil in the silica pores[170]. Similar results were reported by Wang et al. (2006) while preparing SD of nifedipine using porous silica.

23.9.4 *Determination of Drug Content and Dissolution*

The SD of a drug can be dispersed in a suitable solvent in which the drug is freely soluble. The drug is allowed to diffuse with the aid of sonication and filtered to determine the loaded drug.

$$\%\,Drug\,loading = \frac{Amount\ of\ drug\,loaded}{Amount\ of\ SD\,recovered} \times 100$$

This test is useful to judge efficiency of the drug loading method and calculation of the final dosage.

23.10 Final Dosage Form

Oral solid dosage forms are most preferred for MPS-based drug delivery. Tablets and capsules are recommended dosage forms for the delivery of poorly soluble drugs using MPS. Wet granulation of drug-loaded MPS improves the flow behavior and compact ability by increasing the particle size and bulk density, and by making the MPS surface smoother. It should be noted that drug may leach out onto the silica surface if over-wetting happens during granulation resulting in premature drug release. However, controlled amount of moisture is needed and should be used to create agglomerates for granulation. Vialpando et al. (2012) did a study to understand premature drug release from the wet granules of drug-loaded MPS. Poorly water-soluble model compounds, itraconazole, fenofibrate, naproxen, and ibuprofen, were loaded into OMS COK-12, and wet granulated using a polyvinylpyrrolidone (PVP) binder solution. It was observed that premature drug release is compound dependent and can be reduced by decreasing the initial drug load of the material and the binder solution addition rate or by increasing the granulation temperature and binder solution concentration. Premature release has been reported if compaction force is more than 120 MPa which breaks the structure of OMS. The use of a compression aid like microcrystalline cellulose may help to prevent such problems (Vialpando et al. 2012). Similarly, various researchers have studied the tableting parameters of drug-loaded silica. Takeuchi et al. (2005a) studied the compression or die wall pressures with simple formulations composed of diluents and disintegrant together with 9 % of drug-loaded silica particles. Xu et al. (2009) developed controlled release formulations by compressing drug-loaded silica into tablets without excipients followed by a pH-sensitive polymer coating.

23.11 Biocompatibility, Toxicity, and Regulatory Status of MPS

In general, the use of silica-based materials is not considered harmful to humans. Because of its unique properties, MPS has been studied widely for applications in the pharmaceutical industry. Colloidal silicon dioxide and silica gels (MPS) are listed in the Inactive Ingredients Database of US Food and Drug Administration and included in various pharmacopoeias. MPS have been in commercial use for pharmaceutical and food products for many years and there is no reported evidence of adverse reactions.

References

Aerts CA, Verraedt E, Mellaerts R, Depla A, Augustijns P, Van Humbeeck J, Van den Mooter G, Martens JA (2007) Tunability of pore diameter and particle size of amorphous microporous silica for diffusive controlled release of drug compounds. J Phys Chem C 111:13404–13409

Aerts CA, Verraedt E, Depla A, Follens L, Froyen L, Van Humbeeck J, Augustijns P, Van den Mooter G, Mellaerts R, Martens JA (2010) Potential of amorphous microporous silica for ibuprofen controlled release. Int J Pharm 397:84–91

Ahola M, Kortesuo P, Kangasiniemi I, Kiesvaara J, Yli-Urpo A (2000) Int J Pharm 195:219–227

Aiello R, Cavallaro G, Giammona G, Pasqua L, Testa F (2002) Mesoporous silicate as matrix for drug delivery systems of non-steroidal antiinflammatory drugs. Stud Surf Sci Catal 142:1165–1172

Andersson J, Rosenholm J, Areva S, Linden M (2004) Chem Mat 21:4160

Andronis V, Yoshioka M, Zografi G (1997) J Pharm Sci 86:346–351

Attard GS, Glyde JC, Göltner CG (1995) Nature 378:366

Aznar E, Marcos MD, Martinez-Manez R, Sancenon F, Soto J, Amoros P, Guillem C (2009) J Am Chem Soc 131:6833–6843

Bagshaw SA, Prouzet E, Pinnavaia TJ (1995) Science 269:1242

Bahl D, Bogner RH (2006) Amorphization of indomethacin by co-grinding with Neusilin US2: amorphization kinetics, physical stability and mechanism. Pharm Res 23:2317–2325

Beck JS, Vartuli JC (1996) Recent advances in the synthesis, characterization and applications of mesoporous molecular sieves. Curr Opin Solid State Mater Sci 1(1):76–87

Beck JS, Vartuli JC, Roth WJ, Leonowicz ME, Kresge CT, Schmitt KD, Chu CTW, Olson DH, Sheppard EW (1992) A new family of mesoporous molecular sieves prepared with liquid crystal templates. J Am Chem Soc 114(27):10834–10843

Brinker CJ (2005) Colloidal silica: fundamentals and applications. In: Bergna HE, Roberts WO (eds) Taylor and Francis, New York, p 615

Brouwers J, Brewster ME, Augustijns P (2009) Supersaturating drug delivery systems: the answer to solubility-limited oral bioavaiulability? J Pharm Sci 98(8)

Cavallero G, Pierro P, Palumbo FS, Testa F, Luigi P, Aiello R (2004) Drug delivery devices based on mesoporous silicate. Drug Deliv 11(1):41–46

Charnay C, Bégu S, Tourné-Péteilh C, Nicole L, Lerner DA, Devoisselle JM (2004) 1213 Inclusion of ibuprofen in mesoporous templated silica: drug loading and release 1214 property. Eur J Pharm Biopharm 57:533–540

Chen S, Garcia-Bennett AE, Yokoi T, Sakamoto K, Kunieda H, Terasaki O, Tatsumi T (2003) Nat Mater 2:801–805

Chiola V, Ritsko JE, Vanderpool CD (1971) 3556725

Chiou WL (1971) Pharmaceutical applications of solid dispersion system. J Pharm Sci 60(9):1281–1302

Datta S, Grant DJW (2004) Nat Rev Drug Discov 3:42–57

Delle Piane M, Corno M, Ugliengo P (2013) Does dispersion dominate over H-bonds in drug−surface interactions? The case of silica-based materials as excipients and drug-delivery agents. J Chem Theory Comput 9:2404−2415

Doadrio AL, Sousa EMB, Doadrio JC, Perez Pariente J, Izquierdo-Barba I, Vallet-Regi M (2004) Mesoporous SBA-15 HPLC evaluation for controlled gentamicin drug delivery. J Control Release 97(1):125–132

Doadrio AL, Doadrio JC, Sanchez-Montero JM, Salinas AJ, Vallet-Regi M (2010) A rational explanation of the vancomycin release from SBA-15 and its derivative by molecular modelling. Microporous Mesoporous Mater 132(3):559–566

Fernández-Núez M, Zorrilla D, Montes A, Mosquera MJ (2009) Ibuprofen loading in surfactant-templated silica: role of the solvent according to the polarizable continuum model. J Phys Chem A 113:11367–11375

Finnie KS, Jacques DA, McGan MJ, Blackfordb MG, Barbé CJ (2006) Encapsulation and controlled release of biomolecules from silica microparticles. J Mater Chem 16:4494–4498

Foraker AB, Walczak RJ, Cohen MH, Boiarski TA, Grove CF, Swaan PW (2003) Microfabricated porous silicon particles enhance paracellular delivery of insulin across intestinal Caco-2 cell monolayers. Pharm Res 20(1):110–116

Garcia-Bennet AE, Kozhevnikova M, König N, Zhou C, Leao R, Knöpfel T, Pankratova S, Trolle C, Berezin V, Bock E, Aldskogius H, Kozlova EN (2013) Delivery of differentiation factors by mesoporous silica particles assists advanced differentiation of transplanted murine embryonic stem cells. Stem Cells Transl Med 2:906–915

Giraldo LF, López BL, Pérez L, Urrego S, Sierra L, Mesa M (2007) Mesoporous silica applications. Macromol Symp 258:129–141

Giri S, Trewyn BG, Stellmaker MP, Lin VSY (2005) Angew Chem Int Ed 44:5038–5044

Gu J, Huang K, Zhu X, Li Y, Wei J, Zhao W, Liu C, Shi J (2013) Sub-150 nm mesoporous silica nanoparticles with tunable pore sizes and well-ordered mesostructure for protein encapsulation. J Colloid Interface Sci 407:236–242

Gupta MK, Tseng Y-C, Goldman D, Bogner RH (2002) Pharm Res 19:1663

Halamova D, Badanicova M, Zelenak V, Gondova T, Vainio U (2010) Naproxen drug delivery using periodic mesoporous silica SBA-15. Appl Surf Sci 256(22):6489–6494

Heikkilä T, Salonen J, Tuura J, Kumar N, Salmi T, Murzin DY, Hamdy MS, Mul G, Laitinen L, Kaukonen AM, Hirvonen J, Lehto V-P (2007) Evaluation of mesoporous TCPSi, MCM-41, SBA-15, and TUD-1 materials as API carriers for oral drug delivery. Drug Deliv 14:337–347

Hixson AW, Crowell JH (1931) Dependence of reaction velocity on upon surface and agitation I: theoretical considerations. Ind Eng Chem 23:923–931

Ho J, Huang Y, Danquah MK, Wang H, Forde GM (2010) Synthesis of biodegradable polymer-mesoporous silica composite microspheres for DNA prime-protein boost vaccination. Eur J Pharm Sci 39:412–420

Horcajada P, Rámila A, Pérez-Pariente J, Vallet-RegI M (2004) Influence of pore size of MCM-41 matrices on drug delivery rate. Microporous Mesoporous Mater 68(1-3):105–109

Hu Y, Zhi Z, Zhao Q, Wu C, Zhao P, Jiang H, Jiang T, Wang S (2012) 3D cubic mesoporous silica microsphere as a carrier for poorly soluble drug carvedilol. Microporous Mesoporous Mater 147:94–101

Improving solubilty of poorly water soluble drug indomethacin by incorporating porous material in solid dispersion

Izquierdo-Barba I, Martinez A, Doadrio AL, Pérez-Pariente J, Vallet-Regı M (2005) Release evaluation of drugs from ordered three-dimensional silica structures. Eur J Pharm Sci 26:365–373

Izquierdo-Barba I, Sousa E, Doadrio J, Doadrio A, Pariente J, Martinez A, Babonneau F, Vallet-Regi M (2009) Influence of mesoporous structure type on the controlled delivery of drugs: release of ibuprofen from MCM-48, SBA-15 and functionalized SBA-15. J Sol-Gel Sci Technol 50(3):421–429

Jin ZW, Liang H (2010) Effects of morphology and structural characteristics of ordered SBA-15 mesoporous silica on release of ibuprofen. J Disper Sci Technol 31:654–659

k. Finnie mraci c chem, l. Kong, d. Jacques, h-q. Lin, s. Mcniven, s. Calleja, e. Gorissen And c. Barbé. Encapsulation and controlled release from silica particles. Chemistry in Australia

Kalaiselvan R, Mohanta GP, Manna PK (2006) Pharmazie 61:618

Kiekens F, Daems T, Castelein P, Parasote T, Paul J, Martens JA, Van Den Mooter G (2012) High-throughput system to evaluate in-vitro release via ordered mesoporous silica. AAPS poster

Kiekens F, Eelen S, Verheyden L, Daems T, Martens J, Van Den Mooter G. Use of ordered mesoporous silica to enhance the oral bio-availability of ezetimibe, a poorly water soluble compound, in dogs. Poster AAPS

Kinnari P, Mäkilä E, Heikkilä T, Salonen J, Hirvonen J, Santos HA (2011) Comparison of mesoporous silicon and non-ordered mesoporous silica materials as drug carriers for itraconazole. Int J Pharm 414:148–156

Konno T, Kinuno K (1989) Physical and chemical changes of medicinal in mixtures with adsorbents in the solid state: II. Application of reduced pressure treatment for the improvement of dissolution of flufenamic acid. Chem Pharm Bull 37:2481–2484

Konno T, Kinuno K, Kataoka K (1986) Physical and chemical changes of medicinals in mixtures with adsorbents in the solid state: I. Effect of vapor pressure of the medicinals on changes in crystalline properties. Chem Pharm Bull 34:301–307

Kovačič B, Vrečer F, Planinšek O (2011) Solid dispersions of carvedilol with porous silica. Chem Pharm Bull 59(4):427–433

Kozlova SA, Kirik SD (2010) Post-synthetic activation of silanol covering in the mesostructured silicate materials MSM-41 and SBA-15. Microporous Mesoporous Mater 133:124–133

Kresge CT, Leonowicz ME, Roth WJ, Vartuli JC, Beck JS (1992) Ordered mesoporous molecular sieves synthesized by a liquid-crystal template mechanism. Nature 359(6397):710–712

Lai CY, Trewyn BG, Jeftinija DM, Jeftinija K, Xu S, Jeftinija S, Lin VSY (2003) J Am Chem Soc 125:4451–4459
LaMer VK, Dinegar RH (1950) Theory, production and mechanism of formation of monodispersed hydrosols. J Am Chem Soc 72(11)
Li X, Xie QR, Zhang J, Xia W, Gu H (2011) The packaging of siRNA within the mesoporous structure of silica nanoparticles. Biomaterials 32:9546–9556
Liao C-C, Jarowski CI (1984) Dissolution rates of corticoid solutions dispersed on silica. J Pharm Sci 73(3)
Limnell T, Heikkila T, Santos HA, Sistonen S, Hellsten S, Laaksonen T, Peltonen L, Kumard N, Murzin DY, Louhi-Kultanen M, Salonen J, Hirvonen J, Lehto VP (2011a) Physicochemical stability of high indomethacin payload ordered mesoporous silica MCM-41 and SBA-15 microparticles. Int J Pharm 416:242–251
Limnell T, Santos HA, Mäkilä E, Heikkilä T, Salonen J, Murzin DY, Kumar N, Laaksonen T, Peltonen L, Hirvonen J (2011b) Drug delivery formulations of ordered and nonordered mesoporous silica: comparison of three drug loading methods. J Pharm Sci 100(8)
Mai Khanfar a, Mohammad M. Fares b,⇑, Mu'taz Sheikh Salem a,⇑, Amjad M. Qandil c, d
Mehanna MM, Motawaa AM, Samaha MW (2011) Tadalafil inclusion in microporous silica as effective dissolution enhancer: optimization of loading procedure and molecular state characterization. J Pharm Sci 100(5)
Mellaerts R, Aerts CA, Van Humbeeck J, Augustijns P, Van den Mooterc G, Martens JA (2007) Enhanced release of itraconazole from ordered mesoporous SBA-15 silica materials. Chem Commun 1375–1377
Mellaerts R, Mols R, Jammaer JAG, Aerts CA, Annaert P (2008) Increasing the oral bioavailability of the poorly water soluble drug itraconazole with ordered mesoporous silica. Eur J Pharm Biopharm 69:223–230
Mellaerts R, Houthoofd K, Elen K, Chen H, Van Speybroeck M, Van Humbeeck J, Augustijns P, Mullens J, Van den Mooter G, Martens JA (2010) Aging behavior of pharmaceutical formulations of itraconazole on SBA-15 ordered mesoporous silica carrier material. Microporous Mesoporous Mater 130:154–161
Miura H, Kanebako M, Shirai H, Nakao H, Inagi T, Terada K (2010) Enhancement of dissolution rate and oral absorption of a poorly water-soluble drug, K-832, by adsorption onto porous silica using supercritical carbon dioxide. Eur J Pharm Biopharm 76:215–221
Miura H, Kanebako M, Shirai H, Nakao H, Inagia T, Terada K (2011) Stability of amorphous drug, 2-benzyl-5-(4-chlorophenyl)-6-[4-(methylthio)phenyl]-2Hpyridazin-3-one, in silica mesopores and measurement of its molecular mobility by solid-state 13 C NMR spectroscopy. Int J Pharm 410:61–67
Monkhouse DC, Lach JL (1972) J Pharm Sci 61:1435–1441
Moritz M, Laniecki M (2012) Application of SBA-15 mesoporous material as the carrier for drug formulation systems. Papaverine hydrochloride adsorption and release study. Powder Technol 230:106–111
Mortera R, Fiorilli S, Garrone E, Verné E, Onida B (2010) Pores occlusion in MCM-41 spheres immersed in SBF and the effect on ibuprofen delivery kinetics: a quantitative model. Chem Eng J 156(1):184–192
Muoz B, Rámila A, Pérez-Pariente J, Díaz I, Vallet-Regí M (2003) MCM-41 organic modification as drug delivery rate regulator. Chem Mater 15:500–503
Nguyen TD, Tseng HR, Celestre PC, Flood AH, Liu Y, Stoddart JF, Zink JI (2005) Proc Natl Acad Sci U S A 102:10029–10034
Noyes AA, Whitney WR (1897) The rate of solution of solid substances in their own solutions. J Am Chem Soc 19:930–934
Nunes CD, Vaz PD, Fernandes AC, Ferreira P, Romao CC, Calhorda MJ (2007) Loading and delivery of sertraline using inorganic micro and mesoporous materials. Eur J Pharm Biopharm 66(3):357–365

Pan X, Julian T, Augsburger L (2008) Increasing the dissolution rate of a low-solubility drug through a crystalline-amorphous transition: a case study with indomethicin. Drug Dev Ind Pharm 34:221–231.

Park HS, Kim CW, Lee HJ, Choi JH, Lee SG, Yun YP, Kwon IC, Lee SJ, Jeong SY, Lee SC (2010) A mesoporous silica nanoparticle with charge-convertible pore walls for efficient intracellular protein delivery. Nanotechnology 21:225101(9 p)

Park S-Y, Pendleton P (2012) Mesoporous silica SBA-15 for natural antimicrobial delivery. Powder Technol 223:77–82

Patel VI To improve the solubility of poorly water soluble drugs using porous material. Thesis submitted to Pharmacy and healthscience. Long Island University, Brooklyn, NY

Porta F, Lamers GE, Zink JI, Kros A (2011) Peptide modified mesoporous silica nanocontainers. Phys Chem Chem Phys 13:9982–9985

Qian KK, Bogner RH (2012) Application of mesoporous silicon dioxide and silicate in oral amorphous drug delivery systems. J Pharm Sci 101(2):

Qu F, Zhu G, Huang S, Li S, Zhang JSD, Qiu S (2006) Controlled release of Captopril by regulating the pore size and morphology of ordered Mesoporous silica. Micropor Mesopor Mater 92:1–9

Ramila A, Munoz B, Perez-Pariente J, Vallet-Regi M (2003) Mesoporous MCM-41 as drug host system. J Sol-Gel Sci Technol 26(1):1199–1202

Salonen J, Kaukonen AM, Hirvonen J, Lehto VP (2008) Mesoporous silicon in 1443 drug delivery applications. J Pharm Sci 97:632–653

Sarode A, Wang P, Obara S, Worthen DR (2014) Supersaturation, nucleation, and crystal growth during single- and biphasic dissolution of amourphous solid dispersions: polymer effects and implocations for oral bioavailabilty enhamcement of poorly water soluble drugs. Eur J Pharm Biopharm 2–13

Scherer GW, Brinker CJ (1990) Sol-gel science. Academic, New York

Serajuddin ATM (1999) Solid dispersion of poorly water-soluble drugs: early promises, subsequent problems, and recent breakthroughs. J Pharm Sci 88(10):1058–1065

Shen SC, Ng WK, Chia L, Hu J, Tan RB (2011) Physical state and dissolution of ibuprofen formulated by co-spray drying with mesoporous silica: effect of pore and particle size. Int J Pharm 410:188–195

Sheth A, Jarowski CI (1990) Use of powdered solutions to improve the dissolution rate of polythiazide tablets. Drug Dev Ind Pharm 16(5):769–777

Siepmann J, Siepmann F (2013) Mathematical modeling of drug dissolution. Int J Pharm 453:12–24

Sing KSW, Everett DH, Haul RAW, Moscou L, Pierotti RA, Rouquerol J, Siemieniewska T (1985) Reporting physisorption data for gas/solid systems with special reference to the determination of surface area and porosity. Pure Appl Chem 57(4):603–619

Smirnova I, Mamic J, Arlt W (2003) Adsorption of drugs on silica aerogels. Langmuir 19:8521–8525

Song SW, Hidajat K, Kawi S (2005) Functionalized SBA-15 materials as carriers for controlled drug delivery: influence of surface properties on matrix-drug interactions. Langmuir 21(21):9568–9575

Stein J, Fuchs T, Mattern C (2010) Advanced milling and containment technologies for superfine active pharmaceutical ingredients. Chem Eng Technol 33:1464–1470

Sun Y, Sun Y-L, Wang L, Ma L, Yang Y-W, Gao H (2014) Nanoassembles constructed from mesoporous silica nanoparticles and surface-coated multilayer polyelectrolytes for controlled drug delivery. Microporous Mesoporous Mater 185:245–253

Takeuchi H, Nagira S, Yamamoto H, Kawashima Y (2004) Solid dispersion particles of tolbutamide prepared with fine silica particles by the spray-drying method. Powder Technol 141:187–195

Takeuchi H, Nagira S, Tanimura S, Yamamoto H, Kawashima Y (2005a) Tabletting of solid dispersion particles consisting of indomethacin and porous silica particles. Chem Pharm Bull 53:487–491

Takeuchi H, Nagira S, Yamamoto H, Kawashima Y (2005b) Solid dispersion particles of amorphous indomethacin with fine porous silica particles by using spray-drying method. Int J Pharm 293:155–164

Tanev PT, Pinnavaia TJ (1995) Science 267:865
Tang Q, Chen Y, Chen J, Li J, Xu Y, Wu D, Sun Y (2010) Drug delivery from hydrophobic-modified mesoporous silicas: control via modification level and site-selective modification. J Solid State Chem 183:76–83
Tomasi J (2004) Theor Chem Acc 112:184
Vallet-Regi M, Ramila A, del Real RP, Perez-Pariente J (2001) A new property of MCM-41: drug delivery system. Chem Mater 13(2):308–311
Vallet-Regi M, Doadrio JC, Doadrio AL, Izquierdo-Barba I, Perez-Pariente J (2004) Hexagonal ordered mesoporous material as a matrix for the controlled release of amoxicillin. Solid State Ionics 172(1-4):435–439
Van Speybroeck M, Barillaro V, Thi TD, Mellaerts R, Martens J, Van Humbeeck J, Vermant J, Annaert P, Van denMG, Augustijns P (2009) Ordered mesoporous silica material SBA-15: a broad-spectrum formulation platform for poorly soluble drugs. J Pharm Sci 98:2648–2658
Van Speybroeck M, Mellaerts R, Mols R, Thi TD, Martens JA, Van Humbeeck J, Annaert P, Van denMG, Augustijns P (2010) Enhanced absorption of the poorly soluble drug fenofibrate by tuning its releaserate from Ordered Mesoporous Silica. Eur J Pharm Sci 41:623–630
Verraedt E, Van den Mooter G, Martens JA (2011) Novel amorphous microporous silica spheres for controlled release applications. J Pharm Sci 100(10)
Vialpando M, Aerts A, Persoons J, Martens J, Van Den Mooter G (2011) Evaluation of ordered mesoporous silica as a carrier for poorly soluble drugs: influence of pressure on the structure and drug release. Published online 8 March 2011 in Wiley Online Library (wileyonlinelibrary.com). doi:10.1002/jps.22535
Vialpando M, Backhuijs F, Martens JA, Van den Mooter G (2012) Risk assessment of premature drug release during wet granulation of ordered mesoporous silica loaded with poorly soluble compounds itraconazole, fenofibrate, naproxen, and ibuprofen. Eur J Pharm Biopharm 81:190–198
Vivero-Escoto JL, Slowing II, Wu CW, Lin VSY (2009) J Am Chem Soc 131:3462
Vogt M, Kunath K, Dressman JB (2008) Dissolution enhancement of fenofibrate by micronization, cogrinding and spray-drying: comparison with commercial preparations. Eur J Pharm Biopharm 68:283–288
Wang L, Cui FD, Sunada H (2006) Preparation and evaluation of solid dispersions of nitrendipine prepared with fine silica particles using the melt-mixing method. Chem Pharm Bull 54(1):37–43
Wang G, Otuonye AN, Blair EA, Denton K, Tao Z, Asefa T (2009) Functionalized mesoporous materials for adsorption and release of different drug molecules: a comparative study. J Solid State Chem 182(7):1649–1660
Watanabe T, Wakiyama N, Usui F, Ikeda M, Isobe T, Senna M (2001) Stability of amorphous indomethacin compounded with silica. Int J Pharm 226:81–91
Waterman KC, MacDonald BC (2010) Package selection for moisture protection for solid, oral drug products. J Pharm Sci 99(11)
Waters LJ, Hussain T, Parkes G, Hanrahan JP, Tobin JM (2013) Inclusion of fenofibrate in a series of mesoporous silicas using microwave irradiation. Eur J Pharm Biopharm 85:936–941
Xia T, Kovochich M, Liong M, Meng H, Kabehie S, George S et al (2009) Polyethyleneimine coating enhances the cellular uptake of mesoporous silica nanoparticles and allows safe delivery of siRNA and DNA constructs. ACS Nano 3:3273e86
Xia X, Zhou C, Ballell L, Garcia-Bennett AE (2012) In vivo enhancement in bioavailability of atazanavir in the presence of proton-pump inhibitors using mesoporous materials. ChemMedChem 7:43–48
Xie M, Shi H, Li Z, Shen H, Ma K, Li B, Shen S, Jin Y (2013) A multifunctional mesoporous silica nanocomposite for targeteddelivery, controlled release of doxorubicin and bioimaging. Colloids Surf B Biointerfaces 110:138–147
Xu W, Gao Q, Xu Y, Wu D, Sun Y (2009) pH-controlled drug release from mesoporous silica tablets coated with hydroxypropyl methylcellulose phthalate. Mater Res Bull 44:606–612

Xu W, Riikonen J, Lehto V-P (2012) Mesoporous systems for poorly soluble drugs. Int J Pharm xxx:xxx–xxx

Yang KY, Glemza R, Jarowski CI (1979) Effects of amorphous silicon dioxides on drug dissolution. J Pharm Sci 68(5):

Yu H, Zhai QZ (2009) Mesoporous SBA-15 molecular sieve as a carrier for controlled release of nimodipine. Microporous Mesoporous Mater 123(1-3):298–305

Yu C, Yu Y, Zhao D (2000) Chem Commun 575–576

Zhang Y, Zhi Z, Jiang T, Zhang J, Wang Z, Wang S (2010) Spherical mesoporous 1560 silica nanoparticles for loading and release of the poorly water-soluble drug 1561 telmisartan. J Control Release 145:257–263

Zhao D, Feng J, Huo Q, Melosh N, Fredrickson GH, Chmelka BF, Stucky GD (1998a) Triblock copolymer syntheses of mesoporous silica with periodic 50 to 300 Angstrom pores. Science 279(5350):548–552

Zhao D, Huo Q, Feng J, Chmelka BF, Stucky GD (1998b) Nonionic triblock and star siblock copolymer and oligomeric surfactant syntheses of highly ordered, hydrothermally stable, mesoporous silica structures. J Am Chem Soc 120(24):6024–6036

Zhao XS, Lu GQ (1998) Modification of MCM-41 by surface silylation with trimethylchlorosilane and adsorption study. J Phys Chem B 102:1556–1561

Zheng J, Tian X, Sun Y, Lu D, Yang W (2013) pH-sensitive poly(glutamic acid) grafted mesoporous silica nanoparticles for drug delivery. Int J Pharm 450:296–303

Zhuravlev LT (2000) The surface chemistry of amorphous silica. Zhuravlev Model Colloids Surf A 173:1–38

Index

N. Shah et al. (eds.), *Amorphous Solid Dispersions*,
Advances in Delivery Science and Technology, DOI 10.1007/978-1-4939-1598-9